Bartolomé Andreo · Francisco Carrasco
Juan José Durán · Pablo Jiménez
James W. LaMoreaux
Editors

Hydrogeological and Environmental Investigations in Karst Systems

Editors
Bartolomé Andreo
Francisco Carrasco
Pablo Jiménez
Department of Geology
and Centre of Hydrogeology
University of Málaga (CEHIUMA)
Málaga
Spain

James W. LaMoreaux
University Boulevard
Tuscaloosa
USA

Juan José Durán
Geological Survey of Spain
Instituto Geológico y Minero de España
Madrid
Spain

ISSN 1866-6280 ISSN 1866-6299 (electronic)
ISBN 978-3-642-17434-6 ISBN 978-3-642-17435-3 (eBook)
DOI 10.1007/978-3-642-17435-3

Library of Congress Control Number: 2014946748

Springer Heidelberg New York Dordrecht London

Printed on acid-free paper

Springer is part of Springer Science+Business Media (www.springer.com)

Presentation

Karst is the result of climatic and geohydrological processes, mainly in carbonate and evaporite rocks, during geological periods of Earth history. Dissolution of these rock formations over time has generated karst aquifers and environments of significant water and mineral resources. In addition, beautiful landscapes have been created which constitute natural parks, geosites, and caves. Due to their origin and nature, karstified areas require investigation with special techniques and methodology. International collaboration and discussions on advances in karst research are necessary to promote Karst Science.

The International Symposium on Karst Aquifers is one of the worldwide events held periodically to specifically address karst environments. The symposium constitutes an ongoing international forum for scientific discussion on the progress made in research in karst environments. The first and second symposiums were organized in Nerja (near Malaga, Spain), in 1999 and 2002; the third and fourth symposiums were held in Malaga city in 2006 and 2010.

The 5th International Symposium on Karst Aquifers (ISKA5) occurred in Malaga on during October 14–16, 2014. It was organized by the Centre of Hydrogeology University of Málaga (CEHIUMA) and the Spanish Geological Survey (IGME), in cooperation with UNESCO and the International Association of Hydrogeologists (IAH) Karst Commission.

More than 100 contributions were received from 30 countries on five continents. Presentations made during the symposium and published in this book are a compendium of 70 of these manuscripts. Papers submitted by April 2014, were peer-reviewed and subsequently accepted by the Scientific Committee. Contributions are grouped into five sections:

- Methods Utilized to Study Karst Aquifers.
- Karst Hydrogeology.
- Mining and Engineering in Karst media.
- Karst Cavities.
- Karst Geomorphology and Landscape.

A large part of the contributions, 30 %, is related to Methods Utilized to Study Karst Aquifers. Several issues are addressed: methods for groundwater recharge assessment, dye tracer and stable isotope applications, analysis of hydrodynamic data and hydrochemistry, among others.

Most contributions, 40 %, however, are on Karst Hydrogeology. These are primarily in connection with various topics such as numerical modeling in karst, floods, karst groundwater flow, protection of karst aquifers or pollution, and vulnerability in karst.

Five percent of the published papers deal with Mining and Engineering in Karst Media. These papers are about tunnels, hydrogeological risks, and karst risk assessment in mining and civil engineering.

Another section concerning Karst Cavities encompasses 15 % of the contributions. These chapters deal with corrosion and speleogenetic processes, speleothems, CO_2 sources, the global carbon cycle in endokarst, and the study of past climate.

Karst Geomorphology and Landscape constitutes the remaining 10 % of the contributions. These papers are related to karst features, wetlands, hypogene speleogenesis, geodiversity, and karstic geosites.

The results of project work performed by karst specialists worldwide are described in the book. Included in it are experiences from pilot sites, methodologies, monitoring, and data analyses in various climatic, geological, and hydrogeological contexts. Material presented may be utilized for activities such as teaching and technical-professional applications particularly as they apply to the increasingly multidisciplinary nature of karst studies. Information provided may also be useful to decisions makers in making critical decisions regarding development in karst regions. Scientists and engineers and many of the lay public interested in karst environments will benefit from the contents.

Bartolomé Andreo
Francisco Carrasco
Juan José Durán
Pablo Jiménez
James W. LaMoreaux

Acknowledgments

The editors of the book would like to thank the members of the Organizing and the Scientific Committees and the authors and reviewers of the manuscripts of the Fifth International Symposium on Karst Aquifers (ISKA5) for their contributions and collaboration. Members of the Scientific Committee composed of 41 international specialists from various karst domains, along with members of the Organizing Committee are listed accordingly. All the manuscripts were reviewed and only those that met the required criteria and quality were accepted.

ISKA5 held in Malaga, Spain in October, 2014, was organized by the Centre of Hydrogeology at the University of Malaga and the Spanish Geological Survey (IGME) in cooperation with UNESCO and the International Association of Hydrogeologists (IAH) Karst Commission. Support from the organizations that collaborated in organizing the conference is acknowledged and appreciated. Logos of these organizations, Nerja Cave Foundation, Junta de Andalucía (Consejería de Fomento y Vivienda—PROTEKARST project), Club of Groundwater, and Malaga Academy of Science, are included.

Results of contributions published are related to selected projects including DIKTAS, Groundwater Governance, IHP and IGCP 598 of UNESCO; CGL2012-32590 of Spanish Directorate of Research; P10-RNM-6895, P11-RNM-8087 and PROTEKARST, as well as, Research Group RNM 308 of the Andalusian Government. The publication is one of several produced under the auspices of the IAH Karst Commission.

Scientific Committee

Contributions published in this book have been reviewed by members of the Scientific Committee which includes the following professionals:

Bartolomé Andreo (University of Málaga, Spain)
Alice Aureli (UNESCO)
Ralf Benischke (Joanneum Research Institute, Graz, Austria)
Cristina C. Bicalho (University of Brasilia, Brazil)
José Benavente Herrera (University of Granada)
Catherine Bertrand (University of Franche-Comté, France)
Ogden Bonacci (University of Split, Croatia)
Lhoussaine Bouchaou (University of Ibn Zohr, Agadir, Morocco)
Francisco Carrasco (University of Málaga, Spain)
Zhang Cheng (Karst Dynamics Laboratory, Guilin, China)
Yuan Daoxian (Institute of Karst Geology, Guilin, China)
Nathalie Doerfliger (BRGM, France)
Wolfgang Dreybrodt (University of Bremen, Germany)
Juan Jose Durán (Spanish Geological Survey—IGME, Spain)
Christophe Emblanch (University of Avignon, France)
Dave Evans (FloSolutions, Lima, Peru)
Francesco Fiorillo (University of Sannio, Italy)
Stephen Foster (International Association of Hydrogeologists, UK)
Nico Goldscheider (Karlsruhe Institute of Technology)
Chris Groves (Western Kentucky University, USA)
John Gunn (University of Birmingham, UK)
Pierre Yves Jeannin (SISKA, Switzerland)
Cao Jianhua (Kart Dynamic Laboratory, Guilin, China)
Neven Kresic (AMEC, Washington, USA)
Neno Kukuric (IGRAC, The Netherlands)
James W. LaMoreaux (P.E. LaMoreaux and Associates, Inc., USA)
Barbara Mahler (USGS, USA)
Nicolas Massei (University of Rouen, France)

Organizing Committee

Other organizations collaborated in the Symposium: Nerja Cave Foundation, Junta de Andalucía (Consejería de Fomento y Vivienda—PROTEKARST project), Club of Groundwater, and Malaga Academy of Science.

The 5th International Symposium on Karst Aquifers (ISKA5) was organized by researchers from the Centre of Hydrogeology at the University of Málaga (CEHIUMA) and the Spanish Geological Survey (IGME), in the framework of their "Advanced Hydrogeological Studies" partnership. The Symposium was sponsored by UNESCO and the IAH Karst Commission.

Bartolomé Andreo (CEHIUMA)
Juan Antonio Barbera Fornell (CEHIUMA)
Luis Carcavilla Urquí (IGME)
Francisco Carrasco (CEHIUMA)
Juan José Durán (IGME)
Pablo Jiménez (CEHIUMA)
Luis Linares Girela (Malaga Academy of Science)
Cristina Liñán (CEHIUMA and Nerja Cave Foundation)
Ana Isabel Marin Guerrero (CEHIUMA)
Sergio Martos-Rosillo (IGME)
Matías Mudarra (CEHIUMA)
Damián Sánchez García (CEHIUMA)
Iñaki Vadillo (CEHIUMA)

Organized and sponsored by

In collaboration with

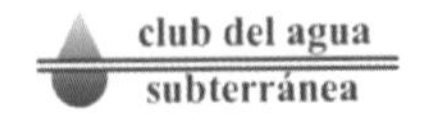

Contents

Introduction

For the most part, the study of karst landforms and the study of caves began as separate subjects. The study of karst dates from the establishment of geomorphology as a science in the middle years of the nineteenth century. Investigations of caves began even earlier with the pioneering explorations of Adolf Schmidl in Austria, Imre Voss in Hungary, Eduard Martel in France and many others. By the middle of the twentieth century, cave studies had evolved into a branch of science called Speleology. Although it was understood that both caves and karst landforms resulted from the dissolution of carbonate (or evaporite) rocks in slightly acidified groundwater, the science was largely observational. The only investigative tools were maps, photographs, and the direct observations of the explorers. Theoretical models were qualitative and deductive as, for example, the theories of cave origin proposed in the United States by W.M. Davis and J.H. Bretz.

The structure and practice of cave and karst science advanced dramatically in the second half of the twentieth century. Many advances were accomplished by "borrowing" both concepts and experimental methods from other sciences. One of the most important borrowings was the physical chemistry of carbonate rock dissolution which allowed a quantitative description of cave and karst landform excavation and of speleothem deposition. The kinetics of carbonate reactions are more complicated but the developing understanding of the rates of carbonate reactions by geochemists was carried over into an explanation for the development of long cave conduits within the rock mass and, more recently, the development of computer models for speleogenesis. Chemical measurements on karst waters, particularly time series data (chemographs) have been a valuable characterization tool.

Early investigations of Karst Geomorphology—mostly dealing with surface landforms and only loosely coupled to cave studies—identified many types of karst but only one karst process: the dissolution of the bedrock by meteoric water acidified by carbonic acid. It is now understood that karst can be formed by different fluid chemistries in fluids moving along a variety of pathways. Thus karst features formed by downward-moving fluids (meteoric water) are epigenetic karst, while karst features (mostly caves) formed by upward-moving fluids are hypogenetic karst. Karst formed by water moving sideways is interstratal karst although there could be some quibbling about this term. Surface landforms are exokarst while caves and related features are endokarst. A further distinction is made between karst developed in dense, well-compacted Mesozoic and Paleozoic limestones and karst developed in more permeable, uncompacted, Cenozoic limestones where diagenetic processes are still underway. These are called telogenetic karst and eogenetic karst respectively.

A major advantage to contemporary karst researchers is the availability of greatly expanded data bases, both regional descriptions of karst and also descriptions and maps of caves. One example is the *International Atlas of Karst Phenomena* presently edited by Karl-Heinz Pfeffer. David Weary and Daniel Doctor of the United States Geological Survey have recently (2014) released a draft version of a new digital karst map of the United States. Many countries have established national cave data bases where cave descriptions and maps are archived. In the United States cave surveys are usually organized by individual states, sometimes privately and sometimes in cooperation with governmental agencies. On a world scale, a most valuable cave data base is the *Berliner Höhlenkundliche Berichte* edited by Michael Laumanns and his colleagues.

Examples of the various sources of information that are fed into the rapidly increasing knowledge of caves and karst are listed schematically in (Fig. 1). Use of X-ray diffraction and the scanning electron microscope revealed many minerals in cave deposits—319 species by 2011. Application of standard engineering fluid mechanics to cave streams and water-filled conduits replaced previous assumptions of Darcy's law behavior in karst aquifers.

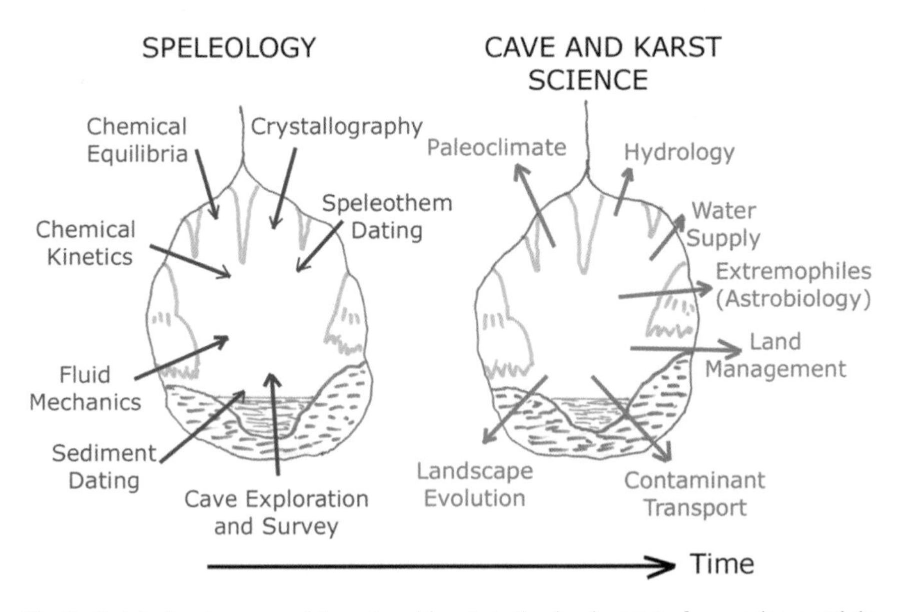

Fig. 1 Sketch showing some of the external inputs to the development of cave science and the outputs that cave and karst science have contributed science at large

A most important advance was the application of isotope geochemistry to providing an absolute chronology for both speleothems and clastic sediments in caves. In the mid-twentieth century, estimates for the age of speleothems and the age of caves were little more than guesses based on estimates of the rates of speleothem growth, on the volume of water needed to excavate a cave passage, and on the relation of large cave passages to nearby river terraces. Measurement of the concentrations of ^{238}U, ^{234}U, and ^{230}Th in speleothems allowed a calculation of their ages. With greatly improved measurement methods using thermal ionization mass spectroscopy, ages can be determined on milligram samples, thus allowing the determination of absolute chronologies for profiles along speleothem growth axes. Identification of magnetic pole reversals in sequences of clastic cave sediments provided a few fixed time markers for sediment deposition. The most recent karst dating breakthrough has been the use of cosmogenic isotopes to provide absolute dates for the deposition of clastic sediments in caves. By measuring the concentrations of ^{26}Al and ^{10}Be in the quartz in cave sediments with an accelerator mass spectrometer, the time of sediment deposition can be calculated. The sediment age is also a minimum age for the passage in which the sediment was deposited. Size, sorting, and distribution of clastic sediments in fossil caves are important clues to the hydrologic evolution of karst drainage basins.

From the beginning, back in the nineteenth century, the study of caves and karst had always been an insular sort of science, a rather minor subdivision of the Earth Sciences with little connection (or importance) to other parts of the subject. That relationship changed dramatically in the latter part of the twentieth century (Fig. 1). There was a transition, as indicated by the arrows in Fig. 1, from a situation where other aspects of the Earth Sciences were contributing to the understanding of caves and karst to one in which information gained from the study of caves and karst contributed to other aspects of the Earth Sciences.

One of the most dramatic of these role reversals was in classical Geomorphology, the evolution of river valleys. The evolution of river valleys had been studied by tracing and mapping terraces left on the valley walls as the rate of downcutting increased in response to climatic or tectonic drivers. The age of the terraces and their residual gravels was a central question to which only rough answers were usually available. Caves were correlated with the terrace levels. With cosmogenic isotope dates for cave sediments, the ages of the caves became more accurately known than the age of the terraces. Instead of terraces being used to date caves, caves can be used to accurately date terraces and thus trace the chronology of valley development.

A relatively new—beginning in the 1990s—and extremely important contribution of cave and karst science is the use of speleothems as paleoclimate archives. Stalagmites, especially uniform cylindrical stalagmites, sliced longitudinally, reveal a microstratigraphy containing the growth history of the stalagmite over the period of its growth. U/Th dating provides a time scale for the profile. Sampling for analysis of $^{18}O/^{16}O$, $^{13}C/^{12}C$, $^{2}H/^{1}H$, $^{86}Sr/^{87}Sr$, and other trace elements, as well as color and luminescence banding produce a series of profiles that become a rich record of climate and vegetation changes on the land surface at least back to the mid-Pleistocene. The records are clear; how to interpret them is the subject of much contemporary research.

Cave biology is an important subject in its own right with a long history. A new aspect is the role of microbiology which forms a bridge between the geological and biological sciences. Microbial processes are central to the development of hypogenetic karst, acting as oxidizing and reducing agents in the sulfur chemistry that is central to hypogenetic processes. Biofilms appear to be everywhere in caves and their importance is only now being recognized.

There is no paradigm shift with respect to Karst Hydrology and Hydrogeology. The movement of water through karstic aquifers has been a central aspect of the subject since the beginning. What is new is the importance of Karst Hydrology to the larger scientific community, to political leaders and regulators, and to the general public. Karst aquifers are major water supplies in many regions of the world so yield, reliability, and freedom from contamination are major management issues. The protection of water supplies and the protection of karst lands are major themes in this 5th International Symposium on Karst. A substantial fraction of the papers address issues in hydrology and water resources.

Those familiar with karst understood that most of the flow was through systems of conduits with only small contributions from the generally low permeability carbonate rock. In the United States, at least, hydrogeologists in the 1950s and 1960s tended to regard karst aquifers as a continuous medium with caves producing, at most, a blip in the flow field. By the 1970s and 1980s the localized turbulent flow in conduits was becoming appreciated and great progress was made in quantitative measurements of spring hydrographs and chemographs, water balance calculations, isotope chemistry, and other means of aquifer evaluation. Various approaches to combining this information into computer models were underway by the 1980s.

Approaches to modeling karst aquifers are based on a range of underlying concepts. Continuous media models separate the aquifer into cells, each with its own hydraulic conductivity. These work best in fractured carbonates with poorly developed conduit systems. Pipe flow models address mainly the conduit system and provide good descriptions of the conduit flow system. Input–output (global or kernal function) models ignore the details of the aquifer completely and construct transfer functions that convert the input hydrograph into the output hydrograph. Most recent attention has been focused on comprehensive (double permeability) models that take account of the exchange between the conduit flow and the storage, recharge, and discharge between the conduits and the matrix. This family of models is undergoing rapid development as, for example, the U.S. Geological Survey's MODFLOW-CFP package.

Issues concerning water supply, contamination, land-use management, and other aspects of karst are often taken as local problems to be solved by local authorities. However, all of these issues have many features in common so that cave and karst science is really a global endeavor. This means that communication, comparing results, comparing problems, and comparing solutions, are also of global importance. As a communication device, international meetings are absolutely invaluable. International Symposium on Karst, in Malaga, is one of the worldwide periodic events on karst. It has been held every four years in Malaga (Spain) since 1992 being an international event for scientific discussion on karst media. The main objective of the 5th International Symposium on Karst is to discuss and disseminate the latest trends in research into karst media on the basis of the results obtained with different methodologies in various karst areas in the world. Thus, this book includes part of the contributions presented at the 5th International Symposium on Karst peer reviewed and accepted by Scientific Committee. It contains contributions from over 20 countries including countries from four continents.

PA, USA William B. White

Comparative Study of the Physicochemical Response of Two Karst Systems During Contrasting Flood Events in the French Jura Mountains

C. Cholet, M. Steinmann, J.-B. Charlier and S. Denimal

Abstract This paper presents preliminary results from two karst systems belonging to the "Jurassic Karst" observatory in the French Jura Mountains. The sites are characterized by localized and diffuse recharge. Physicochemical monitoring was performed at the karst outlet (springs), as well as in the unsaturated zone (cave and epikarstic spring). During two contrasting flood events, water level, temperature, electrical conductivity, dissolved organic carbon, turbidity, and dissolved oxygen were recorded at high frequency and compared. These preliminary results allow to propose a conceptual model for both sites. It was possible to distinguish specific autogenic and allogenic recharge mechanisms and to characterize the respective contribution of the saturated and unsaturated zones.

1 Introduction

High-frequency monitoring of physicochemical parameters contributes to a better understanding of the hydrogeological functioning of karst systems (e.g., Pronk et al. 2006; Fournier et al. 2007; Mudarra et al. 2013; Tissier et al. 2013). In this study, two karst systems were monitored belonging to the "Jurassic Karst" (JK) hydrogeological observatory in the French Jura Mountains (https://zaaj.univ-fcomte.fr/spip.php?article13&lang=en). The aim of this observatory is to characterize (i) the hydrological response of the karst in relation to the hydrodynamic properties of the system and (ii) the hydrogeochemical response in relation to land use in the infiltration zone.

C. Cholet (✉) · M. Steinmann · S. Denimal
Chrono-Environnement, UMR 6249 UFC/CNRS, 16 Route de Gray, 25000 Besançon, France
e-mail: cybele.cholet@univ-fcomte.fr

J.-B. Charlier
BRGM, 1039 Rue de Pinville, 34000 Montpellier, France

B. Andreo et al. (eds.), *Hydrogeological and Environmental Investigations in Karst Systems*, Environmental Earth Sciences 1,
DOI 10.1007/978-3-642-17435-3_1

Both sites of the study have comparable recharge areas of about 40 km^2 and were monitored with an identical equipment installed in the unsaturated zone (cave stream or epikarstic spring), and at the main karst spring at the catchment outlet. Based on two rainfall events, the physicochemical response of each karst system was analyzed in order to get insights into their hydrogeological behavior.

2 Study Sites

2.1 Presentation

The Fourbanne and Lods study sites (40 km^2) are located in the French Jura Mountains (Fig. 1). The Lods site is characterized by dominant autogenic recharge, whereas autogenic and allogenic recharge occurs at Fourbanne.

The Fourbanne site is located in the Doubs valley 20 km NE of Besançon within tabular Dogger limestones crosscut by N to NE-trending normal faults. The Fourbanne spring, situated close to the Doubs river at 260 m a.s.l., is the unique outlet of the Fourbanne system. Tracer tests demonstrated that the system is mainly recharged by stream water losses from the En-Versenne and la Vernoye sinkholes, located about 9 km NNE of the Fourbanne spring (Charmoille 2005). The resulting cave stream is accessible 3 km downstream of the sinkholes, through an artificial pit at Fontenotte (En-Versenne cave).

The Lods site is situated 25 km SE of Besançon within a SW-NE trending depression at the edge of the Ornans plateau and in direct vicinity of the folded and thrusted belt of the Faisceau Salinois. The catchment has two outlets at its south-western end (Grand Bief and Truite d'Or springs), both located in the village of Lods in the Loue valley, at an altitude of 360 m a.s.l. Hydrological connections between the catchment outlet and the recharge area are well documented by numerous tracer tests. The unsaturated zone is not accessible within the recharge area, but the shallow spring of Dahon, located at the northern limit of the recharge area of the Lods site, drains a local, about 30 m deep and 2 km^2 wide area, which can be considered as representative for the unsaturated zone of the Lods catchment.

2.2 Field Monitoring

Both karst systems were monitored at the main outlet and in the unsaturated zone: the Fourbanne spring and the cave stream of Fontenotte for the Fourbanne site, and the Grand Bief spring and the epikarstic Dahon spring for the Lods site, respectively. All sites were monitored since December 2013 with an identical equipment. Stations at the outlet were equipped with multiparameter loggers to measure water level (WL), temperature (Temp.), electrical conductivity (EC), turbidity (Turb.), and dissolved oxygen (DO). Stations in the unsaturated zone were monitored by

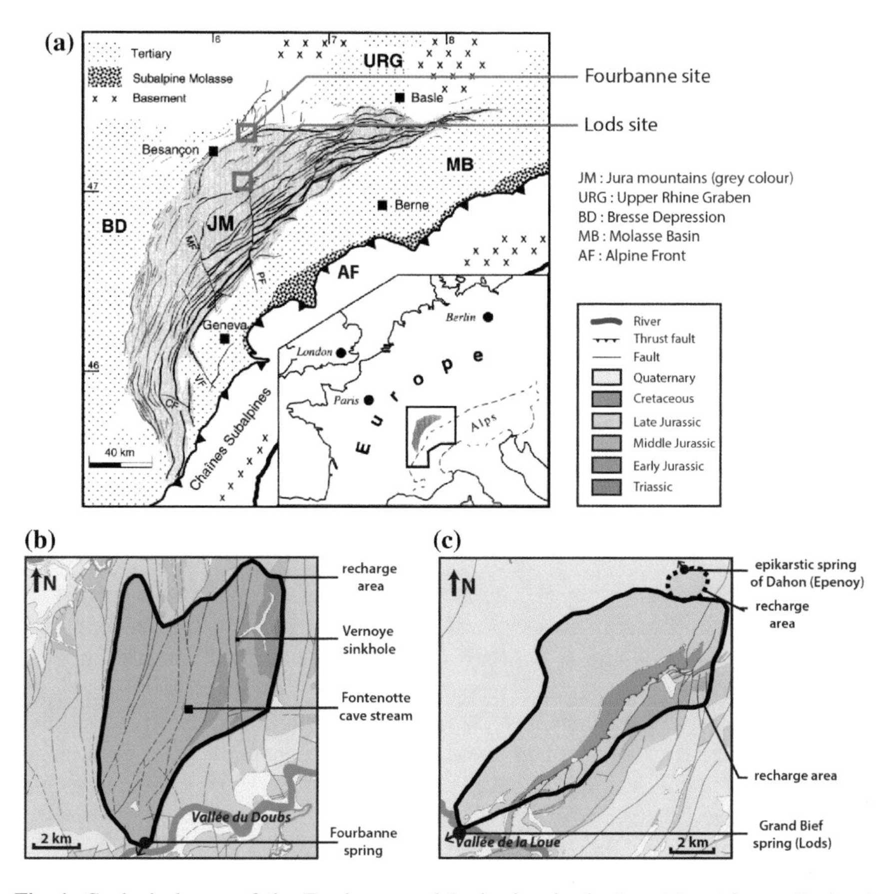

Fig. 1 Geological map of the Fourbanne and Lods sites in the Jura Mountains: **a** Regional situation (modified from Becker 2000), **b** Fourbanne site, **c** Lods site

CTD data-loggers monitoring WL, Temp. and EC. All stations were equipped with GGUN field fluorometers (Schnegg 2003) yield a semi-quantitative estimate for organic carbon (DOC) and Turb (see Charlier et al. (2012) for device calibration). The time step for data acquisition was 15 min for all devices. Hourly precipitation data from Meteo France are derived from the Branne station located 10 km E of the stream cave of the Fourbanne site; and from Epenoy, at less than 1 km from the epikarstic spring of the Lods site.

2.3 Selection of Flood Events

From time series from December 2013 to January 2014, two flood events with contrasting amplitudes were selected: (i) a small flood event corresponding to a single rainfall event on 16 Jan. 2014 (total rainfall of 13 and 15 mm at Fourbanne and

Lods, respectively) and (ii) a large event on 25 Dec. 2013 with 40 mm of rainfalls within 9 h, followed on 26 Dec. 2013 by 13 mm of continuous rainfall over 17 h.

For both events, the amount of precipitation was almost identical for both sites. The selection criteria were a dry pre-event period of at least 1 week in order to compare events with comparable initial hydrological conditions. In that context, it is expected that a contrasting amount of rainfall allows to compare the hydrological response of the karst systems according to the recharge intensity. In the following, these two contrasting events will be referred to as "small event" and "large event", respectively.

3 Results

3.1 Physicochemical Response to a Small Rainfall Event

3.1.1 Fourbanne Site

Figure 2 (left) shows the physicochemical response of the Fourbanne site for the small flood event. Regarding water levels (WL) of the cave stream and the spring, the response was very similar with a response time of 4.5 and 7.5 h for the cave stream and the Fourbanne spring, respectively.

The response time of the other parameters characterizing solute (DOC, DO, EC) and particular transport (Turb.) was much longer, notably at the spring. The response time at the spring was delayed with respect to the cave stream, in particular for EC and Turb. Moreover, the large DOC peak at the spring is surprising because no significant variations occurred in the cave stream, suggesting that localized infiltration is not the main pathway for DOC transport to the karst outlet during small flood events.

3.1.2 Lods Site

Figure 2 (right) shows the physicochemical response of the Lods site during the same small flood event. The dynamic response of the Lods site for the small event was very similar as for the large event. The WL peak appeared 6 and 11 h after the rainfall peak at the epikarstic spring and at the Grand Bief spring, respectively.

It is interesting to observe that globally T, EC, DOC and Turb. peaks were measured before the WL peaks, during increasing discharge. This observation is true for both the epikarstic and the Grand Bief main spring.

The DOC signal at the main spring presented a first peak in phase with Turb. On the contrary to the epikarstic spring, it was followed by a second one, which was highly buffered and associated to a similar Temp. peak (EC was unfortunately not available). It means that the epikarstic signal may be captured at the Grand Bief spring at the beginning of the flood, but that additional processes controlled the global signal at the karst outlet.

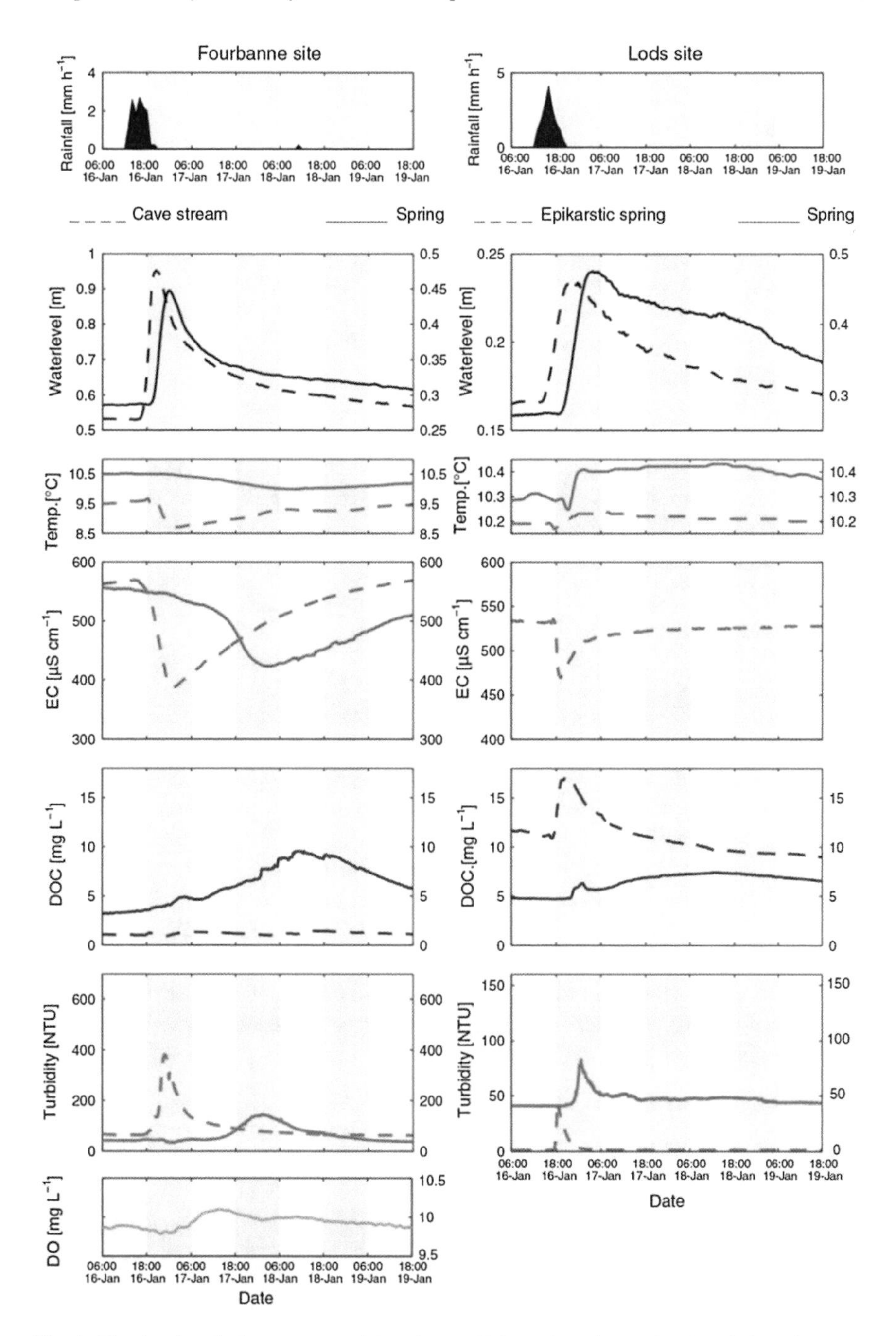

Fig. 2 Physicochemical responses of Fourbanne (*left*) and Lods (*right*) site during the small flood event of the 16 Jan. 2014

3.2 Physicochemical Response to a Large Rainfall Event

3.2.1 Fourbanne Site

Figure 3 (left) shows the physicochemical response of the Fourbanne site during the large flood event. The WL, T, and EC signal (dashed lines) of the Fontenotte cave stream reproduced clearly the two precipitation peaks: WL increased, whereas EC and T decreased, with most extreme values at about 6.5, 8, and 10 h after the first precipitation peak, respectively. In contrast, the Fourbanne spring at the karst outlet presented only a single and buffered WL peak about 34 h after the main rainfall peak.

Temperature signal started to drop 10 h after the first precipitation peak, reaching lowest values 6 h later. Unfortunately, no EC signal was available for the karst outlet. Turb. and DOC signals had contrasted responses with a time lag of 7 h between the two monitoring stations. The DOC response showed an original signal with a strong negative peak prior to a slight positive peak, suggesting that DOC was first diluted by infiltrating rainwater, followed by DOC input from surface soil.

3.2.2 Lods Site

Figure 3 (right) shows the physicochemical responses of Lods site during the large flood event. Globally, the response for this large event was very similar as for the small event but with higher peak values. The WL peak of the epikarstic spring and the main spring occurred 7 and 12 h after the first rainfall peak, respectively.

The EC response at the epikarstic spring showed a fast decrease when WL increased, followed by a fast return close to equilibrium. The negative EC dilution peak and the positive DOC peak appeared for the epikarstic spring and the main spring before the WL peak at 1.5 and 10 h after the first rainfall peak, respectively. It is interesting to note that the EC signal at the karst outlet showed, on the contrary to the Fourbanne site, first a small positive peak prior to strong dilution. This positive peak was well correlated with the increase of WL and Turb. and appeared just before the DO increase.

4 Discussion and Conclusion

The aim of this paper was a comparison of the physicochemical responses of two karst systems with allogenic (Fourbanne site) and a dominant autogenic recharge (Lods site). For both flood events, the sites presented distinct hydrodynamic and hydrochemical behaviors at the karst outlet. The differential analysis of the responses between the stations localized in the unsaturated zone and at the outlet,

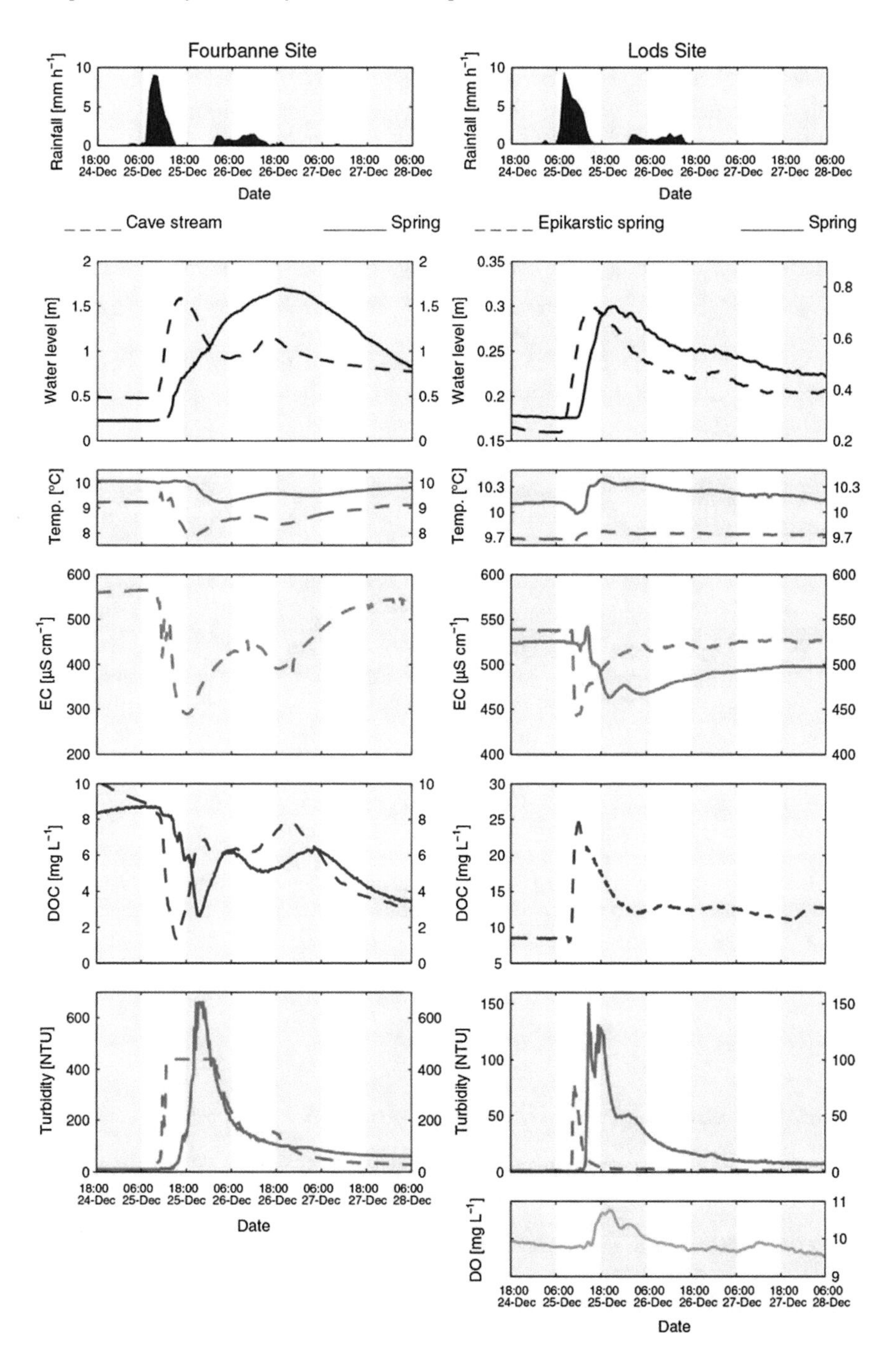

Fig. 3 Physicochemical responses of Fourbanne (*left*) and Lods (*right*) site during the large rainfall event of the 25 Dec. 2013

allowed us to identify some specific features relative to the type of recharge and relative to the role of the saturated zone.

The Fourbanne site was much more sensitive to rapid infiltration through a well-developed conduit network from the surface. Moreover, the variability of EC and Turb. along with the absolute Turb. values in the Fourbanne site, demonstrated the significant role of the allogenic recharge and its high contribution to the outlet. In the hydrodynamic response at the Lods site was slower than the response of EC, DOC, Temp., DO, and Turb. Surprisingly, the highest DOC concentrations occurred at the epikarstic spring. These observations suggest for this site a larger contribution from autogenic recharge.

From our results, a hydrogeological model may be drawn for both sites. Underground transfers in the Fourbanne site may be mainly of piston-type, with fast pressure transfer, and slower mass transfer through conduits within a voluminous saturated zone (mass transport required more than 1 day to run through the 6 km between the cave stream and the Fourbanne outlet for the small event, but only about 5–10 h for the large event). At the Lods site, the preservation of an epikarstic signal at the outlet suggests that the system is horizontally compartmented with a significant contribution of epikarstic water to total runoff. The high DOC and low EC values indicate that solutes coming from the epikarstic compartment were mainly derived from organic surface soils. This contrasting behavior of the Lods and Fourbanne site is supported by the water temperatures, which increased for both events at Lods, whereas they decreased at Fourbanne. In conclusion, this preliminary study furnishes a working hypothesis for the hydrological and physicochemical functioning of the two study sites with allogenic (Fourbanne site) and autogenic (Lods site) recharge mechanisms.

Acknowledgments This work was financed by the *Région Franche-Comté* and the *BRGM*. We thank Bruno Régent for his energy and technical expertise during the installation of the field sites, and the Speleological Group of Clerval Baume-les-Dames for their help during the instrumentation of the Fontenotte cave.

References

Becker A (2000) The Jura Mountains – an active foreland fold-and-thrust beld?. Tectonophysics 321:381–406

Charlier JB, Bertrand C, Mudry J (2012) Conceptual hydrogeological model of flow and transport of dissolved organic carbon in a small Jura karst system. J Hydrol 460–461:52–64

Charmoille A (2005) Traçage hydrochimique des interactions hydrauliques et mécaniques entre les volumes perméables et peu perméables au sein des aquifères fracturés carbonatés. PhD thesis, Université de Franche-Comté, Besançon, France

Fournier M, Massei N, Bakalowicz M, Dussart-Baptista L, Rodet L, Dupont JP (2007) Using turbidity dynamics and geochemical variability as a tool for understanding the behavior and vulnerability of karst aquifer. Hydrogeol J 15(4):689–704

Mudarra M, Andreo B, Barbera JA, Mudry J (2013) Hydrochemical dynamics of TOC and NO_3^- contents as natural tracers of infiltration in karst aquifers. Environ Earth Sci 71(2): 507–523

Pronk M, Goldscheider N, Zopfi J (2006) Dynamics and interaction of organic carbon, turbidity and bacteria in a karst aquifer system. Hydrogeol J 14:473–484

Schnegg P-A (2003) A new field fluorometer for multi-tracer tests and turbidity measurement applied to hydrogeological problems. In: Proceedings of the eight international congress of the Brazilian Geophysical Society, Rio de Janeiro

Tissier G, Perrette Y, Dzikowski M, Poulenars J, Hobléa F, Malet E, Fanget B (2013) Seasonal changes of organic matter quality and quantity at the outlet of a forested karst system (La Roche Saint Alban, French Alps). J Hydrol 482:139–148

Spatial and Temporal Hydrodynamic Variations of Flow in the Karst Vadose Zone (Rustrel, France) in Function of Depth and Fracturing Density

A. Barbel-Perineau, C. Emblanch and C. Danquigny

Abstract The hydrodynamical response at 45 flow points in the gallery of the Low Noise Underground Laboratory—carbonate aquifer (Rustrel, southern France) was monitored in order to determine the relationship between the hydrodynamical functioning of each flow component—slow, intermediate and quick, the depth and the fracturing density. Analysis of the relationship between the distribution of each flow component in function of depth and fracturing density in the karst vadose zone revealed the importance of (1) the variation of flow activation conditions with the depth, (2) the evolution of the flow component distribution with the depth and the fracturing density, and (3) the variability of the vadose zone role in supplying baseflow discharge, in function of carbonate thickness and fracturing state of the study area.

1 Introduction

The study of the flows in the vadose zone of a karst system needs some access to this part of the aquifer. Such kind of flow has been directly monitored since 2004 in the Laboratoire Souterrain à Bas Bruit (LSBB) located in Rustrel (South-East of France), an underground laboratory dug across the vadose zone of the Fontaine-de-Vaucluse karst aquifer.

This artificial gallery (3,800 m length) intersects arbitrarily fault networks from 30 to 500 m depth in carbonated rocks. Spatial and temporal hydrodynamical variations of 3 perennial and 42 temporary flows have been monitored throughout the gallery (Fig. 1), under variable and contrasting climatic conditions from 2004 to 2012 (Barbel-Perineau 2013). Within this period, all of the flow points have

A. Barbel-Perineau (✉) · C. Emblanch · C. Danquigny
UAPV-INRA, UMR 1114 EMMAH, 84914 Avignon, France
e-mail: aurore.perineau@alumni.univ-avignon.fr

B. Andreo et al. (eds.), *Hydrogeological and Environmental Investigations in Karst Systems*, Environmental Earth Sciences 1,
DOI 10.1007/978-3-642-17435-3_2

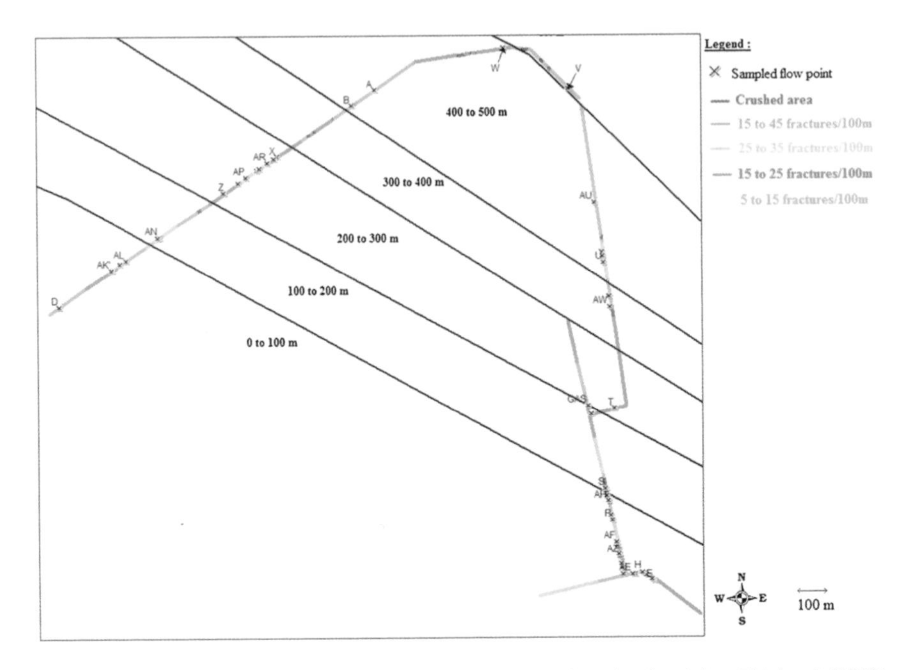

Fig. 1 Flow points location within the gallery and fracturing density (after Thiebaud (2003) modified)

been observed and measured at least once only during the 2008–2009 hydrological cycle. Thus in this study, only the 2008–2009 hydrological cycle is considered.

In previous studies (Perineau et al. 2010; Barbel-Perineau 2013; Perineau-Barbel et al. 2013), results highlight (1) an hydrodynamical flow classification in three components that have been characterized: a slow component, with permanent flows, regardless of the hydrological conditions (amount of rainfall); an intermediate component, with temporary flows, but these flows have some temporal flow continuity when they are active (few days to several months), and are assumed to obey a non linear hysteretic function; a quick component, with temporary flows too, but these flows have no temporal flow continuity (1 day to several days), they run only during important groundwater recharge and when the discharges of the permanent flows (slow component) are about or even reach their upper discharge levels. These three previous flow components are not specific to the study area because it has been shown (Barbel-Perineau 2013) that they match with the three flow components deduced from the numerical model of Tritz et al. (2011) in the vadose zone.

Results also demonstrate flow organization depending on depth and fracturing density; indeed the number of flow points decreases with depth. Moreover, results show that flows become permanent in depth and flows are concentrated in well-fractured and karstified areas, in which flow continuity is ensured (Perineau et al. 2010). It is now necessary to cross these previous results in order to study organization of these flow components in function of depth and geological layout of the vadose zone. Limestone is relatively homogeneous in this geological set, so the geological layout will be studied through fracturing density in the vadose zone.

Using the LSBB hydrodynamical data set, the objective of this paper is to study the relationship between the flow organization and the hydrodynamical functioning of each flow component with depth and fracturing density: which is the most important flow component in function of depth and fracturing density?

2 Data and Methods

The LSBB (http://www.lsbb.eu) is an underground gallery dug for a military purpose and converted into a research laboratory. It is hosted in the vadose zone limestone (Lower Cretaceous) of the Fontaine-de-Vaucluse Mounts, in Rustrel (France). The gallery is 3.8 km long. The rock's cover over the gallery varies from 0 to 519 m due to the topography. As the gallery comes across the karst medium, it intersects also with some flow paths through the vadose zone.

In this underground area, flows are accessible in many natural cavities (speleology) and in galleries of the LSBB which provide a readily access to them.

Between 2004 and 2008, drought period, (Barbel-Perineau 2013; Perineau-Barbel et al. 2013), only three permanents and two temporary flow points are observed and measured (Garry et al. 2008) at different depths. The rest of temporary flow points (40) was observed for the first time in winter 2008 during a heavy rainy period.

All the flow points (permanent and temporary) are located throughout the gallery (Fig. 1). This study is based on hydrodynamical flow measurements in a gallery. Its originality consists of: (i) the number of observation points (45) and (ii) flows that intersect randomly throughout a gallery and not caves, as it is often the case (e.g. Baldini et al. 2006; Fairchild et al. 2006).

To measure discharge rates of each flow point, the wall of the gallery is drilled to reach the rock and flows; then water inflows are concentrated in a spillway with a funnel. Discharge is weekly manually measured at the outlet of the funnel.

Discharge rates are weekly manually measured for many technical reasons: (i) for temporary flow points, it is difficult to set up and look after of around fifty pressure captors (dust, scaling and calibration problems), (ii) the three permanent flow points are equipped with pressure captor but scaling problems complicate the conversion of the water head in discharge with a calibrating curve.

Geological and geotechnical surveys made during the gallery drilling (CEBTP 1968) provide a characterization of lithology, faults, joints, cracks, karstification, and seepages. Moreover, the fracturing frequency (Fig. 1) was calculated by Thiebaud (2003) and the depth of each flow point was also calculated.

To have enough hydrodynamical data in the whole vadose zone, the gallery is divided into five depth classes of 100 m thick (Table 1). Thus for each depth class, the gallery length is large enough to be considered as representative (Table 1).

Table 1 Total linear length of gallery (m) per each depth class

	Total linear length of gallery per each depth class (m)	%
0–100 m	713	22
100–200 m	496	15
200–300 m	492	15
300–400 m	441	13
>400 m	1,138	35
Total	3,280	100

3 Results and Discussions

3.1 Flow Components Distribution in Function of the Depth

Table 2 indicates the flow point's distribution for each flow component in function of depth (five depth classes of 100 m thick) in the vadose zone. Values with a star depict two flow points with peculiar hydrodynamical characteristics:

- Regarding the slow component, the flow point between 0 and 100 m depth (D) is located near the epikarst (30 m depth); however, this flow point shows identical hydrodynamical and hydrochemical characteristics studied in details by Carrière (2014).
- Regarding the intermediate component, (i) the temporary flow point (C), situated between 200 and 300 m depth, is located at boundaries of two flow components, slow and intermediate, because when it is running during wet periods, it is almost permanent (up to 1 year) (ii) the flow point (W) located in depth (>400 m) is a temporary flow point, but when this flow point is running, its hydrodynamical and hydrochemical characteristics are strongly identically to hydrodynamical characteristics of slow component (Barbel-Perineau 2013). Thus, this temporary flow point is also located at boundaries of the two flow's components, slow and intermediate.
- Except the particular case of the D point, the slow component only circulates in depth. Even if this flow point is characterized as a slow component, its peculiar hydrodynamical features are linked to its epikarst position, or a lack of observation linked to a non-representativeness of the LSBB. However, outside the epikarst, it seems logical to find the slow flow component in depth.
- However, the quick component exists throughout the vadose zone, only during strong recharge periods (with large amount of rainfall and strong intensity) and only when the two other flow components are active (Barbel-Perineau 2013). So its hydrodynamical characteristics involve an hydrodynamical flow's behavior within the vadose zone less structured than the slow component. Note that in this study the quick component does not include direct runoffs from surface karst landing to the saturated zone (no doline, even… in the study area).

Table 2 Flow components distribution in function of depth (number of flow points in each depth class)

	0–100 m	100–200 m	200–300 m	300–400 m	>400 m
Quick component	13	7	5	4	1
Intermediate component	6	4	1*	–	1*
Slow component	1*	–	–	–	2
Total	20	11	6	4	4

Values with a star are specific values which are explain in the text

- Finally, the intermediate component shows a different distribution in the vadose zone, this flow component prevails in the firth 200 m depth, and then this component seems to disappear, or rather to become more hydrodynamically organized, i.e., to become permanent flows, typical of the slow component.

3.2 *Flow Components Distribution in Function of Depth and Fracturing Density*

Tables 3, 4 and 5 summarize, for each flow component, the ratio between the number of flow point and 100 m of gallery length, in each depth class and fracturing class, during the 2008–2009 hydrological cycle (within this period all of the flow points have been observed and measured at least once).

The fracturing density is also divided into five classes, from slightly fractured area (5–15 fractures/100 m) to crushed area. The comparison of Tables 3, 4, and 5 indicates that the intermediate component (Table 4) tends to preferentially gather in well-fractured, karstified areas, from the surface (to 200 m depth), on contrary to the two other flow's components. Regarding the peculiar temporary flow point W (see above), theoretically located in the 15–25 fracturing class, in depth (>400 m), more precisely this flow point is located in the gallery within a local small-scale crushed area (in the work of Thiebaud (2003), small-scale crushed areas have been indexed as single fractures). So to have an accurate table, the corresponding value ("0.09*") is written in corresponding cells "crushed areas" versus ">400 m."

The quick component (Table 3) circulates in the whole vadose zone, whatever the depth or fracturing density may be, because important hydraulic connectivity and strong pressure are both necessary to observe this flow's component running in the vadose zone (Barbel-Perineau 2013). Flows corresponding to the quick component circulate throughout well-developed drain paths which can get through the whole vadose zone. Nevertheless, as the slow component flows corresponding to the quick component tend to concentrate in depth in well-fractured areas, in which karstification process is well-developed.

Finally, the slow component (Table 5) is characterized almost exclusively by flow circulations in depth, in preferential flow paths, well fractured (except the flow point D as explained above).

Table 3 Quick component distribution in function of depth and fracturing density

	0–100 m	100–200 m	200–300 m	300–400 m	>400 m
5–15 fractures/100 m	0.14	**	0.00	0.00	0.09
15–25 fractures/100 m	0.98	0.20	0.41	0.00	0.00
25–35 fractures/100 m	0.56	0.81	0.61	0.45	0.00
35–45 fractures/100 m	0.14	0.40	**	0.23	0.00
Crushed areas	0.00	0.00	0.00	0.23	0.00

The two stars mean that the fracturing class is not represented in the depth class and zero corresponds to the lack of flow in depth classes

Table 4 Intermediate component distribution in function of depth and fracturing density

	0–100 m	100–200 m	200–300 m	300–400 m	>400 m
5–15 fractures/100 m	0.28	**	0.00	0.00	0.00
15–25 fractures/100 m	0.56	0.40	0.00	0.00	**
25–35 fractures/100 m	0.00	0.00	0.00	0.00	0.00
35–45 fractures/100 m	0.14	0.60	**	0.00	0.00
Crushed areas	0.00	0.00	0.20	0.00	0.09*

The two stars mean that the fracturing class is not represented in the depth class and a zero corresponds to the lack of flow in depth classes

Table 5 Slow component distribution in function of depth and fracturing density

	0–100 m	100–200 m	200–300 m	300–400 m	>400 m
5–15 fractures/100 m	0.14	**	0	0	0.00
15–25 fractures/100 m	0.00	0	0	0	0.00
25–35 fractures/100 m	0.00	0	0	0	0.00
35–45 fractures/100 m	0.00	0	**	0	0.18
Crushed areas	0.00	0	0	0	0.00

The two stars mean that the fracturing class is not represented in the depth class and a zero corresponds to the lack of flow in depth classes

Each flow component is organized in function of the state of the media, the quick component is a temporary component because flow through drains involve low water storage to supply flows of this component. The functioning of the intermediate component in function of depth and fracturing density tends to show a water supply, a support to the slow component. Intermediate flows could be actually considered as next slow flows, but not enough structured to drain a necessary volume of water storage to supply base flow.

Finally, the role of the vadose zone varies with the rock thickness, (i) the existence and/or the distribution of the three flow's components at the bottom of the vadose zone can be a function of both the karstification state and the vadose zone thickness of studied areas, (ii) permanent baseflow (slow component) within the vadose zone is linked to the thickness of considered vadose zone, which may be several hundred meters. For a long time, vadose zone is supposed to achieve a major role of storage and baseflow discharge supply (e.g., Emblanch et al. 2003; Perrin 2003; White 2006; Mudarra and Andreo 2011). Recently, this hypothesis has been indirectly demonstrated (e.g. Padilla et al. 1994; Charlier et al. 2012) and directly (e.g. Perrin 2003; Garry et al. 2008; Pronk et al. 2009; Barbel-Perineau 2013), with measurements within the vadose zone.

This study puts the role of the vadose zone into perspective. Indeed inside karstic aquifers with an important thickness of vadose zone, the vadose zone has a major hydrodynamic role, even during prolonged low-flow periods. Conversely inside karstic aquifers with a little thick vadose zone, its hydrodynamical role is less important. Thus, in future coming modeling studies of karst aquifers, it would be wise to adjust assigned coefficients of each karstic sub-system, especially the vadose zone, in function of several parameters, as thickness, karstification state, fracturing state, instead of set them arbitrarily (as e.g. Bezes 1976; Fleury et al. 2007).

4 Conclusion

Finally, this study highlights the distribution of flow components within the vadose zone in relation with the depth and the fracturing density: (1) the slow component is characterized almost exclusively by flow circulations in depth, in preferential flow paths, well-fractured areas. (2) The quick component circulates in the whole vadose zone, whatever the depth and the fracturing density may be, because an important hydraulic connectivity and strong pressure are both necessary to observe it. Flows corresponding to the quick component circulate throughout well-developed drain paths, which can get through the whole vadose zone, (3) tends to preferentially gather in well-fractured, karstified areas, from the surface (to 200 m depth), the intermediate component shows a particular distribution in the vadose zone, this flow component prevails in the firth 200 m depth, and then seems to disappear, or rather to become more hydrodynamically organized, i.e., to become permanent flows, typical of the slow component.

Thus, this study allows deducing the importance of (1) the variation of flow activation conditions with the depth and (2) the evolution of the flow component distribution with the depth and fracturing density, (3) the variability of the vadose zone role in supplying baseflow discharge, in function of carbonate thickness and fracturing state of the study area.

Acknowledgments Authors wish to thank the Platform for Fundamental and Applied Interdisciplinary Research, LSBB (www.lsbb.eu). The study is founded by the network of hydrogeological researches sites H+ (www.hplus.ore.fr).

References

Baldini J, McDermott F, Fairchild I (2006) Spatial variability in cave drip water hydrochemistry: Implications for stalagmite paleoclimate records. Chem Geol 235(3–4):390–404

Barbel-Perineau A (2013) Caractérisation du fonctionnement de la zone non saturée des aquifères karstiques. Approche directe par études hydrodynamiques et hydrochimiques sur le BREO de FdV-LSBB. PhD Thesis, Univerity d'Avignon et des Pays de Vaucluse, 223 p

Bezes C (1976) Contribution à la modélisation des sysèmes karstiques: établissement du modèle Bemer: son application à quatre systèmes karstiques du Midi de la France. PhD thesis, University of Montpellier

Carrière SD (2014) Etude hydrogéophysique de la structure et du fonctionnement de la zone non saturée du karst. PhD thesis, Université d'Avignon, 210 p

CEBTP (1968) Mont Ventoux, Rustrel, PCT 1, Etude géotechnique. Dossier SF 67/223.6.039

Charlier J-B, Bertrand C, Mudry J (2012) Conceptual hydrogeological model of flow and transport of dissolved organic carbon in a small Jura karst system. J Hydrol 460:52–64

Emblanch C, Zuppi GM, Mudry J, Blavoux B, Batiot C (2003) Carbon 13 of TDIC to quantify the role of the unsaturated zone: the example of the Vaucluse karst systems (Southeastern France). J Hydrol 279(1–4):262–274

Fairchild IJ, Tuckwell GW, Baker A, Tooth AF (2006) Modelling of dripwater hydrology and hydrogeochemistry in a weakly karstified aquifer (Bath, UK): implications for climate change studies. J Hydrol 321(1–4):213–231

Fleury P, Plagnes V, Bakalowicz M (2007) Modelling of the functioning of karst aquifers with a reservoir model: application to Fontaine de Vaucluse (South of France). J Hydrol 345(1–2):38–49

Garry B, Blondel T, Emblanch C, Sudre C, Bilgot S, Cavaillou A, Boyer D, Auguste M (2008) Contribution of artificial galleries to the knowledge of karstic system behaviour in addition to natural cavern data. Int J Speleol 37(1):75–82

Mudarra M, Andreo B (2011) Relative importance of the saturated and the unsaturated zones in the hydrogeological functioning of karst aquifers: the case of Alta Cadena (Southern Spain). J Hydrol 397(3–4):263–280

Padilla A, Pulido-Bosch A, Mangin A (1994) Relative importance of baseflow and quickflow from hydrographs of karst spring. Ground Water 32(2):267–277

Perineau-Barbel A, Danquigny C, Emblanch C (2013) Hydrodynamic characterization of flows in the vadose zone by direct measurements in karst aquifer. In: Paper presented at the international symposium on hierarchical flow systems in Karst Régions, Budapest, Hungary (4–7 September 2013)

Perineau A, Danquigny C, Emblanch C, Pozzo di Borgo E, Boyer D, Poupeney J (2010) Hydrodynamic organisation of the flows in the unsaturated zone of the Fontaine de Vaucluse karst system. First results. In: Inter-disciplinary underground science and technology (i-DUST) conferences, Apt, France

Perrin J (2003) A conceptual model of flow and transport in a karst aquifer based on spatial and temporal variations of natural tracers. University of Neuchâtel

Pronk M, Goldscheider N, Zopfi J, Zwahlen F (2009) Percolation and particle transport in the unsaturated zone of a karst aquifer. Ground Water 47(3):361–369

Thiébaud E (2003) Etude structurale et hydrogéologique du site du Laboratoire Souterrain à Bas Bruit de Rustrel, Rapport de Master, Université de Franche-Comte, Besançon, 50 p

Tritz S, Guinot V, Jourde H (2011) Modelling the behaviour of a karst system catchment using non-linear hysteretic conceptual model. J Hydrol 397(3–4):250–262

White WB (2006) Conceptual models for carbonate aquifers. Ground Water 7(3):15–21

Characterization of the Functionality of Karstic Systems Based on the Study of the SI_c–Pco_2 Relation

S. Minvielle, N. Peyraube, R. Lastennet and A. Denis

Abstract This paper aims at characterizing the functionality of karstic systems. In order to do so, an extend of a method based on the saturated index with respect to calcite (SI_c) and CO_2 partial pressure (Pco_2) is used. The initial method already applied by Peyraube et al. (2012) uses the Pco_2 of water at equilibrium with atmosphere (Pco_{2_eq}) and the Pco_2 of water for a SI_c equal to zero (Pco_{2_sat}). The Pco_{2_sat} variation gives information on flows conditions within the karstic system. It describes the degree of karstification of the system. Systems with developed saturated zone are characterized by under-saturated waters as well as saturated ones with variable Pco_{2_sat}. At last, fissured system has Pco_{2_sat} which vary only slightly. Moreover, variations of Pco_{2_sat} show homogenization capacities of a karst.

1 Introduction

Degree of karstification can be considered from a functional aspect: from flows of waters. Fissured karst outlets possess waters with few variation of their chemistry. These systems have high capacities to homogenize types of waters. In opposite, highly karstified systems springs are characterized by different types of waters flowing from springs. Then the importance of karst vulnerability can be

S. Minvielle (✉) · R. Lastennet · A. Denis
University of Bordeaux, I2M-GCE - CNRS 5295, Talence, France
e-mail: sebastien.minvielle@u-bordeaux.fr

R. Lastennet
e-mail: roland.lastennet@u-bordeaux.fr

A. Denis
e-mail: alain.denis@u-bordeaux.fr

N. Peyraube
University of Orléans, ISTO, Orléans, France
e-mail: nicolas.peyraube@cnrs-orleans.fr

B. Andreo et al. (eds.), *Hydrogeological and Environmental Investigations in Karst Systems*, Environmental Earth Sciences 1,
DOI 10.1007/978-3-642-17435-3_3

approximated by the study of mass transfers. The aim of this paper is to estimate degree of karstification using SI_c–Pco_2 method (Peyraube 2012), which has not been used to characterize it.

Three karstic systems, whose recharge areas are the same size, are considered (Fig. 1): Notre Dame des Anges, Font Marin and Groseau. These springs are situated around the catchment area of the karstic system of Fontaine de Vaucluse. Notre Dame des Anges and Font Marin springs belong to Vauclusian-karstic systems.

Notre Dame des Anges and Groseau outlets are located in the Barremian and Bedoulian (lower Cretaceous) limestones, characterized by an Urgonian facies. Waters of the Font Marin spring emerge from geologic formations including Barremian limestones and its substratum composed of Valanginian marls and lower Hauterivian clayey limestones. Unlike previous springs, Saint Trinit spring is an epikarstic outlet of a system, which is composed of Barremian and Bedoulian-aged limestones.

2 Method

In this paper, calcium-carbonate equilibrium will be consider using the Henry's constant (K_0) and temperature dependency constants (K_1, K_2, K_c) from Plummer and Busenberg (1982). The SI_c–Pco_2 method is exhaustively described by Peyraube et al. (2012).

The carbon dioxide partial pressure with which water is at atmospheric equilibrium (Pco_{2_eq}) is determined based on activity of bicarbonate ion ((HCO_3^-)), measured pH (pH_m) and equilibrium constants K_0 and K_1:

$$\log\left(Pco_{2\_eq}\right) = \log\left(\left(HCO_3^-\right)\right) - pH_m - \log(K_0 \cdot K_1) \tag{1}$$

In addition, the saturation index with respect to calcite (SI_c) can be determined. This parameter characterizes the tendency of water to precipitate ($SI_c > 0$) or dissolve ($SI_c < 0$) calcite. The SI_c is defined by:

$$SI_c = -\log\left(Pco_{2\_eq}\right) + 2 \cdot \log\left(\left(HCO_3^-\right)\right) + \log\left(\left(Ca^{2+}\right)\right) + \log\left(\frac{K_2}{K_0 \cdot K_1 \cdot K_c}\right) \tag{2}$$

where (X) is the activity of the X element.

Equation (2) denotes a linear relation between SI_c and $\log(Pco_{2\_eq})$. This behavior can be illustrated by a line of slope 1 named gassing-degassing line (G&D line). When G&D line intercepts the x-axis, the Pco_2 value represents the Pco_2 (Pco_{2_sat}) for which water is in equilibrium with respect to calcite ($SI_c = 0$). Pco_{2_sat} for a sample is defined by:

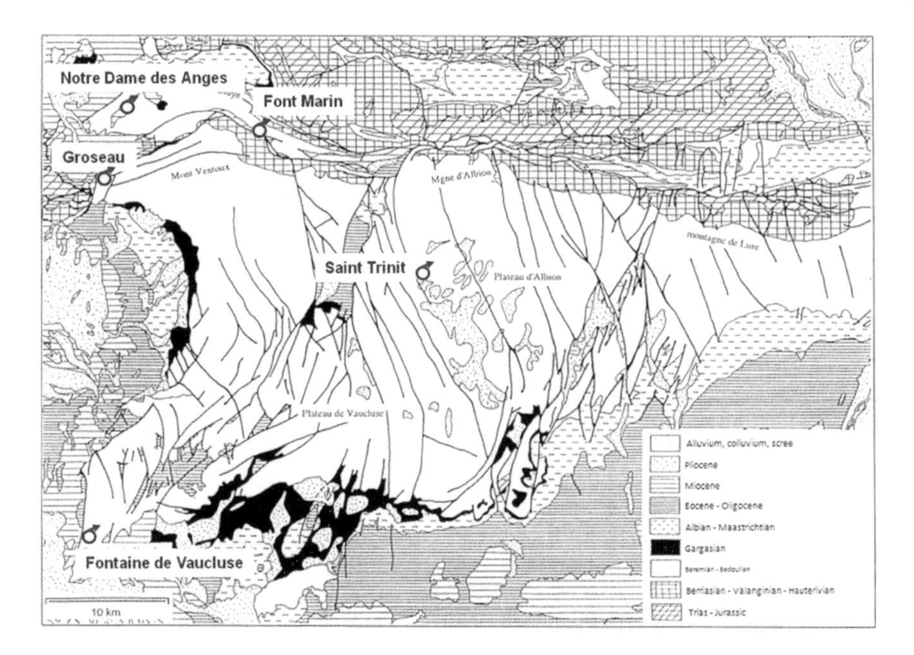

Fig. 1 Geologic context of the study (from Couturaud 1993)

$$\mathrm{Pco}_{2\_\mathrm{sat,sample}} = 10^{\wedge}\left(2 \times \log\left(\mathrm{HCO}_3^-\right) + \log\left(\mathrm{Ca}^{2+}\right) + \log\left(\frac{\mathrm{K}_2}{\mathrm{K}_0\mathrm{K}_1\mathrm{K}_c}\right)\right) \quad (3)$$

Similarly to $\mathrm{Pco}_{2\_\mathrm{sat,sample}}$, a $\mathrm{Pco}_{2\_\mathrm{sat}}$ for a spring $\mathrm{Pco}_{2\_\mathrm{sat,spring}}$ (thereafter only called $\mathrm{Pco}_{2\_\mathrm{sat}}$) can be calculated. It corresponds to the average of $\mathrm{Pco}_{2\_\mathrm{sat}}$ of individual waters of the outlet. This parameter is intrinsic to the spring and allows to compare springs each others and characterize them.

3 Results and Discussion

In order to distinguish types of waters, a PCA is realized based on 175 samples (65 from Notre Dame des Anges spring, 56 from Font Marin spring and 54 from Groseau spring) and six variables (bicarbonate, magnesium, silica and chlorides concentrations, $\mathrm{SI_c}$ and $\mathrm{Pco}_{2\_\mathrm{eq}}$). Variations of Saint Trinit waters on the six variables are high. Analysis of PCA cannot be done if these samples are used to establish PCA since samples dispersions of Font Marin, Groseau and Notre Dame des Anges springs are not visible on factorial planes. Consequently, water samples of Saint Trinit spring are projected on the samples space and do not contribute to the establishment of the PCA. Results of PCA can be seen though the variable space (Fig. 2) and the sample space (Fig. 3). The first factorial plane (F1–F2) is the best-described one for variables. These two principal components explain 73.44 % of the total variance.

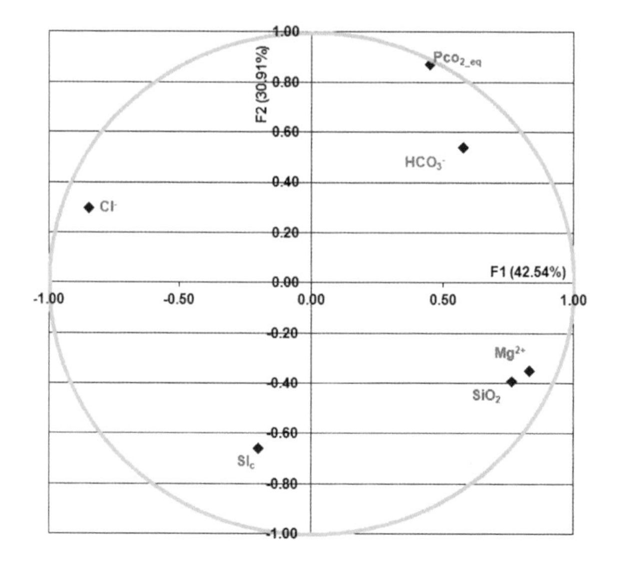

Fig. 2 Variable space in the first factorial plan based on waters spring of Notre Dame des Anges ($N = 65$), Font Marin ($N = 56$) and Groseau ($N = 54$) (1991–1992)

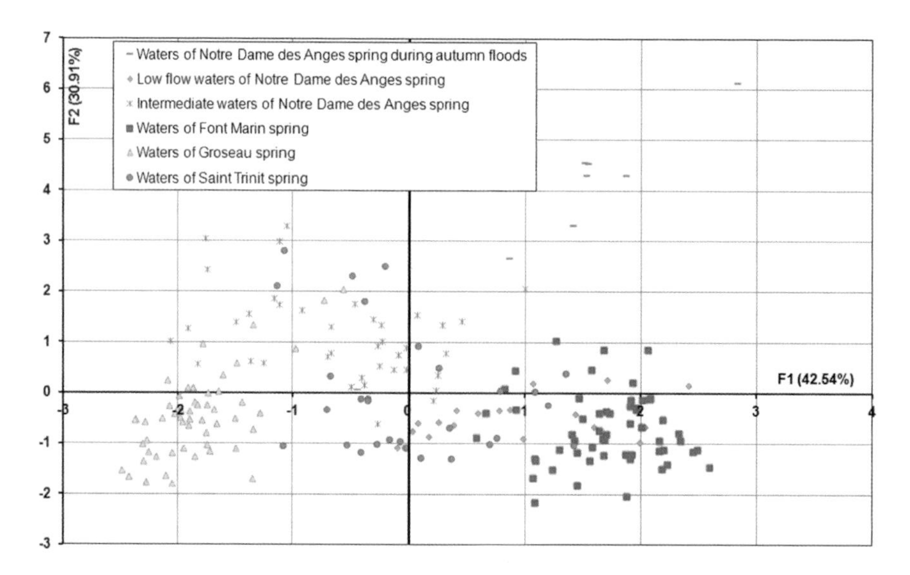

Fig. 3 Sample space in the first factorial plan based on waters springs of Notre Dame des Anges ($N = 65$), Font Marin ($N = 56$) and Groseau ($N = 54$), waters of Saint Trinit spring ($N = 29$) are projected in this space (1991–1992)

The first axis ($F1$) is associated to HCO_3^-, Mg^{2+} and SiO_2 in its positive part, Cl^- in its negative part (Fig. 2). It represents the residence time of water. In its positive part, waters tend to have a longer residence time whereas in its negative part, shallow water participation increases. The second axis ($F2$) exposes the opposition between Pco_{2_eq} in the positive part of the axis and SI_c in the negative part.

Waters of Notre Dame des Anges spring (Fig. 3) is expanded along three poles: Pco_{2_eq}, Mg^{2+} and SiO_2, and Cl^-. During autumn floods, samples shift toward the Pco_{2_eq} pole. A majority of low-flow waters moves to the pole {Mg^{2+}; SiO_2}. This trend characterizes waters of saturated zone with high magnesium and silica contents. Finally, waters with highest negative values on F1-axis showed an important contribution of shallow waters. This participation decreases as waters recede from the Chlorides pole.

Others waters spring do not exhibit several types of waters. Samples of Groseau spring show no variations along {Mg^{2+}; SiO_2} poles. Only few variations of the Pco_{2_eq} pole can be noticed. This is related to gassing or degassing of water. This stability of the chemistry of residence time markers highlights homogenization capacities of the system. Then the Groseau spring tends to be an outlet of a fissured aquifer than a karstic aquifer.

The spring of Groseau (Fig. 4) has the weakest model Pco_{2_sat} value (0.55 %) and the weakest variation range of waters Pco_{2_sat} (between 0.50 and 0.60 %). Waters emerging from Notre Dame des Anges spring have the widest range of variation of Pco_{2_sat} (from 0.74 to 2.08 %) with a model Pco_{2_sat} of 1.06 %. The model Pco_{2_sat} value of Font Marin spring is 0.91 % with an extensive of waters Pco_{2_sat} of 0.50 % (from 0.59 to 1.09 %). The last spring of this site is Saint Trinit spring. It is characterized by waters with a range of variation of Pco_{2_sat} from 1.16 to 1.98 %. The model Pco_{2_sat} of this system is 1.50 %.

Groseau system possesses the lowest and the weakest range of Pco_{2_sat}. PCA indicates that Groseau spring is an outlet of a fissured aquifer than a karstic aquifer. Waters of fissured karsts have low range of variation of Pco_{2_sat} related to their homogenization capacities.

Differences in Pco_{2_sat} are related to soil (nature, type, occupation) and structure of the karstic system. Groseau and Notre Dame des Anges waters springs emerge from the saturated zone of their own system. Nevertheless, Pco_{2_sat} of Groseau spring is smaller than this of Notre Dame des Anges spring. Intake area of Groseau system is located at an altitude of 1,100 m, higher than supply area of Notre Dame des Anges spring. In the Vaucluse county, high uplands are characterized by a vegetation depletion and thin soils. Consequently, in altitude, soil CO_2 is low. As a result, water has a weak Pco_{2_sat}.

However, intake area of Saint Trinit system is higher than those of Notre Dame des Anges and Font Marin springs. The Pco_{2_sat} of Saint Trinit is the highest. Saint Trinit is an epikarstic spring which is then located near the production zone of CO_2. In addition, intake area of this system undergoes agricultural pressures which can increase soil Pco_2 and then Pco_{2_sat} of soil waters. More the water moves away the CO_2 production zone and remains in the transmission zone, more the Pco_{2_sat} in the water decreases. Therefore, Pco_{2_sat} in Saint Trinit waters are high.

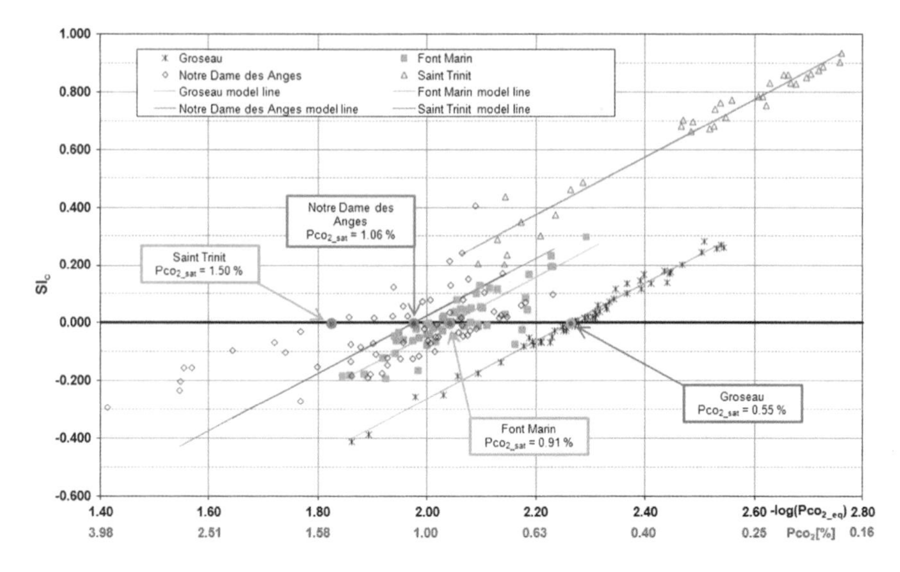

Fig. 4 Representation in a ($-\log(Pco_{2\_eq})$; SI_c) graph of waters spring of Groseau ($N = 54$), Font Marin ($N = 56$), Notre Dame des Anges ($N = 65$) and Saint Trinit ($N = 29$) (1991–1992)

Variability of Saint Trinit Pco_{2_sat} is related to CO_2 production (Cowan et al. 2013) and agricultural supplies.

SI_c are weaker when moisture state of soil/epikarst is high. Consequently, variations of waters Pco_{2_eq} (gassing or degassing) and SI_c in this epikarstic spring are due to moisture content in soil/epikarst.

PCA divides waters of Notre Dame des Anges spring in three water types: one representing low-flow waters, one occurring during autumn floods and the last deriving from a mix between the two previous poles. SI_c–Pco_2 method distinguishes waters of autumn floods too. To complete the description of Notre Dame des Anges, the three groups identified in the PCA are carried over the ($-\log(Pco_{2\_eq})$; SI_c) graph (Fig. 5).

Unlike PCA distinction, sample of the 22/09/1992—with coordinates of (2.09; 0.406) in Fig. 5—has been added to autumn flood waters. Samples of the 11/10/1991 and the 09/09/1992—with coordinates respectively of (2.04; 0.038) and (1.77; −0.268) in Fig. 5—have not been used to determine G&D line of low-flow waters and flood waters due to their remoteness to the others samples. These values are considered outlier, probably due to an error of pH measurement.

Waters of Notre Dame des Anges spring flowing during autumn floods are aligned along G&D model line of Saint Trinit; they have similar Pco_{2_sat}. Then Notre Dame des Anges outlet is feeding by epikarstic supply flowing quickly through the system.

Low-flow waters have the lowest Pco_{2_sat}. Like autumn flood waters, low-flow waters are aligned along a G&D line with a 0.81 % Pco_{2_sat} value. The Pco_{2_sat} indicating by the G&D line of autumn flood waters is 1.85 %.

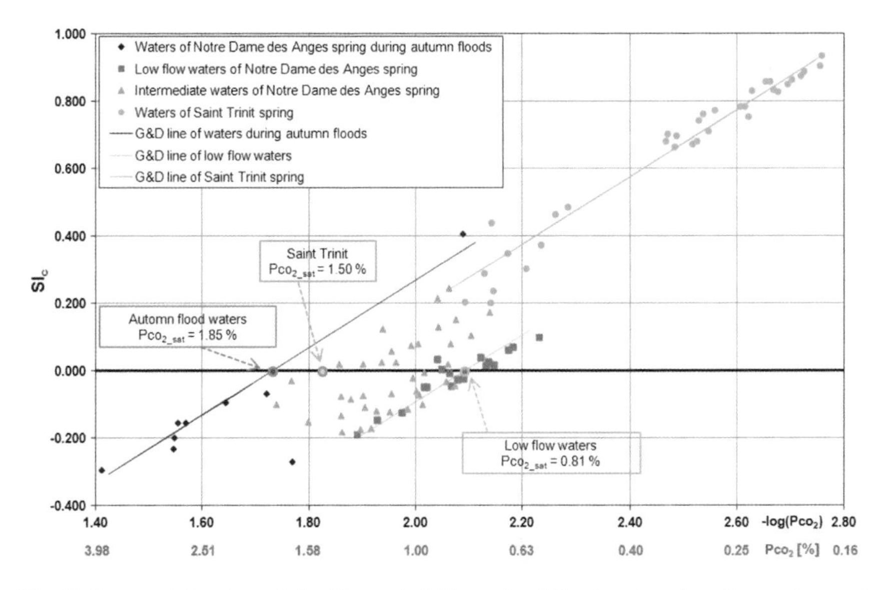

Fig. 5 Representation in a ($-\log(Pco_{2\_eq})$; SI_c) graph of flood waters, low-flow waters and intermediate waters of Notre Dame des Anges spring, and waters of Saint Trinit spring (1991–1992)

Intermediate waters are essentially distributed between these two values. They are mixes between the two previous poles. Their Pco_{2_sat} depend on Pco_{2_sat} of the two end-members. Some of these intermediate waters (hereafter named group A) have a Pco_{2_sat} near those of low-flow waters. Nevertheless, their SiO_2 and Mg^{2+} contents are lower than in low-flow waters (Table 1). Therefore, time residence in contact with the massif is sufficient to equilibrate waters with air Pco_2 of the massif but not enough sufficient to mineralize large amounts of Mg^{2+} and SiO_2.

In flood periods, epikarstic waters replace low-flow waters. Such differentiation implies that waters in the system maintain the chemical signature, which presents during low-flow. Consequently, saturated zone contains stocks of waters which feed the spring during low-flow periods. This supports the fact that the system of Notre Dame des Anges is highly karstified. Two end-members are then distinguishable, characterized by their Pco_{2_sat}. Waters from the transmission zone have a chemical composition which corresponds to a mix between these two poles.

Waters of Font Marin spring are not divided into several groups as in Notre Dame des Anges spring. Then the system of Font Marin has more capacities to homogenize waters. Consequently, more the variation range of Pco_{2_sat} is wide, more the system is characterized by different types of waters. A wide variation range of Pco_{2_sat} described a high karstified system. Such classifications of Notre Dame des Anges and Font Marin systems agree with those of Lastennet and Mudry (1997) and Batiot et al. (2003).

Table 1 Magnesium and silica contents (in mg/L) for group A waters ($N = 12$) and low-flow waters ($N = 17$) (1991–1992)

	Magnesium (mg/L)			Silica (mg/L)		
–	Minimum	Maximum	Average	Minimum	Maximum	Average
Group A	3.68	5.60	4.74	5.15	6.25	5.91
Low-flow waters	4.92	6.18	5.73	6.31	7.76	6.84

4 Conclusions

Degree of karstification can be qualitatively obtained by the SI_c–Pco_2 method. Pco_{2_sat} brings information about origins of waters and flowing conditions in the system. A little range of variations of waters Pco_{2_sat} describes a fissured system or a highly karstified system. Large range of variations of waters Pco_{2_sat} characterizes an evolved system in terms of karstification. Karst is highly compartmentalized with low homogenization capacities. End-members representing epikarst and saturated zone are then identifiable. If Pco_{2_sat} dispersion does not exhibit end-members, system is described as having an average karstification.

References

Batiot C, Emblanch C, Blavoux B (2003) Total organic carbon (TOC) and magnesium (Mg^{2+}): two complementary tracers of residence time in karstic systems. C R Geosci 335:205–214

Couturaud A (1993) Hydrogéologie de la partie occidentale du système karstique de Vaucluse, Thèse doctorale Université d'Avignon

Cowan D, Osborne MC, Banner JL (2013) Temporal variability of cave-air CO_2 in Central Texas. J Cave Karst Studies 75:38–50

Lastennet R, Mudry J (1997) Role of karstification and rainfall in the behavior of heterogeneous karst system. Environ Geol 32(2):114–123

Peyraube N, Lastennet R, Denis A (2012) Geochemical evolution of groundwater in the unsaturated zone of karstic massif, using the Pco_2–SI_c relationship. J Hydrol 430:13–24

Plummer LN, Busenberg E (1982) The solubility of calcite, aragonite and waterite in CO_2–H_2O solutions between 0 and 90 °C, and evaluation of the aqueous model of the system $CaCO_3$–CO_2–H_2O. Geochim Cosmochim Acta 46:1011–1040

Contribution of Hydrogeological Time Series Statistical Analysis to the Study of Karst Unsaturated Zone (Rustrel, France)

C. Ollivier, C. Danquigny, N. Mazzilli and A. Barbel-Perineau

Abstract Characterising the hydrodynamic processes in the unsaturated zone is a prerequisite to efficiently protect and manage the karst water resource. In this study, we use rainfall-discharge cross-correlation analysis to improving the understanding of the hydrodynamic functioning of the karst unsaturated zone. This analysis is based on sparse discharge measurements from five flow points located within the artificial LSBB gallery. Our results are consistent with hydrochemical analyses, and they evidence the influence of inter-annual rainfall variability on the hydrological functioning of the unsaturated zone. They also show that cross-correlation may prove an efficient tool for sparse data analyses.

1 Introduction

This work focuses on the hydrogeological behaviour of preferential water flow pathways through karst unsaturated zone. It benefits from an almost unique experimental facility, the Low-Noise Underground Research Laboratory of Rustrel—Pays d'Apt (LSBB, www.lsbb.eu) which is a 3.7 km artificial gallery dug in the unsaturated zone of the karst system of "Fontaine-de-Vaucluse" (SE France, Fig. 1). Through a typical Cretaceous karstified carbonate platform, this gallery arbitrarily intersects tectonic and karstified features from 0 to 519 m in depth. Since 2003, the water flow rate of up to 53 water flow points has been monitored on a weekly basis. Based on this monitoring, a classification of the water flow points into three groups with (i) diffuse, (ii) intermediate, and (iii) fast flows behaviours, has been proposed (Barbel-Perineau 2013). The hydrodynamic behaviour of the monitored flow points is dependent on both depth and fracturing

C. Ollivier (✉) · C. Danquigny · N. Mazzilli · A. Barbel-Perineau
UMR1114 EMMAH, UAPV, 84914 Avignon, France
e-mail: chloe.ollivier@alumni.univ-avignon.fr

C. Ollivier · C. Danquigny · N. Mazzilli · A. Barbel-Perineau
UMR1114 EMMAH, INRA, 84914 Avignon, France

B. Andreo et al. (eds.), *Hydrogeological and Environmental Investigations in Karst Systems*, Environmental Earth Sciences 1,
DOI 10.1007/978-3-642-17435-3_4

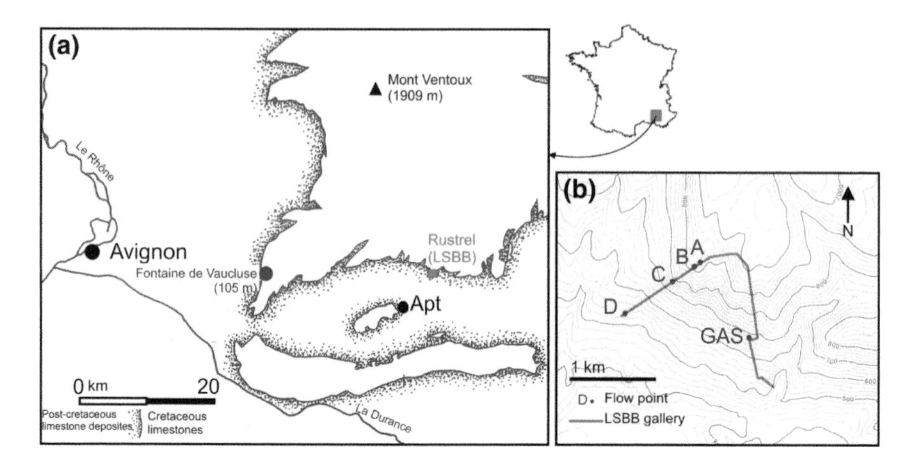

Fig. 1 Situation map for the LSBB (southern France). **a** Site location compared to the recharge area of the "Fontaine-de-Vaucluse" spring. **b** Position of the monitored flow points (*A*, *B*, *C*, *D*, *GAS*) along the LSBB gallery (*red line*). The *solid lines* indicate topographic line (m ASL)

density (Barbel-Perineau et al. 2013). The statistical analysis of spring discharge time series is a classical tool for the hydrogeological investigations of karst aquifers (e.g. Mangin 1984; Padilla and Pulido-Bosch 1995; Labat et al. 2000; Massei et al. 2006). Most statistical approaches require either a high sampling frequency (quasi-continuous sampling), or a regular sampling scheme. However, these requirements are often not fullfilled due to operational constraints and costs, in particular for hydrochemical time series. In this work, we use cross-correlation to analyse the hydrodynamic behaviour of flow points located within the LSBB gallery: (i) in the unsaturated zone, (ii) with a low-frequency, highly heterogeneous sampling scheme and (iii) considering the influence of antecedent rainfall conditions.

2 Materials and Methods

2.1 Basis of Cross-Correlation Analysis

Cross-correlation analysis considers the correlation of the time series, applying a lag time to one of them. In this study, we use Pearson product-moment correlation coefficient R for the analysis of the rainfall-discharge linear relationship (Padilla and Pulido-Bosch 1995):

$$R(t) = \frac{\sum_{i=1}^{n} (Q_{i+t} - \bar{Q})(P_i - \bar{P})}{\sqrt{\sum_{i=1}^{n} (Q_{i+t} - \bar{Q})^2 \sum_{i=1}^{n} (P_i - \bar{P})^2}} \quad (1)$$

where t is the lag time between the two series, Q is the discharge time series, P the rainfall time series, n the sample length.

2.2 Discharge and Rainfall Time Series

In this study, we consider the 2003–2013 discharge time series for the five most sampled water flow points identified in the unsaturated zone along the LSBB gallery: points A, B, C, D, and GAS. All these points have approximately the same elevation, but they are located at different depth (considered from the local topography, Fig. 1): "A" is located at −442 m, "B" at −421 m, "C" at −256 m, "D" at only −33 m, "GAS" at −191 m. The sampling step is an average 11 days but with a very irregular scheme. In particular, measurements frequency is increased during rainy periods. For these points, we consider the relationship with the daily rainfall measured at the nearest station (Saint-Saturnin-Les-Apt, 8 km away from the LSBB field site). A preliminary assessment of the effect of the low-frequency, irregular sampling scheme is performed using the daily discharge time series at the outlet of the karst system (Fontaine-de-Vaucluse spring).

3 Results and Discussion

3.1 Comparison of Different Sampling Schemes for the Fontaine-de-Vaucluse Time Series

Figure 2 shows the discharge time series (Fig. 2a, c, e, and g) and the rainfall-discharge cross-correlation analysis (Fig. 2b, d, f, and h) for the Fontaine-de-Vaucluse spring (2004–2012), based on 4 different sampling schemes. The maximum of the correlation coefficient is reached for a lag time of 6 days, which means that the linear component can be detected at the watershed scale (Fig. 2b). This component is quickly attenuated by more complex components. Applying the sampling scheme from the LSBB flow points to the Fontaine-de-Vaucluse time series leads to a degradation of the cross-correlation; however, both the shape and the average value of the cross-correlation function are conserved (Fig. 2d). A linear reconstruction of the discharge time series (Fig. 2e) reduces the dispersion of the cross-correlation function, but the maximum of the correlation is widened (Fig. 2f). A nearest neighbour reconstruction of the discharge time series (Fig.2g) induces a shift in the maximum of the cross-correlation function (Fig. 2h).

It stems from this comparison that: (i) a low frequency, irregular sampling scheme leads to a degradation of the cross-correlation function, in particular due to a higher dispersion of the correlation coefficients, but it may preserve the shape of the cross-correlation function; (ii) time series reconstruction methods may bias the cross-correlation function; (iii) fast hydrodynamic of karstic flow point may not be reproduced with suggested reconstruction methods with such sparse data. In the following, we use no time series reconstruction methods before the cross-correlation analysis.

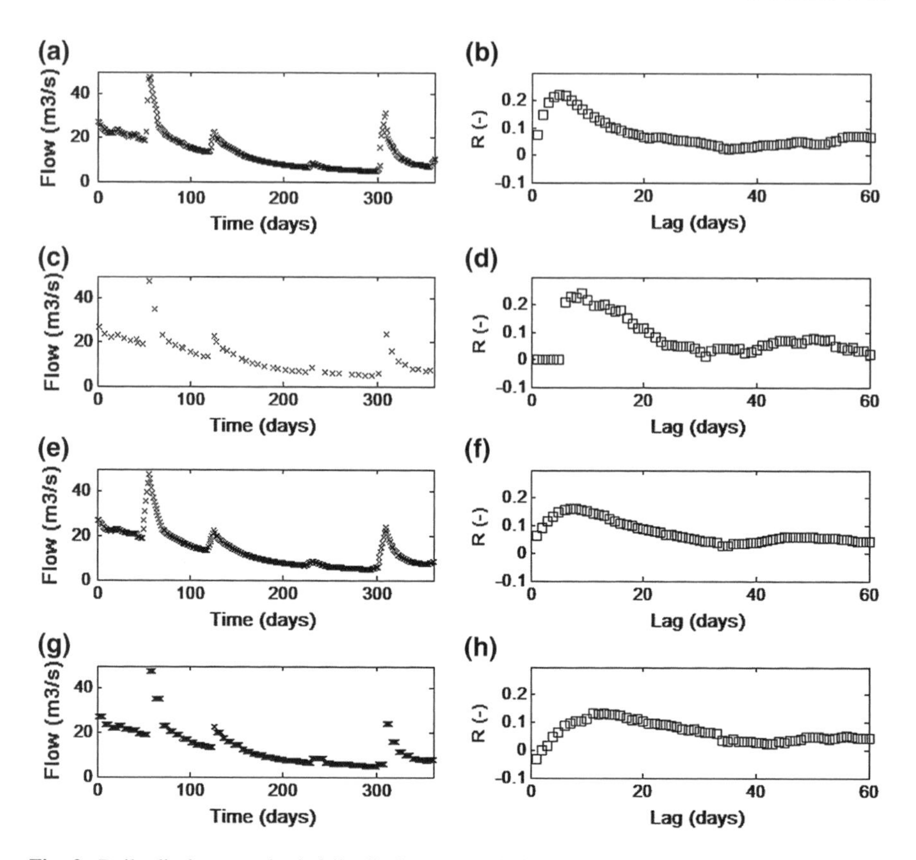

Fig. 2 Daily discharge and rainfall–discharge correlation at the Fontaine–de–Vaucluse, based on different sampling schemes: **a** Daily discharge sampling and **b** the cross–correlation with rainfall–discharge.; **c** LSBB flow points sampling scheme of the flow and **d** associated cross-correlation.; **e** Linear reconstruction of the discharge time series and **f** associated cross-correlation.; **g** Nearest neighbour reconstruction of the discharge time series and **h** associated cross-correlation. For clarity purpose, discharge time series are focused in the year 2004

3.2 Comparison of Rainfall-Discharge Cross-Correlation Functions Under Different Hydrological Conditions

The cross-correlation analysis is repeated for two hydrological cycles (2005–2006 and 2011–2012) that have almost equal cumulative rainfall amount but different wetness antecedent conditions. Figure 3 shows the cumulative rainfall for hydrological years from 2003–2004 to 2012–2013. Both 2005–2006 and 2011–2012 cumulative rainfall is roughly equal to 585 mm. However, the 2003–2005 cumulative rainfall is approximately 1,200 mm whereas the 2009–2011 cumulative rainfall reaches 1650 mm.

Figure 4 shows the cross-correlation functions for the rainfall-discharge relationship at the Fontaine de Vaucluse spring and the LSBB flow points, for the

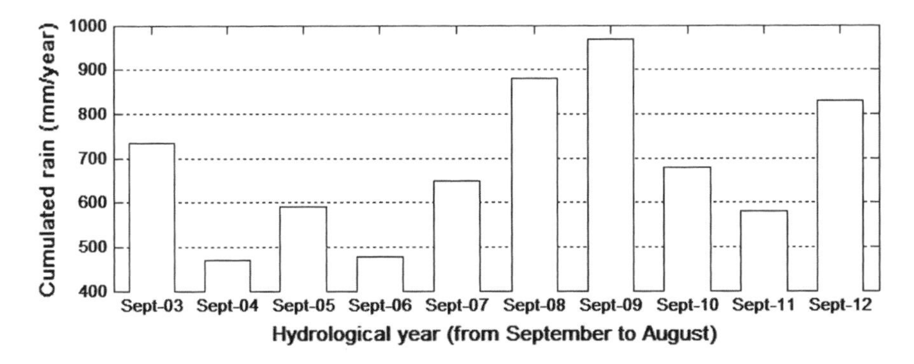

Fig. 3 Cumulative rainfall for the 2004–2005 to 2012–2013 hydrological years

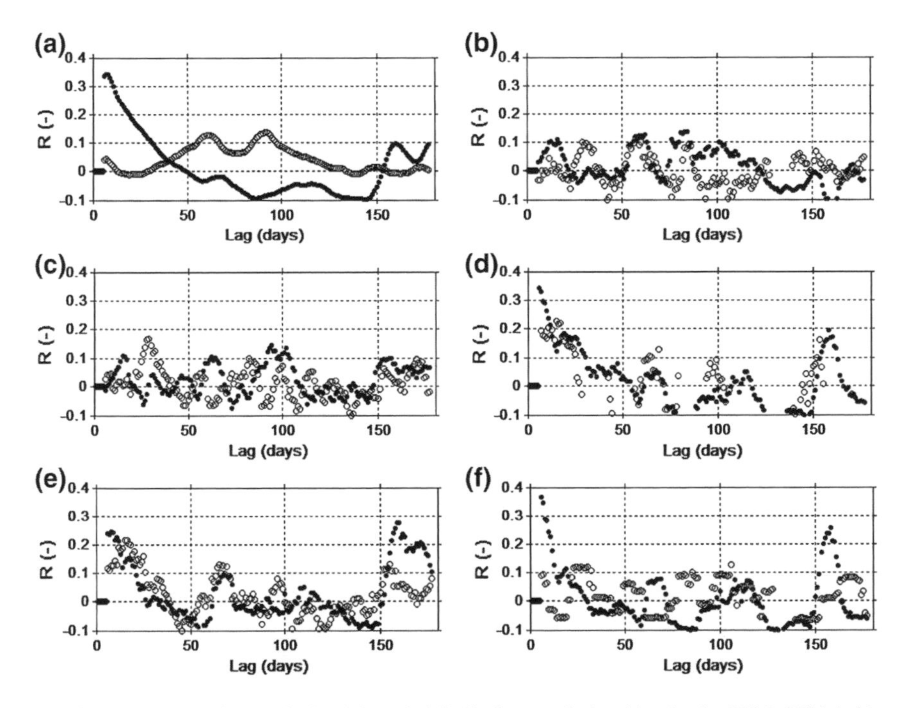

Fig. 4 Cross-correlation analysis of the rainfall-discharge relationship, for the 2005–2006 (*white circles*) and 2011–2012 (*black circles*) hydrological cycles: **a** *Fontaine de Vaucluse*, **b** *Point A*, **c** *Point B*, **d** *Point C*, **e** *Point D*, **f** *Point GAS*

2005–2006 and 2011–2012 hydrological cycles (considered separately). Significant autocorrelation at lag 6 months shows up clearly, due to the seasonal pattern of rainfall. We observe three markedly different responses, with: (i) no significative cross-correlation (points A and B), (ii) significative cross-correlation, similar for both hydrological years (points C and D), (iii) significative cross-correlation for the 2011–2012 hydrological year only or with a very different cross-correlation function for the 2005–2006 year (Fontaine de Vaucluse spring and point GAS).

Points A and B are located ~420 m in depth, with a diffuse flow behaviour according to Barbel-Perineau (2013). However, their reactivity to rainfall events is lower than that of e.g. point D (Fig. 4e). The point D also presents a diffuse flow behaviour, but it is located at −33 m in depth. Based on these observations, we propose that the absence of significative linear rainfall-discharge relationship for points A and B could be related to a relatively high degree of organisation of the karst conduit network.

Point C (−256 m) and point D (−33 m) have, respectively, intermediate and diffuse flow behaviours, according to their classification by Barbel-Perineau (2013). The cross-correlation analysis shows similar hydrodynamic behaviours, with a maximum at ~6 days lag time (Fig. 4d, e). Compared to 2011–2012 results, there is a slight increase in the lag time associated with the 2005–2006 cross-correlation maximum (approx. +12 days).

Based on these elements, we propose that the stability of the linear rainfall-discharge relationship for these points during contrasted hydrological conditions be related to a relatively high water storage upside these flow points. These results are in good agreement with the hydrochemical and hydrogeophysical investigations performed by previous authors: (i) the relation of points C and D with unsaturated zone water storage have been detected by chemical analyses of time tracers (Magnesium and TOC) (Garry 2007; Barbel-Perineau 2013), (ii) unsaturated zone water storage understanding is strengthened by hydrogeophysical investigations performed from the surface, above point D (Carrière 2014).

Point GAS (−191 m) has an intermediate flow behaviour according to Barbel-Perineau (2013). The linear rainfall-discharge relationship is not significative for the 2005–2006 time series, and strong for the 2011–2012 time series (the maximum of the cross-correlation being reached for a lag time lower than 6 days). The absence of reactivity of this point for the 2005–2006 year may be due to the lack of connectivity in the upward rock cover.

The Fontaine de Vaucluse spring shows a significative linear component of the rainfall-discharge relationship for both 2005−2006 and 2011−2012 time series, but with a shift in the cross-correlation maximum from 60 days lag time (2005−2006) to 8 days lag time (2011−2012) and an increase in the absolute value of the cross-correlation maximum. This result evidences the influence of antecedent hydrological conditions on the hydrological functioning of the watershed.

4 Conclusion

This contribution addresses both methodological and conceptual issues.

First, sparse data are a recurrent problem in environmental sciences, and they are generally considered as unfit to statistical analyses. This study shows that data stemming from low-frequency, highly heterogeneous sampling schemes can be further analysed using cross-correlation analysis. These results show that cross-correlation analysis can be successfully used with sparse data time series. The proposed approach could be adapted, for example hydrochemical time series analysis.

Second, characterising the hydrodynamic processes in the unsaturated zone is a prerequisite to efficiently protect and manage the karst water resources. This study provides complementary insights into the hydrodynamic functioning of the unsaturated zone, based on discharge measurements from five flow points located within the artificial LSBB gallery. Our results are consistent with hydrochemical analyses. They also evidence the influence of inter-annual rainfall variability on the hydrological functioning of the unsaturated zone.

Acknowledgments The authors wish to thank the Platform for Fundamental and Applied Interdisciplinary Research, LSBB (www.lsbb.eu) for providing technical support and good field conditions. The CIRAME for providing metrological data (www.agrometeo.fr). The study is founded by the network of hydrogeological researches sites H + (www.hplus.ore.fr).

References

Barbel-Perineau A (2013) Caractérisation du fonctionnement de la zone non saturée des aquifères karstiques. Université d'Avignon, Ph.D. Thesis, p 223

Barbel-Perineau A, Danquigny C, Emblanch C (2013) Hydrodynamic characterization of flows in the vadose zone by direct measurements in karst aquifer. In: Proceeding of KarstFlow 2013, Budapest

Carrière SD (2014) Etude hydrogéophysique de la structure et du fonctionnement de la zone non saturée du karst. Université d'Avignon, Ph.D. Thesis, p 210

Garry B (2007) Etude des processus d'écoulements de la zone non saturée pour la modélisation des aquifères karstiques. Université d'Avignon. Ph.D. Thesis, p 218

Labat D, Ababou R, Mangin A (2000) Rainfall–runoff relations for karstic springs. Part I: convolution and spectral analyses. J Hydrol 238:123–148

Massei N, Dupont JP et al (2006) Investigating transport properties and turbidity dynamics of a karst aquifer using correlation, spectral, and wavelet analyses. J Hydrol 329(1–2):244–257

Mangin A (1984) Pour une meilleure connaissance des systèmes hydrologiques à partir des analyses corrélatoires et spectrales. J Hydrol 67:25–43

Padilla A, Pulido-Bosch A (1995) Study of hydrographs of karstic aquifers by means of correlation and cross-spectral analysis. J Hydrol 168:73–89

Transmissive and Capacitive Behavior of the Unsaturated Zone in Devonian Limestones, Implications for the Functioning of the Epikarstic Aquifer: An Introduction

A. Poulain, G. Rochez, I. Bonniver and V. Hallet

Abstract The hydrogeological behavior of the unsaturated zone in karst limestones is far to be completely understood. However, this part of aquifer systems is the most important for the vulnerability of drinking water resources. In Wallonia, the majority of tap water is provided by Devonian and Carboniferous limestones, which consist of highly karstified and fractured formations. The purpose of this research was to apply an experimental approach in order to assess the functioning of the unsaturated part of karst systems. We choose to follow two drip sites within the Han-sur-Lesse cave system (Rochefort, Belgium), the actual dataset covers the 2008–2013 period and the recording device is still active. We also conduct dye tracer experiment (uranin) from the surface with a fluorometer into the cave coupled with a drip collector. In this paper, we present the contextualization of the study site, the experimental methodology, and the first results.

1 Introduction

The superficial part of karst area is usually referred as the epikarstic area. This zone presents an increased degree of dissolution, weathering, and fracturation and therefore exhibits higher porosity and permeability compared with the underlying bedrock (Mangin 1975; Klimchouk 2004) (Fig. 1). Due to this, it may store a part of the infiltrated water up to a near-saturation state and, then, constituting a perched epikarstic aquifer within the unsaturated zone. Limestone and karst aquifers represent an important part of global groundwater reserves. Their vulnerability is

A. Poulain (✉) · G. Rochez · I. Bonniver · V. Hallet
Department of Geology, University of Namur, 61 Rue de Bruxelles, 5000 Namur, Belgium
e-mail: amael.poulain@unamur.be

B. Andreo et al. (eds.), *Hydrogeological and Environmental Investigations in Karst Systems*, Environmental Earth Sciences 1,
DOI 10.1007/978-3-642-17435-3_5

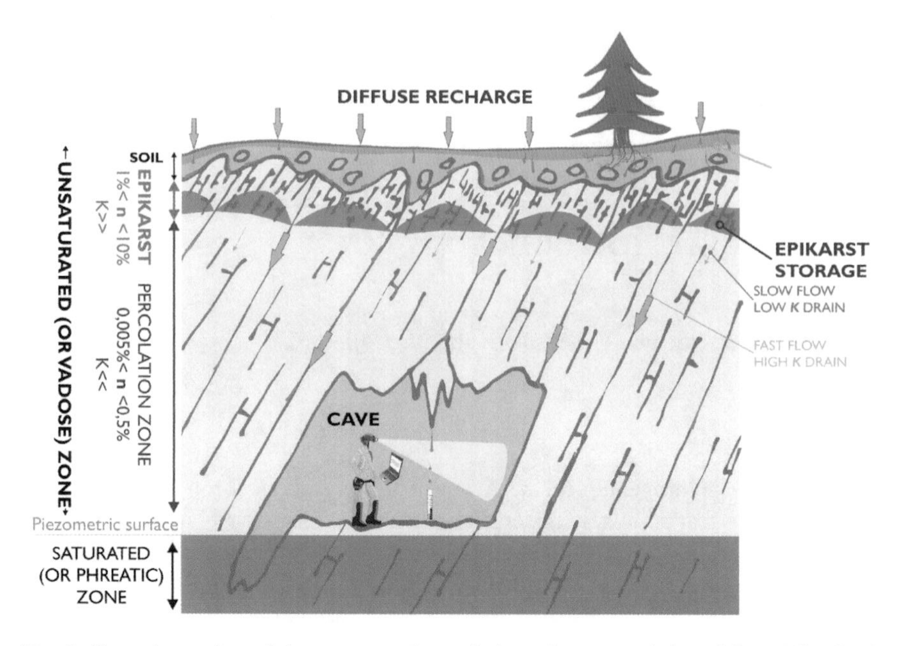

Fig. 1 General overview of the unsaturated zone in karst limestones (adapted from Klimchouk 2004)

important, and therefore the protection of this resource is a major issue for the preservation of the water reserves.

The unsaturated zone (including epikarst) represents a significant transfer path for the aquifer recharge. Transmissive and storage capacity features of the unsaturated zone are poorly understood, mainly due to the lack of direct observational data and the complexity and spatial variability of these processes. One way to understand the behavior of this area, and particularly the role of the epikarst aquifer, is to interpret drip signals into the open karst of the unsaturated zone. Recent studies have already used this methodology in order to assess the various processes inside the vadose zone. Arbel et al. (2010) and Scheffer et al. (2011) used drip rate recording and tracer experiment to estimate the infiltration and recharge processes through the unsaturated zone in Israeli karstified limestones. Jex et al. (2012) provides a summary of published and unpublished sources of cave drip rate monitoring data since more than 15 years. The literature references and the study opportunities based on such data are wide; our first goal was to apply these kind of methodologies at the scale of Belgium. As the Paleozoic limestones of Wallonia are the major reserve of drinkable water in Belgium, it is essential to understand the functioning of such aquifer for the estimation of their vulnerability. In order to collect enough data to allow comparisons with historical data in various places, we chose to install drip rate recording devices in the Han-sur-Lesse cave in 2008. A tracer experiment was also carried out during March 2009.

The first part of the current project was the treatment of datasets from the two drip-sites. The breakthrough curve of the tracer experiment between the surface and this cave has been also compiled and synthesized. This paper presents the geological and karstic context of the study site, our experimental methodologies and the complete results for the 2008–2013 period, with a short analysis.

2 Study Site

The Han-sur-Lesse cave system is located within the Calestienne in south Belgium, near the city of Rochefort (Fig. 2a). The Calestienne is a morpho-structural unit of Belgium which forms a thin band of SW-NE hills between the Fagne-Famenne plain (mean elevation of 200 m) and the Ardenne highlands (300–694 m). This unit is formed by a series of folded and fractured Upper Devonian limestones (Givetian), highly karstified. The south-Belgian Givetian rocks are mainly composed by biostromal limestones with some argillaceous limestones or nodular shales intercalations, which may locally acts as impervious layers within the karst aquifer. The flow of the Lesse River over these lithologies has favored the formation of an underground meander cutoff through the Boine massif which is now known as the Han-sur-Lesse cave system.

The Han-sur-Lesse cave system is the biggest in Belgium with a cumulated length of 13,843 m. The network is composed of an active karst with the underground Lesse river and several access to the local water table and a « fossil » part with multiple galleries within the unsaturated zone.

The first drip site is installed inside the Père Noël Cave (Fig. 2b). This cave is developed along the bedding plane in the upper part of the Givetian series in the Fromelennes Formation. This is a highly active cave in terms of percolation, which has produced one of the most concretionary cave in Belgium. The drip monitoring station is installed under a perennial stalactite flow that ranges from 15 to 70 l/day. This site was chosen by Genty and Deflandre for drip monitoring between 1991 and 1996 (Genty and Deflandre 1998).

The second drip site is in the Salle d'Armes (Fig. 2b), which is located in the Mont d'Haurs Formation within the Han-sur-Lesse cave s.s. The stalactite is a young speleothem with a drip rate ranging from 0 to 50 l/day.

3 Methodology

3.1 Dripping Rate Measurement

The two dripping sites where equipped with a simple self-siphoning device composed by a graduated cylinder combined with a pressure probe (Fig. 1). The cylinder volume is about 1 l (depending of the probe volume and the siphoning tube size) and is emptied by a glass tube when the water reaches a given level. The probe is a CTD-Diver (Eijkelkamp—10 m Diver) with a time-step of 5 min which

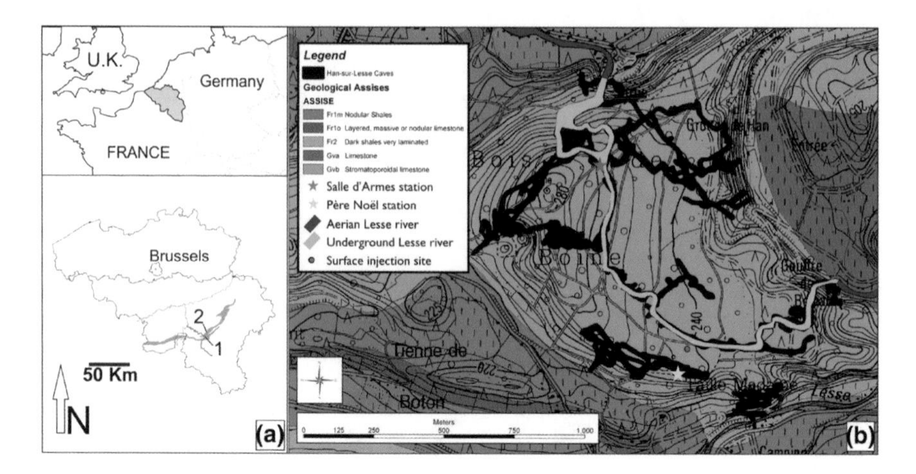

Fig. 2 Location map of Han-sur-Lesse cave system. **a** Location of the Calestienne (*gray*) in Belgium, the study site (*1*) and the city of Rochefort (*2*). **b** Geological map of the study site with the map of the caves

records the pressiometric level into the cylinder (atmospheric pressure + water pressure). As the device is filled/emptied by the dripping flow, we record a saw tooth pattern which allows calculating the dripping rate per day with a simple data treatment. The variation in atmospheric pressure is considered to be too low to influence the results of this methodology.

3.2 Infiltration Calculation

The meteorological data were given by the Royal Meteorological Institute of Belgium (RMI) for the Han-sur-Lesse station which is about 1 km from the Père Noël station and 400 m from the Salle d'Armes station. Daily estimations of temperature (°C), precipitation (mm) and atmospheric pressure are available for the 2008–2013 period, and we consider them to be reliable due to the very short distance. The infiltration water amount was estimated with the water excess (WE) based on the precipitation and the evapotranspiration given by the Thornthwaite formula (Thornthwaite 1954) as suggested by Genty and Deflandre (1998) for the same area. Due to the karst context, runoff is supposed to be negligible.

3.3 Tracer Experiment

The tracer experiment was conducted during March 2009 in the Père Noël cave area. For this particular purpose, the dripping water of the Père Noël station was diverted from the graduated cylinder into a nearby rimstone. This rimstone was large enough to install an automatic fluorometer (GGUN-FL30 by Albillia Switzerland) in order to measure the tracer concentration with a 15 min timestep. The combination of discharge measurement and tracer concentration allows us to calculate the tracer recovery.

We use the fluorescent tracer uranin ($C_{20}H_{10}Na_2O_5$) for this experiment. It is the most used fluorescent tracer due to its very low detection limit (10^{-3} μg/l), low sorptivity, low cost, and nontoxicity at such concentration (Schudel et al. 2002). For the injection point, we choose to project the percolation site on the surface according the local dip of the limestone layers (Fig. 2). We inject 500 g of uranin with approximately 500 l water in this precise location (1 m^2). The distance between the surface and the dripping site is 70 m according to the local dip value, no other cavity is known between these two points.

4 Results and Discussion

4.1 Dripping Rate Chronicles (2008–2013)

Excepted for the beginning of 2008 and the end of 2010, the drip chronicle is generally complete for the 2008–2013 period and shows remarkable inter-annual correlations as well as very different signals between the two sites (Fig. 3). This reflects the variability of the processes and responses of the vadose zone, regarding the recharge water as well as their recurrence during several hydrological years.

4.1.1 Père Noël Site (Red Curve)

The stalactite at the Père Noël station is perennial with drip rate between 15 and 70 l/day. It shows four different stages during the year:

- *Phase 1*: During the periods without infiltration (summer and autumn), a long and slow decreasing is observed until a minimal value around 15–20 l/day. This is interpreted as the epikarst drainage and suggests that the epikarst storage provides most of the discharge of unsaturated percolations during low water level periods.
- *Phase 2*: After the decrease phase, a quick and sharp increase of the discharge is observed, despite the recovery of the infiltration since several days or weeks. This highlights the storage capacity of the epikarst and the fact that a certain infiltration threshold is needed prior to feed the stalactite drip. This has been already observed by many studies in other karst systems (Arbel et al. 2008, 2010; Scheffer et al. 2011).
- *Phase 3*: The drip discharge is highly variable, and correlated to the recharge rate. This could indicate that the storage capacity of the epikarst is reached. This phase lasts for 4–5 months.

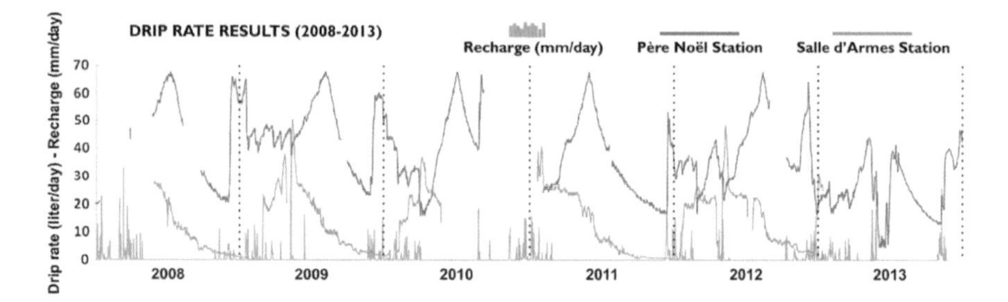

Fig. 3 Drip rate results for 2008–2013 at the two recording stations within the Han-sur-Lesse caves

- *Phase 4*: A slow, but significant increasing of the stalactite discharge until a punctual maximal discharge value, without any infiltration input. This value is higher than the maximal discharge observed at the beginning of the winter. Furthermore, this maximum discharge is the same in 2008, 2009, 2010, 2011 and 2012 (67.7 l/day, with an accuracy of 0.97 l/day).

The three first annual phases at the Père Noël station were already observed by Genty and Deflandre (1998). The fourth phase was not observed, and cannot be explained at this point of our study.

Other new features provided by these data are many sharp decreases of the drip rate (5–10 l in few days). These anomalies are highly correlated to major or prolonged infiltration events. The best example of this phenomenon is observed between January and July 2009, with seven decreasing events. An explanation could be a bypass phenomenon when the infiltration rate is too high and overflows the usual infiltration path, leading to a decrease of the drip rate. This seems to be the best explanation as no peak flow related to the infiltration peak is observed at the stalactite, even after a few days.

On the contrary, during periods with little or without infiltration, the drip rate increases frequently with a linear trend. This is related to the Phase 4 of the hydrological cycle which takes place during the longest period without infiltration water.

4.1.2 Salle d'Armes Site (Blue Curve)

The drip rate chronicle at the Salle d'Armes station is much simpler compared to the previous one. The discharge is in a range between 0 and 50 l/day and shows a seasonal variation in three stages: (i) due to the infiltration lack, a long and slow decease during summer and fall: depending on the year, the stalactite may dry up during a certain period; (ii) the sharp increase when the epikarst storage is filled by the recovery of the infiltration in winter; and (iii) a high discharge stage with quick reactions to the recharge variations.

4.2 Tracer Experiment

The uranin was injected above the *Père Noël* cave the 27th of March at 14:00 and the first appearance of the tracer into the dripping water was observed the 28th at approximately 5:30 (Fig. 4). The maximum water transit speed through the limestone massif (70 m thick) is approximately 4.5 m/h. This first result is quite surprising, while there is no evidence (at the surface and in the cave) for a dissolution enlarged flow path, where quick flow is possible. The maximum tracer concentration (72.2 ppb) is observed 3.5 days after the release, which means a modal velocity of 0.83 m/h. The total breakthrough lasted for more than 150 days, with a return to the background noise after 200 days. The tracer mass recovery is 0.013 %.

During this tracer test, the electrical conductivity of the water shows two significant decreases, corresponding to two periods of negative drip rate anomalies (during high infiltration periods). This means that the drip water is influenced by low-conductive precipitation water. Otherwise, when the discharge is increasing (corresponding to low or absent infiltration periods), the electrical conductivity is also increasing. This particular feature of the drip rate and E.C record reflects an epikarst water drainage caused by lack of infiltration water. We note that even if precipitation water influences the E.C., the water temperature is very stable and does not show any meteoric influence.

4.3 Discussion

The Han-sur-Lesse cave provides a relevant study site for unsaturated zone investigation in Devonian limestones of Belgium.

The drip rate chronicle is very satisfying and shows only few periods of technical issues leading to a lack of discharge data. The particular behavior of the Père Noël stalactite drip has not been explained yet. It is evident that these unknown features of stalactite drip are very interesting, as they reflect particularities of the epikarst functioning. The comparison with the Salle d'Armes drip will be an interesting reference, because they show different behaviors for the same drip rate range and for the same surface infiltration conditions.

The tracer experiment exhibits the duality of the epikarst behavior. On one hand, it shows a very fast percolation within the unsaturated zone. The BTC reflects a high permeable pathway, with no significant dilution along the way as shown by the shape of the curve and the high maximal concentration. On the other hand, the tracer release lasted for more than 150 days with only 0.013 % recovery. The long-term storage capacity of the epikarst is evident, regarding the two last characteristics. Furthermore, no significant clue of the water flush (500 l) was recorded in the drip chronicle after the injection. This indicates that the tracer probably spreads widely around the injection point, and this may be an explanation for a very low tracer recovery, but a high maximum concentration. A quick visual

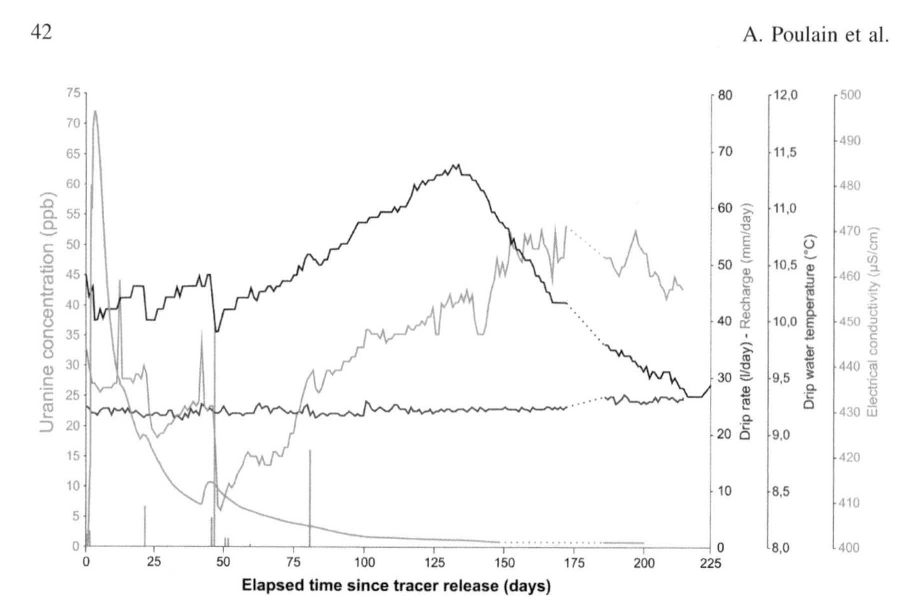

Fig. 4 Breakthrough curve at the Père Noël station during the tracer experiment in March 2009

inspection of the surrounding percolation points was made during the uranin breakthrough (the tracer was visually detectable in the cave) but no other tracer seepage was detected.

5 Conclusions and Research Prospect

The very simple recording methodology for drip rate allowed us to collect 6 years of data for two stalactite sites. The device shows is ability to record precise and reliable data, whereas very little maintenance is required. The different stations are still operational and will record more data during the next few years. The available discharge data highlights recurring processes inside the vadose zone, as well as strong seasonal difference between the two locations. These records offer a promising base to identify the behavior of the epikarst regarding the karst aquifer recharge in the belgian Devonian limestones. A better understanding of these processes will lead to an improved management of drinkable water reserves in Wallonia.

Tracer results confirm the fast vertical drainage inside the unsaturated zone but reflect a potentially large dispersion (low tracer recovery) and a long (>150 days) storage of the tracer.

The following steps of the project are: (i) looking over the existing literature, to drive our analysis perspectives; (ii) building a conceptual model of the unsaturated zone in limestones, regarding discharge and tracer experiment data; (iii) integrating a numerical model able to predict discharge variation cycles with the infiltration data, in order to test the conceptual sketch; and (iv) attempting to validate our models with future data from drip stations, and new tracer experiments.

References

Arbel Y, Greenbaum N, Lange J, Shtober-Zisu N, Grodeck T, Wittenberg L, Inbar M (2008) Hydrologic classification of cave drip in a Mediterranean climate, based on hydrograph separation and flow mechanisms. Isr J Earth Sci 57(3–4):291–310

Arbel Y, Greenbaum N, Lange J, Inbar M (2010) Infiltration processes and flow rate in developed karst vadose zone using tracers in cave drips. Earth Surf Process Landforms 35:1682–1693

Genty D, Deflandre G (1998) Drip flow variation under a stalactite of the Père Noël cave (Belgium). Evidence of seasonal variation and air pressure constraints. J Hydrol 211:208–232

Jex C, Mariethoz G, Baker A, Graham P, Andersen M, Acworth I, Edwards N, Azcurra C (2012) Spatially dense drip hydrological monitoring and infitration behavious at the Wellington Caves, South East Australia. Int J Speleol 41(2):283–296

Klimchouk AB (2004) Toward defining, delimiting and classifying epikarst: its origin, processes and variants of geomorphic evolution. Speleogenesis Evol Karst Aquifers 2(1):1–13

Mangin A (1975) Contribution à l'étude hydrodynamique des aquifères karstiques. Annales de Spéléol 30(1):21–124

Scheffer N, Cohen M, Morin E, Grodek T, Gimburg A, Magal E, Gvirtzman H, Neid M, Isele D, Frumkin A (2011) Integrated cave drip monitoring for epikarst recharge estimation in a dry Mediterranean area, Sif Cave, Israel. Hydrol Process 25:2837–2845

Schudel B, Biaggi D, Dervey T, Kozel R, Müller I, Ross JH, Schindler U (2002) Einsatz kükstlichen Tracer in der Hydrogeologie—Praxishilfe. Berichte des BWG 3:91p

Thornthwaite CW (1954) The measurement of potential evapotranspiration. Mather, Seabroook, 225 p

Feasibility and Limits of Electrical Resistivity Tomography to Monitor Water Infiltration Through Karst Medium During a Rainy Event

S.D. Carrière, K. Chalikakis, C. Danquigny, R. Clément and C. Emblanch

Abstract The common hydrogeological concepts assume that water enters in karst media by preferential pathways. But it is difficult to identify these pathways, particularly if soil or scree covers the karst features. When and where does water enter in the hydrosystem? How fast? A unique large-scale Electrical Resistivity Tomography (ERT) surface-based time-lapse experiment was carried out during a typical Mediterranean autumn rainy episode (230 mm of rain over 17 days). A total of 120 ERT time-lapse sections were measured over the same profile during and after this event (30 days). The main goal was to evaluate efficiency and limits of the ERT to monitor water infiltration, under natural conditions. Apparent (directly measured) and inverted resistivity's variation during the rainy event highlights some interesting zones. They could be interpreted as preferential pathways, where water dynamic seems quicker in term of moistening and drainage. Nevertheless, these results have to be interpreted reasonably because ERT does not provide enough precision to determine exact pathways geometry and functioning. In addition, forward modeling provided relevant data treatment limitations mainly for the deeper parts of the sections.

1 Introduction

Identifying and locating potential preferential pathways such as conduits, faults, or fractures is peculiarly difficult in karst areas. In addition, characterizing and locating water movements through these pathways is probably one of the most

S.D. Carrière (✉) · K. Chalikakis · C. Danquigny · C. Emblanch
UMR1114 EMMAH, UAPV, 84914 Avignon, France
e-mail: simon.carriere@alumni.univ-avignon.fr

S.D. Carrière · K. Chalikakis · C. Danquigny · C. Emblanch
UMR1114 EMMAH, INRA, 84914 Avignon, France

R. Clément
IRSTEA, 1 Rue Pierre-Gilles de Gennes CS 10030, 92761 Antony CEDEX, France

B. Andreo et al. (eds.), *Hydrogeological and Environmental Investigations in Karst Systems*, Environmental Earth Sciences 1,
DOI 10.1007/978-3-642-17435-3_6

important and challenging task in order to improve enhanced understanding of karst water dynamics. Hydrogeologists have been developing several methods mainly related to natural and artificial tracing to access these water pathways. Geophysical tools could help identifying these potential pathways with a suitable methodology (Carrière et al. 2013; Chalikakis et al. 2011; Jardani et al. 2006; Meyerhoff et al. 2012; Valois 2011). However, even if such features were fully identified, effective water flows pathways would probably remain almost impossible to predict and monitor. In karst areas, the link between structure and hydrodynamic functioning remains a bottleneck. This could probably explain also the limited use of physically based gridded flow models for karst hydrodynamic modeling. Ground Penetrating Radar (GPR) is probably the most efficient geophysical technique to image near surface features under adequate surface conditions (Al-fares et al. 2002; Carrière et al. 2013; Chalikakis et al. 2011). However, GPR does not allow identifying if these features impact water dynamic. Electrical Resistivity Tomography (ERT) technique due to its robustness and reliability has been widely used in karst areas to identify karst features (Cardarelli et al. 2006; Valois et al. 2010; Zhou et al. 2000). In porous semihomogenous media, ERT, due to its sensitivity to water content, is a commonly used geophysical technique, fast enough to follow up water infiltration during rainy events (Clément et al. 2009; Descloitres et al. 2008).

A unique large-scale ERT surface-based experiment was carried out during a typical Mediterranean autumn rainy episode (17 days) in the LSBB (Low Noise Underground Research Laboratory, Rustrel, France) within the Fontaine de Vaucluse karst hydrosystem (Fig. 1). This experiment takes place on karstified urgonian limestone described by Masse and Fenerci-Masse (2011) (Fig. 1c and d). There is not developed epikarst and soil cover is thin or absent. A total of 230 mm of rain was registered and 120 ERT time-lapse sections were measured over the same profile during and after the event (30 days). The main goal was to evaluate efficiency and limits of the ERT to monitor water infiltration via previously recognized karst features (Carrière et al. 2013), under natural conditions.

2 Implantation/Acquisition Strategy and Field Constraints

At the experimental site, the slope, the vegetation, and the gravel cover induced a very important preparation work to clear the ERT chosen profile before measurements. In addition, the implantation of electrodes, mainly at limestone outcrops, was an additional difficult task; electrode holes were mechanically dug within the rock; saltwater and mud were used to ensure a good quality ground contact. Bad electrode-ground contact could induce artifacts during data acquisition. These artifacts are comparable to near surface inhomogeneities (Abbas et al. 2004; Ritz et al. 1999) which could deteriorate the final inverted resistivity section (LaBrecque et al. 1996).

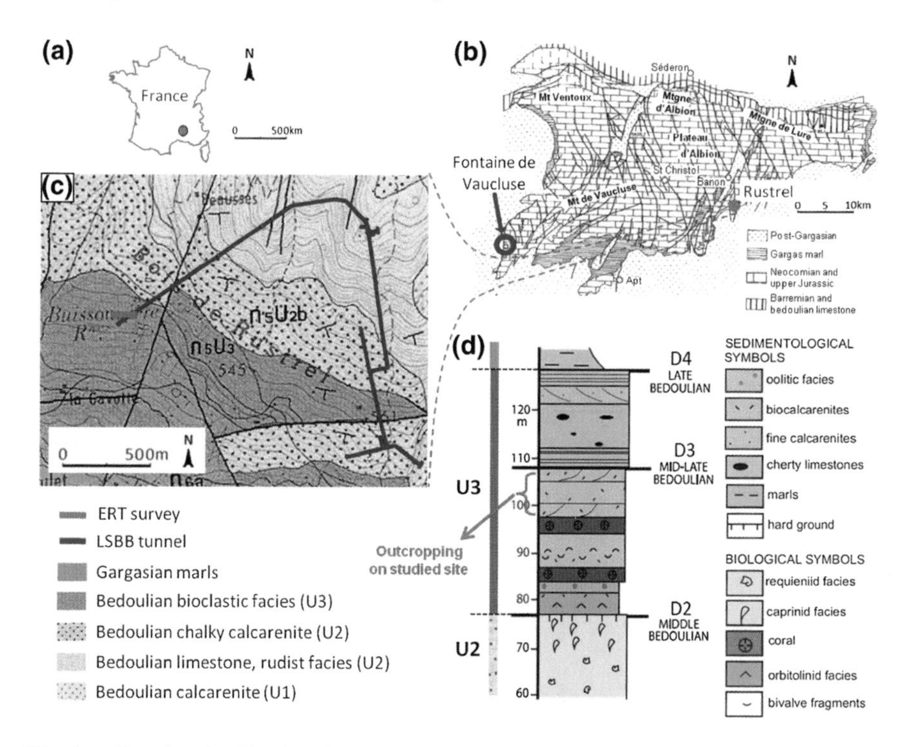

Fig. 1 **a** Fontaine de Vaucluse basin located in France; **b** Test site located in Fontaine de Vaucluse basin (after Puig 1987); **c** Extract of local geological map, n°942 (Blanc et al. 1973); **d** Regional lithostratigraphic log (Masse and Fenerci-Masse 2011 modified)

The chosen ERT profile follows an E–W direction. This direction is perpendicular to the general slope, subperpendicular to one of the main faulting and lineament directions and the most heterogeneous direction, in term of apparent resistivity spatial distribution (Carrière et al. 2013).

The acquisition system used for ERT time-lapse measurements is an ABEM Terrameter SAS 4000 (Dahlin 2001) with 4 channels and 64 electrodes. The profile has 126 m length and a 2 m interelectrode spacing. Electrodes exact position was measured with Real Time Kinematic (RTK) acquisition method every 2 m using a differential GPS (TRIMBLE GPS 5800), with an accuracy around 1 cm for X and Y and 2 cm for Z. The electrodes stayed in the field during the entire campaign.

For the time-lapse measurements, the gradient array was chosen for its robustness and rapidity (Dahlin and Zhou 2004). This protocol totals 1,360 measurement points. For each measurement point, the acquisition time was 0.1 s and the delay time was 0.2 s. Each measurement point cycle spends around 1.8 s. To ensure data quality, during acquisition if a data point presented a repetition Root Mean Square (RMS) >1 %, the measurement was repeated until five times. A 50 Hz filter was also applied to reduce anthropogenic noise. During the 30-day campaign, an ERT time-lapse section was acquired every 3 h during the rainy event (17 days) and every day until 2 weeks after the rainy event.

3 Data Treatment and Results

3.1 Classical Processing

During this monitoring campaign, a total of 120 ERT sections were acquired. When data acquisition quality was not satisfactory (repetition measurement of all the section RMS >2 %) the section was removed. Finally, a total of 106 ERT sections were kept to follow data treatment.

Apparent resistivity values have been averaged for each section. This mean apparent resistivity decreases strongly during the rain event, from 1,750 to 1,050 Ω m (Fig. 2a). These variations do not seem related with temperature variations because air temperature remained stable during the campaign (Carrière 2014). Thus, we can reasonably relate these resistivity variations with water content variation. Analysis of this mean apparent resistivity indicator allows selecting eight critical times step presented in Fig. 2b and c, before, during and after the rainy event.

Several processing strategies and softwares were used for data treatment and analysis; a commonly used software, Res2Dinv (Loke 2012; Loke and Barker 1996) and a research software package, including DC2DinvRes and BERT (Boundless Electrical Resistivity Tomography) (Günther et al. 2006; Günther and Rücker 2013; Rücker et al. 2006). These results were presented and discussed in detail in (Carrière 2014).

The inversion results presented in Fig. 2b were performed using DC2DinvRes. We used a standard time-lapse inversion following the approach proposed by Loke in (Loke 1999). First, the initial time step model was computed. Second, we used it as a reference model for the other time step. Finally, we compared the resulting calculated models (Fig. 2c) using percentage change in calculated resistivity ($\Delta\rho_{calc}$), Eq. 1.

$$\Delta\rho_{inv} = 100 * \left(\frac{\rho_T}{\rho_0} - 1\right). \tag{1}$$

where ρ_0 is the calculated resistivity at the first time step, ρ_T is the calculated resistivity at time step "n".

During the first four selected time-steps, the electrical resistivity was decreasing progressively mainly in the near surface (Fig. 2c). Resistivity decreased until 80 % between the initial time step and the maximum of the rainy event. This strong variation is probably related to the important moistness of the near surface horizons. This is not observed at the deeper part of the sections. Moreover, resistivity seems to increase in several deeper zones. But these increases are calculation artifacts due to the inversion process. They appear commonly in ERT time-lapse sections under electrical conductive zones (Clément et al. 2009, 2010). In this case, deeper horizons are not truly investigated because the ERT sensibility decreases quickly with depth. This low sensitivity is due to the near surface

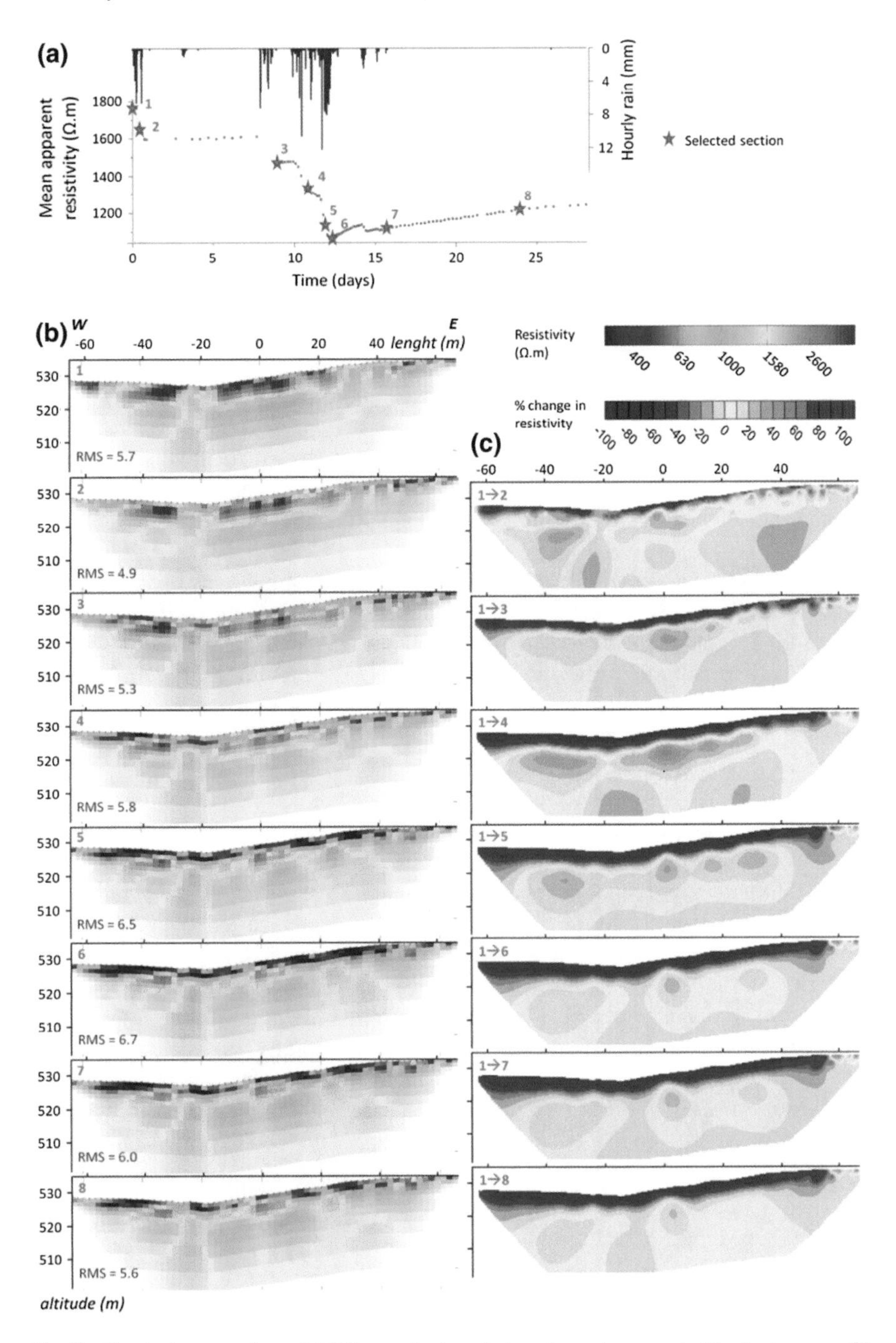

Fig. 2 Classical processing of ERT monitoring during the rainy event, Gradient array, 64 electrodes. **a** Evolution of mean apparent resistivity during monitoring versus rain. Each *brown point* represents one ERT section. **b** Resistivity models processed using DC2DinvRes, color attenuation represents coverage index. **c** Percentage change in resistivity from initial model

conductive horizon which concentrates the majority of the electrical current. This incapacity of ERT time-lapse to image the deeper part of the profile is demonstrated in the following part by direct modeling.

3.2 Inversion Process Limitations

There are several possibilities to evaluate the quality of resistivity models obtained by data inversion. Such as the sensitivity and resolution matrix (Loke 1999), the Depth Of Investigation (DOI) index (Oldenburg and Li 1999) and the coverage (Günther 2004). Some of them were presented and discussed in detail by Carrière in (2014) for this monitoring campaign.

In this paper, we chose to evaluate ERT time-lapse data inversion process limitations by direct modeling (theoretical simulation of the observed measurements). This nonautomatic solution is very explicit and relevant for a nonspecialist user.

A resistivity model with near surface conductive horizon was used like reference model (Fig. 3a). The resistivity of two blocks in depth (5 × 5 m and 5 × 3 m) is gradually decreased from 20 until 95 % of their initial resistivity (Fig. 3b) simulating the water presence in depth. For each model, a new apparent resistivity dataset was calculated by direct modeling. The same way as for the field measurements, each data set is inverted and variations between this model and reference model (Fig. 3a) are calculated (Eq. 1) and presented in Fig. 3c.

The results of these tests, presented in Fig. 3 by direct modeling, highlight the incapacity of the inversion process to image deep resistivity variations during a rainy event. It is necessary to reach a deep resistivity variation close to 90 %, in order to obtain a measurable resistivity variation at the surface. In other words, ERT surveys do not have the sensibility to detect water variation in depth. This low sensitivity is due to the near surface conductive zone which concentrates the majority of the electrical injected current. Then, there are not enough current propagated in depth to characterize the medium.

In this way, we can conclude that with actual technology, surface-based ERT is not able to detect deep water variation in karst media during a rainy event. If interpretation of monitoring using inversion process is limited, we will see in the following part that rough result (apparent resistivity) could provide additional information.

3.3 Change in Apparent Resistivity

During ERT campaigns apparent measured resistivity (ρ_α) analysis is usually neglected. However, these rough results could provide, without calculation artifacts, complementary information to the inverted resistivity.

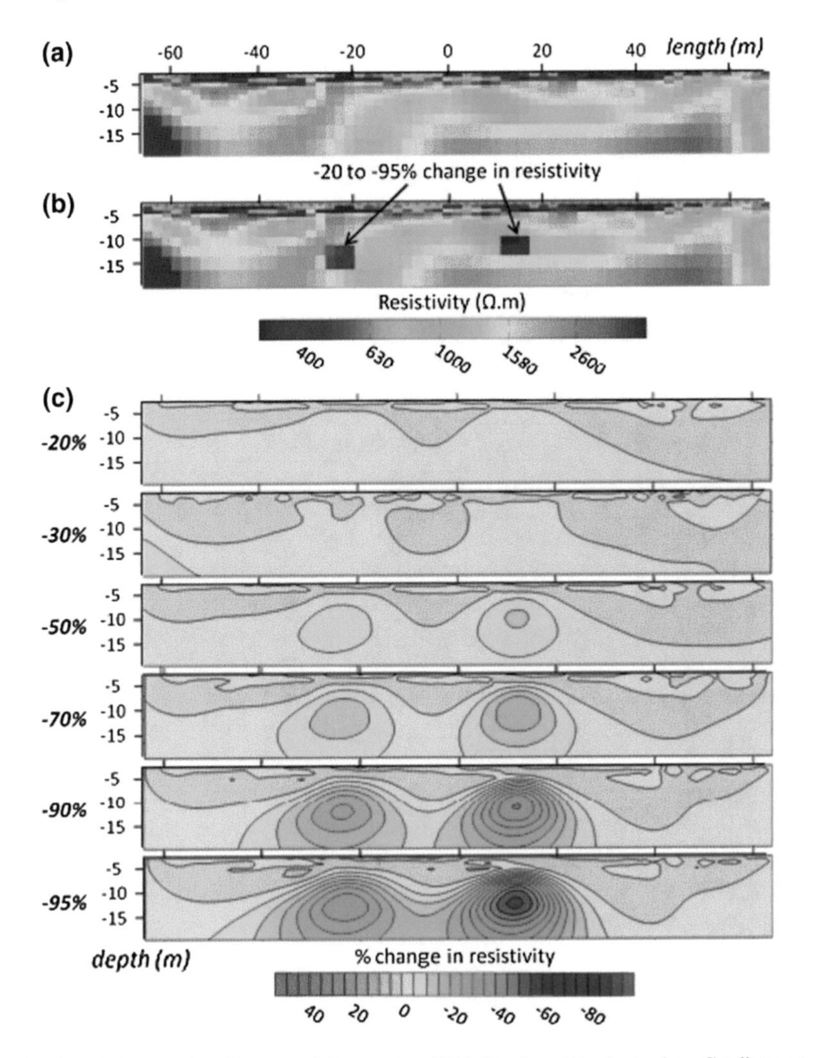

Fig. 3 Sensitivity test by direct modeling using DC2DinvRes, 64 electrodes, Gradient array. **a** Resistivity model during rainy event, with near surface conductive zone; **b** Same model than "A" with two blocs where resistivity is decreased from 20 to 95 %; **c** Change in resistivity between reference model "A" and inverted model from "B"

For this monitoring campaign, we analyzed apparent resistivity variations ($\Delta\rho_a$) between two consecutive time steps. These variations are normalized by the time (Δ_T) between both measurement (ρ_n and ρ_{n-1}) using the following equation (Eq. 2).

$$\Delta\rho_{app} = \left(\frac{\rho_n}{\rho_{n-1}} - 1\right) * \frac{100}{\Delta T} \quad (2)$$

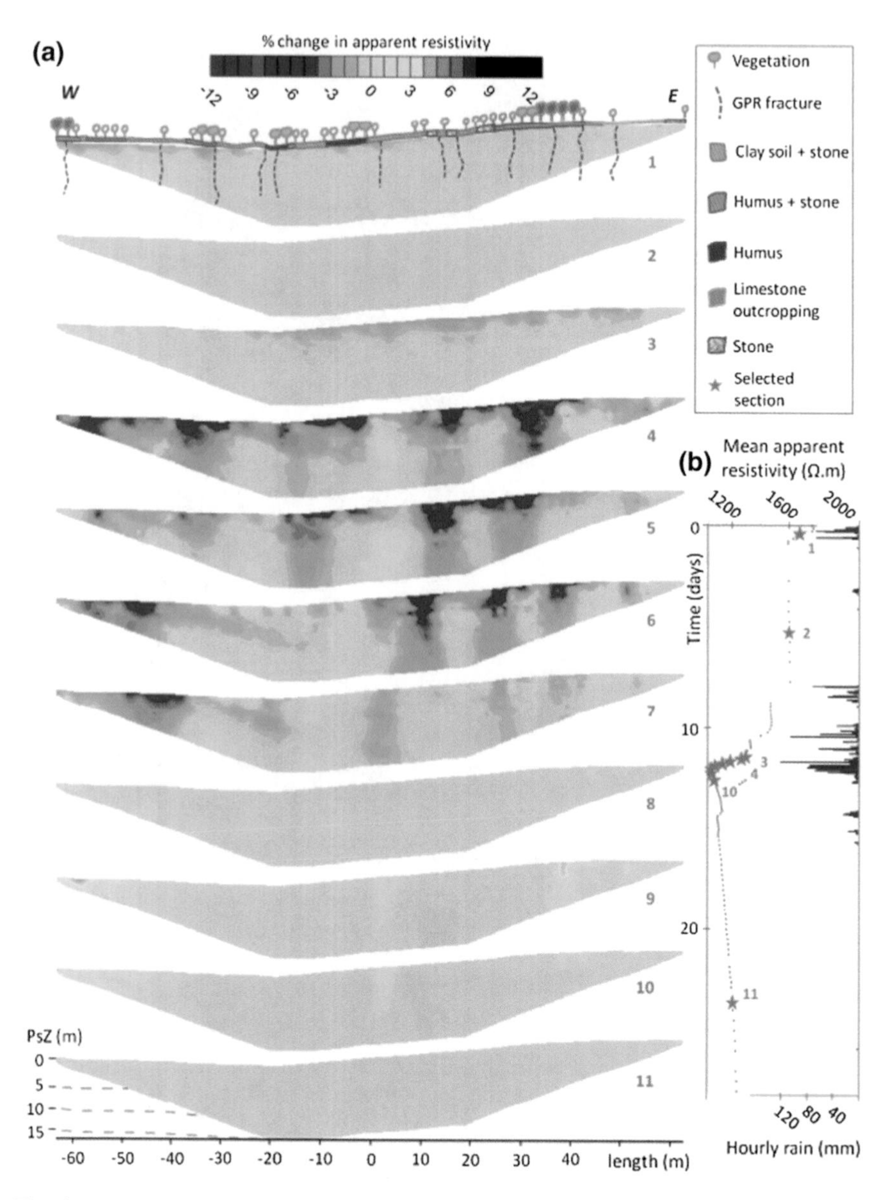

Fig. 4 **a** Hourly change in apparent resistivity between two consecutive time steps. Positioning of fracture detected by GPR and basic representation of vegetation and soil cover is shown (PsZ is pseudo-depth). **b** Evolution of mean apparent resistivity during monitoring versus rain. Each *brown point* represents one ERT section

Thus, the results are presented (Fig. 4) in hourly percentage change in ρ_α with a basic representation of vegetation and soil cover of the profile. Previously recognized karst features (Carrière et al. 2013) are also presented.

At the beginning of the rainy event (Fig. 4a/section 1), ρ_α decreases moderately and homogeneously along the section. This resistivity decrease could be related to a moistening of the near surface horizons. After the first rainy event (Fig. 4a/section 2), ρ_α is quickly stabilized.

During the following strong rain episode (Fig. 4a/section 4 to 7), moistening process appears very heterogeneous and some zones look like preferential pathways. However, it is impossible to link observed ρ_α variations in depth with deep moistening process. These observed variations can be directly influenced by near surface variations. Just after the rain (Fig. 4a/section 9 and 10), ρ_α increases in some zones. That could be related with drainage process. These zones fit with some zones identified before like preferential pathway. This second observation reinforced the hypothesis of preferential pathways in some zones but remains impossible to precise the geometry of these pathways. Other zones where drainage process is not identifiable could be related to zones where soil is thicker and remains moist after the rain.

The preferential pathways identified do not seem related to variations of vegetation density or soil cover. However, these zones seem related with fractures previously detected by GPR (Carrière et al. 2013). It is important to point out that we do not image the pathways but we probably image the effect of pathways above the soil.

4 Discussion and Conclusion

A unique large-scale ERT experiment was carried out during a typical Mediterranean autumn rainy episode (30 days). A total of 230 mm of rain was registered and more than 100 ERT time-lapse sections were measured over the same profile during the event. The main goal was to evaluate efficiency and limits of the ERT to monitor water infiltration via previously recognized karst features, under natural conditions. With the first rain, the near surface electrical resistivity decreases strongly due to progressive moistening of soil by the rain. The sensibility of the ERT decreases quickly with depth due to this conductive near surface horizon which concentrates the majority of electrical current. Forward modeling (theoretical simulation of the observed measurements) provided relevant information about data treatment limitations mainly for the deeper parts of the profile.

Consequently, the classic ERT data processing is limited because the inversion process induces important artifacts. Thus, it is necessary to analyze the lateral and vertical variation of the electrical resistivity. Combined observation of apparent and inverted resistivity variations during and after the rainy event highlights some interesting zones. They could be interpreted as preferential pathways, where water dynamic seems quicker in term of moistening and drainage. Nevertheless, these results have to be interpreted reasonably because ERT does not provide enough precision to determine pathways geometry and functioning.

References

Abbas AM, Atya MA, Al-Sayed EA, Kamei H (2004) Assessment of groundwater resources of the Nuweiba area at Sinai Peninsula, Egypt by using geoelectric data corrected for the influence of near surface inhomogeneities. J Appl Geophys 56(2):107–122

Al-fares W, Bakalowicz M, Guérin R, Dukhan M (2002) Analysis of the karst aquifer structure of the Lamalou area (Hérault, France) with ground penetrating radar. J Appl Geophys 51(2–4):97–106

Blanc MM, Masse J-P, De Peyronnet P, Roux M, Weydert P, Rouire J (1973) Notice explicative et carte géologique, France (1/50 000), feuille Sault-de-Vaucluse (942). BRGM, France, Orléans: 15

Cardarelli E, Di Filippo G, Tuccinardi E (2006) Electrical resistivity tomography to detect buried cavities in Rome: a case study. Near Surf Geophys 4(6):387–392

Carrière SD, Chalikakis K, Sénéchal G, Danquigny C, Emblanch C (2013) Combining electrical resistivity tomography and ground penetrating radar to study geological structuring of karst unsaturated zone. J Appl Geophys 94:31–41

Carrière SD (2014) Etude hydrogéophysique de la structure et du fonctionnement de la zone non saturée du karst. Thèse de Doctorat. Université d'Avignon et des Pays de Vaucluse: 210

Chalikakis K, Plagnes V, Guerin R, Valois R, Bosch FP (2011) Contribution of geophysical methods to karst-system exploration: an overview. Hydrogeol J 19(6):1169–1180

Clément R, Descloitres M, Günther T, Ribolzi O, Legchenko A (2009) Influence of shallow infiltration on time-lapse ERT: experience of advanced interpretation. CR Geosci 341(10–11):886–898

Clément R, Descloitres M, Günther T, Oxarango L, Morra C, Laurent JP, Gourc JP (2010) Improvement of electrical resistivity tomography for leachate injection monitoring. Waste Manag 30(3):452–464

Dahlin T (2001) The development of DC resistivity imaging techniques. Comput Geosci 27(9):1019–1029

Dahlin T, Zhou B (2004) A numerical comparison of 2D resistivity imaging with 10 electrode arrays. Geophys Prospect 52(5):379–398

Descloitres M, Ribolzi O, Le Troquer Y, Thiébaux J-P (2008) Study of water tension differences in heterogeneous sandy soils using surface ERT. J Appl Geophys 64:83–98

Günther T (2004) Inversion methods and resolution analysis for the 2D/3D reconstruction of resistivity structure from DC measurements. Freiberg, Germany, University of Mining and Technology, Ph.D. 160

Günther T, Rücker C, Spitzer K (2006) Three-dimensional modelling and inversion of dc resistivity data incorporating topography—II inversion. Geophys J Int 166:506–517

Günther T, Rücker C (2013) Boundless electrical resistivity tomography—BERT 2—the user tutorial. www.resistivity.net

Jardani A, Dupont JP, Revil A (2006) Self-potential signals associated with preferential groundwater flow pathways in sinkholes. J Geophys Res 111(B9)

LaBrecque DJ, Miletto M, Daily W, Ramirez A, Owen E (1996) The effects of noise on Occam's inversion of resistivity tomography data. Geophysics 61(2):538–548

Loke MH (1999) Time-lapse resistivity imaging inversion. In: 5th meeting of the environmental and engineering society section. Budapest, Hungary

Loke MH (2012) Geoelectrical imaging 2D&3D. Geotomo software manual. www.geotomosoft.com

Loke MH, Barker RD (1996) Rapid least-squares inversion of apparent resistivity pseudosections by a quasi-Newton method. Geophys Prospect 44(1):131–152

Masse J-P, Fenerci-Masse M (2011) Drowning discontinuities and stratigraphic correlation in platform carbonates. The late Barremian-early Aptian record of Southeast France. Cretac Res 32(6):659–684

Meyerhoff SB, Karaoulis M, Fiebig F, Maxwell RM, Revil A, Martin JB (2012) Visualization of conduit-matrix conductivity differences in a karst aquifer using time-lapse electrical resistivity. Geophys Res Lett 39(24):L24401

Oldenburg DW, Li YG (1999) Estimating depth of investigation in dc resistivity and IP surveys. Geophysics 64(2):403–416

Puig JM (1987) Le système kasrtique de la Fontaine de Vaucluse. Avignon, Université d'Avignon et des Pays de Vaucluse. PhD: 207

Ritz M, Robain H, Pervago E, Albouy Y, Camerlynck C, Descloitres M (1999) Improvement to resistivity pseudosection modelling by removal of near-surface inhomogeneity effects: application to a soil system in South Cameroon. Geophys Prospect 47(2):85–101

Rücker C, Günther T, Spitzer K (2006) Three-dimmensional modelling and inversion of dc resistivity data incorporating topography—I modelling. Geophys J Int 166:495–505

Valois R (2011) Caractérisation structurale de morphologies karstiques superficielles et suivi temporel de l'infiltration à l'aide des méthodes électriques et sismiques. Paris, Univ. Paris VI. PhD: 244

Valois R, Bermejo L, Guerin R, Hinguant S, Pigeaud R, Rodet J (2010) Karstic morphologies identified with geophysics around saulges caves (Mayenne, France). Archaeol Prospection 17(3):151–160

Zhou W, Beck BF, Stephenson JB (2000) Reliability of dipole-dipole electrical resistivity tomography for defining depth to bedrock in covered karst terranes. Environ Geol 39(7):760–766

Role of the Soil-Epikarst-Unsaturated Zone in the Hydrogeological Functioning of Karst Aquifers. The Case of the Sierra Gorda de Villanueva del Trabuco Aquifer (Southern Spain)

M. Mudarra and B. Andreo

Abstract Temporal evolutions of discharge, water chemistry (electrical conductivity, temperature, alkalinity, Cl^-, SO_4^{2-}, Ca^{2+}, Na^+, Mg^{2+}) and carbonate controlling variables (PCO_2 and $SI_{calcite}$), together with natural tracers of infiltration (TOC and NO_3^-), monitored in two karst springs of the Sierra Gorda de Villanueva del Trabuco aquifer (Southern Spain), reflect the greater relative importance of the soil-epikarst-unsaturated zone in the hydrogeological functioning of this aquifer. Each recharge event provoked an increase of discharge rates and water mineralization, the latter being due to an increase of alkalinity, and Ca^{2+}, Cl^- and SO_4^{2-} contents, while Mg^{2+} content decreased. Moreover, these variations were mostly accompanied by the rise of TOC content, while concentration of NO_3^- only rose during the first flood episodes (normally in autumn), and progressively fell during winter and spring times. Water temperature varied annually in a similar way to changes in air temperature.

1 Introduction

Hydrogeological responses of karst springs are widely investigated in recent decades with the aim of obtaining basic information about the behavior of the aquifers they drain. In this respect, environmental soil tracers contribute to understand (qualitatively) the hydrogeological functioning of carbonate aquifers

M. Mudarra (✉) · B. Andreo
Department of Geology and Centre of Hydrogeology,
University of Málaga (CEHIUMA), 29071 Málaga, Spain
e-mail: mmudarra@uma.es

B. Andreo
e-mail: andreo@uma.es

B. Andreo et al. (eds.), *Hydrogeological and Environmental Investigations in Karst Systems*, Environmental Earth Sciences 1,
DOI 10.1007/978-3-642-17435-3_7

and some of the hydrogeochemical processes (dynamic of infiltration, transit time, flow conditions, transport, etc.) taking place within them (Batiot et al. 2003; Perrin et al. 2003; Hunkeler and Mudry 2007; Mudarra et al. 2014). Moreover, a detailed and widely monitoring of environmental tracers, jointly with other natural responses, provide insights into the structure and dynamics of karst aquifers, especially regarding to the location of storage (soil-epikarst, unsaturated or saturated zones) and the degree of participation of these in the functioning of the system (Hunkeler and Mudry 2007; Mudarra and Andreo 2011).

In this work, data series of discharge, water temperature, and hydrochemistry (including environmental tracers) from two karst springs located in southern Spain, have been coupled and used with the aim of characterizing the hydrogeological processes (infiltration, residence time, functioning, etc.) that take place within the Sierra Gorda de Villanueva del Trabuco aquifer (Fig. 1). This permits to define the role of the saturated and the unsaturated zones (including soil and epikarst) in the hydrogeological functioning of this aquifer.

2 Study Site

The Sierra Gorda of Villanueva del Trabuco aquifer (6.3 km^2) is located approximately 30 km north of the city of Málaga, Southern Spain (Fig. 1). The altitudes ranging from 800 to 1,450 m a.s.l. in this steep area. The prevailing climate is temperate Mediterranean, with a marked seasonal pattern in the annual distribution of rainfalls (mainly in autumn and winter). The mean annual precipitation recorded during the study period (September 2006–March 2009) was 678 mm (Mudarra 2012), with differences in the quantity year by year: 624 mm (2006/2007), 564 mm (2007/2008) and 847 mm (2008/2009). The mean annual air temperature for the study period has varied between 11.1 °C, in high areas, and 17.1 °C in aquifer borders (Mudarra 2012).

The Sierra Gorda of Villanueva del Trabuco aquifer consists of 400–450 m thick of fractured and karstified Jurassic dolostones and limestones (Martín-Algarra 1987), being bounded at the base by Upper Triassic clays and evaporite rocks (mainly gypsum), while at the top there are Cretaceous-Paleogene marly-limestones and marls (Fig. 1). The geological structure is characterized by a normal limb, with moderate dip (30°), of an anticline fold with vergence toward S-SE, from which overthrust has developed with similar vergence over Cretaceous-Paleogene marls (Fig. 1b). This overthrust individualizes the aquifer of Sierra Gorda de Villanueva del Trabuco of another located to the southeast (aquifer of Sierras de Camarolos and del Jobo). At the north and western borders of aquifer, Flysch-type clays and sandstones outcrops. Aquifer recharge takes place through direct infiltration of rainwater, while discharge is mainly produced through springs located at the northern border of carbonate outcrops (Fig. 1a): Pita spring (24 L/s annual mean discharge rate) situated at 825 m a.s.l., and Eulogio spring (55 L/s) located at 835 m a.s.l.

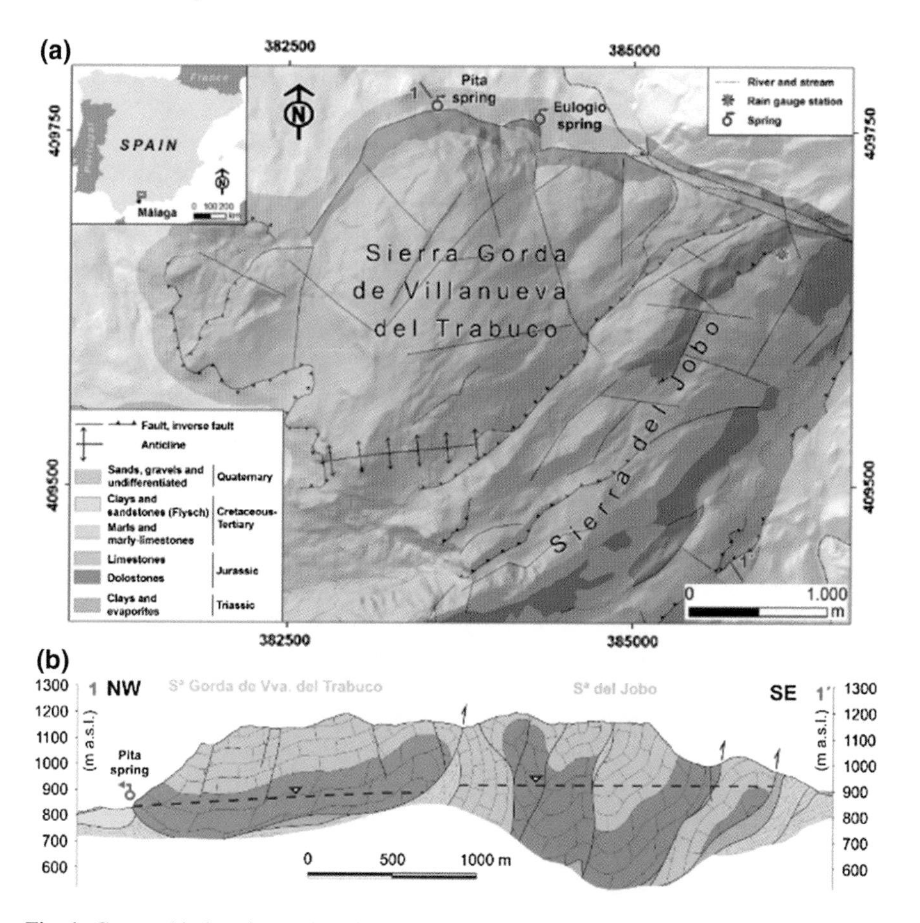

Fig. 1 Geographic location and geological-hydrogeological sketch (**a**) and cross-section (**b**) of Sierra Gorda of Villanueva del Trabuco aquifer. Altitude in meters above sea level

3 Methodology

From September 2006 to March 2009, records of discharge, electrical conductivity (EC) and water temperature, together with the most common hydrochemical parameters (including Total Organic Carbon–TOC-), were monitored in the water drained by Pita and Eulogio springs. In addition, precipitation was also recorded daily at a rainfall station located on the carbonate outcrops (1,130 m a.s.l., Fig. 1a). During the study period, sampling periodicity was adapted to the different hydrological conditions that affected to the aquifer (recharge, recession, depletion): daily in high flow and every 2 weeks during periods of depletion (average sampling periodicity was weekly). The total number of samples obtained was 96 at Pita spring and 87 at Eulogio spring (in 500 ml PVC bottles, stored at 4 °C, analysis within 48 h). In the field, and at the same time as the samples were obtained, water temperature and EC were measured in situ using portable equipment WTW, Cond 315i with a precision of ±0.1 °C and ±1 μS/cm, respectively. In addition to hand measurements, Pita spring was monitored with data logger devices (WTW, Cond 340i), providing an hourly record of water temperature and EC.

The hydrochemical parameters considered were analyzed at the laboratory of the Centre of Hydrogeology of University of Malaga. Alkalinity (TAC) was measured by volumetric titration using 0.02 N H_2SO_4 to pH 4.45. Major components were performed using high pressure liquid ion chromatography (HPLC, Metrohm 791-Basic IC model), with an accuracy of ±0.01 mg/L (previously filtered in line and precolumn-filter). TOC content was measured by combustion (after HCl treatment) of the organic matter present in the samples using a Shimadzu V-TOC carbon analyser, calculated as total carbon (TC) less inorganic carbon (IC). The CO_2 partial pressure (PCO_2) and the saturation index with respect to calcite ($SI_{calcite}$) were calculated using the EQ3NR code (Wolery 1992).

4 Results

Figures 2 and 3 show the temporal evolution of discharge, EC and water temperature values, together with the hydrochemical parameters analyzed in the water drained by Pita and Eulogio springs in the present research; this figure also shows the rainfall recorded in the area for the same time period. Analysis of data obtained in both springs permits to observe relatively quick variations in flow rates as response to precipitation events. These variations are more rapid and accused in Eulogio spring (ranges from 1 to 650 L/s) than in Pita spring (from 7 to 219 L/s). In both cases, increases in discharge rates are accompanied by a sympathetic rise in EC (Figs. 2 and 3), which later falls during the periods of low water conditions. The magnitudes of these increases were proportional to the quantity and intensity of precipitation events. Thus, the maximum variation observed in Pita spring was of 60 μS/cm (autumn 2008), whereas it was of 90 μS/cm (February 2009) in case of Eulogio spring.

The temporal evolution of water temperature at Pita and Eulogio springs presented a seasonal variation mainly influenced by monthly air temperature changes in the area (Figs. 2 and 3): the lowest values (12.6 °C at Pita spring; 12.2 °C at Eulogio spring) occur with the winter minimums of the air temperature (coinciding with high water conditions), whereas the maximum values (13.3 °C at Pita and Eulogio springs) were recorded in summer, during periods of depletion. At Eulogio spring, this seasonal evolution is incomplete due to a lack of measures during the summer months. Additionally, some significant decreases in temperature can exist at this spring (0.4 °C), as consequence of the different recharge events that took place in autumn and winter. Nevertheless, rainfalls had little or no influence on hydrothermal response of Pita spring.

Variations in water mineralization are caused by corresponding changes in Alkalinity and Ca^{2+} content (Figs. 2 and 3), which increase progressively at the same time as the flow rate. During or immediately after recharge events, an

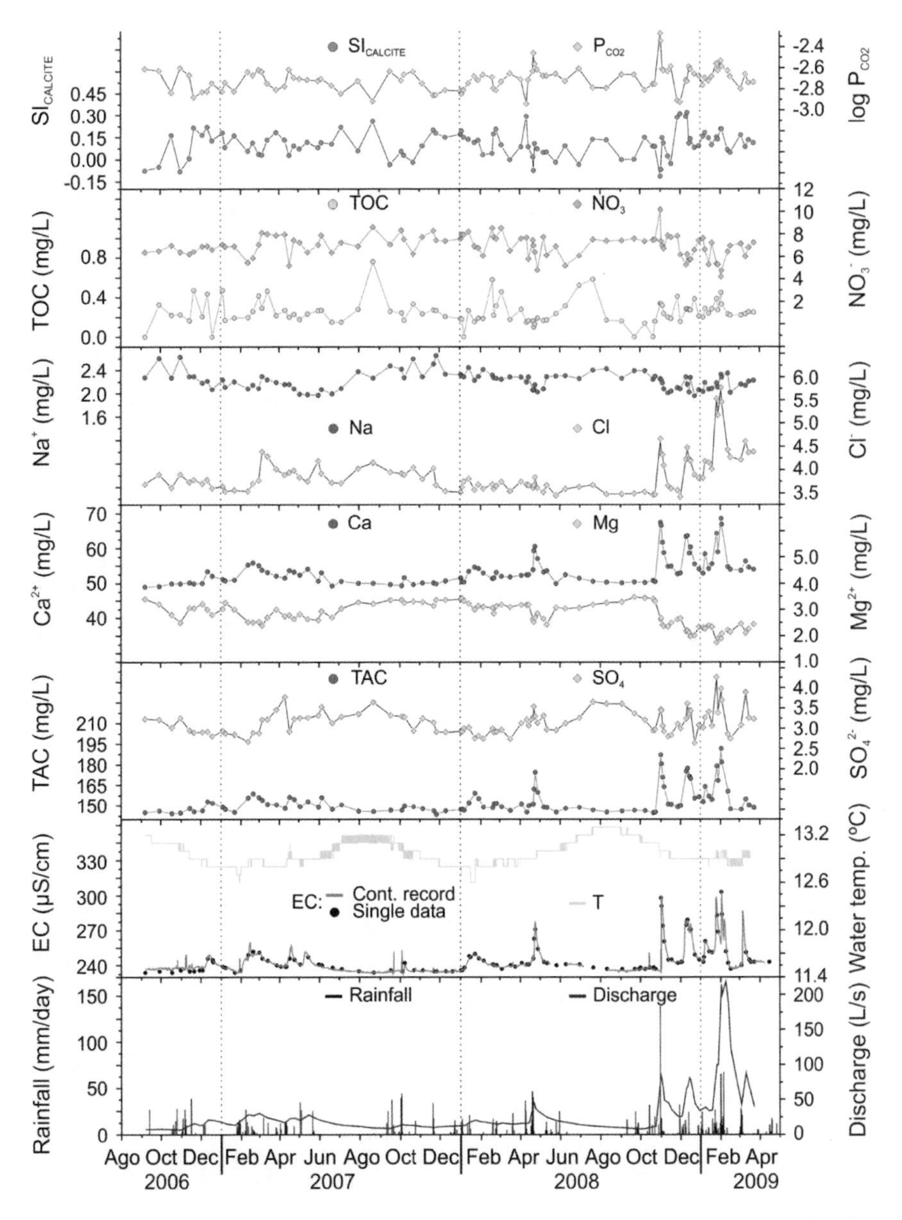

Fig. 2 Temporal evolution of discharge rates, water temperature and the principal chemical components of the water from Pita spring, with respect to precipitation events

increase in Cl^- and, in lesser extent, in SO_4^{2-} contents took place. Variations of all these chemical parameters are slightly higher in the water of Eulogio spring than in that of Pita spring. On the other hand, the greatest concentrations of Mg^{2+} were detected in the spring waters during depletion periods, especially at the

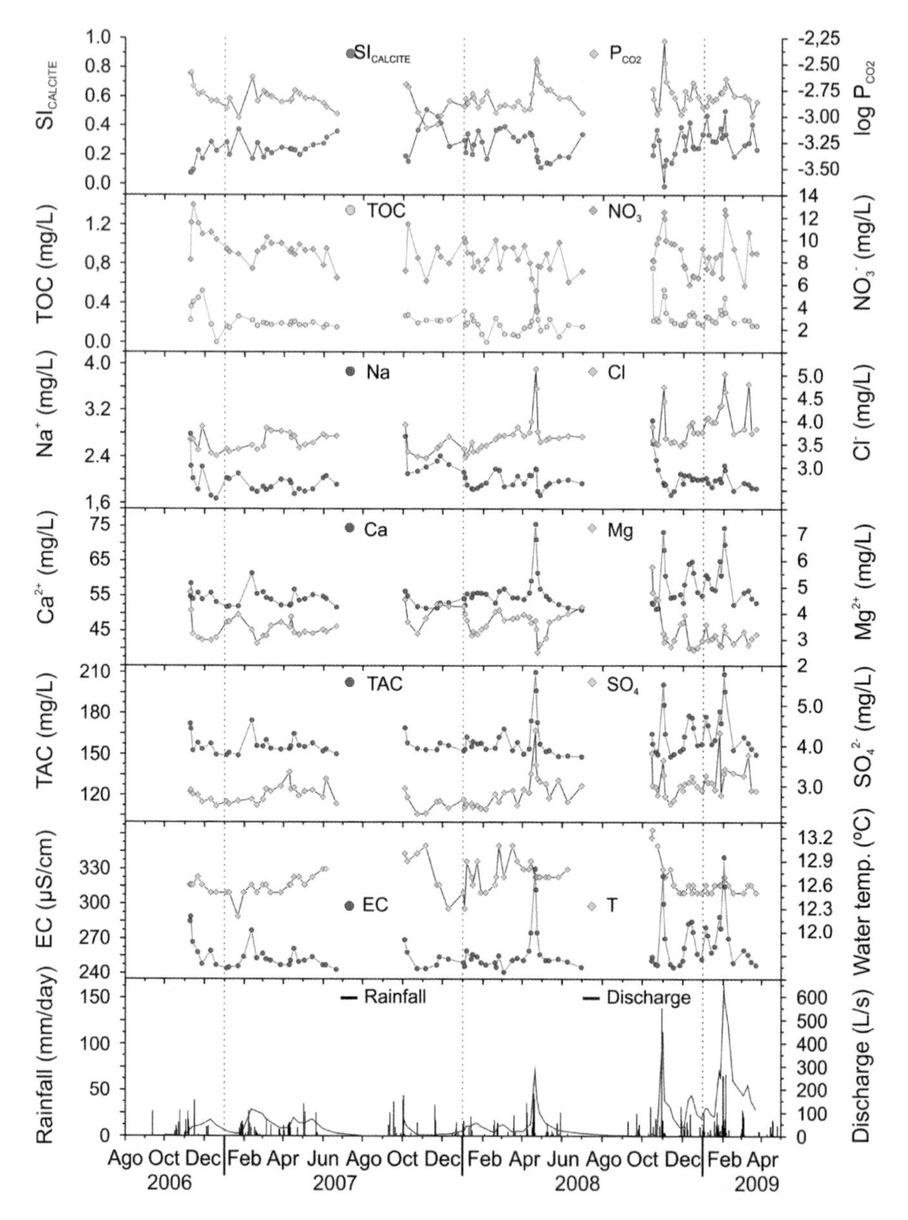

Fig. 3 Temporal evolution of discharge rates, water temperature and the principal chemical components of the water from Eulogio spring, with respect to precipitation events

beginning of the hydrological year, with falls during or immediately after recharges. Na^+ content did not follow the general pattern of EC; on the contrary, they show a general trend similar to Mg^{2+} content, which could indicate, at least partly, the same origin for both in the saturated zone of the aquifer. In fact, temporal evolutions of Na^+ and Cl^- are not always coincident.

With respect to TOC and NO_3^- contents, both natural tracers of recently infiltration, they trended to be higher at the beginning of the hydrologic year (specially at Eulogio spring), when the first recharge events occurred, and fell during the subsequent periods of recession. Other less accentuated increases in TOC content could be observed at springs as consequence of most abundant precipitation events recorded at winter and spring-time. In opposition, the concentration of NO_3^- fell in both springs during the same periods, except at Eulogio spring in February 2009, when the values of this parameter again increased (Fig. 3). During summer of 2007 and, above all, during the summer months of 2008, there was an increase and subsequent decrease in TOC at Pita spring (Fig. 2), which was not caused by precipitation events, but probably by single-point livestock concentrations in the surroundings of the spring (Mudarra 2012).

PCO_2 values calculated in the water of both springs showed similar temporal evolutions and ranges of variation (Fig. 3): they rose during recharge periods and suddenly decreased after few days. This can be specially observed in the precipitation event of November, 2008 (Figs. 2 and 3), which coincided with an increase in TOC and NO_3^- content, coming from the edaphic layer. In general, $SI_{calcite}$ behaves an opposite ways to PCO_2, with decreases in recharge events and increases after these. During the most abundant rainfall, water of both springs are near to equilibrium or slightly subsaturated respect to calcite, whereas in flow recession and in low water condition they tended to be over-saturated. Changes on $SI_{calcite}$ and PCO_2 indicate that, during recharge periods, springs drained very recent infiltrated water, with high CO_2 content from the soil and epikarst. These aggressive waters provoke rapid calcite dissolution (increases in TAC values and Ca^{2+} contents), which favor conditions near to equilibrium in the outflows.

5 Discussion

According to previous authors (Lastennet and Mudry 1997; Emblanch et al. 1998; Batiot et al. 2003; Perrin et al. 2003), the hydrochemical variations recorded at the water drained by Pita and Eulogio springs reflect the greater relative importance of the soil-epikarst-unsaturated zone in the hydrogeological functioning of the sierra Gorda de Villanueva del Trabuco aquifer (Mudarra and Andreo 2011). Each increase of springs flow rates during periods of recharge is followed by a rise in EC values and of the components most influencing it (TAC, Ca^{2+}, Cl^- and SO_4^{2-}), simultaneously to a fall in Mg^{2+} content; the latter indicator of residence time of the water within the aquifer (Batiot et al. 2003), especially in the saturated zone. In addition, these variations are usually accompanied with increases of diverse magnitude in TOC contents. This pattern reveals the arrival to the springs of water that was previously stored in the soil and epikarst, and within the fractures and conduits of the unsaturated zone, where it was affected by evaporation processes and dissolution of carbonate minerals during periods without rainfalls. Later, during high flow conditions, this water was rapidly pushed and mobilized toward springs by new infiltration water (piston effect in the unsaturated zone; Mudarra and Andreo 2011). Likewise, it is possible that a fraction of TAC and Ca^{2+} variations in spring waters would also be a consequence of rapid calcite dissolution in the unsaturated zone of the aquifer by rapid infiltration of water with high CO_2 content (Bakalowicz 1979; Dreybrodt 1988). Nevertheless, aggressive infiltration water, with higher PCO_2 values, would provoke a decrease in $SI_{calcite}$ values during high flow conditions (Figs. 2 and 3).

The above-mentioned hydrochemical variations, although of slight magnitude in some cases (Figs. 2 and 3), took place relatively quickly, which suggests that, in response to rises in flow rate, the water that was previously stored within the fractures of the unsaturated zone is rapidly mobilized toward the springs. During the dry year 2006/2007, the first water arriving at springs was the water from the lower part of the unsaturated zone, which was of greater TAC and had a higher Ca^{2+} content (Fig. 2), followed by the water coming from soil and epikarst (with higher contents of TOC, Cl^- and, in some recharges, also NO_3^-). The lag between the maximum of TAC and Ca^{2+} content, on the one hand, and those of TOC and Cl^- on the other, could be indicative of the transit time of the water through the unsaturated zone, from the soil and epikarst to the spring (exceeding 2 days; Mudarra and Andreo 2011). In opposition, during the most abundant pluviometric events of 2008 and 2009, no lags between these components were detected (Figs. 2 and 3), which could be interpreted as a result of a homogenization of the infiltration water within the unsaturated zone (from soil and epikarst until the lower part of unsaturated zone), previously to the spring discharge.

PCO_2 variations were, in some cases, similar to those of TOC contents, overall at Eulogio spring (Fig. 3). Normally, a simultaneous enrichment of CO_2 and organic matter take place in the edaphic layer. However, an increase of PCO_2 values at the water of Pita spring was observed during some floods, with no changes (or small) in TOC contents (Fig. 2). This might be the result, on the one hand, to the poorly organic matter production in the soil and, on the other hand, to a limited degassing in the unsaturated zone, where low PCO_2 values obtained in soil would be kept.

During low water conditions, the saturated zone also plays a role in the hydrogeological functioning of the aquifer. The greater residence time of the water within the aquifer causes small increases in Mg^{2+} content. However, the very slight increase in the concentration of this cation could be also related to the groundwater flow through the unsaturated zone (Batiot et al. 2003; Mudarra and Andreo 2011). This explanation is consistent with the fact that Pita and Eulogio springs present the lowest mean temperature and EC values of all springs located in the region (Mudarra 2012), being presumably associated with superficial groundwater flows within the aquifer. The time lag observed during some flood events between the maximum of TAC, Ca^{2+} and Cl^- content, and the minimum of Mg^{2+}, especially at Eulogio spring (Fig. 3), could be also associated to the transit time of water from the surface to the unsaturated zone (lower than 1 day), minor than from the surface to the saturated zone (greater to 1 day).

6 Conclusions

Pita and Eulogio springs respond to the most important precipitation events with significant increases in groundwater flow (faster and higher in Eulogio spring), sometimes slightly delayed in time. In this case, the mineralization of the water increases rather than decreasing because it contains larger quantities of most of its chemical components (piston effect of the unsaturated zone), except Mg^{2+}, which is indicative of the residence time of the water within the saturated zone of this site. This hydrogeological behavior indicates that the aquifer drained by these springs has a moderate degree of functional karstification, even within the unsaturated zone, which seems to affect its functioning more than does the saturated zone (especially under recharge situations) and provokes a less attenuated response to rainfall events.

Acknowledgments This work is a contribution to the projects CGL2008-06158 and CGL-2012-32590 of DGICYT and IGCP 598 of UNESCO, and to the Research Group RNM-308 of the Junta de Andalucía.

References

Bakalowicz M (1979) Contribution de la géochimie des eaux à la connaissance de l'aquifère karstique et de la karstification. Ph.D. These. Sci. Nat. University Pierreand Marie Curie

Batiot C, Liñán C, Andreo B, Emblanch C, Carrasco F, Blavoux B (2003) Use of TOC as tracer of diffuse infiltration in a dolomitic karst system: the Nerja cave (Andalusia, Southern Spain). Geophys Res Lett 30(22):2179. doi:10.1029/2003GL018546

Dreybrodt W (1988) Processes in karst systems. Springer, Berlin

Emblanch C, Blavoux B, Puig JM, Mudry J (1998) Dissolved organic carbon of infiltration within the autogenic karst hydrosystem. Geophys Res Lett 25:1459–1462. doi:10.1029/98GL01056

Hunkeler D, Mudry J (2007) Hydrochemical methods. In: Goldscheider N, Drew DP (eds) Methods in karst hydrogeology. Taylor & Francis, London, pp 93–121

Lastennet R, Mudry J (1997) Role of karstification and rainfall in the behavior of a heterogeneous karst system. Environ Geol 32:114–123

Martín-Algarra M (1987) Evolución geológica alpina del contacto entre las Zonas Internas y Externas de la Cordillera Bética. Ph.D. Thesis, University of Granada

Mudarra M (2012) Importancia relativa de la zona no saturada y zona saturada en el funcionamiento hidrogeológico de los acuíferos carbonáticos. Caso de la Alta Cadena, sierra de Enmedio y área de Los Tajos (provincia de Málaga). Ph.D. Thesis, University of Málaga

Mudarra M, Andreo B (2011) Relative importance of the saturated and the unsaturated zones in the hydrogeological functioning of karst aquifers. The case of Alta Cadena (Southern Spain). J Hydrol 397(3–4):263–280. doi:10.1016/j.jhydrol.2010.12.005

Mudarra M, Andreo B, Barberá JA, Mudry J (2014) Hydrochemical dynamics of TOC and NO3- contents as natural tracers of infiltration in karst aquifers. Environ Earth Sci 71:507–523. doi:10.1007/s12665-013-2593-7

Perrin J, Jeannin PY, Zwahlen F (2003) Implications of the spatial variability of infiltration-water chemistry for the investigation of a karst aquifer: a field study at Milandre test site, Swiss Jura. Hydrogeol J 11:673–686

Wolery TJ (1992) EQ3NR, a computer program for geochemical aqueous speciation-solubility calculations: theoretical manual, user's guide and related documentation (Version 7.0.). Report UCLR-MA-110662 Pt III. Lawrence Livermore National Laboratory, Livermore, USA

Significance of Preferential Infiltration Areas for Groundwater Recharge Rate Estimated with APLIS in the Mountain Karst Aquifer System of Sierra de las Nieves (Southern Spain)

C. Guardiola-Albert, S. Martos-Rosillo, J.J. Durán, E. Pardo-Igúzquiza, P.A. Robledo-Ardila and J.A. Luque Espinar

Abstract Researchers have been troubled with finding a reliable technique for estimating groundwater recharge in carbonate aquifers. Many studies have recognized the importance of preferential flow in karst systems. Despite this evidence, preferential infiltration has no effect on estimating recharge with a classical soil water budget. The present research aims to determine the significance of the correct location of preferential infiltration areas in estimating recharge with the APLIS method. The study was carried out in Sierra de las Nieves aquifer (Southern Spain). The effect of correctly estimating the preferential infiltration areas is studied through different levels of complexity in the geomorphological information: (i) a classical geomorphologic map with pothole databases available, (ii) a reviewed geomorphologic map, and (iii) fracture density and epikarst cartography obtained with field work and remote sensing interpretation. The obtained results provided a mean difference of more than 10 % of recharge in the whole aquifer, and up to 20 % when the pixel scale is considered at.

Keywords APLIS · Preferential infiltration · Recharge estimation · Sierra de las Nieves

C. Guardiola-Albert (✉) · S. Martos-Rosillo · J.J. Durán · E. Pardo-Igúzquiza · P.A. Robledo-Ardila · J.A. Luque Espinar
Geological Survey of Spain, Instituto Geológico y Minero de España (IGME), C/Ríos Rosas, 23, 28003 Madrid, Spain
e-mail: c.guardiola@igme.es

B. Andreo et al. (eds.), *Hydrogeological and Environmental Investigations in Karst Systems*, Environmental Earth Sciences 1,
DOI 10.1007/978-3-642-17435-3_8

1 Introduction

Groundwater recharge is the amount of rainwater entering to the aquifer during a given period of time. However, recharge may also come from surface water sources or from the groundwater of adjacent aquifers. Recharge may be defined as a percentage of annual precipitation, and is then known as the rate of recharge or of effective infiltration. Knowledge of aquifers recharge is important because it allows identifying the water inputs or resources entering the aquifer, that is, the amount of water potentially available for sustainable human use. This knowledge is fundamental for appropriate hydrologic planning and water management.

Recharge is an issue addressed in different manuals of Hydrogeology. However, the spatial distribution of recharge is a less studied aspect, especially in the karst aquifers due to their special characteristics in regard to its hydrogeological behavior and modalities of infiltration. During the last few years, several studies have considered this important aspect with different approaches and degrees of success (Andreo et al. 2008; Andreu et al. 2011). In Andreo et al. (2008), the APLIS method developed by IGME-GHUMA (2003) was presented. APLIS evaluates the mean recharge and the spatial distribution of the carbonate aquifers, from investigations in aquifers in the South of Spain, located in different geological and climatic conditions. This method was modified and improved by Marín (2009) adding one correction factor that depends on the hydrogeological characteristics of the materials outcropping on the surface. Lately, the method has been effectively applied in several karst Spanish aquifers and Cuban aquifers. APLIS provides a way of calculating long-term spatial average recharge. It is not recommended to quantify short-term recharge because the recharge process can be influenced by several factors such as the degree of saturation of the soil and unsaturated zone or rainfall intensity.

The experience in using this method suggests that most of the input variables in APLIS are easy to calculate from available data bases or previous reports. However, a preferential infiltration factor is typically estimated based on field works or interpreting geomorphologic maps to identify zones where infiltration through exokarstic forms is predominant. Additional sources of information can also be used: topographic maps, aerial and satellite images, speleological data, spring hydrographs, tracer tests results, and hydrochemical data. It is not easy to join all this material and often databases are not exhaustive or complete. Moreover, contradictory interpretation is sometimes possible depending on the karst expert opinion.

As a part of a bigger hydrogeological research project (KARSTINV), it was necessary to estimate the recharge in the karst Sierra de las Nieves aquifer. An unexpected finding came out when it was decided to improve the preferential infiltration map with actual and more complex sources of data. Frequently, this information is difficult to translate into a preferential infiltration factor without being subjective. This work aims on investigating the effect of different kinds of information that hydrogeologists can use to develop preferential infiltration maps on the APLIS annual recharge percentage.

2 Study Area

Sierra de las Nieves is an important relief of the Málaga province (southern Spain) with outcrops of dolostones and limestones of Triassic and Jurasic age, which are permeable by fracturation and karstification. Karst features are abundant in this aquifer. The drainage of the aquifer takes place at its southern edge through several springs. Response to precipitation is very rapid such as occurs in karst aquifers with conduit flow behavior. The mean annual precipitation is 1,000 mm/year. The values of average annual temperature of the air and the evapotranspiration calculated with the method of Hargreaves are 15 °C and 980 mm/year, respectively. The aquifer is recharged by infiltration of rainwater, and, occasionally, by snow. Using the water balance method a coefficient of recharge of 55 % was found by IGME-GHUMA (2003).

3 Methodology

3.1 APLIS Method

The APLIS method allows to estimate the mean rate of annual recharge in carbonate aquifers, expressed as a percentage of precipitation, on the basis of the characteristics of these aquifers (Andreo et al. 2008) such as altitude (A), slope (P), lithology (L), preferential infiltration layers (I), soils (S), and a correction factor (F_h) that depend on the hydrogeologic characteristics of the materials outcropping on the surface (Marín 2009). Its application requires developing a map of that data, in a geographic information system. For each of these layers of information, it applies a score system to create a map of each variable. Scores range from 1 to 10, following an arithmetic progression with a step width of 1, with the aim to easily equate to aquifer recharge rates. The value 1 indicates minimal incidence of the variable in the recharge of the aquifer, whereas the value 10 expresses the maximum influence on recharge. Ranks and scores are described in Andreo et al. (2008) and Marín (2009).

The following expression is used to estimate the recharge rate (Marín 2009):

$$R = [(A + P + 3 \cdot L + 2 \cdot I + S) / 0.9] F_h \quad (1)$$

The weight of each variable in the above expression is intended to represent its importance in determining the recharge rate, in accordance with the results obtained from the prior analysis of the variables that influence recharge. By dividing by 0.9, recharge rates are obtained that range from a minimum of 8.88 % to a maximum of 88.88 % of the rainfall onto the surface of the aquifer.

The layers of information corresponding to each variable are taken from different sources. The digital elevation model map, scale 1:10,000, was used for the mapping of altitude and slope. The geological map was grouped in lithological sets of similar hydrogeological characteristics taking into account the geomorphologic forms. The discretization used for all the maps in this work is of 5 × 5 m.

3.2 Preferential Infiltration

The APLIS method was already applied in Sierra de las Nieves (IGME-GHUMA 2003), obtaining an annual recharge rate of 54 %. In this previous work, preferential infiltration areas were assessed limiting the sectors with karst forms of favorable infiltration and slopes of less than 8 %. Geomorphological maps were also consulted for this task. In this previous study, a 12 % of the total aquifer area was assigned to have preferential infiltration distributed in two zones (Fig. 1): the Nava paleopolje and the port of Los Pilones, with low slope and many dolines and simes.

For the present study, this information has been significantly improved by satellite images, interpretation of aerial images and exhaustive field work. The evaluation of this material resulted in different levels of complexity and types of geomorphologic indicators. These maps had to be transformed into *I* scores and three different preferential infiltration maps were constructed depending on the data used (Fig. 1): *Case 1.* Geomorphological map and cells that include a polje (closed depression covering a considerable surface area) as well as the ones containing at least one significant pothole are included. Finally areas with a slope less than the 3 % are added. *Case 2.* An improved geomorphologic map of Sierra de las Nieves aquifer has been developed from exhaustive terrain work and aerial photo interpretation. In this case it was decided to create two different ranges of scores to obtain the *I* map. *Case 2a*, with just 1 (scarce exokarstic forms) or 10 (abundant exokarstic forms) values as in Andreo et al. (2008), resulting in the map shown in Fig. 1. And *Case 2b*, a new proposed score system to improve the APLIS method of grading the scores between 1 and 10 with the ranking shown in Table 1. *Case 3.* Fracture density and epikarst cartography from field work and remote sensing interpretation give the preferential infiltration areas.

4 Results and Discussion

Figure 2 shows the calculation results for each of the studied cases. The results shown are based on the application of APLIS Eq. (1) applying each of the four different maps shown in Fig. 1, and the rest of the variables remain the same. Each of the figures displays the annual recharge rate for each of the combined maps. All the figures show different values reflecting the differences of areas of preferential

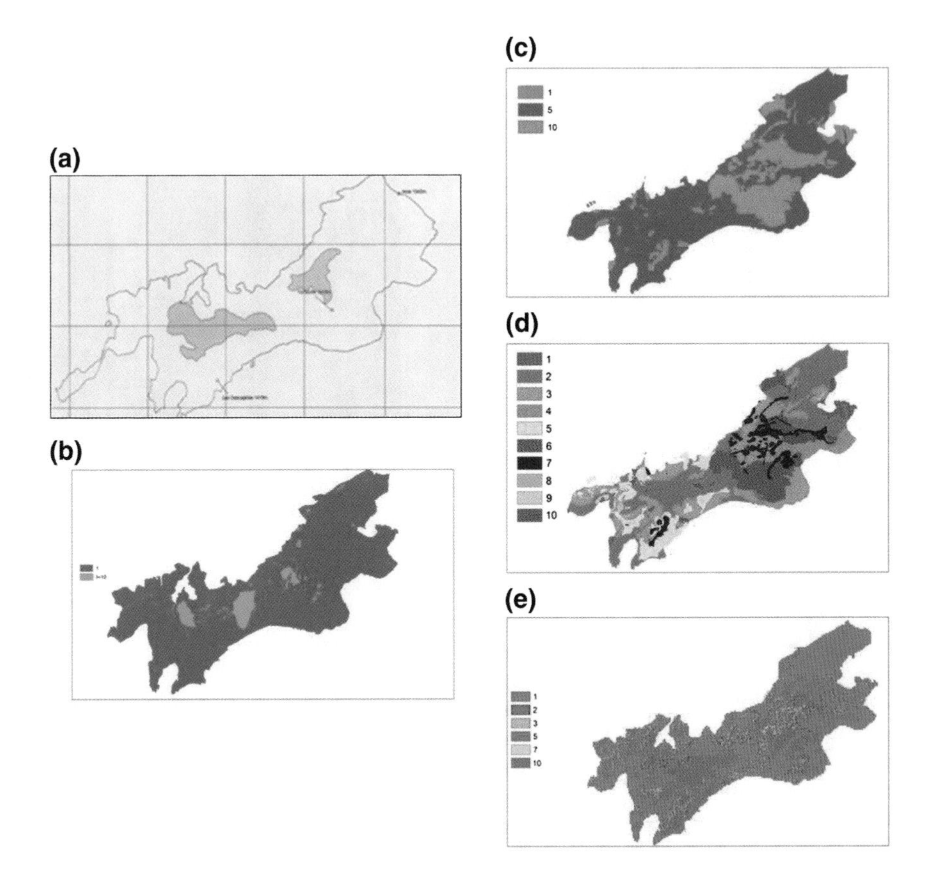

Fig. 1 Preferential infiltration areas used in the different studied cases. **a** I areas from IGME-GHUMA (2003). **b** I areas Case 1. **c** I areas Case 2a. **d** I areas Case 2b. **e** I areas Case 3

Table 1 Proposal of new *I* scores for preferential infiltration variable in APLIS

Exokarst form	*I* score
Doline	10
Polje	10
Doline with crops	9
Karst flatted	8
Karren field nuede with high development and fractures	7
Karren field nuede with high development and microformes	6
Karren field low covered and fractured	6
Quarry	5
Karren field nuede with low development	5
Karren field low covered	4
Karren field covered and fractured	3
Karren field covered	2

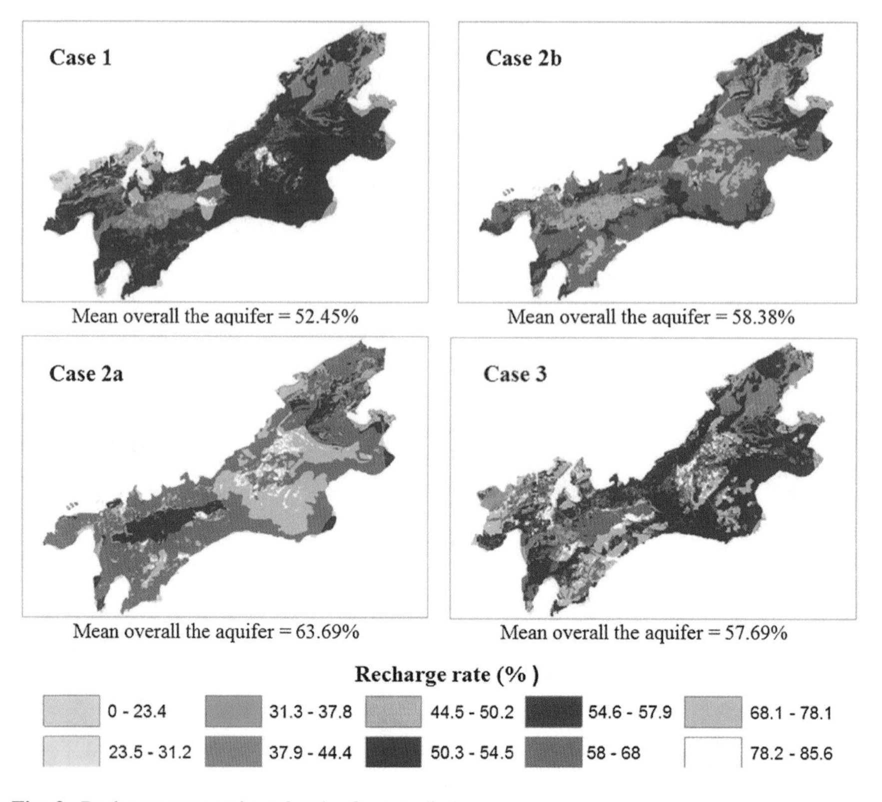

Fig. 2 Recharge rates values for the four studied cases

infiltration. *Case 1* is the one that results in the smallest values (mean of 52.45 %) and *Case 2a* in the highest (mean of 63.69 %). *Cases 2b* and *3* have very similar mean values (58.38 and 57.69 %, respectively). On the other hand *Case 2b* has less dispersed values (10.71 % of standard deviation) than *Case 3* (12.41 % of standard deviation).

The minimum recharge value (i.e., $I = 1$ in all the aquifer) is computed for comparison purposes between the studied cases. This minimum value is subtracted from the four resulted maps. Mean values of these differences are 0.9 for *Case 1*, 11.3 for *Case 2a*, 6.0 for *Case 2b* and 5.8 for *Case 3*. These numbers, again, show that *Cases 2b* and *3* are very similar. Figure 3 reflects the spatial distribution of differences and the variations among all the methods. *Cases 1*, *2a* and *3* tend to concentrate high preferential infiltration rate in some areas, depending on the expert interpretation and the data available. *Cases 1* and *3* give very low preferential infiltration rates in the rest of the aquifer, while *Case 2a* gives a medium value. For *Case 2b*, a new score table for preferential infiltration has given, with a more graduate recharge rate. New *I* scores are reflected in the results and in the

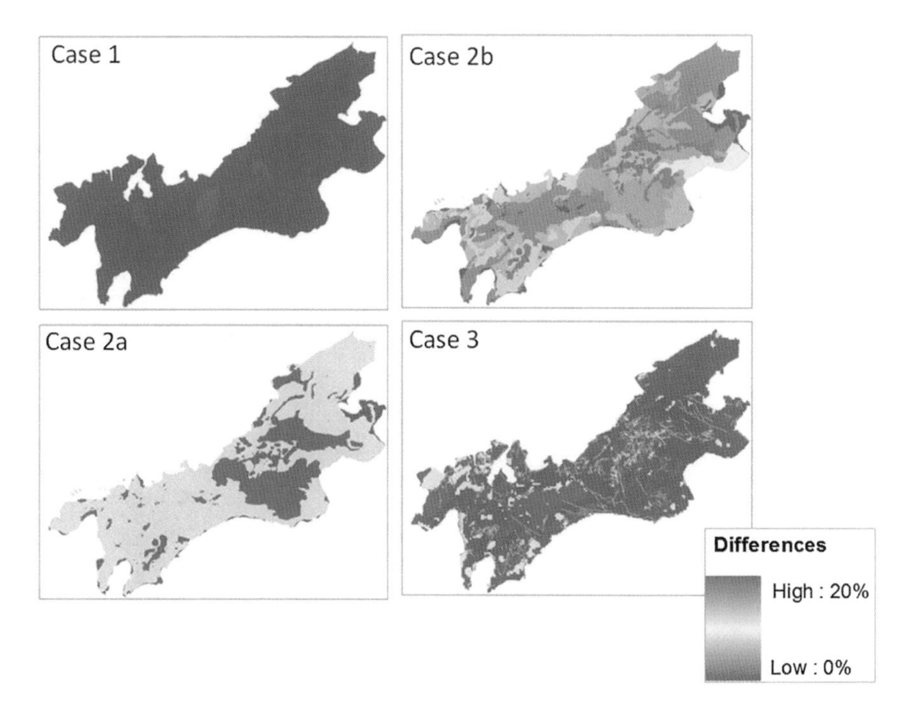

Fig. 3 Spatial distribution of differences of recharge rate values (%) for the four studied cases and the map resulted from using I = 1 all over the aquifer

differences with the map of $I = 1$ in the whole aquifer (Fig. 3). This proposal results congruent with the ones computed in IGME-GHUMA (2003) and appear to be more realistic following the recharge process in karst aquifers.

5 Conclusions

The study was able to evaluate differences introduced by the preferential infiltration factor in the recharge rate estimation. The results obtained provided a mean difference of more than 10 % of recharge in the whole aquifer, arriving to differences of 20 % when the pixel scale is looked at. The study of karst and epikarst has allowed developing a realistic estimation of the actual aquifer recharge. The proposal of using a new score for preferential infiltration gives congruent values introducing an improvement in the APLIS method.

Acknowledgments This work was supported by research project CGL2010-15498 from the Spanish National Ministry of Science and Innovation.

References

Andreo B, Vías JM, Durán JJ et al (2008) Methodology for groundwater recharge assessment in carbonate aquifers: application to pilot sites in Southern Spain. Hydrogeol J 16:911–925

Andreu JM, Alcalá FJ, Vallejos A et al (2011) Recharge to mountainous carbonated aquifers in SE Spain: different approaches and new challenges. J Arid Environ 75(12):1262–1270

IGME-GHUMA (2003) Estudios metodológicos para la estimación de la recarga en diferentes tipos de acuíferos carbonáticos: aplicación a la Cordillera Bética. 3 tomos

Marín AI (2009) Los Sistemas de Información Geográfica aplicados a la evaluación de recursos hídricos y a la vulnerabilidad a la contaminación de acuíferos carbonatados. Caso de la Alta Cadena (Provincia de Málaga). Tesis de Licenciatura, Universidad de Málaga

A Method for Automatic Detection and Delineation of Karst Depressions and Hills

E. Pardo-Igúzquiza, J.J. Durán, P.A. Robledo-Ardila, J.A. Luque-Espinar, A. Pedrera, C. Guardiola-Albert and S. Martos-Rosillo

Abstract Karst depressions of decametric scale (dolines, uvalas, poljes, and other endorheic basins) play an important role in the hydrogeology of karst aquifers. They are traps of sediment and when their detritic filling has an important thickness they can retain a large amount of water delaying their percolation towards the water table or towards the networks of conduits. Many times the delineation of the depressions may be difficult because the study area may be very large, or inaccessible or hidden by vegetation. In those circumstances, it is of great help to have an automatic method of depression detection and delineation. The proposed procedure uses the digital elevation model, a geographical information system, an algorithm of pit removal and basic operations of map algebra. The method provides the depth of each detected depression measured from its rim. This fact can be used to detect the center of maximum depth as well as for calculating

E. Pardo-Igúzquiza (✉) · J.J. Durán · P.A. Robledo-Ardila · J.A. Luque-Espinar · C. Guardiola-Albert · S. Martos-Rosillo
Geological Survey of Spain, Instituto Geológico y Minero de España (IGME), C/Ríos Rosas, 23, 28003 Madrid, Spain
e-mail: e.pardo@igme.es

J.J. Durán
e-mail: jj.duran@igme.es

J.A. Luque-Espinar
e-mail: ja.luque@igme.es

C. Guardiola-Albert
e-mail: c.guardiola@igme.es

S. Martos-Rosillo
e-mail: s.martos@igme.es

A. Pedrera
Instituto Andaluz de Ciencias de la Tierra (CSIC), Avda. de Las Palmeras 4, 18071 Granada, Spain
e-mail: pedrera@ugr.es

B. Andreo et al. (eds.), *Hydrogeological and Environmental Investigations in Karst Systems*, Environmental Earth Sciences 1,
DOI 10.1007/978-3-642-17435-3_9

morphometric parameters using depth. The final map of depressions can be characterized by altitude in order to have morphometric parameters related with elevation. The algorithm has been extended for detection and delineation of karst hills. The methodology is illustrated with the Sierra de las Nieves karst aquifer in the province of Malaga, Southern Spain, where the depressions and hills show a strong structural control.

Keywords Dolines · Pits · Hills · Geomorphology · Automatic cartography

1 Introduction

Karst depressions are one of the most typical forms of the karst landscape (also known as exokarst) at the scale from several meters to several hundreds or thousands of meters (Figs. 1 and 2). Among these depressions, one has dolines, uvalas, poljes, and other endorheic depression in karst terrains. Karst depressions are very important for different reasons. First of all, these depressions are trap of sediments which study can be used for inferring the evolution of the karst massif as well as for paleoclimatic studies. Second, these depressions have an important role in the hydrogeology of the karst, that is, in the balance of water, recharge and hydrodynamics of water flow. On the other hand, karst hills also give information of the karst evolution and on the evolution of slopes.

The mapping of both, depressions and hills, is of great importance and always must be based on field work. However, there are situations were it may be of great help to have an automatic method of identification and delineation of depressions and hills that can help to complement field work. Among those special situations are when the study area is very large, when the area is covered by vegetation or when the area is not accessible. In this work, there is a brief description of the methodology used for automatic detection and delineation of karst depressions. The methodology is extended for the detection and delineation of karst hills and there is an application to the Sierra de las Nieves karst aquifer in the province of Málaga in Southern Spain.

2 Methodology

The methodology for the automatic identification and delineation of depressions was proposed for the first time in Durán et al. (2012) and applied in detail in Pardo-Igúzquiza et al. (2013). The methodology is general for the identification of any terrain depression but in the previous two works the method was applied to karst terrains, that is, for the detection of karst depressions. The methodology is extended here for the detection of positive reliefs, that is, karst hills in the case of

Fig. 1 Examples of karst depressions in the Sierra de las Nieves karst aquifer. **a** High altitude depression; **b** Small polje with ponor in the *center* of the *image*; **c** Captured doline in the *center* of the *image*, and **d** Small circular doline with ponors (potholes) in the *border*

karst terrains. The method of karst depression identification and delineation is based on the well-known algorithm of Jenson and Domingue (1988). These authors developed an advanced method for removing the pits of digital elevation models (DEM) in order to obtain pit-free DEMs. Thus, by a simple operation of map algebra, where the difference between the original DEM and the pit-free DEM is taken, a map of depressions is obtained where there is the following information: (1) identification of the depression and delineation of its limits and (2) a map of depth of the depressions from their rims. The method gives very high resolution maps of depth if a high resolution DEM is used as shown in Fig. 3. The extension of the methodology to detection and delineation of hills is very simple. It is enough to change the sign of the DEM (so that positive reliefs become depressions) and to apply the same methodology as for detection of depressions. The maps obtained can be used in morphometric analysis and geomorphologic mapping among other uses.

Fig. 2 Ponors of different dolines of the Sierra de las Nieves aquifer. **a** Ponor with collapse. **b** Ponor as pothole on the cabonatic rock. **c** Ponor working after a rainfall event. **d** Ponor with a collapse that has cover the swallet

3 Case Study

The study area is the Sierra de las Nieves karst aquifer in the province of Málaga, Southern Spain. The karst massif consists mainly of a succession of carbonate rocks: Triassic marbles and dolostones, Jurassic limestones, and a Tertiary carbonatic breccia. The Mesozoic sequence is folded by an NE–SW trending overturned syncline with a vergency toward the NW (Liñán 2005). The carbonatic breccia unconformably lies over the Mesozoic sucession but is also deformed by the fold. The method described in the previous section has been applied to a DEM of the Sierra de las Nieves aquifer with a resolution of 5 m and provided by the Instituto Geográfico Nacional. A detail of the binary map of dolines is shown in Fig. 4 that has been obtained from a map of dolines dephs as the one in Fig. 3. Using the binary map, it is possible to calculate the density of dolines as the map shown in Pardo-Igúzquiza et al. (2013) and by applying a mask (Fig. 5) it is possible to detect the three main recharge areas that discharge at the three main karst springs. It is also of great help in the delineation of the hydrogeological basins of each karst spring and in defining areas of high probability of conduit

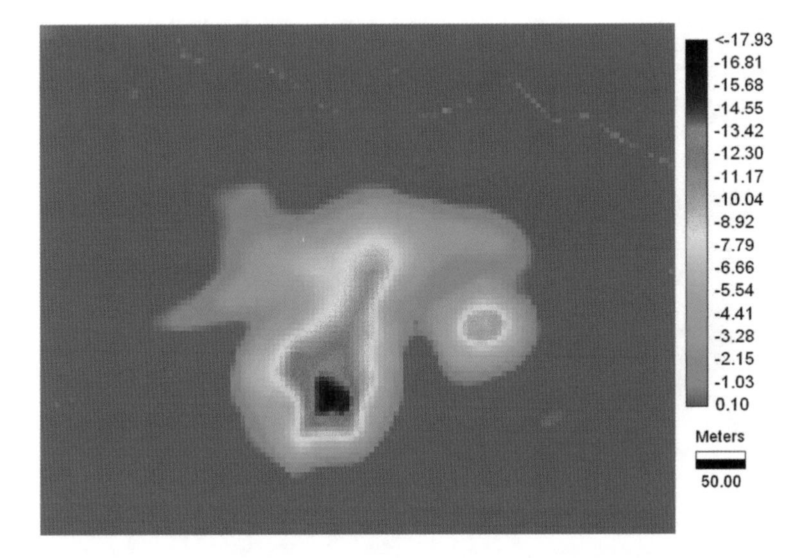

Fig. 3 High resolution dolines mapped with the proposed methodology. The *colors* represent the depth of the doline from the rim. The deepest depth of 32 m corresponds to the ponor and is reached by the doline on the *left hand side* that also shows a clear elongation along fractures. However, the doline of the *right-hand side* is very rounded (Fig. 1d) and the ponor is on the *border* of the doline as pothole in the rock (similar to Fig. 2b). The streams of depressions in the north part are artifacts along a stream and are eliminated by applying an area threshold

development. With respect to the mapping of karst hislls, Fig. 6 shows a map of the detail of an area where the hills and the depressions have been represented. The outline of hills is more related with the evolution of karst hills and shows a more rounded area or a very linear limit when associated to a fracture or fault.

4 Discussion and Conclusions

The delineation of karst depressions can be done automatically by the procedure described in this work. The results are of great application in karst hydrogeology and geomorphology. The map provides information of the depth of depression from the rim, and this can be used to evaluate many morphometric parameters using the fact that the altitude is known by their location in the digital elevation model. In addition, the number of dolines and their area can be evaluated. More morphometric parameters can be obtained from this information and maps of density of dolines can be obtained. The map of dolines can be used in geomorphology maps and the map of density of dolines can be used as shown in Figs. 4 and 5 for making hydrogeological conjectures in relation to the recharge areas, hydrogeology basin delineation and high probability areas for the development of karst conduits. Karst hills can be delineated in the same way by changing the sign

Fig. 4 Clusters of dolines (*green*) in the central part of the Sierra de las Nieves karst aquifer. The *white line* represents the border of the aquifer

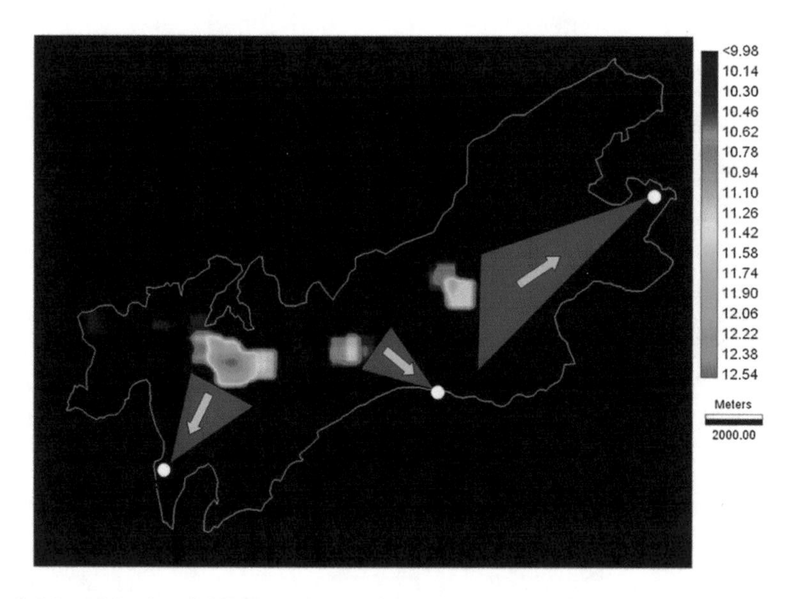

Fig. 5 Map of density of dolines. After applying a mask, there are clearly shown the three main areas that recharge the three main springs. The *triangles* show the areas of highest probability for active karst conduit drainage

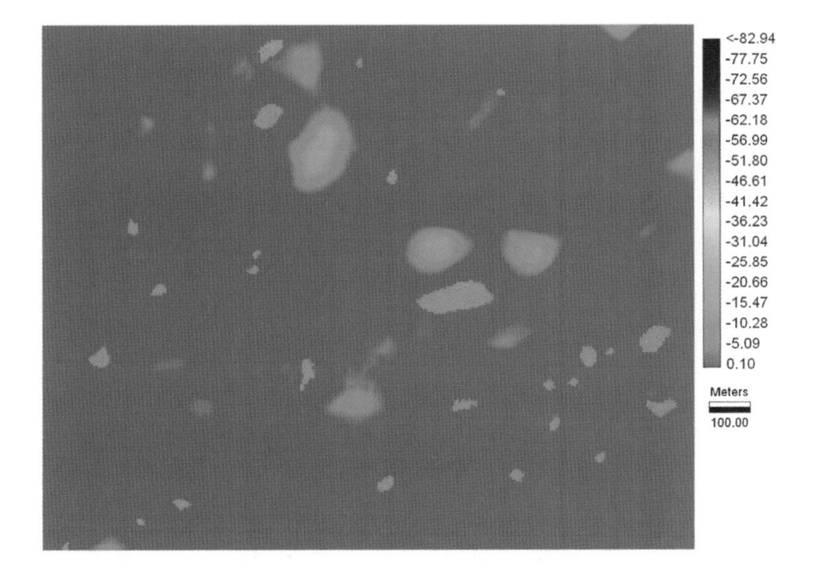

Fig. 6 Karst hills (*shadows of green*) and karst depressions (*red*) for a particular area of the Sierra de las Nieves karst aquifer. Although the disposition may look like random, there is a clear structural control when seen at a smaller scale

of the digital elevation model and applying the same methodology of delineation of depressions. The procedures presented in this work for karst depression is completely general and can be applied in other context where depressions can be lakes or craters for example.

Acknowledgments This work was supported by the research project CGL2010-15498 from the Ministerio de Economía y Competitividad of Spain. We would like to thank the groups of speleology that work in the Sierra de las Nieves: Grupo de Exploraciones Subterráneas de la Sociedad Excursionista de Málaga, Interclub Sierra de las Nieves (Ronda), Centro Excursionista del Sur Escarpe, MAINAKE Sociedad Espeleo-Excursionista y el Grupo de Exploraciones Subterráneas de Tolox. Also we would like to thank to the managers of the Parque Natural de la Sierra de las Nieves.

References

Durán JJ, Pardo-Igúzquiza E, Robledo PA (2012) Detección automática de depresiones utilizando el modelo digital del terreno y su aplicación a la cartografía geomorfológica. Avances de la Geomorfología en España (2010–2012) Actas de la XII Reunión Nacional de Geomorfología Santander, Universidad de Cantabria

Jenson SK, Domingue JO (1988) Extracting topographic structure from digital elevation data for geographic information system analysis. Photogramm Eng Remote Sens 54(11):1593–1600

Liñán C (2005) Hidrogeología de acuíferos carbonatados en la unidad Yunquera-Nieves (Málaga). Publicaciones del Instituto Geológico y Minero de España. Serie: Hidrogeología y Aguas Subterráneas 16, p 322, Madrid

Pardo-Igúzquiza E, Durán JJ, Dowd PA (2013) Automatic detection and delineation of karst terrain depressions and its application in geomorphological mapping and morphometric analysis. Acta Carsologica 42(1):17–24

Comparison of the APLIS and Modified-APLIS Methods to Estimate the Recharge in Fractured Karst Aquifer, Amazonas, Peru

K. Espinoza, M. Marina, J.H. Fortuna and F. Altamirano

Abstract Estimates of groundwater recharge were made for a karst aquifer in the Amazonas Region of northern Peru based on two methods: the APLIS method, and the modified-APLIS method. The study area is a 9.6 km^2 section of the upper La Florida Catchment, located on the eastern slope of the Andes mountain range at an elevation of approximately 2500 m. Annual precipitation is over 1,500 mm/year, with high intensity storms during the rainy season (November–April). The study area is characterized by high relief, including deep canyons, and is underlain by limestone-dolomite of the Upper Triassic Pucara Group, which contains karst and epikarst features. These conditions make it difficult to accurately estimate the magnitude and distribution of runoff, infiltration and groundwater recharge. The APLIS and the modified-APLIS methods were applied to estimate the spatial distribution and the annual rate of groundwater recharge in the study area. These methods were developed for carbonate aquifers under Mediterranean conditions, and some modifications were required to apply them to the study area in Peru. The average annual recharge rate estimated with the APLIS method is 48 % of the total rainfall, with high recharge areas occurring in canyons where porous dolomite outcrops. The average recharge with the modified-APLIS method is 24 %. The difference in the recharge rate by these methods is considerable, while the areas of greatest recharge are similar mainly due to the application of aquifer and non-aquifer designations common to both methods. The recharge distributions estimated with these methods were introduced into a numerical groundwater flow

K. Espinoza (✉) · M. Marina · J.H. Fortuna · F. Altamirano
Water Resources Department, Mining Environmental Group, Pedro de Osma Avenue 418, Barranco-Lima (Lima 04), Peru
e-mail: kespinoza@klohn.com

M. Marina
e-mail: mrojo@klohn.com

J.H. Fortuna
e-mail: jfortuna@klohn.com

F. Altamirano
e-mail: faltamirano@klohn.com

B. Andreo et al. (eds.), *Hydrogeological and Environmental Investigations in Karst Systems*, Environmental Earth Sciences 1,
DOI 10.1007/978-3-642-17435-3_10

model for the study area. The calibration process indicates that the modified-APLIS method provides a more reasonable representation of the recharge rates for the study area.

Keywords Carbonate aquifer • APLIS method • APLIS modified method • Recharge and karst

1 Introduction

There are different mechanisms of assessing groundwater recharge under various geological and climate conditions. Estimating groundwater recharge in karst environments presents unique problems, due to the inherent characteristics of karst landforms and the dual nature of recharge. The APLIS method (Andreo et al. 2008) and the modified-APLIS method (Marín 2009) were developed to estimate the mean annual recharge in carbonate aquifers under Mediterranean conditions. The APLIS method was subsequently applied to a karst aquifer terrain in Cuba (Farfán et al. 2010), and produced acceptable results. In this study, was applied these techniques for recharge estimation in karstic areas in the Amazonas Region of Peru.

The aim of this study is (a) to estimate the autogenic recharge into the carbonate aquifer trough the APLIS and modified-APLIS methods in a Peruvian aquifer, and (b) to evaluate the suitability of these methods to our study site.

2 Hydrogeological Characteristics

The study area is a 9.6 km^2 section of the upper La Florida Catchment, located in the Amazonas Region of northern Peru. The site sits on the eastern slope of the Andes mountain range at an elevation of approximately 2,500 m. The site has high relief, steep-topography, and high cliff faces exposed in canyons.

The study area is an exploration project, and the site of a planned underground Pb–Zn mine. Estimates of recharge are necessary to constrain potential groundwater inflows to a proposed underground mine, and assist with mine water management planning. Data from the exploration project was used to develop a conceptual hydrologic and hydrogeologic model for the site, which was supplemented by spring surveys, hydrologic mapping, soil infiltration tests, and underground hydrogeologic mapping and flow measurements.

The site is underlain by the Triassic-Jurassic Pucará Group, which includes three formations from oldest to youngest: the Chambará Fm (bottom), the Aramachay Fm (middle), and the Condorsinga Fm (upper). The Chambará formation directly underlies most of the project site and contains three members simply classified as Chambara 1 (basal member), Chambara 2, and Chambara 3 (Fig. 1).

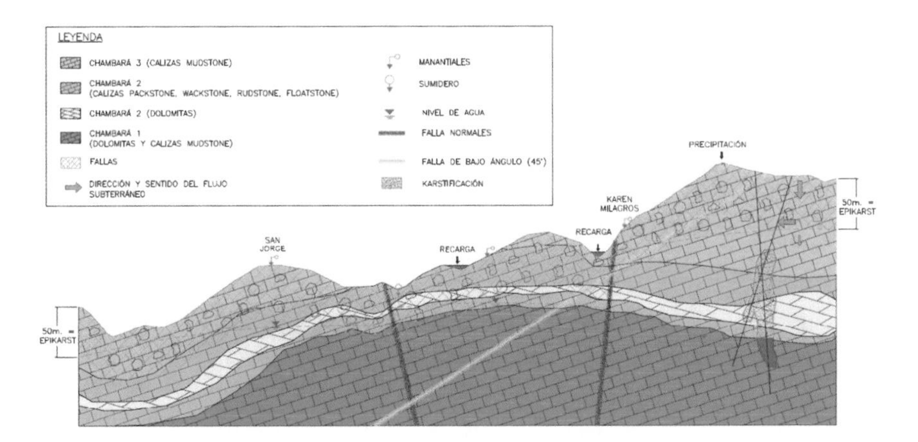

Fig. 1 Schematic hydrogeological conceptual model of Florida catchment

Limestone and dolomite of these formations form a carbonate aquifer system. The middle member contains an altered, porous dolomite layer hosts the orebody.

The La Florida catchment is characterized by small epikarst features in the upper 10–50 m, especially in the fault-controlled tributary canyons where fractured bedrock is prevalent. Exploration drilling and tunneling have encountered caves and karstic features at greater depths of up to 500 m below ground, but these features are rarely expressed at ground surface.

Despite high annual precipitation of 1,556 mm/year, some surface water streams lack flow except during intense rainfall events, and shallow infiltration appears to be rapid in many areas. In other streams, flow originates mid-valley from karst features. In the higher elevation valleys, thin soil cover with low vertical permeability is present.

In general, in all locations the limestone layers have very low primary permeability where unfractured, and field evidence and observations indicate that subsurface water migration is predominantly sub-horizontal along bedding planes. Shallow infiltration from the highlands typical remerges as seeps and springs along fractures and bedding planes high in canyon walls.

These conditions make it difficult to accurately estimate the magnitude and distribution of runoff, infiltration, and groundwater recharge.

3 Application of the APLIS Method

The APLIS method (Andreo et al. 2008) and the modified-APLIS method (Marín 2009) were developed to estimate the mean annual recharge in carbonate aquifers under Mediterranean conditions. They are GIS-based methods that incorporate the intrinsic variables that influence recharge: Altitude (A), Pendiente (P) or slope, Lithology (L), Infiltration (I), and Soil type (S).

The study area rises from an elevation of 1,800 m a.s.l in the southwest to more than 2,700 m a.s.l in the northeast. Based on elevation, only 4 of the 10 possible APLIS altitude categories (A) are presented.

Study area values for the slope term (P) are high, reflective of the steep slopes in the area. Approximately 58 % of the area has slopes at angles greater than 46 %. Only 6 % of the study area has slopes less than 21 %.

The lithology (L) has been categorized based on the existing local geological, geomorphological and hydrogeological maps, as well as field inspections and observations. The Aramachay Fm, which consists of shales, limestones, and turbidites, has been assigned as category 3, "gravel and sand" due to a lack of evidence of fissuring or karstification. The Chambara 1 Fm has been assigned to category 4, because the dolomite, limestone and mudstone of this formation show no evidence of karstic features. The limestones of Chambara 2 Fm have been assigned to category 7, "limestones and fractured dolomites, somewhat karstified", while the porous dolomite layer of Chamabra 2 Fm has been classified as category 9 because of the higher degree of fracturing and karstification. The Chambara 3 Fm, which consists of dolomite, limestone, mudstone, shales, and turbidites has been assigned to category 5, "limestones, cracked dolomites". This formation directly underlies over 90 % of the study area.

The infiltration landform (I) term was set by assigning preferential infiltration forms to areas observed during field works, including the main faults and fault controlled valleys, dry stretches of streams, and the shallow karst cavities with a depth below surface of less than 10 m. However, these features make up a very small percentage of the study area.

Soil maps based on the FAO classification (Farfán et al. 2010) normally used with the APLIS method were not available for the study area. The Soil (S) parameter values were assigned based on a site soils map using USDA Soil Taxonomy classifications (1999), and modified based on field infiltrometer tests undertaken using the Guelph Permeameter to assess relative soil permeability.

Three types of soils (S) were classified in the study area, Mollisols Udolls Typic Hapludols, Ultisols Udults Typic Paleudults and Inceptisols Udepts Humic Dystrupets.

Areas mapped as Mollisols Udolls Typic Hapludols were assigned an S value of 4 based on soil characteristics and infiltration tests. Udults Typic Paleudults soils were assigned a value of 2, similar to "planasols" in the Farfán et al. classification. The Inceptisols Udepts Humic Dystrupets were assigned value of 1, as they are similar to "vertisols". The quaternary soils located in the areas around streams are assigned value 8 "calcareous regosols and fluvisols".

For both the APLIS and APLIS-modified methods, the recharge rates were calculated using ArcGIS Desktop 10.

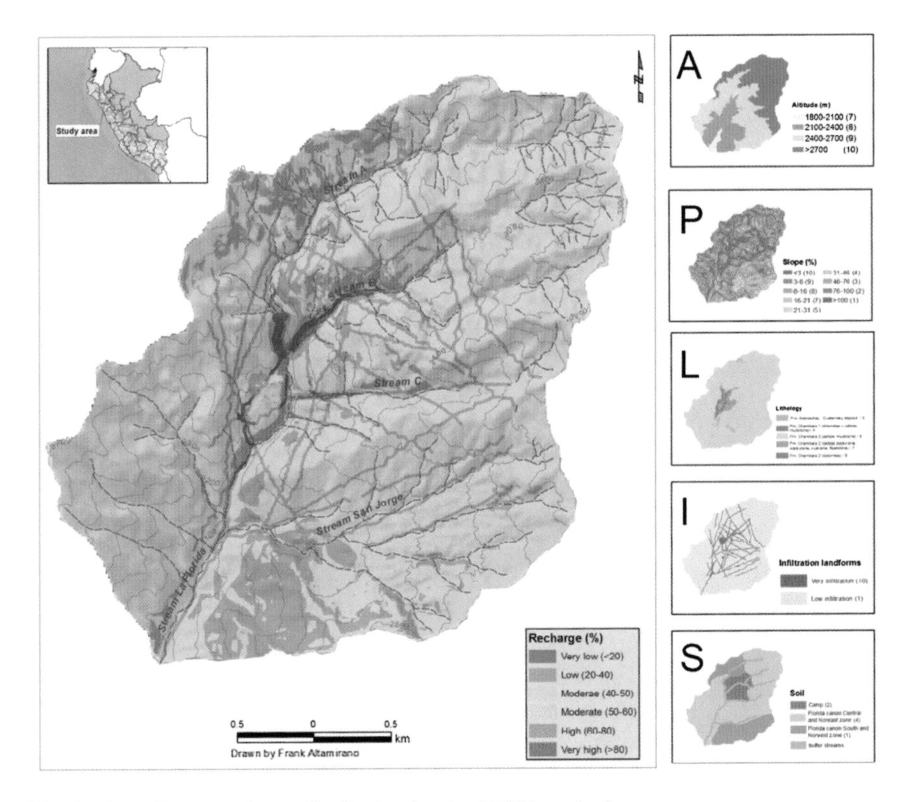

Fig. 2 Groundwater recharge distribution by the APLIS method

The recharge rates for the APLIS method are shown in Fig. 2. The majority of the study area has estimated infiltration rates in the "moderate" category (40–50 %). Very high recharge zones are located at the confluence of the main streams, where there are zones of preferential infiltration. The overall infiltration rate for the study area using the APLIS method is 48 %.

4 Application of the Modified-APLIS Method

Marín proposed modifications to the APLIS method (Marín 2009). These modifications include additional slope classes, refinement of the infiltration term, and a new factor, termed the correction coefficient of recharge (Fh), that allows for designations of exposed bedrock as "aquifer" or "non-aquifer".

The intent of the "non-aquifer" classification using the modified-APLIS method is to allow for classification of non-carbonate rock units in a study area where the APLIS method is applied (Marín 2009). In our study, field evidence indicates that limestones of the Chambara 3 Fm, overlying the dolomite aquifer are relatively impermeable, and therefore we have used the non-aquifer designation in our application of the modified-APLIS method as described below.

The modified-APLIS method was applied to the study area, and the main differences were in the infiltration map and in the coefficient of recharge. In the original APLIS method, classification of the infiltration term is assigned a score of 10 (for preferential infiltration areas) or 1 (for other areas). The modified version adds an intermediate category (5) for areas with some preferential landforms. The infiltration landform (I) term was set by assigning 10 for the preferential infiltration forms as in the APLIS method, but further evaluation was done to identify intermediate recharge areas. The conceptual site model and associated data - including borelogs and rock quality designations from exploration drilling, identified caves, and fault maps (local, regional) -were used to classify infiltration areas and features. Areas with flowing stretches of streams, main and secondary faults, karst cavities that start at a depth more than 10 m below ground surface, and areas with poor rock quality were scored with 10 in the infiltration map. The areas with karst cavities which not reached the surface, and areas with medium rock quality are categorized as moderate development of preferential infiltration forms and rating with 5.

Dolomites, faults, shallow karst cavities and areas with poor rock quality are considered as materials with aquifer characteristics, and are assigned correction coefficient of recharge (Fh) 1. More than half of the study area is underlain by limestone of the Chambara 3 Fm, which has very low primary permeability. These areas were generally assigned an Fh score of 0.1. In areas where this formation is faulted, and in other areas where caves, faults and karst features reach the surface and may have serve as preferential infiltration corridors, an Fh score of 1 was assigned as these areas are assumed to have connection to the underlying dolomite aquifer based on the conceptual site model.

The mean recharge rate estimated with the modified-APLIS method is 24 % (Fig. 3), with most areas classified as having very low (<20 %) infiltration rates. The highest recharge rates (high and very high class) are associated with outcropping dolomites of the Chambara 2 Fm and other areas with preferential infiltration, such dry streams and epikarst located along major fault valleys that make up the tributary canyons.

Table 1 shows the estimated by APLIS method, modified-APLIS method and the ground water model.

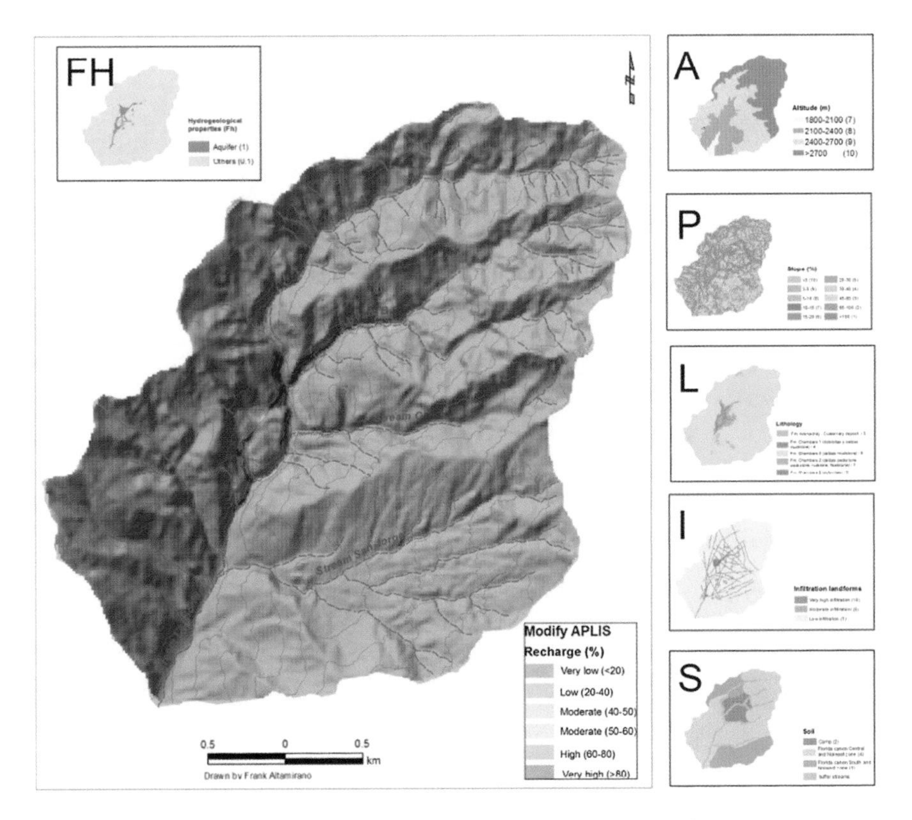

Fig. 3 Groundwater recharge distribution by the Modified-APLIS method

Table 1 Aquifer recharge rate estimated by APLIS method and modified-APLIS method and the ground water model

Recharge rate (%)	Recharge class	Percentage of the study area in recharge class	
		APLIS method	Modified-APLIS method
≤20	Very low	0	12.4
20–40	Low	7	0.2
40–60	Moderate	15	0.3
50–60	Moderate	15	2.2
60–80	High	48	40.7
>80	Very high	15	44.1
Mean recharge rate for the study area		48.0	24.0

5 Conclusions

The APLIS method and the modified-APLIS method have been applied to estimate the mean annual recharge (%) and spatial distribution in the Florida catchment, and resulting in values of 48 and 24 % respectively. The key difference in the recharge rate estimated by these methods is determined by whether exposed bedrock is designated as "aquifer" or "non-aquifer".

References

Andreo B, Vías J, Durán JJ, Jiménez P, López-Geta JA, Carrasco F (2008) Methodology for groundwater recharge assessment in carbonate aquifer: application to pilot sites in southern Spain. Hydrogeol J 16(5):911–925

Andreo B, Vías J, López-Geta JA, Carrasco F, Durán JJ, Jiménez P (2004) Propuesta metodológica para la estimación de la recarga en acuíferos carbonatados. Boletín Geológico y Minero 115(2):177–186

Farfán H, Corvea JL, Bustamante I (2010) Sensitivity analysis of APLIS method to compute spatial variability of karst aquifers recharge at the National Park of Viñales (Cuba), Advance in research in karst media. Springer, Berlin, pp 19–24. doi:10.1007/978-3-642-12486-0

Marín AI (2009) Los sistemas de información geográfica aplicados a la evaluación de recursos hídricos y a la vulnerabilidad a la contaminación de acuíferos carbonatados. Caso de la Alta Cadena (Provincia de Málaga). Bachelor Thesis, University of Malaga

USDA (1999) Soil taxonomy, a basic system of soil classification for making and interpreting soil surveys. United States Department of Agriculture Natural Resources Conservation Service, Agriculture handbook. USDA, Washington, DC, p 436

Synthesis of Groundwater Recharge of Carbonate Aquifers in the Betic Cordillera (Southern Spain)

S. Martos-Rosillo, A. González-Ramón, P. Jiménez, J.J. Durán, B. Andreo and E. Mancera-Molero

Abstract This paper presents a synthesis of the results and the evaluation methods of recharge in 51 carbonate aquifers of the Betic Cordillera. The average infiltration coefficient is 38 %, with a standard deviation of 12 %. The method of evaluation of recharge most applied is soil water balance, which served to take the first steps in groundwater management in Spain. Other widely used methods are water balance of the aquifer, chloride mass balance, and empirical methods such as APLIS. In eastern areas of the Cordillera, where the semiarid conditions are more overt, distributed models are used to assess recharge and calibrate it with data from the piezometric level evolution. In general, the annual recharge rates obtained appear higher when correlated with the annual rainfall. Thus, the data presented in this work contribute to a correct evaluation of renewable resources associated with the carbonate aquifers of the region. The high capacity of recharge and good quality of water for different uses, and especially to supply the population, makes these aquifers essential in the face of strong demand. Moreover, the data presented should be of special interest for future comparisons involving recharge assessments and different scenarios of climate change and changes in land use.

S. Martos-Rosillo (✉) · A. González-Ramón · J.J. Durán
Geological Survey of Spain, Instituto Geológico y Minero de España (IGME), C/Ríos Rosas, 23, 28003 Madrid, Spain
e-mail: s.martos@igme.es

A. González-Ramón
e-mail: antonio.gonzalez@igme.es

J.J. Durán
e-mail: jj.duran@igme.es

P. Jiménez · B. Andreo · E. Mancera-Molero
Department of Geology and Centre of Hydrogeology, University of Málaga (CEHIUMA), 29071 Málaga, Spain
e-mail: pgavilan@uma.es

B. Andreo
e-mail: andreo@uma.es

B. Andreo et al. (eds.), *Hydrogeological and Environmental Investigations in Karst Systems*, Environmental Earth Sciences 1,
DOI 10.1007/978-3-642-17435-3_11

1 Introduction

Groundwater in southern Spain, where semiarid conditions prevail, is essential to supply the population, to maintain associated ecosystems, and to support the agricultural and tourism sectors. The area investigated in this work—which coincides with the outcrops of the Betic Cordillera, extending from the province of Alicante at the SE of Spain, to the province of Cádiz in the SW—has a length of some 600 km and a half-width of the order of 130 km. Although the total surface outcrop of this rock type is not very important, aquifers consisting of limestone, dolomite, and marble provide significantly higher recharge rates than detrital aquifers and the low permeability hard rocks of the Betic Cordillera. This fact, coupled with the good quality of their water and the important capabilities of storage, has led to a greater awareness of the significance of these aquifers and the need to protect them. Water resources associated with the carbonate aquifers of the Betic Cordillera, in addition to covering current demand, prove essential to ensure demand peaks that recur during the long and frequent drought periods inherent in the Mediterranean region. In this sense, correct evaluation of the recharge of these strategic reservoir systems is the key to determining the flow of sustainable yield in order to achieve adequate groundwater management.

The recharge in carbonate aquifers is more complex to assess than in detritic aquifers. The high variability of porosity and permeability of carbonate rock bear an impact on (1) infiltration (diffuse and/or concentrated), (2) flow along the unsaturated zone (matrix, fractures and the karstic conduit network), and (3) discharge (diffuse and/or concentrated) (Bakalowicz 2005). For these reasons, scientific literature on the evaluation of recharge in carbonate materials is scarce, and very few articles deal specifically with southern Spain. In the Betic Cordillera, most recharge studies of carbonate aquifers have been carried out in the past two decades, and the first regional study was in 2003 (IGME-GHUMA 2003). However, a substantial number of reports and articles have been published in hydrogeological congresses in Spanish, making them somewhat more difficult to access. Here we review the most rigorous works involving recharge assessment, comparing recharge values obtained by means of several methods, possibly independent ones. At the end of the document, Appendix I shows some characteristics of the aquifers, including annual recharge, annual rainfall, coefficient infiltration, the method applied to evaluate recharge, and corresponding bibliographic references.

Thus, the aim of this paper is to present a synthesis of the rates of recharge occurring in carbonate materials of S Spain and of the methods applied for evaluation. This compilation is of particular interest because the assessment of groundwater recharge is one of the key challenges in determining the sustainable yield of aquifers. Furthermore, such data serve for comparative future work regarding the impact of climate change and changes in land use on groundwater reserves.

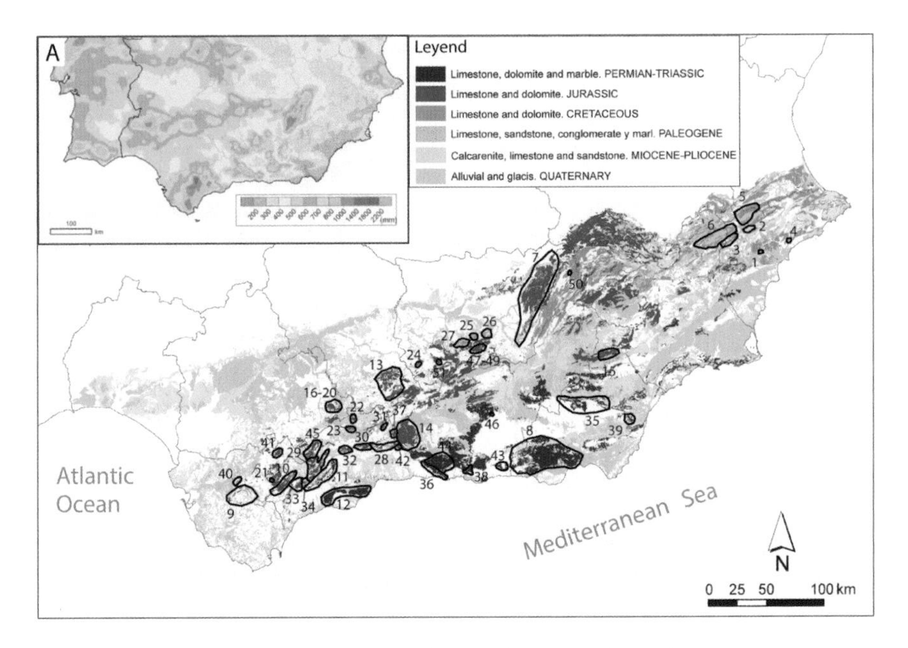

Fig. 1 Map of aquifers in Betic Cordillera. The aquifer number is listed in Appendix I. *A* Mean annual precipitation in the southern Iberian Peninsula

2 Study Sites

The Betic Cordillera is an extensive Alpine mountain range (Fig. 1), formed during the Miocene. Subdivided into the External and Internal Zones, the former integrating Mesozoic and Tertiary materials that were deposited on the continental margin of the Iberian microplate; these materials were folded and detached from their basement during the lower and middle Miocene. The Internal Zone is an allochthonous tectonic element of a higher order. It is a fragment of a subplate that originally occupied a position within the current Mediterranean, and during the lower Miocene it was disaggregated, ejecting materials to the W until they collided with the Iberian microplate.

The older sediments appear in the Internal Zone. In the areas where the most important aquifers are located, they consist of shales at the bottom, overlain by phyllites and quartzites, and thick sections of dolomite, limestone or dolomitic marble, sometimes with metapelitic intercalations. The shale could be metamorphosed Paleozoic sediments, while the phyllites come from sediments of the lower Triassic and the carbonates have been dated as middle-upper Triassic (Sanz de Galdeano 1997). Aquifers tend to be large and many are close to the Mediterranean coast, which permits they can be exploited.

The External Zone of the Betic Cordillera is divided, in turn, into the Prebetic and Subbetic domains. The substrate in both domains consists of clay and evaporite sediments of Triassic origin. The Jurassic comprises mainly carbonates, whereas the Cretaceous is of a marly nature in the Subbetic and calcareous in the Prebetic. The carbonate aquifers therefore prevail in both domains. The Prebetics have large Cretaceous aquifers in the north-eastern part of the range. The Subbetic aquifers often show great division and on the other hand extensive spatial distribution, which makes them ideal for urban supply.

From meteorological stand point, rainfall generally decreases from the W to E in the Betic Cordillera, with exceptions such as the Cazorla and Segura mountains (Fig. 1). Evapotranspiration increases in the same sense, as does the mean annual temperatures. Aquifer recharge rates are, therefore, conditioned by the geographic location. Other factors affecting recharge are the mean height of the outcropping permeable surfaces, the vegetal cover and the degree of surface karstification, which also decreases from W to E.

3 Overview of Methods

In the Betic Cordillera, carbonate aquifer recharge has been estimated using different approaches. The soil water balance method (SWB) is the one most commonly applied. In most carbonate aquifers, groundwater management is based on recharge estimates made by SWB, often contrasted with the aquifer water balance (AWB) after several years monitoring the inputs and outputs of the system. Although empirical methods such as that of Kessler have been applied in different aquifers (see Appendix I), the most used is the APLIS method (Andreo et al. 2008). The chloride mass balance (CMB) (Eriksson and Khunakasem 1969) has also been applied in continental Spain (Alcalá and Custodio 2014). On the other hand, Contreras et al. (2008) used a satellite-based model to estimate potential recharge in the Sierra de Gádor. Numerical modeling techniques can also be applied to estimate recharge in carbonate aquifers, the lumped models being more widely used than the three-dimensional models based on numerical code distribution. Accordingly, lumped models with ERAS (Murillo and De la Orden 1996), based on water table fluctuations to estimate recharge in overexploited aquifers, are most common in recent years (Aguilera and Murillo 2009; Martos-Rosillo et al. 2009, 2013; Andreu et al. 2011; Martínez-Santos and Andreu 2010). Others lumped models utilized are the so-called VISUAL BALAN (Guardiola-Albert et al. 2012), VENTOS (Bellot et al. 2001) and TRIDEP (Padilla and Pulido-Bosch 2008). Recently, Pardo-Igúzquiza et al. (2012) developed a distributed model of evaluation of recharge in carbonated aquifers which was applied to the Sierra de las Nieves aquifer. In the Betic Cordillera, MODFLOW has been used on occasion to evaluate recharge. Martinez-Santos and Andreu (2010) compared recharge results from ERAS and MODFLOW codes in different aquifers from the south-eastern Betic Cordillera. In turn, Pulido-Velázquez et al. (2014) have modeled the flow and recharge of the Serral-Salinas aquifer (SE Spain) combining VISUAL BALAN and MODFLOW codes in order to simulate the aquifer response to different climate change scenarios.

Table 1 Statistical values of surface outcrop (*S*), annual precipitation (*P*), annual recharge (*R*) and infiltration coefficient measured (*R/P*) of the 51 aquifers studied

	S (km^2)	P (mm/year)	R (mm/year)	R/P (%)
Mean	83	648	262	38.1
SD	156	216	146	11.8
Median	26	586	237	38.3
Percentile 10	4	414	115	25.6
Percentile 90	196	964	469	54
Max	823	1,299	720	67.2
Min	1	242	7	2.9

Legend Standard deviation (*SD*), Maximum (*Max*), Minimum (*Min*)

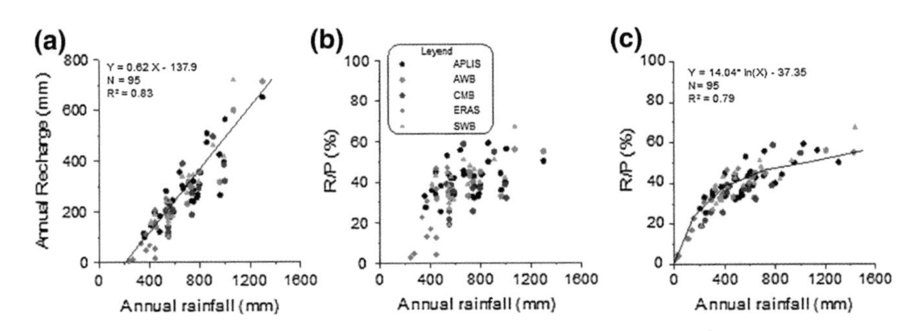

Fig. 2 **a** Annual rainfall versus annual recharge, **b** Infiltration coefficient (*R/P*) versus annual rainfall and **c** Infiltration coefficient (*R/P*) versus annual recharge

4 Overview of Results

The bibliographic data collected in this work reflect the small size of the aquifers, with a median value of 26 km^2. Mean annual rainfall and its associated standard deviation amount to 648 $\pm$ 216 mm/year. The average annual recharge is 262 mm/year, with a standard deviation of 146 mm/year. The median value of recharge, and percentiles 10 and 90, are 237, 469, and 115 mm/year, respectively. The average coefficient of recharge is 38 %, with a standard deviation close to 12 % (Table 1).

Figure 2a represents annual rainfall against annual recharge deduced from different methods used. The diagram shows up a high correlation coefficient ($R^2 = 0.83$) between annual rainfall and recharge in carbonate aquifers of the Betic Cordillera. Regardless of the method used, Fig. 2b reveals important differences in R/P with the same annual rainfall (they may be greater than 40 %). Finally, Fig. 2c shows that most of the points indicate recharge coefficients falling between 30 and 60 % of annual rainfall; moreover, the relationship between the infiltration coefficient measured and annual recharge is not linear, so that when annual recharge is higher the infiltration coefficient tends to stabilize, making a coefficient over 60 %.

5 Conclusions

The bibliographic data used in this study correspond to 51 carbonate aquifers of the Betic Cordillera indicate mean annual recharge and mean coefficient infiltration of 262 mm/year and 38 %, respectively. Selected aquifers are representative of a broad range of meteorological and geologic conditions of this region.

A high correlation between annual rainfall and annual recharge in all aquifers has been obtained. However, even with a same annual rainfall recharge, the infiltration coefficient can be very different. This reflects the significant differences in the degree of surface karstification and the degree of development of the vegetal cover-soil-epikarst system in the Betic Cordillera.

Acknowledgments This report was produced in the framework of the Associate Unit "Advanced Hydrogeologic Studies" between the IGME and the Hydrogeology Group at the University of Málaga. It is a contribution to the projects CGL-2010-15498, CGL-2012-32590 of DGICYT and IGCP 598 of UNESCO, and to the Research's Groups RNM-126 and 308 of the Junta de Andalucía.

Appendix I. Recharge Estimating Using Variety of Methods

Number	Aquifer	S (km^2)	Method	P (mm/year)	R (mm/year)	R/P (%)	References
1	*Ventós-Castellar*	7	ERAS	242	7	3,0	Andreu et al. (2001)
			ERAS	272	13	5,0	Martinez-Santos and Andreu (2010)
2	*Peñarrubia*	41,5	ERAS	372	114	31,0	Murilo and Roncero (2005)
			ERAS	446	163	37,0	Aguilera and Murillo (2009)
3	*Serral-Salinas*	198	ERAS	372	49	13,0	Corral et al. (2004)
			ERAS	446	18	4,0	Aguilera and Murillo (2009)
4	*Cabeçó d'Or*	15	ERAS	400	68	17,0	Murilo and De la Orden (1996)
5	*Solana*	118	ERAS	333	76	23,0	Murilo et al. (2004)
			ERAS	446	208	47,0	Aguilera and Murillo (2009)
6	*Jumilla-Villena*	320	ERAS	446	56	13,0	Aguilera and Murillo (2009)

(continued)

(continued)

Number	Aquifer	S (km^2)	Method	P (mm/year)	R (mm/year)	R/P (%)	References
7	*Sierra de Cazorla*	823	AWB	740	297	40,0	Moral (2005)
			CMB	740	190	26,0	Benavente et al. (2004)
			APLIS	740	244	33,0	Andreo et al. (2008)
			CMB	740	274	37,0	Alcalá and Custodio (2014)
8	*Sierra de Gadór*	670	SWB	550	150	27,0	Pulido-Bosch et al. (2000)
			CMB	550	120	22,0	Alcalá et al. (2011)
			SWB	550	164	30,0	Contreras et al. (2008)
			SWB,CMB	550	105	19,0	Alcalá et al. (2011)
			CMB	550	142	26,0	Alcála and Custodio (2014)
9	*Sierra de las Cabras*	34	AWB	964	318	33,0	IGME-GHUMA (2003)
			APLIS	964	347	36,0	Andreo et al. (2008)
10	*Sierra de Líbar*	89	AWB	1299	714	55,0	IGME-GHUMA (2003)
			APLIS	1299	652	50,2	Marín (2009)
11	*Sierra de las Nieves*	129	AWB	1004	552	55,0	IGME-GHUMA (2003)
			APLIS	1004	564	56,2	Marín (2009)
12	*Sierras Blanca and Sierra de Mijas*	173	AWB	712	320	45,0	IGME-GHUMA (2003)
			APLIS	712	309	43,4	Marín (2009)
13	*Sierra de Cabra and*	196	AWB	800	360	45,0	IGME-GHUMA (2003)
	Rute-Horconera		APLIS	800	310	38,7	Andreo et al. (2008)
14	*Sierra Gorda de Loja*	334	SWB, Kessler	861	474	50,0	IGME-GHUMA (2003)
			APLIS	861	510	59,2	Marín (2009)
15	*Sector Sierra de María Sector*	33	AWB	450	144	32,0	Gónzalez-Ramón et al. (2013); IGME-IARA (1990)

(continued)

(continued)

Number	Aquifer	S (km²)	Method	P (mm/year)	R (mm/year)	R/P (%)	References
	Sierra de María	70	APLIS	484	185	38,2	Andreo et al. (2008)
16	*Becerrero*	26,2	SWB	586	185	31,6	Martos-Rosillo et al. (2013)
			APLIS	586	246	42,0	
			CMB	586	208	35,5	
17	*Mingo*	0,6	SWB	534	240	44,9	
			APLIS	534	283	53,0	
			ERAS	534	252	47,2	
18	*Hacho*	1,6	SWB	543	183	33,7	
			APLIS	543	239	44,0	
			ERAS	543	188	34,6	
19	*Águilas*	4,4	SWB	564	240	42,6	
			APLIS	564	237	42,0	
			CMB	564	217	38,5	
20	*Pleites*	2,9	SWB	571	183	32,0	
			APLIS	571	183	32,0	
21	*Sierra de la Silla*	10	CMB	995	386	38,8	
			SWB	995	418	42,0	
			APLIS	995	398	40,0	
22	*Sierra de Mollina-La Camorra*	11,8	SWB	443	192	43,2	Ruiz-Gonzalez (2012)
			APLIS	443	199	44,9	
23	*Sierra de Humilladero*	4,9	SWB	443	192	43,2	
			APLIS	443	199	44,9	
24	*Ahíllo*	8,1	ERAS	544	178	32,7	De Mingo (2013)
25	*Torres-Jimena (Jimena)*		SWB	413	161	39,0	In preparation (IGME-DPJ)
			APLIS	413	149	36,0	
26	*Bedmar-Jódar*	17	SWB	415	156	37,6	Mancera (2013)
			APLIS	415	150	36,0	
27	*Mancha Real-Pegalajar*	22,9	CMB	795	294	37,0	González-Ramón (2007)
			SWB	795	318	40,0	
28	*La Alta Cadena*	76	APLIS	665	303	45,5	Mudarra (2012)
			SWB	665	339	51,0	
			CMB	665	390	58,7	
29	*Serrania de Ronda Oriental*	104	APLIS	908	497	54,7	Barberá (2014)
			SWB	908	463	51,0	
			CMB	908	498	54,8	
30		28,7	APLIS	641	357	55,7	

(continued)

(continued)

Number	Aquifer	S (km^2)	Method	P (mm/year)	R (mm/year)	R/P (%)	References
	Torcal de Antequera						CEHIUMA (2010)
31	*Sierra de Archidona*	7,6	APLIS	590	199	33,8	
32	*Sierra del valle de Abdalajís*	40,5	APLIS	591	203	34,3	
33	*Sierra de Jarastepar*	45,2	APLIS	960	425	44,3	
34	*Dolomias de Ronda*	18,2	APLIS	810	355	43,8	
35	*Sierra de los Filabres*	130,4	APLIS	362	100	27,6	
36	*Sierra de las Alberquillas*	117,1	APLIS	750	242	32,2	
37	*Sierra de Gibalto-Arroyo Marin*	7,7	APLIS	787	268	34,0	
38	*Sierra de Escalate*	20,7	APLIS	442	150	33,9	
39	*Bédar-Alcornia*	20,1	APLIS	354	117	33,1	
40	*Sierra Valleja*	5,9	APLIS	780	257	32,9	
41	*Sierra de Líjar*	24,2	APLIS	670	293	43,7	
42	*Sierra de En medio-Los Tajos*	27,8	APLIS	752	286	38,0	
43	*Albuñol*	26,1	APLIS	482	123	25,5	
44	*Sierra Almijara*	70,6	APLIS	750	322	42,9	
45	*Sierra de Cañete Sur*	40,7	APLIS	588	201	34,1	
46	*Carcabal*	1,9	CMB SWB	530 751	207 346	39,1 46,1	González-Ramón et al. (2011); ITGE-COPTJA (1998)
47	*Mágina NE-1*	6,5	AWB	548	210	38,3	Gollonet et al. (2002)
			SWB	548	181	33,0	
48		17	AWB	717	310	43,2	

(continued)

(continued)

Number	Aquifer	S (km²)	Method	P (mm/year)	R (mm/year)	R/P (%)	References
	Mágina SW-1		SWB	717	344	48,0	
49	*Mágina SW-2*	13,4	AWB	548	193	35,3	
			SWB	548	210	38,3	
50	*Sierra del Espino*	2	AWB	1071	600	56,0	Moral (2005)
			SWB	1071	720	67,2	
51	*Montesinos*	7,3	AWB	702	233	33,2	ITGE-COPTJA (1996)
			SWB	702	274	39,0	

References

Aguilera H, Murillo JM (2009) The effect of possible climate change on natural groundwater recharge based on a simple model: a study of four karstic aquifers in SE Spain. Environ Geol 57:963–974

Alcalá FJ, Custodio E (2014) Spatial average aquifer recharge through atmospheric chloride mass balance and its uncertainty in continental Spain. Hydrol Process 28(2):218–236

Alcalá FJ, Cantón Y, Contreras S, Were A, Serrano-Ortiz P, Puigdefábregas J, Solé-Benet A, Custodio E, Domingo F (2011) Diffuse and concentrated recharge evaluation using physical and tracer techniques: results from a semiarid carbonate massif aquifer in southeastern Spain. Environ Earth Sci 62(3):541–557

Andreo B, Vías J, Durán JJ, Jiménez P, López-Geta JA, Carrasco F (2008) Methodology for groundwater recharge assessment in carbonate aquifers: application to pilot sites in southern Spain. Hydrogeol J 16:911–925

Andreu JM, Alcalá FJ, Vallejos A, Pulido-Bosch A (2011) Recharge to mountainous carbonated aquifers in SE Spain: different approaches and new challenges. J Arid Environ 75(12):1262–1270

Andreu JM, Delgado J, García-Sánchez E, Pulido-Bosh A, Bellot J, Chirino E, de Ortiz Urbina JM (2001) Caracterización del funcionamiento y la recarga del acuífero del Ventós-Castellar (Alicante). Revista Sociedad Geológica de España 14(3–4):247–254

Bakalowicz M (2005) Karst groundwater: a challenge for a new resources. Hydrogeol J 13:148–160

Barberá JA (2014) Investigaciones hidrogeológicas en los acuíferos carbonáticos de la Serranía de Ronda Oriental. Doctoral Thesis. University of Malaga, Malaga, Spain

Bellot J, Bonet A, Sanchez JR, Chirino E (2001) Likely effects of land use changes on the runoff and aquifer recharge in a semiarid landscape using a hydrological model. Landsc Urban Plan 55:41–53

Benavente J, Hidalgo MC, Izquierdo A, Mabrouki K, Rubio JC (2004) Contenido en cloruros y en isótopos estable (^{18}O y D) de las precipitaciones en un área montañosa (Alto Guadalquivir, Jaén). Geogaceta 36:111–114

CEHIUMA (2010) Trabajos de mejora del conocimiento y protección contra la contaminación y deterioro del estado de las masas de agua subterránea de las Demarcaciones Hidrográficas Andaluzas de carácter intracomunitario, conforme a lo establecido en las Directivas 2000/60/CE y 2006/118/CE. Unpublished report

Contreras S, Boer M, Alcalá FJ, Domingo F, García M, Pulido-Bosch A, Puigdefabregas J (2008) An ecohydrological modelling approach for assessing long-term recharge rates in semiarid karstic landscapes. J Hydrol 351:42–57

Corral MM, Murillo JM, Rodriguez L (2004) Caracterización del funcionamiento de la unidad hidrogeológica de Serral-Salinas (Alicante). VIII Simposio de Hidrogeología, vol XXVI. Zaragoza, pp 53–62

De Mingo B (2013) Evaluación de la recarga del acuífero carbonático de Ahíllo (Jaén). Licenciature Thesis. Pablo de Olavide University, Sevilla, p 63

Eriksson E, Khunakasem V (1969) Chloride concentrations in groundwater, recharge rate and rate of deposition of chloride in the Israel coastal plain. J Hydrol 7:178–179

Gollonet J, González-Ramón A, Rubio JC (2002) Nuevas aportaciones sobre el funcionamiento hidráulico del sistema kárstico de Sierra Mágina. Karst and Environment. Nerja. Málaga, pp 211–217

González Ramón A, Peinado Parra T, Delgado Huertas A, Cifuentes Sánchez VJ (2013) Características hidrológicas, hidroquímicas e isotópicas del acuífero Orce-María (Almería). Aportaciones al modelo conceptual de funcionamiento hidrogeológico. X Simposio de Hidrogeología. Hidrogeología y Recursos Hidráulicos. Tomo XXX:91–102

González-Ramón A (2007) Hidrogeología de los acuíferos kársticos de las Sierras de Pegalajar y Mojón Blanco (Jaén). Doctoral Thesis, Universidad de Granada

González-Ramón A, Delgado A, Mudarra M (2011) Análisis de la respuesta a la recarga en carbonatos alpujárrides mediante el estudio hidrodinámico, hidroquímico e isotópico del manantial del Carcabal (La Peza, Granada). Boletín Geológico y Minero 122(1):93–108

Guardiola-Albert C, Martos-Rosillo S, Jiménez P, Liñán C, Pardo-Igúzquiza E, Cerezuela R, Pulido D, Luque-Espinar JA, Durán JJ, Robledo-Ardila PA (2012) Comparación de distintos métodos de evaluación de la recarga en el Sector Occidental del acuífero kárstico de la Sierra de las Nieves (Málaga). En: El Agua en Andalucía. Retos y avances en el inicio del milenio. IGME. Madrid

IGME-IARA (1990) Investigación hidrogeológica de los acuíferos de Sierra de Orce y Cúllar Baza. Unpublished report

IGME-GHUMA (2003) Estudios metodológicos para la estimación de la recarga en diferentes tipos de acuíferos carbonáticos: aplicación a la Cordillera Bética. Unpublished report

ITGE-COPTJA (1996) Reconocimiento hidrogeológico y ejecución de sondeos de investigación en el sector Quiebrajano-Víboras (Jaén). Unpublished report

ITGE-COPTJA (1998) Plan de Integración de los recursos hídricos subterráneos en los sistemas de abastecimiento público de Andalucía. Sector de Acuíferos de Padul-La Peza y Albuñuelas. Unpublished report

Molina JL, Pulido-Velázquez D, García-Aróstegui JL, Pulido-Velázquez M (2013) Dynamic bayesian networks as a decision support tool for assessing climate change impacts on highly stressed groundwater systems. J Hydrol 479:113–129. doi:http://dx.doi.org/10.1016/j.jhydrol.2012.11.038

Mancera E (2013) Evaluación de la recarga en un acuífero carbonático sometido a explotación intensiva. El acuífero de Bedmar-Jodar (Jaén). Master Thesis, p 78

Marín AI (2009) The application of GIS to evaluation of resources and vulnerability to contamination of carbonated aquifer. Test site Alta Cadena (Málaga province). Degree thesis, University of Malaga

Martínez-Santos P, Andreu JM (2010) Lumped and distributed approaches to model natural recharge in semiarid karst aquifers. J Hydrol 388:389–398

Martos-Rosillo S, Rodríguez-Rodríguez M, Pedrera A, Cruz-Sanjulián J, Rubio J (2013) Groundwater recharge in semi-arid carbonate aquifers under intensive use: the Estepa Range aquifers (Seville, southern Spain). Environ Earth Sci 70:2453–2468

Martos-Rosillo S, Rodríguez-Rodríguez M, Moral F, Cruz-Sanjulián JJ, Rubio JC (2009) Analysis of groundwater mining in two carbonate aquifers in Sierra de Estepa (SE Spain) based on hydrodynamic and hydrochemical data. Hydrogeol J 17:1617–1627

Moral F (2005) Contribución al conocimiento de los acuíferos carbonáticos de la Sierra de Segura (Alto Guadalquivir y Alto Segura). Thesis Doctoral. Univ. Pablo de Olavide. Dep. de Ciencias Ambientales, p 580, Sevilla

Mudarra M (2012) Importancia relativa de la zona no saturada y zona saturada en el funcionamiento hidrogeológico de los acuíferos carbonáticos. Caso de la Alta Cadena, sierra de Enmedio y área de Los Tajos (provincia de Málaga) Doctoral Thesis. University of Malaga, Malaga, Spain

Murillo JM, De la Orden JA (1996) Sobrexplotación, alternativas de gestión y evaluación del efecto del cambio climático en la recarga natural del acuífero Kimmeridgiense de Cabezón de Oro (Alicante). Recursos Hídricos en Regiones Kársticas, Vitoria, pp 73–88

Murillo JM, De la Orden JA, Roncero FJ (2004) El modelo "ERAS" una herramienta sencilla para estimar la recarga a los acuíferos que tienen una respuesta rápida. Congreso XXXIII IAH- 7° ALHSUD. Groundwater flow understanding: from local to regional scales. Zacatecas City, Méjico. http://www.igeograf.unam.mx/aih

Murillo JM, Roncero FJ (2005) Natural recharge and simulation of the management using the model ERAS. Application to Peñarubia aquifer (Alicante). Boletín Geológico y Minero 116:97–112

Padilla A, Pulido-Bosch A (2008) A simple procedure to simulate karstic aquifers. Hydrol Process 22:1876–1884

Pulido-Bosh A, Pulido-Lebouf P, Molina L, Vallejos A, Martín-Rosales W (2000) Intensive agricultura, wetlands, quarries and water management. A case study (Campo de Dalías, SE Spain). Environ Geol 40(1–2):163–168

Pardo-Igúzquiza E, Durán JJ, Dowd PA, Guardiola-Albert C, Liñán C, Robledo-Ardila PA (2012) Estimation of spatio-temporal recharge of aquifers in mountainous karst terrains: application to Sierra de las Nieves (Spain). J Hydrol 470–471:124–137

Pulido-Velázquez D, García-Aróstegui JL, Molina JL, Pulido-Velázquez M (2014) Assessment of future groundwater recharge in semi-arid regions under climate change scenarios (Serral-Salinas aquifer, SE Spain). Could increased rainfall variability increase the recharge rate? Hydrol Process. doi:10.1002/hyp.10191

Ruiz-González, P.(2012). Evaluación de la recarga de los acuíferos carbonáticos de las Sierras de Mollina - La Camorra y Humilladero (Málaga). Proyecto Fin de Carrera. Universidad Pablo Olavide, Sevilla, p 72

Sanz de Galdeano C (1997) La Zona Interna Bético-Rifeña. Monográfica Tierras del Sur. Universidad de Granada, p 316

Recharge Processes of Karst Massifs: Examples from Southern Italy

F. Fiorillo and M. Pagnozzi

Abstract Recharge of karst aquifers occurs when rainfall (or snowmelt) infiltration crosses the soil mantle and percolates through the vadose zone. In karst environments, the infiltration can occur in both concentrated and diffuse forms. In several areas of the Mediterranean, karst massifs are important sources of drinking water. In southern Italy, karst massifs are generally characterized by wide endorheic basins with seasonal lakes, which constitute large parts of the spring catchments. The origin of these endorheic basins is related to tectonism during the upper Pliocene-Pleistocene epochs and subsequent erosion and karstification. These endorheic basins constitute the most important recharge areas of karst massifs in central-southern Italy, and have been designated as groundwater protection areas. This study focuses on the karst massifs of the Picentini Mountains, which is characterized by rugged, steep landscape, and comprised of mainly dolostone and limestone. These karst massifs feed many basal karst springs with discharges up to thousands of liters for second, and constitute the main water resource in the region of Campania. The hydrological processes in these basins are simulated using a Geographic Information System (GIS)-based model on an annual scale. The results of the annual scale model have been used to successfully calibrate a daily time step model of infiltration and run off.

Keywords Karst massif • Recharge • Southern Italy • Spring discharge

1 Introduction

Recharge of aquifers occurs when rainfall (or snowmelt) infiltration crosses the soil mantle and percolates through the vadose zone. In karst environments, the infiltration can occur in both concentrated and diffuse forms. The first is connected to

F. Fiorillo (✉) · M. Pagnozzi
Dipartmento di Scienze e Tecnologie, University of Sannio, Benevento, Italy
e-mail: francesco.fiorillo@unisannio.it

B. Andreo et al. (eds.), *Hydrogeological and Environmental Investigations in Karst Systems*, Environmental Earth Sciences 1,
DOI 10.1007/978-3-642-17435-3_12

sinkholes and shafts, which allow rapid drainage of surface runoff and rapid transmission through the vadose zone and to the saturated zone. Diffuse recharge occurs by slower drainage from the soil mantle and fractured limestone. The percolation time can take weeks to months to reach the water table depending on the thickness and hydraulic conductivity of the vadose zone (Fiorillo and Doglioni 2010).

The hydraulic characteristics of the Terminio and Cervialto massifs were analyzed. Each of these massifs is characterized by large endorheic basins with high discharge basal springs. Several decades of spring flow and chemistry data exists thus allowing to accurately estimate the recharge values for the spring catchments. In particular, the role of the endorheic areas has been distinguished from the other zones of the spring catchment. The endorheic areas do not allow the escape of the run off from the spring catchment, whereas in the latter, here namely "open areas," the run off processes allow the escape of water from the spring catchment, especially during intense storms.

The GIS-based hydrological model first estimates the recharge on an annual scale and is subsequently used to calibrate the daily time step recharge-runoff model (Fiorillo et al. 2014).

2 Geological and Hydrogeological Framework

The northern sector of the Picentini mountains (Fig. 1) contains the Cervialto and Terminio massifs which were formed by Quaternary tectonic uplift. These steep, rugged mountains have several fault-scarps in the carbonaceous rocks with high slope angles. The endorheic areas of the Terminio and Cervialto massifs occupy 39 and 25 % of the total spring catchment area, respectively. The Cervialto massif has the highest elevation and highest precipitation of the two.

The ground elevation reaches 1,809 and 1,806 m a.s.l. for the Mt. Cervialto and Terminio massifs, respectively (Fig. 2). Limestone and limestone–dolomite (Late Triassic-Miocene) dominates these massifs with thicknesses ranging between 2,500 and 3,000 m. Recent pyroclastic deposits from the Somma-Vesuvius volçanoes cover the Picentini mountains with thicknesses between a few centimeters on the steep slopes to several meters within the flatter areas. These deposits play an important role in the infiltration of water into the karst substratum.

The Terminio and Cervialto karst massifs feed several basal karst springs (Fig. 1) with discharges up to thousands liters per second, and constitute the principal water resource in Southern Italy.

The Serino group of springs, located in the valley of the Sabato River along the north-western boundary of the Picentini massif, includes the Acquaro-Pelosi springs (377–380 m a.s.l.) and the Urciuoli spring (330 m a.s.l.). These springs are fed by the Terminio massif (Civita 1969; Fiorillo et al. 2007), with an overall mean annual discharge of 2.25 m^3/s. Roman aqueducts (first century A.D.) were supplied by these springs and the Urciuoli spring was re-tapped between 1885 and 1888 by the Serino aqueduct, which is comprised of a gravity channel followed by

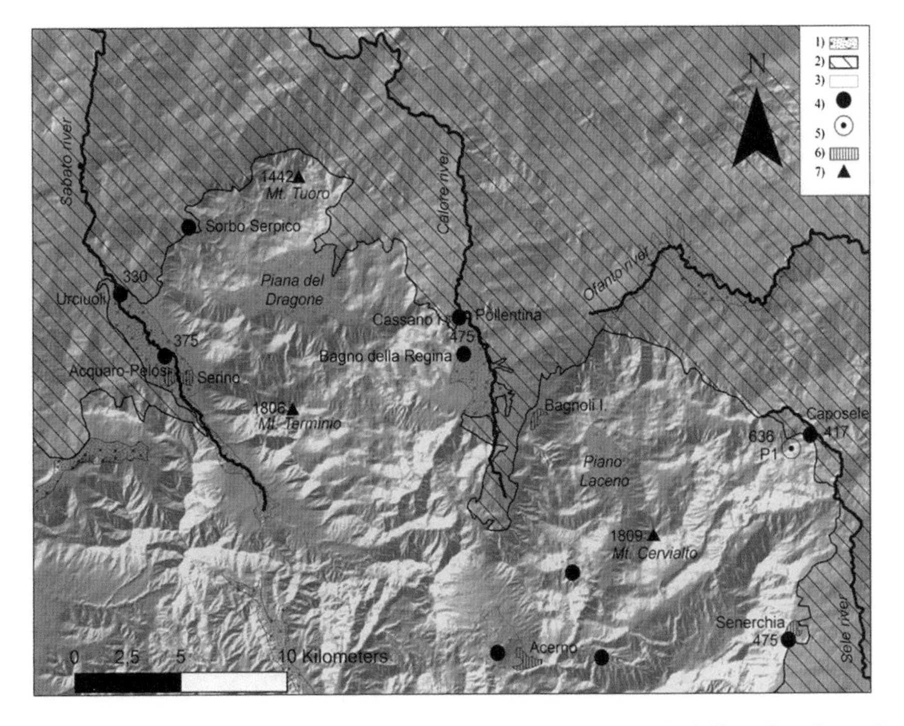

Fig. 1 Sketch of north-eastern sector of Picentini Mountains; Legend: *1* Slope breccias and debris, pyroclastic, alluvial and lacustrine deposits (Quaternary); *2* flysch sequences (Paleogene–Miocene); *3* calcareous-dolomite series (Jurassic–Miocene); *4* main karst spring; *5* monitoring well; *6* village *7* mountain peak

a system of pressured conduits that is used to supply water to the Naples area. Additionally, the Aquaro and Pelosi springs were also re-tapped in 1934 by the Serino aqueduct.

The Cassano group of springs is located in the Calore river basin along the northern boundary of the Picentini Mountains, and is formed by the Bagno della Regina, Peschiera, Pollentina, and Prete springs (473–476 m a.s.l.). Also these springs are primarily fed by the Terminio massif (Civita 1969), with an overall mean annual discharge of 2.65 m^3/s. In 1965, these springs were tapped to supply the Puglia region with water, and a gravity tunnel was joined to the Pugliese aqueduct.

The Caposele group of springs is formed by the Sanità spring (417 m a.s.l.), which is located at the head of the Sele river basin along the north-eastern boundary of the Picentini Mountains. This spring, which is primarily fed by the Cervialto mountain (Celico and Civita 1976), has a mean annual discharge of 3.96 m^3/s. The spring was tapped in 1920 by the Pugliese aqueduct, which passes through the Sele-Ofanto divide via a tunnel and supplies the Puglia region with

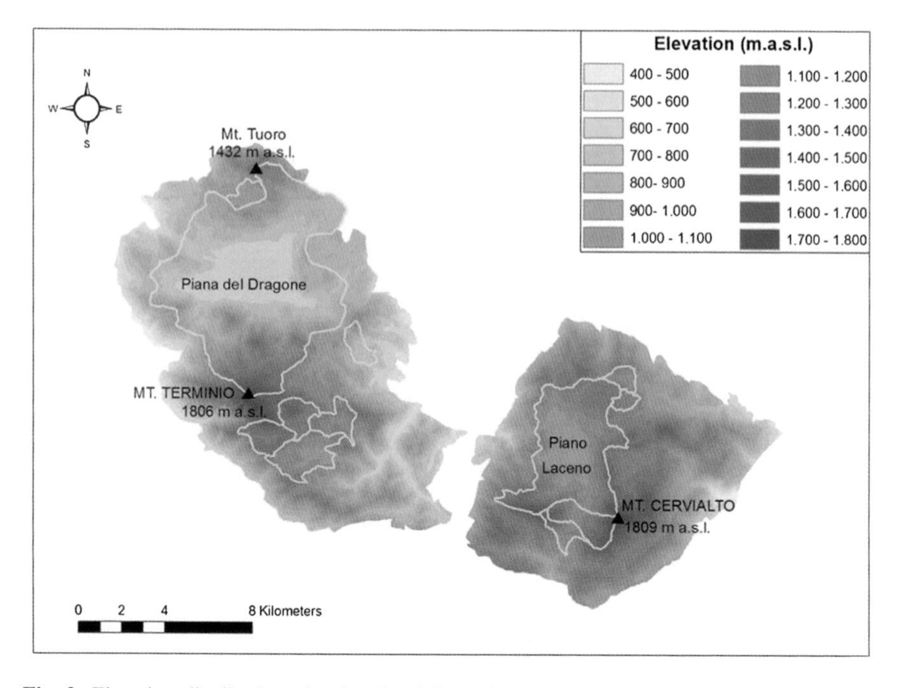

Fig. 2 Elevation distributions for the Terminio and Cervialto catchments; endorheic areas are outlined by *yellow line*

water. On the basis of geological and hydrogeological features, the Cervialto and Terminio massifs can be considered large spring catchments, with areas of 110 and 163 km^2, respectively.

These massifs are characterized by large internally drained (endorheic) basins (Fig. 2) with high recharge coefficients. The origin of these endorheic basins is related to tectonism during the upper Pliocene-Pleistocene epochs and subsequent erosion and karstification. The Terminio massif is characterized by several endorheic basins (Fig. 3) with the largest being the Piana del Dragone (55.1 km^2). Several sinkholes drain this endorheic basin however their combined capacities are not sufficient to prevent flooding in the area. Drainage was improved by engineering works at the *Bocca del Dragone* sinkhole and limit wet season flooding. Tracer tests have confirmed hydraulic connection between this sinkhole and the Cassano springs (Celico et al. 1982).

The Cervialto massif is characterized by the several endorheic basins with the largest being Piano Laceno (20.5 km^2). A permanent lake exists in this basin which is surrounded by several sinkholes which limit the extent of the lake during the wet season.

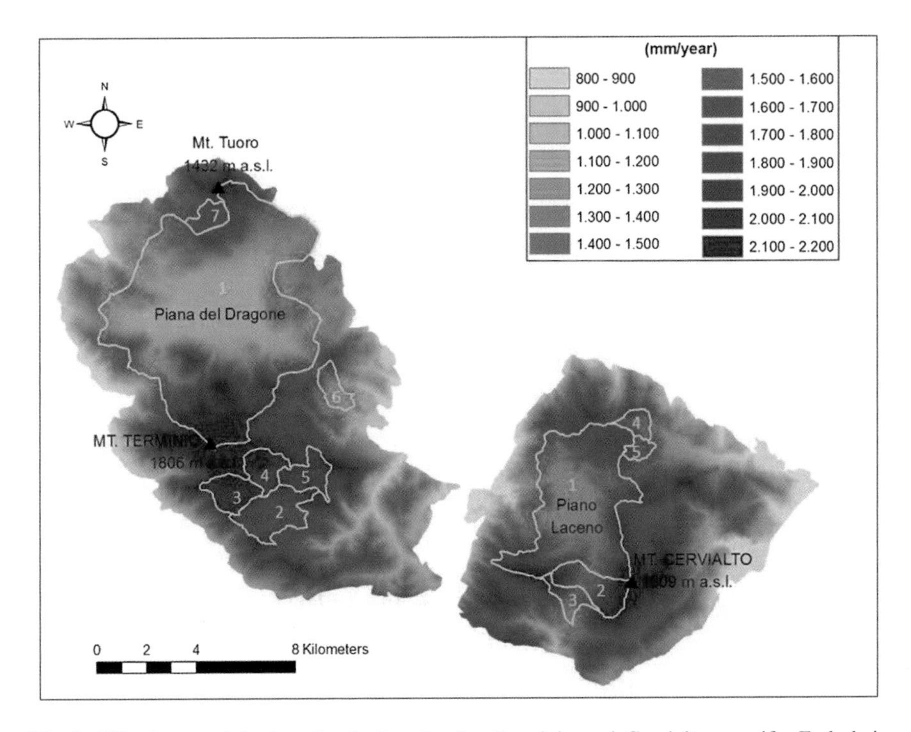

Fig. 3 Effective precipitation distribution for the Terminio and Cervialto massifs. Endorheic areas are outlined in *yellow*; and associated with a specific number in Table 1 (modified from Fiorillo et al. 2014)

3 Groundwater Recharge Model

Recharge can be defined as the downward flow of water reaching the water table, adding to the groundwater reservoir (De Vries and Simmers 2002). Following this definition, we estimate the amount of precipitation (rainfall and snow) which is not lost by evapotranspiration processes, by run off, or accumulated into soil without percolating downward. The difference between precipitation P, and the actual evapotranspiration AET, provides an estimation of the amount of precipitation which is free to infiltrate into soil or to develop run off. This difference can be evaluated at daily, monthly or annual scales, and is defined as effective precipitation, P_{eff} (Fig. 3):

$$P_{\mathrm{eff}} = P - \mathrm{AET} \tag{1}$$

At daily or monthly scale, most of procedures simply evaluate the effective precipitation as the difference between precipitation P, and potential evapotranspiration, PET:

Table 1 Main hydrological parameters obtained by the analysis of Terminio and Cervialto endorheic areas; *F*, Afflux; F_{eff}, Effective Afflux; *RO* Run-off; Q_p water amount pumping from the endorheic areas; *R* recharge; C_R, effective recharge coefficient; C'_R, recharge coefficient (modified from Fiorillo et al. 2014)

Springs			Massif	Area	Item	Elevation (minimum) m a.s.l.	km^2	F m^3 × 10^6	F$_{eff}$ m^3 × 10^6	RO m^3*10^6	Q$_p$ m^3 × 10^6	R m^3 × 10^6	C$_R$	C'$_R$
Group	m^3/s	m^3 × 10^6												
Caposele	3.96	128.5	Cervialto	Piano Laceno	1	1047	20.5	43.8	32.9	0.0	0.0	32.9	1	0.75
				Piano Acernese	2	1168	3.3	7.4	5.8	0.0	0.0	5.8	1	0.78
				Piano dei Vaccari	3	1164	1.4	3.0	2.3	0.0	0.0	2.3	1	0.77
Bagnoli Irpino	0.07			Valle Rotonda	4	1156	1.1	2.3	1.8	0.0	0.0	1.8	1	0.78
				Raia dell'Acera	5	1246	0.7	1.5	1.2	0.0	0.0	1.2	1	0.80
				Closed areas, A$_E$	–	–	27.0	58.0	44	0.0	0.0	44.0	1	0.76
Others	0.10			Open areas, A$_O$	–	–	83	172.4	128.3	43.8	0.0	84.5	**0.66**	0.49
				Springs catch., A$_C$	–	–	110	230.4	172.3	43.8	0.0	128.5	0.75	0.56
Serino	2.25	170.5	Terminio	Piana del Dragone	1	668	55.1	103.9	71.6	0.0	6.3	65.3	0.91	0.63
				AcquadellePietre	2	1061	4.3	8.9	6.7	0.0	0.0	6.7	1	0.75
Cassano Irpino	2.65			Campolaspierto	3	1279	2.3	5.1	4	0.0	0.0	4.0	1	0.78
				Piani d'Ischia	4	1210	2.1	4.6	3.5	0.0	0.0	3.5	1	0.76
				Piano di Verteglia	5	1177	2.1	4.4	3.4	0.0	0.0	3.4	1	0.77
Sorbo Serpico	0.43			Piana di Cetola	6	752	1.4	2.6	1.8	0.0	0.0	1.8	1	0.69
				Piana Sant'Agata	7	1047	1.3	2.6	2.1	0.0	0.0	2.1	1	0.81
				Closed areas, A$_E$	–	–	68.6	132.1	93.1	0.0	6.3	86.8	0.93	0.66
Montella	0.15			Open areas, A$_O$	–	–	94.3	180.2	125.0	41.4	0.0	83.7	**0.67**	0.46
				Springs catch., A$_C$	–	–	162.9	312.3	218.1	41.4	6.3	176.8	0.81	0.57

$$P_{\text{eff}} = P - \text{PET} \tag{2}$$

assuming $P_{\text{eff}} = 0$ when $P < \text{PET}$.

In GIS environment, the spatial distribution of the precipitation allows to estimate the total amount of the afflux, F, in a specific area, A, by:

$$(F)_A = \frac{\sum_1^n P}{n} \times A \tag{3}$$

where n is the number of cells in the area A. If the actual evapotranspiration AET, is subtracted from the rainfall (Eq. 1), the effective afflux, F_{eff}, in a specific area A, is:

$$(F_{\text{eff}})_A = \frac{\sum_1^n P_{\text{eff}}}{n} \times A \tag{4}$$

In endorheic areas, A_E, the recharge amount, R, can be considered equal to effective afflux.

Table 1 shows results obtained in this study for each massif, where values of each endorheic area are also reported.

The Cervialto and Terminio massifs constitute a useful hydrogeological condition to evaluate the recharge, because the groundwater of these massifs is almost completely drained by spring outlet. In fact, these karst systems are bounded by impervious terrains, and only along the western boundary a limited groundwater drainage toward alluvial of Sabato river exists (Fig. 1).

The estimation of effective afflux F_{eff}, and the spring output measurements, provides a gross estimate of the recharge coefficients. The recharge coefficient, $C'_R = R/F$, and the effective recharge coefficient, $C_R = R/F_{\text{eff}}$, are computed at an annual scale and refer to a long term period of time; if these coefficients are computed for a specific year, they can vary from a year to another. This occurs because the spring discharge may have a memory effect of the previous years, especially after dry years (Fiorillo 2009). Also, the monthly rainfall distribution may have an important role, so that the same total annual rainfall on a specific spring catchment could provide different values of spring discharge outlet, as noted by Bonacci (2001).

The long-term period used to estimate C_R allows to smooth the influence of a specific year, and provides useful tools to find the amount of rainfall which feeds the spring outlet.

The coefficient C'_R is more easily evaluated, but it depends on the temperature and rainfall distributions. Literature provides several examples, even if they were computed without GIS support. For the karst aquifers of central Italy, Boni et al. (1982) found a rough estimation of C'_R around of 0.7; for a Dinaric karst aquifer Bonacci (2001) a value around 0.56; for a Greek karst aquifer Soulios (1991)

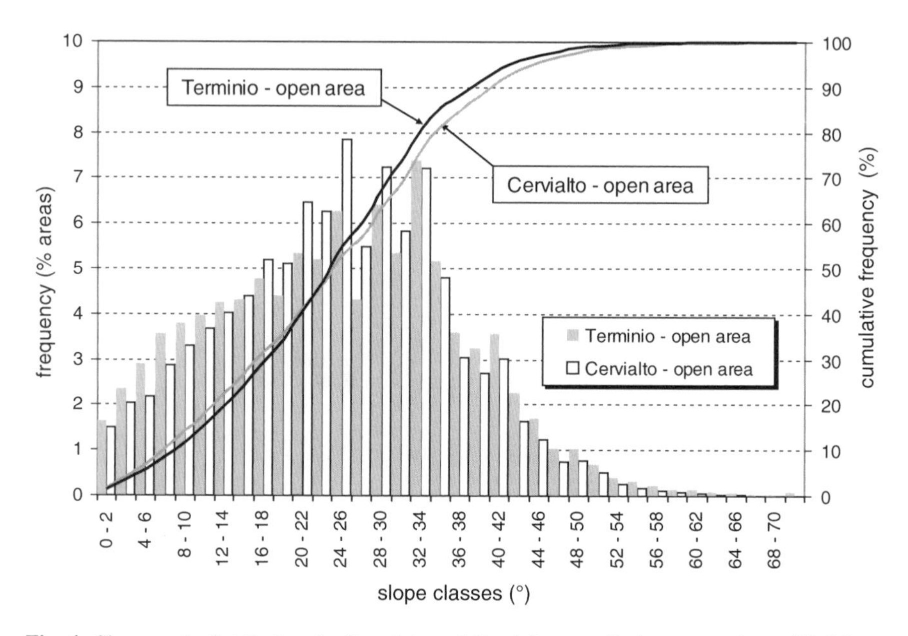

Fig. 4 Slope angle distribution for Terminio and Cervialto massifs (*open areas*), modified from Fiorillo et al. 2014

found a value almost 0.5. Allocca et al. (2014) estimated the annual mean recharge of aquifers of central southern Italy based on regional approach; for the Terminio and Cervialto massifs they found values smaller than that of Table 1, due to difference on evaluating the afflux and definition of the catchment boundaries.

4 Hydrologic Parameters Estimation on Annual Mean Scale and Daily Mean Scale

The effective recharge coefficient computed for the open areas, $(C_R)_{Ao}$, appears to have comparable values for the Cervialto (0.66) and Terminio (0.67) massifs; the possible explanation can be found in the similar slope angle distribution (Fig. 4) and in a similar karst conditions. The Annual Mean Scale (ASR) model is based on the estimation of some fundamental hydrological parameters: the afflux and the evapotranspiration.

The Daily Mean Scale (DSR) model has allowed to fix a threshold for the daily recharge, and considers a single point to estimate the recharge/run off (1D model), located approximately in the middle of the Cervialto catchment. This could be a limitation of the model since it does not consider elevation and the related spatial distribution of recharge and runoff. A high resolution daily time step model for the entire catchment would need very strong computational power.

5 Conclusion

A GIS-based model was developed to assess groundwater recharge of two major karst aquifers located in southern Apennines (Terminio and Cervialto massifs), using two distinct time scales: an Annual Scale Recharge (ASR) which was used to calibrate the Daily Scale Recharge (DSR). The ASR model provides a useful approach for assessing long-term groundwater recharge, especially for large areas with strong morphological irregularities, and with limited hydrological data. In particular, the afflux, run off, and recharge are computed using GIS, allowing an estimation of the recharge and runoff coefficients, distinguished for open and endorheic areas, while the DSR model has been calibrated by the results of the ASR model, and allows to split the amount of daily rainfall which cause run off from that recharging the aquifer. The daily estimation of recharge is a useful tool for water management, allowing the control of recharge condition of aquifers, especially during dry years. The application of this method could help to improve the design of appropriate management models for groundwater and surface resources of karst aquifers as well as the elaboration of accurate strategies to mitigate the effects of climate and land-use change.

References

Allocca V, Manna F, De Vita P (2014) Estimating annual groundwater recharge coefficient for karst aquifers of the southern Apennines (Italy). Hydrol Earth Syst Sci 18:803–817

Bonacci O (2001) Monthly and annual effective infiltration coefficient in Dinaric karst: example of the Gradole karst spring catchment. Hydrol Sci J 46(2):287–300

Boni C, Bono P, Capelli G (1982) Valutazione quantitativa dell'infiltrazione efficace in un bacino dell'Italia centrale: confronto con analoghi bacini rappresentativi di diversa litologia. Geologia Applicata e Idrogeologia, Bari 17:437–452

Celico P, Civita M (1976) Sulla tettonica del massiccio del Cervialto (Campania) e le implicazioni idrogeologiche ad essa connesse. Boll Soc Nat Naples 85:555–580

Celico P, Magnano F, Monaco L (1982) Prove di colorazione del massiccio carbonatico del monte Terminio-monte Tuoro. Notiziario Club Alpino Italiano, Napoli 46(1):73–79

Civita M (1969) Idrogeologia del massiccio del Terminio-Tuoro (Campania). Memorie e Note Istituto di GeolApplUniv di Napoli 11:5–102

De Vries JJ, Simmers I (2002) Groundwater recharge: an overview of processes and challenges. Hydrogeol J 10:5–17

Fiorillo F, Esposito L, Guadagno FM (2007) Analyses and forecast of the water resource in an ultra-centenarian spring discharge series from Serino (Southern Italy). J Hydrol 36:125–138

Fiorillo F (2009) Spring hydrographs as indicators of droughts in a karst environment. J Hydrol 373:290–301

Fiorillo F, Doglioni A (2010) The relation between karst spring discharge and rainfall by the cross-correlation analysis. Hydrogeol J 18:1881–1895

Fiorilllo F, Pagnozzi M, Ventafridda G (2014) A model to simulate recharge processes of karst massifs. Hydrological Processes (in press)

Soulios G (1991) Contribution à l'ètude des coubes de recession des souces karstiques: example du pays hellènique. J Hydrol 124:29–42

Use of Tracing Tests to Study the Impact of Boundary Conditions on the Transfer Function of Karstic Aquifers

L. Duran, M. Fournier, N. Massei and J.-P. Dupont

Abstract The impact of the variations of multiple environmental parameters on the response of karstic systems was investigated after a campaign of tracing tests acquired in very different hydrologic conditions. Principal components analysis and hierarchical clustering were applied on both environmental variables and karstic system response variables (parameters of the RTD curves). Equations between the RTD parameters and the most relevant variables were established using a symbolic regression algorithm. This model giving RTDs parameters in function of boundary conditions is more accurate than the PCA analysis since it takes into account the nonlinearity of the relations between variables. It appeared that the variations of the RTD parameters depend mainly on the piezometric level downstream of the aquifer, the cumulated rainfall preceding the injection, and on the tide coefficient (suggesting sensitivity to the annual variations of tide, in this case of a karstic system under marine influence). So the RTDS parameters are controlled by the hydraulic conditions downstream of the system, including tide. The dispersivity was found to be very sensitive to the precipitation and tides variations at a daily scale.

1 Introduction

Karstic aquifers represent 25 % of the water resources worldwide (Ford and Williams 2007), and are vulnerable to pollution, especially in regard of the high flow velocities in the conduits. Therefore, comprehension of those systems and the study of their vulnerability is a major stake in the management of the water resource. Tracing tests are a privileged way to understand the transport mechanisms within a karst system and to characterize its flow dynamics (Bakalowicz 2005). Commonly, one or two tracing tests are carried out. Still, the response of

L. Duran (✉) · M. Fournier · N. Massei · J.-P. Dupont
M2C, UMR 6143, CNRS, Morphodynamique Continentale et Côtière, Université de Rouen, Bât. IRESE a, 76821 Mont Saint Aignan, France
e-mail: lea.duran1@univ-rouen.fr

B. Andreo et al. (eds.), *Hydrogeological and Environmental Investigations in Karst Systems*, Environmental Earth Sciences 1,
DOI 10.1007/978-3-642-17435-3_13

the system (characterized by the breakthrough curve (BTC) and the residence time distribution (RTD)) changes significantly depending on the hydrologic conditions during the tracing test. For example, the study of the apparent velocity of the tracer can give some clues about the structure of the system (Dörfliger 2010). In order to fully characterize the functioning of a karstic system, tracing tests should ideally be carried out in different hydrologic conditions such as: low, medium, and high seasons, during rainfall events, and with variations of the water table. The transport variations in low flows and high flows have been investigated (Göppert and Goldscheider 2008; Larocque et al. 1998), and the impact of other components of the surface system like vegetation, nature of soils has been studied by simulation (Doummar et al. 2012). The variations of the BTCs according to the tracer used have been modeled (Geyer et al. 2007). Nevertheless, the impact of multiple boundary conditions (including various downstream controls) on the response of a karstic aquifer has not been fully assessed yet.

This study is based upon a campaign of tracing tests acquired on the same karstic system, but in different hydrologic conditions. Some statistical analyses have been carried out on the responses of this karstic system in order to assess the relative importance of all of the environmental variables and to establish relations between the parameters the karstic system response and those of the boundary conditions.

2 Methods: PCA, HC, Symbolic Regression

The principal components analysis method (PCA), commonly used to interpret hydrogeological data (Bakalowicz 1997; Fournier et al. 2007; Helena et al. 2000; Moore et al. 2009), was used in this study to assess the links between the response of the aquifer and the boundary conditions of the system. The individuals of the PCA are the tracing tests conducted on the study site. The selected variables are of two types. The first type corresponds to environmental variables (or boundary conditions of the karst aquifer): upstream boundaries with the precipitation, downstream boundaries, with the piezometric level in the aquifer, the tide coefficient, the level of the river (see description of the site). The second type is composed by variables related to the tracing test: recovery rate, characteristic times, dispersivity, and RTD parameters.

Hierarchical clustering was performed on the data: an agglomerative method with Ward algorithm (maximizing inter-class inertia, in order to obtain compact, spherical clusters) has been used on Euclidean distance matrix. In order to find the optimal number of clusters, Partitioning around medoids has been performed on HC results. The Kruskal Wallis post hoc test and Tukey's post hoc tests were then used to determine which groups were significantly different (Saporta 2011).

Once the more relevant variables were identified and the partitioning of the data studied, the next step was to find some relations between the environmental variables and the RTD variables. For that purpose, we used a software able to test a high number of possible relations between several variables: Eureqa (Version 0.98 beta).

Its functioning is based on a machine learning technique called Symbolic Regression to unravel the intrinsic relationships in data and explain them as simple equations. Using Symbolic Regression, Eureqa can create predictions (Schmidt and Lipson 2009). But even before obtaining the relationships between the variables, one of the interests is to identify which variables are relevant in the estimation of one particular parameter, as the algorithm calculates a lot of equations changing the nature and the number of variables implied.

3 Study Site and Tracing Tests Campaign

The Norville karst system is located in Normandy (France), near the Seine River. The site has been studied since 1999 and is described in various publications (Fournier et al. 2008; Massei 2001). It is a part of a national observation network on karstic systems (SNO karst).

The geology and geomorphology of the site are characteristic of the upper Normandy: the Seine River cuts deeply into chalk plateaus. An aquifer takes place in those formations than can be variously karstified. On the top of this Mesozoic chalk layer, the weathered chalk has formed a clay with flint layers, quite impervious. Swallow holes are penetrating the clay and chalk layers: the infiltration of water into the karstified chalk aquifer can be very quick. More locally, a N70E fault affects the site with a net slip of 120 m. South to the fault, the formations are the ones described before, while North to the fault, cretaceous layers of clay alternate with sand formations. Norville study site is a sinkhole-spring system: upstream, the small Bébec river drains a watershed of 10 km^2, before infiltrating into the ground when reaching the Triquerville fault (Fig. 1). The Bébec discharge has an important variability, from 5 l/s during low flow up to 400 l/s after storms events. Downstream, the perennial Hannetôt spring, at the bottom of the chalk cliffs, has been proved to be the resurgence of the Bébec River

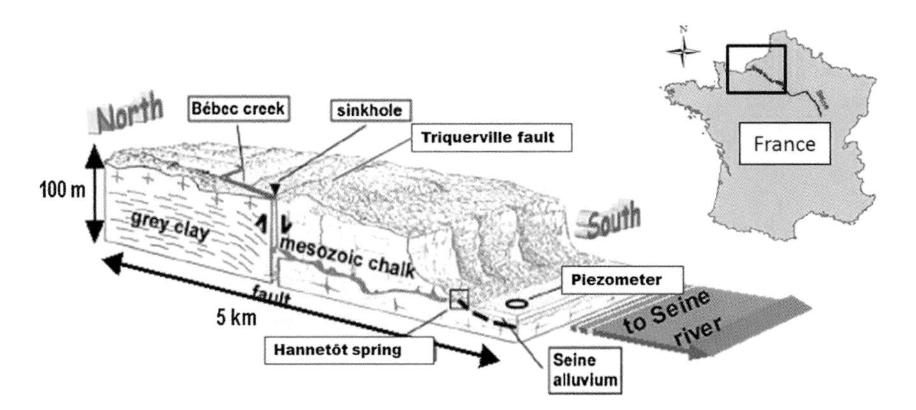

Fig. 1 Norville system (modified from Massei 2001)

with multiple tracing tests (Massei 2001). Karstic conduits are also likely to be found within the chalk aquifer between the spring and the Seine River, underneath alluvial aquifer; hence a hydraulic connection between the karstic aquifer and the Seine River cannot be excluded (Fournier et al. 2008). The variations of the level of a piezometer located south of the Seine, in the same chalk layer, are correlated to the variations at the spring.

The tracing tests are one of many tools available to assess the functioning of a karst system (Lepiller and Mondain 1986). In this study, 14 tracer tests were performed between 1999 and 2013, under various boundary conditions, including the precipitation, the piezometric level, the time within the hydrological cycle, and the tide conditions (being in the context of a karstic system under coastal influence). The tracer was injected into a perennial flow, in a sinkhole. Some of those tracing tests were made at less than 10 h interval in order to assess the role of the tide causing the BTCs to overlap. For that reason, it was sometimes necessary to separate these complex overlapping BTCs before interpreting them: we used PeakFit 4.0 (SPSS Inc.). The BTCs were then analyzed with the TRAC software, released by the BRGM (Klinka et al. 2012). The characteristic parameters of the BTCs were calculated, as well as the RTDs, and the dispersivity was estimated using the simulation tool. Through RTDs, different tracing tests within the same system or in different systems can be easily compared; the parameters of the RTD are linked to the dispersive parameters of the studied system. Here, the normalized RTDs (Fig. 2) present some important variations (area, time of the peak, tailing).

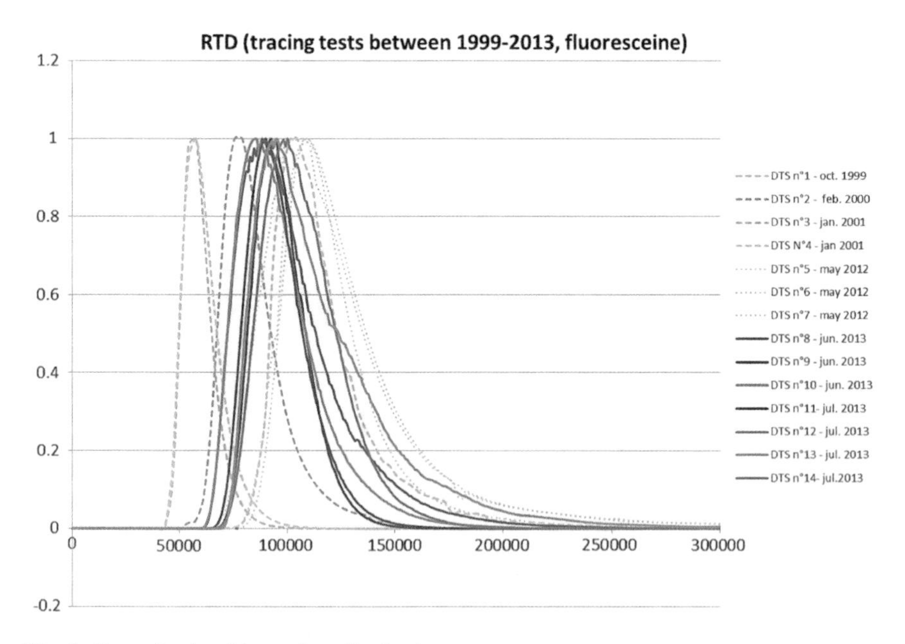

Fig. 2 Normalized residence time distribution curves

Table 1 Main results of the PCA (variables contributing positively and negatively)

Main results of PCA	Axis I	Axis II	Axis III	Axis IV	Axis V
Inertia (%)	53.7	14.17	10.34	6.7	6.05
Variables +	RTD parameters, BTC times	–	–	–	mar
Variables –	pf, qmoy, vmod, p7j, %rest	$p + 24$, $p - 24$	–	–	–

pf piezometric level within the chalk aquifer; *mar* tide coefficient; *P7j* cumulated rainfall during the week before the injection; $P - 24$, $P + 24$, rainfalls the day before and after the injection; Qmoy, mean discharge (l/s)

4 Results and Discussion

4.1 PCA Analysis

The results of the PCA analysis are presented in Table 1 and Fig. 3.

The percent of inertia explained by the different axis as well as the information given by the different components conducted to keep axis I, II, and V, explaining 73.89 % of the inertia. On axis I, several environmental variables (piezometric level, rainfall) contribute strongly negatively, when response parameters contribute positively. The main information on axis II is that the rainfalls of the 24 h before and after the injection contribute negatively. The tide coefficient (mar) contributes to axis V for 75 %. The fact that most of the environmental variables

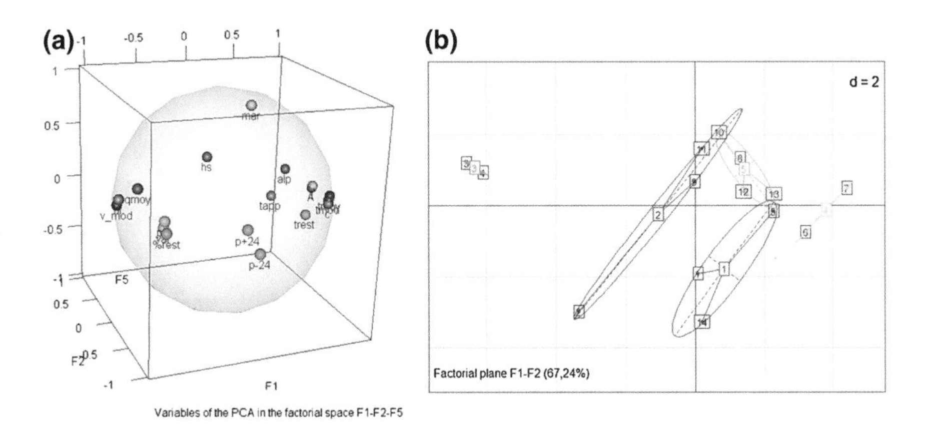

Fig. 3 Variables of the PCA **a** and tracing tests **b** in the factorial space F1–F2–F5. (*hf* piezometric level within the chalk aquifer; *mar* tide coefficient; *P7j* cumulated rainfall during the week before the injection; *P − 24*, rainfall the day before the injection; *P + 24*, rainfall the day after the injection; *Qmoy*, mean discharge (l/s); A, c and l: area, center and width of RTD; *vmod* modal velocity; *alp* dispersivity, trest, *%rest* recovery rate, *trest* time of recovery, *tapp* time of apparition)

are opposite to the RTD parameters on axis I suggests that they are strongly related; the objective is to characterize those links. In the individuals' space, some clear distinctions between the tracing tests appear: tracing tests number 3 and 4 contribute negatively to axis I, when tracing tests 1, 2, and 14 contribute negatively to axis II. In order to understand the structure of the data, some clustering and partitioning have been conducted.

4.2 Clustering and Interpretation of the Groups

The partitioning with medoids has been tested with a number of groups k from 2 to 10. The best partitioning is the one with the highest average silhouette width; here we obtained a width of 0.51 for $k = 5$. The next step is the hierarchical clustering with $k = 5$ (Fig. 4, left). The resulting groups contain from two elements to four elements, and can be drawn in the factorial plane (Fig. 4, right).

In order to test the significance of those groups, the Kruskal-Wallis test was applied. The null hypothesis was rejected for variables hf, %rest, tapp, tmod, tmoy, qmoy, alp, a, c, and l. This implies that for those variables, at least one group was significantly different from the other. The simple ANOVA test gave the same results. The ANOVA with regression on component gave a p-value of 0.0014, indicating that the partitioning was significant. Then the Tukey post hoc test was conducted to identify more precisely the groups which are significantly different. Independently, we attempted to assess the common conditions for the tracing tests in each group. The first group corresponds to tracing tests 3 and 4, with very high level in the piezometric level (extreme conditions) and precipitation conditions. Group 2 (tracing tests 2, 9, and 11) could be characterized by a rather low tide and a medium piezometric level. Group 3 (tracing tests 5, 1, and 14), would also concern tracing tests at a medium piezometric level, but with high tide conditions. Group 4 (tracing tests 6 and 7), corresponds to a low piezometric level. And at last, group 5 (tracing tests 8, 10, 12 and 13) would gather tracing tests with a high tide coefficient, or at default a high tide condition.

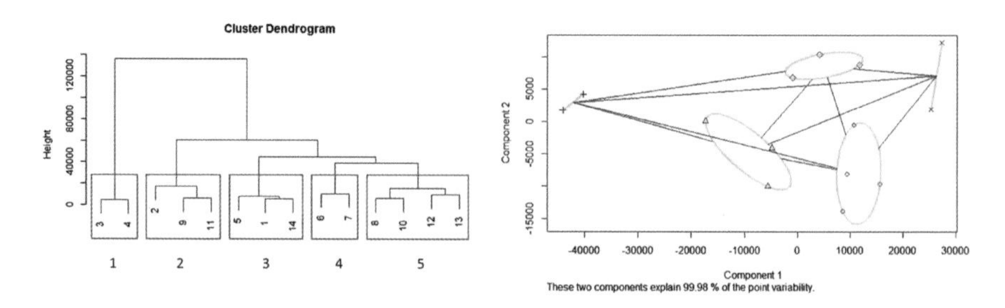

Fig. 4 Partitioning around medoids and cluster dendogram

4.3 Role of the Environmental Parameters

The clustering is particularly significant for all the variables characterizing the response to the karstic systems, like the RTD parameters. The PCA highlights the importance of the environmental parameters: the piezometric level in the chalk aquifer (downstream of the spring), as well as the spring discharge, are the major contributors to the first component of the PCA, and are anti-correlated to the parameters of the RTD curves. On axis II, the cumulated rainfall during the 24 h before the injection of tracer and the 24 h after the injection are the main contributors, and therefore it can be assumed that they play an important role in the response of the system. At last, the tide coefficient is by far the most important contributor to axis V; the tide is likely to be an important downstream boundary condition of the system.

4.4 Relations Between Environmental Variables and RTD Parameters

A lot of different configurations have been tested, by changing the number of variables included in the algorithm, and the forms of the mathematical equations. Each parameter of the RTD curve was expressed in function of all the environmental parameters; and the dispersivity in function of those same parameters. The possibilities of relations between A, the area of the RTD, were investigated. The piezometric level appears in 17 models over the 18 models calculated; it indicates that this variable is relevant to characterize the area, which is consistent with the results of the PCA. The second variable that appears in most models is the tide coefficient (in 14 models over 18). Then come the accumulated rainfall on the week before the injection and the rainfall 24 h before the injection. The level within the Seine River and the mean discharge are not well represented. The type of model obtaining the best R^2 Goodness of Fit and correlation coefficient ($R^2 = 0.96$ and $C = 0.987$) is:

$$A = a + b * \text{qmoy} + d * P7j - c * \text{mar} - f * \text{Pant} - g * pf - h * P7j * P_{\text{post}} - i * hs^2$$

The same investigation was conducted for the center of the RTD curve C. The piezometric level was also the most represented variable (in 19 over 21 models and 19 occurrences). But the second most recurrent variables were the precipitation of the week before and the day before and the mean discharge. The best fitted model obtained a R^2 goodness of fit of 0.992 and a correlation coefficient of 0.996. The corresponding equation is:

$$C = a + b * \text{Pant} + d * P7j^2 - \text{qmoy} - c * hs - f * pf - g * \text{Pant}^2$$

The piezometric level was the principal variable in most of the models for the width of the RTD. The second one was the tide coefficient, followed by the mean discharge and precipitation preceding the injection. The equation obtaining the best R^2 goodness of fit and correlation coefficient (0.97 and 0.99) is:

$$l = a + b * \mathrm{qmoy} - d * pf - c * \mathrm{mar} + i * \mathrm{mar} * \mathrm{qmoy} + g * \mathrm{mar} * hs - h * \mathrm{Pant} - i * hs - j * P7j + k * pf * P7j$$

As for the dispersivity, the most recurrent variable in all models was the cumulated rainfalls during the week preceding the injection (in 22 models over 24, with 42 occurrences). The second one was the level of the Seine River. The tide coefficient and the piezometric level in the chalk aquifer have an intermediary importance, appearing respectively in 14 and 15 models. The rainfalls at 24 h and the mean discharge were less significant with few occurrences. With a R^2 goodness of fit of 0.98 and correlation coefficient of 0.99, the selected equation for alpha is:

$$\alpha = a + b * P7j + c * hs + d * pf + e * \mathrm{mar} + hs * (f * \mathrm{Post} - g * pf) + h * \mathrm{qmoy}$$

4.5 Discussion

The fact that the majority of models for A, l, and c contain the piezometric level pf as a key variable is consistent with the results of the PCA. The tide coefficient is essential to calculate the area and the width of the RTD. The importance of the precipitation was also highlighted, especially influencing the center of the RTD. The mean discharge appears as an important contributor to axis I in the PCA, whereas the algorithm used to estimate the relations between the environmental parameters and the parameters of the RTD does not enhance this discharge as essential. As for the dispersivity, the results of the algorithm indicate that the precipitations during the week preceding the injection contribute strongly to its variations. Unlike the parameters of the RTD, the dispersivity is sensitive to the variations of the level of the Seine River that is to say of the tide, but at a daily scale, not in its annual variations.

The equations obtained are to be taken with caution, since they are calculated on a relatively small sample (14 tracing tests). These equations have to be tested on later tracing tests. Moreover, even though the tracing tests have been conducted in order to cover the maximum diversity of environmental conditions, some are not well represented. For example, the tracing test $n°$ 14 is the only one with an important rainfall occurring during the injection. That explains its position in the factorial plane F1–F2, contributing negatively to axis II, which corresponds to $P + 24$ (rainfall in the day following the injection). So this variable could have a non-negligible effect, but could be underestimated because of the tracing tests sample.

Nevertheless, this is a first model of the response of karstic system (RTDs) in function of environmental parameters. Moreover, it takes into account the non-linearity of the relations between the variables, by opposition to the PCA which indicates only the possible linear relations between them. Another important result is that the shape of the RTDs is controlled by downstream conditions, including tide.

5 Conclusion

In this study, we investigated the links between various environmental parameters and some characteristic parameters of the karstic system response. A campaign of tracing tests was conducted in different hydrologic conditions (high flows, low flows, tide variations, rainfalls…) on the same karstic system. For each tracing test, the parameters of the BTC and the RTD curves and the dispersivity were assessed. A PCA indicated that most relevant variables were the piezometric level downstream of the karstic system, the cumulated rainfall, and the mean discharge. These were anti-correlated to the parameters of the normalized RTD curve. The precipitation around the injection was also relevant, as well as the tide coefficient. We studied the structuration of the data by using the partitioning around medoids and hierarchical clustering. These groups were linked to corresponding environmental conditions.

We established a model giving the expression of RTDs parameters in function of environmental variables. It appears that the downstream control is essential: the piezometric level and the tide coefficient are the most relevant variables. The cumulated rainfall and the mean discharge were considered as necessary variables in the estimation of the RTD parameters. It suggests that the mean discharge alone cannot provide enough information to estimate the RTD parameters and that the precipitation preceding the tracing tests is essential in the form of the RTD. The results also suggest that the dispersivity is very sensitive to the precipitation, and to the tide. The piezometric level was also an important variable, unlike the mean discharge. Equations were selected for each variable, but need to be cross-checked and validated through other tracing tests.

As perspectives, these analyses could be completed with more tracing tests, in order to confirm the trends brought out (especially investigating the rainfall parameter, with tracing tests during storms). The results could be compared to other karstic system. In the case of coastal karstic systems, it would enable to validate the hypothesis of the influence of annual variation of the tide on the response of the system, as well as the daily component of the tide on the dispersivity. For continental aquifers, it would be interesting to assess the joint role of the piezometric level and the precipitations in the response.

References

Bakalowicz M (1997) Water geochemistry: water quality and dynamics. In: Standford J, Gibert J, Danielopol D (eds) Ground-water ecology. Academic Press, San Diego, pp 97–127

Bakalowicz M (2005) Karst groundwater: a challenge for new resources. Hydrogeol J 13(1):148–160. doi:10.1007/s10040-004-0402-9

Dörfliger N (2010) Guide méthodologique: Les outils de l'hydrogéologie karstique pour la caractérisation de la structure et du fonctionnement des systèmes karstiques et l'évaluation de leur ressource

Doummar J, Sauter M, Geyer T (2012) Simulation of flow processes in a large scale karst system with an integrated catchment model (Mike She)—Identification of relevant parameters influencing spring discharge. J Hydrol 426–427:112–123. doi:10.1016/j.jhydrol.2012.01.021

Ford D, Williams P (2007) Karst Hydrogeology and Geomorphology. (John Wiley & Sons, Ed.). MacMaster, Canada; Wiley, Aukland, New Zealand, p 578. doi:10.1002/9781118684986

Fournier M, Massei N, Bakalowicz M, Dupont J (2007) Use of univariate clustering to identify transport modalities in karst aquifers. C. R. Geosci 339(339):622–631. doi:10.1016/j.crte. 2007.07.009

Fournier M, Massei N, Mahler BJ, Bakalowicz M, Dupont JP (2008) Application of multivariate analysis to suspended matter particle size distribution in a karst aquifer, 2345 (October 2007), 2337–2345. doi:10.1002/hyp

Geyer T, Birk S, Licha T, Liedl R, Sauter M (2007) Multitracer test approach to characterize reactive transport in karst aquifers. Ground Water 45(1):36–45. doi:10.1111/j.1745-6584. 2006.00261.x

Göppert N, Goldscheider N (2008) Solute and colloid transport in karst conduits under low- and high-flow conditions. Ground Water 46(1):61–68. doi:10.1111/j.1745-6584.2007.00373.x

Helena B, Pardo R, Vega M, Barrado E, Fernandez JM, Fernandez L (2000) Temporal evolution of groundwater composition in an alluvial aquifer (Pisuerga river, Spain) by principal components analysis. Water Resour 34(3):807–816

Klinka T, Gutierrez A, Thiéry D (2012) Validation du logiciel TRAC: Aide à l ' interprétation de traçages en milieu poreux Rapport final Validation du logiciel TRAC: Aide à l ' interprétation de traçages en milieux poreux Rapport final. BRGM/RP-59425-FR, p. 58

Larocque M, Mangin A, Razack M, Banton O (1998) Contribution of correlation and spectral analyses to the regional study of a large karst aquifer (Charente, France). J Hydrol 205(3–4):217–231. doi:10.1016/S0022-1694(97)00155-8

Lepiller M, Mondain P (1986) Les traçages artificiels en hydrogéologie karstique. Mise en oeuvre et interprétation. Hydrogéologie 1:33–52

Massei N (2001) Transport de partcules en suspension dans l'aquifère crayeux karstique et à l'interface craie/alluvions. University of Rouen

Moore PJ, Martin JB, Screaton EJ (2009) Geochemical and statistical evidence of recharge, mixing, and controls on spring discharge in an eogenetic karst aquifer. J Hydrol 376(3–4):443–455. doi:10.1016/j.jhydrol.2009.07.052

Saporta, G. (2011). Probabilités, analyse des données et statistique (Editions T), pp 243–266

Schmidt M, Lipson H (2009) Distilling free-form natural laws from experimental data. Science (New York, N.Y.) 324(5923):81–85. doi:10.1126/science.1165893

A Computer Method for Separating Hard to Separate Dye Tracers

P.-A. Schnegg

Abstract Tracer tests are an irreplaceable tool for hydrogeologists. They are used to determine the paths of water flow between two spots in a catchment below the surface of the earth. Usually, hydrogeologists carry out tracer tests with only one tracer at a time, but sometimes two or more fluorescent substances are simultaneously injected into different spots and collected in a spring. Then, the resulting cocktail is analyzed by optical methods (fluorescence spectrometer) to separate the tracers and calculate their concentrations. Molecules with sufficiently different excitation spectra are easily separated. But two among the most frequently used tracers, uranine (Na fluoresceine) and eosine, are very close in this respect. Their separation is well-known to be difficult. Other examples are sodium naphthionate and amino G acid, two very useful tracers since they are colorless and therefore unnoticed in surface waters. The eluent of charcoal bags (fluocapteurs) is another example. Beside the released tracer, there is a very high fluorescence background of dissolved organic matter (DOM) from which it must be optically separated. The shape of the excitation spectrum of a fluorescent tracer can be approximated by a Gaussian curve. This curve is completely described by three parameters: peak wavelength, height, and width. The spectrum of a cocktail of two tracers is the sum of two such Gaussian curves. To separate these two curves, we use an algorithm based on the steepest descent in the parameter space to find the best set of $2 \times 3 = 6$ parameters of the model that best fits the measured curve. We achieve good separation even with a concentration ratio smaller than 1:10.

Keywords Separation · Fluorescence spectrometer · Tracer test

---

P.-A. Schnegg (✉)
University of Neuchâtel & Albillia Co, Neuchâtel, Switzerland
e-mail: pierre.schnegg@unine.ch
URL: http://www.albillia.com

B. Andreo et al. (eds.), *Hydrogeological and Environmental Investigations in Karst Systems*, Environmental Earth Sciences 1,
DOI 10.1007/978-3-642-17435-3_14

1 Introduction

Groundwater tracing with dye tracers is of common usage in karst hydrology. This is a method of choice for identifying preferential flow paths in the underground. After the tracer solution has been injected into the ground, a fraction as small as one part per billion collected downstream is still detectable through its fluorescence. Sometimes in multitracer tests, two or three different dye tracers are simultaneously injected at neighboring sites. Often, collected water samples result in a mixture of these tracers. It is necessary to resolve overlapping peaks to establish the concentration of each tracer. This operation is carried out in the laboratory with the fluorescence spectrometer, or directly in the field with the portable filter fluorometer (Schnegg 2003). The success of this analysis depends on the choice of tracers. Their optical properties (wavelength of the excitation and emission bands) must be sufficiently different. Good examples are the following cocktails:

- Uranine—rhodamine (any rhodamine: sulfo B, amino G, WT). Spectral distance: 60–70 nm
- sodium naphthionate—uranine or rhodamine. Spectral distance: 120–180 nm

Unfortunately, a very good tracer such as eosine is not so easy to separate from other dye tracers because its spectrum displays intermediate wavelengths (spectral distance to uranine and rhodamine: 25–48 nm). Colorless tracers are also of great interest, but the spectral distance between two of them, sodium naphthionate and amino G acid (7-Amino-1,3-naphtalene disulfonic acid, monopotassium salt hydrate), is only 39 nm. For this reason, almost all tracer tests avoid these two last combinations.

Noting that the spectra (excitation, emission, or synchroscan spectra) can be represented by single Gaussian curves or the sum of two such curves, we developed a computing method for the separation of two tracers. A Gaussian curve is completely described by three parameters: its peak amplitude, peak position, and width. The spectrum of a mixture of two tracers is therefore the sum of two Gaussian curves (sometimes three if a tracer has two peaks). The separation problem is solved if the six parameters can be determined. As the involved mathematical relationships are highly nonlinear, an Algebraic solution is not available. We use the algorithm of steepest descent to extract the parameters. The program automatically varies the six parameters in turn and stops when the difference between the calculated and the measured response has reached the absolute minimum. Although the software of curve fitting for resolving overlapping peaks is commercially available (PeakFit, Eigenvector, etc.), there is no mention of such method in the area of tracer tests.

2 Minimization Routine MINDEF

The routine MINDEF was written at the Institute of Physics of the University of Neuchâtel (Beiner 1970) and used since then in various optimization problems. The following example (Rosenbrock's valley) illustrates this routine with only two parameters $x1$, $x2$ to have a good grasp of the principle. The function is

$$f(x_1, x_2) = 100\left(x_2 - x_1^2\right)^2 + (1 - x_1)^2 \tag{1}$$

The function has its minimum at point $[x_1 = 1; x_2 = 1]$ and its value is zero (Fig. 1).

Any pair of values can be assigned to $x1$, $x2$ as start point. Routine MINDEF varies the parameters and after some iteration finds its way to the lowest point, the minimum of the function. As an example, starting at point $[x_1 = -1.2; x_2 = 1]$ the minimization ends after 126 steps at point [1.015; 1.014] with a misfit of 2.96×10^{-2}. CPU time on a PC is a few seconds. Each parameter can be assigned a variation weight, so that if its weight is set to zero, the parameter is not allowed to vary. The minimization algorithm is extremely robust.

3 Application to the Separation of Two Tracers

In this article, the method was applied to the separation of two different tracer cocktails, each of them difficult to separate into individual tracer concentrations: (1) uranine with eosine and (2) sodium naphthionate with amino G acid (colorless

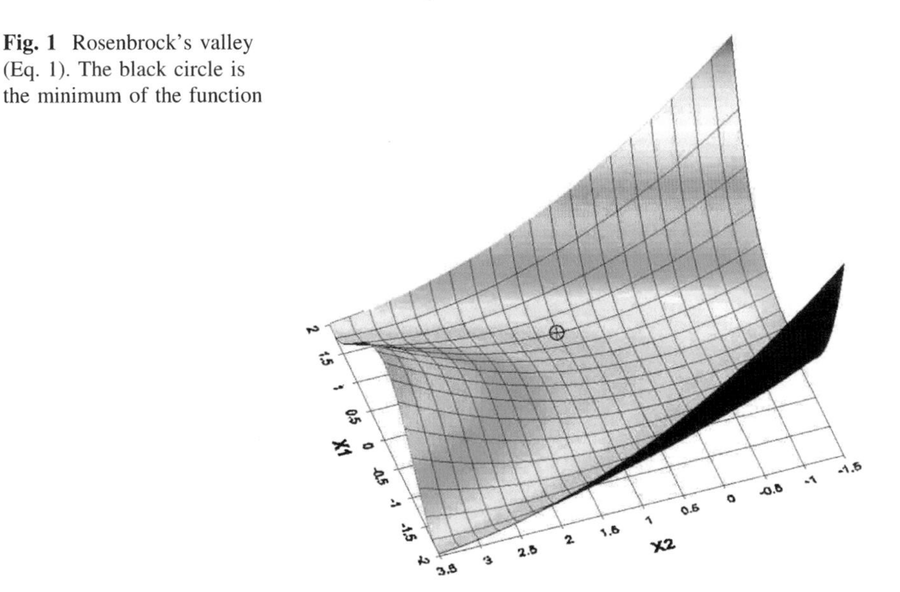

Fig. 1 Rosenbrock's valley (Eq. 1). The black circle is the minimum of the function

dyes). To check the efficiency of the method, we prepared known mixtures of the tracers. The starting solutions were prepared at a concentration of 10 μg/L (10 ppb).

The different cocktails were analyzed on a Perkin Elmer LS50B luminescence spectrometer, with excitation and emission slits of 10 nm and a scan speed of 500 nm/mn.

Figure 2 shows synchroscan spectra of uranine and eosine at 10 μg/L concentration along with their mixtures (from 9 μg/L uranine/1 μg/L eosine to 1 μg/L uranine/9 μg/L eosine). Figure 3 shows synchroscan spectra of sodium naphthionate and amino G acid at 10 μg/L concentration, as well as their mixtures (from 7.5 μg/L sodium naphthionate/2.5 μg/L amino G acid to 2.5 μg/L sodium naphthionate/7.5 μg/L amino G acid). The curves of single tracers show perfect Gaussian shapes (except for amino G acid whose curve is well approximated by the sum of two Gaussians). Therefore, the separation of uranine and eosine is a problem of finding $2 \times 3 = 6$ parameters. We could deduce that the number of parameters for the separation of sodium naphthionate and amino G acid would be higher (3×3). Fortunately, this is not the case since the secondary Gaussian curve of amino G acid is proportional to the main one. In addition, the total number of unknown parameters can be lowered since the center and width data can be measured with good precision by scanning the one-tracer solutions. Then, in both examples, the separation routine will handle only two parameters, the peak amplitudes.

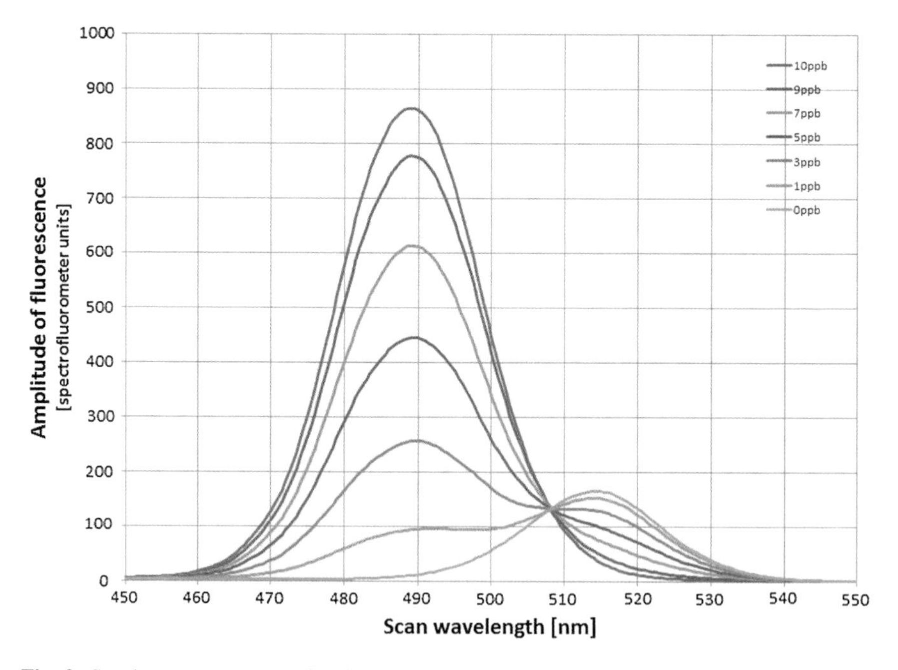

Fig. 2 Synchroscan spectra of mixtures of uranine and eosine 10 μg/L solutions. The concentration legend is for uranine. Corresponding eosine concentrations were 10 μg/L minus the concentration of uranine

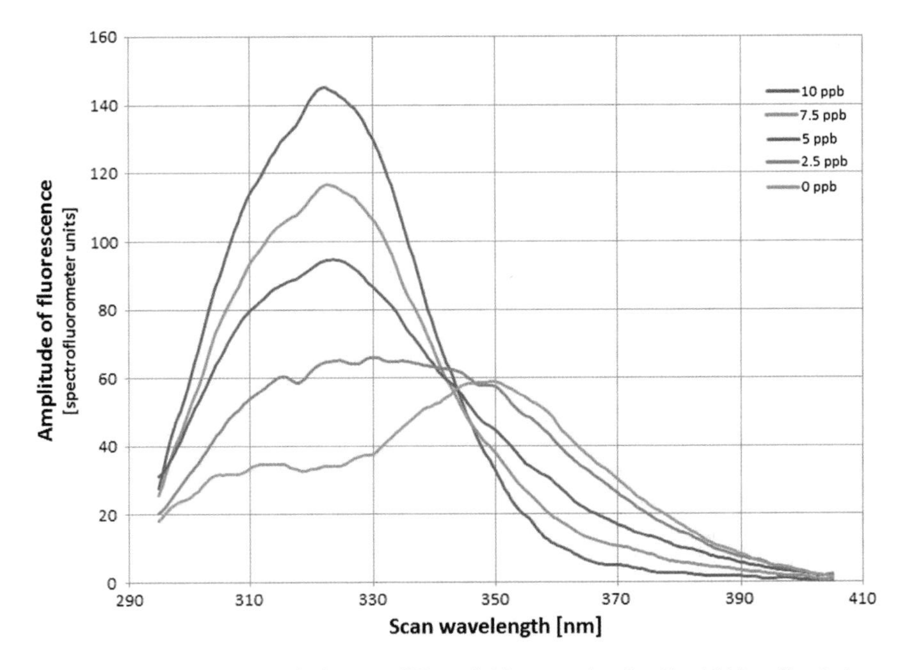

Fig. 3 Synchroscan spectra of mixtures of Na naphthionate and amino G acid 10 μg/L solutions. The concentrations legend is for naphthionate. Corresponding amino G acid concentrations were 10 μg/L minus the concentration of naphthionate

Table 1 Results of the separation with MINDEF of uranine and eosine from mixtures prepared with 10 μg/L solutions

% uranine (true)	% uranine (MINDEF)	% eosine (MINDEF)	% eosine (true)
0	0.69	100.00	100
10	10.45	89.63	90
30	29.05	72.56	70
50	49.02	53.66	50
70	70.49	34.76	30
90	89.55	17.07	10
100	100.00	7.93	0

Table 1 shows the results of the separation by MINDEF of uranine and eosine from mixtures prepared with 10 μg/L solutions. For uranine, the error is less than 1 % regardless of the concentration ratio, but for eosine it strongly depends on its concentration. The error is below 10 % for eosine concentrations in excess of 30 %, but it is close to 100 % when eosine concentration is lower than 10 %.

The case of the naphthionate-amino G acid cocktail is illustrated in Fig. 4. Suppose that all we know is that our water sample contains these two tracers, but at unknown ratio. We will run the program so that the response of our model best fits

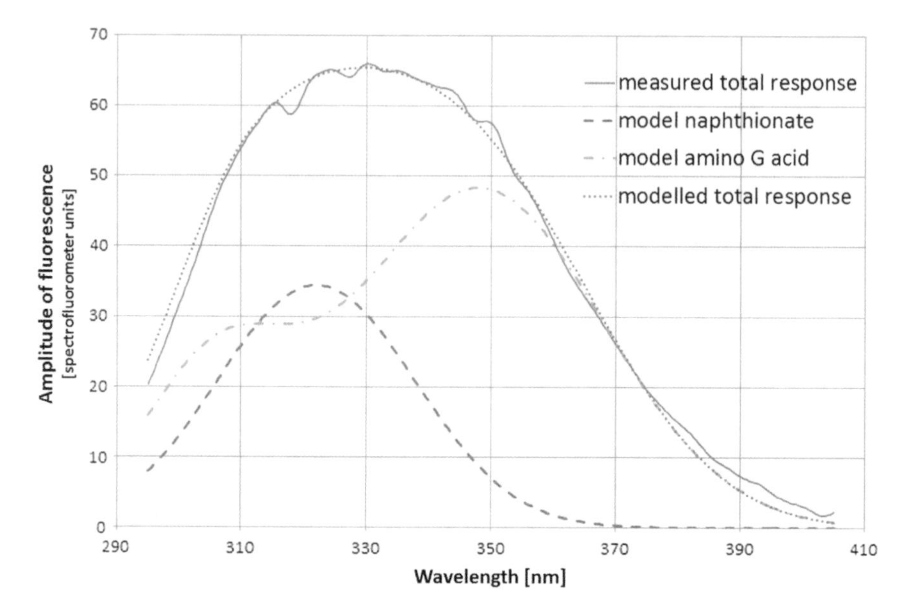

Fig. 4 Separation of a mixture of 2.5 μg/L naphthionate and 7.5 μg/L amino G acid

the curve measured with the spectrofluorometer. The total number of unknown parameters is six, since each Gaussian curve is fully determined by three parameters (the secondary peak of amino G acid is proportional to the main peak and therefore no additional parameter is required). But this number of parameters can be easily reduced to two: peak widths and wavelengths of both tracers can be determined once and for all by applying MINDEF on synchroscan spectra of one-tracer solutions. Then, these parameters are kept fixed in the final modeling phase of the tracer cocktail (variation weight set to zero) and only two unknown parameters, the peak amplitudes are varied and finally determined. Note that the common baseline was automatically subtracted by the luminescence spectrometer before applying MINDEF.

Table 2 shows the results of the separation by MINDEF of Na naphthionate and amino G acid from mixtures prepared with 10 μg/L solutions. From the values of the Tables 1 and 2, we see in both cases that the tracer with the highest quantum yield (peak amplitude) also displays the smallest error in the determination of its concentration.

The case of charcoal bags (fluocapteurs) is not illustrated in this work. However, we have also checked that the method is well suited for separating the tracer response from the background (luminescence of the dissolved organic matter (DOM)). In this problem, one Gaussian curve represents the response of the tracer and a second Gaussian curve the response of the DOM. Since DOM fluorescence appears in the UV to blue region of the spectrum, i.e., at wavelengths shorter than the tracer emission, only the long wavelength tail of the second curve (DOM) is normally represented in the spectrum of the eluent. However, to obtain a good fit

Table 2 Results of the separation by MINDEF of Na naphthionate and amino G acid from mixtures prepared with 10 μg/L solutions

% naphthionate (true)	% naphthionate (MINDEF)	% amino G acid (MINDEF)	% amino G acid (true)
100	100.00	7.02	0
75	75.52	28.07	25
50	54.55	50.88	50
25	23.78	84.21	75
0	0.00	100.00	100

of the measured curve, it is necessary to allow the variation of the three parameters associated with the DOM Gaussian curve, but at the end we are only interested in the amplitude of the tracer peak.

4 Discussion

We used synchroscan luminescence spectra of the tracers for the separation, but other spectra such as excitation or emission spectra can be used as well. However, the best result is always achieved when the peak separation is a maximum.

The assumption behind this separation method is the linearity of the synchroscan signals with regard to the concentration of the tracers. For excitation wavelengths above 500 nm, linearity is almost the rule. But for shorter wavelengths, a nonlinear term should be added. This is particularly true if secondary maxima located between 200 and 300 nm are used in the modeling instead of those of Fig. 4.

To improve the quality of the separation, some more parameters should be included in the MINDEF search, to account for the nonlinearity. The full polynomial dependency of the signal versus concentration could be measured and included in the optimization method.

The MINDEF routine is nice because it does not require the analytic expression of the derivative of the function to be known. But any other similar routine can be used for this work, such as MINUIT (James 2004), designed for particle physics in the 1970s by physicists of the CERN, and many others.

References

Beiner J (1970) FORTRAN Routine MINDEF for Function Minimization. Institute of Physics, University of Neuchâtel, Switzerland, p 5

James F (2004) MINUIT Tutorial. Function Minimization. In: Reprinted from the proceedings of the CERN computing and data processing school (CERN 72–21), Pertisau, Austria, Geneva, 10–24 Sept 1972

Schnegg P-A (2003) A new field fluorometer for multi-tracer tests and turbidity measurement applied to hydrogeological problems. In: Eighth international congress of the Brazilian geophysical society, Rio de Janeiro, Brazil. (Available on the ResearchGate page of the author)

Standardized Approach for Conducting Tracing Tests in Order to Validate and Refine Vulnerability Mapping Criteria

M. Sinreich and A. Pochon

Abstract An approach for conducting tracer tests in karst systems is proposed in order to assess differing degrees of groundwater vulnerability. It consists of (i) a standardized artificial recharge scenario, (ii) the selection of conservative tracers, and (iii) application to contrasting vulnerability situations within a catchment. Results from multi-tracer testing at a karst site in Switzerland provided breakthrough curves that were significantly different in terms of mass recovery, which is considered the key parameter for defining a quantitative protection effect. The presented approach may provide a better basis for both punctual validation of vulnerability maps and refinement of associated assessment methods.

1 Introduction

Tracing experiments are recognized as an essential tool for validating groundwater vulnerability mapping in karst terrains (Perrin et al. 2004). However, data interpretation is limited for many experiments as only the connection between highly vulnerable zones in the catchment area and related springs is proven. The contrast in vulnerability within well-developed karst systems, as deduced from mapping methods such as EPIK (Doerfliger et al. 1999), is rarely tested, although it is crucial for applying protection measures in the catchment (intra-basin comparison) or for transferring results to other systems (inter-basin comparison). This is particularly important, for example, in large areas characterized by low natural protection, as is typical of karst aquifers in Switzerland and many other alpine and peri-alpine regions.

M. Sinreich (✉)
Hydrology Division, Federal Office for the Environment FOEN, 3003 Bern, Switzerland
e-mail: michael.sinreich@bafu.admin.ch

A. Pochon
Hydro-Geol Sàrl, CP 2430, 2001 Neuchâtel, Switzerland
e-mail: alain.pochon@hydro-geol.ch

B. Andreo et al. (eds.), *Hydrogeological and Environmental Investigations in Karst Systems*, Environmental Earth Sciences 1,
DOI 10.1007/978-3-642-17435-3_15

2 Conceptual Approach

In general, the minimum transit time derived from tracing tests is not sufficient to characterize the protective function of overlaying layers, due to the frequent presence of preferential flow paths, even in the case of significant topsoil and subsoil layers. Tracer concentration attenuation and reduced mass recovery over the duration of the main breakthrough may be more representative in this context. While transit time reflects a distance effect in large catchments or is an indicator of the existence of preferential flow paths through potentially protective layers, the main breakthrough recovery is dependent on the storage capacity of the unsaturated zone, including the epikarst. Transient storage in the soil and subsoil is thought to play the most important role in relation to intrinsic vulnerability in many karst systems.

Considering the influence of storage (Smart and Friederich 1986) requires the application of a well-defined and targeted experimental setup. Unlike conventional localized tracer injections, i.e. at swallow holes or into a trench, irrigation through water sprinkling that creates a realistic recharge over a defined area of land surface is an appropriate scenario, which is also much easier to control and interpret than tracer testing under natural rainfall conditions. Such an approach consists of three steps: (i) development of a standardized artificial recharge scenario, which can estimate transient storage and is applicable in a uniform manner, (ii) testing the conservative behavior of several dye tracers under given conditions in order to be able to use them concurrently during multi-tracer experiments, and (iii) application to several karst systems and to diverse settings in order to assess differing degrees of vulnerability.

3 Field Application

3.1 Experimental Setup and Tracer Selection

The proposed procedure includes initial surface irrigation over an area of about 10 m^2 until complete water saturation of the uppermost layers (topsoil, subsoil, epikarst) can be assumed, followed by diffuse tracer application. This area is then irrigated with a volume of water corresponding to a 100 mm recharge over a duration of several hours to simulate unfavorable yet realistic scenarios in terms of maximum daily rainfall (in the order of a 10-year return period) or diffuse contamination events. Tracer retention in the diverse subsystems over the irrigation duration results in reduced mass recovery and tracer concentration attenuation. In addition, a reference injection should be performed in a nearby zone, which is extremely vulnerable (i.e., swallow hole or bare karrenfield) and connected to the sampled spring. The concurrent use of different tracers permits the definition of a setting-specific protection effect for each injection location.

The nonreactive behavior of several tracers, which is a prerequisite to quantitatively determine contrasts in intrinsic vulnerability, has to be tested (Sinreich and Flynn 2011). Figure 1 shows the results of one of these tests conducted in the vadose zone of a karst aquifer overlain by a 0.5 m thick soil layer. Simultaneous tracer pulse injection was performed with a constant irrigation rate, corresponding to maximum soil infiltration capacity as evidenced by runoff generation during initial surface irrigation. An automatic sampler was installed at the main water outlet in a cave 35 m below the surface, which allowed high temporal resolution monitoring of tracer concentration. Breakthrough and mass recovery were similar for three of the tracers, i.e. uranine, duasyne, and naphthionate. Their recovery rates, which varied between 41 and 46 %, are close to the 55 % of injected water recovered, suggesting that these tracers behave in a conservative manner, whereas sulforhodamine B (35 %)—due to its sorption tendency—did not fulfill these requirements under the given conditions despite the relatively thin soil cover.

3.2 Multi-Tracer Test

The described approach was applied on several alpine and peri-alpine karst systems in Switzerland. An example from the Swiss folded Jura Mountains is presented (Lionne spring at l'Abbaye). The catchment of the site extends to some 16 km^2 and is characterized by karstified Jurassic limestone locally covered by glacial deposits of moderate to low permeability with a thickness up to several meters. Karst landforms as well as the conduit network are highly developed. Groundwater is assumed to be of a vadose flow type in the majority of the catchment. Phreatic conditions occur in the vicinity of the spring, which discharges between 20 and 10,000 L/s with an annual mean value of approx. 500 L/s.

A multi-tracer test was performed concurrently using the three dyes considered conservative as illustrated by Fig. 1. All injection points are situated within a radius of 50 m at a distance of about 1,600 m to the spring suggesting similar lateral pathways. However, each injection location had different topsoil/subsoil characteristics in terms of thickness and permeability, which are assumed to provide contrasting vulnerability (Fig. 2):

(1) Karrenfield with only scarce and aerated forest soil patches (most vulnerable)
(2) Limestone covered by about 0.3 m of more compacted pasture soil
(3) Limestone covered by about 0.5 m of soil and 1 m of silty moraine (least vulnerable).

High data quality was necessary to allow a quantitative interpretation of the results. This was achieved by using up-to-date monitoring technologies installed at the spring, which included a field fluorometer (real-time data via Web server), two automatic samplers and continuous discharge measurement. Furthermore, the fluorescence properties of the selected dyes facilitated their analytical differentiation in the laboratory and provided precise tracer concentrations.

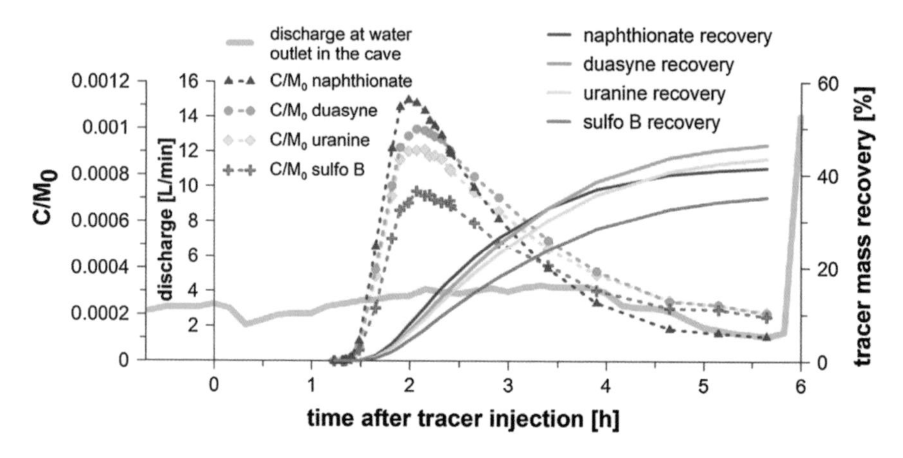

Fig. 1 Breakthrough and cumulative recovery curves for a short-distance comparative tracer test. Uranine, duasyne and naphthionate were considered conservative given the similar form of their breakthrough curves and mass recovery close to irrigation water recovery

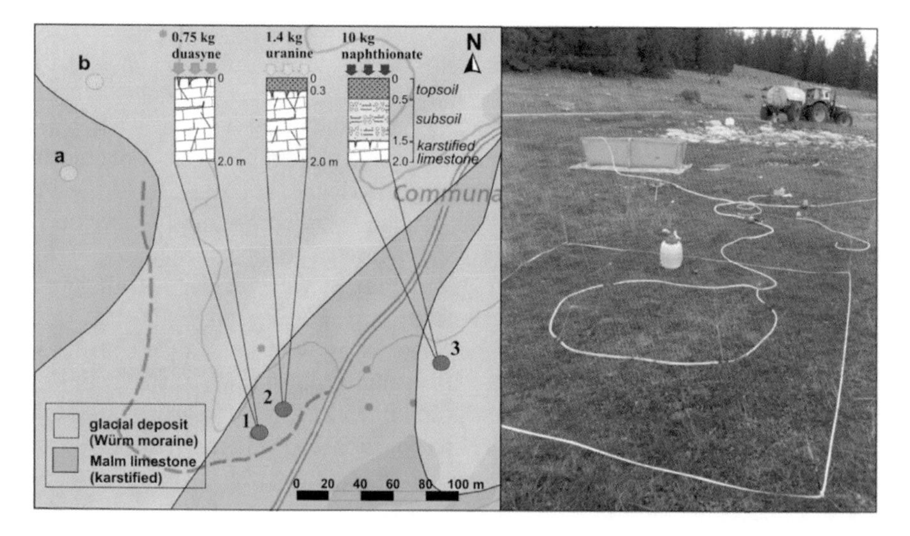

Fig. 2 Injection points (*1*, *2*, *3*) and geological setting for the multi-tracer experiment with logs highlighting the contrasts of vulnerability tested (*left*). The two additional points (*a*, *b*) make reference to tracer injections during a previous study (Perrin et al. 2004). Tracer injection was performed according to the described standardized scenario displayed for injection point 2 (*right*)

Tracer recovery at the spring reflects the contrasting vulnerability evaluated in the field, with the highest value (32 %) for the karrenfield, lower recovery (21 %) for the karstified limestone featuring greater soil cover, and finally drastically reduced recovery (3 %) for the tracer injected into the glacial subsoil deposits (Fig. 3). The dominant transit time for each tracer, however, was almost the same,

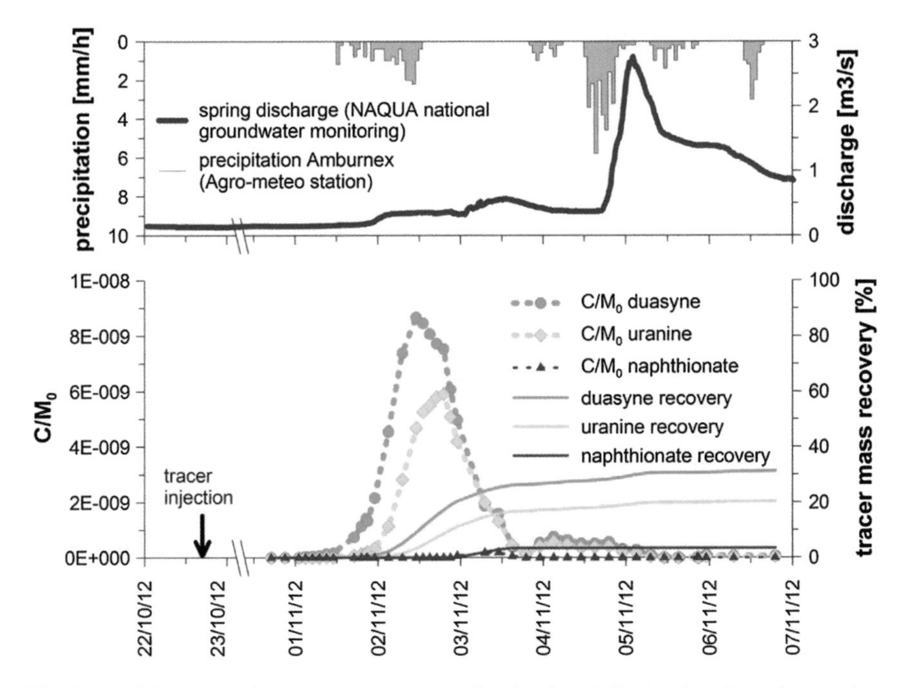

Fig. 3 Breakthrough and mass recovery curves for the three injection locations, i.e. uranine, duasyne and naphthionate tracers. The hydrological conditions remained stable during the 10 days between injection and tracer arrival (only 15 mm of precipitation, most of which was deposited as snow) and a pronounced flood event occurred only towards the end of the breakthrough

with peak velocities between 150 m/day (duasyne and uranine) and 140 m/day (naphthionate). While first detection of both duasyne and uranine tracers was clearly unaffected by natural recharge events, a moderate discharge rise could be observed before naphthionate arrival. Thus, it cannot be excluded that such an additional volume of water was necessary to push the tracer into the karst network, and that it had remained stored in the subsoil with the 100 mm during post-injection irrigation.

A similar degree of karstification is assumed for all injection locations, such that differences in attenuation must be essentially related to the topsoil and subsoil. The presented experiment thus addresses the protective effect of the overlaying layers, which in many assessment methods is considered a crucial criterion influencing the vulnerability of highly karstified systems (Doerfliger et al. 1999; Daly et al. 2002; Zwahlen 2004). The two injection locations covered solely by pedologic soil (topsoil) exhibited only slight differences in mass recovery, although the highest quantity of tracer was recovered from the location with lowest soil cover. In contrast, the moraine cover retained 90 % of the tracer with respect to the karrenfield injection.

Using tracer mass recovery as the key parameter to quantitatively determine the vulnerability is confirmed by comparison with the result of a previous tracer experiment featuring injections in close proximity to the present locations (Perrin et al. 2004). Two pulse injections each corresponding to a net recharge of about 1000 mm were performed on a karrenfeld (Fig. 2, point a) during high and low water stages. The experiment revealed dramatically contrasting transit times varying between 1.5 and 11 days, whereas tracer mass recoveries did not differ significantly (40 and 50 %). This is consistent with the results of the present study suggesting that transit time—which is very dependent on hydrological conditions—is not reliable for vulnerability validation in developed karst systems. Another dye injected into more than 2 m of glacial deposit was not detected at the spring (Fig. 2, point b).

4 Conclusions and Outlook

The presented approach sets out to provide a better quantitative basis for both punctual validation of vulnerability maps and a refinement of associated assessment methods. It highlights the fact that quantitative field evaluation of karst groundwater vulnerability is only reliable if related tracing tests are conducted in a standardized and well-controlled manner. This includes the use of conservative tracers, a defined injection and irrigation scenario through water sprinkling, high data quality in relation to tracer concentrations and hydrological conditions, as well as a high level of understanding of the functioning of the karst setting under consideration.

Existing methods generally use point count and rating systems to describe vulnerability, even though mapping criteria are not related to quantitative information on the extent of contaminant attenuation. Standardized tracing tests, however, allow the attribution of a protection effect—i.e., a degree of mass attenuation—to the situation encountered in the field. Breakthrough curves obtained from tracer injections at the studied catchment differed significantly, both in terms of maximum relative concentration and mass recovery, permitting quantification of the natural protective effect. In this context, mass recovery was found to be the key parameter for evaluating contrasts in vulnerability, which is virtually independent of hydrological conditions.

Mass recovery accurately determined from tracer testing can also be linked to the influence of the individual vulnerability mapping criteria (e.g., soil thickness or epikarst development). Multi-tracing experiments, where each injection setting differs only by a single criterion or class, thereby represent the most efficient means of deducing their protective effect from tracer breakthrough curves. This is the case for example for two injections into adjacent locations with different classes of soil thickness, though with all other criteria remaining equal. Based on the interpretation of numerous tracer experiments, the next step will consist of defining a protection unit for a quantitative classification of each mapped criterion.

The protection provided between a certain location at the land surface in the catchment and the related spring can then be determined from the cumulative effect of each subsystem (topsoil, subsoil, and epikarst), i.e., the total number of protection units describing the groundwater vulnerability in a quantitative manner.

References

Daly D, Dassargues A, Drew D, Dunne S, Goldscheider N, Neale S, Popescu IC, Zwahlen F (2002) Main concepts of the "European approach" to karst-groundwater-vulnerability assessment and mapping. Hydrogeol J 10:340–345

Doerfliger N, Jeannin PY, Zwahlen F (1999) Water vulnerability assessment in karst environments: a new method of defining protection areas using a multi-attribute approach and GIS tools (EPIK method). Environ Geol 39(2):165–176

Perrin J, Pochon A, Jeannin PY, Zwahlen F (2004) Vulnerability assessment in karstic areas: validation by field experiments. Environ Geol 46(2):237–245

Sinreich M, Flynn R (2011) Comparative tracing experiments to investigate epikarst structural and compositional heterogeneity. Speleogenesis Evol Karst Aquifers 10:60–67. www.speleogenesis.info

Smart PL, Friederich H (1986) Water movement and storage in the unsaturated zone of a maturely karstified carbonate aquifer, Mendip Hills, England. Environmental problems of Karst Terrains and their solutions. National Water Well Association. Dublin, Ohio, pp 59–87

Zwahlen F (ed) (2004) Vulnerability and risk mapping for the protection of carbonate (karst) aquifers. COST Action 620. European Commission, Brussels/Luxembourg, pp 297

Hydrogeological Characterization of Karst Tributaries of the San Franciscan Depression, River Corrente, West Bahia, Brazil

C.C. Bicalho, M. Berbert-Born and E. Silva-Filho

Abstract Brazilian economy has experienced substantial growth in the last decade. Balancing economic development and the associated increase of water consumption with environmental sustainability is a challenge for both society and government. Karst aquifers are important freshwater resources for the growing population in some regions of Brazil. The west region of the state of Bahia is known for its abundant water resources. Corrente River (basin surface: 42,732 km^2), provides ca. 30 % of the total water flow of the Sao Francisco River (basin surface 631,133 km^2). During recent years, Bahia's state has undergone a marked process of economic growth driven by agricultural modernization. Important transformations of soil occupation can be observed, and water resource exploitation is often disorganised and predatory. The karst aquifers on the San Franciscan depression are located downstream of an intensively exploited region, the Urucuia sandstone aquifer. Karst aquifers are characterised by high vulnerability to contamination and low capacity of self-purification. This study aims to: (i) perform a preliminary characterization of the hydrogeological behaviour of the karst system in relation with the whole basin context by characterising water hydrochemistry, and (ii) to verify the existence of anthropogenic pollution in order to delineate the aquifer vulnerability. The first results showed a great heterogeneity of aquifer waters, denoted by water chemistry. This heterogeneity was not only related to a large basin surface and different recharge conditions, but also to aquifer compartmentalisation in depth. Results denoted the possible existence of redox processes associated to organic deposits in depth.

Keywords Karst · Hydrogeology · Hydrochemistry · Natural tracers

C.C. Bicalho (✉) · E. Silva-Filho
Departamento de Geoquímica, Instituto de Química, UFF, Niterói, RJ 24020-150, Brazil
e-mail: ccbicalho@gmail.com

M. Berbert-Born
CPRM – Serviço Geologico Do Brazil, Avenida SGAN-Quadra 603 - Conjunto J, Parte a - 1° Andar, Brasília, DF 70830-030, Brazil

B. Andreo et al. (eds.), *Hydrogeological and Environmental Investigations in Karst Systems*, Environmental Earth Sciences 1,
DOI 10.1007/978-3-642-17435-3_16

1 Introduction

Considering the significant growth in the Brazilian economy during the previous decade, the development of efficient water resources management is a crucial challenge for both society and government. Karst aquifers store important freshwater resources for the growing population in some regions of Brazil. The state of Bahia in Brazil has undergone noticeable economic growth driven by agricultural modernisation. Significant transformation of land and soil occupation can be observed and water resource exploitation is often disorganised and predatory. Recently, a massive landslide occurred on the Urucuia plateau carrying down an extensive mass of soil towards karst creeks located downstream, causing a serious ecological accident (Moraes 2013).

The west region of the state of Bahia is known for its abundant water resources (Gaspar 2006). The karst aquifers on the San Franciscan depression are located downstream of an intensively exploited region, the Urucuia sandstone aquifer (Fig. 1). Corrente River (basin surface: 42,732 km^2), provides ca. 30 % of the total water flow of the Sao Francisco River (basin surface 6,31,133 km^2). Like most of the karst aquifers, the San Frascican unities are characterised by high contamination vulnerability enhanced by low residence times and low self-depuration capability. The present study region is particularly more vulnerable due to its location directly downstream to very erodible unit, the Urucuia plateau (CPRM/UFBA 2007).

The present study aims to: (i) perform a preliminary characterisation of the water hydrochemistry in order to understand the behaviour of water circulation and water chemical evolution along the basin and its main compartments (e.g. the karstic sub-basins); (ii) improve the knowledge of the karst aquifer behaviour, in a local and a whole basin scale and (iii) verify the existence of anthropogenic pollution in order to delineate the aquifers' vulnerability.

2 Materials and Methods

The São Francisco River (6,31,133 km^2) flows from south to north, i.e. from Serra da Canastra (Minas Gerais) towards an Atlantic estuary (Fig. 1).

Corrente river has a catchment area of 42,732 km^2, an average flow-rate of 36 m^3/s. The average annual rainfall in the basin during the last 74 years is 995, 41 mm. First recharge events of the hydrological cycle occur in September/October. The period from May to September is nearly dry: it rains 4 % of the total annual precipitation. The rainy season take place from October to April, it rains 96 % of the annual precipitation (HIGESA 1995).

The geology on the Corrente basin area is characterised by the presence of two distinct geological provinces: the Pelitic-carbonate sequence of Late Proterozoic, which constitute the Bambuí Limestones Group and the Sandstone plateau of the

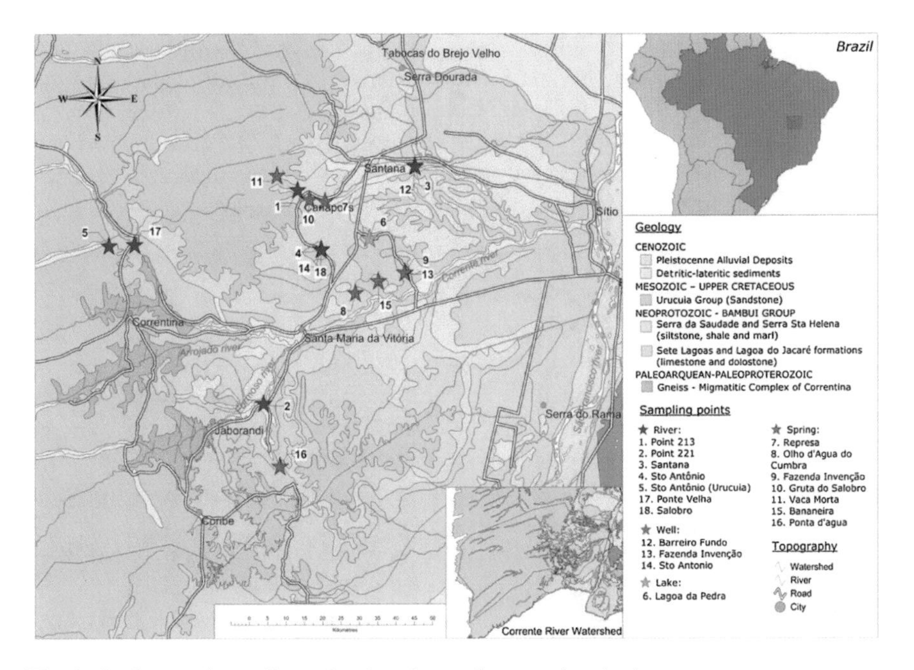

Fig. 1 Geology and sampling points location at Corrente river basin

Urucuia Formation, which is the predominant formation in Corrente River basin (50 %). The study area of main interest is the Sanfranciscan Depression, which is a long stretch of the west bank of São Francisco River, which extends longitudinally in Bahia among the sandstone plateau of Urucuia until the river channel (CPRM/UFBA 2007; HIGESA 1995).

The Urucuia sandstones comprises a very large Phanerozoic intracratonic basin extended in N-S (Sanfranciscan Basin), which is one of the most important granular aquifer in Brazil. It is known as a free aquifer with local confinement due to silicified horizons. The lower strata, lying above the pelitic-carbonatic basement (Bambuí Group) is a clayey unit (Geribá Formation) recognised as the substratum of the aquifer.

Sampling campaigns were undertaken on August-2011, November 2011 and March 2012. Eighteen water points were sampled during this period, including: seven springs, seven streams, three wells and one lake were sampled along the basin, from upstream to downstream (Fig. 1). The sampling points were chosen along the Corrente River network from the Urucuia sandstone massif until the proximities of Corrente river base level. Urucuia Plateau has a very large surface, but is geologically very uniform, consequently, few sampled points could represent the water chemistry. On a longitudinal profile from the Urucuia Plateau until the outlet, three different karsts were monitored (Karst 1, 2 and 3) (Fig. 1).

Temperature, pH, and Electrical conductivity were measured in situ before sampling. Total alkalinity was measured within a day, by acid titration. The samples were filtered in cellulose acetate, acidified with HNO_3. Major elements (Br, Cl, NO_3, SO_4, Ca, Mg, Na and K) were analysed by ionic chromatography and Trace elements (Li, B, Al, V, Cr,Mn, Co, Ni, Cu, Zn, As, Rb, Sr, Mo, Cd, Ba, Pb and U) were analysed using ICP-MS (Thermo Scientific). Multivariate statistical analysis of collected data was used to characterise hydrochemically the different water types along the basin.

3 Results and Discussion

Great discrepancies are observed for the hydrochemistry data presented along the study area, e.g. the wide variation of Electric Conductivity, which values vary from 0 up to 1400 μS/cm. This reflects the known lithological heterogeneity of the study area and gives insights to the significance that hydrochemical evolution plays on waters fingerprinting along its path.

Discriminating Factorial Analysis (DFA) was used to identify the main variables responsible for the general hydrochemistry of the studied dataset. The Discriminant Factorial Analysis for the entire basin samples showed clearly three basic groups (Fig. 2): the Urucuia Group, Lake and Karst group (Karst 1, Karst 2 and Karst 3).

The Urucuia group, formed by the samples collected up in the plateau (Santo Antonio River—Urucuia) and the spring located on the plateau basis (Vaca Morta Spring). This group is basically characterised by an extremely low mineralisation. The group of samples collected on the Lake (Lagoa da Pedra), which is characterised by an evaporation signature on stable isotopes. The group of the karst

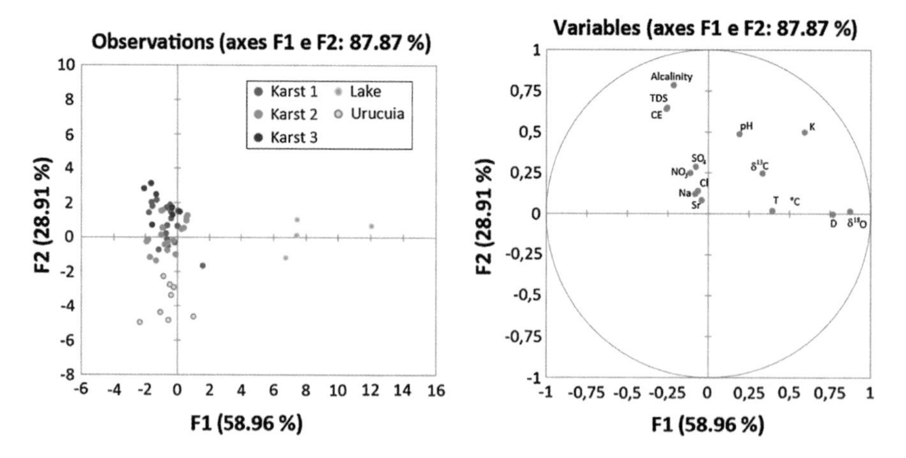

Fig. 2 Discriminant factorial analysis—entire basin

samples, which is formed by three sub-groups: Karsts 1 Karst 2 and Karst 3. Those groups were characterised separately.

The group of the karst samples is characterised by high heterogeneity on samples properties, which is expected on such complex and karstified limestone terrain. The DFA applied for the karst samples (Fig. 3), considered the discriminating variables as three distinct groups of karst.

Karst 1 is composed by the samples located on the upstream part of the basin, right downstream to the Urucuia Plateau. This sub-system is characterised by a rapid and more surficial groundwater circulation. This water has a more marked anthropogenic fingerprint, with higher concentrations in K and NO3. The samples of Karst 1 group are: *Córrego Ponto 213, Riacho Santana, Corrego Santo Antônio, Nascente da Represa, Nascente gruta do Salobro, Poço Santo Antônio and Ressurgência da Bananeira.*

Karst 2 is composed by a group of samples that is located in the lower part of the basin, which level seems to be underneath the Corrente river base level. It is characterised by high mineralised waters (EC > 2000 μs/cm) that are also enriched on concentrations for elements like Cl, Sr, SO4 and Na. This group probably corresponds to high residence-time groundwaters, its water mineralisation is probably influenced by other minerals than calcite (e.g.: gipsita and pirita). More complex geochemical reactions should also be existent for these water chemical evolution (e.g: redox reactions), which need to be more studied in-depth. The samples of Karst 2 group are: *Nascente Olho d'Água do Cumbra, Poço Barreiro Fundo, Riacho Salobro and Rio Ponte Velha.*

Karst 3 is composed by an intermediary karst system, which is located on the right-bank of the Corrente River. It is an autochthone karst system, apparently less subjected to anthropogenic influence than Karst 1, and with a more simple chemistry, i.e. waters facies are magnesium- calcium-carbonic. The samples of Karst 3 group are: *Nascente Fazenda Invenção, Nascente Ponta d'àgua, Corrego ponto 221 and Poço Fazenda Invenção.*

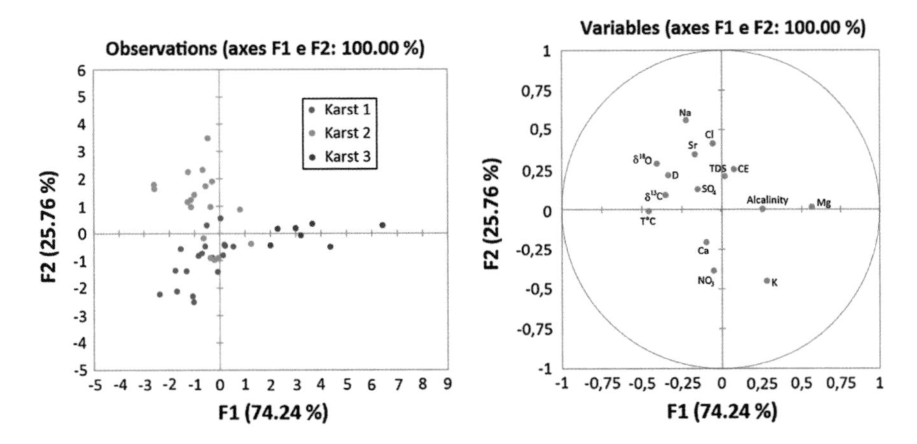

Fig. 3 Discriminant factorial analysis—only karsts

4 Conclusion: Conceptual Model

The groundwaters' chemical characterisation allowed decrypting the general aquifer structuring and groundwater circulation, which is represented through a conceptual model (Figs. 4 and 5). The model presents water circulation from upstream to downstream on the longitudinal profile W-E (Fig. 4).

The karst system 1 is located directly downstream to the sandstones; it is the first uncovered limestone from upstream to downstream (in grey on Fig. 4). This sub-system is recharged by the sandstones located directly upstream. This karst is a shallow-free-fissured aquifer. Recharged water acquires carbonated facies in a relatively short flow-path. In this region flat accumulation surfaces were observed ("*Patamares do Chapadão*" geomorphological unit) denoting higher availability of water resources; consequently most of the settlements are located in this area.

The karst system 2 is also relatively shallow and fissured, but in a context of a very irregular landscape (notched relief with large canyons). There are some residual sandstone covers. This region has few inhabitants and suffers from a lack of surface water.

The karst system 3 is composed by high mineralised waters, and has a likely deep circulation flow-path, which is located in a low hydraulic gradient context. Its

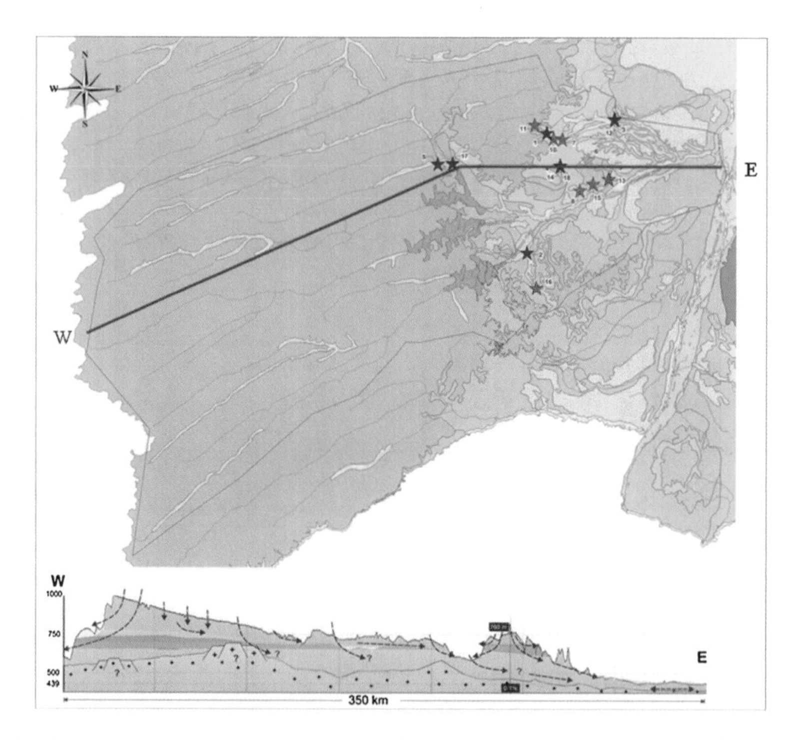

Fig. 4 Conceptual model of groundwater circulation along the longitudinal profile indicated by the line WE on the plan

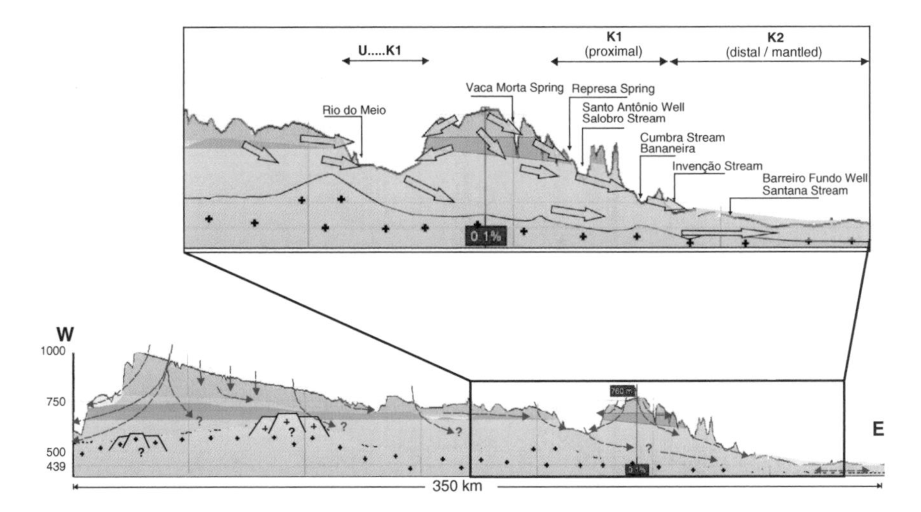

Fig. 5 Conceptual model in detail of the more downstream portion of the profile (karst 3)

water properties denote the presence of organic matter and/or sulphides and chlorides. This system seems to be hydrogeologically controlled by limestones situated underneath the Urucuia Plateau. Karst 3 groundwaters are chemically fingerprinted by high residence-time waters issued from limestones located underneath the sandstones. This karst sub-system is the most complex case of the studied area.

This preliminary study allowed a global understanding of the system behaviour, including recharge and groundwater circulation. The first results showed a great heterogeneity of aquifer waters denoted by water chemistry. This heterogeneity was not only related to a large basin surface and different recharge conditions, but also to aquifer compartmentalisation.

The relationship between the granular Urucuia aquifer and the granular-fissured Bambuí aquifer beneath Urucuia Plateau is not still clearly elucidated. There are unconformities pointing the existence of structural lineaments and different compartments that will be the object of supplementary investigation in order to better understand this system and elucidate groundwater origins.

References

CPRM/UFBA (2007) Hidrogeologia da Bacia Sedimentar do Urucuia: Bacias Hidrográficas dos Rios Arrojado e Formoso. Meta B, Caracterização Geológica e Geométrica dos Aquíferos—Revisão Geológica e Levantamento Geofísico (Comportamento das Bacias Sedimentares da Região Semi-Árida do Nordeste Brasileiro). CPRM/UFBA: Rede Cooperativa de Pesquisa, p 72

Gaspar MTP (2006) Sistema Aqüífero Urucuia: Caracterização Regional e Propostas de Gestão, in Instituto de Geociências. Universidade de Brasília, Brasília

HIGESA (1995) Plano Diretor de Recursos Hídricos - Bacia do Rio Corrente: Plano Setorial de Saneamento. Salvador. 1995. p. Paginação irregular

Moraes C (2013) Terra Ronca sofre danos irreversíveis 2013. O Hoje. Available in http://www.ohoje.com.br/noticia/10621/terra-ronca-sofre-danos-irreversiveis. Accessed 11 Dec 2013

Middle Term Evolution of Water Chemistry in a Karst River: Example from the Loue River (Jura Mountains, Eastern France)

J. Mudry, F. Degiorgi, E. Lucot and P.-M. Badot

Abstract Plotting multiyear chemographs in a karst river may display evolutional trends in water quality. In recent decades, different factors could explain for instance phosphate decrease and nitrate increase. In the Jura karst area, different springs and rivers not only exhibit variation of anthropogenic molecule concentrations, but also evolution of major element concentrations, that results in a change in electrical conductivity. Over a 30-year period, average electrical conductivity of the Loue River, which is totally supplied by karst springs, has increased from 260 to 470 μS/cm. Such an 81 % variation is only explainable by increases in major components, i.e. calcium and hydrogenocarbonate ions, which are the almost exclusive by-products of karstification processes in the Jurassic limestones of the Jura Mountains. Indeed, no direct anthropogenic cause may be invoked for such an evolution. An increase of dissolution throughout the system is necessarily correlated to an increase in the carbon dioxide transfer, resulting from an increase in dissolution and/or production rate.

1 Introduction

Generally, hydrochemical datasets of springs and rivers in karst context are used to display either pollutions or global changes. The Loue River, situated in the Jura Mountains (Eastern France) is a well-known river for sport fishery (trout and grayling). Recently, it has been subjected to several severe fish mortality episodes, despite the fact that water meets the chemical quality standards.

These problems led to examine finely the historical records of ionic concentrations and concentrations of various pollutants. These physical and chemical data

J. Mudry (✉) · F. Degiorgi · E. Lucot · P.-M. Badot
Chrono-Environnement, UMR 6249/CNRS University of Franche Comté, 16 Route de Gray, 25000 Besançon, France
e-mail: jacques.mudry@univ-fcomte.fr

B. Andreo et al. (eds.), *Hydrogeological and Environmental Investigations in Karst Systems*, Environmental Earth Sciences 1,
DOI 10.1007/978-3-642-17435-3_17

have been collected at the drinking water production plant of the city of Besançon, located at Chenecey-Buillon, in the middle valley of the Loue. No significant pollution episodes, but only gentle incremental variations were displayed. This fact does not argue for a massive pollution, but conversely for a progressive worsening of multi-factor environmental quality.

One of the major historical changes in physical-chemical parameters is an increase in electrical conductivity (Teleos 2002). Fitting a 40 year conductivity dataset with time displays a rough linear relationship (Fig. 1).

2 Global Reasons

Several trends in karst water balance and in subsequent chemical change can be ascribed to global climate change, such as North Atlantic Oscillation (NAO, Knight et al. 2006). But throughout the interested period, no significant trend can be highlighted in the discharge of the river (Villeneuve et al. 2012).

On the other hand, in the period 1970–2006, atmospheric carbon dioxide concentration increased by 17 % from 325 to 380 ppm. This difference cannot explain the synchronous change in electrical conductivity by 81 %, from 260 to 470 μS/cm. Nevertheless, combined with the correlated global climate change, it could be involved in soil modification trends (Jackson et al. 2009).

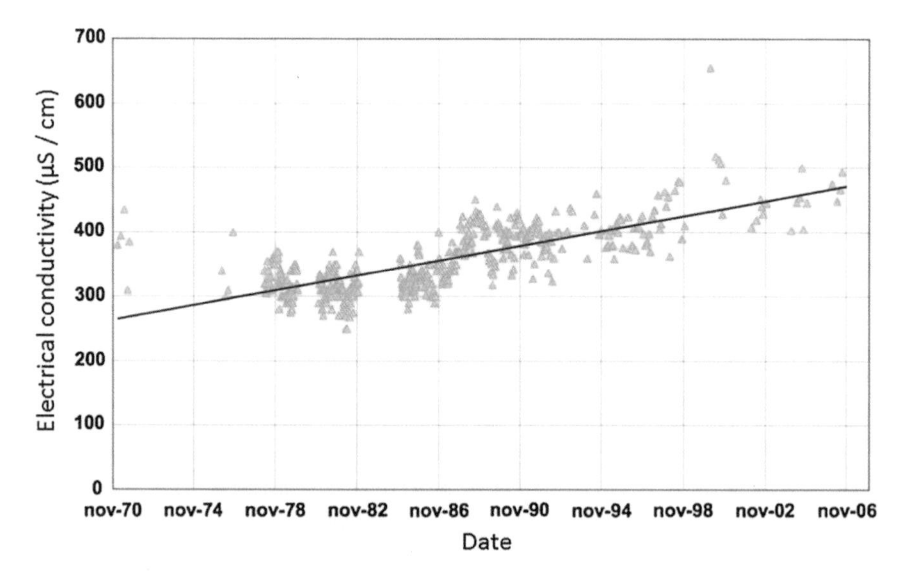

Fig. 1 Evolution of electrical conductivity (Loue River, Chenecey-Buillon). Cond = 0.0156 $t_{(j)}$ − 139.19; R = 0.74 (HS)

3 Pollution Reasons

As a whole, the main indicators of human activities in a watershed show contrasted evolutions in 40 years (Fig. 2).

Nitrate concentration, despite a wide range of variation due to leaching episodes, displays an increasing trend. On average, nitrates increase by 66 %. This increase could be linked to a (moderate) increase of the cattle density in the recharge area, and to a (significant) increase of milk productivity *per capita* (Lambert 2007; Lambert et al. 2010) through the period of interest (Fig. 3).

Indeed, milk production was 5,387 kg per year in 1979 and 7,303 kg in 2000. This increase by 36 % of milk productivity is arithmetically correlated to a 36 % increase of nitrogen in excreta. But passing from 4 to 8 mg/L of nitrogen cannot explain 81 % of conductivity increase: doubling this nitrate concentration is increasing electrical conductivity up to 5 μS/cm only (versus 210 μS/cm observed in the river).

Phosphorus, originating in sewage treatment plants and in manure spreading, displays a trend towards reduction, mainly peaks, because of the improvement of urban sewage collection and treatment. The cases of ammonium and chlorophyll are similar: descending trend for maxima.

No analysed pollutant can be invoked to explain the high conductivity increase of the last 40 years.

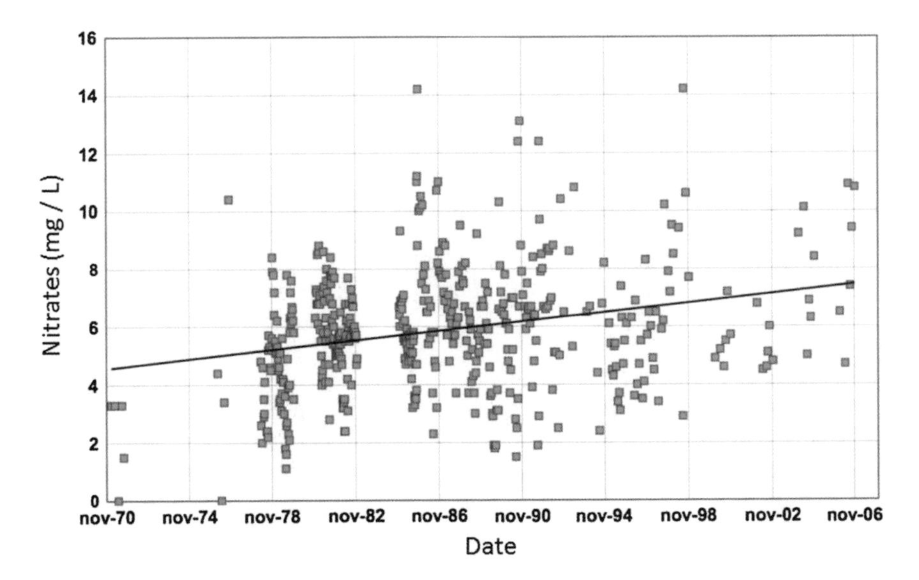

Fig. 2 Evolution of nitrate concentration (Loue River, Chenecey-Buillon)

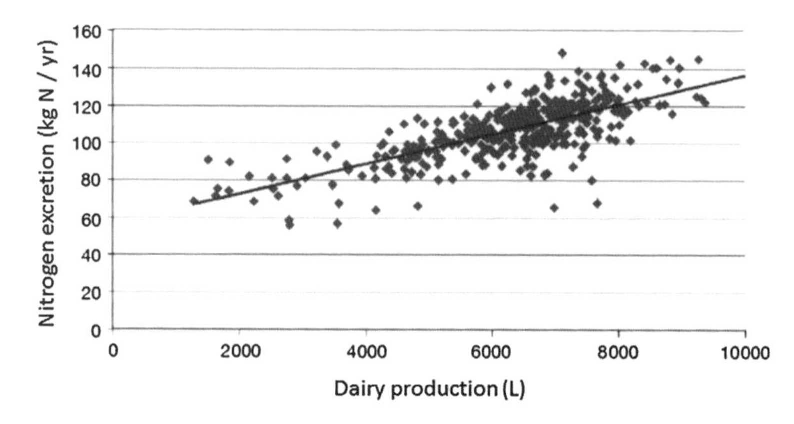

Fig. 3 Estimation of the dairy cow's nitrogen excretion versus milk production (Lambert 2007)

4 Land Use Reasons

What has actually changed in the recharge area during the last 40 years?

Population and correlatively traffic have increased. Sewage collection and treatment are now more efficient. Wood treatment in sawmills and also within the forest use more insecticides and fungicides, but these toxic pollutants cannot explain any calcic-carbonic mineralisation change. But the land use management is quite different today. Grasslands have no more a natural growth; they are deeply ploughed, with a significant consequence: increased reaction surface between soil aggregates, solution and atmosphere. This practice promotes mineralisation of organic carbon, therefore CO_2 production, enabling solid carbonate dissolution.

5 Conclusions

A dramatic increase of electrical conductivity in a karst context with moderate population and economic activity growths cannot be assigned to a sole direct pollution effect. It requires a spatial phenomenon involving massive additional CO_2 production. The present hypothesis is that the soil itself undergoes a loss of organic carbon owed to a change in agricultural practices, involving deep ploughing. This process could also be magnified by global climate change and/or increased atmospheric Carbon dioxide. A balance in soil carbon will support these hypotheses.

References

Jackson RB, Cook CW, Pippen JS, Palmer SM (2009) Increased below-ground biomass and soil CO_2 fluxes after a decade of carbon dioxide enrichment in a warm-temperate forest. Ecology 90(12):3352–3366

Knight JR, Folland CK, Scaife AA (2006) Climate impacts of the Atlantic multidecadal oscillation. Geophys Res Lett 33:L17706. doi:10.1029/2006GL026242

Lambert R (2007) Rejets azotés de la vache laitière en région wallonne. *In*: 14ème rencontre autour des recherches sur les ruminants, 5–6 décembre, Paris, 14

Lambert R, De Toffoli M, Dufrasne I, Hornick JL, Stilmant D, Seutin Y (2010) Towards a revision of the dairy cow's standard for nitrogen production: Justification and what are consequences for soil link rate of dairy farms. Biotechnologie Agron Soc et Environ 14:67–71. http://www.pressesagro.be/base/text/v14ns1/67.pdf

TELEOS (2002) Study of ecological aquatic potentiels écologiques of the 'Natura 2000' sites of the Loue and Lison rivers, DIREN of Franche—Comté, 87p + annexes + atlas. Unpublished (in French). https://www.yumpu.com/fr/document/view/16903417/etude-des-potentiels-ecologiques-aquatiques-des-sites-natura-2000. Accessed 15 March 2014

Villeneuve A, Humbert JF, Berrebi R, Devaux A, Gaudin Ph, Pozet F, Massei N, Mudry J, Trevisan D, Lacroix G, Bornette G, Verneaux V (2012) Expertise report about fish mortalities and cyanobacteria blooms of the Loue river. Study of the functionning of the Loue River and watershed. Report ONEMA, p 32. Unpublished (in French). http://www.esrl.noaa.gov/gmd/ccgg/trends/ Assessed 5 March 2014

Hydrologic Influences of the Blanco River on the Trinity and Edwards Aquifers, Central Texas, USA

B.A. Smith, B.B. Hunt, A.G. Andrews, J.A. Watson, M.O. Gary, D.A. Wierman and A.S. Broun

Abstract The Blanco River of central Texas provides an important hydrologic link between surface and groundwater as it traverses two major karst aquifer systems—the Trinity and Edwards Aquifers. The Blanco River is characterized by alternating gaining and losing stretches due to the presence of springs that discharge water into the river and swallets that drain water from the river. The region consists primarily of Lower Cretaceous limestone, dolomite, and marls. One of the more significant springs along the Blanco River is Pleasant Valley Spring. During below-average flow conditions, Pleasant Valley Spring becomes the headwaters of the Blanco River even though the headwaters, under wet conditions, are about 50 km upstream. Water that enters the Edwards Aquifer from the Blanco River can eventually discharge at both San Marcos Springs to the south and Barton Springs to the north. During periods of extreme drought, when other recharging streams are dry, the Blanco River can provide enough water to the Edwards Aquifer that will help maintain flow at Barton Springs where endangered species of salamanders need sufficient flow of high-quality groundwater. In the western part of the study area, increasing rates of pumping from the Trinity Aquifer, combined with impact from drought, are reducing heads in the aquifer and are subsequently reducing springflows (such as from Pleasant Valley Spring) that sustain the Blanco River. Decreasing flow in the Blanco River can lead to less recharge to the Edwards Aquifer and less discharge from San Marcos and Barton Springs. A better understanding of these aquifer systems and how they are influenced by the Blanco River is important for management of groundwater in an area undergoing significant population growth.

B.A. Smith (✉) · B.B. Hunt · A.G. Andrews · J.A. Watson
Barton Springs/Edwards Aquifer Conservation District, 1124 Regal Row, Austin, TX, USA
e-mail: brians@bseacd.org

M.O. Gary
Edwards Aquifer Authority, San Antonio, TX, USA

D.A. Wierman · A.S. Broun
Independent Consultant, Dripping Springs, TX, USA

B. Andreo et al. (eds.), *Hydrogeological and Environmental Investigations in Karst Systems*, Environmental Earth Sciences 1,
DOI 10.1007/978-3-642-17435-3_18

1 Introduction

This study was conducted to better understand the relationship between the Blanco River and the underlying Trinity and Edwards Aquifers. The Blanco River extends across about 110 km of Hill Country terrain in central Texas. Only under very wet conditions does the river have continuous flow of water across its entire length. Under extreme drought conditions the Blanco River flows over only about half of its length. Karst features related to faults and fractures, stratigraphy, and lithology determine whether water is entering or leaving the river. This study interprets recently-collected flow and potentiometric data and results of dye trace studies to describe the interaction of the Blanco River with the underlying aquifers. The Trinity Aquifers and the Edwards Aquifer provide water to millions of people in central Texas. There are no others sources of water over much of this area.

2 Geography and Hydrogeology

The study area (Fig. 1) consists of the Edwards Plateau to the west and the Balcones Fault Zone (BFZ) to the east. The boundary between the Edwards Plateau and the BFZ is approximately the line on Fig. 1 between the Edwards Aquifer and the Trinity Aquifer. The climate of the study area is classified as subtropical humid, with mild winters and hot summers, although during some periods it may be more accurately considered semi-arid, especially in the western portions in the Hill Country. Average annual rainfall for this area is about 770 mm (Hunt et al. 2012). The entire area is prone to drought, which may be severe and persist for months to many years. Virtually every decade has had one or more significant drought periods that has lasted for a substantial part of a year or more. The most extreme drought in recorded history in this area was the decadal drought from 1947 to 1956 (BSEACD 2014). During the hottest and driest periods most of the rain will evaporate with very little left to run off the surface of infiltrate the subsurface. Some storm events of up to 200 mm of rain have occurred in the area with no appreciable flow developing in the creeks (Smith et al. 2011).

The geologic units that make up the Edwards Aquifer are mostly limestone and dolomite. The Trinity Aquifers are composed of (from stratigraphically highest to lowest) the Upper Glen Rose Limestone, Lower Glen Rose Limestone, Hensell Sand, Cow Creek Limestone, and the Hammett Shale (Fig. 2). The Upper and Lower Glen Rose Limestones consist mostly of limestone, dolomite, shale, and marl. Some units of the Upper and Lower Glen Rose Limestones contain evaporites.

Studies of structures along the BFZ (Collins and Hovorka 1997) indicate that much of this area consists of southeast dipping, *en echelon*, normal faults with throws of as much as 260 m. Some of these faults are continuous over many kilometers, while others extend only a few kilometers or less. Fault blocks between the points where fault displacement decreases to zero are called relay ramps. The relay ramps transfer

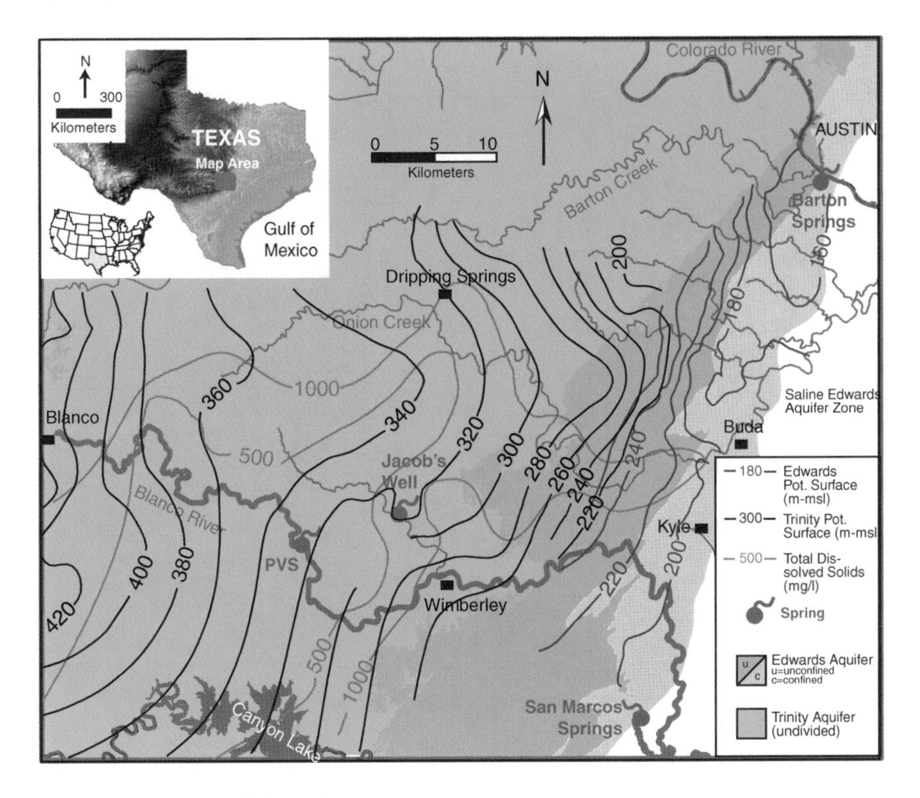

Fig. 1 Location map of the study area

displacement from one fault to an adjacent fault. Faulting and fracturing along the BFZ play a significant role in the development of the Edwards and Trinity Aquifers.

Since the Edwards and Trinity Aquifers consist largely of limestone and dolomite, karst features, such as caves, sinkholes, and solutionally enlarged fractures, direct water into the subsurface. In the upland areas, runoff from rain events might flow into these recharge features for only a few hours. The majority of water recharging the Edwards Aquifer in the study area comes from flow in the creeks and rivers. Slade (2014) estimates that 75 % of the water entering the Edwards Aquifer in the area between Kyle and Barton Springs (Fig. 1) comes from infiltration of water in the stream beds, with the remaining 25 % infiltrating in the uplands between the streams.

3 Blanco River Flow

Under average-to-dry conditions, the Blanco River (Fig. 2) consists of long stretches of a dry streambed and intermittent stagnant pools with no flow. Other stretches of the river maintain flow under all but the most extreme drought

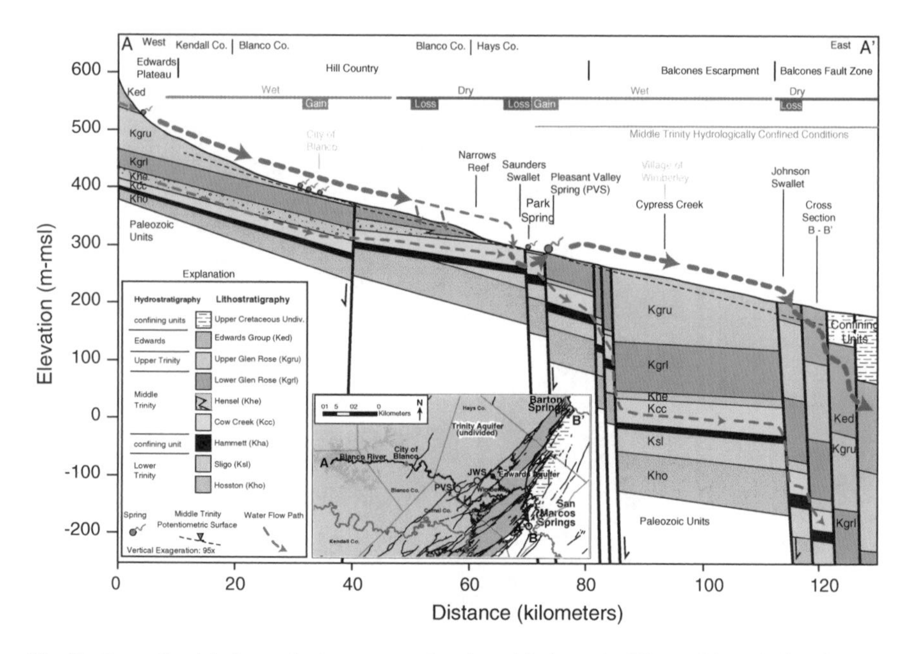

Fig. 2 Generalized hydrogeologic cross section A to A' along the Blanco River during drought conditions

conditions (Wierman et al. 2010). What controls flow, or the lack of flow, in the river, in addition to periods of rainfall or drought, are karst features that allow water to enter the subsurface or to discharge to the river. West of the City of Blanco (Figs. 1 and 2), rain falls on the semi-arid landscape of the Texas Hill Country. Some of that water penetrates the thin soil and passes through the vadose zone where it recharges the Upper Trinity Aquifer. Some of the water falling on the surface might run off to a normally-dry creek bed, and then flow into the Blanco River. Near the City of Blanco some of the water that recharges the shallow aquifer in the upgradient areas discharges into the Blanco River through a series of small springs. About 20 km downstream of the City of Blanco, the Blanco River becomes a losing stream as water flows into the subsurface through numerous small recharge features. At Saunders Swallet in Hays County (Fig. 2), the river is dry most of the time, other than periods of high rainfall. Less than 1 km downstream of Saunders Swallet, groundwater discharges from several springs (Park Springs) and the Blanco River becomes a flowing river again. Another 2 km downstream is Pleasant Valley Spring (Fig. 3) from which discharge has been measured at up to 0.8 m^3/s. Along this stretch of the river, groundwater discharges at multiple locations though gravels on the stream bed. Pleasant Valley Spring occurs where a northeast-trending fault and associated fractures cross the Blanco River. During drought conditions, Pleasant Valley Spring is the headwaters of the Blanco. Even under average conditions, Pleasant Valley Spring provides the majority of water flowing in the river. From this point, the Blanco River continues

Fig. 3 View of Pleasant Valley Spring on the Blanco River. Groundwater discharges from the aquifer through a gravel bar that cuts diagonally across the river bed. Inset photograph is an upstream view of the Blanco River where groundwater discharges from a gravel bar beneath the large, *red* tree to the *left*

as a flowing stream for another 40 km where recharge features, such as Johnson Swallet (Fig. 2), in the Edwards recharge zone divert enough water to cause the river to go dry again under moderate drought conditions. When enough water is flowing in the river, some of that water will continue to flow past the recharge zone of the Edwards Aquifer. From that point on, the geologic units underlying the river have low permeability and the Blanco River continues to where it joins the San Marcos River.

4 Recharge to the Trinity Aquifers

Described earlier is the pathway of water where it either flows in the river or where it is interchanged between the river and the shallow aquifer. However, there are other pathways for the water to flow further away from the river. One of these pathways is where the water enters the Middle Trinity Aquifer and moves deeper into the subsurface (Fig. 2). The geologic units in the study area are encountered at greater depths to the east owing to normal faulting with downthrown blocks to the east of the faults. One area for recharge to the Middle Trinity Aquifer is near

Saunders Swallet (Fig. 2) where the Hensel Sand and Cow Creek Limestone are exposed at the surface. Another area is about 8 km upstream of the Narrows. Some of the water entering the Cow Creek Limestone might discharge into the Blanco River again at Pleasant Valley Spring where it migrates upward about 10–20 m along faults and fractures. In this area, the Hensel consists mostly of dolomite with low permeability and acts as an aquitard. The Hammett Shale acts as a aquitard below the Cow Creek. Any water in the Cow Creek that does not exit the aquifer at this point will follow a deep pathway as faulting takes the Cow Creek to depths of 400 m within the study area and considerably deeper as the water migrates deeper into the subsurface to the east. This pathway is not directly across the faults, but lateral flow is thought to occur along relay ramps where the amount of throw along a fault decreases to zero (Collins and Hovorka 1997). At these depths, evaporite minerals are common in the Upper and Lower Glen Rose formations. These evaporite minerals were initially present in the same sediments to the west, but they have largely been removed by infiltration of meteoric water. Therefore, these units have very low permeability, and faults that cut across these units have low permeability (Smith and Hunt 2010). A multiport monitor well was installed near Onion Creek (Fig. 4) that penetrated the Cow Creek Limestone in the deepest sampling port in the well at a depth of 410 m (Wong et al. 2013). Water samples collected from the Cow Creek are of moderately low total dissolved solids (TDS) values (about 900 mg/L)(Wong et al. 2013). Considering the potentiometric and TDS contours shown in Fig. 1, it is likely that the source of the water encountered in the Onion Creek monitor well comes from the Blanco River or from that general area. Where this Cow Creek groundwater ultimately surfaces is unknown. Some of that water might flow north toward the Colorado River, finding pathways along permeable faults to discharge into the river. Or, the water might continue moving deeper into the basin where other faults could allow it to reach the surface and discharge into rivers on the Texas coastal plain.

5 Recharge to the Edwards Aquifer

Water in the Blanco River that reaches the recharge zone of the Edwards Aquifer will either enter karst features along that stretch to recharge the Edwards, or will flow downstream to join the San Marcos River. Flow data indicate that up to 0.6 m^3/s of Blanco River water may enter the Edwards Aquifer along this stretch (USGS 2014, unpublished data). From here, the water can either flow south to discharge at San Marcos Springs, or it may flow north to discharge at Barton Springs, or it may flow in both directions. A dye trace study conducted in 2009 showed that dye injected into Johnson Swallet on the Blanco River (Fig. 2) arrived at both springs (Johnson et al. 2012).

The main factor that determines the direction of flow of groundwater beneath the Blanco River is the amount of water recharging the Edwards from Onion Creek, which is about 4 km north of the Blanco River (Fig. 1). About 5.7 m^3/s can

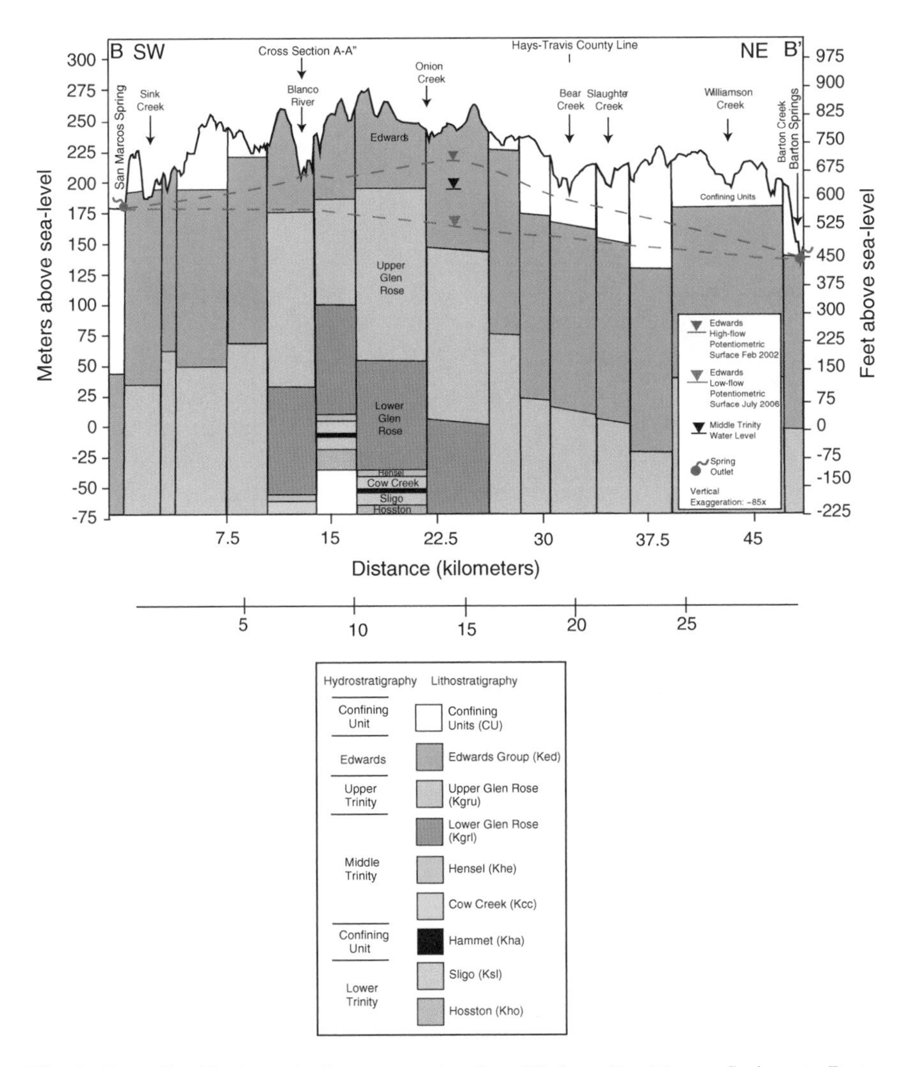

Fig. 4 Generalized hydrogeologic cross section B to B' from San Marcos Springs to Barton Springs

recharge the Edwards Aquifer from Onion Creek (Smith et al. 2011). The Blanco River, with fewer large recharge features and a shorter stretch over the recharge zone can only provide about 0.6 m^3/s of recharge. The large amount of recharge from Onion Creek allows a significant groundwater mound to develop beneath Onion Creek during wet conditions (Fig. 4). Groundwater flowing south from the Onion Creek mound dominates the smaller mound beneath the Blanco River allowing water from Onion Creek to flow south and eventually discharge at San Marcos Springs (Figs. 1 and 4) (Smith et al. 2012; Johnson et al. 2012). Under dry conditions, flow ceases in Onion Creek, while the Blanco River continues to flow.

The Onion Creek mound slowly dissipates, eventually allowing water from the Blanco River to flow north to Barton Springs. Figure 4 shows a cross section with the Onion Creek groundwater mound under wet conditions and another potentiometric surface during drought conditions with hydraulic heads as much as 50 m lower.

6 Conclusions

Water flows from the headwaters of the Blanco River to the outlets at San Marcos and Barton Springs or into the deep subsurface along various convoluted pathways. Some of the water flows in the river for much of its journey and some flows mostly in the subsurface. But, some of that water is following pathways that can take it from the river to the subsurface and back to the river again multiple times. A complex combination of geology, structures, lithologies, and karst features determines the actual pathways. A better understanding of the relationships between the Blanco River and the Trinity and Edwards Aquifers is needed to allow for better management of the water resources of central Texas. Considering the implications of water supply for the people and endangered species that live in San Marcos and Barton Springs, further studies need to be made to provide for optimum management of the river and the aquifers.

References

BSEACD (2014) Regional habitat conservation plan for groundwater use and management of the Barton Springs segment of the Edwards Aquifer. Barton Springs/Edwards Aquifer conservation district, prepared for U.S. fish and wildlife service, Austin, TX, Feb 2014, p 191, plus appendices

Collins EW, Hovorka SD (1997) Structure map of the San Antonio segment of the Edwards Aquifer and Balcones fault zone, south-central Texas: structural framework of a major limestone aquifer: Kinney, Uvalde, Medina, Bexar, Comal and Hays Counties. The University of Texas at Austin, Bureau of economic geology, miscellaneous Map No. 38, scale 1:250,000, p 14

Hunt BB, Smith BA, Slade R, Gary RH, Holland WF (2012) Temporal trends in precipitation and hydrologic responses affecting the Barton Springs segment of the Edwards Aquifer, central Texas. In: 62nd annual convention on gulf coast association of geological societies transactions, Austin, TX, 21–24 Oct 2012

Johnson S, Schindel G, Veni G, Hauwert N, Hunt B, Smith B, Gary M (2012) Tracing groundwater flowpaths in the vicinity of San Marcos Springs, Texas. Edwards Aquifer Authority, Report No. 12-03, p 139, San Antonio, TX

Slade R (2014) Documentation of a recharge-discharge water budget and main streambed recharge volumes, and fundamental evaluation of groundwater tracer studies for the Barton Springs segment of the Edwards Aquifer. Tex Water J Tex Water Resour Inst 5(1):12–23

Smith BA, Hunt BB (2010) Hydraulic interaction between the Edwards and Trinity Aquifers. In: Wierman D, Broun A, Hunt B (eds) Hydrogeologic atlas of the Hill Country Trinity Aquifer, Barton Springs/Edwards Aquifer, USA, 17 plates, July 2010

Smith BA, Hunt BB, Johnson SB (2012) Revisiting the hydrologic divide between the San Antonio and Barton Springs segments of the Edwards Aquifer: insights from recent studies. In: 62nd annual convention on gulf coast association of geological societies journal vol 1. Austin, TX, pp 55–68, 21–24 Oct 2012

Smith BA, Hunt BB, Beery J (2011) Final report for the onion creek recharge project, northern Hays County, Texas. Barton Springs/Edwards aquifer conservation district, Austin, TX, prepared for the Texas Commission on Environmental Quality, Aug 2011, p 47, plus appendices

USGS (2014) Unpublished data, U.S. Geological Survey. www.waterdata.usgs.gov/nwis/uv?08171000

Wierman, DA, Hunt BB, Broun AS, Smith BA (2010) Recharge and groundwater flow. In: Wierman D, Broun A, and Hunt B (eds) Hydrogeologic atlas of the Hill Country Trinity Aquifer, Barton Springs/Edwards Aquifer, USA, 17 plates, July 2010

Wong C, Kromann J, Hunt B, Smith B, Banner J (2013) Investigating groundwater flow between Edwards and Trinity Aquifers in central Texas. Ground Water Journal, National Ground Water Association

Chemical, Thermal and Isotopic Evidences of Water Mixing in the Discharge Area of Torrox Karst Spring (Southern Spain)

J.A. Barberá and B. Andreo

Abstract Time analysis of the chemical, thermal and isotopic characteristics of spring waters was conducted to infer the different flow types feeding the Torrox spring (Teba-Peñarrubia carbonate aquifer, southern Spain). The results suggest that the relatively high mineralized and thermal groundwater drained by the outlet is conditioned by the mixing among recently infiltrated waters through the carbonate outcrops, runoff infiltration in La Venta river and (not yet proved) deep groundwater coming from neighbouring aquifers. This example comprises a hydrogeological characterization of the groundwater discharge in karst aquifers from the analysis of the spring natural responses, which is crucial to understand its functioning and, consequently, to plan a suitable management and protection of karst waters.

1 Introduction

The sources of waters flowing within carbonate (karst) aquifers are often difficult to determine, in particular if the contrast of the hydrogeological parameters (among them hydrochemical and isotopic tracers) is scarce. In some cases, as occurs in binary karst aquifers, the allogenic recharge may be significant, providing additional chemical and isotopic heterogeneity that facilitates the identification of the water origin. The former variability can result even higher if background levels of rock-dissolution-derived parameters are unusually elevated and/or anthropogenic activities exist.

J.A. Barberá (✉) · B. Andreo
Department of Geology and Centre of Hydrogeology, University of Málaga (CEHIUMA), 29071 Málaga, Spain
e-mail: jabarbera@uma.es

B. Andreo
e-mail: andreo@uma.es

B. Andreo et al. (eds.), *Hydrogeological and Environmental Investigations in Karst Systems*, Environmental Earth Sciences 1,
DOI 10.1007/978-3-642-17435-3_19

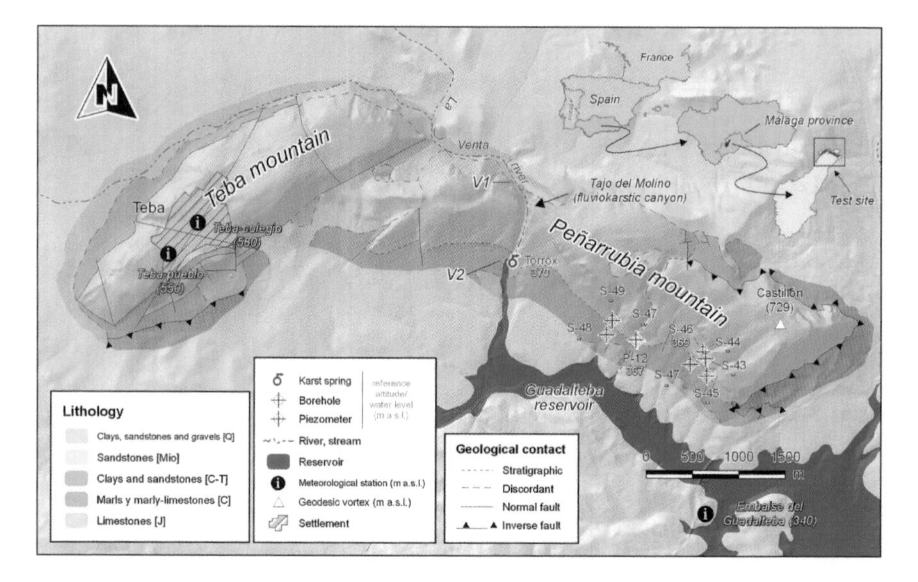

Fig. 1 Location and geological and hydrogeological characteristics of the test site. Age of the geological formations: *J* Jurassic, *C* Cretaceous, *C-T* Cretaceous-Terciary, Mio-Miocene and *Q* Quaternary

This work highlights the applicability of the commonly used hydrogeological methods (water temperature, hydrochemistry and water stable isotopes) to characterize the flow components participating in a single discharge area of a small karst aquifer from southern Spain.

Test site comprises carbonate outcrops (Teba and Peñarrubia massifs) covering a surface of 7.2 km^2, which are located to the N of the Málaga province (Fig. 1), in southern Spain. These permeable rocks are made up by Jurassic limestones with a thickness of several hundreds of meters, being the aquifer basement a Triassic formation of clayey materials with evaporites. The geological structure is defined by an anticline form and divided in two carbonate massifs: the western sector (Teba mountain) is NE-SW oriented and the eastern one (Peñarrubia mountain) has ESE-WNW direction.

The main hydrological feature in the study site is La Venta river, that intersects the aquifer sector of the Peñarrubia mountain (from the N to the S), where a fluviokarstic canyon 100 m deep has been developed (Tajo del Molino area; Fig. 1). This river gathers the runoff generated on its watershed plus the effluent of waste waters from little settlements and farms. The discharge of the Teba-Peñarrubia aquifer occurs by the Torrox spring and by pumping in the boreholes drilled in the eastern sector of the aquifer (Peñarrubia mountain).

A regular field measurement campaigns (discharge, electrical conductivity EC, water temperature and pH) and water sampling were performed between August 2007 and May 2010 in the Torrox spring (Fig. 1): daily during the recharge periods and fortnightly in the dry season. Major ions, Alkalinity, TOC and $\delta^{18}O$ were analyzed in the Centre of Hydrogeology of the University of Málaga using a high liquid pressure chromatograph (potentiometric method in the case of Alkalinity), carbon analyzer and a cavity ring down spectrometer (CRDS), respectively.

2 Results and Discussion

A preliminary analysis of the Torrox spring hydrograph (Fig. 2) shows a flood event per year as response to the rainfall infiltration, as well as smooth flow variations with a significant delay between the rainfall and the maximum peak discharge (e.g. December 2009–March 2010). This hydrodynamic behaviour seems to be coherent with a low degree of the karstification in the aquifer sector that drains, although it could also be influenced by the pumping and/or by the infiltration of surface waters. The latter has been demonstrated from point flow measurements done in the *V1*–*V2* sections (Fig. 1) of the nearby La Venta river (Barberá 2014).

The mean temperature of the groundwater during the studied period is 22.2 °C (Barberá, 2014), 4.8 °C higher than the average air temperature (17.4 °C) recorded at the meteorological station of Embalse del Guadalteba (Fig. 1). Therefore, the Torrox spring has a thermal behaviour that suggests the drainage of deep flows (probably from neighbouring aquifers) in which groundwater reaches the thermal equilibrium with the host rock.

Thermograph of the Torrox spring (Fig. 2) displays temperature values almost invariant (22.3–22.5 °C) in low flow conditions, excepting in the flood event of May 2008, which diminish during the recharge periods. The decreases of the groundwater temperature occur progressively and may last during several months (e.g. December to March–April 2009 and 2010), until the minimum annual temperature value is recorded. However, this thermal pattern can be preceded by a single increasing (of several tenths of °C), as it was observed during November 2008 and December of 2009 (Fig. 2). Thus, the time response and the amplitude of the groundwater temperature changes depend on the distribution, intensity and quantity of the recharge episodes. Once depletion in the spring flow is established, groundwater drained tends to recover the pre-event temperature values ($\sim$22.3 °C).

The study of the chemographs (Fig. 2) reveals a positive trend of the EC values during the study period. Spring waters are less mineralized in low flow conditions and present higher EC values during the recharge periods. Therefore, the time response and the magnitude of EC variations, which practically consist in increases followed by progressive decreases, also depend on the recharge conditions in the aquifer.

The increasing of EC values observed in the waters drained by the Torrox spring (Fig. 2) suggests the mixing, in different proportions, of the groundwater flowing through the aquifer (800–900 μS/cm; see hourly EC record in late December 2009 in Fig. 2) with the surface waters of La Venta river (more

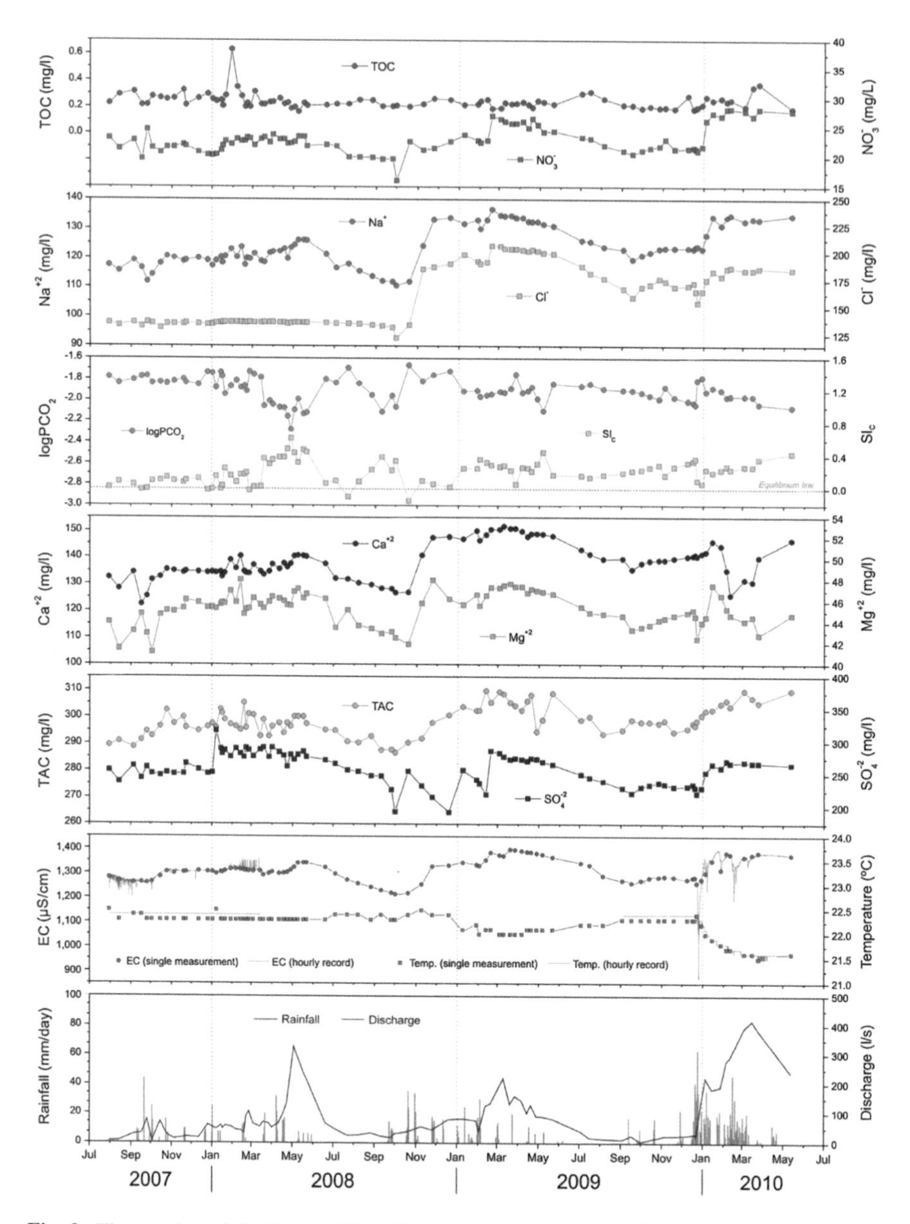

Fig. 2 Time series of discharge, EC and temperature data and major ions and TOC contents measured in the Torrox spring waters

mineralized: 2,000–3,000 μS/cm), which infiltrates along with the fluviokarstic canyon of the Tajo del Molino area (Fig. 2).

Groundwater mineralization is conditioned by the variations of Alkalinity values and the contents of Ca^{+2}, Mg^{+2}, Na^{+} and Cl^{-} (Fig. 2). These hydrochemical components show a general trend toward higher values during the sampling period, as occur with the EC data (Fig. 2). The concentrations of all parameters increase, each hydrological year, as consequence of the aquifer recharge and diminish as the spring flow depletion is established. This hydrochemical behaviour was more evident during 2008/2009 water year. The latter observations suggest a common origin of almost all chemical components, associated to La Venta river waters, due to the high concentrations of the major ions (particularly those related with the rock dissolution) and the higher hydrochemical variability (Barberá 2014). However, this interpretation may also be coherent with the groundwater transference from other hydrogeological systems implying longer and deeper flows (probably of evaporite origin), highly mineralized and enriched in SO_4^{-2}, Ca^{+2}, Mg^{+2}, Na^{+} and Cl^{-} and with higher Alkalinity values. NO_3^{-} contents show a similar time evolution with respect to that of EC, Alkalinity, Ca^{+2}, Mg^{+2}, Na^{+} and Cl^{-} (Fig. 2), with minimum values in the dry season and maximum ones during the recharge period. The maximum peaks of NO_3^{-} concentrations are progressively higher during the research period (23.9 mg/l in 2007/2008; 27.1 mg/l in 2008/2009 and 28.4 mg/l in 2009/2010). These peaks coincide with the flood peaks in the spring discharge and maximum runoff in the river, which suggest that the NO_3^{-} enrichment in the groundwater drained by the Torrox spring are due to the surface waters (plus leachates from farm activities) infiltration in La Venta river.

The spring waters display low contents of TOC (0.2–0.3 mg/l), scarce variability and a decreasing trending of this parameter during the sampling period (Fig. 2). However, several rapid and single TOC peaks (of about 0.6 mg/l) are observed which seem to be associated to the most intense recharge episodes.

The lower contents of the organic matter dissolved in the groundwater, in comparison with the surface waters of La Venta river (1–5 mg/l), can be explained by different and complex biogeochemical processes affecting TOC and NO_3^{-} concentrations (Sánchez-Monedero et al. 2001; Toran and White 2005). On one hand, the hydrochemical responses, generally buffered respect to the input signal rainfall could promote the mineralization of TOC (which is transformed in HCO_3^{-}), along with the groundwater flow path within the aquifer (Batiot et al. 2003a, b). On the other hand, dilution, favoured by the mixing with groundwater poor in NO_3^{-}, and denitrification processes, probably due to the anoxic conditions in depth zones of the aquifer (Panno et al. 2001; Vesper et al. 2001), should exist.

Torrox spring waters are in equilibrium or oversaturated with respect to calcite (Fig. 2). SI_C values became progressively higher during the sampling period. Conversely to the former parameter, the partial pressure of CO_2 ($logPCO_2$) evolves with a decreasing trending. This evolution is congruent with the progressive mixing between groundwater and surface waters from La Venta river, which are equilibrated with the atmosphere ($logPCO_2$ of −3.5). Nevertheless, single peaks of the CO_2 contents, with a certain magnitude, and coinciding with the first flood

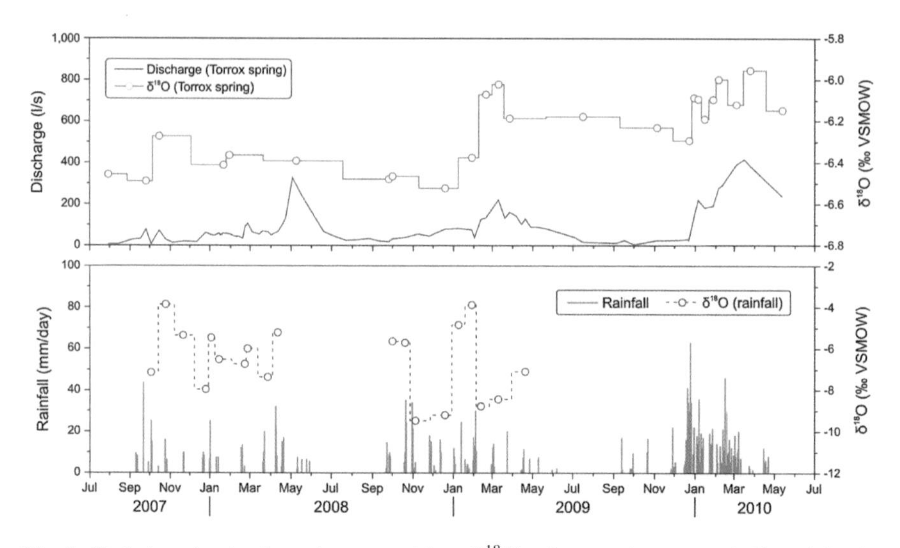

Fig. 3 Variations in the isotopic composition ($\delta^{18}O$) of meteoric waters (collected in the Peñarrubia mountain) and the groundwater drained by the Torrox spring

events on each water year are observed in the spring waters, probably as consequence of the rapid rainfall infiltration through the carbonate exposures.

Figure 3 shows the time evolution of the $\delta^{18}O$ data measured in the groundwater samples of Torrox spring. In this figure two periods with different average $\delta^{18}O$ values are distinguished: from July 2007 to February 2009, with a mean $\delta^{18}O$ value close to −6.4 % and from February 2009 to May 2010, in which the average value of this parameter is higher, around −6.1 %. The less negative isotopic composition of the spring waters in the second period seems to reflect the mixing of evaporated surface waters, enriched in $\delta^{18}O$, and the groundwater of the aquifer, isotopically depleted. Thus, surface waters contribute in a major proportion to the spring flow during the rainiest periods (2008/2009 and 2009/2010 water years), when the river flow is higher, being the effects of the allogenic recharge in the isotopic composition (but also in the hydrochemical and thermal natural responses) of the groundwater more evident during these periods. The higher participation of the surface water component in the spring flow during recharge periods has been demonstrated from point flow measurements done in the La Venta river (Barberá 2014).

However, the Torrox spring responds to the main recharge episodes with single increases of $\delta^{18}O$ values, which reproduce, in a certain extent, the isotopic variations measured in meteoric waters (e.g. October 2007 and December 2008–March 2009; Fig. 3). The isotopic responses recorded, as consequence of the recharge, show a low variability ($\Delta\delta^{18}O < 0.4$ %), which is in accordance with a scarce hierarchization of the drainage system, probably developed in an aquifer sector with a low development of inner karstification.

3 Conclusions

The periodic monitoring of the chemical, thermal and isotopic characteristics of groundwater drained by a karst spring have permitted to characterize the different flow types which participate in the discharge of a small carbonate aquifer (Teba and Peñarrubia mountains, southern Spain). Groundwater drained by the Torrox spring is a mixing of three water types with different origin. The diffuse recharge trough the carbonate outcrops is represented by lower mineralized waters, poor in the majority of the hydrochemical parameters and isotopically depleted. The all-ogenic component consists in cooler waters with high EC and Alkalinity values and enriched in SO_4^{-2}, Ca^{+2}, Mg^{+2}, Na^+, Cl^-, NO_3^-. The third mixing component that hypothetically contributes to the spring discharge comprises thermal and highly mineralized waters of deep origin. Despite of the qualitative characterization of the different water types participating in the Torrox spring flow, further research are needed to precise the end-members components (e.g. using dye tracers and specific natural tracers as $\delta^{13}C$, $\delta^{34}S$, $^{87}Sr/^{86}Sr$ and geothermometers) and to quantify the mixing proportions (by mass and heat balance).

Acknowledgments This work is a contribution to the research projects P06-RNM 2161 of Junta de Andalucía, CGL2008-06158 BTE and CGL2012-32590 of DGICYT and IGCP 513 of UNESCO, and to the research group RNM-308 of Junta de Andalucía.

References

Barberá JA (2014) Hydrogeological research in the carbonate aquifers of the Serranía oriental de Ronda (Málaga). Ph.D. thesis (in Spanish)

Batiot C, Emblanch C, Blavoux B (2003a) Carbone organique total (COT) et Magnésium (Mg^{2+}): Deux traceurs complémentaires du temps de séjour dans laquifère karstique. CR Geosci 335:205–214

Batiot C, Liñán C, Andreo B, Emblanch C, Carrasco F, Blavoux B (2003b) Use of TOC as tracer of diffuse infiltration in a dolomitic karst system: the Nerja Cave (Andalusia, southern Spain). Geophys Res Lett 30(22):2179

Panno S, Hackley K, Hwang H, Kelly W (2001) Determination of the sources of nitrate contamination in karst springs using isotopic and chemical indicators. Chem Geol 179:113–128

Sánchez-Monedero MA, Roig A, Pareded C, Bernal MP (2001) Nitrogen transformation during organic waste composting by the Rutgers system and its effect on pH, EC and maturity of the composting mixtures. Bioresour Technol 78:301–308

Toran L, White WB (2005) Variation in nitrate and calcium as indicators of recharge pathways in Nolte spring, PA. Environ Geol 48(7):854–860

Vesper DJ, Loop C, White WB (2001) Contaminant transport in karst aquifers. Theor Appl Karstology 13–14:101–111

Characterization of Carbonate Aquifers (Sierra de Grazalema, S Spain) by Means of Hydrodynamic and Hydrochemical Tools

D. Sánchez, B. Andreo, M. López, M.J. González and M. Mudarra

Abstract Hydrodynamic and hydrochemical monitoring of springs has been largely used to study carbonate aquifers and to determine their hydrogeological functioning. In this work, temporal evolutions regarding flow discharge, major components, and natural soil tracers (Total Organic Carbon and NO_3^-) of four springs draining karst aquifers located in Sierra de Grazalema Natural Park (Southern Spain) have been analyzed. Results show the existence of aquifers with a high degree of karstification in which recharge water rapidly infiltrates and causes sharp water dilutions and steep flow increases. These aquifers coexist with others characterized by a lower development of karstification processes, but higher natural attenuation capacities.

D. Sánchez (✉) · B. Andreo · M. Mudarra
Department of Geology and Centre of Hydrogeology, University of Málaga (CEHIUMA), 29071 Málaga, Spain
e-mail: dsanchez@uma.es

B. Andreo
e-mail: andreo@uma.es

M. Mudarra
e-mail: mmudarra@uma.es

M. López
Consejería de Agricultura Pesca Y Medio Ambiente de La Junta de Andalucía, Ctra. N-IV, Km. 637, 11407 Jerez de La Frontera, Cádiz, Spain
e-mail: manuel.lopez.rodriguez@juntadeandalucia.es

M.J. González
FULCRUM Aldapa Kalea, 9-13. Apdo. 89, 48940 Leioa, Bizkaia, Spain
e-mail: mjgonzalez@fulcrum.es

B. Andreo et al. (eds.), *Hydrogeological and Environmental Investigations in Karst Systems*, Environmental Earth Sciences 1,
DOI 10.1007/978-3-642-17435-3_20

1 Introduction

Karstic aquifers present a high degree of heterogeneity in their drainage due to the internal hierarchization of karstic elements (conduits, fractures, and fissures), enlarged via dissolution processes within carbonate rocks. Temporal evolutions concerning the hydrochemical, hydrothermal, and hydrodynamic characteristics of the spring water, especially during recharge periods, have been largely used to characterize the hydrogeological functioning of karstic aquifers (Shuster and White 1971; Mudry 1987; Genthon et al. 2005). The analysis of these evolutions provides information on the volumes of water transferred from the recharge area to the discharge one, as well as the contribution of the different zones of the aquifer (saturated and unsaturated zones, epikarst, and soil) to the spring water (Emblanch et al. 1998; Batiot et al. 2003; Mudarra and Andreo 2011).

Sierra de Grazalema Natural Park is located in the province of Cádiz (Southern Spain, Fig. 1). It covers a total area of 260 km^2 from which 150 km^2 are carbonate permeable rocks. Sierra de Grazalema represents one of the wettest areas in Spain with mean annual values of precipitation as high as 1,800 mm. However, some towns located in this region have problems to supply drinking water to population during drought periods.

In this work, we analyze the hydrodynamic and hydrochemical temporal series of four springs draining karstic aquifers of Sierra de Grazalema. The monitoring was developed from November 2012 to March 2014 and included in situ measurements of electrical conductivity (EC), temperature, and pH as well as the collection of water samples. The frequency of monitoring was every 1–3 weeks depending on the precipitation pattern. Analytical results of water samples from December 2013 and forward were not available at the moment of writing this manuscript, and consequently temporal evolutions for major components are limited to that date.

The aquifers are constituted by Jurassic limestones and dolostones around 500 m thick. The outcrops that constitute the study area are hydrogeologically separated by Flysch materials (Boyar Corridor, Fig. 1). The structure is made up of folds lying NE-SW that generate anticlines where limestones and dolostones outcrop and synclines formed by Cretaceous marls.

Four springs have been selected to carry out this study: Cornicabra, Nueve Caños, Benamahoma, and Esparragosilla (Fig. 1). Cornicabra and Nueve Caños drain Jurassic limestones from Sierra de Ubrique and Sierra Alta; Esparragosilla drains dolostones of Sierra de la Silla and Benamahoma is located at the western edge of Sierra del Pinar dolomitic aquifer.

The hydrological year 2012/2013 was wet with an annual rainfall value higher than the average (up to 50 %) in all weather stations. The autumn of hydrological year 2013/2014 was dry and the first significant precipitations took place at the end of December, January and February.

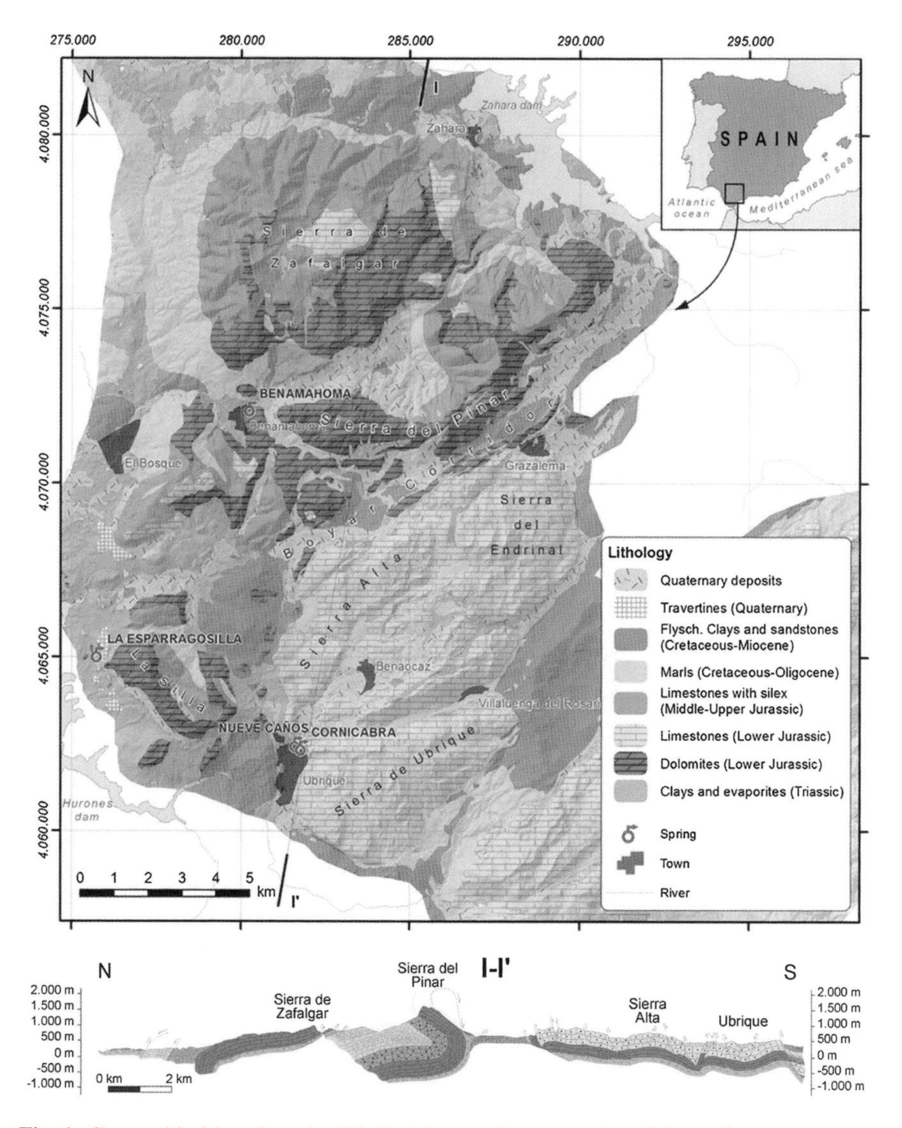

Fig. 1 Geographical location, simplified geology and cross section of the study area

2 Results and Discussion

2.1 Cornicabra

Figure 2 shows the hydrodynamic and hydrochemical evolution of Cornicabra spring. The flow discharge presents a large variability with values ranging from less than 50 l/s in summer to 2,460 l/s after the main precipitation events. This

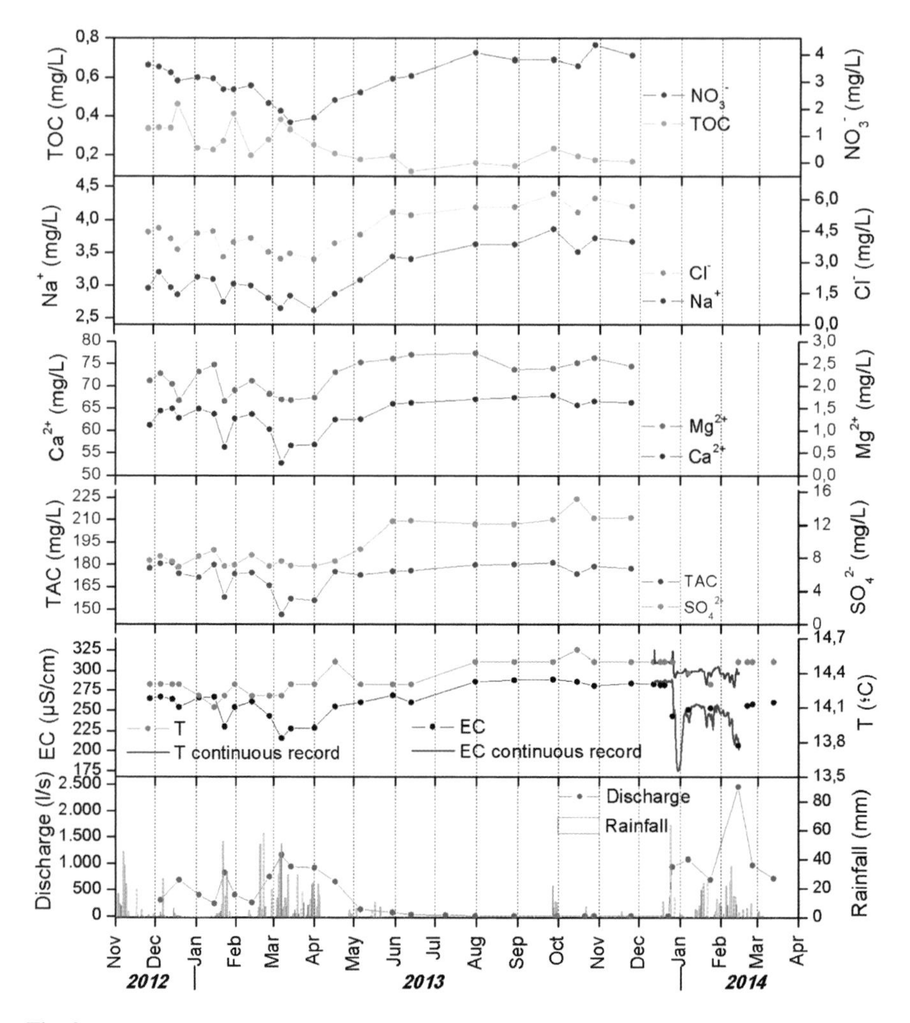

Fig. 2 Hydrodynamic and hydrochemical time series recorded at Cornicabra spring

flow pattern is characteristic of karstic systems, with sudden and rapid changes in response to rainfall events and a low natural attenuation capacity.

The hydrochemical evolution shows rapid dilutions following rainfall events. The magnitudes of these dilutions are proportional to that of the recharge and they tend to recover their initial state once the precipitation is finished. The effects of rainfall on the water mineralization are noticeable only 1 day after. The highest EC and temperature values were recorded in low water conditions.

In general terms, each precipitation episode was associated with an increase in groundwater flow and TOC, and a decrease of Alkalinity (TAC), SO_4^{2-}, Ca^{2+}, Mg^{2+}, Na^+ and Cl^- contents. This suggests a direct mixing of rapidly infiltrating water with that stored at the saturated zone of the aquifer, which is characterized by greater residence time and higher concentrations of major components.

The aquifer sector drained by Cornicabra spring shows a well developed functional karstification. Rain water infiltrates into the aquifer and circulates rapidly by conduits and fractures, causing increases in flow discharge and decreases in EC and major components concentrations.

2.2 Nueve Caños

The temporal evolution of flow discharge at Nueve Caños spring (Fig. 3) also shows high variability, with values ranging from less than 15 l/s in low water conditions to 1,610 l/s after high recharge periods. Both the increase and decrease of flow respond quite rapidly to the rainfall regime, reflecting a low attenuation capacity of the sector drained by this spring.

The hydrochemical evolution shows a gradual dilution effect of water caused by recharges occurred from November 2012 to April 2013, which is followed by an increased trend that leads to initial values. Although some rainfall events caused sudden EC falls, the general trend is gradual. The minimum EC values are recorded some weeks after the peak of discharge.

The continuous record of EC started in October 2013 shows stable values during low water conditions (309–311 μS/cm). The first significant recharge event of the hydrological year, registered at the end of December 2013, caused a slight increase of mineralization (317 μS/cm). The absence of analytical data for this period does not permit a proper interpretation of this event; however, the stability of temperature values suggests that water may come from the saturated zone of the aquifer. Rainfall recorded since middle January 2014 onward do not have an effect on EC until the first week of February, when the mineralization fell to 299 μS/cm. During this dilution process water temperature stayed the same, which might be explained by the drainage of infiltration water that has reached the aquifer temperature (Liñán et al. 2009).

Each dilution episode was associated with a decrease in TAC, Ca^{2+} and Mg^{2+} contents and an increase in Cl^-, Na^+ and TOC. This reflects the dilution of the water from the saturated zone of the aquifer by the mixing with water from the epikarst and the soil (Cl^-, Na^+ and TOC).

Nueve Caños drains an aquifer sector with a relatively developed degree of karstification and a rather low natural attenuation capacity. Flow discharge shows a large variability and EC evolution is characterized by a relatively small range of variation (28 μS/cm during the studied period) and delayed responses to the main recharge events.

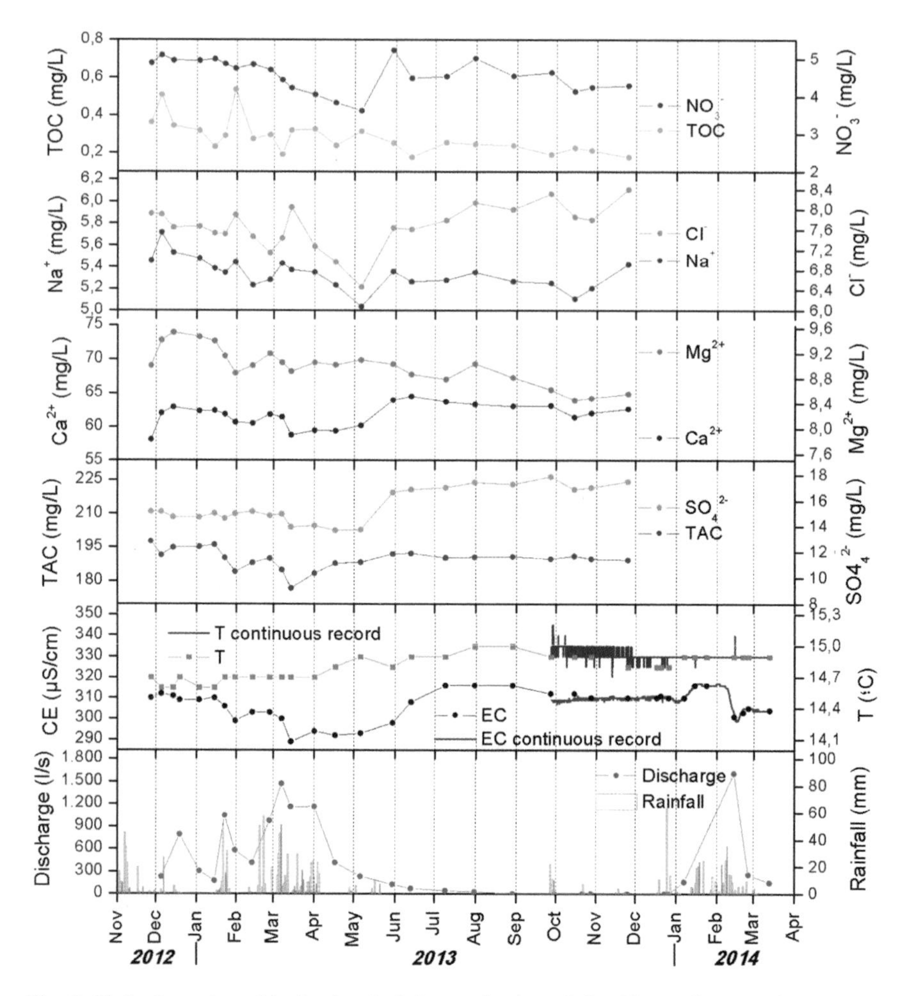

Fig. 3 Hydrodynamic and hydrochemical time series recorded at Nueve Caños spring

2.3 Benamahoma

Figure 4 illustrates the hydrodynamic and hydrochemical temporal evolution of Benamahoma spring. The hydrograph is characterized by steep rising and recession limbs, and a high variability as it is shown by the large range of values for the control period: from 150 l/s in low water conditions to more than 9,000 l/s after intense recharge (February 2014).

As results of natural recharge on Benamahoma spring, rapid water dilutions occur which tend to recover the initial state of mineralization once the precipitation is finished. Dilutions are associated with fast and intense falls of EC values such as the recorded after the first significant rainfall of the hydrological year

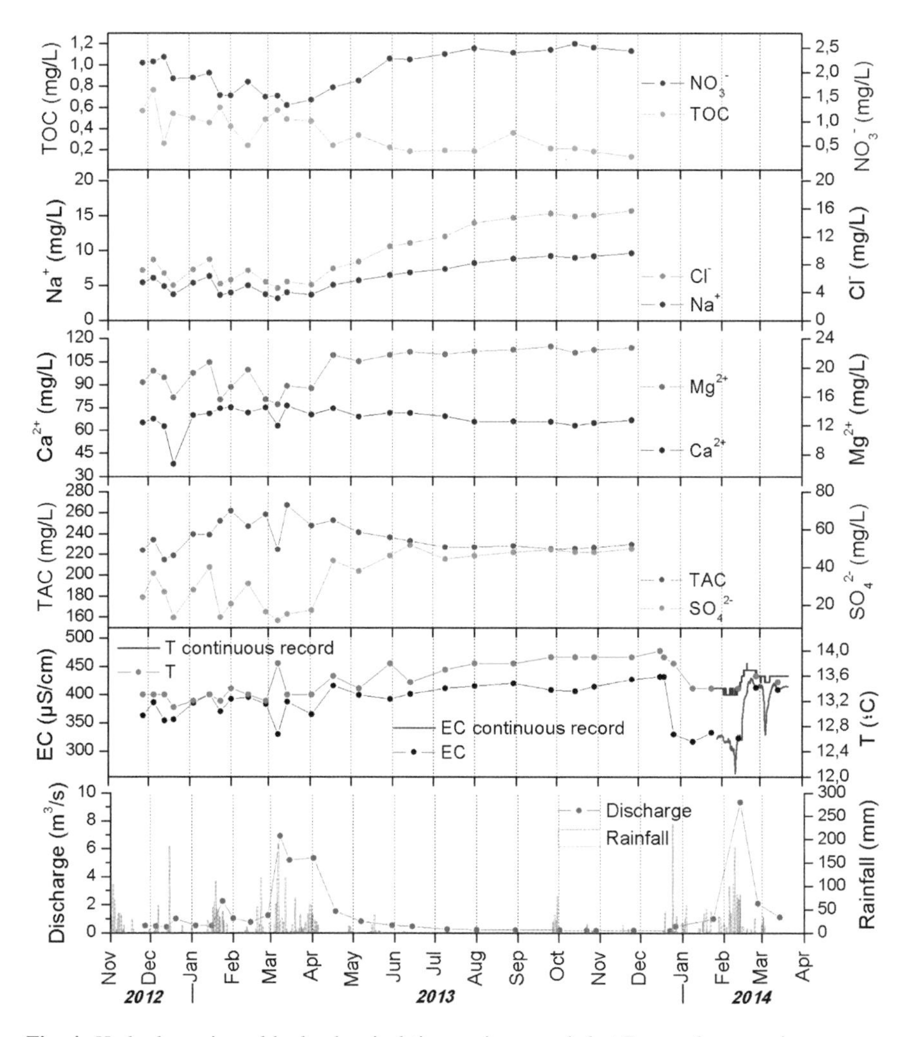

Fig. 4 Hydrodynamic and hydrochemical time series recorded at Benamahoma spring

2013/2014 (25/12/2013), which caused a decrease of 100 μS/cm. Subsequent precipitation events generated new dilution effects (February, March 2014) which were followed by rapid upward trends to initial EC values. The range of variation of EC for the monitoring period was 171 μS/cm (261–432 μS/cm).

Water temperature was colder during dilution episodes and reached the maximum values at the end of the low water period. Water dilutions were associated with decreased concentrations of TAC and all major components (Ca^{2+}, Mg^{2+}, SO_4^{2-}, Cl^-, Na^+) and peaks of flow discharge and TOC.

The Mg^{2+} content, which is indicative of the residence time of the water (Batiot et al. 2003; Mudarra and Andreo 2011), together with the SO_4^{2-}, an indicator of the aquifer's evaporitic substratum, present increasing trends during low water conditions. These evolutions characterize the contribution of the saturated zone to the karstic spring during low water conditions.

Benamahoma spring drains a large sector of the Sierra del Pinar aquifer with a well developed karstification through which rapid infiltrated water reaches the discharge points, causing water dilutions and flow discharge increases. The effect of recharge on the spring mineralization and flow usually lasts only for a few days. This is caused by rapid circulation of water from the recharge areas to the discharge points. However, the drainage of relatively high flows (>150 l/s) even after nearly 9 months without significant rainfalls, suggests the coexistence of smaller fractures/fissures.

2.4 Esparragosilla

The maximum flow recorded during the monitoring period was 447 l/s and the minimum 23.5 l/s (Fig. 5). The hydrograph shows a relatively gradual increase of flow as response to rainfall until March 2013, when the maximum flow discharge occurred. No intermediate peaks are identified during the flood period, even after intense precipitation events, and the discharge peak is followed by a gentle recession limb. This hydrodynamic evolution is characteristic of aquifers with a low degree of karstification whose output signal (flow discharge) tends to smooth the input signal (rainfall).

Esparragosilla has the hottest and most mineralized water among the monitored springs, with mean EC and temperature values of 2,250 μS/cm and 17.5 °C, respectively. This is due to the contribution of Triassic evaporitic rocks (halite and gypsum), which constitute the substratum of the aquifer.

The effect of recharge on Esparragosilla spring is a progressive increase of water mineralization. The highest EC values were registered when flow discharge was the maximum, whereas in low water conditions the EC decreased. Water mineralization is mainly controlled by SO_4^{2-}, Ca^{2+} and Mg^{2+} contents, which are representative of the saturated zone of the aquifer (Mg^{2+}) and the evaporitic substratum (SO_4^{2-} and Ca^{2+}). Alkalinity concentration remained stable during the whole monitoring period except for one sample (April 2013).

Esparragosilla has the greatest TOC contents among the springs analyzed, with maximum values of up to 0.9 mg/l. This may be caused by the biological activity that exists in the discharge point, which is not found in the other springs, or by livestock developed in nearby areas.

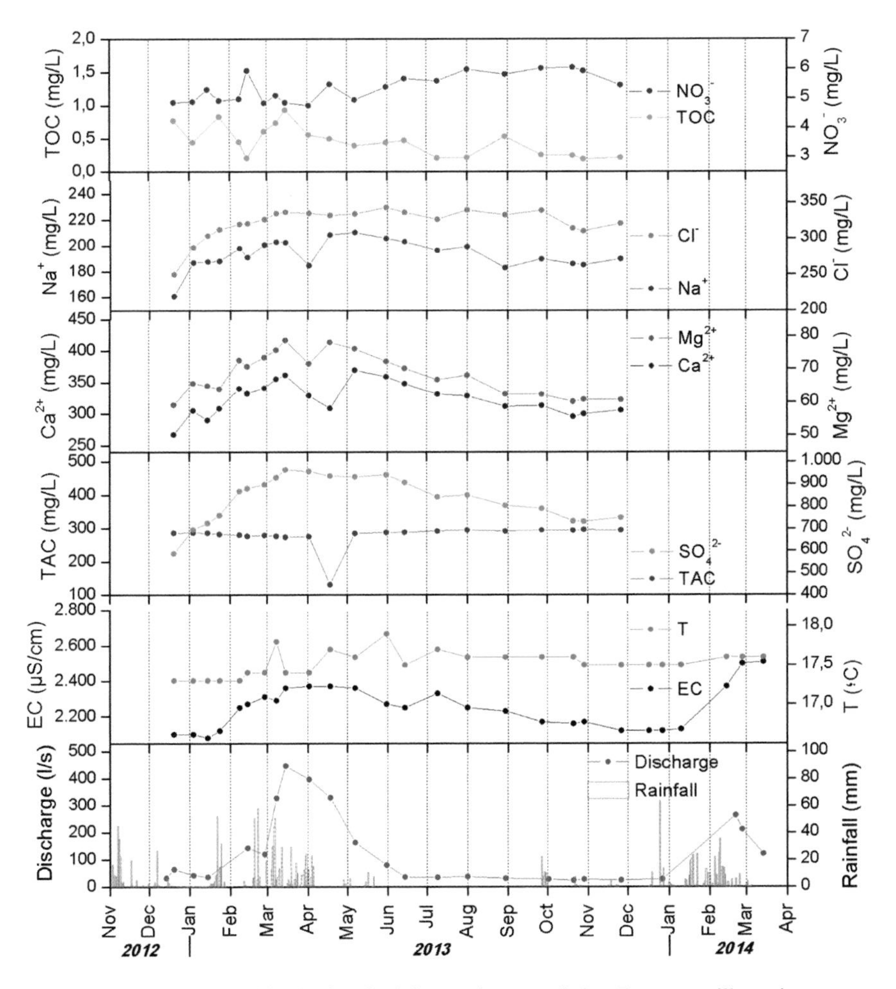

Fig. 5 Hydrodynamic and hydrochemical time series recorded at Esparragosilla spring

The aquifer drained by Esparragosilla spring shows a certain natural attenuation capacity, which is characteristic of diffuse flow systems. The effect of recharge into the aquifer is the mobilization of water stored in the saturated zone, which is characterized by greater residence time and presents higher mineralization.

3 Conclusions

The comparative analysis of the hydrodynamic and hydrochemical temporal evolutions of four discharge points of the Sierra de Grazalema carbonate aquifer has permitted the identification of different hydrogeological responses to natural recharge.

Cornicabra and Benamahoma springs showed the most karstic functioning based on rapid and sudden water dilutions and steep flow discharge increases after each important precipitation event. Dilutions are caused by decreasing concentrations of all major components. This pattern and the increase of TOC concentrations are indicative of the rapid infiltration of recharge water into the aquifer.

On the contrary, Esparragosilla spring drains a diffuse flow system. Its hydrograph displays progressive flood and recession curves, and the depletion part of the curve tends to be almost horizontal. This is indicative of a higher natural attenuation capacity. The effect of recharge is not the dilution of water, but an increase of mineralization mainly due to a piston effect from the saturated zone (increase of Mg^{2+}) which is in contact with the Triassic substratum (SO_4^{2-} and Ca^{2+}).

Finally, Nueve Caños spring drains an aquifer sector with a moderate degree of karstification. Flow discharge variations are large and relatively rapid. However, the hydrochemical response to rainfall shows smoother variations. The first significant recharge of the hydrological year causes the mobilization of more mineralized water, probably coming from the saturated zone of the aquifer; nevertheless, subsequent rainfall events produce the dilution of the spring water.

The hydrogeological characterization of aquifers functioning is necessary to manage groundwater resources, especially in areas with water supply difficulties during drought periods such as the sector where Sierra de Grazalema aquifers are located.

Acknowledgments This work is a contribution to the projects CGL2008-06158 and CGL-2012-32590 of DGICYT and IGCP 598 of UNESCO, and to the Research Group RNM-308 of the Junta de Andalucía. The authors wish to thank the Agencia de Medio Ambiente y Agua de Andalucía for its support in the development of this research.

References

Batiot C, Emblanch C, Blavoux B (2003) Total Organic Carbon (TOC) and magnesium (Mg^{2+}): two complementary tracers of residence time in karstic systems. CR Geosci 335:205–214

Emblanch C, Blavoux B, Puig JM, Mudry J (1998) Dissolved organic carbon of infiltration within the autogenic karst hydrosystem. Geophys Res Lett 25:1459–1462

Genthon P, Bataille A, Fromant A, D'Hulst D, Bourges F (2005) Temperature as a marker for karstic waters hydrodynamics. Inferences from 1 year recording at La Peyrére cave (Ariège, France). J Hydrol 311:157–171

Liñán C, Andreo B, Mudry J, Carrasco F (2009) Groundwater temperature and electrical conductivity as tools to characterize flow patterns in carbonate aquifers: The Sierra de las Nieves karst aquifer, south Spain. Hydrogeol J 17:843–853

Mudarra M, Andreo B (2011) Relative importance of the saturated and the unsaturated zones in the hydrogeological functioning of karst aquifers: the case of Alta Cadena (Southern Spain). J Hydrol 397:263–280

Mudry J (1987) Apport de traçage physico-chimique naturel à la connaissance hydrocinématique des aquifères carbonatés. Thèse Sciences Naturelles, Université de Franche-Comté, Beçanson, 378 pp

Shuster ET, White WB (1971) Seasonal fluctuations in the chemistry of limestone springs: a possible means for characterizing carbonate aquifers. J Hydrol 14:93–128

In Situ Study of Hydrochemical Response of a Fractured-Layered Carbonate Regional Aquifer: Comparative Analyses of Natural Infiltration and Artificial Leakage of a Large Dam Lake (Vouglans, Jura, France)

C. Bertrand, Y. Guglielmi, S. Denimal, J. Mudry and G. Deveze

Abstract Leakage detection and prediction of fractured rocks is an important question in hydropower engineering. Some of large water reservoirs are built in karstic carbonate areas. In order to understand underground circulations in the limestone/dolomite foundation of Vouglans dam (Jura, France), analysis of groundwater chemistry, according to geological conditions, is used. Statistical analyses (PCA and DFA) are used (i) to characterise more precisely the low contrast in chemical composition resulting from the interaction between surface water, groundwater and carbonate environment, and (ii) to reclassify individuals in homogeneous groups with respect to the variables studied. Three hydrodynamic behaviours were determined in the sector. The area of influence of the rapid transit of the lake water varies with the seasons, and particularly in response to changes of the water level of the lake. The hydrostatic pressure of the water column has an influence on the opening and closing of cracks at the bottom of the dam.

C. Bertrand (✉) · S. Denimal · J. Mudry
Chrono-Environnement, UMR 6249/CNRS University of Franche Comté, 16 Route de Gray, 25000 Besançon, France
e-mail: catherine.bertrand@univ-fcomte.fr

Y. Guglielmi
CEREGE UMR7330, CNRS - Aix-Marseille Université, 3 Place Victor Hugo, 13331 Marseille, France

G. Deveze
DIN/CEIDRE - Département TEGG, Service Géologie Géotechnique, 905 Avenue Du Camp de Menthe, 13097 Aix En Provence Cedex 02, France

B. Andreo et al. (eds.), *Hydrogeological and Environmental Investigations in Karst Systems*, Environmental Earth Sciences 1,
DOI 10.1007/978-3-642-17435-3_21

1 Introduction

Leakage detection and prediction of fractured rocks is a relevant question in hydropower engineering, CO_2 sequestration in geological reservoirs/cap rock systems, tunnels and storage caverns, and in general fluid manipulations in the subsurface. Most of the dams have been experiencing these problems with respect to degrees of security and related sustainability in their lives. It can be useful to survey dam seepage by environmental isotopes and groundwater chemical analyses, according to geological condition. Some abundant information is provided by change of chemical and environmental isotope compositions after surface water seeping into stratums (Käss 1998; Mudry et al. 2008; Plata and Araguas 2002). In many cases, leakage passage existing in dam foundations can be detected by analysing hydrochemistry and isotope of groundwater in boreholes or springs (Kendall and McDonnel 1998). Large water reservoirs are sometimes built in karstified carbonate areas (De Waele 2008). The understanding and evaluation of environmental impacts of such human activities in karst are important to find a balance between development and preservation of these complex hydrogeological systems (Milanovic 2002, 2004). Moreover, in this managed system, water flows are constrained by varying hydraulic and mechanical limit conditions. In this latter case, these variations are conditioned by the water level in the reservoir, and induce increases of the stress applied on the dam's abutment (water weight) and pressure increase of the fluid, through the discontinuity network. These cumulative variations lead to significant changes in hydraulic properties of fractures, which convey water fluxes through the massif. Artificial tracer methods using purposefully drilled wells have been widely applied to detect the bypass routes of reservoir water in dam foundations built in carbonate terrain (Silar 1988). We propose to use natural tracing to constrain the hydrodynamic behaviour in the dam at two different hydrological contrasted seasons.

2 Study Site

2.1 Presentation

The Vouglans dam (Jura, France) is situated on the Ain course, in the outer area of the Jura. It is positioned on a large tabular structure oriented north–south. Runoff on the massif supporting the dam is guided by open discontinuities in the rock. These discontinuities are in the form of karstic long sub-horizontal lines, parallels to the stratification plans. They are interconnected by aligned vertical wells, following fault planes. The foundations of the dam have been concerned by two types of treatment: injections to form a grout curtain, and a drainage system consisting of tunnels and open drill holes. Drains allow to reduce water pressure in rocky foundation and piezometers allow the control of this water pressure, especially to

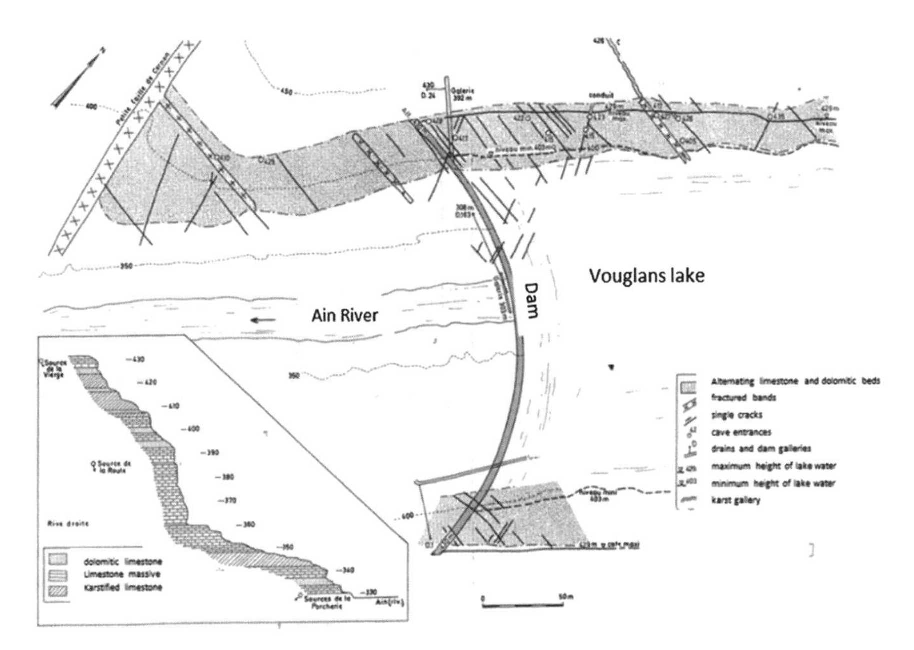

Fig. 1 Location and geological map of the Vouglans dam (Jura, France)

avoid uplift of the flood reception of slab spillways. The Vouglans massif is drained by a series of stepped sources (source de la Vierge, 430 m a.s.l.; source de la route, 385 m a.s.l.; source de la Porcherie, 330 m a.s.l.). The dam is an arch dam (Fig. 1) with a development peak of 427 m and a height of 103 m above the bed of the Ain. An "upstream" fissure is also observed, corresponding to a decollement zone between concrete and foundation upstream of the dam. Its average thickness is 25 m at the base and 6 m at the crest. In the dam, stepped galleries allow the control of the dam's deformation. These galleries continue in rock support, where drains allow the control of lake leakage. The reservoir is located in the present valley of Ain river. The lake shows 1,600 ha area, and contains 605 mm^3 of water volume. Its maximum depth is 100 m. The lake shows an amplitude variation of water level of about 34 m.

2.2 *Measurements*

In the massif and in the lake, physical and chemical parameters (pH, temperature, electrical conductivity, major cations and anions) have been measured in low water (June 2002) and high water (March 2004) periods, corresponding to two different levels of the lake water: 427 and 420.5 m a.s.l..

A total of 34 drains distributed in the entire dam have been sampled: on the right bank, six samples (numbers 1–6) have been taken in three galleries (348, 370 and 392 m a.s.l.). On the left side of the dam, two drains have been sampled (numbers 7 and 8) in two galleries (respectively, 365 and 338 m a.s.l.). At the base of the dam: 26 samples (numbers 9–34) have been collected in the perimeter gallery (303 m a.s.l.). Two points representative of the lake water are called 35 and 36. A sampling on three springs has also been performed at the 'source de la Porcherie' (n° 38; 330 m a.s.l.), the 'source de la Route' (n° 39; 385 m a.s.l.), and 'source de la Vierge' (n° 40; 430 m a.s.l.).

The pH, electrical conductivity and temperature were measured in the field with a WTW (LF30) apparatus. pH and electrical conductivity probes were calibrated on a regular basis with standard buffer solutions. Two depth profiles of temperature (°C), electrical conductivity (μS/cm), and pH were obtained, using a specific probe (''Seabird'' SBE19 probe CDT probe, accuracy: T $\pm$ 0.1 °C, electrical conductivity $\pm$0.5 %, pH $\pm$ 0.01).

Water samples were collected in polyethylene bottles and were filtered at 0.45 μm. Analyses of Na^+, Ca^{2+}, K^+, Mg^{2+} were performed by atomic absorption spectrometry (AA 100 Perkin–Elmer) with detection limits of 0.01; 0.5; 0.1 and 0.1 mg/L, respectively. Anion analyses of SO_4^{2-}, NO_3^-, Cl^- were performed by high pressure ion chromatography (Dionex DX 100) with detection limits of 0.1; 0.05 and 0.1 mg/L, respectively. The HCO_3^- concentrations were measured by acidic titration with a N/50 H_2SO_4, maximum 48 h after sampling(), with 1 % accuracy. For the Vouglans hydrochemical conditions (pH between 6 and 8.5), total and carbonate alkalinity can be considered as equalling HCO_3^- ion concentration. The calculated charge balance error for the reported analyses was performed with the PHREEQC code (Parkhurst and Appelo 1999). Analyses which display a charge balance lower than 10 % were only taken into account. Chemical analyses were performed at the Chrono-Environnement laboratory at the University of Franche-Comté in France.

3 Results

3.1 *Different Water Source*

Statistical analyses are used to characterise more precisely the low contrast in chemical composition, resulting from the interaction between surface water, groundwater and infiltration in a carbonated medium (PCA). The discriminant factor analysis is used in order to reclassify individuals in homogeneous groups in term of the variables studied (DFA).

3.1.1 Hydrochemical Groups

A principal component analysis (PCA) with the XLstat software has been performed, using all samples collected in the site, that is to say 11 physical and

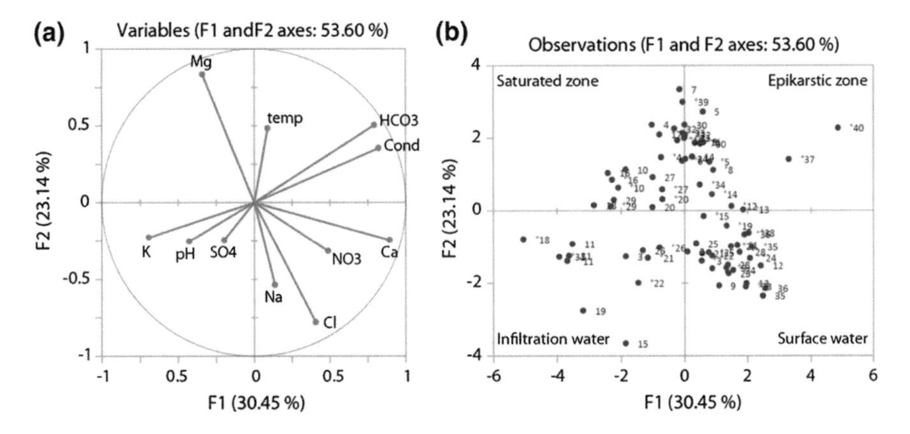

Fig. 2 Principal component analysis of the samples taken into the abutments and the drains of the dam and the lake. **a** Diagram variables. **b** Diagram of individuals (°1: June 2002 campaign, 1: March 2004 campaign)

chemical variables and 75 observations for the 2002 and 2004 data analysis. Averages values were used for both chemical profiles of the lake.

Figure 2 presents the PCA variables and observations. Regarding variables (Fig. 2a), bicarbonates are well correlated with electrical conductivity and are negatively correlated with potassium and sulphate. These variables outline a first axis on Factor 1 of surface water infiltration: potassium represents a shallow water origin, potassium being stored mainly in plants and partially leached during the decomposition of organic matter (Chaudhuri et al. 2007). Another origin is the human activity as manure application on calcareous soils (Griffioen 2001; Charlier et al. 2012). Bicarbonates are in turn essentially acquired during the interaction between the soil and the epikarst. On axis 2 Magnesium is anti-correlated to chlorides, nitrates and sodium. All of them are independent tracers defined by axis 1. Magnesium is a residence time tracer of carbonate reservoirs. He opposes anthropogenic markers that characterise superficial waters. Epikarst springs display high bicarbonates values and high electrical conductivity, while lake waters are characterised by high values of chloride, sodium and nitrates.

Regarding PCA observations, Fig. 2b shows the distribution of sampling points around these four poles. Points representative of the saturated zone are located in the galleries, on the right and left sides of the dam, as well as some central drains of the perimeter gallery. The springs 'source de la Vierge' and 'source de la Route' are draining epikarst waters. Surface waters are grouped around the lake waters and are found in some of the drains of the perimeter gallery. Infiltration of surface water can be found in the drains of the right side and in the perimeter gallery.

Table 1 Assignment of supplementary elements with respect to hydrological season (low water to high water periods)

Point	Group
3	3 → 1
6	2 → 4
12	4 → 3
13	4 → 3
15	4 → 1
19	4 → 1
21	1 → 4
22	1 → 3
25	3 → 4
26	4 → 1
29	1 → 2

Group 1 infiltration water; *group 2* saturated zone water, *group 3* shallow water, *group 4* epikarstic water

3.1.2 Seasonal Effect

DFA enables displaying, analyzing and above all predicting, using a set of standardised variable set, belonging to a group of individuals to predefined groups. It intends to compare, within a statistical population separated in a priori separated groups, intergroup variance to intragroup variance. In the framework of our study, 75 samples have been dispatched into principal and supplementary individuals. In order to have equally weighted belonging groups, only four principal individuals have been defined in each pole (Mudry 1991). The 59 other samples have been declared as supplementary, with the aim of classifying them into the different poles (Group 1 infiltration water; group 2: saturated zone water, group 3: surface water, group 4: epikarstic water), according to the hydrological season (Table 1) The supplementary individuals which have been assigned are mainly situated in the perimeter gallery (points 12–29). In a general way, points declared in the 'epikarstic water' group have be reassigned either in the 'infiltration water' or 'shallow water' groups for the winter samples, which indicates inflows from different origins and transfer time, with respect to seasons. For any other point, there is no assignment.

The variable diagram (Fig. 3a) displays an opposition along axis 2 between a groundwater pole (Mg) and a surface water one (Na, Cl and NO_3). Along axis 1, infiltration pole is opposed to epikarst one. Individuals are classified according to these poles (Fig. 3b), with a very noticeable dispersion around the pole for the points 'supp./epikarst', 'supp./infiltration superficial water' and part of points 'supp./saturated zone water', which expresses influence of the other poles. DFA evidences mixtures between various poles. Epikarst pole is the lowest classified one, because bicarbonates are present in each pole.

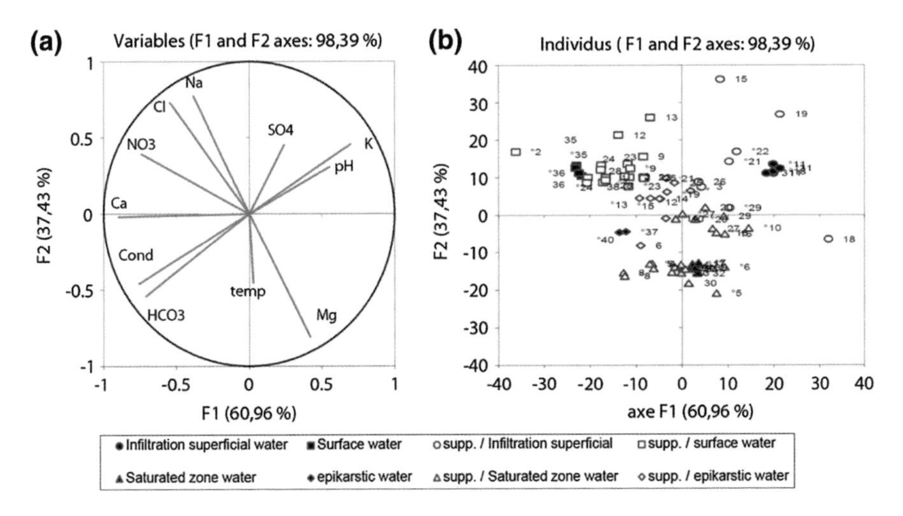

Fig. 3 Discriminant analysis (DFA) of the samples taken into the abutments and the drains of the dam and the lake. **a** Diagram of variables. **b** Diagram of individuals (°1: June 2002 campaign, 1: March 2004 campaign). N = 4*4, 59 supplementary elements

3.2 Hydrodynamic Behaviours

The correlation between concentrations of chloride versus magnesium is used to distinguish three groups of drains according to their flow type (Fig. 4):

- drains that are directly influenced by the lake (chloride concentrations above 3.8 mg/L, and magnesium concentrations below 5 mg/L). These drains are characterised by circulation of water from the lake, with short residence time;
- drains that are influenced by water of the saturated zone (chloride concentrations below 2 mg/L, and magnesium concentrations above 17 mg/L), which are typical of groundwater circulation with a long residence time;
- drains that have chloride and magnesium concentrations in between the previous values, which correspond to water mixtures between lake and groundwater along the drains.

4 Effect of Lake Level Variation

Both in winter and in summer, we differentiate three cases in this area:

- drains directly influenced by rapid transit of the lake;
- drains directly influenced by a slow water transit from deep reservoir, bearing the signature of water-rock carbonate interactions with a long residence time;
- drains influenced by mixing between lake and deep reservoir waters.

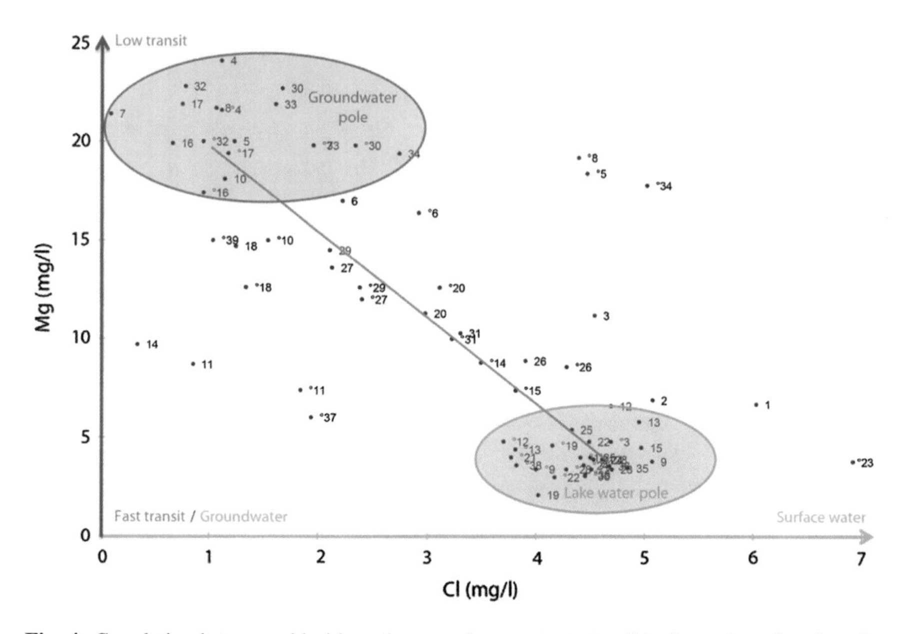

Fig. 4 Correlation between chloride and magnesium contents (mg/L) of samples taken into the abutments and the drains of the dam and the lake

In summer (Fig. 5a), only five drains are influenced by rapid transit of the lake waters. In this season, the area of influence of the rapid transit of the lake waters affects mainly drains upstream of the dam, and the closest one. The mixing zone between lake water and groundwater is close to the foot of the arch dam. Only drains beneath the stilling basin are not influenced by the lake. In winter (Fig. 5b), the area of influence of the rapid transit of the lake extends to drains a little more distant than the previous one, but still very close to the foot of the arch. Then the waters of the lake are mixed in the deep aquifer, and only drains beneath the stilling basin are not influenced by the lake (similar to winter results).

So the difference between winter and summer hydrodynamic behaviour comes from the downstream propagation of the influence zone of lake waters. When the reservoir is full and the dam is cold (high water period), arch moves downstream and the ‘crack open’, spreading the lake water in the basin.

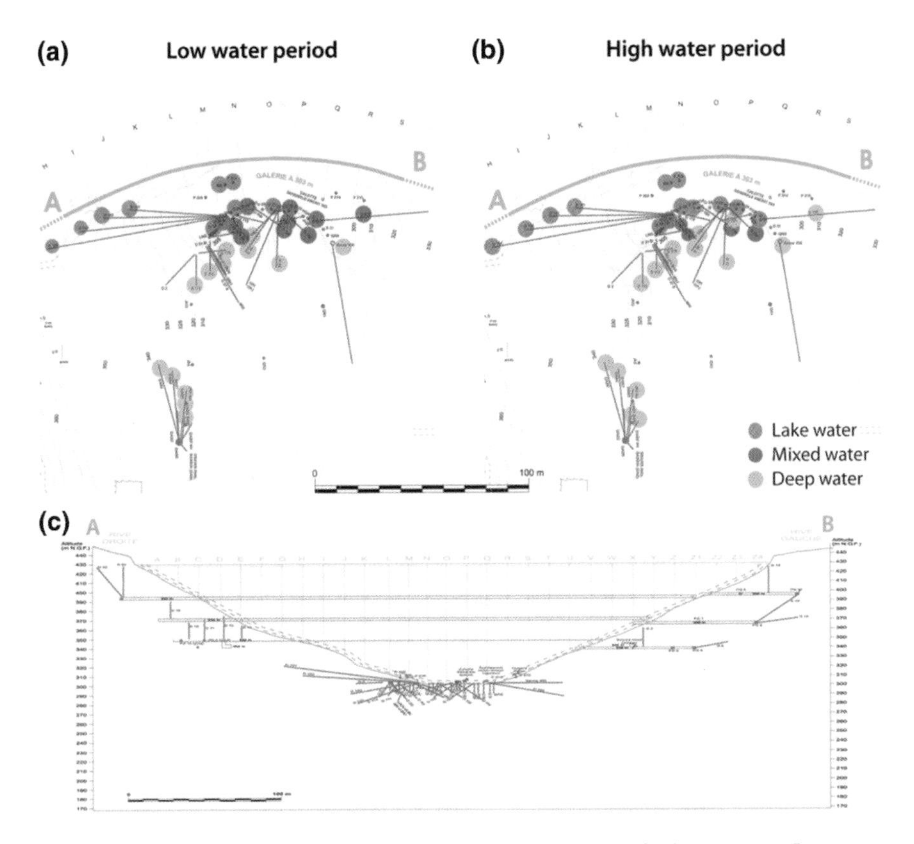

Fig. 5 Identification of drains sensitive to the zone of influence of lake waters. **a** Low water period (June 2002); **b** High water period (March 2004), **c** cross section of the dam and localization of a part of drain network

5 Conclusion

This hydrochemical investigation enables defining water typologies in a lowly contrasted environment, due to the common carbonated context, and providing a conceptual model of water fluxes within the dam.

Tracers highlighted by CPA and DFA enable to classify water types previously described in three groups: (i) recent waters, with a short residence time (dam reservoir water), (ii) 'old' waters, with a large residence time (deep groundwater) and (iii) intermediate waters, originating from mixtures between both components. Water infiltrations in the Vouglans dam's abutments originate predominantly from the lake upstream of the dam. On both banks, the shallowest perched groundwaters, especially the saturated layers situated between 395 and 410 m a.s.l. right bank are partly recharged by surface infiltrations, and partly from the lake leakages. Deeper groundwaters, and groundwater of the bottom of the valley are recharged only by the lake or by underground inflows from the perched groundwaters. Hydraulic circulations through the foundations are well characterised. The significant seasonal effect observed at the wall footer or the designation of drains 'intercepting the upstream fissure in any season' or 'only cold arch /full reservoir' is made possible by hydrochemical measurements.

Acknowledgments This work was financed by EDF. We gratefully acknowledge G. Castanier (Service Géologie Géotechnique DIN/CEIDRE—Département TEGG Aix en Provence Cedex).

References

Charlier JB, Bertrand C, Mudry J (2012) Conceptual hydrogeological model of flow and transport of dissolved organic carbon in a small Jura karst system. J Hydrol 460–461:52–64. doi:10.1016/j.jhydrol.2012.06.043

Chaudhuri S, Clauer N, Semhi K (2007) Plant decay as a major control of river dissolved potassium: a first estimate. Chem Geol 243:178–190

De Waele J (2008) Interaction between a dam site and karst springs: The case of Supramonte (Central-East Sardinia, Italy). Eng Geol 99:128–137

Griffioen J (2001) Potassium adsorption ratios as an indicator for the fate of agricultural potassium in groundwater. J Hydrol 254:244–254

Käss W (1998) Tracing techniques in hydrogeology. A.A. Balkema, Rotterdam, p 518

Kendall C, McDonnell JJ (eds) (1998) Isotope tracers in catchment hydrology. Elsevier Science B.V., Amsterdam, 839p

Milanovic P (2002) The environmental impacts of human activities and engineering constructions in karst regions. Episodes 25(1):13–21

Milanovic P (2004) Water resources engineering in karst. CRC Press, Boca Raton, p 328

Mudry J (1991) L'analyse discriminante, un puissant moyen de validation des hypothèses hydrogéologiques. Revue des Sciences de l'Eau (Québec - Paris) 4:19–37

Mudry J, Andreo B, Charmoille A, Liñan C, Carrasco F (2008) Some applications of geochemical and isotopic techniques to hydrogeology of the caves after research in two sites (Nerja cave-S Spain, and Fourbanne spring system-French Jura) internat. J Speleo 37(1):67–74

Parkhurst DL, Appelo CAJ (1999) User's guide to PHREEQC (version 2), a computer program for speciation batch reaction, one dimensional transport and inverse geochemical calculations. Water resources research investigations report 99-4259, 321p

Plata Bedmar A, Araguas Araguas L (2002) Detection and prevention of leaks from dams. A.A. Balkema Publishers, Rotterdam, p 419

Silar J (1988) Can water losses from reservoirs in karst be predicted? In: Yuan D (ed) Proceedings of the international association of hydrogeologists, vol 2. IAHS Publication No. 176, Wallingford, pp 1148–1152

Spatiotemporal Variations of Soil CO_2 in Chenqi, Puding, SW China: The Effects of Weather and LUCC

R. Yang, M. Zhao, C. Zeng, B. Chen and Z. Liu

Abstract Long-term monitoring of soil CO_2 dynamic was applied under different weather conditions and land uses to investigate the influences of weather and LUCC on soil CO_2 variations as well as its potential carbon sink. Observation results demonstrate that seasonal variations of soil CO_2 are mainly controlled by temperature-water combined effect, rising to the peak in wet–hot season and declining to the valley during dry–cold time. In spatial scale, soil CO_2 concentration is largely regulated by LUCC and follows the descending order of forest, shrubbery, dry land, and paddy land except for rice growing season, but with an ascending sequence for coefficient of variations (C.V.s), among which the highest C.V. and the abrupt changes of CO_2 in paddy land are mainly due to the alternate cultivation, flooding irrigation, and draining mechanism. Furthermore, by means of modeling calculations, wet–hot weather conditions and land uses as forests are provided with higher dissolved inorganic carbon (DIC) equilibrium concentration than the other sites, reflecting stronger karst process as well as larger potential carbon sink.

Keywords Terrestrial carbon cycle · Land use/cover change · Weather conditions · Soil CO_2 · Karst process-related carbon sink

---

R. Yang · M. Zhao · C. Zeng · B. Chen · Z. Liu (✉)
State Key Laboratory of Environmental Geochemistry, Institute of Geochemistry, Chinese Academy of Sciences, 46 Guanshui Road, Guiyang 550002, China
e-mail: liuzaihua@vip.gyig.ac.cn

R. Yang
e-mail: yangrui@vip.gyig.ac.cn

R. Yang · M. Zhao · C. Zeng · B. Chen · Z. Liu
Puding Comprehensive Karst Research and Experimental Station, Institute of Geochemistry, CAS and Science and Technology Department of Guizhou Province, Puding 562100, China

B. Andreo et al. (eds.), *Hydrogeological and Environmental Investigations in Karst Systems*, Environmental Earth Sciences 1,
DOI 10.1007/978-3-642-17435-3_22

1 Introduction

Accompany with the increased concentration of greenhouse gasses in the atmosphere, including carbon dioxide, methane, nitrous oxide, etc., the energy equilibration will be seriously affected, and thus the global climate (Manabe and Stouffer 1979; Jenkinson et al. 1991; Melillo et al. 2002). Since the industrial revolution, the emissions of greenhouse gasses, particularly CO_2, which is responsible for 60–80 % of total greenhouse effect (Lashof and Ahuja 1990; Rodhe 1990), is highly raised due to fossil fuels combustion, land use/cover change (LUCC), and the development of agriculture, forestry, animal husbandry, etc., leading to a series of problems on global climate change, especially continuous global warming (Vitousek 1994; Lasco et al. 2002; Upadhyay et al. 2005; Cihlar 2007). To cope with this problem and cut CO_2 emissions, in December 1997, Kyoto Protocol was formulated by the United Nations Framework Convention on Climate Change (UNFCCC), requiring joint implementation, emissions trading, and clean development mechanism (Steffen et al. 1998). Due to the growing concern on global warming as well as the accompanying political and economic pressure on CO_2 emission reduction, studies on global carbon cycle have attracted an increasing attention (Steffen 1999).

For terrestrial carbon cycle, LUCC is considered as the main artificial factor apart from fossil fuel combustion (Lambin et al. 2001; Levy et al. 2004). Previously, researches about LUCC-related carbon cycle mainly focus on biomass change and soil carbon sequestration (Del Galdo et al. 2003; Albani et al. 2006; Chen et al. 2006; Batlle-Bayer et al. 2010). While improving soil management practices is recognized as an important measure for implementing clean development mechanism (Pingoud et al. 1999; Lal 2004), soil respiration is widely studied in almost all ecosystems to evaluate the effects of soil management practices on global warming and carbon cycle (Liu et al. 2013). However, due to the regional diversity, in karst area, LUCC also affects underground carbon migration and conversion, causing the variations of CO_2 in the soil and the consequent differences of carbonate erosion, thus regulates the intensity of karst process and the capacity of karst process-related carbon sink. In terrestrial environments, soil is one of the main paths where CO_2 can be added into water. While some of CO_2 originating from soil inevitably escapes into atmosphere as soil respiration, part of it will dissolve in soil water and potentially recharge the underlying aquifer through a downward flux in the unsaturated zone (Tsypin and Macpherson 2012; Yang et al. 2012). It is considered that water and CO_2 are the major chemical driving forces for carbonate dissolution and have significant influences on karst process-related carbon cycle (Liu et al. 2007, 2010). Therefore, monitoring of CO_2 in the soil under different land uses is conducive to reveal the intensity of karst process as well as its potential carbon sink and contribute to guide LUCC managements in karst region.

In this research, long-term monitoring of soil CO_2 dynamic was applied under different weather conditions and land uses to investigate the influences of weather and LUCC on soil CO_2 and its potential carbon sink. On the basis of field survey, four representative sites in Chenqi, Puding, SW China were chosen to monitor the concentration of CO_2 in the soil per day during May 2011–May 2012, covering a complete hydrologic year. It was found that the concentration of soil CO_2 presents large diversity in different land uses and shows frequent response to vegetation alterations as well as weather condition changes, implying the disparate capacities of karst carbon sink under LUCC managements, and thus the role of karst process in global carbon cycle is considerable and need to be reappraised.

2 Materials and Methods

2.1 Study Area

The research sites were set in Chenqi epikarst system, which is located in Puding County, Guizhou Province, SW China. The climatic conditions here are characterized by humid subtropical monsoon climate with an annual average precipitation of 1,315 mm (over 80 % precipitation occurs from May to October) and the mean air temperature of about 15.1 °C. The main lithology is limestone and dolomite of the Guanling Formation of middle Triassic, intercalated with gypsum strata (Zhao et al. 2010).

The main geomorphology of Chenqi is a typical karst valley, covering with loose Quaternary deposits, most of which have been reclaimed for farmland. The primary land uses here are secondary forest (mainly consisting of pear trees), shrubbery, and tilled land, including paddy land and dry land. The corresponding area proportion of each land use has been investigated (Zhao et al. 2010) and was presented in Table 1. As observation sites to monitor soil CO_2, we chose four plots for total land uses, among which, shrubbery and secondary forest were hardly influenced by human activities, while vegetation and usage were altered around the year in paddy land and dry land by tillage and husbandry.

2.2 Monitoring and Analysis

To observe the variations of CO_2 in the soil under different land uses and depths, 24 polyvinylchloride (PVC) pipes were designed as gas measuring tubes and divided into eight groups equally (8×3). Every two groups of the tubes were embedded into soil at the depths of 30 and 60 cm, respectively by drilling in each sample plot. At the top of each tube, a removable cap was covered to avoid gas exchange between the tube and atmosphere. A portable CO_2 analyzer GM70 (made by VAISALA in Finland with the probe measuring range from 0 to 10 %

Table 1 Land uses in Chenqi (Zhao et al. 2010)

Land uses	Secondary forest	Shrub land	Dry land	Paddy land	Total
Area (km^2)	0.087	0.306	0.729	0.189	1.31
Percentage (%)	6.61	23.35	55.65	14.39	100

and the resolution of 1 ppm) was applied to measure soil CO_2 concentration in each tube from 11:00 am to 12:00 am per day.

Local rainfall data was logged through the rain gage, a sub-module of CTDP300 multi-parameter water quality meter (made by Greenspan Corporation in Australia with the rainfall resolution of 0.5 mm). Soil temperature was recorded by AV-10T (temperature sensor produced by AVALON in America with a resolution of 0.1 °C).

3 Results

We recorded the dynamic data of rainfall, soil temperature, and soil CO_2 concentration during May 2011–May 2012, covering a complete hydrologic year, and presented the data graphically in time series (Figs. 1, 2, 3, and 4) to allow visual assessment alongside the statistical text (Table 2). The graphical results about the variations on these points are analyzed as follow.

3.1 Dynamic Weather Conditions

Precipitation data was logged by a rain gauge and inspected according to the nearest meteorological station record. During the research period, the total rainfall is about 783 mm, only approximate two-thirds of the annual average precipitation, implying relative drought. The precipitation is distributed mainly from May to October and concentrated in six periods (Figs. 1, 2, rain 1–6). Beyond this, two arid periods exist in July and August during rainy season.

Soil temperature is observed synchronously and shows obvious seasonal variations, high in summer and low in winter (Figs. 1, 2). However, the temperature sensor was only set in shrubbery. Due to the absence of temperature data in other land uses, soil temperature in shrubbery was presented to simulate the general variation tendency of temperature in all land uses year-round.

According to different weather conditions, the study period was divided into three parts, including two wet–hot growing seasons and a dry–cold dormant season from November 2011–March 2012 (Figs. 1, 2).

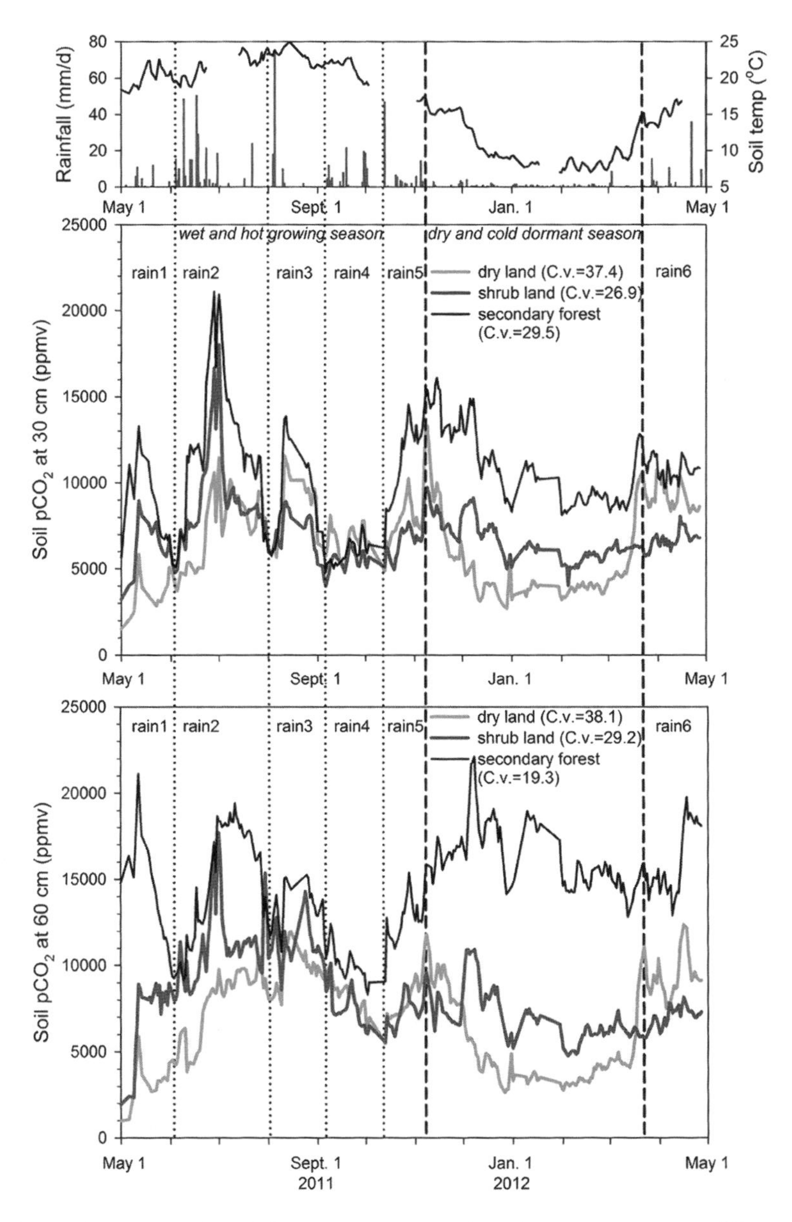

Fig. 1 Continuous data of rainfall, soil temperature and soil CO_2 concentration in different land uses during May 2011–May 2012. Soil temperature was recorded by single sensor which was buried in shrub at the depth of 30 cm. Though soil temperature did not exactly match the one in each site, it could approximate the tendency of soil temperature variation year-round

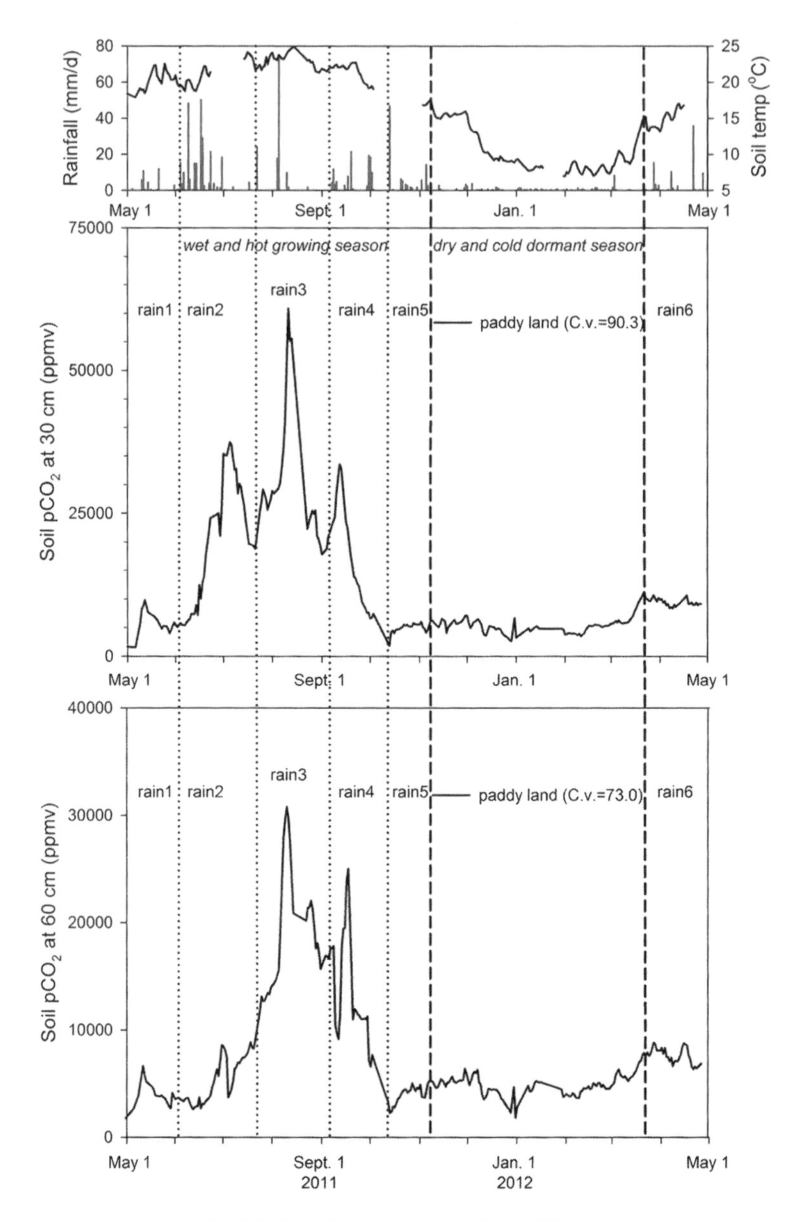

Fig. 2 Continuous data of rainfall, soil temperature and soil CO_2 concentration in paddy land during May 2011–May 2012. Temperature data was taken from shrub land due to the absence of sensor in paddy land

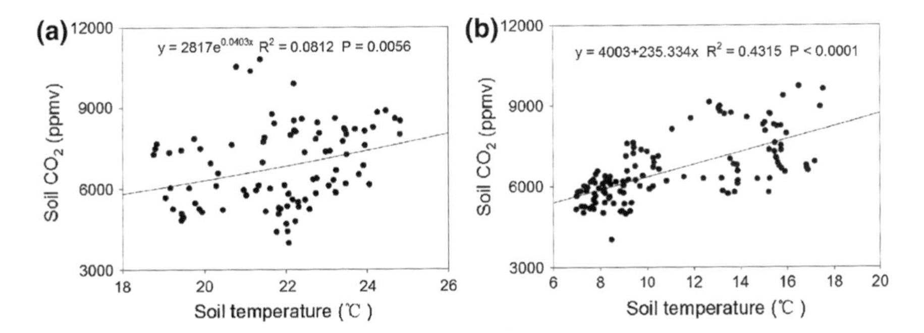

Fig. 3 Correlations between soil temperature and soil CO_2 in different seasons in shrub land. **a** Exponential correlation in wet–hot growing season. **b** Linear correlation in dry–cold dormant season

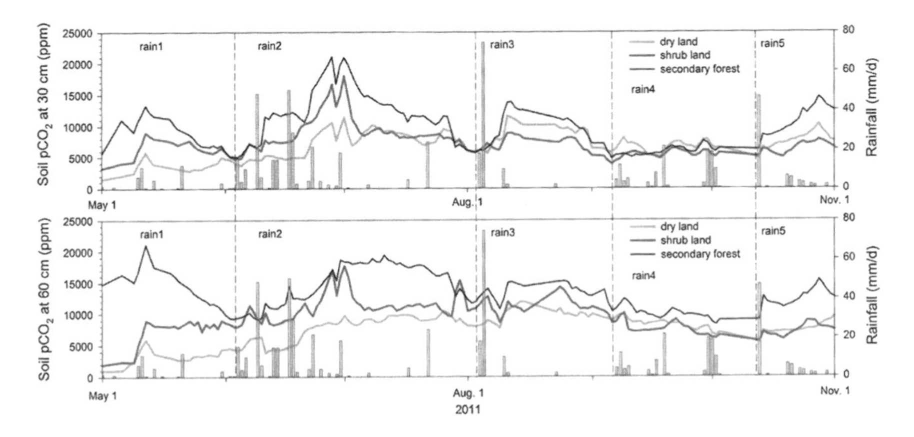

Fig. 4 Continuous data of rainfall and soil CO_2 concentration in different land uses during rainy season in 2011

Table 2 Minimum, maximum and annual mean values of soil CO_2 and the relevant coefficient of variation in different land uses

Land uses	Secondary forest		Shrub land		Dry land		Paddy land	
	30 cm	60 cm	30 cm	60 cm	30 cm	60 cm	30 cm	60 cm
Min. (ppmv)	4,807	8,307	3,230	1,930	1,513	983	1,407	1,783
Max. (ppmv)	21,087	22,100	18,013	17,695	13,280	12,363	60,645	30,810
Mean (ppmv)	10,471	14,734	6,999	8,241	6,574	6,956	12,096	7,507
C.v. (%)	29.5	19.3	26.9	29.2	37.4	38.1	90.3	73.0
N	306	306	306	306	306	306	306	306

C.v. or coefficient of variation = (standard deviation/mean) %, *N* means number of samples

3.2 Soil CO_2 Variations

Seasonal cycles of soil CO_2 concentration are remarkable in time series (Fig. 1). Soil CO_2 in all land uses except secondary forest show in-phase changes with soil temperature, rising to peaks in summer and decreasing to troughs in winter, indicate that temperature is an important influence factor on soil CO_2 generation. In general, with higher soil temperature, more soil CO_2 is produced. For a detailed discussion later, soil temperature vs. soil CO_2 was presented to assess their correlations (Fig. 3a, b). As soil temperature is logged just in shrubbery at a soil depth of 30 cm, lacking of data in other land uses, the figures can only represent the correlations in shrubbery. Graphic results illustrate that the correlations are significant and positive both in wet–hot growing season (Fig. 3a) and dry–cold dormant season (Fig. 3b). Meanwhile, higher P value and lower R value in wet–hot growing season suggest a preponderant water-temperature combined effect on soil CO_2.

During rainy season, soil CO_2 fluctuates according to both soil temperature and rainfall. In detail, every abrupt change of soil CO_2 almost corresponds to each rainfall periods (Fig. 4, rain 1–5). When there is a rain, there appears a rising trend in soil CO_2 and can last for several days, then fall back. Similar changes are constantly repeated until the end of rainy season. It could be explained as that rainfall (and/or soil moisture) plays a significant role in soil CO_2 generation and diffusion.

In different land uses, the concentration of soil CO_2 presents large diversity, generally highest in secondary forest and lowest in dry land (Figs. 1, 2). The annual variation coefficients of soil CO_2 follow the descending order as: paddy land, dry land, shrubbery, and secondary forest (Table 2, Figs. 1, 2). Besides, soil CO_2 in paddy land is particularly high during rice growing season and shows frequent fluctuations accompany with flooding and draining alternation. The distinct concentration and variations of soil CO_2 are likely responded to different vegetation, cultivation, soil physicochemical properties, etc.

Soil CO_2 in different depths varies from each other (Figs. 1, 2) and maybe owing to the vertical distribution of plant roots and soil structure. In dry land and shrubbery, there is an inconspicuous rising of soil CO_2 from 30 to 60 cm in the soil profile. A similar trend is significantly presented in secondary forest around the whole year, while an inverse change is shown in paddy land during rice growing season.

4 Discussion

In general, soil CO_2 originates mainly from root respiration, oxidative decomposition of organic matter, and microbial respiration (Reardon et al. 1979; Hanson et al. 2000) and diffuses in three directions, upward, downward, and lateral.

The loss of CO_2 in the soil mainly consists of the upward diffusion to atmosphere and the downward migration to groundwater as gas phase (CO_{2g}) directly or dissolved inorganic carbon (DIC) indirectly. In addition, carbonate dissolution in unsaturated area will cause soil CO_2 consumption, which could be reflected in groundwater hydrochemical characteristics by downward migration (Tsypin and Macpherson 2012; Yang et al. 2012).

When CO_2 in the soil is more generated than consumed, the concentration is risen correspondingly. Conversely, the lower concentration is caused. The factors that lead to the variations of soil CO_2 could be weather conditions and LUCC.

4.1 Effects of Weather Conditions

Temperature is considered to be an important factor on soil CO_2 generation. Generally, a higher soil temperature will result in a stronger root respiration (Atkin et al. 2000; Burton et al. 2002) and a faster microbial decomposition (Raich and Schlesinger 1992; Atkin et al. 2000), thus soil CO_2 is more generated. However, the influences of temperature on soil CO_2 are distinct from each season. Therefore, the time series was separated into wet–hot growing season and dry–cold dormant season to discuss temperature effects in both seasons.

According to the analytical results (Fig. 3), the correlations between soil temperature and CO_2 in both seasons are significant at the 0.01 level. The lower P value and higher R value (Fig. 3b) indicate that temperature effect on soil CO_2 is more preponderant in dry–cold dormant season than in wet–hot growing season. This is mainly because that, during dry–cold dormant season, soil moisture is relatively stable due to less precipitation and evaporation, thus soil temperature dominates CO_2 variations. On the contrary, temperature effect is weak due to frequent rainfall intervention and strong evaporation, which could change soil moisture and regulate the generation, dissolution, and diffusion of soil CO_2.

Soil moisture is a restrictive factor on soil CO_2 production, low water content can inhibit CO_2 generation in soil, sufficient soil moisture could result an abundant CO_2 production by root and microbe (Almagro et al. 2009). Besides, water content also affects the diffusion rate of CO_2 in soil profile (Davidson and Trumbore 1995), thus controlling the accumulation of soil CO_2. Unfortunately, due to frequent instrument malfunctions, detailed soil moisture data is missed during research period. For this reason, we only discuss the effects of rainfall.

In rainy season, the influences of precipitation on soil CO_2 can be divided into long-term and short-term effects. Long-term effects are usually positive to CO_2 accumulation. Rainfall will increase soil water content, which could promote microbial activity and the metabolism of plants, then boost the rate of root respiration and organic matter decomposition (Orchard and Cook 1983; Davidson et al. 1998) and thus improve CO_2 production. Moreover, rain water will seal the soil pores and reduce gas diffusion rate (Linn and Doran 1984; Skopp et al. 1990), hence decrease CO_2 upward loss as soil respiration. On the other side, the interstitial water may also restrict oxygen diffusion through pore spaces into soil and thus slow down oxidative decomposition of organic matter (Linn and Doran 1984; Liu et al. 2013), resulting a low CO_2 production. This negative effect on CO_2 accumulation may be less, compared with the positive ones we mentioned earlier. Therefore, the ultimate expression of rainfall long-term effects is a rising CO_2 production and decreasing soil respiration, thus the increasing soil CO_2 concentration. As we have presented in Fig. 4, soil CO_2 increased corresponding to each rainfall period. The rising processes may last for a long time, generally several days, and will fall back accompanied with the evaporation of soil moisture.

Short-term effects of rainfall may cause a rapid decline in soil CO_2 within a short time. However, the effects could be show up only under high-resolution monitor strategy, for instance, with 15 min interval due to the short responsive time. As we recorded only a set of data per day, the discussion is mainly based on the previous work (Yang et al. 2012). At the beginning of rainfall, CO_2 in the shallow soil could be released into atmosphere due to the replacement of CO_2 in the soil pores by rain water (Chen et al. 2005); thus, soil respiration is increased instantaneously and CO_2 in the soil is decreased. After that, some of soil CO_2 dissolve in soil water and react with carbonate or infiltrate downward into groundwater directly, leading to a large consumption of CO_2 and the relevant response of groundwater chemical features (Yang et al. 2012), the responsive time mainly depend on rainfall intensity and groundwater recharge mode. Though carbonate erosion also exist in dry season, precipitation in rainy season will enhance the effect and shorten the responsive time into a few hours. Consequently, the short-term effects could result a faster CO_2 consumption than production in a short time, presenting a decline in soil CO_2 concentration.

In brief, seasonal variations of soil CO_2 can be explained as a result of temperature-water combined effects. Wet–hot weather conditions are optimum for soil CO_2 generation and accumulation, showing higher soil CO_2 concentration during summer than winter (Figs. 1, 2).

4.2 Influences of LUCC

In spatial scale, as sample plots were chosen under the same climate condition, variations of soil CO_2 are mainly reflected by LUCC, which can diversify vegetation, organic matter content, and soil physicochemical property. In fallow season, concentration of soil CO_2 declined in the order of secondary forest, shrubbery, paddy land, and dry land (Figs. 1, 2). In general, the highest biomass and organic matter content in forest land (Zhang 2011) lead to the strongest root respiration and microbial decomposition, hence the maximum of soil CO_2 concentration. However, in farming season, soil CO_2 in paddy land rise to the highest due to the special irrigation.

In secondary forest, the seasonal variation trend of soil CO_2 is weak (Table 2, Fig. 1), especially at 60 cm depth. The main reason could be that, in winter the forest soil is covered by a thick litter layer, which could hold soil temperature and moisture (MacKinney 1929) and replenish organic matter (Berg 2000), thus conducive to the generation of soil CO_2. Besides, forest soil structure is relatively stable due to the absence of tillage, distinct from tilled land where the soil would be turned over after harvest (usually in autumn). The stable structure of forest soil may slow down soil CO_2 from upward diffusion, and thus stabilize its concentration. On the other hand, root respirations of trees contribute large proportion to soil CO_2 production (Ewel et al. 1987; Behera et al. 1990; Bowden et al. 1993), especially in forest, root respiration is the major contributor of soil CO_2 (Edwards 1975). While most of the trees are perennial in forest, root respiration is uninterrupted and may be strong even in winter because that leaves are withered and roots almost take up the total plant breathe, showing the inconspicuous seasonal variation trend of soil CO_2 in woodland. This can also explain the second-lowest variation coefficient in shrub land (Table 2, Fig. 1), but due to the absence of litter layer cover and the less biomass, the annual variation coefficient is higher than forest.

In tilled lands (including paddy land and dry land), however, the seasonal cycle of soil CO_2 differ from woodland and is remarkable (Table 2, Figs. 1 and 2), which is owing to the seasonal crops cultivation and plowing strategy. In farming season (usually in spring and summer), crops flourish, and thus more CO_2 is generated from root respirations. Also, the rising temperature will boost microbial decomposition and cause a high soil CO_2. In fallow season, due to the plowing after harvest, the absence of vegetation and the reducing temperature, soil CO_2 is less generated and diffused faster, result in low concentration of soil CO_2.

In paddy land, as a result of rice cultivation and flooding irrigation, the concentration of soil CO_2 is far higher than the other land uses and has several abrupt changes (Fig. 2). During rice growing season, a water layer is necessary for covering over the soil surface (flooding irrigation), and could cut down most of the upward diffusion of soil CO_2 (Liu et al. 2013). Although some soil CO_2 may be dissolved in water, but the water is stagnant and will not cause the migration of CO_2. Furthermore, a high-density soil layer at the depth of 30–40 cm (a plow pan as aquitard) will hinder gas downward diffusion and liquid infiltration (Forrer et al. 1999), and thus accumulates a large amount of CO_2 in the soil air. On the contrary, draining of the paddy land will release CO_2 to atmosphere as soil respiration, corresponding to the troughs of soil CO_2 during the relative dry period (Fig. 2). That is, when the water layer is dried up without rainfall and irrigation, CO_2 in the soil reduced rapidly.

Generally, soil CO_2 concentration increases slightly with the soil depth (Reardon et al. 1979). Observation result shows that CO_2 concentration at 60 cm is higher than 30 cm (Figs. 1, 2), especially in forest due to the deep roots system. However, an inverse trend was found in paddy land, providing additional evidence for the blocking effect on CO_2 diffusion by plow pan at 30–40 cm depth.

In conclusion, LUCC significantly affect the spatiotemporal variations of soil CO_2. Forest is conducive to the production and conservation of CO_2 in the soil, presenting the highest concentration and the lowest variation coefficient of CO_2.

Table 3 Soil CO_2 partial pressure and the calculated equilibrium values of DIC under different land uses and weather conditions according to (Dreybrodt 1988) (16 °C)

Land uses	Soil CO_2 partial pressure (ppm)			DIC equilibrium concentration (mmol/L)		
	Wet–hot season	Dry–cold season	Annual	Wet–hot season	Dry–cold season	Annual
Secondary forest	12,177	13,186	12681.5	4.750	4.931	4.839
Shrub land	8,477	6,578	7527.5	4.036	3.599	3.817
Dry land	7,966	4,043	6004.5	3.924	2.945	3.464
Paddy land	13,481	4,642	9061.5	4.990	3.121	4.152

4.3 Implications for Carbon Cycle in Karst Area

Karst process-related carbon cycle, as a result of the water–rock–soil–gas–organism interaction, is a complex process, and significantly affects global carbon budget (Yuan 1997; Gombert 2002; Liu et al. 2010). CO_2 and water, as the indispensable driving factors, play an important role in karst process. In general, a higher soil CO_2 and water content will lead to a faster carbonate dissolution rate, and thus a greater capacity of carbon sink.

Karst process-related carbon sink is mainly considering of CO_2 consumption caused by carbonate dissolution in $CaCO_3$–CO_2–H_2O system. This process can be described in the following equation:

$$CaCO_3 + CO_2 + H_2O = Ca^{2+} + 2HCO_3^- \quad (1)$$

As a result of this reaction, in karst area, the amount of CO_2 absorbed by water is significantly larger than in non-karst system. CO_2 dissolves as Eq. (2) and establishes an equilibrium mixture of carbonic acid, bicarbonate, and carbonate ions, which formed most of DIC (Liu et al. 2010).

$$CO_2 + H_2O \Leftrightarrow H_2CO_3 \Leftrightarrow H^+ + HCO_3^- \Leftrightarrow 2H^+ + CO_3^{2-} \quad (2)$$

According to Dreybrodt, DIC equilibrium concentration in karst system under different soil CO_2 partial pressure could be calculated out (Dreybrodt 1988), indicating CO_2 consumption and reflecting the capacity of karst carbon sink relatively (Table 3).

As we have mentioned earlier, changes in soil CO_2 concentration under different land uses and weather conditions will lead to different DIC equilibrium concentration. Calculation result shows that wet–hot weather conditions and land uses as forests are provided with higher DIC equilibrium concentration due to higher soil CO_2 partial pressure, and thus optimum for strengthening karst process and the potential carbon sink.

However, it is difficult to quantify the influence of LUCC on karst carbon sink because of the complex factors, for instance, soil CO_2 concentration, soil moisture, rainfall infiltration coefficient, organic matter content, etc. Therefore, further studies need to be developed combining with some simulative and/or modeling experiments under controlled conditions.

5 Conclusions

We have investigated spatiotemporal variations of soil CO_2 in epikarst systems. Four sites in Chenqi, Puding, SW China, representing different land uses, were chosen to monitor the concentration of CO_2 in the soil during May 2011–May 2012, covering a complete hydrologic year. Observation results demonstrate that, (1) seasonal variations of soil CO_2 are controlled by temperature-water combined effects, high in wet–hot growing season and low in dry–cold dormant season. (2) In spatial scale, soil CO_2 concentration is regulated by LUCC and follows a descending order of forest, shrubbery, dry land, and paddy land apart from rice growing season. (3) The abrupt variations of CO_2 in paddy land are caused by rice cultivation, flooding irrigation, and draining mechanism. (4) Wet–hot weather conditions and land uses as forests are optimum for the production, accumulation, and conservation of CO_2 in the soil, thus strengthen karst process and the potential carbon sink.

To sum up, this study has showed spatiotemporal variations of soil CO_2 and the potential capacity of carbon sink in karst area under different land uses, implying that the role of karst process in global carbon cycle is considerable and needs to be reappraised based on LUCC managements.

Acknowledgments This work was supported by the National Basic Research Program of China (Grant No. 2013CB956703) and Strategic Priority Research Program (XDA05070400) of Chinese Academy of Sciences, and National Natural Science Foundation of China (41172232 and 41103084).

References

Albani M, Medvigy D, Hurtt GC et al (2006) The contributions of land-use change, CO_2 fertilization, and climate variability to the Eastern US carbon sink. Global Change Biol 12:2370–2390

Almagro M, López J, Querejeta JI et al (2009) Temperature dependence of soil CO_2 efflux is strongly modulated by seasonal patterns of moisture availability in a Mediterranean ecosystem. Soil Biol Biochem 41:594–605

Atkin OK, Edwards EJ, Loveys BR (2000) Response of root respiration to changes in temperature and its relevance to global warming. New Phytol 147:141–154

Batlle-Bayer L, Batjes NH, Bindraban PS (2010) Changes in organic carbon stocks upon land use conversion in the Brazilian Cerrado: a review. Agr Ecosyst Environ 137:47–58

Behera N, Joshi SK, Pati DP (1990) Root contribution to total soil metabolism in a tropical forest soil from Orissa, India. Forest Ecol Manag 36:125–134
Berg B (2000) Litter decomposition and organic matter turnover in northern forest soils. Forest Ecol Manag 133:13–22
Bowden RD, Nadelhoffer KJ, Boone RD et al (1993) Contributions of aboveground litter, belowground litter, and root respiration to total soil respiration in a temperate mixed hardwood forest. Can J Forest Res 23:1402–1407
Burton A, Pregitzer K, Ruess R et al (2002) Root respiration in North American forests: effects of nitrogen concentration and temperature across biomes. Oecologia 131:559–568
Chen D, Molina JAE, Clapp CE et al (2005) Corn root influence on automated measurement of soil carbon dioxide concentrations. Soil Sci 170:779–787
Chen H, Tian H, Liu M et al (2006) Effect of land-cover change on terrestrial carbon dynamics in the southern United States. J Environ Qual 35:1533–1547
Cihlar J (2007) Quantification of the regional carbon cycle of the biosphere: policy, science and land-use decisions. J Environ Manage 85:785–790
Davidson EA, Belk E, Boone RD (1998) Soil water content and temperature as independent or confounded factors controlling soil respiration in a temperate mixed hardwood forest. Global Change Biol 4:217–227
Davidson EA, Trumbore SE (1995) Gas diffusivity and production of CO_2 in deep soils of the eastern Amazon. Tellus B 47:550–565
Del Galdo I, Six J, Peressotti A et al (2003) Assessing the impact of land-use change on soil C sequestration in agricultural soils by means of organic matter fractionation and stable C isotopes. Global Change Biol 9:1204–1213
Dreybrodt W (1988) Processes in karst systems, physics, chemistry, and geology. Springer, Heidelberg
Edwards NT (1975) Effects of temperature and moisture on carbon dioxide evolution in a mixed deciduous forest floor1. Soil Sci Soc Am J 39:361–365
Ewel KC, Cropper WP, Gholz HL (1987) Soil CO_2 evolution in Florida slash pine plantations. II. Importance of root respiration. Can J Forest Res 17:330–333
Forrer I, Kasteel R, Flury M et al (1999) Longitudinal and lateral dispersion in an unsaturated field soil. Water Resour Res 35:3049–3060
Gombert P (2002) Role of karstic dissolution in global carbon cycle. Global Planet Change 33:177–184
Hanson P, Edwards N, Garten C et al (2000) Separating root and soil microbial contributions to soil respiration: a review of methods and observations. Biogeochemistry 48:115–146
Jenkinson D, Adams D, Wild A (1991) Model estimates of CO_2 emissions from soil in response to global warming. Nature 351:304–306
Lal R (2004) Soil carbon sequestration impacts on global climate change and food security. Science 304:1623–1627
Lambin EF, Turner BL, Geist HJ et al (2001) The causes of land-use and land-cover change: moving beyond the myths. Global Environ Change 11:261–269
Lasco RD, Lales JS, Arnuevo M (2002) Carbon dioxide (CO_2) storage and sequestration of land cover in the Leyte geothermal reservation. Renew Energ 25:307–315
Lashof DA, Ahuja DR (1990) Relative contributions of greenhouse gas emissions to global warming. Nature 344:529–531
Levy P, Friend A, White A et al (2004) The influence of land use change on global-scale fluxes of carbon from terrestrial ecosystems. Clim Change 67:185–209
Linn DM, Doran JW (1984) Effect of water-filled pore space on carbon dioxide and nitrous oxide production in tilled and nontilled soils. Soil Sci Soc Am J 48:1267–1272
Liu Y, K-y Wan, Tao Y et al (2013) Carbon dioxide flux from rice paddy soils in central China: effects of intermittent flooding and draining cycles. PLoS One 8:e56562
Liu Z, Dreybrodt W, Wang H (2010) A new direction in effective accounting for the atmospheric CO_2 budget: considering the combined action of carbonate dissolution, the global water cycle and photosynthetic uptake of DIC by aquatic organisms. Earth-Sci Rev 99:162–172

Liu Z, Li Q, Sun H et al (2007) Seasonal, diurnal and storm-scale hydrochemical variations of typical epikarst springs in subtropical karst areas of SW China: soil CO_2 and dilution effects. J Hydrol 337:207–223

MacKinney A (1929) Effects of forest litter on soil temperature and soil freezing in autumn and winter. Ecology 10:312–321

Manabe S, Stouffer RJ (1979) A CO_2-climate sensitivity study with a mathematical model of the global climate. Nature 282:491–493

Melillo JM, Steudler PA, Aber JD et al (2002) Soil warming and carbon-cycle feedbacks to the climate system. Science 298:2173–2176

Orchard VA, Cook FJ (1983) Relationship between soil respiration and soil moisture. Soil Biol Biochem 15:447–453

Pingoud K, Lehtilä A, Savolainen I (1999) Bioenergy and the forest industry in Finland after the adoption of the Kyoto protocol. Environ Sci Policy 2:153–164

Raich JW, Schlesinger WH (1992) The global carbon dioxide flux in soil respiration and its relationship to vegetation and climate. Tellus B 44:81–99

Reardon EJ, Allison GB, Fritz P (1979) Seasonal chemical and isotopic variations of soil CO_2 at Trout Creek, Ontario. J Hydrol 43:355–371

Rodhe H (1990) A comparison of the contribution of various gases to the greenhouse effect. Science 248:1217

Skopp J, Jawson MD, Doran JW (1990) Steady-state aerobic microbial activity as a function of soil water content. Soil Sci Soc Am J 54:1619–1625

Steffen W (1999) The IGBP global carbon project: understanding the breathing of the planet. Global Change Newsletter 37:1

Steffen W, Canadell J, Apps M et al (1998) The terrestrial carbon cycle: implications for the Kyoto Protocol. Science 280:1393–1394

Tsypin M, Macpherson GL (2012) The effect of precipitation events on inorganic carbon in soil and shallow groundwater, Konza Prairie LTER Site, NE Kansas, USA. Appl Geochem 27:2356–2369

Upadhyay TP, Sankhayan PL, Solberg B (2005) A review of carbon sequestration dynamics in the Himalayan region as a function of land-use change and forest/soil degradation with special reference to Nepal. Agr Ecosyst Environ 105:449–465

Vitousek PM (1994) Beyond global warming: ecology and global change. Ecology 75:1861–1876

Yang R, Liu Z, Zeng C et al (2012) Response of epikarst hydrochemical changes to soil CO_2 and weather conditions at Chenqi, Puding, SW China. J Hydrol 468–469:151–158

Yuan D (1997) The carbon cycle in karst. Z Geo N F 108:91–102

Zhang C (2011) Carbonate rock dissolution rates in different land uses and their carbon sink effect. Chinese Sci Bull 56:3759–3765

Zhao M, Zeng C, Liu Z et al (2010) Effect of different land use/land cover on karst hydrogeochemistry: a paired catchment study of Chenqi and Dengzhanhe, Puding, Guizhou, SW China. J Hydrol 388:121–130

Characterization and Dynamics of Two Karst Springs in a Soil-Covered Karst Area, Lagoa Santa, Southeastern Brazil

P.F.P. Pessoa and A.S. Auler

Abstract Hydrogeological studies were performed on two karst springs located in a mining area of soil-covered karst in the Lagoa Santa region, in subtropical southeastern Brazil. Although they are located close together, the Tadinho and Cafundó springs exhibit distinct hydrogeological behaviors. Groundwater flow routes were determined through quantitative and qualitative rhodamine tracer tests through artificial injection in swallow holes. A 2-year discharge monitoring demonstrated that there is a 4-month delay between pluvial and discharge peaks at Tadinho spring due to the presence of a constriction, which causes the retention of the dye and a delay in the discharge response. Tadinho spring also displays an average discharge that is smaller than the total injection flow feeding the spring, as indicated by an only 20 % dye recovery. Conversely, Cafundó Spring displays a closer response to pluvial peaks because it is located within a much larger groundwater system, which major outlet is Tadinho Spring. Tadinho is characterized as an underflow karst spring (sensu Worthington 1991) because its discharge displays a constant depletion coefficient. Cafundó spring is interpreted as an overflow spring, situated in a higher topographic position, being constrained by the geometry and porosity of the aquifer system. A 4-year hydrochemical monitoring program showed sharp variations in calcite saturation indexes for Tadinho spring, with a negative correlation between rainfall and aggressiveness. The denudation rates for Tadinho spring are, on average, 22.5 mm/ka, in agreement with other studies in the Brazilian karst. Water budget calculations and spring hydrograph analysis indicate that the catchment area of the springs is much larger than determined by surface divides, with the Tadinho catchment area comprising significant areas of mantled karst.

P.F.P. Pessoa (✉) · A.S. Auler
Instituto Do Carste, Rua Brasópolis 139, Belo Horizonte, MG 30150-170, Brazil
e-mail: ppessoa@hidrovia.com.br

B. Andreo et al. (eds.), *Hydrogeological and Environmental Investigations in Karst Systems*, Environmental Earth Sciences 1,
DOI 10.1007/978-3-642-17435-3_23

1 Introduction

Two karst springs, Tadinho and Cafundó, were analyzed to characterize their groundwater dynamics and hydrochemical behavior. Both springs are located in the covered karst area of Lagoa Santa, southeastern Brazil, in the vicinity of a limestone quarry owned by ECL—Cimentos LIZ. Due to the thick soil cover, karst expression is limited, compared with that of the more exposed Lagoa Santa karst located elsewhere in the area between Velhas River and Mata Creek (Fig. 1).

The thickness of soil cover has a direct influence on the groundwater pattern. Subsurface dissolution features are well developed, as suggested by a large number of voids detected by borehole data.

Two karst springs were studied in detail, and both are located close to the quarry. Cafundó spring lies only 250 m from the open pit mine, while Tadinho Spring is located 1,600 m further south. A third spring, Carrapato spring, was not studied due to access restrictions. From tracer tests, it was possible to infer that only 20 % of the rhodamine WT mass injected in a swallow hole located upstream in the quarry was detected at Tadinho spring, and none was detected at Cafundó spring. In the same way, only 20 % of the total discharge injected (100 l/s) was measured at Tadinho spring, with no variations at Cafundó spring.

This study intends: (i) to characterize the karst hydrogeology in the vicinity of ECL limestone quarry, focusing on the identification of hydrogeological constraints responsible for controlling the groundwater behavior in the catchment area of the Tadinho and Cafundó karst springs; and (ii) to define the aquifer compartments, based on systematic discharge monitoring of the springs and boreholes.

The study area involves carbonate and pelite rocks of the Bambuí group, deposited over a gneiss-migmatite basement. Stratigraphy comprises four major sequences. The topmost Serra de *Santa Helena Formation* contains pelite rocks represented by siltites, mudstone with some sandstone, as well as carbonate lenses of argillaceous limestone and fine-grained calcarenites. These rocks show well developed weathering and restricted groundwater circulation. This domain is classified as an aquitard.

The *Lagoa Santa member* of the *Sete Lagoas Formation* contains pure micritic limestone with intercalations of siltite, breccia, stromatolites, and mylonitic zones. Subhorizontal laminations displaying calcite and quartz lenses are frequent, and represent aquifers with well-developed conduit flow.

The *Pedro Leopoldo member* of the *Sete Lagoas Formation* contains calcisiltites, argillaceous limestone with sparitic to micritic textures, stromatolites with some intercalations of fine-grained calcarenites, and milonites. Shearing zones show graphite and quartz veins. There is an occurrence of a basal marble. Overall, this sequence shows limited groundwater flow, as indicated by a fractured aquifer with karstified horizons.

The *Basal complex* contains migmatites and granitoids and is classified as aquiclude.

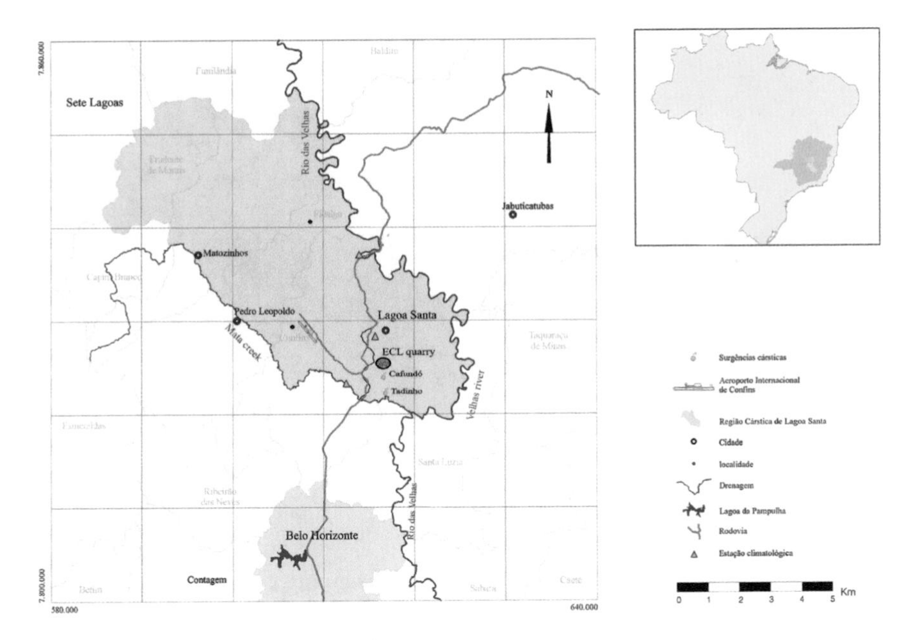

Fig. 1 Location of the study area and springs

2 Hydrogeology

Tadinho spring represents the most important outlet in the area, with an average discharge of 70 l/s. It is located in the southern portion of the study area, 400 m away and 25 m above the Mata creek river bed.

A discharge measurement was performed through a calibration curve and staff gauge, based on current meter measurements with a maximum error of 5 %. Curve calibration included discharge measurements using Rhodamine WT.

Discharge measurement in Cafundó spring occurred between May 2002 and July 2003 and relied upon periodic readings of a Parshall flowmeter located 100 m downstream.

2.1 Tadinho Spring

Discharge variations at Tadinho spring were subjected to systematic measurement over a 14-month period (Fig. 2), to determine the water input into the spring catchment area.

Following tracer confirmation of a connection between the swallet—that receives the dewatered volume from the mine—and the Tadinho spring, a 20 % increase in discharge was observed. Specifically, of the 100 l/s pumped to the

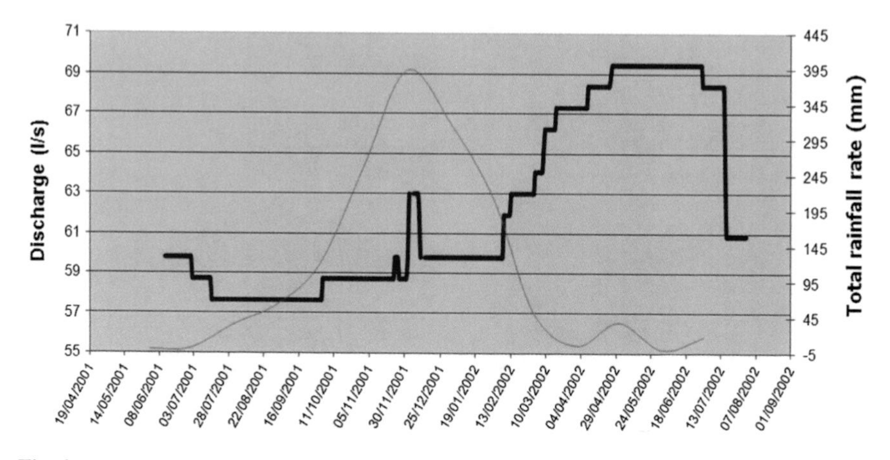

Fig. 2 Hydrograph for Tadinho spring, based on mean monthly discharge (*blue line*). The rainfall rate over the same period is shown in *pink*

swallet, only 20 l/s reached the spring, with the remaining volume flowing towards undetermined groundwater routes.

Due to this artificial discharge increment, the pumped volume was not considered during the construction of the hydrograph.

The interpretation of the hydrograph shows a 4-month delay between rainfall peaks and changes in discharge. During the dry season of 2001 (June–August), the discharge was quite stable, displaying low values ($\sim$57 l/s). During the same interval in the following year, the discharges were higher ($\sim$80 l/s), finally decreasing to a minimum of 58 l/s in mid-July.

These variations in the basal flow discharge are controlled by precipitation, because during the hydrological year 2000–2001, the total precipitation was abnormally low, reaching only 888 mm between October and June; in the same time period in the following year, the total rainfall was 1,354 mm.

Discharge variations are characterized by a period of low discharge values (57 l/s) that start at the beginning of the measurements (June 2001) and last approximately until the end of December 2001, reflecting not the onset of the rainy season between September and October. Discharge increases only in January 2002, when it experiences a linear increase until reaching a somewhat constant discharge at approximately 82 l/s for about 3 months, from mid-April to mid-July 2002. A 4-month delay between rainfall peaks and spring discharge is observed in the period 2001–2002, similar to what is observed between the rainfall minima and the beginning of the period marked by discharge decrease. The start of the period of discharge increase occurs when there is a decrease in precipitation levels, showing a lack of direct correlation. Maximum discharge is association with the end of the rainy season and decreases rapidly afterwards.

It is possible to infer that the period between December and March is characterized by discharge rise, due to increased aquifer recharge. Depending on rainfall

Table 1 Types of springs according to discharge correlations (after Worthington 1991)

Types of springs	Ratios Q_x/Q_n	Periods with $Q > 0$
Full-flow	High	Constant
Underflow	Low	Constant
Overflow	∞ ($Q_n = 0$)	Constant/periodic
Base flow and overflow (underflow–overflow)	Low to ∞	Constant/periodic

Table 2 Hydrological parameters for Tadinho spring

Discharge characteristics	Discharge values
Maximum discharge (Q_x)	82.12 l/s
Minimum discharge (Q_n)	47.27 l/s
Mean discharge (Q_m)	64.5 l/s
Ratio (Q_x/Q_n)	1.7

levels, the decrease starts between May and July and continues until November, with a low, but constant, base flow discharge.

According to Worthington (1991), the simplest procedure to analyze discharge regime of karst springs involves the correlation between maximum annual discharge (Q_x) and minimum annual discharge (Q_n) taking into account the time period over which the measurements are performed ($Q > 0$). Four main types of springs can be defined (Table 1). Hydrological parameters for Tadinho spring are shown in Table 2.

Based on the classification by Worthington (1991), the Tadinho spring can be characterized as an *underflow spring* with constant discharge ($Q > 0$). However, it also belongs to the subtype *losing or high-stage underflow*, with a regime associated with the loss or absence of fast flow and discharge both constant and smaller than the discharge released by the catchment zone, suggesting loss or divertion of water along the route.

Several studies refer to the high level of complexity associated with the water budget of karst systems (Milanovic 1976; Bonacci 1987; Worthington 1991). Nevertheless, one of the situations described by Worthington (1991) fits well with the Tadinho karst system, in which the hydrogeological pattern can be associated with the existence of a constriction upstream from the spring.

Assuming the existence of a constriction, the low or high stage underflow of this aquifer is able to stabilize the water level upstream from the constriction until an overflow route is activated.

Figure 3 displays a linear log-normal pattern, in which the components associated with the recession are controlled by a constant ($Q_n = q_u$) or slightly decreasing ($Q_n > q_u$) trend.

Tracer tests have demonstrated that there is no hydrogeological connection between the swallet that receives the pumped water and Cafundó spring. However, it is possible that the distinct Cafundó system may work as an overflow route

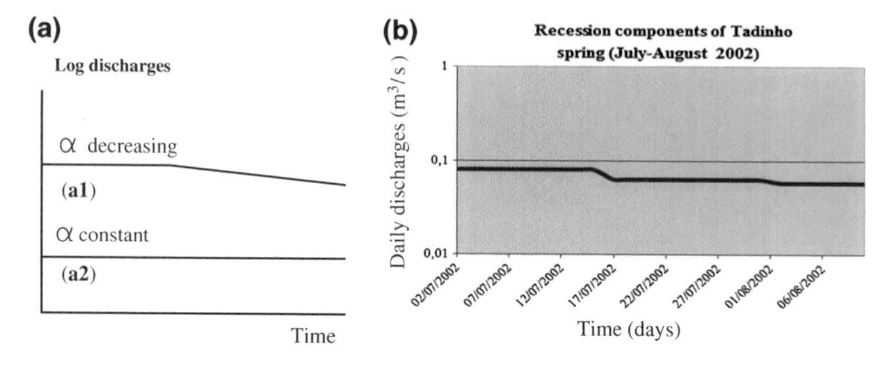

Fig. 3 **a** Recession curve pattern for Tadinho spring, according to the classification of Worthington (1991), where *a* is the depletion exponent. **b** Depletion components from Tadinho spring

outlet, in a distributary pattern in which the smaller Cafundó spring is responsible for partially draining Tadinho spring high levels. The catchment area of Cafundó spring comprises soil-covered karst, in which delayed storage may enable a constant supply of water, making this spring a permanent one with an overflow component because it occurs at a higher elevation than Tadinho spring.

At Tadinho spring, the periods of high discharge are characterized by variations that depend on rainfall intensity as well as the geometry and porosity of the aquifer, both in the vadose and phreatic zones. Discharge can decrease very quickly over the period of few days, as observed by the end of July 2002, suggesting the existence of fast flow routes in this underflow spring.

Water level analysis in piezometers in the same system shows that the difference in recession time is due to recharge time, effective porosity, and rainfall intensity (Pessoa et al. 2007).

However, according to Torbarov (1976), basic hydraulic parameters in any karst aquifer, such as permeability and effective porosity, vary in time and space. During the recession period, these parameters also depend on other variables, such as the precipitation regime and its surface distribution, and on the water level at the moment when the recession curve starts.

Thus, the fastest discharge decrease is related to the largest volume of rainfall, which is able to saturate a larger portion of the aquifer comprised of conduits enlarged by dissolution, as well as associated fractures. Smaller rainfall periods are only able to saturate the conduits. A rapidly decreasing recession curve reflects water delivered from these larger conduits, while a slower decrease is associated with the fractured media.

According to Worthington (1991) when the discharge of a spring is higher than the discharge related to its catchment area, it is characterized by a decreasing recession coefficient (a) (Fig. 3a1). However, the flow regime at Tadinho spring is characterized by a discharge equal to or less than its aquifer catchment, suggesting the presence of constant recession coefficient (Fig. 3a2).

2.2 Cafundó Spring

Cafundó spring was monitored over a 15-month period, from May 2002 to July 2003. Daily discharge measurements are presented in the hydrograph (Fig. 4).

Cafundó spring displays a regime quite different from Tadinho spring, especially during the dry season. This portion of the hydrograph is characterized by constant recession coefficients, but with a discharge that presents a decreasing component over time, except for a small portion related to fast flow (a = 0.02) before the inception of base flow.

Analyzing the rainfall pattern, the responses in discharge variation tend to be rapid over the entire monitored period, showing a good correlation between precipitation and discharge. Because the Cafundó spring catchment area is quite small (<1.0 km^2) the fast response may be due to water that quickly infiltrates in the limestone outcrops close to the quarry, as shown in Fig. 4.

Cafundó spring hydrological regime tends to follow the rainfall peaks in a direct way, which is in agreement with the basal discharge pattern when analyzed using the recession criteria suggested by Worthington (1991).

Figure 5 shows the recession trends for Cafundó spring, and Table 3 display the coefficients (a) for this spring. If one considers the pattern shown in Fig. 5 and the proposed classification of springs according to the recession coefficient (Worthington 1991), Cafundó spring can be classified as an underflow–overflow spring because it lies topographically between the higher Carrapato spring (upstream) and the major discharge zone represented by the lower Tadinho spring (downstream).

The coefficient values show considerable variation. Although they may suggest well-developed karst aquifers, the presence of a quarry upstream from the spring may exert major control over its hydrological behavior. The spring discharge pattern remained approximately constant throughout the monitoring period.

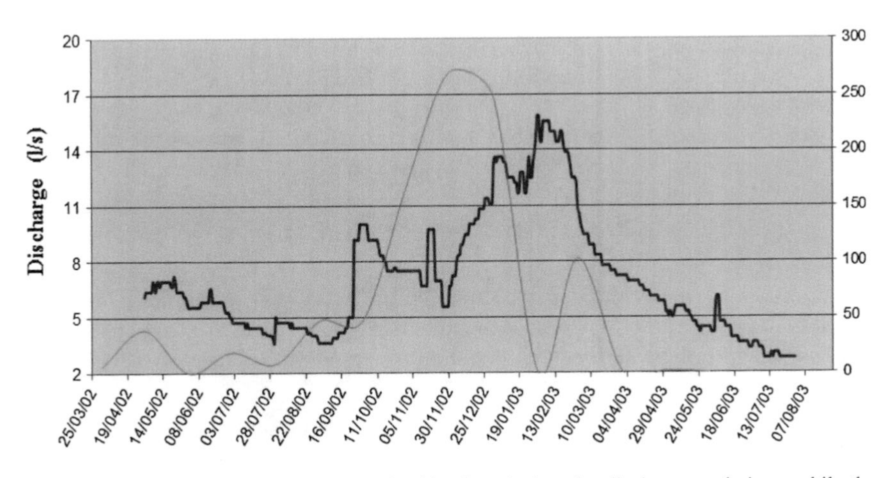

Fig. 4 Hydrograph of Cafundó spring. The *blue line* depicts the discharge variations, while the *pink line* shows the precipitation

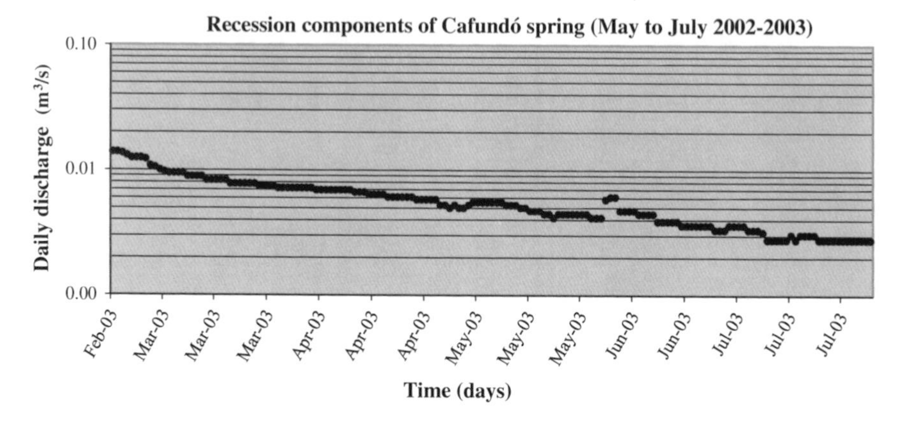

Fig. 5 Recession hydrographs for Cafundó spring

Table 3 Recession coefficients (a) based on discharge hydrograph for Cafundó spring

Measurement interval		Time (days)	Q_0 (m^3/s)	Q_f (m^3/s)	a (day^{-1})
1	10/05/02–04/09/02	117	0.0069	0.0036	0.006
2	09/02/03–06/03/03	25	0.0156	0.0108	0.020
3	07/03/03–31/07/03	146	0.0106	0.0028	0.009

Q_o initial discharge; Q_f final discharge

Table 4 Discharge hydrological parameters for Cafundó spring

Discharge characteristics	Discharge values
Maximum discharge (Q_x)	16.7 l/s
Minimum discharge (Q_n)	2.5 l/s
Mean discharge (Q_m)	7.6 l/s
Ratio (Q_x/Q_n)	7

Although the north pit mine is important, it does not play a major role in the variation of discharge. Other variables must be considered, such as the relief over the catchment area, variable thickness of the soil cover, and especially the configuration and location of the fracture system near the south pit mine. The highest recession coefficient (a = 0.02) is related to the fastest discharge over a period of 25 days, which according to Milanovic (1976), may be due to accumulated water in interconnected fissures and to joints enlarged by dissolution. A similar pattern was found by Karmann (1994) as part of the base flow fed by waters accumulated in the interconnected joints and enlarged karstified fractures in the Pérolas–Santana system in southern Brazil.

However, the relationship between the maximum and minimum discharge (Q_x/Q_n) at Cafundó spring is not very high. The main parameters related to discharge measurements are shown in Table 4, and recharge areas are displayed in (Fig. 6).

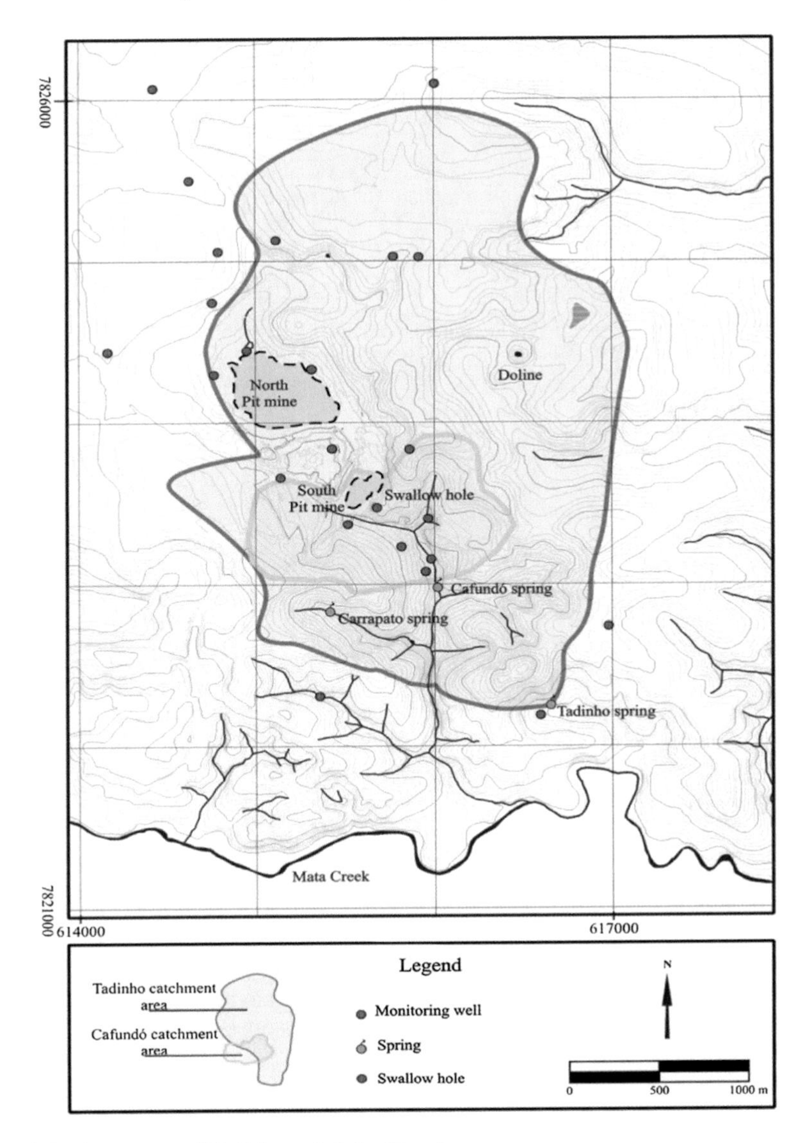

Fig. 6 Recharge area of Tadinho and Cafundó springs

3 Hydrochemistry

Hydrochemical monitoring shows a small difference in concentrations for hardness, bicarbonate, calcium, and specific conductance for both springs. However, considering the calcite saturation index, it was observed that the groundwater at Cafundó and Carrapato springs is undersaturated with respect to calcium carbonate, mostly due to a difference in pH, while for Tadinho, the SI indicates that

Table 5 Hydrochemistry results for 12 sampling sites (January/2002–June/2004)

Parameters *Δ* ¦ *Max* *Min* ¦ *vc*	**Recharge zone** ***ZONE 1*** ***(n: 48)***	**Intermediate flow** ***ZONE 2*** ***(n: 23)***	**Discharge zone** ***ZONE 3*** ***(n: 24)***
Total hardness (mg/L) related to $CaCO_3$	44.5 ¦ 124.5 8.6 ¦ 51.4	179.3 ¦ 214.5 112.7 ¦ 13	151.5 ¦ 155.8 143.4 ¦ 7
Specific conductance (µS/cm)	120 ¦ 333 21 ¦ 67.6	*328* ¦ *680* *60* ¦ *48.6*	254 ¦ 439 65 ¦ 33.7
Alkalinity (mg/L HCO_3^-)	*46* ¦ *120* *8* ¦ *51*	179.8 ¦ 220 111 ¦ 12.2	149.3 ¦ 192 131 ¦ 7.2
Ca^{++} (mg/L)	15.7 ¦ 44.9 3.2 ¦ 47.6	61.8 ¦ 78.9 42 ¦ 14	55.5 ¦ 72 62.6 ¦ 9.8
Mg^{++} (mg/L)	1.3 ¦ 9 0.09 ¦ 89.7	6 ¦ 18.9 0.55 ¦ 81.5	3.2 ¦ 10 0.67 ¦ 64.8
Ca/Mg (molar)	14.9 ¦ 71.9 1 ¦ 82.5	15.8 ¦ 79.5 1.4 ¦ 94	17.9 ¦ 61.4 2.9 ¦ 71.7
Temperature (°C)	23.5 ¦ 28.4 21 ¦ 4.9	23.38 ¦ 26.2 21 ¦ 4.2	22.7 ¦ 24.2 20 ¦ 4.5
pH	6.25 ¦ 7.49 5.29 ¦ 7.3	7.32 ¦ 7.97 6.77 ¦ 4.6	7.29 ¦ 7.97 6.50 ¦ 4.2
Saturation Index calcite	−2.63 ¦ −0.58 −4.92 ¦ 31	−0.17 ¦ +0.51 −0.89 ¦ *	−0.29 ¦ +0.40 −1.23 ¦ *

vc variation coefficient; *not calculated

the water is supersaturated with respect to calcium carbonate. The periods of higher discharge at Tadinho spring are related to increased water aggressiveness.

Following 4 years of sampling (2001–2004), Tadinho spring shows denudation rates of 22.5 mm/ka, which is similar to the values found in non-covered karst aquifer systems in Brazil.

As shown by Langmuir (1971), differences in the saturation levels and the alkalinity of bicarbonates (and, most likely, the partial pressure of CO_2) sampled between the karst resurgences suggests basic differences in the chemical evolution of its waters. The sampling sites of Carrapato, Tadinho, and Cafundó refer to the discharge zones of the system, while the other sites refer to recharge areas and the intermediate zones of the flow system (Table 5).

4 Final Considerations

The main features of the Tadinho and Cafundó spring system can be described as follows:

- because the geometry and total porosity of the aquifer in both vadose and phreatic zones are constant, the varying volumes of water drained by the karst system as shown by the recession data are due to the precipitation patterns;

- the period that displays higher discharge will depend not only on the amount of rainfall in the humid period but also on the thickness and intensity of saturation of the aquifer below the epikarst zone;
- decreasing discharge is associated (as in the period July–August 2002) to the fast dewatering of the vadose portion of conduits;
- slow decrease in discharge is most likely associated with a less intense saturation of bedrock, in which the smaller rainfall levels were only able to promote small rises in the hydraulic head of the aquifer. This volume is released in a diffuse way, resulting in slower and more prolonged decreasing discharge trends;
- following Worthington (1991), the hydrological parameters indicate that Tadinho spring can be classified as an underflow spring containing constrictions due to conduit obstruction, resulting in constant hydraulic heads and discharge over much of the year. The discharge is characterized by constant and very low recession coefficients, except during periods in which the spring discharge is higher than the discharge associated with its catchment area, implying a temporary decrease in the recession coefficient;
- Cafundó spring has a different pattern than does Tadinho spring. Due to its location over the total catchment area, it represents a possible overflow system, with the highest recession coefficient being part of the base flow fed by waters accumulated in the interconnected joints and conduits that exist around the south pit mine border;
- the values of recession coefficient for Cafundó spring are greater than Tadinho spring, and the differences are related to its small catchment areas and the time dependence of rainfall rapid recharge at sites connected with its northern portions, which includes south pit mine;
- hydrochemical data show a good temporal-spatial correlation, with the parameters being controlled by climate seasonality, aquifer geometry, and available water sources.

References

Bonacci O (1987) Karst hydrology, with special reference to the dinaric karst. Springer, Berlin 184p

Karmann I (1994) Evolução e dinâmica atual do sistema cárstico do Alto vale do Rio Ribeira de Iguape, sudeste do Estado de São Paulo. São Paulo: Instituto de Geociências da Universidade de São Paulo. 228p (Doctorate Thesis)

Langmuir D (1971) The geochemistry of some carbonate ground waters in central Pennsylvania. Geochimica et Cosmochimica Acta 35:1023–1045

Milanović PT (1976) Water regime in deep karst. Case study of the Ombla spring drainage area. In: Yevjevich V (ed) Karst hydrology and water resources. Water Resource Publications, Fort Collins, pp 165–191

Pessoa PFP, Loureiro CO, Auler AS (2007) Using borehole data to assess the dynamics of the epikarst zone in a mantled karst area, Lagoa Santa, Brazil. In: Ribeiro L, Chambel A, Condesso de Melo MT (eds) XXXV Congress of International Association of Hydrogeologists—IAH, abstract book. International Association of Hydrogeologists, Lisbon

Torbarov K (1976) Estimation of permeability and effective porosity in karst on the basis of recession curve analysis. In: Yevjevich V (ed) Karst hydrology and water resources. Water Resource Publications, Fort Collins, pp 121–136

Worthington SRH (1991) Karst Hydrogeology of the Canadian Rocky mountains. Hamilton, 227 p. (Ph.D. thesis, Department of Geography/Mc Master University)

The Karst Hydrostructure of the Mount Canin (Julian Alps, Italy and Slovenia)

L. Zini, G. Casagrande, C. Calligaris, F. Cucchi, P. Manca, F. Treu, E. Zavagno and S. Biolchi

Abstract The Mt. Canin massif, from a hydrogeological and geomorphological point of view, is a unique structure, being an independent part of the Italian Julian Alps (north east Italy) bounded on all sides by impressive karst springs. Extensive outcropping limestones go from the top (2587 m a.s.l.) to the bottom of the valleys (about 500 m a.s.l.) creating an hydrostructure subdivided between two countries originating two transboundary watersheds: the Mediterranean one to the South and the Black Sea to the North. The aim of this paper is to define the dynamic and the characteristics of the groundwaters and to identify the superficial and deep watersheds in order to elaborate the aquifer vulnerability.

1 Introduction

From the political point of view, Canin massif is divided by the national border line (Italy-Slovenia) and even if it is not any more an iron curtain as it was in the past, it separates languages, laws, and different cultures (Fig. 1). Despite this, for the waters it is not the same, in fact the hydrogeological watershed is not coincident with the catchment, but it is on the Italian side (Casagrande and Cucchi 2007a, b).

The highly karstified carbonatic hydrostructure (mainly Triassic Dachstein) is well-bounded, in the valleys, by impermeable layers (Eocene Flysch) or rather less karstifiable ones (Triassic Dolomia Principale) (Casale and Vaia 1972).

The massif, strongly and diffusely karstified is recognized as very important geosite due to the presence of impressive epigean and hypogean glaciokarst and

L. Zini (✉) · C. Calligaris · F. Cucchi · F. Treu · E. Zavagno · S. Biolchi
Dipartimento di Matematica e Geoscienze, Università di Trieste, Trieste, Italy
e-mail: zini@units.it

G. Casagrande · P. Manca
Direzione Ambiente ed Energia - Regione Friuli Venezia Giulia, Trieste - Udine, Italy

B. Andreo et al. (eds.), *Hydrogeological and Environmental Investigations in Karst Systems*, Environmental Earth Sciences 1,
DOI 10.1007/978-3-642-17435-3_24

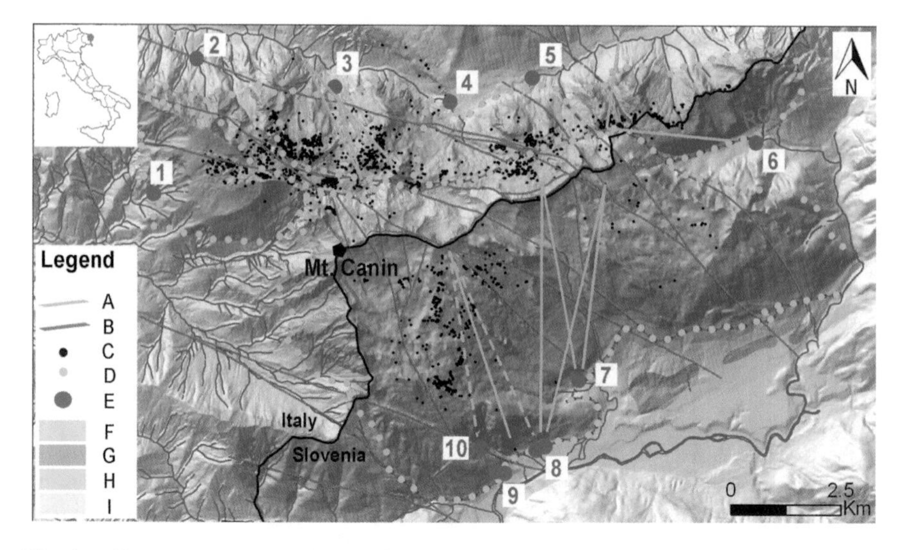

Fig. 1 **a** Tracer tests connected sites (*dotted*, the uncertainties); **b** main faults and overthrusts (RC, Val Resia—Val Coritenza overthrust); **c** caves; **d** groundwater watersheds; **e** main springs (*1* Fontanone del Mt. Sart, *2* Tamaroz, *3* Fontanon di Goriuda, *4* Sella Nevea, *5* Rio del lago, *6* Mogenza, *7* Glijun, *8* Zvikar, *9* Bocic, *10* Boka); **f** Quaternary deposits; **g** Flysch; **h** Calcari del Dachstein; **i** Dolomia Principale

karst landforms (Cucchi and Finocchiaro 2009). For this reason, since 1996 it was included in the Julian Prealps Nature Park and it is frequented all over the year by tourists and speleologists.

More than 2,000 caves, explored by speleologists since more than 50 years, are present inside the massif (Casagrande et al. 1999). The explorations permitted to discover very extensive and intricate networks of meaningful galleries developed for more than 60 km with pits of exceptional size (Cucchi et al. 2002). Six caves are deeper than 1,000 m. The massif also includes the deepest single shaft of the world, deep 643 m.

The hydrodynamic of the aquifer is highly dependent from the geological and structural settings, characterized by a very fast circulation, highlighted also by a quick increase in the spring discharge where values for all the springs, considering also the effluent character of the rivers, are reaching more than 500 m^3/s (Casagrande et al. 2011). This behavior can be identified also through geochemical analyzes.

To reach the aim of the present research, a geomorphological and hydrogeological description of the area of the massif is given, analyses were performed on the lithological and structural conditions that gave rise to the epigean and hypogean karst morphologies. A further review has been also realized on the hydrochemical characteristics, on the groundwaters behavior and on the spring systems.

2 Karst Geomorphology

Mt. Canin massif is one the most beautiful example of karst environment in the world recognized as geosite. Surface karst landforms as hypogean ones attracted speleologists and researchers since nineteenth century discovering and studying more than 2000 caves included the deepest vertical shaft of the world, 643 m.

The stately karst plateau that we are able to enjoy nowadays is the result of the glacial shaping action that till several thousands of years ago was covering the whole massif. Glacial morphologies, result of the last ice age, ended about 10,000 years ago, are still well visible on the steep slopes and in the valleys as the wide basin where Sella Nevea is (Cucchi et al. 2000a, b). On the northern side of the Canin Mt. are well visible the signs left by the glacial erosion: the clear color change is meaningful of the actual ice retreat. The plateau is the result of two jointed actions: a physical one due to the glacial erosion and the chemical one due to the carbon dioxide present in the water and dissoluting the carbonate rocks. The plateau infact is worldwide known for its rich karst superficial morphologies responsible of the huge extension of the deep caves favored by the cool-temperate climate conditions where the solvent potential is higher (Ford and Williams 2007). On the carbonatic surface, it is possible to observe all the different shapes typical of the high-mountain karst environment. Among the macro-landforms were recognized wide collapse dolines and solution ones, sinkholes, crevasses and elongated depressions identifiable also through GIS techniques (Telbisz et al. 2011). A huge variety of small marvelous micro-karst features as small solution pits, grooves and runnels termed karren and kamenitze can also be observed. Morphological features are, in this high altitude plateaus environment, particularly rough and intense due the vegetation free surface. Exogenous agents are contributing also in the considerable differences in elevation among cave entrances, 90 % of them are in fact between 1,600 and 2,100 m a.s.l. in correspondence of lower steepness plateau acting as wide absorption areas. At lower elevations, cave entrances drastically decrease due to slope steepness and high-vegetation degree. To this is necessary to add that on the Italian side between 800 and 1,000 m.a.s.l. and on the Slovenian one at 400 m a.s.l. dolostones and silico-clastic rocks are outcropping. This variation of hydraulic conductivity favors the presence of spring cave entrances.

3 Hydrogeological Framework and Hydrochemical Characteristics

Canin massif represented always a challenge for the researchers studying the karstic hydrostuctures. The massif represents a highly karstified structure with quick transit times. From the plateau, waters reach the springs in few hours with high increasing discharge values linked to the precipitations. Several tracer tests permitted to identify the different recharge sectors.

Canin Mt. represents a unique structure, well defined and separated within the Julian Alps being surrounded by huge springs where wide limestone outcrops to the valley floors (Calcari del Dachstein) (Ponton 2011). Precisely, the hydrostructure is bounded by contacts with impermeable lithologies (Flysch) or with low karstifiable ones (Dolomia Principale). At high altitude, single persistent discontinuities are very important not only from a geomorphological point of view, but also from the hydrogeological one, being the main responsible of the permeability and hydraulic conductivity for the unsaturated zone (Cucchi and Vaia 1987). The structural constrains are not only the vertical ones, but meaningful is the backthrust (Val Resia–Val Coritenza) of the Dolomia Principale over the Calcare del Dachstein that divide the hydrostructure into two different subunits. The northern one (comprehending the main part of the Italian side) is dipping toward north and the southern one (higher northern Italian side) that south dips have direct consequences to the groundwater flow direction. The percolation zone is the highest part of the hydrostructure with a very fast drainage Casagrande and Cucchi (2007a, b), Casagrande et al. (2011). The net is heavily influenced by the local geological and structural characteristics. Vertical features are prevailing which enlargement is favored by discontinuities, circulation, and high karstifiable layer thicknesses. These drainage systems, in few hours, convey the waters into the depths of the massif forming sudden streams and waterfalls. The saturated zone (or phreatic zone) instead, is made of always-flooded highly effective drainage systems and moderate effective ones. Sited at the toe of the massif, the phreatic conduits have a prevailing horizontal development (acquired during speleological surveys), which conveys the waters to the spring areas. The distribution and the shape of the systems are relied to the Dolomia Principale and to the overthrust forming a permeable limit (Fig. 1). The hydrostructure has 10 perennial springs and some others have an intermittent character (Cucchi et al. 2000a, b). In the northern mountainside, the geological and structural situation take to the occurrence of free draining springs, while at the toe of the southern side of the massif springs are mainly dammed. Some springs are outflowing from explorated caves as Fontanone del Monte Sart, Fontanon di Goriuda, Tamaroz, Boka, Srnica, Bocic Jama, and Mala Boka. In some other cases, waters are flowing out from opened fractures, between layers or from the quaternary deposits that during rainy seasons can generate a real spring belt. In the Italian territory only Fontanon di Goriuda spring was relevant, but, during the latest explorations in August 2012, a new outflow was discovered and named Tamaroz (from the speleologists of the S.A.S. Trieste). Also springs have a high discharge variability with very different values between low water regime and high ones. This is due mainly to the heavy rainfalls (estimated to be 3,000 mm/y) in the area that, connected to the high karstification of the massif. The total discharge evaluation is extremely complex also due to logistic difficulties, to the presence of overflow springs and to the effluent character of the streams. Casagrande and Cucchi (2007a, b), estimated that, during low water regime, the outflow from the Canin hydrostructure is of about 8 m^3/s increasing up to 500 m^3/s during high water regime. The min and max discharges evaluated for the main springs are: Fontanone del Monte Sart (30 l/s–>10 m^3/s), Tamaroz (30 l/s–>1 m^3/s), Fontanon di Goriuda

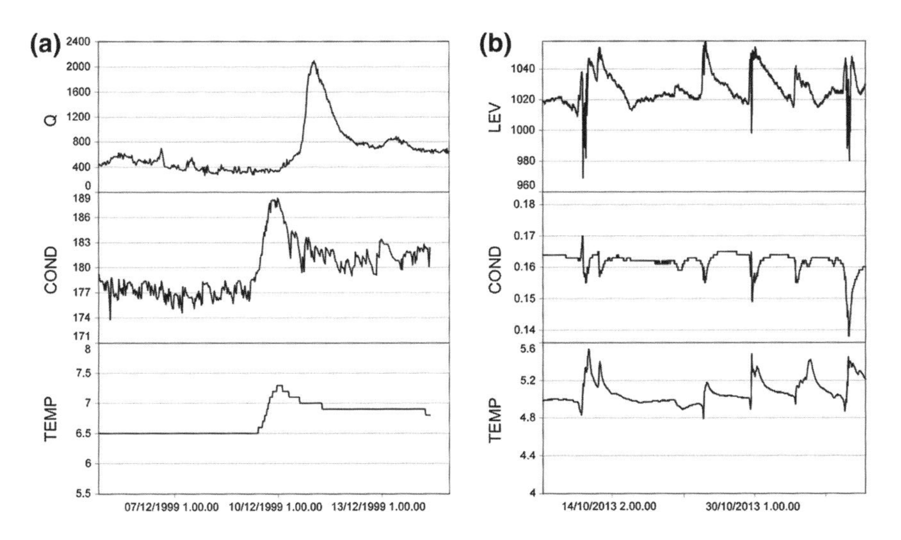

Fig. 2 **a** Monitored parameters (TEMP [°C], COND [μS/cm] and Q [l/s]) for the Glijn spring; **b** monitored parameters (TEMP [°C], COND [mS/cm] and LEV [mm]) for the Tamaroz spring where TEMP is the temperature, COND is the EC, Q the discharge and LEV the water level

(10 l/s–10 m^3/s), Sella Nevea springs (10 l/s–2 m^3/s), Rio del lago springs (0.5 l/s–0.2 m^3/s), Moznica (0.25 l/s–>25 m^3/s), Glijun (250 l/s–>80 m^3/s), Zvica (150 l/s–0.8 m^3/s), Bocic (200 l/s–>0.5 m^3/s), Boka (25 l/s–>40 m^3/s) and Srnica (0 l/s–2 m^3/s). A quarter of these resources are used for hydroelectric purposes and only a small amount for drinking ones. To characterize the outflowing waters, three springs (Tamaroz, 25.01.13–13.11.13; Glijun and Zvikar, 5.11.99 – 5.9.00) were monitored through devices that measure in continuous, water level, temperature and Electrical Conductivity (EC). On some of the emerging waters, over the years, geochemical analyses were done. In this paper, to understand the groundwater circulation, two main springs are considered, one on the Italian side, Tamaroz, and one on the Slovenian side, Glijun. The monitoring surveys realized in the Glijun and Tamaroz springs (Fig. 2a, b) are representative of the two different flow behavior present in the south and north sides. Almost all the Canin hydrostructure is characterized by a short and quick circulation where the groundwater rising can be detected only few hours after the rainfalls with discharge values one order of magnitude greater. The best example of this behavior is represented by the cave springs of Fontanon di Goriuda and Boka that by wide conduits are linked to the recharge plateau areas. As result of this, the geochemical analyses confirm that generally, Canin groundwaters have a low mineralization, peculiar of the quick circulation. Absolute major ion values, EC and hardness are confirming the thesis: most part of the springs have a Ca value between 26.4 and 59.3 mg/l, Mg content between 0.61 and 13.1 mg/l and a variable HCO_3 value between 71 and 200 mg/l (Cucchi et al. 2000a, b). The other ion values are subordinate. During floods, bicarbonates drop due to the rapid disposal of the recharged waters having a short residence time, while a sulfate increasing is observed. The peak represents a marker

in the arrival of neoinfiltration waters. From the realized chemical analyses, all the waters present a calcium bicarbonate facies but the springs in the northern side show a Mg enrichment due to the dolomite rock crossing.

Despite the general context of a very fast circulation in karst conduits, some springs testify the possibility of reserves formation (Cucchi et al. 2000a, b). This is the case of the southern springs and especially of Glijun one, representing the most important spring of the whole massif. It is the only one having a constant discharge value of about 500 l/s (Fig. 2a). Studying its hydrographs, it is possible to understand that the circulation behind it is very wide and complex, characterized by a high effective drainage system where well developed karst conduits are linked to open fractures and only secondly to slightly karstified areas being an important renewable resource. This behavior is clearly identifiable with the spring hydrograph analysis. The increased water level value while flooding, linked to the synchronous EC and temperature fluctuation, is attributable to the so called "piston effect" due to the resident waters mobilization replaced by neoinfiltration ones. It is possible to note that after the event, the parameters are not going back quickly to the base level but showing a certain inertia due to the presence of huge renewable reserves. This behavior is not the same in the northern sector (Fig. 2b). The hydrograph representing Tamaroz spring highlight a prompt water level increase while raining with a consequent increasing in temperature and a decreasing in EC values. After the peak, the curve go quickly back to the base level highlighting a highly effective drainage system, typical of an aquifer where a huge water reserve is not present (Galleani et al. 2011). Stimulated by the charming option to link with a unique conduit the plateau to the valley springs, over the years, since 1968, several tracer tests were realized (Casagrande et al. 2011). Initially qualitative tests permitted to identify the connections between hypogean waters and the related springs. Only during the last 10 years, the use of captors and laboratory analyses took to semiquantitative evaluations highlighting a groundwater complex circulation. Most part of the tests were realized tracing at the bottom of the caves where flowing waters were present. Through tracer tests (Cucchi et al. 1997) it has been possible to identify three different sectors. The northeastern one that drains the waters towards Rio del Lago spring, the northwestern one that drains the waters towards Fontanon di Goriuda spring and the central and southern draining waters towards Glijun and Boka springs.

4 Conclusions

Through all the investigations realized arised that an important role in the hydrodynamics is played by the Val Resia–Val Coritenza backthrust that separates the hydrostructure in two different sectors. The top of Mt. Canin ridge do not correspond to a water divide effectively represented by the back thrust itself demonstrated by the traced waters.

The Mt. Canin massif has been studied since the end of nineteenth century when speleologists and researchers started to identify and connect the hundreds of caves present in the area. The hydrostructure so became an open-air laboratory where physical and geochemical parameters were collected and analyzed. As the result of the studies, it was defined that an important role in the hydrodynamics is played by the Val Resia-Val Coritenza backthrust that separates the hydrostructure in two different sectors. The top of Mt. Canin ridge infact, do not correspond to a water divide effectively represented by the back thrust itself. Traced waters in the Italian caves outflow in the Slovenian springs. Both sectors are intensively karstified, but thanks to the geological settings, in the southern side more water reserves are present than in the northern one as highlighted by the different behavior recorded by the more representative springs. The transition between Dolomia Principale dolostone and Calcari del Dachstein limestones is gradual and can generate an undefined permeability limit. For its peculiar karstic characteristics, the transboundary Canin aquifer can be recognized as a high vulnerable one and its exploitation for hydroelectric and drinking purposes needs to be carefully evaluated in order to avoid any overexploitation and pollution.

References

Casagrande G, Cucchi F (2007a) L'acquifero carsico del Monte Canin, spartiacque tra Adriatico e Mar Nero. In: L'acqua nelle aree carsiche in Italia, a cura di F. Cucchi, P. Forti & U. Sauro, Mem. Ist. It. Spel., serie II, 19:57–64, Bologna

Casagrande G, Cucchi F (2007b) L'acquifero carsico del Monte Canin, spartiacque tra Adriatico e Mar Nero. In: L'acqua nelle aree carsiche in Italia, a cura di F.Cucchi, P.Forti & U.Sauro, Mem. Ist. It. Spel., serie II, 19:57–64, Bologna

Casagrande G, Cucchi F, Manca P, Zini L (1999) Deep hypogean karst phenomena of Mt. Canin (Western Julian Alps): a synthesis of the state of present research. Acta Carsologica 28(1):57–69

Casagrande G, Cucchi F, Manca P, Zini L (2011) L'idrostruttura del Monte Canin. In: Il fenomeno carsico delle Alpi Giulie. Mem. Ist. It. Spel. S.II 24:155–180

Casale A, Vaia F (1972) Prima segnalazione della presenza del Giurassico sup. nel gruppo del Monte Canin (Alpi Giulie). Studi Trentini di Sc. Naturali 49(1):14–26

Cucchi F, Finocchiaro F (2009) Geositi del Friuli Venezia Giulia. Regione Autonoma Friuli Venezia Giulia

Cucchi F, Vaia F (1987) Nota preliminare sull'assetto strutturale della Val Raccolana (Alpi Giulie). Gortania—Atti Museo Friul. Storia Nat. 8:5–16

Cucchi F, Gemiti F, Manca P, Semeraro R (1997) Underground water tracing in the east part of the karst of Canin massif (Western Julian Alps). Ipogea 2:141–150

Cucchi F, Casagrande G, Manca P (2000a) Le forme glacio-carsiche. In: Carulli GB, editore, (2000)—Guida alle escursioni. Società Geologica Italiana—80° Riunione Estiva, Trieste, 6-8 settembre 2000 (eds.) Edizioni Università di Trieste, pp 90–96

Cucchi F, Casagrande G, Manca P (2000b) Chimismo ed idrodinamica dei sistemi sorgivi del Monte Canin (Alpi Giulie Occidentali). Atti e Memorie della Commissione Grotte "E. Boegan", 37:93–123

Cucchi F, Casagrande G, Manca P (2002) Il contributo della speleologia alle conoscenze geologiche ed idrogeologiche del Massiccio del M. Canin (Alpi Giulie, ITA-SLO). Mem Soc Geol It 57:471–480

Ford D, Williams P (2007) Karst hydrogeology and geomorphology. Wiley, Chichester
Galleani L, Vigna B, Banzato C, Lo Russo S (2011) Validation of a vulnerability estimator for spring protection areas: the VESPA index. J Hydrol 396:233–245
Ponton M (2011) Note geologiche sulle Alpi Giulie Occidentali. In: Il fenomeno carsico delle Alpi Giulie. Mem. Ist. It. Spel. S.II, 24:57–79
Telbisz T, Mari L, Szabo L (2011) Geomorphological characteristics of the Italian side of Canin massif (Julian Alps) using digital terrain analysis and field observations. Acta Carsologica 40(2):255–266

Analysis of Groundwater Pathways by High Temporal Resolution Water Temperature Logging in the Castleton Karst, Derbyshire, England

J. Gunn

Abstract Temperature is the easiest water quality parameter to measure and a variety of robust, waterproof and relatively inexpensive temperature loggers are available that enable high-resolution data to be collected from remote locations. It has long been known that in regions where there is a seasonal variation in surface air temperature the water temperature at karst springs can be used to distinguish between those systems that are fed partly by sinking streams (wider annual temperature range) and those fed only by autogenic percolation water (narrower annual temperature range). However, there has been little analysis of short-term temperature changes within open and flooded conduit systems as well as at springs, which can provide significant additional information. The potential utility of water temperature logging is demonstrated in the Peak-Speedwell cave system, Derbyshire, England where high temporal resolution (2-min) water temperature data have provided information on internal geometry, residence times and velocities.

1 Introduction

The temporal variability of water temperature in a cave or at a spring provides an indication of the routes followed by water to the sampling point. Where there is a long transit time between surface and sampling point heat exchange with the surrounding rock will dampen any signal from the surface. However, as the transit time decreases there is less time for heat exchange and more of the surface signal will be present. Conduit diameter and the volume of water will also influence the extent of signal retention. Temperature is an easy parameter to measure and for many years mercury thermometers were used routinely to record spot water

---

J. Gunn (✉)
Limestone Research Group, School of Geography, Earth and Environmental Sciences, University of Birmingham, Edgbaston, Birmingham B15 2TT, UK
e-mail: j.gunn.1@bham.ac.uk

B. Andreo et al. (eds.), *Hydrogeological and Environmental Investigations in Karst Systems*, Environmental Earth Sciences 1,
DOI 10.1007/978-3-642-17435-3_25

temperature at sinking streams, springs, and, less commonly, in caves. The frequency of measurements was rarely better than daily and commonly much less regular, but the data were sufficient to draw simple conclusions on groundwater flow systems. For example, Pitty et al. (1979) sampled at approximately monthly intervals and used the temperature standard deviation to distinguish springs dominated by autogenic recharge from those dominated by allogenic recharge. Gunn (1981) measured water temperature at surface and underground sites in the Waitomo karst, New Zealand, weekly over a 59 week period, and used the data to distinguish between throughflow, subcutaneous flow, shaft flow and vadose seepage. Crowther and Pitty (1982) measured water temperatures of surface streams and groundwaters in three Malaysian karst area at 3- or 6-week intervals and found that they provided a sound basis for characterising groundwater flow and identifying groundwater components in surface streams. More recently, Beddows et al. (2007) used spot measurements of water temperature and specific electrical conductance as natural tracers to characterise the interactions occurring between the freshwater lens, the underlying saline water zone, ocean water and the Yucatan carbonate aquifer.

Significant constraints on earlier work were an inability to log temperature accurately at short intervals and an inability to measure temperature at more than a few centimetres below the water surface. However, there are now a variety of robust, relatively inexpensive commercially available waterproof temperature loggers that have accuracies of ± 0.5 °C or better and can record tens of thousands of data points thereby allowing high temporal resolution data collection. Surprisingly it seems that karst scientists are making little use of these loggers as there are few published papers that discuss temporal variability of water temperature, the most notable being those of Martin and Dean (1999) and Luhmann et al. (2011). Martin and Dean (1999) demonstrated that temperature, logged at 1.5- and 3-min intervals, provided a good natural tracer of short residence time groundwater in the Floridan karst. Luhmann et al. (2011) review previous work, provide a thorough discussion of processes and draw examples from an extensive monitoring programme in which the thermal patterns at 23 different springs and two cave streams in southeastern Minnesota were observed. All of their sites showed one or more of four temperature patterns: *Pattern 1 is an event-scale temperature fluctuation that occurs over hours or days; Pattern 2 is a seasonal fluctuation in temperature that is in phase with surface temperatures; Pattern 3 is a seasonal fluctuation in temperature that is out of phase with surface temperatures and Pattern 4 is long-term temperature stability, over timescales of weeks to years* (Luhmann et al. 2011, p. 327). Luhmann et al. (2011) also suggest that a distinction can be made between localised recharge (directly into a conduit), which is thermally ineffective and hence retains part of the surface signal, and distributed recharge which has moved through more thermally effective parts of groundwater system and hence retains little, if any, of the surface signal.

In this paper, I demonstrate the thermal complexity that is present in karst groundwaters using high temporal resolution water temperature data.

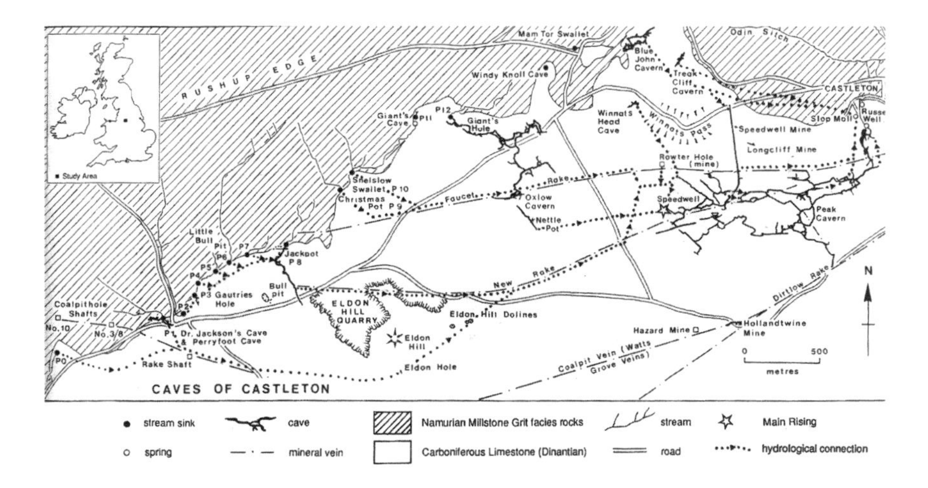

Fig. 1 The Castleton karst

2 Study Area

The Castleton karst (Derbyshire, UK) lies close to the northeastern boundary of the 'White Peak' district, has developed on Carboniferous limestones, and contains over 25 km of 'input' and 'output' cave linked by presently inaccessible conduits (Fig. 1). The input has two components, allogenic recharge from streams that sink at 15 separate points (catchment area ~ 5 km^2) and autogenic recharge onto a ~ 8.4 km^2 area in which the limestones are overlain by soils of loessic origin. Output is from three springs: Peak Cavern Rising (PCR), Slop Moll (SM) and Russet Well (RW) which combine to form the Peakshole Water (PW). Drainage is via the multi-level Peak-Speedwell cave system, which contains more than 16 km of active and relict cave passages, and has around 50 inlet streams, 20 of which flow from permanently water-filled 'sumps'. Cave divers have explored over 1000 m of these sumps one of which (Main Rising, MR) descends to a depth of at least 76 m. The modern Speedwell conduit has developed on a lower inception horizon than that on which the modern Peak conduit has formed and under low to moderate flows all allogenic water, augmented by some autogenic water, flows through the Speedwell conduit and is discharged by RW and SM, while PCR discharges water of solely autogenic origin. Under conditions of high recharge the lower part of the Speedwell conduit is unable to transport all the flow and water backs up and enters the Peak conduit at a number of discrete points, eventually emerging from PCR. Over 50 water-tracing experiments have revealed the broad outline of the underground hydrology (Figs. 1, 2; Gunn 1991) but there is a great deal of internal complexity that is yet to be fully understood and is discussed further in this paper.

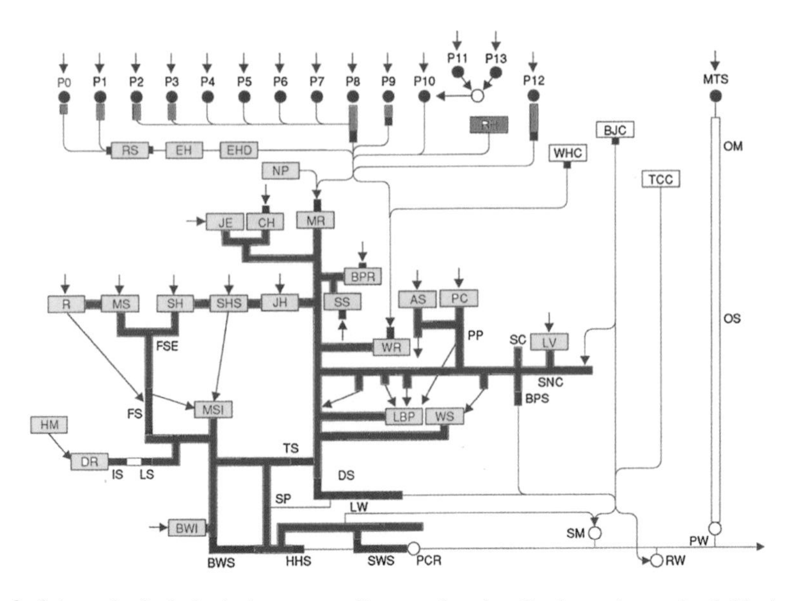

Fig. 2 Schematic hydrological systems diagram for the Castleton karst. [*red blocks* are accessible input cave; *blue blocks* are accessible output cave passage; *black blocks* are flooded conduit accessed by divers; *thin lines* are links proven by water tracing. For full key to sites see Gunn (1991)]

3 Temperature Logging

Table 1 lists the key characteristics of the five types of temperature logger deployed in the Castleton karst: Schlumberger Mini-Diver and Micro-Diver TD loggers and Tinytag TG-4100, TGP-4017 and TGP-4020 data loggers. Loggers were cross-calibrated and response to temperature change assessed. The TGP-4020 loggers have an external temperature probe, and hence have the most rapid response when placed in water at a different temperature, commonly reaching equilibrium in <1-min). The other loggers have an internal sensor and although they respond rapidly to temperature change they may take 10–20 min to stabilise.

In July 2012, mini-Divers were deployed at the system outlet (PW) and the springs (RW, SM & PCR) and micro-Divers at the two main flooded inlets to Speedwell Cavern (MR & WR). Logging was at 1-min (micro) and 2-min (mini) intervals. In December 2013 a TG-4100 logger was co-deployed with the WR Diver to investigate a suspected instrument malfunction and additional TG-4100 loggers were deployed at the outlets to the BPR and SS flooded conduits and at two points along the Speedwell Cavern stream way, the Bung Hole (BH) and Windows Inlet (WI). Logging was at 2-min intervals. TGP-4017 and TGP-4020 loggers were deployed at other locations not discussed in this paper.

Table 1 Characteristics of data loggers deployed in the Castleton karst (logging hours/days assume a 1-min sampling interval)

	Memory	Hours	Days	Accuracy (°C)	Resolution (°C)	Depth
mini-Diver	24,000	400.00	16.67	±0.1	0.01	10 m+
TGP-4017	32,000	533.33	22.22	±0.5	0.01	15 m
TGP-4020	32,000	533.33	22.22	±0.35	0.02	15 m
TG-4100	32,000	533.33	22.22	±0.5	0.01	500 m
micro-Diver	48,000	800.00	33.33	±0.1	0.01	10 m+

Table 2 Descriptive statistics for 2-min data series 18/12/2013–5/3/2014

	WR	MR	BH	WI	RW	SM	PCR	PW
Mean	7.88	7.54	7.54	7.70	7.79	7.97	8.22	7.95
Standard error	0.0013	0.0015	0.0013	0.0010	0.0013	0.0010	0.0005	0.0010
Median	7.96	7.61	7.59	7.72	7.85	8.00	8.24	7.98
Standard deviation	0.32	0.34	0.30	0.16	0.29	0.15	0.11	0.22
Coefficient of variation (%)	4.0	4.6	3.9	2.1	3.8	1.8	1.4	2.8
Minimum	6.36	5.80	6.05	7.20	6.48	7.48	7.38	6.90
Maximum	8.33	8.19	8.31	8.33	8.40	8.51	8.48	8.44
Range	1.97	2.39	2.26	1.13	1.92	1.03	1.10	1.54
Data points analysed	55,757	55,757	55,757	25,834	53,849	20,877	53,837	53,863

4 Results

The project has generated a very large quantity of data most of which remains to be analysed. This paper presents data collected over the period 18 December 2013 to 5 March 2014. Summary statistics based on 2-min data are shown in Table 2 and the hourly average temperatures (calculated from the 1 or 2-min data) are shown on Fig. 3. Overall, the three springs exhibit a small temperature range (<2°C) and low temperature variability (CV < 4 %), but in detail there is remarkable event-related variability with complex temperature troughs related to recharge. Summer data show similar but inverse events with temperature peaks following recharge. Figure 4 illustrates an event that starts about midnight on 24th December. At MR, the temperature falls increasingly rapidly reaching a low point from 06:10 to 06:34 after which it slowly recovers. In contrast, the temperature at WR remains essentially constant until 03:50 when it rises by 0.2 °C before starting to fall at 04:14 and reaching a low point at 06:28. Water tracing experiments have shown that the sinking streams drain to both MR and WR, but the temperature data illustrate a complex conduit geometry.

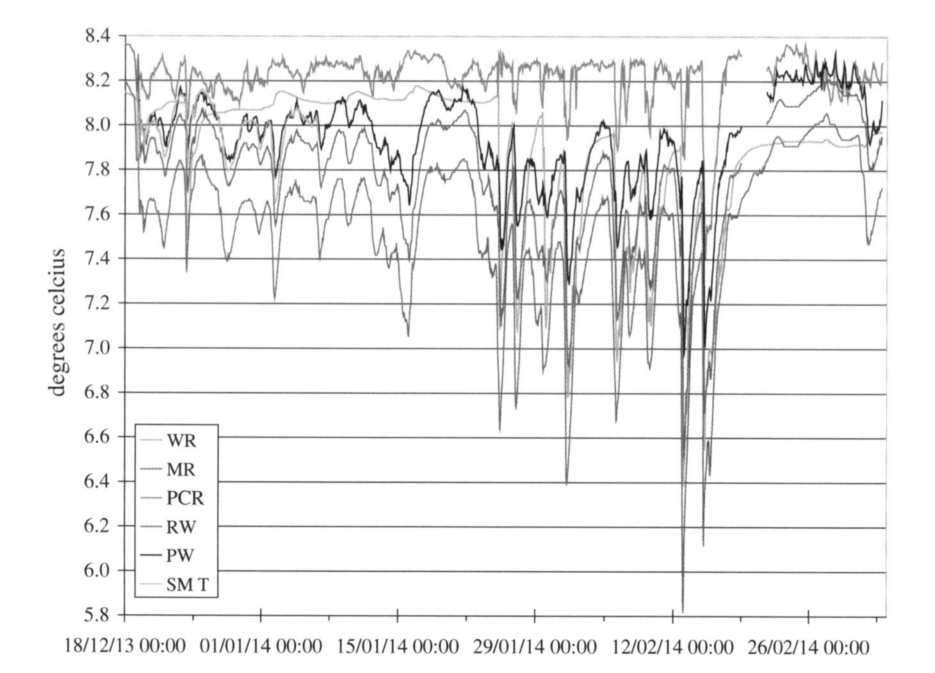

Fig. 3 Hourly average temperature based on 2-min data series

The MR and WR streams flow through open conduit and combine at 'the Whirlpool' which is about 290 m downstream of each rising. The BH logger is a further 140 m downstream and here the lowest temperature was recorded at 07:20. This suggests that the water is moving at a rate of 430 m in 50 min (0.14 m/s).

The WI logger is a further 350 m downstream, but the temperature low was only 10 min after the BH low giving a much higher velocity of 0.58 m/s. This is consistent with the passage gradient, which is markedly steeper and with fewer obstructions to flow. From WI, the open passage continues for a further 325 m to a flooded conduit that drains to RW and SM (straight line distance 550 m), but overflows to PCR when surcharged. If it is assumed that the velocity from WI to the sump is similar to that from BH to WI, then the lowest temperature at the start of the sump would have been at about 07:40. The lowest temperatures at RW and SM were at 07:58 and 08:06, respectively, giving a velocity of >0.46 m/s.

Most of the other events analysed have broadly similar behaviour, but an event on 18–19 December 2013 was markedly different (Fig. 5). At 20:38 on the 18th the temperature at PCR fell by 0.78 °C over a 6 min period and then rose over the next 40 min. Initially, it was thought this might be due to an instrument malfunction, but the temperature at the PW logger, about 140 m downstream, fell by 0.49 °C over a 16 min period commencing at 20:40. There is no obvious explanation for this sharp drop, which is not seen at any other site and takes place during a period of steady water depth. At MR the temperature declines slowly from

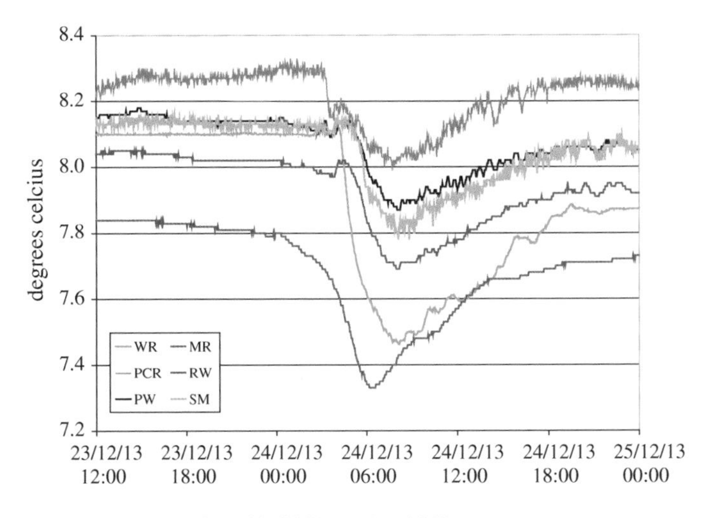

Fig. 4 Two-min temperature data, 23–25 December 2013

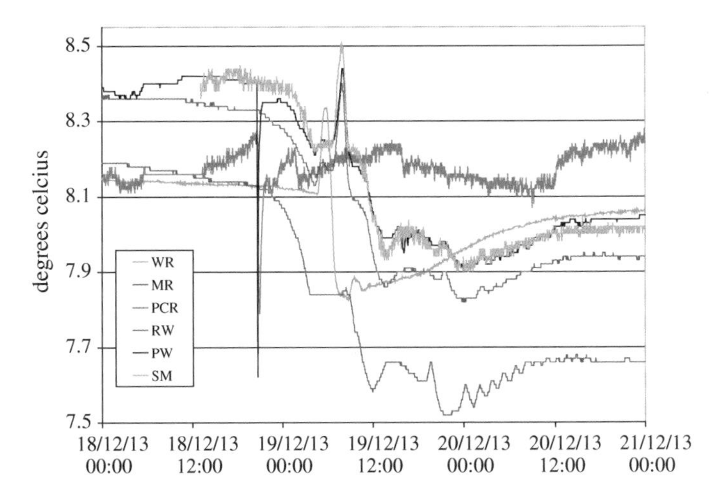

Fig. 5 Two-min temperature data, 18–21 December 2013

8.19 °C at midnight on the 17th to 8.10 °C at 22:38 on the 18th and then decreases more rapidly to 7.84 °C at 03:36 on the 19th. It remains at exactly this temperature until 09:00 after which there is a short small increase followed by a further decrease to 7.58 °C at 12:00. Thereafter, there are further temperature pulses and a period of temperature oscillations with reducing amplitude. At WR there is a very gradual fall in temperature to 8.11 °C at 04:44 on the 19th. There is then a very marked rise of 0.225 °C in 56 min, a 12 min plateau and then a sharp fall of 0.508 °C over 186 min after which there is a short pulse followed by a slow but steady recovery. The temperature pulse at WR is seen at RW after 132 min and at

SM after 136 min, exactly the same delay as was seen in the 23–25 December event. At PCR a water temperature trough at 02:10 coincides with a small peak in water depth, but there is no change in depth or temperature during the MR and WR events. It can therefore be concluded that under the conditions that pertained at that time no allogenic water was overflowing from the Speedwell conduit into the Peak Conduit.

The start of the temperature trough at MR coincided with a sharp fall in water depth, from 227.1 cm at 03:36 to 19.8 cm at 04:16, an average fall of 5.2 cm/min. The water depth then remained essentially static for 184 min before rising 167.1 cm in 18 min (9.3 cm/min). During this period the water depth at WR showed a similar behaviour to that exhibited in other events, with short-period changes of up to 45 cm. The RW and SM springs showed a response that combined the water depth fall and trough seen at MR with the water temperature pulse at WR, and the signal at the two springs was added to the signal from PCR event and transmitted to the PW weir. The shape of the resulting stage hydrograph at PW is remarkably similar to that modelled by Bottrell and Gunn (1991) who only had water depth data from a pen and ink recorder. They hypothesised that the cause was likely to be a mobile sediment blockage behind the MR sump that caused water to back-up and flow from WR. Flow returned when the pressure was sufficient to re-mobilise the sediment. The temperature data from MR and WR provide some support for this hypothesis although further analysis is needed.

5 Conclusion

The springs discharging from the Castleton karst clearly demonstrate event-scale variability and on that basis fall into Thermal Pattern 1 of Luhmann et al. (2011). However, Luhmann et al. also suggest that groundwater systems fed by perennial sinking streams are likely to show seasonal temperature variability and fall into Thermal Pattern 2. This is not the case at Castleton where most of the sinking streams are perennial. It is notable that a clear surface signal is maintained at the springs despite the water circulating to depths in excess of 75 m below the conceptual water table and over 250 m below the ground surface. The temperature and discharge of the three springs, and the surface river fed by them, demonstrate a complex non-linear response to recharge and this would be even more difficult to explain in the absence of data from within the conduit system. Further data collection and analysis is planned and it is suggested that high temporal-resolution temperature measurements at springs, and within the underground conduit system where possible, should form a part of all karst hydrogeological studies.

Acknowledgments This study would not have been possible without the support of Nigel Ball and Nick Coward who undertook most of the underground data logging. The Diver loggers were purchased using a grant from the British Cave Research Association and the Tinytag loggers were loaned by the University of Birmingham.

References

Beddows PA, Smart PL, Whitaker FF, Smith SL (2007) Decoupled fresh–saline groundwater circulation of a coastal carbonate aquifer: spatial patterns of temperature and specific electrical conductivity. J Hydrol 346:18–32

Bottrell S, Gunn J (1991) Flow switching in the Castleton Karst aquifer. Cave Sci 18(1):47–49

Crowther J, Pitty AF (1982) Water temperature variability as an indicator of shallow-depth groundwater behaviour in limestone areas in West Malaysia. J Hydrol 57:137–146

Gunn J (1981) Hydrological processes in karst depressions. Z Geomorph 25:313–331

Gunn J (1991) Water tracing experiments in the Castleton Karst, 1950–1990. Cave Sci 18(1):43–46

Luhmann AJ, Covington MD, Peters AJ, Alexander SC, Anger CT, Green JA, Runkel AC, Alexander EC (2011) Classification of thermal patterns at karst springs and cave streams. Ground Water 49(3):324–335

Martin JB, Dean RW (1999) Temperature as a natural tracer of short residence times for groundwater in karst aquifers. In: Palmer AN, Palmer MV, Sasowsky ID (eds) Karst modelling, vol 5. KWI Special Publication, pp 236–242

Pitty AF, Halliwell RA, Ternan JL, Whittel PA, Cooper RG (1979) The range of water temperature fluctuations in the limestone waters of the central and Southern Pennines. J Hydrol 41:157–160

Oxygen Isotope Composition Snapshot of Spring Waters in a Karstified Plateau

P. Malík, I. Slaninka, J. Švasta and J. Michalko

Abstract In the period of September 10–21, 2002, 317 km^2 of Muránska planina karstic plateau (Slovakia) was mapped for springs in great detail. Within this short time interval, 295 springs were documented and also sampled for $\delta^{18}O$. Although only freshwater of apparently similar origin was sampled, $\delta^{18}O$ was in a range between −10.90 and −7.32 ‰. Plateau is moderately vertically exaggerated (~400 to ~1400 m a.s.l.), but altitudinal differences do not explain this wide oxygen isotope span. Patterns of $\delta^{18}O$ distribution were examined for the possible influence of documented springs' parameters (temperature, discharge, electric conductivity, geology, aquifer circulation type, morphology around spring's orifice) without finding any significant dependency. The main reason for very different $\delta^{18}O$ values in nearby groundwater sources was identified in individual groundwater circulation regimes. According to IAEA GNIP stations' records, a significant contrast in precipitation $\delta^{18}O$ values was found within 14 months previous to the sampling. Heavier oxygen isotopes in the sampled set (around −8 ‰) probably reflect quick circulation influenced by enriched July–August 2002 precipitation (−5.28 ‰). Depleted $\delta^{18}O$ in springs (around −10 ‰), corresponding to the impact of winter 2001/2002 precipitation (−18.4 to −14.7 ‰), suggests longer residence times.

P. Malík (✉) · I. Slaninka · J. Švasta · J. Michalko
Štátny geologický ústav Dionýza Štúra – Geological Survey of Slovak Republic,
Mlynská Dolina 1, 81704 Bratislava 11, Slovakia
e-mail: peter.malik@geology.sk

I. Slaninka
e-mail: igor.slaninka@geology.sk

J. Švasta
e-mail: jaromir.svasta@geology.sk

J. Michalko
e-mail: juraj.michalko@geology.sk

B. Andreo et al. (eds.), *Hydrogeological and Environmental Investigations in Karst Systems*, Environmental Earth Sciences 1,
DOI 10.1007/978-3-642-17435-3_26

1 Introduction

Muránska planina (approximately E20°00′/N48°45′; Fig. 1) is a karstic Plateau built mainly of karstified Middle and Upper Triassic limestones and dolomites belonging at least to two different tectonic units thrusted one over the other, thus forming a complicated geological structure (Švasta et al. 2004). In spite of its complexity, the presence of carbonate rocks in the mentioned units allowed creating of one relatively uniform, hydrogeologically interconnected karstified body. Here, the surface of outcropping limestones and dolomites is 120 km^2 with average altitude of 957 m a.s.l. (Malík et al. 2008). Surrounded by less permeable crystalline granites and metamorphites, as well as by Paleozoic shales, carbonates form a closed unit from the groundwater balance point of view (Fig. 1). All groundwater in the Muránska planina Plateau is supposed to be formed by recharge from precipitation. The only exception is surface water input via swallow holes from the two small (5.5 km^2) adjacent crystalline watersheds on the NE rim (Malík et al. 2008).

2 Materials and Methods

During the fieldwork undertaken for constructing hydrogeological map of the Muránska planina Plateau (Švasta et al. 2004), all natural groundwater outlets had to be identified, visited and documented on the maps at 1:10,000 scale. The basic documentation included manual measurements of spring discharges, water and air temperature, water-specific electric conductivity, and the description of rock environment, supposed type of groundwater circulation and geomorphology around spring orifices. Additionally, samples of water were taken from each of the documented springs. Due to the availability of qualified personnel, all the documentation process was completed in a relatively short period of time. Altogether, 295 springs were documented and sampled by six teams of co-workers trained in the same methodology of field documentation. The mapping and sampling was completed within 12 days, under meteorologically stable and dry conditions in the period from September 10–21, 2002. The total area covered by detailed hydrogeological mapping was 317 km^2 (Fig. 1) and water samples were taken from groundwater springs also for identification of the abundance of stable oxygen isotopes ($\delta O^{18}/O^{16}$).

3 Results and Discussion

Within the 295 samples, the $\delta^{18}O$ values ranged between -10.90 and -7.32 ‰, with median value of -9.48 ‰. The arithmetical mean was -9.41 ‰ and standard deviation value of the set was -0.74 ‰. In spite of short sampling period

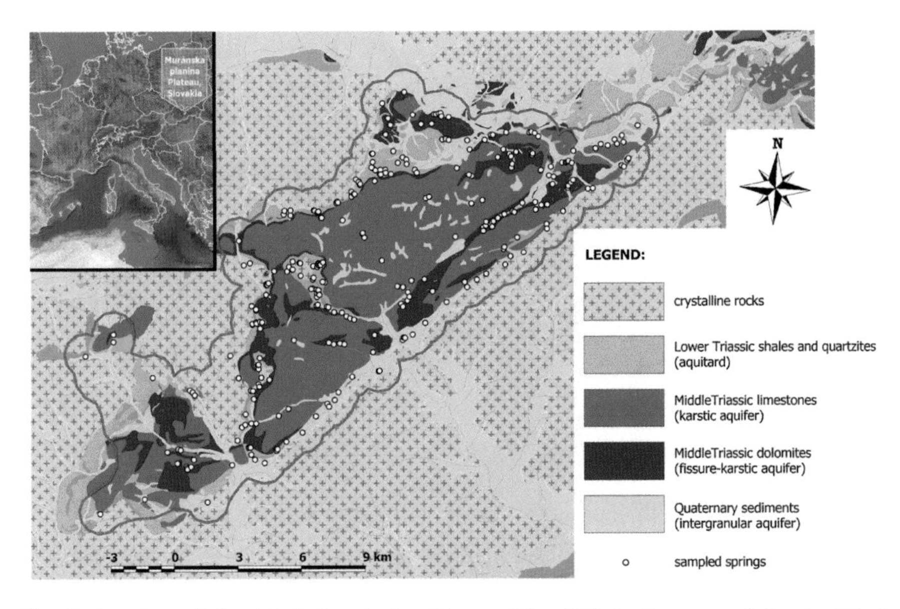

Fig. 1 Location of the Muránska planina Plateau (Slovakia) in Europe and its simplified geological sketch, with position of springs sampled during hydrogeological mapping

(12 days) and similar hydrogeological settings of all sampled groundwater sources, the whole dataset $\delta^{18}O$ values were rather scattered. Although only freshwater of apparently similar origin was sampled, such a range of values cannot be explained by simple impact of altitudinal effect. Muránska planina Plateau is moderately vertically exaggerated, the range of spring altitudes was between 390 and 1170 m a.s.l., the highest point in the area is Kľak with 1408 m a.s.l. These altitudinal differences cannot explain such a wide oxygen isotope span: altitude gradient of $\delta^{18}O$ calculated by Malík et al. (1996) for groundwater in karst springs in mountain ranges of northern Slovakia was 0.1 ‰/100 m and the mean annual altitude gradient of $\delta^{18}O$ in precipitation was 0.21 ‰/100 m (Holko et al. 2012). If so, the $\delta^{18}O$ values in the Muránska planina Plateau should be ranging within an approximate width of interval from 1.0 to 2.1 ‰, and not 3.58 ‰ as were measured. Various factors that might be responsible for such a variegated $\delta^{18}O$ values in the water of springs were subsequently studied.

Possible dependencies of oxygen isotope composition on springs' physical characteristics were studied. Measurable parameters such as discharge, altitude of spring's orifice, air and water temperature and EC were statistically correlated with $\delta^{18}O$ values. However, the correlation in nearly all cases was very poor: for $\delta^{18}O$ versus discharge, the correlation coefficient was only −0.01, for air temperature −0.02, for EC values 0.02; −0.20 for longitude and −0.43 for the latitude of the springs' cartographical position. Surprisingly, but in accordance with the previous paragraph, the altitudinal dependency of $\delta^{18}O$ values (Fig. 2) was also relatively weak (correlation coefficient of −0.36), but together with groundwater temperature (correlation coefficient 0.29) these were the most "significant"

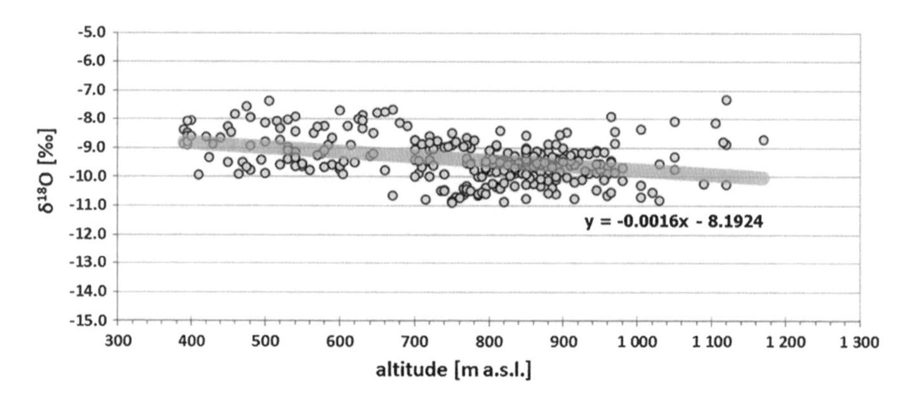

Fig. 2 Muránska planina Plateau—$\delta^{18}O$ values in groundwater versus altitudes of the springs sampled in the period of September 10–21, 2002

relations of $\delta^{18}O$ with measurable numeric values. On the other hand, mutual correlation of groundwater temperature and altitude gave the −0.63 value of correlation coefficient, which means that the measured $\delta^{18}O$ values were probably also driven by other factors than altitudinal effect. Although the $\delta^{18}O$ altitudinal dependency seems to be weak, its overall trend demonstrated in the altitude gradient of 0.16 ‰/100 m (Fig. 2) is in accordance with the gradients previously described in the region (Malík et al. 1996; Holko et al. 2012).

Possible influence of descriptive parameters such as aquifer lithology, estimated aquifer circulation type or morphology of spring's orifice on $\delta^{18}O$ values was studied by comparison of selected groups' statistical values. The role of aquifer lithology and stratigraphy is compared in Table 1.

It seems that relatively high ranges of $\delta^{18}O$ values of 2.0–3.5 ‰ found in nearly all groups and datasets (Table 1) should be caused by differences in individual groundwater circulation patterns. Different mean transit times are influenced by different storage and transmissivity as intrinsic aquifer characteristics and also different hydraulic gradients and flow pathways in infiltration areas of individual springs.

If regularly recorded in time with sufficient frequency, the changes in $\delta^{18}O$ can be used for calculation of the mean transit time. At temperate latitudes, a common simplification used to estimate transit time using the lumped parameter model takes advantage of the strong seasonal changes in the composition of stable isotopes in precipitation, e.g. by computationally simple sine-wave method (McGuire and McDonnell 2006), but for a period usually not longer than 4 or 5 years (DeWalle et al. 1997).

In our dataset, this was not the case: for each spring only one $\delta^{18}O$ value was available. On the other hand, all the data were acquired within short time period and cover all groundwater outputs on the area of 317 km^2. The results of $\delta^{18}O$ in precipitation in precipitation taken from the Global Network of Isotopes in Precipitation (GNIP) database, coordinated by International Atomic Energy Agency (IAEA/WMO 2014) were used as input dataset (Table 2). The station situated

Table 1 $\delta^{18}O$ values in spring types according to aquifer lithology and stratigraphy, estimated aquifer circulation type, morphology of spring's orifice and different discharge categories

	n	Median	Average		*n*	Median	Average
Aquifer lithology and stratigraphy				*Discharge category*			
Granitoids and metamorphic rocks	15	−9.46	−9.45	<0.1 L · s^{-1}	43	−9.16	−8.98
Lower Triassic quartzites	62	−9.54	−9.58	0.1–0.3 L · s^{-1}	75	−9.43	−9.35
Permian and Lower Triassic shales	35	−9.98	−9.87	0.3–1.0 L · s^{-1}	93	−9.55	−9.50
Middle Triassic limestones	88	−9.30	−9.21	1.0–3.0 L · s^{-1}	41	−9.66	−9.68
Middle Triassic dolomites	87	−9.39	−9.29	3.0–10.0 L · s^{-1}	17	−9.74	−9.55
Tectonites of carbonate rocks: rauwacks	8	−9.42	−9.44	10.0–30.0 L · s^{-1}	11	−9.79	−9.53
				>30.0 L · s^{-1}	15	−9.33	−9.37
Aquifer circulation type				*Morphology of spring's orifice*			
Karst-fissure	39	−9.59	−9.43	Point	174	−9.48	−9.41
Fissure	98	−9.57	−9.52	Areal	65	−9.56	−9.50
Fissure-debris	82	−9.43	−9.30	Countour-linear	23	−9.16	−9.27
Debris	76	−9.40	−9.36	Slope-linear	33	−9.31	−9.29
All aquifer types	295	−9.48	−9.41	All aquifer types	295	−9.48	−9.41

n—number of samples

Table 2 Values of $\delta^{18}O$ in precipitation [‰] on Vienna, Krakow and Ondrašová GNIP stations in the period 1992–2002 (IAEA/WMO 2014)

WMO code, station name, altitude	Altitude [m a.s.l.]	Median	Average	Min	Max	σ—standard deviation
1103500 Vienna (Hohe Warte)	203	−8.56	−9.55	−20.74	−2.69	3.52
1256500 Krakow (Wola Justowska)	205	−9.70	−9.96	−19.50	−4.60	3.13
1187800 Liptovský Mikuláš (Ondrašová)	570	−10.04	−10.38	−21.46	−4.44	3.89

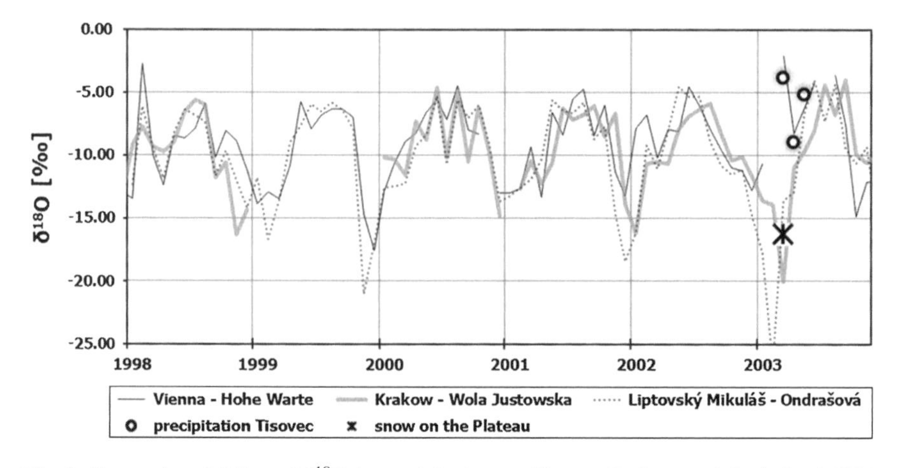

Fig. 3 Seasonal variability of $\delta^{18}O$ in precipitation on Vienna, Krakow and Ondrašová GNIP stations in the period 1992–2002 (IAEA/WMO 2014) together with precipitation data from spring of 2003 directly measured within the area—precipitation in Tisovec, snow samples from the top of the Plateau

260 km to WSW with the longest observation record since 1961 is Vienna (Hohe Warte). Krakow (Wola Justowska), observed since 1975, is placed 135 km towards north. The nearest GNIP station placed on the Slovak territory just 40 km to NWN, observed since 1992 is Liptovský Mikuláš (Ondrašová). Both continental and altitudinal effects (Rozanski et al. 1993) are recognizable in the mean values of Table 2. However, the seasonal variability of $\delta^{18}O$ in precipitation remains similar at all three sites (Fig. 3). Precipitation data directly measured within the area (precipitation in Tisovec, snow samples from the top of the Plateau) during the next winter and spring of 2003 were in accordance with the records of the three previously mentioned GNIP station (Fig. 3).

The direct influence of precipitation on groundwater during the year is controlled also by the evapotranspiration cycle and the precipitation data ought to be filtered by calculated evapotranspiration values (Malík et al. 2008) to receive monthly totals of effective precipitation. To understand the strength of the isotope signal brought by effective precipitation that is supposed to reach the groundwater table, both parameters (monthly amount of precipitation and measured monthly composite $\delta^{18}O$ values) were multiplied (Fig. 4). For reaching a "standard zero value" there, before the multiplication the value of −10 ‰ (close to the precipitation median at Ondrašová, Table 2) was subtracted from the measured monthly $\delta^{18}O$ values. A significant contrast in precipitation $\delta^{18}O$ values is then visible within a period of 14 months previous to the sampling: intensive and depleted during winter time (especially December 2001), and isotopically enriched heavy rainfalls in July and August 2001. The major springs on the SE facing margin of the Muránska planina Plateau (Fig. 5), with "heavier" $\delta^{18}O$ values (around −8 ‰) should be then influenced by enriched July–August 2002 precipitation, with low mean transit time/quick groundwater circulation. On both the western

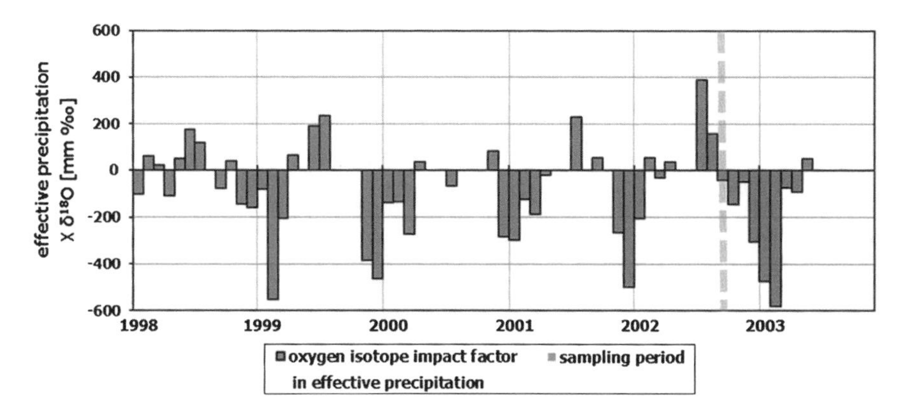

Fig. 4 Intensity of the isotope signal brought by the effective precipitation as multiplication of monthly effective precipitation total and $\delta^{18}O$ measured monthly composite

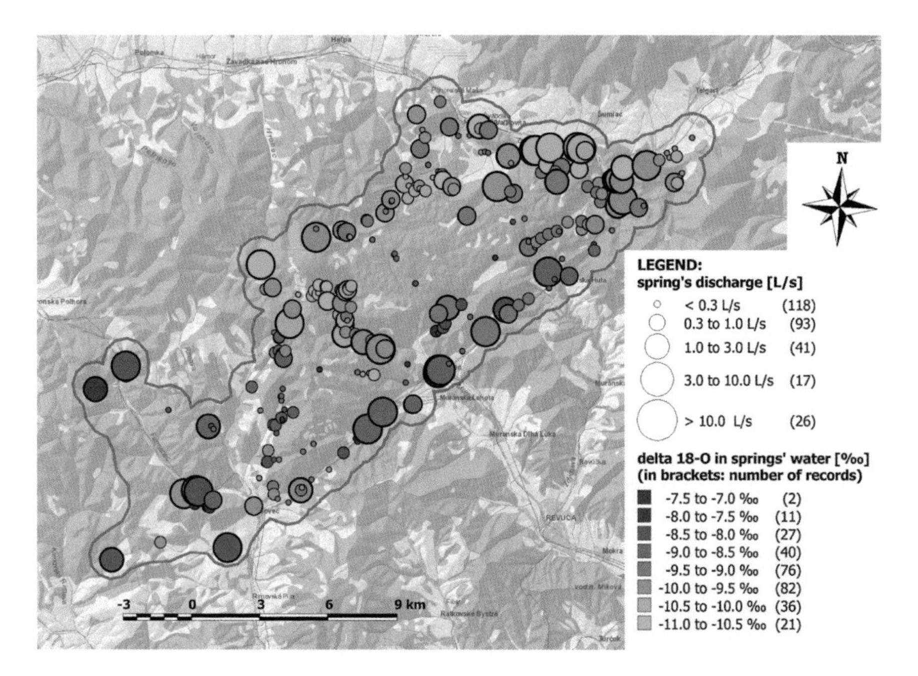

Fig. 5 $\delta^{18}O$ values in the springs of the Muránska planina Plateau sampled in the period of September 10–21, 2002. Diameter of *circles* depends on spring's discharge; its *colour* corresponds to $\delta^{18}O$

and northern rim of the Plateau, big springs with $\delta^{18}O$ of approximately -10 ‰ were possibly still influenced by depleted winter 2001/2002 precipitation and, therefore, with higher groundwater residence times.

4 Conclusion

The wide interval 3.58 ‰ of $\delta^{18}O$ values (between −10.90 and −7.32‰) that was found in the groundwater of 295 springs of the Muránska planina Plateau sampled within a few days of September 10–21, 2002 cannot be an exclusive consequence of altitudinal effect. On the other hand, any significant correlation of $\delta^{18}O$ with groundwater temperature, discharge, electric conductivity, geology, aquifer circulation type, morphology around spring's orifice was found. Differences in individual groundwater circulation regimes are therefore supposed to be the main reason for high $\delta^{18}O$ variation in even adjoining springs. According to IAEA GNIP stations' records, a significant contrast in precipitation $\delta^{18}O$ values was found within 14 months previous to the sampling. In this case, heavier oxygen isotopes in the sampled set probably reflect quick circulation influenced by enriched July–August 2002 precipitation, and low $\delta^{18}O$ in springs suggests longer residence times (impact of winter 2001/2002 precipitation). In this sense, the first group of springs should be more vulnerable (sensitive to potential pollution) than the second (Fig. 5). Monitoring of Cl^- performed in the first half of the next year on the three springs with possibly short mean transit time showed its response to chlorides originated from the winter road maintenance in the period between 60 and 90 days.

References

DeWalle DR, Edwards PJ, Swistock BR, Aravena R, Drimmie RJ (1997) Seasonal isotope hydrology of three Appalachian forest catchments. Hydrol Process 11:1895–1906

Holko L, Dóša M, Michalko J, Kostka Z, Šanda M (2012) Isotopes of oxygen-18 and deuterium in precipitation in Slovakia. J Hydrol Hydromech 60(2012)4:265–276. doi:10.2478/v10098-012-0023-2

IAEA/WMO (2014) Global network of isotopes in precipitation. The GNIP database. Accessible at: http://isohis.iaea.org

Malík P, Michalko J, Mansell SJ, Fendeková M (1996) Stable isotopes in karstic groundwaters of Veľká Fatra mountains Slovakia. In: Proceedings of international symposium on isotopes in water resources management IAEA Vienna, pp 191–192

Malík P, Švasta J, Baroková D (2008) Pokus o rekonštrukciu a rozšírenie hydrologickej bilancie Muránskej planiny na obdobie rokov 1971–1985 [Attempt of reconstruction and extension of the Muránska planina Plateau water balance for the period of 1971–1985, in Slovak with English summary]. SAH Bratislava ISSN 1335-1052, Podzemná voda XIV/1:71–87

McGuire KJ, McDonnell JJ (2006) A review and evaluation of catchment transit time modelling. J Hydrol 330(2006):543–563

Rozanski K, Araguás-Araguás L, Gonfiantini R (1993) Isotopic patterns in modern global precipitation. Climate change in continental isotopic records (Swart PK, Lohmann KL, McKenzie J, Savin S (eds). American Geophysical Union, Washington DC) Geophys Monogr 78:1–37

Švasta J, Slaninka I, Malík P, Vojtková S, Vojtko R (2004) Základná hydrogeologická mapa Muránskej planiny v mierke 1:50 000 [Basic hydrogeological map of the Muránska planina in the 1:50,000 scale, in Slovak]. Final report, manuscript, archive of the Geofond ŠGÚDŠ Bratislava, No. 92456

Groundwater Isotopic Characterization in Ordesa and Monte Perdido National Park (Northern Spain)

L.J. Lambán, J. Jódar and E. Custodio

Abstract The Ordesa and Monte Perdido National Park constitutes the largest calcareous mountain range of Western Europe, where the highest altitude karst of Europe is found. No previous studies regarding groundwater isotopic characterization in this area are known. This work presents the results of two preliminary campaigns carried out during July 2007 and April 2012. The water stable isotopes ($\delta^{18}O$, $\delta^{2}H$) show that the oceanic fronts from the Atlantic are responsible for the high levels of precipitation. In autumn, winter, and spring time, a deuterium excess is found in recharge water, which could be related to snow sublimation and its later condensation on the snow surface. The recharge zones are between 2,500 m and 3,200 m asl. The water tritium content points to short groundwater transit times.

Keywords Environmental isotopes • Groundwater • Deuterium excess • Snow sublimation • Ordesa and Monte Perdido National Park

1 Introduction

The Ordesa and Monte Perdido National Park (PNOMP) is the greatest carbonate rock mountain range of Western Europe. Situated on the Pyrenean mountain range, it is characterized by glacier landscape, with topography strongly conditioned by the erosion of its calcareous materials, which conditions the karstic hydrogeological system found in the PNOMP.

L.J. Lambán (✉)
Geological Institute of Spain (IGME), Madrid, Spain
e-mail: javier.lamban@igme.es

J. Jódar · E. Custodio
Department of Geotechnical Engineering and Geosciences,
Technical University of Catalonia, Barcelona, Spain
e-mail: jorge.jodar@upc.edu

E. Custodio
e-mail: emilio.custodio@upc.edu

B. Andreo et al. (eds.), *Hydrogeological and Environmental Investigations in Karst Systems*, Environmental Earth Sciences 1,
DOI 10.1007/978-3-642-17435-3_27

Karst aquifers have characteristics that make them very different from other aquifers, such as the high heterogeneity generated by the endokarstic network, high velocities of water flow, and short residence times (Kiraly 1997; Motyka 1998). When these aquifers are located in high altitude mountain zones, their hydrogeologic response is influenced by other variables, such as important vertical temperature gradients, abrupt topography that produces extremely short response times to precipitation events, and snow melting dynamics that controls aquifer recharge during winter and spring seasons. Water environmental isotopes are extremely useful to characterize such hydrogeological systems (Collins and Gordon 1981). Recharge zones can be defined by using $\delta^{18}O$ and $\delta^{2}H$ (Araguás-Araguás et al. 2000; Gonfiantini et al. 2001). Tritium content can either be used to separate base flow from runoff due to snow melt or estimate groundwater transit times (Boronina et al. 2005; Gustafson 2007).

Despite its significance as a large karst system and the place where important spring discharges appear, there is not any previous geochemical or isotopic characterization of groundwater in the PNOMP. So, this work presents the first groundwater isotopic ($\delta^{18}O$ and $\delta^{2}H$ and tritium) characterization in the PNOMP. The study focuses on characterizing the general hydrogeological isotopic features such as the seasonal isotopic content variations and the average groundwater transit time.

2 Study Area

The PNOMP is located in the central sector of the Pyrenees, the most important mountain chain of the Iberian Peninsula (Fig. 1). The maximum altitude corresponds to the Monte Perdido Peak, 3,355 m asl (above sea level), one of the highest peaks of the entire Pyrenean range.

From a geological point of view, the most important tectonic characteristic is the presence of two great nappes: Gavarnie (west of the Ara river) and Monte Perdido (to the east) (Fig. 1). The first one includes the Sierra Tendeñera outcroppings (Cretaceous to lower Eocene) and the base of the Ordesa valley (Paleocene). It is characterized by a strong plunging of these materials toward the south, under the Eocene flysch. The second one includes the Eocene outcroppings. It extends to the south, reaching the Ara River. This nappe presents scarce internal deformation, which prevents the local hydraulic connection between the different permeable levels (CHE 1998). The limestones, dolomites and calcarenites of the Upper Cretaceous and the lower Paleocene-Eocene constitute the most important surficial permeable levels, and contain the most significant springs and water discharges. The main karst system is located on the lower Paleocene-Eocene materials (CHE 1998; Ríos-Aragüés 2003).

From a climatic point of view, according to the Köppen-Geiger classification (Peel et al. 2007), the PNOMP has a cold climate with a dry season, with mild and cool summers and significant altitudinal variations (AEMET/IM 2011).

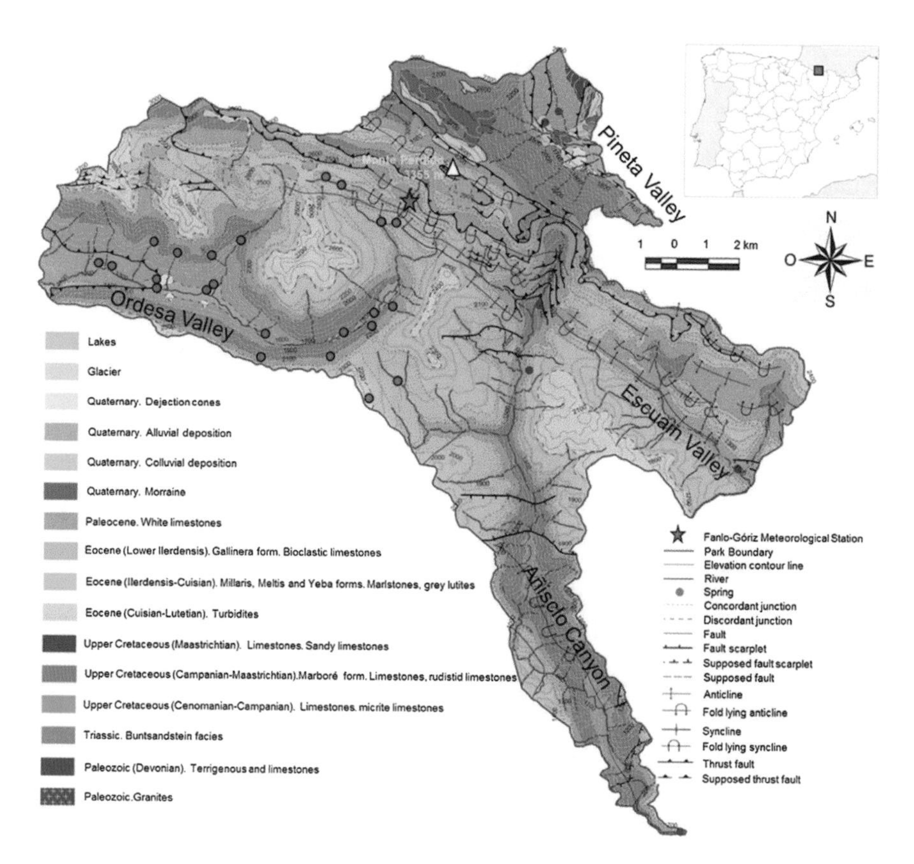

Fig. 1 Location and geological map of the Ordesa and Monte Perdido National Park. The *red points* located in the Ordesa Valley correspond to the sampled springs

The average monthly precipitation presents a maximum in autumn, with a secondary peak in spring, and a minimum in winter (Fig. 2). Rainfall spatial variability shows a W-E gradient (Benito Alonso 2006) owing the oceanic low pressure fronts arriving from the Atlantic. They are responsible for the main precipitation volumes registered in the PNOMP.

3 Materials and Methods

During the period between July/2007 and April/2012, several groundwater sampling campaigns were conducted. The objective was twofold: on the one hand to examine if any seasonal rainfall isotopic variation was observed, and on the other hand to measure the tritium content of groundwater in the most important springs located in the PNOMP (Fig. 1).

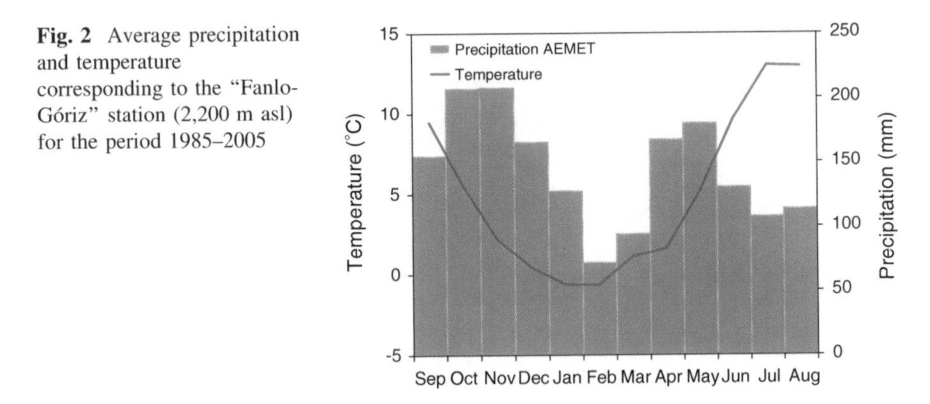

Fig. 2 Average precipitation and temperature corresponding to the "Fanlo-Góriz" station (2,200 m asl) for the period 1985–2005

$\delta^{18}O$ and δ^2H (relative to the V-SMOW standard) in water was analyzed at the IACT-CSIC in Granada by gas mass spectrometry and tritium (in tritium units TU) at the Isotope Hydrology Laboratory of CETA-CEDEX in Madrid by liquid scintillation counting in enriched samples.

4 Results Discussion

4.1 Stable Isotopes

The groundwater samples obtained in PNOMP present an isotopic content (Fig. 3c) that ranges between −7.4 and −12.3 ‰ for $\delta^{18}O$ and between −45.3 and −81.5 ‰ for δ^2H, with coefficients of variation of 0.091 and 0.093 for for $\delta^{18}O$ and δ^2H, respectively.

Figure 3b shows the seasonal averages of $\delta^{18}O$ and δ^2H in rainfall for the PNOMP and for three additional sampling stations located on the Pyrenean axis (Fig. 3a): Santander (IAEA-GNIP), at sea level and under Atlantic influence; (2) Estopiñán Syncline (Pérez 2013), 30 km away towards SE, at 750 m asl, and (3) Barcelona (IAEA-GNIP), at sea level and under Mediterranean influence.

The seasonally averaged rainfall isotopic composition in the PNOMP is lighter than what is observed in the other precipitation sampling stations (i.e., Estopiñán, Santander and Barcelona). This is due to the higher altitude and lower temperature in the PNOMP. Additionally, even though the geographical position of the PNOMP and Estopiñán are equivalent locations from a synoptic point of view, the seasonal averaged isotopic groundwater composition in the PNOMP shows a deuterium excess ($d = 8\delta^{18}O - \delta^2H$ ‰) not observed in Estopiñán. The deuterium excess coincides with the seasonal presence of snow in the study area, where the snow covers almost the whole PNOMP between October and June.

Snow sublimation is an important mass loss mechanism in high mountain systems. In extreme cases it can exceed 3 mm/d (Vuille 1996). Sublimation is

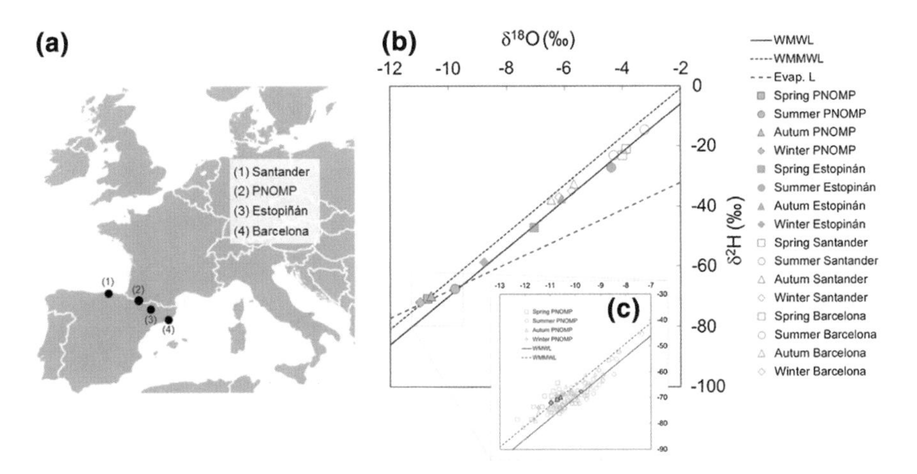

Fig. 3 **a** Geographic location of the PNOMP and the precipitation isotope sampling stations of Santander, Estopiñán and Barcelona and **b** $\delta^{18}O$ and $\delta^{2}H$ seasonal averages. The *solid line* and the densely *dashed lines* are the meteoric *lines of slope* (m) of 8 and deuterium excess of +10 and +15 ‰ respectively, and the *dashed line* (in *blue*) is the evaporation *line for a slope* m = 4.5. **c** Measured and seasonally averaged yearly values of $\delta^{18}O$ and $\delta^{2}H$ (*open* and *solid symbols*, respectively) in the PNOMP

produced in the upper layers of the snow cover, the maximum during the hours of highest sun irradiation in warmer times, provided the snow cover persists. As a consequence of diurnal sublimation, the remaining snow becomes isotopically enriched and with a lower deuterium excess, while the sublimated water vapor becomes lighter and with a higher deuterium excess. The incorporation and recycling of this vapor in the surrounding atmosphere increase the atmospheric deuterium excess at local scale. The low night temperatures cause the atmospheric water vapor to condensate on the surface of the snow and as a result the snow cover increases its deuterium excess (Stichler et al. 2001; Froehlich et al. 2008). The cycle of diurnal snow sublimation and later nocturnal condensation would explain why in the PNOMP recharge water produced under snow cover in autumn, winter, and spring has a higher deuterium excess when the balance of old and new snow is in favor of a condensation excess.

$\delta^{18}O$ shows a linear relationship (Fig. 4) with the sampling point height for the highest altitude springs (i.e., above 1,900 m). A vertical isotopic gradient of −0.32 ‰ $\delta^{18}O$/100 m is obtained, which is coherent with the gradients reported by Arce et al. (2001) and Iribar et al. (1996) for the western areas of the Pyrenees. This result suggests that all the highest altitude springs have a similar altitude difference between their corresponding sampling and the recharge areas. The height differences for all the sampling points can be easily obtained by taking the above vertical $\delta^{18}O$ gradient and the rainfall $\delta^{18}O$ content for a given location ($\delta^{18}O$ = −8.48 ‰ measured at 2,500 m asl).

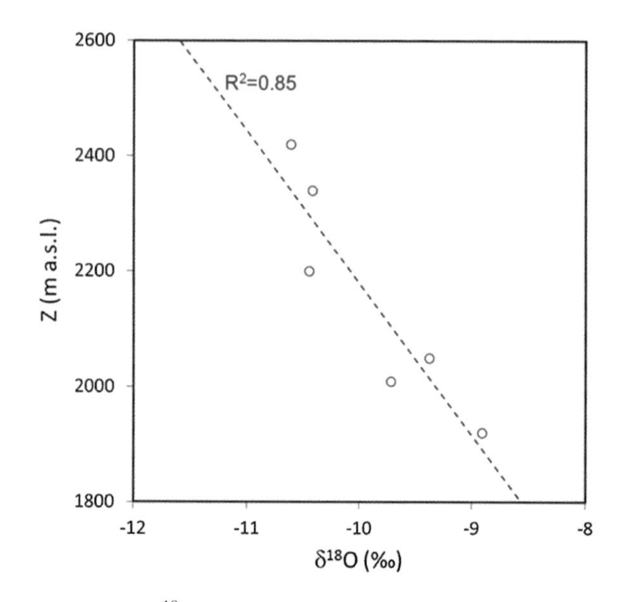

Fig. 4 Relationship between $\delta^{18}O$ and the elevation for the samples taken from springs located above 1,900 m during the 2007 field campaign. The altitudinal isotopic gradient is −0.32 ‰ $\delta^{18}O/100$ m

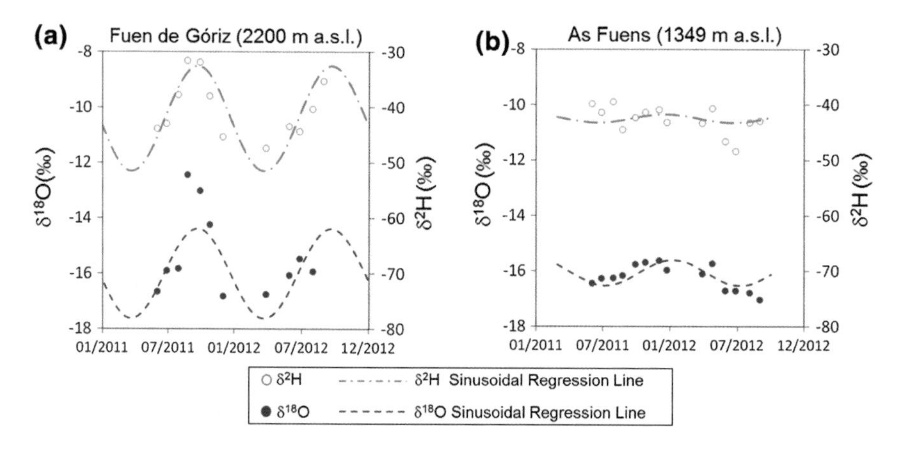

Fig. 5 Temporal evolution of the $\delta^{18}O$, $\delta^{2}H$ concentrations during the 2011–2012 field campaign at Fuente de Góriz (*left*) and as Fuens (*right*) springs

All sampling spring points show a sinusoidal time evolution of isotopic content of both $\delta^{18}O$ and $\delta^{2}H$. This result agrees with the known relationship between the isotopic fractionation and temperature (Clark and Fritz 1997), which shows heavier isotopic compositions in summer and lighter compositions in autumn and winter. Moreover, springs located at higher altitudes (Fig. 5a) present a larger amplitude oscillation than the springs located at lower levels (Fig. 5b).

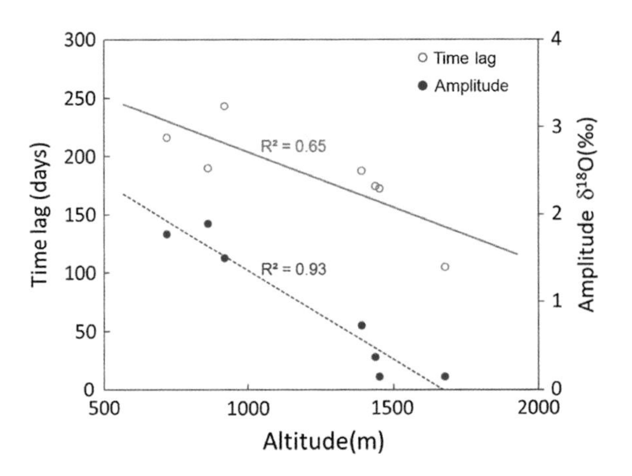

Fig. 6 Time lag and $\delta^{18}O$ amplitude versus altitude difference between recharge and discharge zones of each sampling of the 2011–2012 field campaign

A sinusoidal function has automatically calculated to fit the measured isotopic time series of spring water, using a minimum square criterion (Press et al. 1992). The sinusoidal function is given by $\delta C = \mu + A \cdot \sin(\omega t + \phi)$, where δC is isotopic content, μ is average isotopic content, t is time, A *is* the amplitude, ϕ the angular shift and ω the angular frequency. The estimated parameters are A and ϕ.

The estimated parameters show a linear relationship with altitude difference between recharge and discharge levels of the sampled springs (Fig. 6). This shows how the input isotopic signal is delayed and progressively buffered as the altitude difference increases.

4.2 Tritium

Groundwater tritium content was analyzed only in water samples collected during the July/2007 field campaign. It ranges between 6.1 ± 0.6 TU and 4.6 ± 0.6 TU, with a variation coefficient of 0.093.

There is no any available measurement of tritium content in rainfall at the PNOMP. The tritium content in rainfall is regularly measured at the REVIP (Monitoring Network of Precipitation Isotopes) stations in Spain. Results show small tritium variations during the last 15 years (Díaz-Tejeiro et al. 2009). In fact, the rainfall tritium concentration in the closest stations to the study area show values near those measured in the PNOMP groundwater (Fig. 7). The latter points out quite short transit times of recharge water through the aquifer.

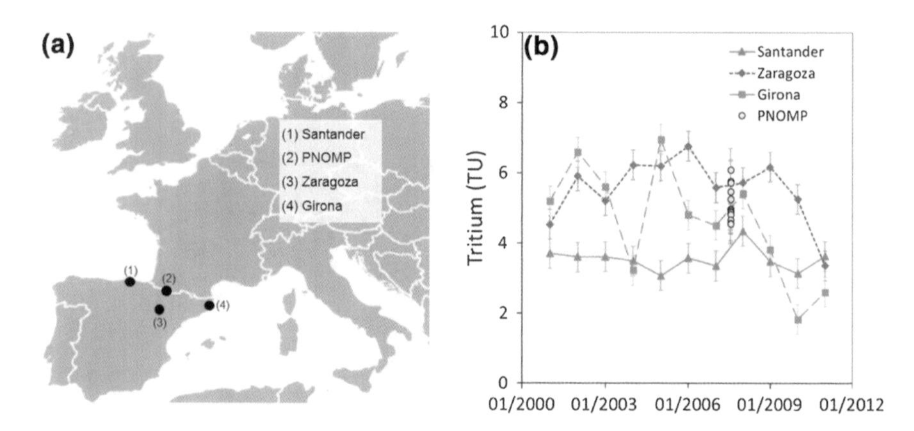

Fig. 7 **a** Location of the PNOMP and the closest precipitation sampling station for tritium (Santander, Zaragoza and Girona) REVIP stations and **b** time series of tritium concentration measured at the PNOMP and at the selected REVIP stations

5 Conclusion

In all the sampled springs of the PNOMP the isotopic groundwater shows a clear seasonal behavior. The isotopic content ranges between −7 and −12 % for $\delta^{18}O$ and between −45 and −81 ‰ for $\delta^{2}H$. Additionally, during autumn, winter, and spring there is a deuterium excess, which could be related to snow sublimation and its later condensation on the snow surface.

The local altitudinal isotopic gradient (−0.3 ‰ $\delta^{18}O$/100 m) agrees with bibliographical values for the Pyrenees. Recharge altitudes are between 2,500 m and 3,200 m asl The amplitude of the isotopic sinusoidal signal of precipitation water decreases, and is delayed as recharge water flows through the aquifer. In the case of the PNOMP, both amplitude and delay show a linear dependency with the difference between recharge and discharge altitudes.

The tritium content measured in groundwater samples is similar to measured tritium concentration in rainfall for close stations, which shows that groundwater in the PNOMP has a quite short residence time, not longer than a few years.

Acknowledgments This research was undertaken in the framework of the project "Hydrological behavior analysis of groundwater dependent wetlands" funded by the Geological Institute of Spain (IGME). The authors would like to thank the Ordesa and Monte Perdido National Park Direction (Gobierno de Aragón) and the Pyrenean Institute of Ecology (IPE-CSIC) by their collaboration. Meteorological data have been provided by the Spanish Meteorological Agency (AEMET). The isotopic rainfall data have been provided by the Centro de Estudios de Técnicas Aplicadas of the "Centro de Estudios y Experimentación de Obras Públicas" (CEDEX), which, together with AEMET, manage the Spanish Monitoring Network for Isotopes in Precipitation (REVIP).

References

AEMET/IM (2011) Atlas climático ibérico [Iberian climate atlas]. AEMET & IM., Madrid

Benito Alonso JL (2006) Vegetación del Parque Nacional de Ordesa y Monte Perdido. Publ. Consejo Protección Naturaleza Aragón, no 50. Zaragoza

Arce M, García MA, Arqued V (2001) Caracterización del oxígeno 18, deuterio y tritio en las aguas del Pirineo. In: Medina A, Carrera J (eds) Las Caras del Agua Subterránea Barcelona, pp 387–393

Araguás-Araguás L, Froehlich K, Rozanski K (2000) Deuterium and oxygen-18 isotope composition of precipitation and atmospheric moisture. Hydrol Processes 14:1341–1355

Boronina A, Renard P, Balderer W, Stichler W (2005) Application of tritium in precipitation and in groundwater of the Kouris catchment (Cyprus) for description of the regional groundwater flow. Appl Geochem 20:1292–1308

CHE (1998) Catalogación de los acuíferos de la Cuenca del Ebro. Oficina de Planificación Hidrológica. Confederación Hidrográfica del Ebro, Zaragoza

Clark ID, Fritz P (1997) Environmental isotopes in hydrogeology. Lewis, Boca Raton

Collins DN, Gordon JY (1981) Meltwater hydrology and hydrochemistry in snow and ice-covered mountain catchments. Nord Hydrol 12:319–334

Díaz-Teijeiro MF, Rodríguez-Arévalo J, Castaño S (2009) The Spanish network for surveillance of isotopes in precipitation: spatial isotopic distribution and contribution to the knowledge of the hydrological cycle. Ingeniería Civil 155:87–97

Froehlich K, Kralik M, Papesch W, Rank D, Scheifinger H, Stichler W (2008) Deuterium excess in precipitation of Alpine regions: moisture recycling. Isot Environ Health Stud 44(1):1–10. doi:10.1080/10256010801887208

Gonfiantini R, Roche MA, Olivry JC, Fontes JCh, Zuppi GM (2001) The altitude effect on the isotopic composition of tropical rains. Chem Geol 181:147–167

Gustafson JR (2007) Snowmelt water isotope fractionation in high elevation seasonal snowpacks: implications for isotope hydrograph studies. http://www.sahra.arizona.edu/unesco/casestudies/WestUS. December 2007. Accessed 25 Feb 2013

Iribar V, Antigüedad I (1996) Definición de zonas de recarga de manantiales kársticos mediante técnicas isotópicas ambientales. Simp.Rec.Hidr. en regiones Kársticas. Gob Vasco/AIH Vitoria-Gasteiz 1:271–280

Kiraly L (1997) Modelling karst aquifers by the combined discrete channel and continuum approach. In: 6th conference on limestone hydrology and fissured aquifers: modelling karst aquifers. Université de Franche-Comté, Sciences et Techniques de l'Environnement, La Chaux-de-Fonds, pp 1–26

Motyka J (1998) A conceptual model of hydraulic networks in carbonate rocks, illustrated by examples from Poland. Hydrol J 6:469–482

Peel MC, Finlayson BL, McMahon TA (2007) Updated world map of the Köppen-Geiger climate classification. Hydrol Earth Syst Sci 11:1633–1644. doi:10.5194/hess-11-1633-2007

Pérez C (2013) Funcionamiento hidrogeológico de un humedal hipogénico de origen kárstico en las Sierras Marginales Pirenáicas: Las Lagunas de Estaña (Huesca). Ph.D. thesis. Universidad Complutense de Madrid

Press WH, Teukolsky SA, Vetterling WT, Flannery BP (1992) Numerical recipes in C. Cambridge University Press

Ríos-Aragüés LM (2003) Introducción al mapa geológico del Parque Nacional de Ordesa y Monte Perdido. Sociedad Española de Espeleología y Ciencias del Karst (SEDECK) Boletín 5:84–99

Stichler W, Schotterer U, Fröhlich K, Ginot P, Kull C, Gäggeler H, Pouyaud B (2001) Influence of sublimation on stable isotope records recovered from high-altitude glaciers in the tropical Andes. J Geophys Res 106(D19):22,613–22,620

Vuille M (1996) Zur raumzeitlichen Dynamik von Schneefall und Ausaperung im Bereich des südlichen Altiplano, Südamerika. Geogr Bernensia G45:1–118

Methodological Procedure for Evaluating Storage Reserves in Carbonate Aquifers Subjected to Groundwater Mining: The Solana Aquifer (Alicante, SE Spain)

A. Ruiz-Constán, C. Marín-Lechado, S. Martos-Rosillo, C. Fernández-Leyva, J.L. García-Lobón, A. Pedrera, J.A. López-Geta, J.A. Hernandez Bravo and L. Rodríguez-Hernández

Abstract A methodological procedure is proposed for evaluating the groundwater reserves in carbonate aquifers. It is applied to the Solana-Onteniente-Volcadores aquifer (Alicante, SE Spain), subjected to groundwater mining exploitation. The large amount of geological and geophysical information available (361 km^2 of geological mapping, 84 lithological columns of boreholes, 47 km of seismic reflection profiles, 68 vertical electrical soundings [VES], 22 MT soundings and 1,600 gravity measurements) was integrated using two different 3D geological modelling codes: 3D Geomodeller and GOCAD. In addition, a petrophysical study of the rock matrix was accomplished with 39 rock samples, aiming to determine their storage capacity and calculate the total groundwater reserves. Information provided by the 3D model stands as a great advance in our hydrogeological knowledge of the region.

Keywords Reserve curve • Porous system • Carbonate aquifer • 3D modelling

A. Ruiz-Constán (✉) · C. Marín-Lechado · S. Martos-Rosillo · C. Fernández-Leyva · J.L. García-Lobón · J.A. López-Geta
Instituto Geológico Y Minero de España (IGME), C/Ríos Rosas, 23, 28003 Madrid, Spain
e-mail: a.ruiz@igme.es

A. Pedrera
Instituto Andaluz de Ciencias de la Tierra (CSIC), Avda. de Las Palmeras, 4, 18100 Granada, Spain

J.A. Hernandez Bravo · L. Rodríguez-Hernández
Diputación de Alicante, Avda. de la Estación 6, 03005 Alicante, Spain

B. Andreo et al. (eds.), *Hydrogeological and Environmental Investigations in Karst Systems*, Environmental Earth Sciences 1,
DOI 10.1007/978-3-642-17435-3_28

1 Introduction

The assessment of storage reserves is essential for prediction of the water-level response in aquifers subjected to groundwater mining. This calls for the development of geologically well-constrained flow models. During the past 30 years, the technology applied in researching deep geological formations has drastically progressed at the expense of the oil industry (e.g. Ahr 2008). In this regard, the recent application of 3D geological models to hydrogeological aims has greatly contributed to the 3D knowledge, evaluation and management of aquifers (e.g. Wu et al. 2008).

The aim of this study is twofold: (i) to establish a methodological procedure to obtain an accurate 3D geological model of deep aquifers integrating previous geological and geophysical data and (ii) to use this information to calculate groundwater reserves for a sustainable groundwater management. We focus the methodology on a natural example: the Solana-Onteniente-Volcadores aquifer.

2 The Solana-Onteniente-Volcadores Aquifer

The Solana aquifer (Fig. 1), located by the NW border of Alicante province (SE Spain), is one of the most important aquifers in the region (280 km^2). However, the area modelled in this study also includes the aquifers of Onteniente (33 km^2) and Volcadores (53 km^2) (IGME-DPA 2013), because the geological model suggests geological continuity. For this reason, it is referred to herein as the Solana-Onteniente-Volcadores aquifer (SOV).

The study area is located in the External Zones of the Betic Cordillera. It is deformed by NW-vergent open folds with reverse listric faults in the NW limb of the antiforms. At the western border, the folds are cut by the Triassic chaotic rocks of the Villena diapir (De Ruig 1992). The stratigraphic sequence starts with Triassic clays and evaporitic rocks superposed by Jurassic limestones, partially dolomitized. Over them, the sequence continues with low-permeable marls, clays and sands (Lower Cretaceous).The stratigraphic sequence of the aquifer is constituted by Upper Cretaceous dolostones and limestones. Over them, there is a thick sequence of marls of Paleogene to Quaternary age.

The SOV aquifer has a total extension of 366 km^2, with 173 km^2 of permeable exposed surfaces. The boundaries of the aquifer are considered closed to groundwater flow, although the Solana eastern limit is controversial. Transmissivity fits a log-normal distribution with a median of 467 m^2/day, and values ranging between 89 and 2.442 m^2/day (IGME-DPA 2006). In natural conditions the groundwater flow, currently subjected to continued storage depletion, is

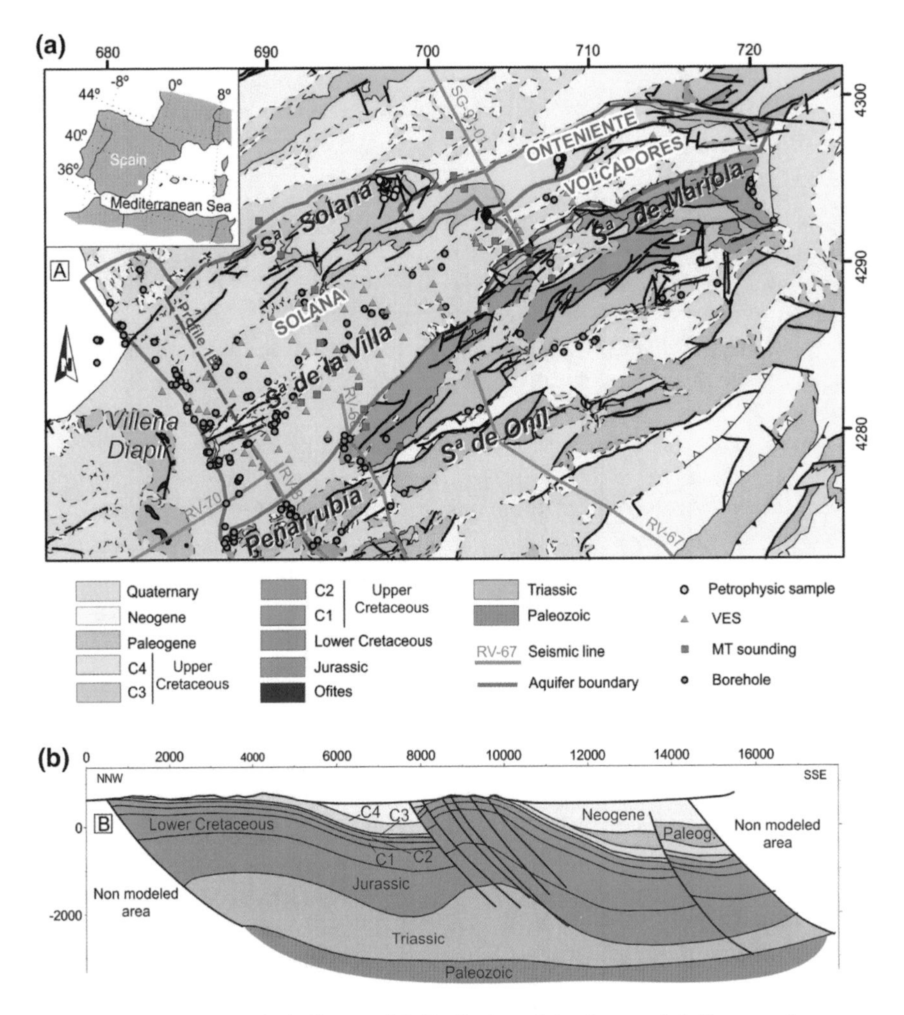

Fig. 1 **a** Geological map including spatial distribution of the data used. **b** Cross-section

towards the SW. The average spring flows were 300 l/s at elevations around 502 m.a.s.l. and with an average hydraulic gradient of 1 ‰. Nowadays, the piezometric level is in between 390 and 340 m.a.s.l. (IGME-DPA 2013). The intensive exploitation of the aquifer since the 60s (315 hm^3 in the period 1958–2008) has led to the disappearance of springs and an accentuated decrease of the piezometric level (122 m in 40 years). At present, recharge is mainly caused by rain infiltration and returns from irrigation; meanwhile the discharge of the aquifer is the exploitation of the numerous pumping wells.

3 Methodology

3.1 3D Modelling: Model Elements, Reference Surfaces and Faults

The basis of an accurate 3D model is the careful establishment of reference surfaces and the homogeneous distribution of input data (Figs. 1, 2). The topographic surface was extracted from a 10 m Digital Elevation Model (http://www.ign.es/ign/layoutIn/modeloDigitalTerreno.do). The lithological contacts, sedimentary beds (756 measurements) and faults were digitalised and georeferenced from previous geological maps, locally modified after a detailed photo interpretation study.

A total of 84 lithological columns and the interpreted seismic horizons of 47 km of seismic reflection lines were also integrated. The seismic reflectors interpreted are the bottom of the Triassic, Jurassic, Lower Cretaceous, Upper Cretaceous (C4-C1), Paleogene and Neogene units (Fig. 1). The time-to-depth conversion was accomplished with the normal move out (NMO) velocities generated during the seismic processing.

Geological cross-sections were developed to facilitate interpretation of the 3D surfaces. Afterwards, a 3D geometric model was performed with 3D Geomodeller. Other geophysical data helped to constraint the 3D model: 68 Vertical Electric Soundings (VESs; http://cuarzo.igme.es/sigeco/default.htm) and 22 new broadband magnetotelluric (MT) sites acquired along two NW-SE profiles.

Gravity data were used as an additional test to the model, performed through a 3D gravimetric inversion using 3D Geomodeller. 883 gravity stations (http://cuarzo.igme.es/sigeco/default.htm) were processed to calculate the Bouguer anomaly. The next steps involve: (i) testing and successively modifying the geological model by forward modelling to locate areas of mismatch and (ii) performing the 3D inversion to fit the density values and the geometry of the contacts.

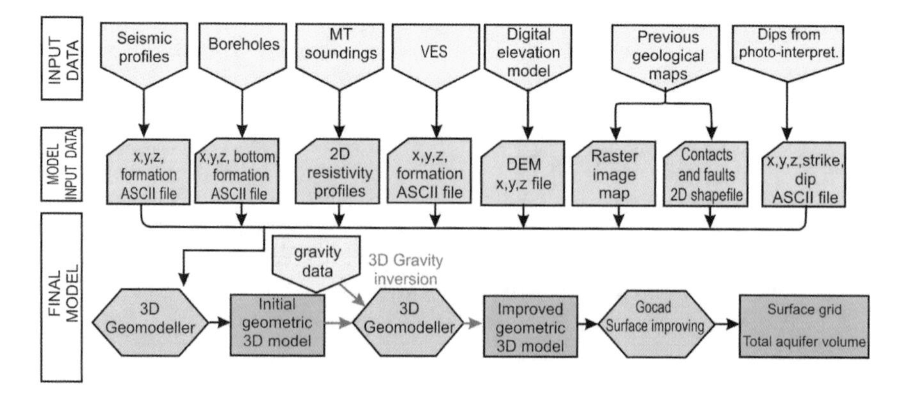

Fig. 2 Diagram of the methodological procedure

3.2 Porous System Characterization

Thirty-nine samples were analysed to determine the porous system of the permeable rocks of the SOV aquifer. This study allows us to establish the correlation between the carbonate rock fabrics and their physical properties, identify and classify samples based on their textural elements (Lucia 1999), and finally determine morphology and size of pores as well as fracture distribution. We calculated the interconnected matrix porosity (p_0) of the samples from core cubes to quantify storage. To avoid problems induced by core sampling, fresh samples without large vugs and without fractures were selected. All the air was extracted from the sample and replaced by water.

3.3 Total Storage Capacity

We assume that almost all the storage capacity of the aquifer is located in the rock matrix. This feature is evident in detritic aquifers, and recently also accepted for carbonate ones (Ahr 2008). Thus, total storage capacity was obtained based on the determination of the saturated volume of the aquifer between the piezometric surface (at different heights) and the top and the bottom of each rock formation distinguished in the 3D model. The next step was to build a 3D grid of the different stratigraphic layers to obtain the aquifer volume using GOCAD software. The groundwater reserve curves, representing the storage volume of an aquifer below a given height, were then calculated for the 25th and 75th percentiles of the field samples.

4 Results

4.1 3D Modelling: Model Elements, Reference Surfaces and Faults

The 3D geological model of the SOV aquifer (Fig. 1b) is formed by nine geological units laterally limited by thrust faults and diapirs. The Neogene-Quaternary unit (average thickness of 650 m) has a NW-vergent synform geometry that diverges into two synforms towards the SW. On the other hand, the Paleogene unit crops out only at the SW border of the 3D model with a maximum thickness of 300 m. Downwards, the Upper Cretaceous unit (subdivided into four sub-units: C4-C1; Fig. 1) is mainly formed by limestones and dolostones, and constitutes the SOV aquifer. The average thicknesses of the sub-units are 250, 70, 135 and 170 m, respectively. Geometrically, the unit is deformed by NW-vergent open folds with thrust faults at the NW flank of the antiforms. Finally, the Lower Cretaceous,

Jurassic and Triassic units have a cylindrical folded geometry and an average thickness (estimated through 3D gravity inversion) of 600, 990 and 750 m, respectively.

4.2 Porous System Characterization

Petrographic studies were carried out on 39 samples: 21 dolomite samples and 18 limestones and partially dolomitized limestones. The statistical sample does not fit a normal distribution (Fig. 3). The interconnected matrix porosity has a median of 4 %, while the 25th and 75th percentiles are, respectively, 1.85 and 6.95 %. The median value is greater than the average values for limestones and dolostones (3 %) of the Betic Cordillera (Pulido-Bosch et al. 2004). Several samples with anomalous porosities (22 %) are associated with large vugs and open microfissures within the core samples due to local dissolution processes.

4.3 Total Storage Capacity

The total storage reserves were calculated with the higher piezometric surface generated in the 3D model. This level corresponds to 502 m.a.s.l at the Solana aquifer, and 525 m.a.s.l. at the Volcadores and Onteniente ones. In the most unfavourable situation, the interconnected matrix porosity value and the storage coefficient used (25th percentile and 10^{-7}, respectively) provide a total storage capacity of 1,476 hm^3, whereas in the most favourable situation (75th percentile and 10^{-4}, respectively) the value is 10,406 hm^3.

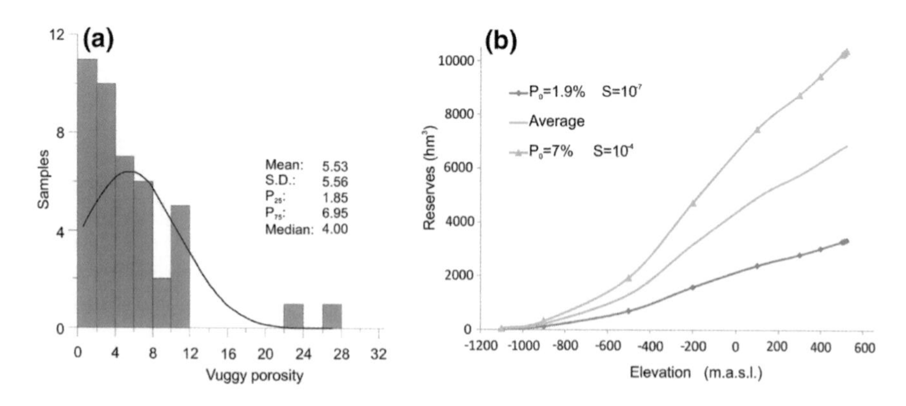

Fig. 3 *Histogram* of interconnected porosity (**a**) and reserve curve (**b**)

5 Hydrogeological Implications

The 3D geological model of the SOV aquifer reveals the continuous geological structure of the system. There is no evidence pointing to the compartmentalization of the aquifer formations into three aquifers with independent hydrogeological behaviours. Nevertheless, the main aquifer units (C3 and C4) have different hydraulic potential than the C1 due to the aquitard behaviour of the C2 unit located between them. Hence a new hydrogeological conceptual model is required, one that differentiates the piezometric information depending on the geological unit reached in each pumping sounding and the effect of faults.

The integration of all the geological and geophysical data enabled us to establish a well constrained model with several reference surfaces (base and top of the units, faults, etc.). This georeferenced information can be exported to generate new and accurate flow and transport groundwater models.

6 Conclusion

A methodological procedure for evaluating groundwater reserves in carbonate aquifers is proposed based on 3D geological modelling and the estimation of the matrix porosity. It was applied in the SOV aquifer (Alicante, SE Spain), which has been altered by groundwater mining exploitation. This methodology is easily applicable to similar aquifers with low to moderate structural complexity, whenever a variety of geological and geophysical data are available.

The 3D geological model successfully reveals the structural geometry of the aquifer and suggests it be analysed as a whole, not as three different aquifers. Additional data, such as piezometric surfaces, geological cross-sections or isobar maps, provide further valuable information that could be used for new analysis. Therefore, 3D geological models appear essential to continue advancing in terms of hydrogeological knowledge.

Acknowledgments This work was financed by the Diputación Provincial de Alicante.

References

Ahr WM (2008) Geology of carbonate reservoirs: the identification, description, and characterization of hydrocarbon reservoirs in carbonate rocks editorial. John Wiley & Sons Inc, Hoboken, p 277

De Ruig MJ (1992) Tectono-sedimentary evolution of the prebetic fold belt of Alicante (SE Spain): a study of stress fluctuations and foreland basin deformation. Ph.D. thesis, Vrije Universiteit, Amsterdam

IGME-DPA (2006) Estudio del funcionamiento hidrogeológico y simulación numérica del flujo subterráneo en los acuíferos carbonatados de Solana y Jumilla-Villena. (Alicante y Murcia), p 117

IGME-DPA (2013) Atlas hidrogeológico de la provincia de Alicante. Instituto Geológico y Minero de España-Diputación Provincial de Alicante
Lucia FJ (1999) Carbonate reservoir characterization. Springer, Berlin, pp 1–226
Pulido-Bosch A, Motyka J, Pulido-Leboeuf P, Borczak S (2004) Matrix hydrodynamic properties of carbonate rocks from the Betic Cordillera (Spain). Hydrol Process 18(15):2893–2906
Wu Q, Xu H, Zhon W (2008) Development of a 3D GIS and its applications to karst areas. Environ Geol 54:1037–1045

Structural Characterization of a Karstic Aquifer Based on Gravity and Magnetics: Los Chotos-Sazadilla-Los Nacimientos (Jaén, SE Spain)

A. Ruiz-Constán, J.P. González de Aguilar, A. Pedrera, S. Martos-Rosillo, J. Galindo-Zaldívar and C. Martín-Montañés

Abstract The geological structure of Los Chotos-Sazadilla-Los Nacimientos carbonate aquifer (S of Jaén) and its hydraulic connection with the La Serreta-Gante-Cabeza Montosa carbonate aquifer have been established through geophysical prospecting–gravity and magnetics–structural measurements and study of piezometric levels. The scarce hydrogeological data, the complexity of the tectonic structure, and the presence of Plio-Quaternary rocks covering the permeable carbonate rocks make it difficult to establish a robust conceptual hydrogeological model. This study focuses on an area where hydraulic disconnection between the two aquifers was traditionally assumed, given the diapiric emplacement of low-permeable rocks between them. A connection between aquifers implies greater groundwater reserves than supposed up to present. These results are valuable for efficient and sustainable groundwater management.

Keywords Gravity data · Magnetic data · Geological structure · Carbonate aquifer

1 Introduction

Over the past two decades and in parallel to regional economic development, the demand and exploitation of aquifers has increased considerably in Spain (Molinero et al. 2011). In view of the decreasing piezometric levels, knowledge of water

A. Ruiz-Constán (✉) · J.P. González de Aguilar · S. Martos-Rosillo · C. Martín-Montañés
Instituto Geológico Y Minero de España (IGME), C/Ríos Rosas, 23, 28003 Madrid, Spain
e-mail: a.ruiz@igme.es

A. Pedrera · J. Galindo-Zaldívar
Instituto Andaluz de Ciencias de La Tierra (CSIC), Avda. de Las Palmeras 4, 18100 Granada, Spain

J. Galindo-Zaldívar
Departamento de Geodinámica, Universidad de Granada. Campus Fuentenueva S/N, 18071 Granada, Spain

B. Andreo et al. (eds.), *Hydrogeological and Environmental Investigations in Karst Systems*, Environmental Earth Sciences 1,
DOI 10.1007/978-3-642-17435-3_29

resources has become a top priority when designing management policies. A multi-method approach—integrating information from geology, hydrology, and diverse geophysical techniques—implies significant reductions in uncertainty (e.g., Duque et al. 2008), when determining the storage, flow path, and evolution of a groundwater system.

In this study, we jointly analyze magnetic and gravity data to characterize the geological structure of a carbonate aquifer in SE Spain: the Los Chotos-Sazadilla-Los Nacimientos aquifer (LSL). These data also allow us to define the hydrogeological continuity between different carbonate outcrops and two adjacent aquifers known as Cabra del Santo Cristo (CSC) and La Serreta-Gante-Cabeza Montosa (LGC). A hydrogeological inventory of piezometric levels was developed to better define the storage water reserves.

2 Los Chotos-Sazadilla-Los Nacimientos and Adyacent Aquifers

The Los Chotos-Sazadilla-Los Nacimientos (LSL) aquifer is located in SE Spain (Fig. 1). It comprises a series of carbonate outcrops (Ruiz Reig et al. 1988a, b): 9 km^2 of Jurassic dolostones and 20 km^2 of Miocene bioclastic calcarenites. The substrate is constituted by Triassic clays and marls with low permeability (Pérez-López 1991). Triassic rocks crop out to the N, W, and S, closing these borders to groundwater flow. However, the nature of the E-SE border, covered by semi-permeable Plio-Quaternary rocks, remains controversial. Renewable resources were estimated at 2.4 hm^3/year (IGME-DPJ 2011).

The area may be split up into three different sectors: (i) Los Chotos-Cortijo Hidalgo sector, in the eastern part, is formed by Jurassic dolostones of Cerro de los Peones and Cerro Los Chotos. (ii) Sazadilla sector, formed by Jurassic dolostones superposed by oolitic limestones; and (iii) Los Nacimientos sector, to the SW, an intensely fractured aquifer made up of Miocene sediments with impermeable red marls at the base.

Information about the hydraulic parameters of the aquifer is very limited. The pumping test of Cerro de los Peones (203930022) has an average transmissivity value of 9,950 m^2/day and a storage coefficient of 2×10^{-2} (ITGE 1990). Figure 1 shows the hydrogeological information available. Los Nacimientos spring, the main drainage point of the western sector, shows an altitude of 996 m.a.s.l., considerably higher than the rest of the aquifer. On the other hand, the piezometric level at Cerro de Los Peones (944 m.a.s.l. in July 2013) and Molino del Barranco spring (932 m.a.s.l.) suggests that the eastern sector has the same altitude as Molino de Gante spring (939 m.a.s.l.).

The cumulated piezometric decrease between April 2004 and August 2009 was 34 meters (Fig. 2). During the period of intense rain, the piezometric level was over the drainage level of the Molino del Barranco spring. Recovery was very fast

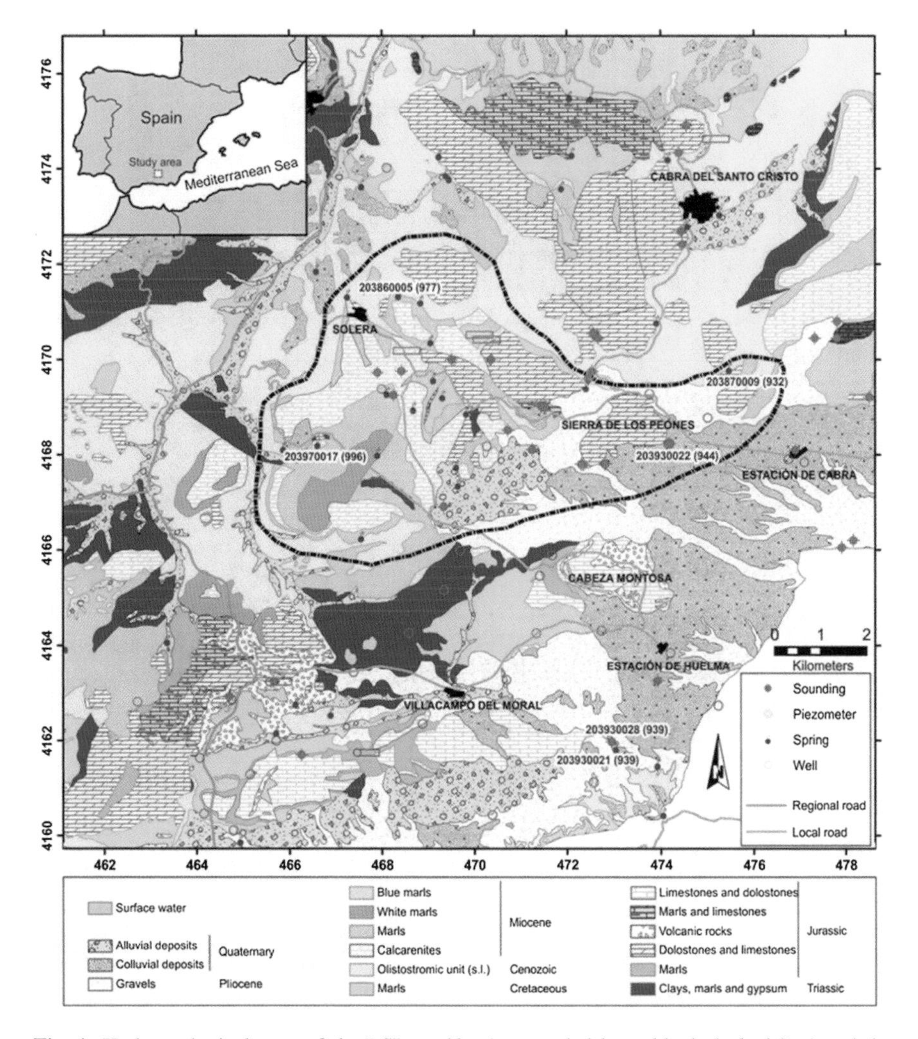

Fig. 1 Hydrogeological map of the LSL aquifer (surrounded by a *black dashed line*) and the adjacent aquifers of CSC and LGC. The discharge points with piezometric levels (in July 2013) are marked as follows: Fuente de las Negras spring (*203860005*), Los Nacimientos spring (*203970017*), Molino del Barranco spring (*203870009*), Cerro de Los Peones sounding (*203930022*), Molino de Gante spring (*203930021*) and Molino de Gante sounding (*203930028*). The map shows, in parentheses, the piezometric level in July 2013

as a result of intense rain for years, suggesting the existence of important reserves. In dry periods, the piezometric levels drop rapidly. This significantly decreasing rate could be justified by the presence of impermeable borders that accelerate the depression level of the pumping cone around the Cerro de los Peones well.

Two adjacent aquifers are related to the LSL aquifer. The CSC aquifer, to the north, is a perched aquifer formed by high permeable Jurassic dolostones and

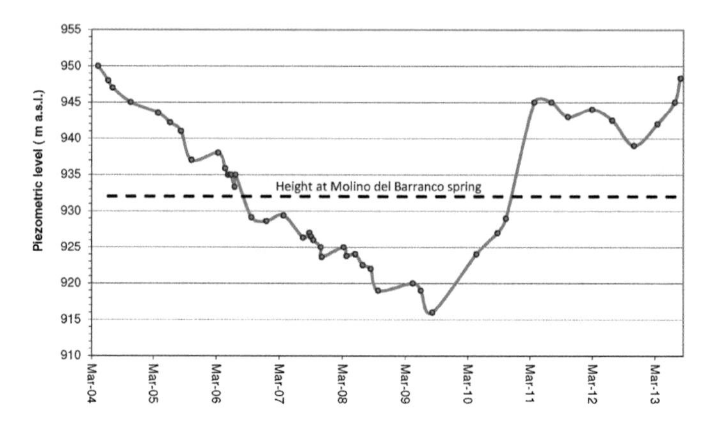

Fig. 2 Piezometric evolution of the cerro de los peones pumping well

limestones (11.4 km^2) over low permeable Triassic clays. The most important discharge point (20 l/s) is El Nacimiento at a height of 960 m.a.s.l. The second aquifer, the LGC aquifer, is formed by Jurassic carbonate rocks with interstratified volcanic rocks (ITGE 1991). Along its SE border there are two discharge points: the Molino de Gante pumping well and spring, at 939 m.a.s.l. Hidden discharge toward the detritic Plio-Quaternary sediments with a moderate permeability is reported around the eastern border (IGME-DPJ 2011).

3 Methodology

A total of 210 gravity stations (Fig. 3a), spaced at an average of 250-300 m, were acquired. They were integrated with 99 additional measurements from the IGME database (http://cuarzo.igme.es/sigeco/default.htm). Gravity data were acquired using a ScintrexAutograv CG-5 gravity meter with a maximum accuracy of 0.001 mGal. The station height was afterward fixed using a 5 m Digital Elevation Model (IGN; http://centrodedescargas.cnig.es/CentroDescargas/index.jsp). Measurements were referenced to the Granada base station of the IGN gravimetric network (http://www.ign.es/ign/main/index.do). The terrain correction was obtained by means of Hammer circles (Hammer 1982). Complete Bouguer anomaly was determined with a standard density of 2.6 g/cm^3 and considering the GRS67 Geodetic Reference System.

Simultaneously, 209 total magnetic field measurements (Fig. 3b) were gathered using a GSM 8 proton precession magnetometer with an accuracy of 1 nT. Susceptibility was measured by means of an Exploranium KT-9 kappameter. Total field magnetic anomalies were calculated through a standard procedure including reduction to the IGRF 2010 (IAGA 2010). Diurnal variations were

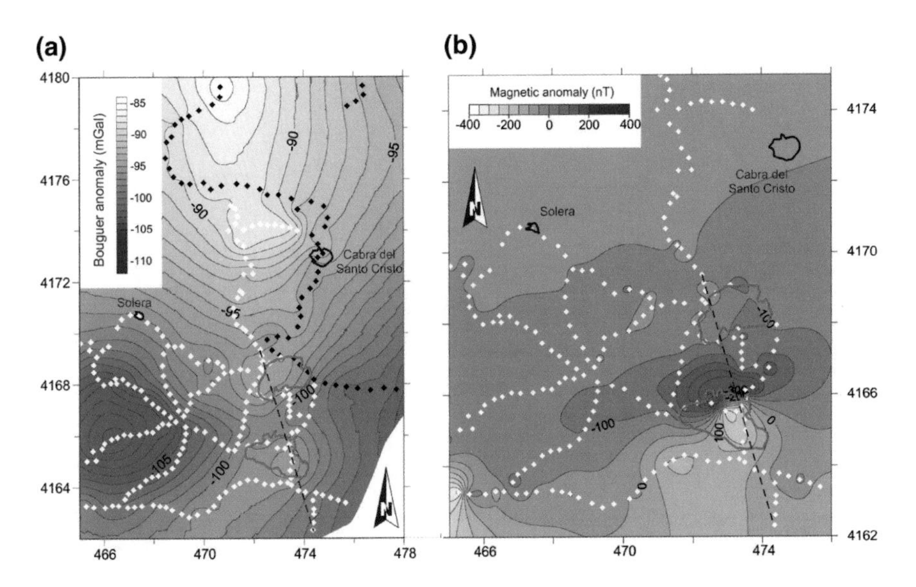

Fig. 3 Bouguer anomaly map (**a**) and Total Field magnetic anomaly map (**b**). *White* symbols are new data; *Black* symbols correspond to the IGME database. The profile modeled is marked. *Black* polygons are the villages and *red* contours the main carbonate outcrops

corrected using the continuous recording of the nearest Intermagnet observatory (www.intermagnet.org). The 2D integrated gravity and magnetic models (Fig. 4) were calculated using GRAVMAG V.1.7 software (Pedley et al. 1993).

4 Results

The tectonic structure of the LSL aquifer was studied, with special attention to its limit with the LGC aquifer. Cabeza Montosa is a N100°E N-vergent antiform with a southern limb dipping at 25–50°. In the northern one, interstratified submarine basalts crop out as erosion has removed part of the Jurassic limestones. The western border of Cabeza Montosa is the periclinal end of the antiform. To the north, the Cerro de los Peones is formed by Jurassic dolostones constantly dipping S. They, in turn, are superposed by Triassic clays, sandstones and gypsum with ductile deformation. Neogene and Quaternary sediments crop out between these two carbonate outcrops.

The Bouguer anomaly (Fig. 3a) shows a southward decreasing trend related to the regional continental crust thickening. The gravity maximum is located in the eastern border of Sierra Mágina (−85 mGal), while the minimum (−111 mGal) is found 5 km south of Solera. Bouguer anomaly values at the main hills of Cabeza Montosa and Cerro de los Peones are −95 and −97 mGal, respectively. Between them, there is another relative minimum of −100 mGal.

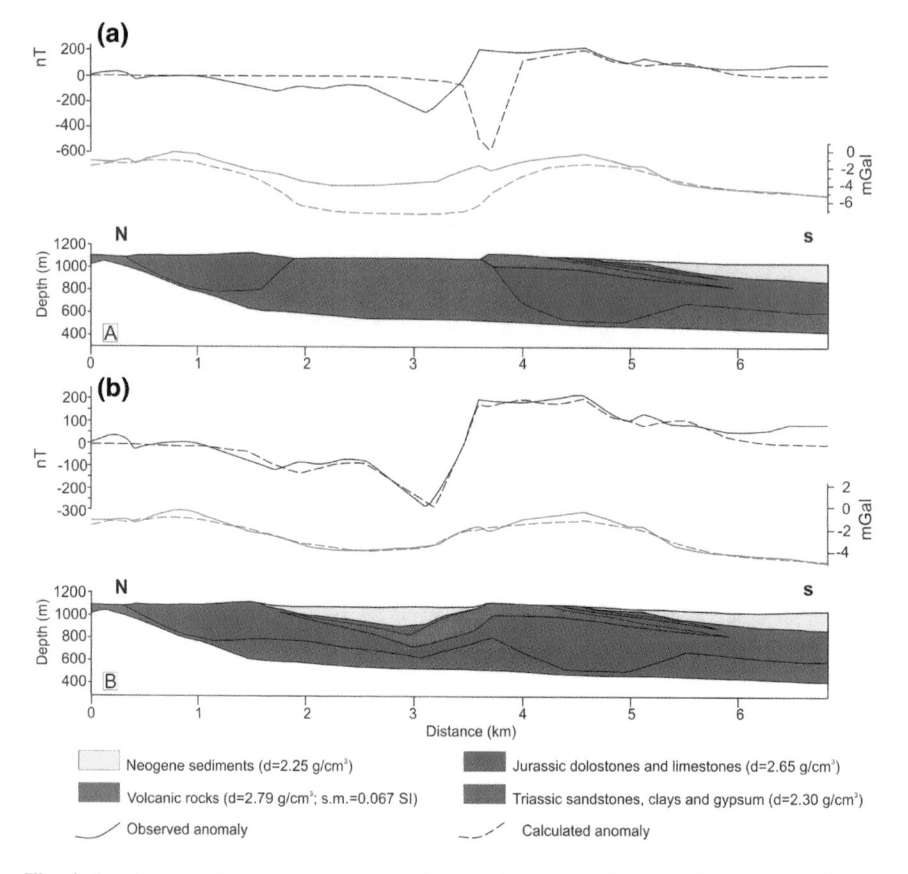

Fig. 4 2D integrated gravity and magnetic models, with (**a**) and without (**b**) hydraulic connection (**d**, density; m.s.,magnetic susceptibility)

The most remarkable feature in the magnetic anomaly map (Fig. 3b) is the magnetic dipole whose minimum is located to the N of Cabeza Montosa (−389 nT) and maximum (+223 nT) to the south of the same hill. Furthermore, a second magnetic dipole is observed 8 km to the W of Cabeza Montosa, although the minimum is poorly defined due to the low coverage of stations in this area. Both dipoles are related to the presence of Jurassic pillow lavas.

A joint gravity and magnetic 2D model (Fig. 4) was performed in a NNW-SSE profile. The average density assigned to each geological unit is related to the main lithology observed (Telford et al. 1990). For modeling purposes, the reference level of the magnetic anomaly was displaced 40 nT to fit the zero value, as a consequence of overestimation of the IGRF 2010 in the study area.

We considered two 2D models, with (Fig. 4b) and without (Fig. 4a) hydraulic connection between the two carbonate outcrops. It was not possible to fit either the gravity anomaly or the magnetic one with the presence of a diapiric body between them (Ruiz Reig et al. 1988b). The best fit was obtained with a synformal geometry of the limestones and the volcanic rocks. The maximum thickness of the main body of volcanic rocks is in the range of 150–200 m, and it thins along the northern limb of the synform. Its interstratified character points to the hydraulic connection hypothesis. Field measurements of magnetic susceptibility in the pillow lavas give a mean value of 0.016 SI, which is in the range proposed for these rocks (Telford et al. 1990). However, the best fit of the model is obtained considering 0.067 SI, slightly higher than field measurements, which implies the existence of remnant magnetization.

5 Discussion and Conclusions

Spatial distribution of the piezometric levels suggests a certain degree of hydrogeological compartmentalization in the LSL aquifer. On the one side, springs and wells of the Los Nacimientos sector have piezometric levels around 1,000 m.a.s.l. On the other hand, data from Molino del Barranco spring (932 m.a.s.l.) and the Cerro de Los Peones well (944 m.a.s.l.), in the Los Chotos-Cortijo Hidalgo sector, suggest there is no hydrogeological connection between the two sectors.

Likewise, the hydrogeological continuity of LSL and LGC aquifers is suggested by the fast recovery of the piezometric levels in the water supply well of Cerro de los Peones and the similarity of the springs heights. This information is relevant to determine the storage reserves available for water supply. Cumulated decrease in dry periods of the piezometric level at Cerro de los Peones can be explained by the presence of lateral impermeable borders of null flow that accelerate the decrease of the pumping cone. In wet periods, in contrast, the recharge of a larger aquifer means faster recovery of the groundwater level in this sector.

In addition, the joint modeling of gravity and magnetic data together with geological information points to the absence of a prominent Triassic diapir disconnecting the carbonate outcrops of Cerro de los Peones and Cabeza Montosa. Such evidence suggests a need to redefine the limits of these formerly disconnected aquifers, and would explain the increase in the volume of groundwater reserves in the area.

Acknowledgments This work was financed by the Diputación Provincial de Jaén and through the project CGL-2010-21048, and the Junta de Andalucía group RNM148.

References

Duque C, Calvache ML, Pedrera A, Martín-Rosales WM, López-Chicano M (2008) Combined time domain electromagnetic soundings and gravimetry to determine marine intrusion in a detrital coastal aquifer (Southern Spain). J Hydrol 349:536–547

Hammer S (1982) Critique of terrain corrections for gravity stations. Geophysics 47:839–840

IGME-DPJ (2011) Atlas hidrogeológico de la provincia de Jaén, p 142–151

ITGE (1990) Informe final de la perforación y aforo realizados para abastecimiento en el término municipal de Cabra del Santo Cristo (Jaén), pp 20

IAGA, Working Group V-MOD (2010) International geomagnetic reference Field: the eleventh generation. Geophys J Int 183(3):1216–1230

Molinero J, Custodio E, Sahuquillo A, Llamas MR (2011) Groundwater in Spain: Legal framework and management issues. In: AN Findikakis and K Sato (eds) Groundwater Management Practices. CRC Press/Balkema, Leiden, pp 123–137

Pedley RC, Busby JP, Dabeck ZK (1993) Gravmag User Manual- Interactive 2.5D gravity and magnetic modelling.British Geological Survey, Technical Report, WK/93/26/R, pp 75

Pérez-López A (1991) El trías de facies germánica del sector central de la cordillera bética. Tesis doctoral. Universidad de Granada, Granada, p 400

Ruiz Reig P, Álvaro López M, Hernández Samaniego A, del Olmo Zamora P (1988a) Mapa Geológico Nacional. Escala 1:50.000. Hoja 948 (Torres). IGME

Ruiz Reig P, Díaz de Neira Sánchez JA, Enrile Albir A, López Olmedo F (1988b) Mapa Geológico Nacional. Escala 1:50.000. Hoja 970 (Huelma). IGME

Telford WM, Geldart LP, Sheriff RE (1990) Applied geophysics, Cambridge University press, Cambridge, pp 770

Fractal Modeling and Estimation of Karst Conduit Porosity

E. Pardo-Igúzquiza, J.J. Durán, P.A. Robledo-Ardila and C. Paredes

Abstract The three-dimensional distribution of karst conduits in a karst aquifer is the main source of its heterogeneity and anisotropy. It also has a strong effect on peak spring discharge, and in the form of the karstic hydrograph, chemograph, and thermograph. The direct access to the conduits is only possible by speleological exploration and cave mapping, which provides a very valuable information for karst modeling. However, this information is biased because the speleologists can only explore a limited part of the conduit network: they can only map conduits of a minimum diameter, the speleological exploration can take many years in order to explore all the leads that appear in a network system and parts of the network are not accessible. In order to estimate conduit porosity we can take advantage of the fractal character of nature. The volume of conduits larger than a given diameter can be estimated in a particular volume of rock. This is done by extrapolating a power law that has been fitted to experimental data. The power law distribution of the volume of conduit voids larger than a particular diameter is related with the fractal behavior of a network of karst conduits. This fact was also used for simulating a three-dimensional network of karst conduits by a stochastic process of diffusion limited aggregation. The procedure is illustrated in the Sierra de las Nieves karst aquifer in the province of Malaga, Southern Spain. Karst conduit

E. Pardo-Igúzquiza (✉) · J.J. Durán · P.A. Robledo-Ardila
Geological Survey of Spain, Instituto Geológico y Minero de España (IGME), C/Ríos Rosas, 23, 28003 Madrid, Spain
e-mail: e.pardo@igme.es

J.J. Durán
e-mail: jj.duran@igme.es

P.A. Robledo-Ardila
e-mail: pa.robledo@igme.es

C. Paredes
Departamento de Matemática Aplicada, Escuela Técnica Superior de Ingenieros de Minas de Madrid (ETSIM), Ríos Rosas 21, 28003 Madrid, Spain
e-mail: cparedes@dmami.upm.es

B. Andreo et al. (eds.), *Hydrogeological and Environmental Investigations in Karst Systems*, Environmental Earth Sciences 1,
DOI 10.1007/978-3-642-17435-3_30

porosity can be used for generation of numerical models of karst systems, and for the mathematical modeling of flow in karst aquifers.

Keywords Karst conduits · Fractals · Power law · Porosity

1 Introduction

The intuitive idea of a fractal is that of a complex object that looks similar at many scales of observation (Mandelbrot 1967). In general, they are very irregular objects that partially fill a space of topological dimension higher than their proper topological dimension. The fractal dimension describes statistically in a single number the complexity of the object (Mandelbrot 1983). It can be considered an index of complexity or irregularity. Fractals are ubiquitous in geosciences: the coastline, the landscape, the seafloor topography, fractures, stream networks, etc. It is very well known the fact that maps or when taking photographs in the field there is the need for a scale. In karst terrains fractals are also ubiquitous (Fig. 1): the karst landscape, the surface of dissolution features, porosity, caves geometry, the network of

Fig. 1 Examples of fractal features in karst terrains: **a** landscape; **b** fractures; **c** karren and kamenitsas, and **d** caves and karst conduits. Examples from the Sierra de las Nieves karst aquifer, Southern Spain

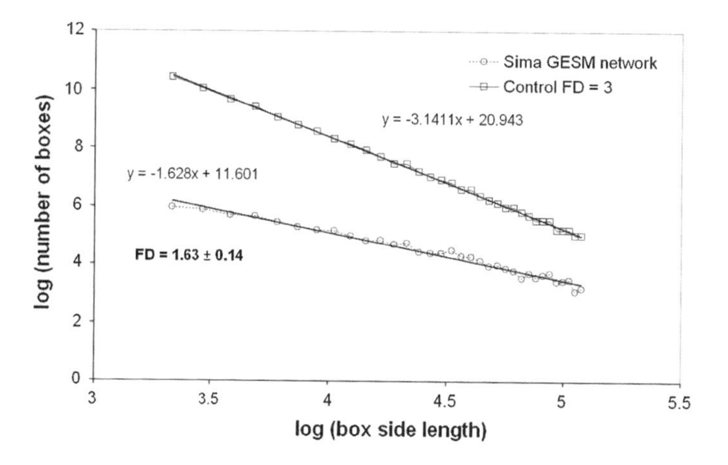

Fig. 2 Fractal dimension of Sima GESM in Sierra de las Nieves. GESM is the deepest cave system of Southern Spain, with more than 1,000 m depth

karst conduits, volumes of karst spring discharge, etc. Directly related with the fractal geometry it is the fact that many objects that occur in nature over different scales show a power-law distribution of relative abundance. That is particularly true for phenomena related to rock fracture (Pickering et al. 1999): earthquakes, fault displacements, fault and fracture trace lengths, fracture apertures, etc. Karst conduit volumes can be considered as fractures or other rock discontinuities (like bedding or the contact between different lithologies) widened by carbonate dissolution under the action of flowing water. A three-dimensional network of karst conduits can be considered a fractal with a typical geometry dimension of 1.67 (Jeannin et al. 2007). Apart from the experimental verification by calculating the fractal dimension of experimental data (Fig. 2), experimentally derived objects resemble simulated objects obtained from fractal models like the conduit networks (Fig. 3) obtained by diffusion-limited aggregation that resembles karst networks (Pardo-Igúzquiza et al. 2012). The purpose of this paper is to study the fitting of a power law of relative abundance to the volume of conduits that are larger than a given diameter. It can be considered an extension of Pardo-Igúzquiza et al. (2014) taking account more suitable methods for fitting a power law to a limited number of the experimental data. The methodology is applied to the network of conduits of the Sierra de las Nieves karst aquifer in the province of Málaga (Southern Spain).

2 Methodology

An application of fractal modeling is the description of scale-invariant distributions and the extrapolation to smaller and larger scales beyond the measured range. The hypothesis for scale invariance using a frequency-size distribution is that the

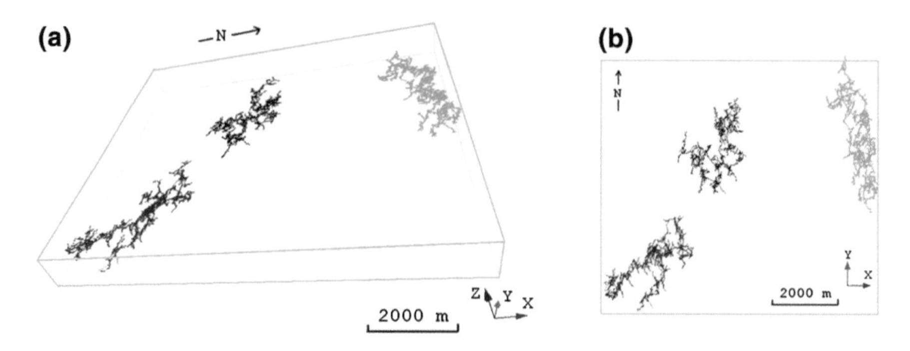

Fig. 3 Three-dimensional networks of karst conduits simulated by a diffusion limited aggregation process that resembles three-dimensional karst networks

cumulative volume of conduits (V) that have a diameter greater than (D) follows a power law of the form:

$$V(D) = kD^{-FD}, \quad (1)$$

where FD is the fractal dimension and k is an empirical fitting parameter. If the distribution is scale invariant, the experimental data will plot as a straight line in a log-log plot (Fig. 5).

The volume is calculated by using the speleological data and assuming that the conduits are cylindrical. The fractal behavior, in Fig. 5, may be seen by the straight-line part in the center of the graph, while at both ends the behavior of the graph is not linear because two kinds of artifacts. On the left hand part there is an artifact due to truncation, that is, measurement limited resolution leads to incomplete sampling of smaller features and underestimation of smaller features. While on the right hand side there is an artifact due to censoring, that is, limited size of the sampled volume (or area) leads to censoring of larger features and underestimation of larger features.

3 Case Study

The study area is the Sierra de las Nieves karst aquifer in the province of Málaga, Southern Spain. The karst massif consists mainly of a succession of carbonate rocks: Triassic marbles and dolostones, Jurassic limestones and a Tertiary carbonatic breccia. The Mesozoic sequence is folded by an NE–SW trending overturned syncline with a vergency toward the NW (Liñán 2005). The carbonatic breccia unconformably lies over the Mesozoic succession, but is also deformed by the fold. After five decades of speleological exploration, more than 26 km of conduits have been mapped including the deepest cave of Southern Spain (Sima GESM or GESM pothole).

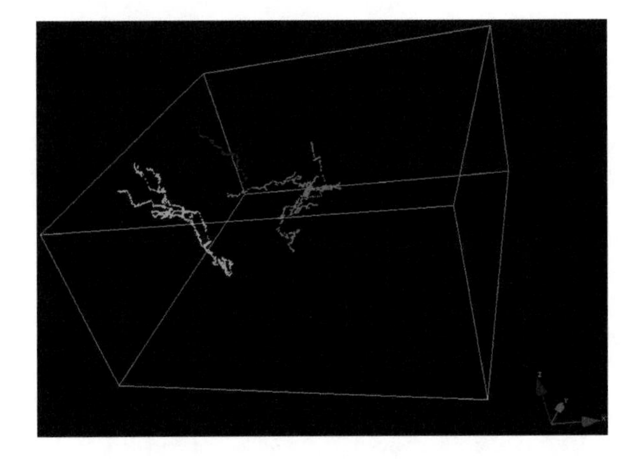

Fig. 4 Conduit networks in the sierra de las nieves aquifer. Sima GESM (*green*) Sima del Aire (*light blue*) and Sima Prestá (*dark blue*). The lenght of the side of the cube is 2 km

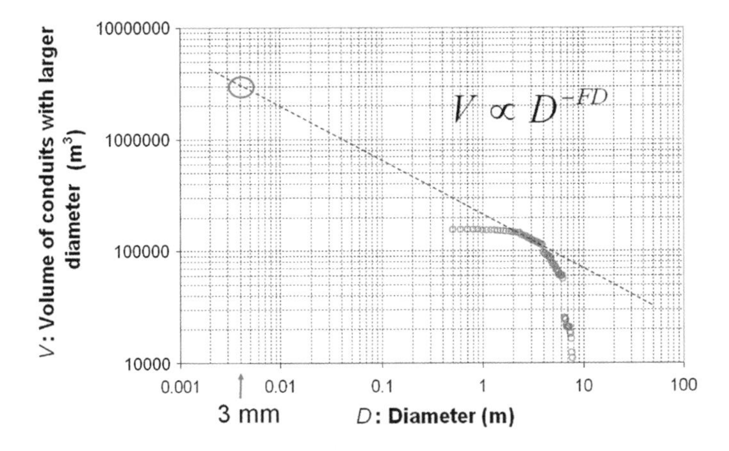

Fig. 5 Log-log plot of the karst conduits in the Sierra de las Nieves aquifer. From the figure it is found that the volume of conduits larger than 3 mm is 3 millions of cubic meters

The three main karst networks (Sima GESM, Sima del Aire, and Sima Prestá) are shown in Fig. 4 and the log-log plot of the distribution of volume of karst conduits has been represented in Fig. 5. A model has been fitted to the power law and extrapolation is used to calculate that the volume of conduits larger than 3 mm is of 3×10^6 m^3 in a total volume of 1.28×10^9 m^3, which implies a conduit porosity of 0.23 %. This value can be used to extrapolate to the rest of the study area draining to the Rio Grande spring, and to simulate networks of karst conduits with the required specifications.

4 Discussion and Conclusions

The spatial distribution of conduits in a karst system introduces a huge heterogeneity in a karst aquifer. The conduit porosity is defined here as the ratio of the volume of conduits larger than 3 mm in diameter with respect to the total volume. Conduit porosity has been calculated for the Sierra de las Nieves aquifer using extrapolation from a power law fitted to experimental cave mapping data. The conduit porosity has been estimated as 0.23 % that should be added to rock matrix and fracture porosity.

Acknowledgments This work was supported by the research project CGL2010-15498 from the Ministerio de Economía y Competitividad of Spain. We would like to thank the groups of speleology that work in the Sierra de las Nieves: Grupo de Exploraciones Subterráneas de la Sociedad Excursionista de Málaga, Interclub Sierra de las Nieves (Ronda), Centro Excursionista del Sur Escarpe, MAINAKE Sociedad Espeleo-Excursionista y el Grupo de Exploraciones Subterráneas de Tolox. Also we would like to thank the managers of the Parque Natural de la Sierra de las Nieves.

References

Jeannin P-Y, Groves C, Häuselaman P (2007) Chapter 3: Speleological investigations. In: Goldscheider N, Drew D (eds) Methods in karst hydrogeology. Taylor and Francis, London, pp 264

Liñán C (2005) Hidrogeología de acuíferos carbonatados en la unidad Yunquera-Nieves (Málaga). Publicaciones del Instituto Geológico y Minero de España, Madrid. Serie: Hidrogeología y Aguas Subterráneas 16, pp 322

Mandelbrot BB (1967) How long is the coast of Britain? Statistical self-similarity and fractional dimension. Science 156:636–638

Mandelbrot BB (1983) The fractal geomety of nature. W. H. Freeman and Co, New York, p 495

Pardo-Igúzquiza E, Dowd PA, Xu C, Durán JJ, Rodriguez-Galiano V (2012) Stochastic simulation of karst conduit networks. Adv Water Resour 35:141–150

Pardo-Igúzquiza E, Durán JJ, Robledo P, Guardiola C, Antonio Luque J, Martos S (2014) Fractal modelling of karst conduits. In: Pardo-Igúzquiza E, Guardiola Albert C, Heredia J, Moreno-Merino L, Durán JJ, Vargas-Guzmán JA (eds) Mathematics of planet earth, Springer, Berlin, Lecture Notes in Earth System Sciences, pp 217–220

Pickering G, Bul JM, Sanderson DJ (1999) Fault populations and their relationship to the scaling of surface roughness. J Geophys Res Solid Earth 104(B2):2691–2701

Integral Porosity Estimation of the Sierra de Las Nieves Karst Aquifer (Málaga, Spain)

E. Pardo-Igúzquiza, J.A. Luque-Espinar, J.J. Durán, A. Pedrera, S. Martos-Rosillo, C. Guardiola-Albert and P.A. Robledo-Ardila

Abstract Karst aquifers are very complex and heterogeneous systems because of the presence of three kinds of porosity (matrix rock porosity, fracture porosity, and conduit porosity) that generally have a large spatial variability. In order to have realistic karst models the three kinds of porosity and their spatial variability must be taken into account. A quantitative model of a karst aquifer is proposed by integration of the three kinds of porosity in a three dimensional numeric model. Nevertheless, the main task of this work is restricted to the proposal of methods for their evaluation. Matrix rock porosity has been measured in the laboratory from samples collected in the field. Matrix rock porosity is well correlated with the lithology and with the structural position of the rock. Fracture porosity has been estimated from fracture mapping and field measurements. A geostatistical method is used to obtain a continuous field of fracture porosity. Conduit porosity has been

E. Pardo-Igúzquiza (✉) · J.A. Luque-Espinar · J.J. Durán · S. Martos-Rosillo · C. Guardiola-Albert · P.A. Robledo-Ardila
Geological Survey of Spain, Instituto Geológico y Minero de España (IGME), C/Ríos Rosas, 23, 28003 Madrid, Spain
e-mail: e.pardo@igme.es

J.A. Luque-Espinar
e-mail: ja.luque@igme.es

J.J. Durán
e-mail: jj.duran@igme.es

S. Martos-Rosillo
e-mail: s.martos@igme.es

C. Guardiola-Albert
e-mail: c.guardiola@igme.es

P.A. Robledo-Ardila
e-mail: pa.robledo@igme.es

A. Pedrera
Instituto Andaluz de Ciencias de La Tierra (CSIC), Avda. de Las Palmeras 4, 18071 Granada, Spain
e-mail: pedrera@ugr.es

B. Andreo et al. (eds.), *Hydrogeological and Environmental Investigations in Karst Systems*, Environmental Earth Sciences 1,
DOI 10.1007/978-3-642-17435-3_31

calculated from a power model fitted to speleologic cave mapping data. However, because of the scarcity of conduit data, probabilistic models must be conjectured. The integration of the three kinds of porosity gives a three dimensional numerical model that can be used in vulnerability mapping, recharge estimation, and mathematical modeling of flow and transport in karst systems. The approach is illustrated with the Sierra de las Nieves karst aquifer in the province of Málaga in Southern Spain.

Keywords Matrix porosity · Fractures · Karst conduits · Fractals

1 Introduction

Karst aquifers are very complex and heterogeneous systems because of the presence of three kinds of porosity: rock matrix porosity, fracture porosity, and conduit porosity (Ford and Williams 2007). The three kinds of porosity may have a large spatial variability both across the horizontal and along the vertical of the karst massif. The presence of conduit porosity is indicative of a well developed karst, while the porous and fracture porosity are very important for the early evolution and development of the system flow (Kaufmann and Braun 2000). While rock matrix porosity is more or less homogeneous, at least for a given lithology and fractures show a large heterogeneity with zones with a high fracture density, the system of karst conduits is the factor that introduces the highest heterogeneity in the karst massif. Also while matrix rock porosity may have certain permanence in the vertical dimension, the number of fractures decreases generally with depth and the system of karst conduits has a large variability in the vertical dimension as shown by the Z-histogram (Pardo-Igúzquiza et al. 2011). Any numerical model of the karst system must take into account these three kinds of porosity and its evaluation is paramount for hydro-dynamical modeling and vulnerability mapping. In this paper, there is a description of the methodologies for the modeling of the three kinds of porosity, and it shows the results obtained in the Sierra de las Nieves karst aquifer in Southern Spain.

2 Methodology

Rock matrix porosity was measured in the laboratory from samples taken from rock outcrops. This field sampling was done methodically in order to sample the different lithologies and to cover the area homogeneously. The rock matrix porosity may be considered a spatial variable, and different interpolation techniques can be used to obtain a continuous map from a limited number of representative measurements. Fracture porosity may be estimated in a two stage process. In the first stage, a map of fracture density is generated from field work and remote sensing images (satellite images and aerial photographs). A continuous map is generated by geostatistical spatial interpolation (Olea 1999). In the second stage, fractal porosity is calculated from karst rock exposures (Fig. 1) in the field and from photographs (Mace et al. 2005). A correspondence is done between the fracture porosity and fracture density in order to have the map of fracture porosity, where the highest values of fracture porosity is associated with the largest values of fracture density. Finally, conduit porosity is calculated by using a fractal approach (Pardo-Igúzquiza et al. 2014) where a power law distribution is fitted to the experimental distribution of the volume of conduits larger than a given diameter (caves accessible by speleologists). Then, the power law is used to extrapolate the volume of conduits to a chosen diameter that is considered the limit between a conduit and a fracture. The methodology is illustrated with data from the Sierra de las Nieves karst aquifer.

3 Case Study

The study area (Fig. 2) is the Sierra de las Nieves karst aquifer, which geology map is shown in Fig. 3a. The karst massif consists mainly of a succession of carbonate rocks: Triassic marbles and dolostones, Jurassic limestones, and a Tertiary carbonatic breccia. The Mesozoic sequence is folded by an NE-SW trending overturned syncline with a vergency toward the NW (Liñán 2005). The carbonatic breccia unconformably lies over the Mesozoic succession, but is also deformed by the fold. Twenty four rock samples were taken across the region, in order to sample the different lithologies and the area homogeneously. The matrix rock porosity was calculated in the laboratory and the results are given in Table 1 where the mean of the porosities have been calculated for each lithology. The geology map of Fig. 3a may be considered a map of rock matrix porosity by assigning the mean porosities of Table 1 to the lithologies in Fig. 3a. The interpolation by a geostatistical approach has not been considered because it has been found the lack of a spatial continuity in the variability of the porosity. Different porosities are confined to each lithology.

The map of fracture porosity is shown in Fig. 3b. It has been calculated following the two steps described in the methodology section: the combination of a map of density of fractures calculated using field data and remote sensing data (Luque et al. 2012) and fractal fracture porosity (Fig. 1) where a value of 7 % has been assigned to the maximum fracture density and a linear decrease with the density of fractures. This simple procedure is shown for illustration purposes only and a more careful treatment can provide more realistic results. Additionally, from a map of dolines (more than 300 dolines mapped in Fig. 3c) allows a map of density of dolines to be calculated where to the largest doline density has been assigned a porosity (of the detrital filling that can reach more than 30 m of thickness) of 6 % and a linear decrease with the density of dolines has been assumed.

Fig. 1 Examples of fracture porosity (**a** 10 %; **b** 3.5 %; **c** 5.5 % and **d** 7 %) in Sierra de las Nieves karst massif calculated along transects (*red line*) in the epikarst

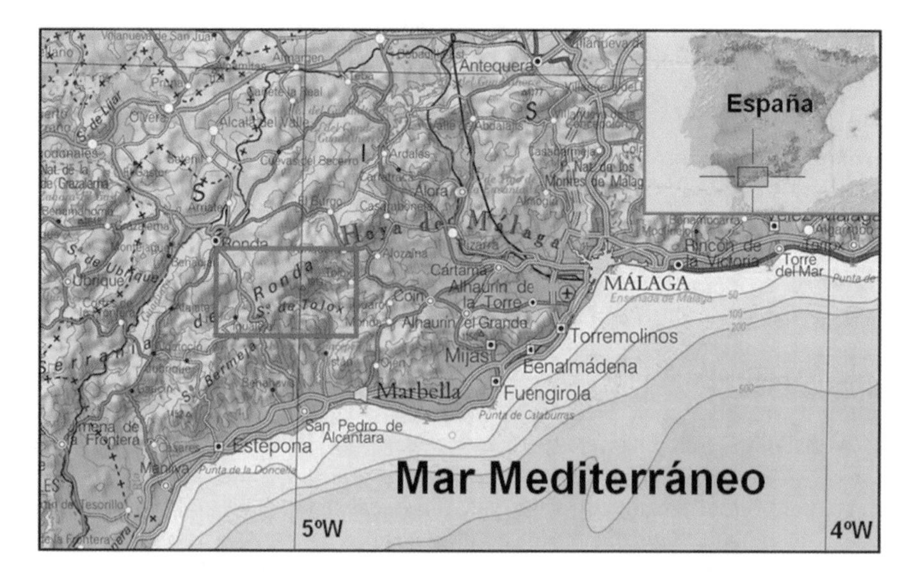

Fig. 2 Geographycal location of the Sierra de las Nieves aquifer in Southern Spain

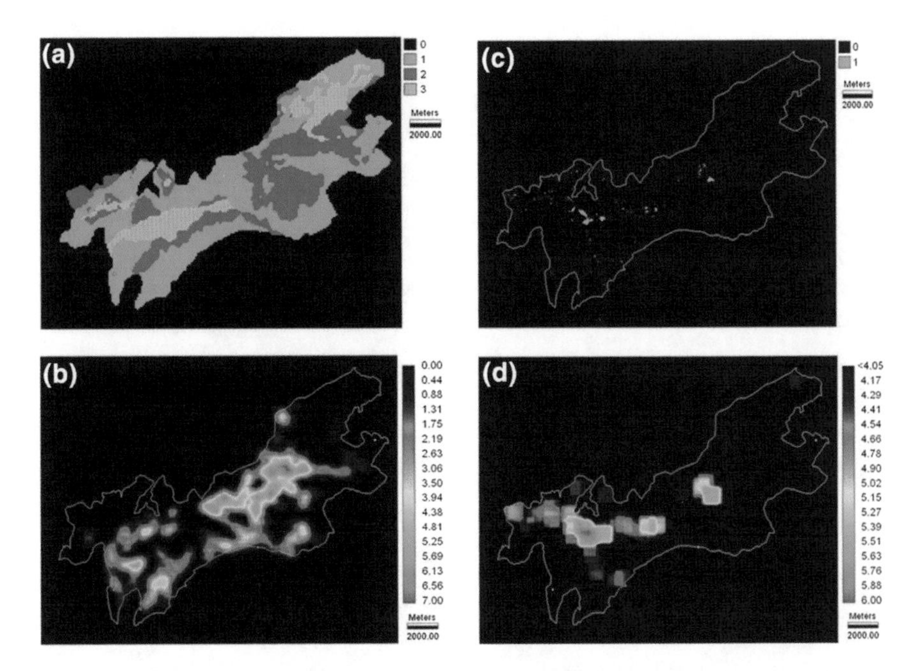

Fig. 3 **a** Simplified geology map of the aquifer (*1* dolostone; *2* limestone; *3* breccia). This is a map of rock matrix porosity by taking into account porosities of Table 1. **b** Map of fracture porosity. **c** Map of karst depressions. **d** Map of porosity of the detritic filling of depressions

Table 1 Matrix rock porosity calculated in the laboratory for the different lithologies

Lithology	Number of data	Mean (%)	Standard deviation (%)
Dolostone	11	0.945	0.758
Limestone	7	1.457	0.896
Breccia	6	2.766	1.659

With respect to the estimation of conduit porosity, the fractal procedure is described in detail in Pardo-Igúzquiza et al. (2014). The actual estimated value is of 0.25 %. However, the conduit porosity has a very high spatial variation. Figure 4a shows the surface density of potholes, while Fig. 4b shows the density of karst conduits projected on the surface. Furthermore, there is a large variation of density of conduits along the vertical as shown by the Z-histogram of Fig. 5. The location of the conduit porosity can be done probabilistically considering Figs. 4 and 5.

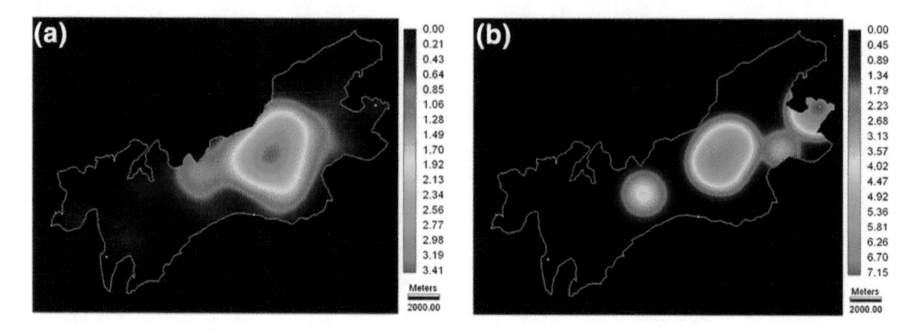

Fig. 4 **a** Density of potholes; **b** Density of the karst conduits (3D network in the interior of the massif) projected on the surface

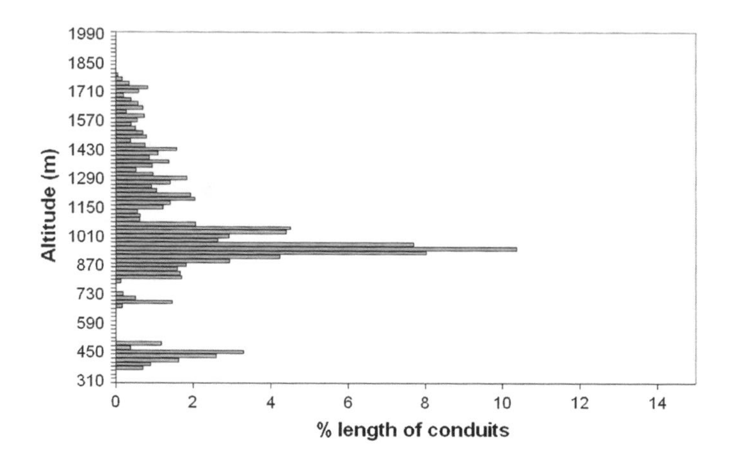

Fig. 5 Z-histogram of the mapped conduits in the Sierra de las Nieves aquifer. That is, the density along vertical of the mapped conduits

4 Discussion and Conclusions

The distribution of porosity in a karst aquifer must consider the three main kinds of karst porosity: porosity of the rock matrix, fracture porosity, and conduit porosity. The three kinds of porosity have a large spatial variability both along the horizontal and along the vertical. The porosity of the rock matrix is more homogeneous across the aquifer, and can be resolved by assigning a value to each lithology, but keeping in mind that there is variability inside each lithology and depending of the structural location of the rock. Fracture porosity is the easiest to calculate on the surface by mapping fracture density and by estimating fractal facture porosity in rock exposures. Fracture porosity decreases with depth, and can be estimated by using some mathematical function that describes its decrease with depth. It could be also correlated with a map of the epikarst in order to evaluate a layer with significant fracture porosity close to the surface, although this is not shown in this work. Finally, conduit porosity is the most difficult to quantify unless it is done in probabilistic terms. By using the data available on the number of potholes and caves mapped on the surface as well as a speleology mapping of karst conduit development and the Z-histogram, it will be possible to build a probabilistic model of occurrence of conduits. A limiting value will be the conduit porosity calculated by extrapolating a power law fitted to the conduit data mapped by the speleologists. A final conclusion is that the close cooperation with the speleological community must be considered in order to succeed in a numerical modeling of the hydrogeology of a karst aquifer. In this work, we have considered just a single hydrogeological variable, porosity, and we have shown the way to its quantification.

Acknowledgments This work was supported by the research project CGL2010-15498 from the Ministerio de Economía y Competitividad of Spain. We would like to thank the groups of speleology that work in the Sierra de las Nieves: Grupo de Exploraciones Subterráneas de las Sociedad Excursionista de Málaga, Interclub Sierra de las Nieves (Ronda), Centro Excursionista del Sur Escarpe, MAINAKE Sociedad Espeleo-Excursionista y el Grupo de Exploraciones Subterráneas de Tolox. Also we would like to thank the managers of the Parque Natural de la Sierra de las Nieves.

References

Ford D, Williams P (2007) Karst hydrogeology and geomorphology. Wiley, Chichester, p 562

Kaufmann G, Braun J (2000) Karst aquifer evolution in fractured, porous rocks. Water Resour Res 36(6):1381–1391

Liñán C (2005) Hidrogeología de acuíferos carbonatados en la unidad Yunquera-Nieves (Málaga). Publicaciones del Instituto Geológico y Minero de España, Madrid. Serie: Hidrogeología y Aguas Subterráneas 16, p 322

Luque JA, Pedrera A, Martos-Rosillo S, Pardo-Igúzquiza E, Durán JJ, Pedro Agustín Robledo-Ardila y Carolina Guardiola-Albert (2012) Cartografía geoestadística de la densidad de fracturación del macizo kárstico de Sierra de las Nieves (Málaga). In El Agua en Andalucía. Retos y avances en el inicio del milenio. In: López-Geta JA, González GR, Rubio RF, Fernández DL (eds) Publicaciones del Instituto Geológico y Minero de España. Serie Hidrogeología y Aguas Subterráneas No. 30 Tomo 2, pp 1501–1510, ISBN: 978-84-7840-863-4

Mace RE, Marrett RA, Hovorka SD (eds) (2005) Fractal scaling of secondary porosity in karstic exposures of the Edwards Aquifer. Sinkholes and the engeneering and environmental impacts of karst. ASCE, Geotechnical Spatial Publication 144, Proceedings of the Tenth Multidisciplinary Conference, pp. 1–10

Olea R (1999) Geostatistics for engineers and earth scientists, Springer, Berlin, p 322

Pardo-Igúzquiza E, Durán JJ, Rodríguez-Galiano V (2011) Morphometric analysis of three-dimensional networks of karst conduits. Geomorphology 132(1–2):17–28

Pardo-Igúzquiza E, Durán JJ, Robledo P, Guardiola C, Luque JA, Martos S (2014) Fractal modelling of karst conduits. In: Pardo-Igúzquiza E, Albert CG, Heredia J, Moreno-Merino L, Durán JJ, Vargas-Guzmán JA (eds) Mathematics of planet earth, Springer, Berlin. Lecture Notes in Earth System Sciences, pp 217–220, ISBN: 978-3-642-32407-9

A Three-Dimensional Karst Aquifer Model: The Sierra de Las Nieves Case (Málaga, Spain)

E. Pardo-Igúzquiza, J.J. Durán and P.A. Robledo-Ardila

Abstract The mathematical modeling of karst aquifers has been limited to black box models many times where the input (recharge) is related to the output (spring discharge) by using a transfer function from system theory and the spectral analysis of time series. These so-called black box models have a limited use because they do not provide information on the local spatial characteristics of the karst aquifer. In order to have a mathematical spatio-temporal model of the karst hydrogeology, several issues must be addressed, including a three-dimensional numeric model of the karst system, a spatial evaluation of recharge, and a consideration of the epikarst, karst depressions, fracture zones, thick vadose zones, and the simulation of conduit flow and karst spring discharge. All these characteristics have been taken into account in the Sierra de las Nieves karst aquifer in the Malaga province, Southern Spain. Furthermore, this aquifer has three hydrogeologic basins, with three main discharge points. It is concluded that spatially distributed karst models have following requirements: the integration of all the available information, the availability of speleologic cave mapping, the development of numerical karst system simulation methods and karst models of flow in the karst system. The problems encountered are discussed and the proposed solutions are described.

Keywords Karst conduits · Spring · Vadose zone · Hydrodynamics · Modeling

E. Pardo-Igúzquiza (✉) · J.J. Durán · P.A. Robledo-Ardila
Geological Survey of Spain, Instituto Geológico y Minero de España (IGME), C/Ríos Rosas, 23, 28003 Madrid, Spain
e-mail: e.pardo@igme.es

J.J. Durán
e-mail: jj.duran@igme.es

P.A. Robledo-Ardila
e-mail: pa.robledo@igme.es

B. Andreo et al. (eds.), *Hydrogeological and Environmental Investigations in Karst Systems*, Environmental Earth Sciences 1,
DOI 10.1007/978-3-642-17435-3_32

1 Introduction

Karst aquifers are very complex and heterogeneous systems because there are three types of porosity (Fig. 1) that usually have a very large spatial variability. Matrix rock porosity is mainly related with lithology. Fracture porosity is related with the density of fractures and the opening of the fractures, and usually decreases with depth. Conduit porosity is more heterogeneous and anisotropic than the other types and may be important in depth as there may be a large network of phreatic conduits.

Furthermore, there may be a well developed epikarst (Fig. 2a), areas of preferential infiltration where recharge is very quick (Fig. 2b), a thick vadose zone (Fig. 2c), karst depressions with an important thickness of detrital filling, and all the previous features can have a very large spatial variability. Additionally, the discharge takes places at important karst springs (Fig. 2d), which are associated with the network of karst conduits. Modeling the hydrogeology of a karst system is challenging (Scanlon et al. 2003). Any mathematical model of a spatial distributed model of karst hydrogeology must take into account those two factors (Fig. 3): to simulate the karst geologic media and to simulate the physical flow processes: recharge, flow along the non-saturated zone, flow along conduits, flow along the saturated zone and discharge at springs. This work describes an approximation for a three-dimensional model of a karst aquifer.

2 Methodology

The previous considerations about the particularities of a karst aquifer must be taken into account when building a mathematical model of a karst system distributed in time and space. Distribution in time is always considered, but distribution in space is a challenge that still has not been satisfactorily solved. In order to solve it, recharge, epikarst, porosity, hydraulic conductivity, non-saturated thickness, and conduits distribution are estimated spatially. The flow along the vadose zone is considered one-dimensional (in the vertical direction) using the gravity flow described by the Richards equation. This one-dimensional flow is coupled with a two-dimensional saturated flow that follows Darcy law. Coupled with the two previous models there is a discrete model of channel flow that follows an open channel flow equation. The model has been applied to the Sierra de las Nieves karst aquifer.

3 Case Study

The study area is the Sierra de las Nieves karst aquifer in the province of Málaga, Southern Spain. The karst massif consists mainly of a succession of carbonate rocks: Triassic marbles and dolostones, Jurassic limestones, and a Tertiary

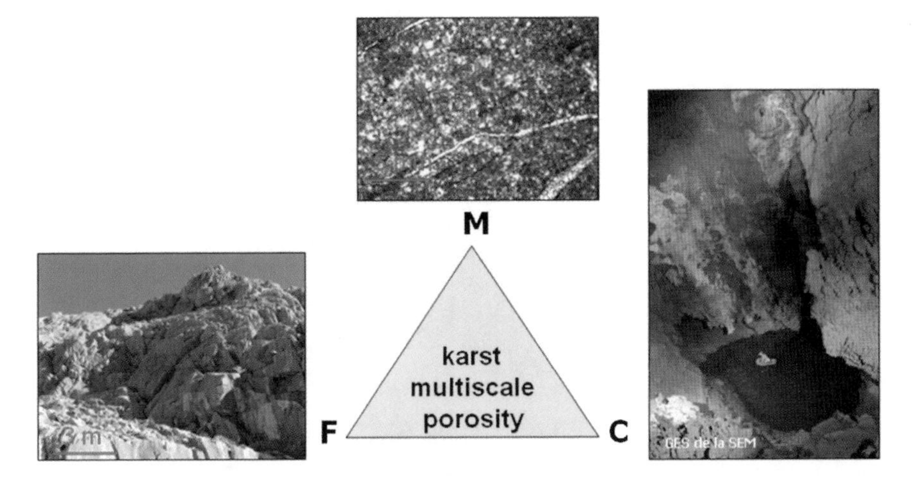

Fig. 1 The karst multiscale porosity introduces a large heterogeneity and anisotropy in a karst system. (*M*) matrix rock porosity; (*F*) fracture porosity; and (*C*) conduit porosity. Examples from the Sierra de las Nieves karst aquifer

Fig. 2 Some of the typical features that must be considered in a mathematical model of a karst aquifer. **a** the epikarst; **b** zones of preferential infiltration; **c** a thick non-saturated zone and **d** discharge at important karst spring associated with karst conduits. Examples from the Sierra de las Nieves karst aquifer

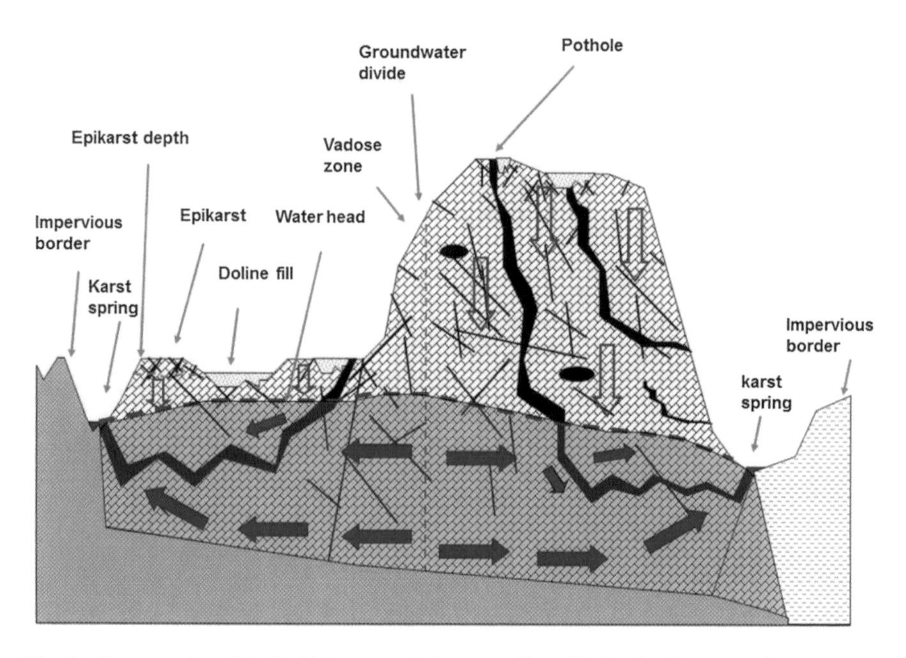

Fig. 3 Conceptual model of a high-mountain karst aquifer with the development of super-caves. This is the model assumed for the Sierra de las Nieves aquifer

carbonatic breccia. The Mesozoic sequence is folded by an NE–SW trending overturned syncline with a vergency toward the NW (Liñán 2005). The carbonatic breccia unconformably lies over the Mesozoic succession, but is also deformed by the fold. The conceptual model of the aquifer (Fig. 3) is that there are two distinctive blocks; a tectonically uplifted block (with respect to the other) that has a thick non-saturated zone that reaches 1 km and that has developed super-caves (right part of Fig. 3). The other block (left part of Fig. 3) has a non-saturated thickness of around 300 m and the most important landforms are karst depressions. Recharge has been estimated spatio-temporally on a temporal basis of 1 day and on a grid of square cells of 100 m on a side. The procedure (Fig. 4a) is described in detail in Pardo-Igúzquiza et al. (2012).

The epikarst has also been evaluated using remote sensing and field work according to a supervised classification method described in (Rodríguez-Galiano et al. 2012). The aquifer has three main hydrogeological basins that discharge by the three main karst springs that drain the aquifer (Fig. 3). The Río Grande river basin (Fig. 4c) is the one that is considered here. Figure 4c shows the thickness of the vadose zone that can reach up to 1,070 m, and Fig. 4d shows the water head table in recession as may be during the dry months of July and August. The hydrograph of Río Grande may be seen in Fig. 5a together with the recharge in one

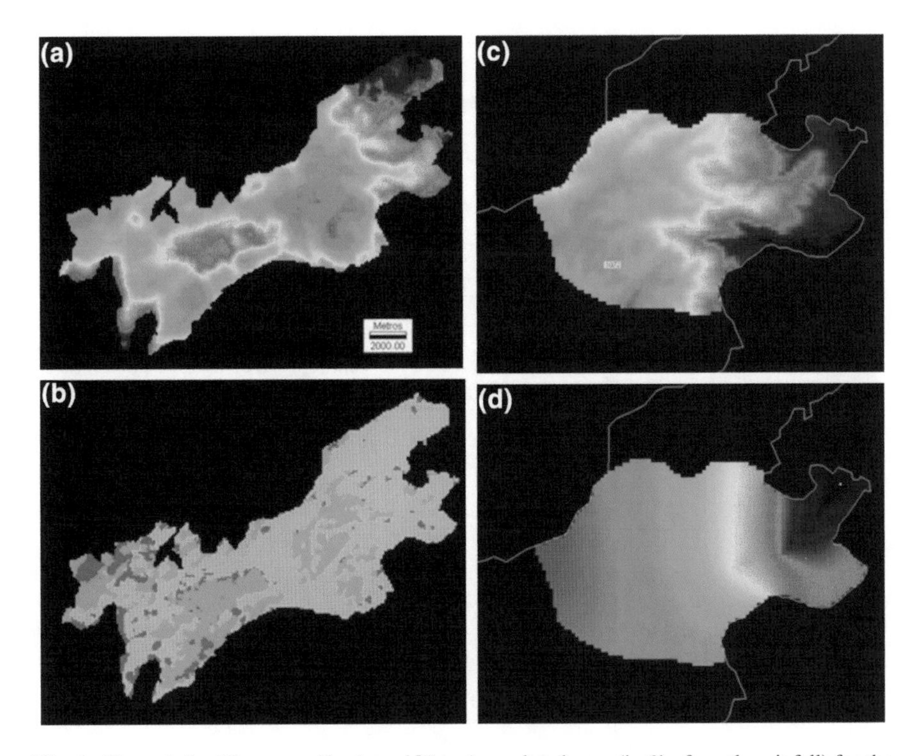

Fig. 4 Sierra de las Nieves aquifer (**a** and **b**). **a** Annual recharge (in % of yearly rainfall) for the hydrologic year 1995/1996. The values range from 40 % (*dark blue*) to 84 % (*dark green*). **b** Map of epikarst development (*green* is well developed, *blue* is medium development and *orange* is bad developed). Río Grande hydrogeological basin (**c** and **d**). **c** Non-saturated thickness, that reaches 1,072 m in the Torrecilla area (*green*). **d** Water level at recession with water table that reaches the 640 m a.s.l. at the *dark green* area

given cell in the Torrecilla recharge area. It can be seen how the fast flow has a delay of 1 day with respect to rainfall. The fast flow represents the 19 % of the total recharge (Fig. 5b), while the slow flow represents the 81 % of the recharge, and has a delay of 14 days (Fig. 5c). Río Grande has the largest discharge of the aquifer (73 hm^3) in 1995/1996, but it also has the fastest recession (Fig. 5d).

The model of the flow along the non-saturated zone (Fig. 6a) shows how the water percolates as an elevator. However, the real picture becomes very complicated because of the orography (Fig. 6b). The recharge is coupled with a two-dimensional Darcy flow model, and Fig. 6c shows the response of the karst sprint to 1 day event when quick flow is not considered, and Fig. 6c shows the spring discharge with fast flow taken into account. The final coupling of all the models is the subject of current research.

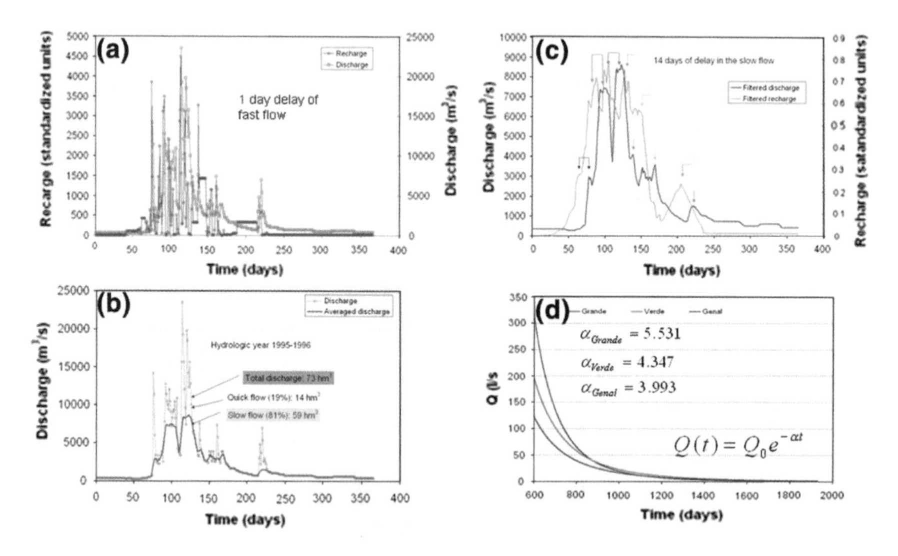

Fig. 5 **a** Hydrograph (*red line*) of the Río Grande karst spring, together with the recharge at one cell in the recharge area close to the Torrecilla (*blue line*). **b** Hydrograph (*red line*) of the Río Grande karst spring, together with the hydrograph of slow flow (*blue line*). **c** By filtering the recharge (*red*) and the discharge (*blue*) it is possible to estimate the delay (14 days) of the slow flow. **d** Models of the recession curves fitted to the observed recessions of the three main karst springs

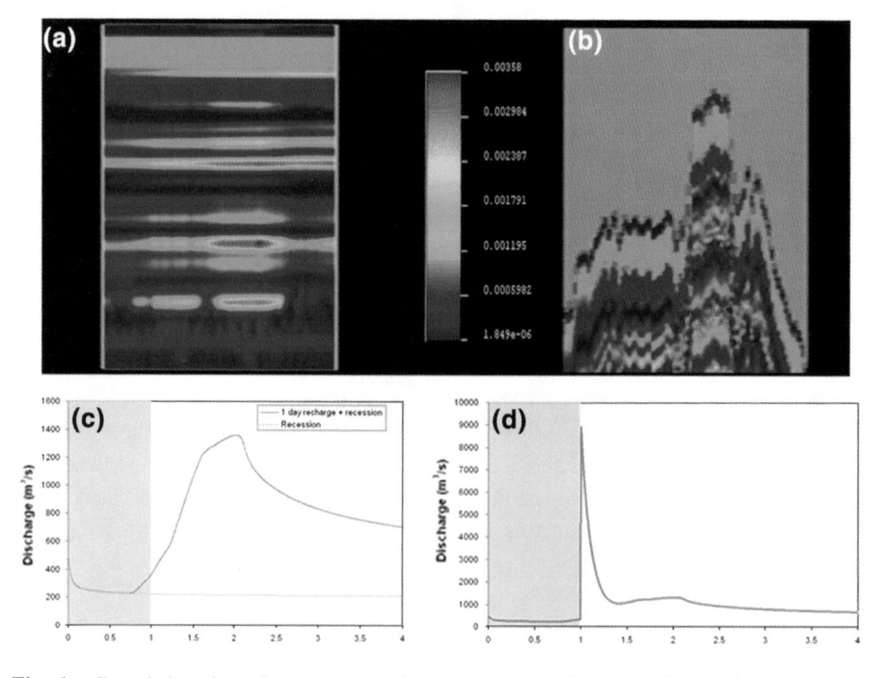

Fig. 6 **a** Percolation along the non-saturated zone across an SW–NE aquifer profile in the case of flat topography. **b** Similar to (**a**) but with the real topography. **c** Spring response to 1 day rainy even. **d** as C but with quick flow along a conduit

4 Discussion and Conclusions

The mathematical model of a karst aquifer is a very challenging task because of both the large anisotropy of the aquifer and the simultaneous fast and slow flow hydrodynamics. A three-dimensional model has been considered by coupling several models: a 1D (vertical) flow along the vadose zone, flow along discrete conduits, and a 2D flow for the saturated zone. In essence, the karst has a dual character in recharge (diffused and concentrated), in the transit flow (fast and slow), and in the discharge (floods and recession). The first results have been discussed with simple configurations of conduits. More research is needed to complete the task satisfactorily.

Acknowledgments This work was supported by the research project CGL2010-15498 from the Ministerio de Economía y Competitividad of Spain. We would like to thank the groups of speleology that work in the Sierra de las Nieves: Grupo de Exploraciones Subterráneas de la Sociedad Excursionista de Málaga, Interclub Sierra de las Nieves (Ronda), Centro Excursionista del Sur Escarpe, MAINAKE Sociedad Espeleo-Excursionista y el Grupo de Exploraciones Subterráneas de Tolox. Also we would like to thank the managers of the Parque Natural de la Sierra de las Nieves.

References

Liñán C (2005) Hidrogeología de acuíferos carbonatados en la unidad Yunquera-Nieves (Málaga). Publicaciones del Instituto Geológico y Minero de España, Madrid. Serie: Hidrogeología y Aguas Subterráneas 16, p 322

Pardo-Igúzquiza E, Durán JJ, Dowd PA, Guardiola-Albert C, Liñan C, Robledo-Ardila PA (2012) Estimation of spatio-temporal recharge of aquifers in mountainous karst terrains: application to Sierra de las Nieves (Spain). J Hydrol 470–471:124–137

Scanlon BR, Mace RE, Barrett ME, Smith B (2003) Can we simulate regional groundwater flow in a karst system using equivalent porous media models? Case study, Barton Springs Edwards aquifer, USA. J Hydrol 276(1–4):137–158

Rodríguez-Galiano V, Pardo-Igúzquiza E, Durán JJ, Chica-Olmo M, Luque-Espinar JA, Guardiola-Albert C, Martos-Rosillo y PS, Robledo-Ardila A (2012) Cartografía del epikarst integrando información de campo, geología e imágenes de satélite: caso de Sierra de las Nieves (Málaga). In: López-Geta JA, Ramos González G, Fernández Rubio R, Lorca Fernández D (eds) El Agua en Andalucía. Retos y avances en el inicio del milenio. Publicaciones del Instituto Geológico y Minero de España. Serie Hidrogeología y Aguas Subterráneas, vol 30, 2, pp 1611–1620

How Karst Areas Amplify or Attenuate River Flood Peaks? A Response Using a Diffusive Wave Model with Lateral Flows

J.-B. Charlier, R. Moussa, V. Bailly-Comte, J.-F. Desprats and B. Ladouche

Abstract This paper investigates the role of karst aquifers on flood generation and propagation using the Hayami Diffusive Wave (DW) model accounting for uniformly distributed lateral flows. The inverse model was applied on the main channel reaches of the Tarn basin at Millau (2,400 km^2) in southern France to assess lateral inflows from karstic springs as well as lateral outflows from river losses. Results show that the DW model, which is simple, parsimonious, and easy-to-use, is able to quantify lateral flows avoiding difficult parameterisation. Surface/groundwater exchanges were characterised on several reaches along the stream, showing a highly variable attenuation/amplification influence of flood peak by karst units during a single flood event. We showed that the upstream part of the karst area have a dominant attenuation role by re-infiltrating part of runoff from the head-water basin in hard-rock areas, while the downstream part have a dominant amplification role due to high contributions of karst groundwater. These results improved the conceptual hydrogeological model of the Grands Causses region.

1 Introduction

The complexity of surface/groundwater exchanges in karstic basins drives their hydrological response. Karst aquifers may re-infiltrate totality of surface flows or amplify up to 80 % the flood peak according to groundwater level (Bailly-Comte et al. 2009; De Waele et al. 2010). Thus, investigating the role of karst aquifers on

J.-B. Charlier (✉) · V. Bailly-Comte · J.-F. Desprats · B. Ladouche
BRGM, 1039 Rue de Pinville, 34000 Montpellier, France
e-mail: j.charlier@brgm.fr

R. Moussa
INRA, Laboratoire D'étude Des Interactions Sol – Agrosystème – Hydrosystème (LISAH), UMR SupAgroM – INRA – IRD, 2 Place Pierre Viala, 34060 Montpellier Cedex 1, France

B. Andreo et al. (eds.), *Hydrogeological and Environmental Investigations in Karst Systems*, Environmental Earth Sciences 1,
DOI 10.1007/978-3-642-17435-3_33

flood generation requires the use of a modelling approach accounting for high lateral inflows and outflows.

To our knowledge, less physically-based flood routing models were used in this topic. A first approach was proposed by Bailly-Comte et al. (2012) using the kinematic wave model coupled with a linear underground reservoir to account for concentrated lateral flows. In the case of river reach having numerous inflows and outflows not precisely localised, a new parsimonious modelling approach accounting for diffusivity is needed, which allows both distributed lateral inflow and lateral outflow to be simulated.

For that, this paper aims to assess the influence of karst aquifers on flood generation investigating the Hayami (1951) diffusive wave model accounting for uniformly distributed lateral inflows/outflows proposed by Moussa (1996), which is an analytical solution, stable and parsimonious. The inverse model was applied on the Tarn basin at Millau in southern France where two superposed karstic aquifers may influence flood routing. This model improves our understanding of the hydro(geo)logical functioning of the karstic rivers in the Grands Causses area.

2 The Hayami Diffusive Wave Model with Lateral Flows

To account for lateral flow contributions between two hydrographic stations, we used an analytical resolution of the Diffusive Wave (DW) equation with lateral flow uniformly distributed over a channel network using the Hayami assumptions: constant celerity C [L T^{-1}] and diffusivity D [L^2 T^{-1}], and no physical downstream boundary condition, as proposed by Moussa (1996):

$$O(t) = \Phi(t) + [I(t) - \Phi(t)] * K(t) \text{ with } \Phi(t) = \frac{C}{L}\int_0^t [Q_A(\lambda) - Q_A(0)]d\lambda \quad (1)$$

where $I(t)$ and $O(t)$ are the upstream (Inflow) and downstream (Outflow) storm flows, respectively, and L is the channel length. The symbol * represents the convolution relation. The Hayami kernel function $K(t)$ is expressed as follows:

$$K(t) = \frac{L}{2(\pi D)^{1/2}} \frac{\exp^{\frac{CL}{4D}\left(2-\frac{L}{Ct}-\frac{Ct}{L}\right)}}{t^{3/2}} \quad (2)$$

$Q_A(t)$ is the corresponding hydrograph of the uniformly distributed lateral flow component per length unit $q(x,t)$ [L^2·T^{-1}]:

$$Q_A(t) = \int_0^L q(x,\ t) \cdot dx \quad (3)$$

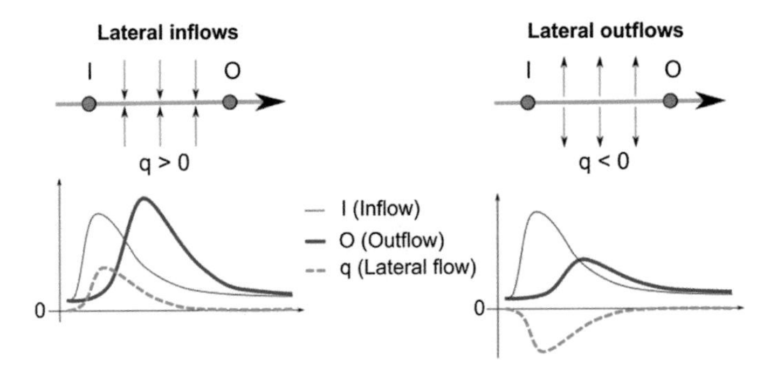

Fig. 1 Diffusive wave model with uniformly distributed lateral flows (from Moussa 1996)

According to hydrological conditions, lateral flow component $q(x,t)$ may be positive or negative when lateral inflow or outflow from the channel occurred, respectively (Fig. 1).

The inverse problem concerns the identification of lateral inflow or outflow between two gauging station on the basis of observed hydrographs at these two stations. The problem is to identify $Q_A(t)$ knowing the two functions $I(t)$ and $O(t)$. According to Moussa (1996) from Eq. (1), we obtain:

$$A(t) = O(t) - I(t) * K(t) \text{ with } \Phi(t) - \Phi(t) * K(t) = A(t) \tag{4}$$

The resolution of Eq. (4) needs, firstly, the identification of $K(t)$ using Eq. (2), and of $Q_A(t)$ as follows:

$$Q_A(t) = Q_A(0) + \frac{L}{C}\frac{d\Phi}{dt} \tag{5}$$

As a first approximation, flood-routing parameters C and D can be estimated from the hydrograph's characteristics when $I(t)$ and $O(t)$ are known (Moussa 1996), avoiding any calibration procedure. Then, the DW model is applied estimating $O(t)$ knowing $I(t)$ and $q(t)$ from Eq. (1). The inverse model is finally used to assess lateral flows knowing $I(t)$ and $O(t)$ from Eq. (4).

3 Study Case

The study site is the Tarn river at Millau (2,400 km^2) in southern France (Fig. 2), which extend from the Cévennes Mountains to the Grands Causses karstic region. The climate is Mediterranean with a mountainous influence. Highest rainfalls occur mainly in autumn, and are usually of short duration and of higher intensity in the Cévennes Mountains compared to the Grands Causses area (Charlier et al. 2012). The Tarn River has two main tributaries: Dourbie and Jonte Rivers. Head

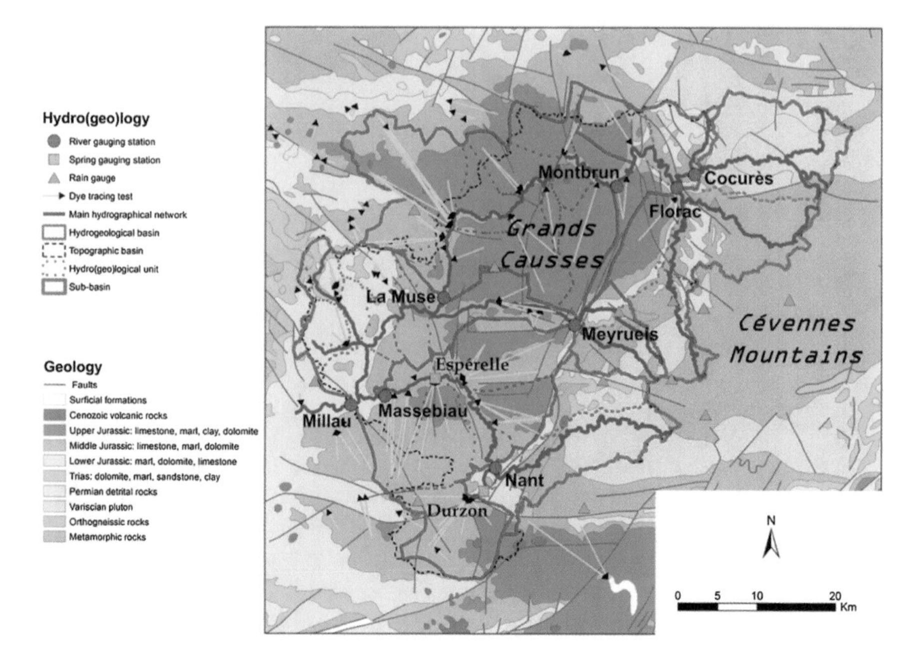

Fig. 2 Hydrogeological map of the Tarn basin at Millau

catchments for these three main streams are located in the hard-rock mountainous areas. In their intermediate and downstream part towards the outlet located at Millau, the three rivers cross over the Grands Causses plateau in deep canyons.

Recharge area of karst systems were delimited from results of dye tracing tests, which give information on existing connections between infiltration points and karst outlets (see Charlier et al. (2012) for the compilation and complete references of dye tracing tests). Rainfall was recorded by *METEO France*, streamflow hydrographs were obtained from national French services *SPC Garonne-Tarn-Lot*, *DDT48*, and *DREAL Midi-Pyrénées*, and karstic spring's discharges were obtained from the *Parc Naturel Régional des Grands Causses.*

4 Results

First, we illustrate the methodology based on a flood event to assess the ability of the modelling approach to simulate karst lateral flows on a reach where lateral spring flows are measured (flood event of the 02 Nov. 2008 in Sect. 4.1). Then, we extend the methodology on two other events having homogeneous and contrasted spatially distributed rainfalls (flood events of the 10 Apr. 2009 and of the 20 Oct. 2006, respectively in Sect. 4.2). These events can be used to quantify the spatial variability of lateral inflows and outflows.

4.1 Illustration of the Modelling Approach

The event of 02 Nov. 2008 on the Dourbie River between Nant and Massebiau stations (Fig. 2) was selected to compare the calculated lateral flow hydrograph using the inverse model with the measured one, i.e. corresponding to the sum of the Espérelle and Durzon spring's hydrographs. During this event, the calculated lateral flow volume is almost equal to the measured one.

Figure 3 presents (i) the simulated outflow hydrograph (dashed blue line) from the observed inflow hydrograph (solid black line) using the DW model accounting for observed lateral flow (solid green line), as well as (ii) the simulated lateral flow hydrograph (dash-dot green line) from the observed inflow hydrograph (solid black line) and the observed outflow hydrograph (solid blue line) using the inverse model. The performance criteria give very good results for the outflow simulation (Nash $Q_{\text{outflow}} = 0.9$) and acceptable results for the lateral flow simulation (Nash $Q_{\text{lateral flow}} = 0.7$), meaning that for this event, spring flows explain the main lateral flow. Globally, regarding the calculated lateral flow by the inverse model, hydrograph is well simulated during recession but over-estimated for peakflow periods. At the beginning of the flood, it is interesting to note a decrease of the baseflow of about 8 $m^3 s^{-1}$ in the channel between the two stations, even if springs are flowing at a rate of 5 $m^3 s^{-1}$. This shows the complexity of lateral contributions in the same channel reach where lateral inflows (localised springs) and outflows (river losses) occurred concomitantly along the reach. Our modelling approach is able to calculate the global lateral hydrograph, which is the sum of all lateral inflows and outflows. This analysis justifies also the use of a model accounting for uniformly—rather than concentrated—distributed lateral flows when concentrated inflows (springs) and outflows (river losses) are not known, and thus, not precisely localised.

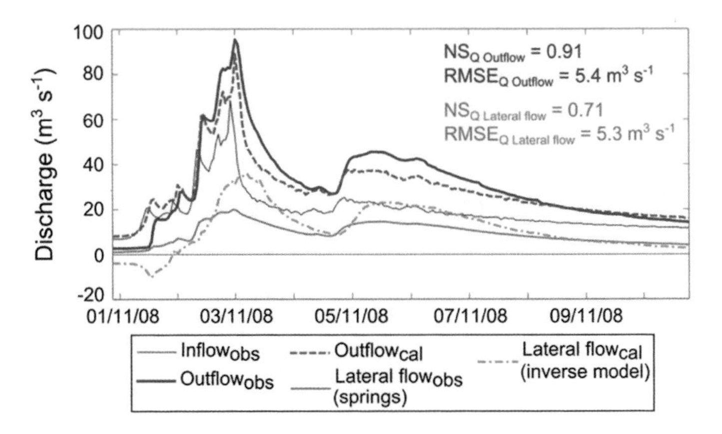

Fig. 3 Observed and simulated lateral hydrographs of the Dourbie River between Nant (*Inflow*) and Massebiau (*Outflow*) stations for flood event of the 02 Nov. 2008

4.2 Spatio-Temporal Variability of Lateral Flows

This section aims at assessing the effect of spatial distribution of rainfall on the hydrological response, and especially between the headwater hard rock basin and the downstream karst basin. We selected two events with similar initial hydric conditions (based on baseflow level): the first flood event (10 Apr. 2009) occurred under spatially homogeneous rainfalls, while the second one (20 Oct. 2006) occurred under spatially contrasted rainfall distribution with up to four-fold higher rainfall intensities and amount in the hard-rock area (headwater basin).

Figure 4 presents measured inflow and outflow, and calculated lateral flow hydrographs for each channel reach using the inverse model. First, under a homogeneous spatial distribution of rainfall (top of Fig. 4), we observe lateral inflows for all reaches, with an increase of the lateral flow peaks from the upstream reach to the downstream one. Between Montbrun and La Muse stations, the calculated lateral hydrograph fits the measured spring's hydrograph multiplied by 100. The same pattern is also observed on the Dourbie River, showing that karst

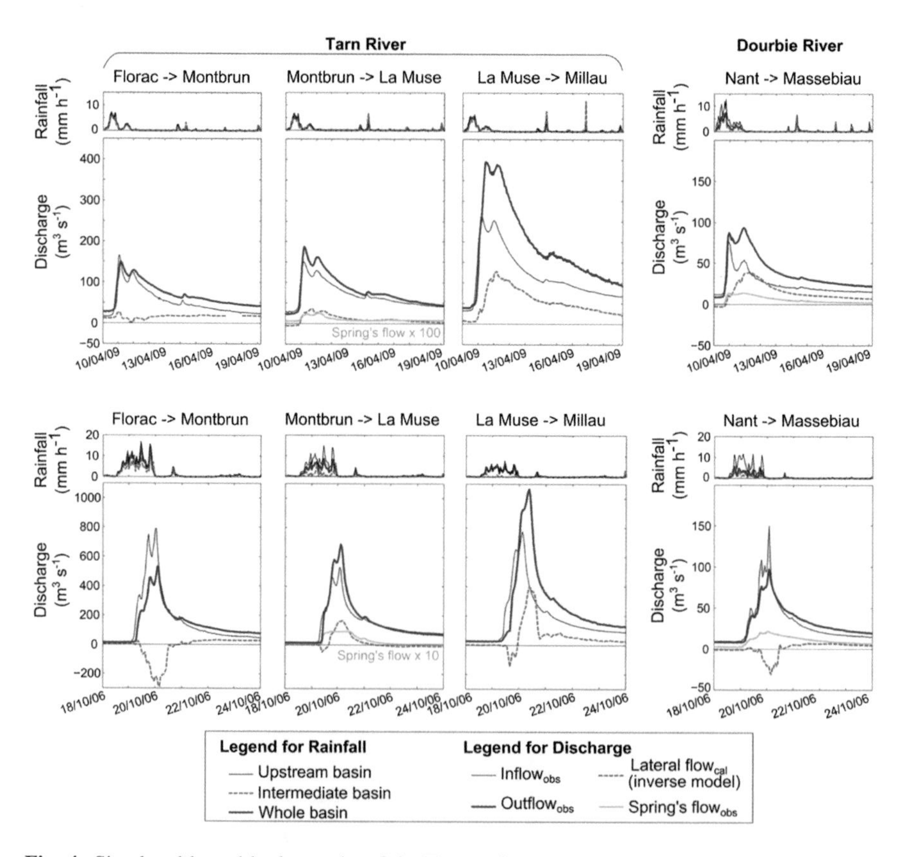

Fig. 4 Simulated lateral hydrographs of the Tarn and Dourbie Rivers for flood events of the 10 Apr. 2009 (*top*) and the 20 Oct. 2006 (*bottom*)

aquifers may explain the global lateral flows in the intermediate and downstream zone of the basin.

Second, under a contrasted spatial distribution of rainfall (bottom of Fig. 4), we observe a various hydro(geo)logical response on the reaches. In the intermediate part of the basin, high lateral outflows are estimated on the two reaches located downstream headwater basins at the hard-rock/karst boundary (cf. Fig. 2) between Florac+Cocurès and Montbrun stations on the Tarn River, and between Nant and Massebiau stations on the Dourbie River. Up to 280 $m^3 s^{-1}$ of instantaneous outflows are estimated on the Tarn River. Downstream, the dominant lateral contributions are inflows, but lateral outflows in the river bed are simulated at the beginning of the event. This shows that the flood wave is partly (25 % of the flood peak) re-infiltrated in the river bed when the stream joins the karst area.

5 Discussion and Conclusion

The aim of this paper was to assess the influence of karst areas on surface flood routing using an inverse modelling approach of the Diffusive Wave (DW) accounting for lateral flows. First, results showed that the DW model is able to simulate the output hydrograph (at a downstream station) from the input hydrograph (at an upstream station) and additional lateral flows in an heterogeneous media such as karstic basin where complex surface/groundwater exchanges occurred: high inflows from large springs as well as strong river losses. In order to assess lateral contributions from karst systems along a channel reach, we used the inverse model, which simulates the uniformly distributed lateral hydrograph from observed input and output hydrographs. To our knowledge, this is the first physically-based modelling approach, which allows both distributed lateral inflow and lateral outflow to be simulated, in order to quantify the flood amplification or attenuation, respectively. Advantages of the DW model are a parsimonious parameterization using only two parameters (celerity C and diffusivity D). Moreover, the Hayami model offers an unconditionally stable analytical solution of the DW equation. Results showed that our model, which is simple, parsimonious and easy-to-use, is able to quantify lateral flows where river losses and karstic spring's inflows occurred.

Second, our results give insights of the conceptual hydro(geo)logical model of flood routing in the Grands Causses area. The inverse model showed that outflows and inflows may occur during the same flood event. This highlights the influence of the groundwater level on lateral flows. Following previous works in Mediterranean karstic area (Bailly-Comte et al. 2009), we may hypothesize that river losses occurred during low water levels inducing karst saturation. However, in our case study, it appears more complex because the analysis in a Dourbie River reach shows that river losses occurred while large springs—feeding the stream—are flowing at the same time. This implies the presence of two superposed aquifers having various saturation levels relative to the river bed altitude, as presented in Fig. 5: (i) an

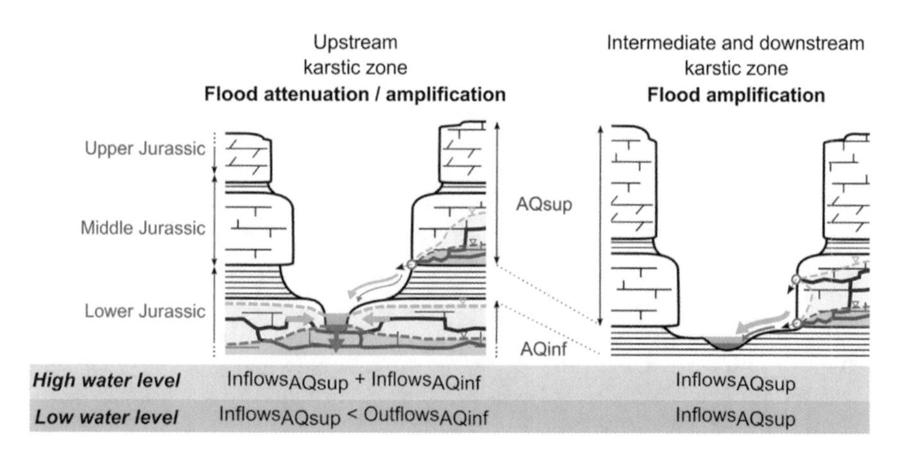

Fig. 5 Hydro(geo)logical scheme of the main karstic rivers in the Grands Causses area

inferior aquifer of which the main outlet is the stream via multiple springs localised in the river bed, and (ii) a superior aquifer of which the outlets of the different units are localised at the bottom of the cliffs above the river bed. Following the classical hydrogeolocial scheme of the Grands Causses area, we may hypothesize that the inferior aquifer is the one developed in the Lower Jurassic outcropping in upstream karstic zones, whereas the superior aquifer is the one developed in Middle and Upper Jurassic in downstream zone, above Toarcian marls.

A last point is that these results help us to better understand flood generation in Mediterranean karstic areas where headwater basins—often located in hard-rock areas (e.g. Cevennes or Pyrenean Mountains in France)—have the highest rainfall intensities. The role of surface/groundwater exchanges varies greatly with the distance from the hard-rock/karst boundary (i.e. where allogenic recharge from the hard-rock area occurred). Under contrasted rainfall distribution (up to four-fold higher rainfall intensities in the headwater basin compared to the karst area) the upstream reach of the karst area have a dominant peakflow attenuation role by re-infiltrating part of runoff. In downstream karst area, reaches have a dominant amplification role: smallest stream losses occurred at the beginning of the flood, but are followed during the peakflow by high inflows from karst aquifer. Under homogeneous rainfalls, stream losses are not visible, and the flood is amplified all over the channel reach towards the outlet. We highlight a highly variable attenuation/amplification influence of flood peak by karst units along the stream channel during a single flood event, complicated the forecast of karst influence on flood generation.

Acknowledgments This work was financed by the *SCHAPI* and the *BRGM*. We thank Laurent Danneville from the *Parc Naturel des Grands Causses* for providing hydrogeological data.

References

Bailly-Comte V, Jourde H, Pistre S (2009) Conceptualization and classification of groundwater-surface water hydrodynamic interactions in karst watersheds: Case of the karst watershed of the Coulazou river (southern France). J Hydrol 376:456–462

Bailly-Comte V, Borrell-Estupina V, Jourde H, Pistre S (2012) A conceptual semidistributed model of the Coulazou river as a tool for assessing surface water–karst groundwater interactions during flood in Mediterranean ephemeral rivers. Water Resour Res 48:W09534

Charlier J-B, Bailly-Comte V, Desprats J-F (2012) Appui au SCHAPI 2012—Module 1—Mise en place d'un indicateur d'aide à la décision pour la prévision de crue en milieu karstique : le bassin du Tarn à Millau. BRGM report. RP-61816-FR, p 88

De Waele J, Martina MLV, Sanna L, Cabras S, Cossu QA (2010) Flash flood hydrology in karstic terrain: Flumineddu Canyon, central-east Sardinia. Geomorphology 120:162–173

Hayami S (1951) On the propagation of flood waves. Disaster Prev Res Inst Bull 1:1–16

Moussa R (1996) Analytical Hayami solution for the diffusive wave flood routing problem with lateral inflow. Hydrol Process 10:1209–1227

Comparison Between Hydrodynamic Simulation and Available Data in a Karst Coastal Aquifer: The Case of Almyros Spring, Crete Island, Greece

A. Archontelis and J. Ganoulis

Abstract The Almyros spring is the outlet of the largest karst coastal aquifer in Crete Island, Greece. The spring is important for drinking water supply of Heraklion, the capital city of Crete. However, because of sea water intrusion, most of the time and especially during the dry season, the water is brackish. Since the 1960s, despite repeated efforts to collect the freshwater resources of the aquifer, only few practical results have been obtained. In order to investigate alternative measures for collecting the aquifer's freshwater a mathematical model was developed for simulating the saturated groundwater flow in the karst porous medium. Although the local use of the Darcy's law for flow simulation in discontinuous porous media such as karst aquifers is problematic, it can give useful results when applied in larger scales. For studying the hydrodynamic characteristics of the Almyros/Heraklion aquifer, the numerical model MODFLOW was used and after calibration, it is shown that useful results can be obtained in practice.

1 Introduction

The Almyros aquifer system is the largest karst aquifer in Crete Island, Greece. Its outlet, the Almyros spring is located near the sea, few kilometers west of the city of Heraklion, which is the capital city of Crete (Fig. 1). Its annual mean water flow rate is of about 250 million m^3, but most of the time the spring water is brackish and despite repeated efforts to halt seawater intrusion, the freshwater resource remains untapped.

A. Archontelis (✉)
Department of Civil Engineering, Aristotle University of Thessaloniki, Thessaloniki, Greece
e-mail: archontelis@civil.auth.gr

J. Ganoulis
UNESCO Chair and Network INWEB, Aristotle University of Thessaloniki, 54124 Thessaloniki, Greece

B. Andreo et al. (eds.), *Hydrogeological and Environmental Investigations in Karst Systems*, Environmental Earth Sciences 1,
DOI 10.1007/978-3-642-17435-3_34

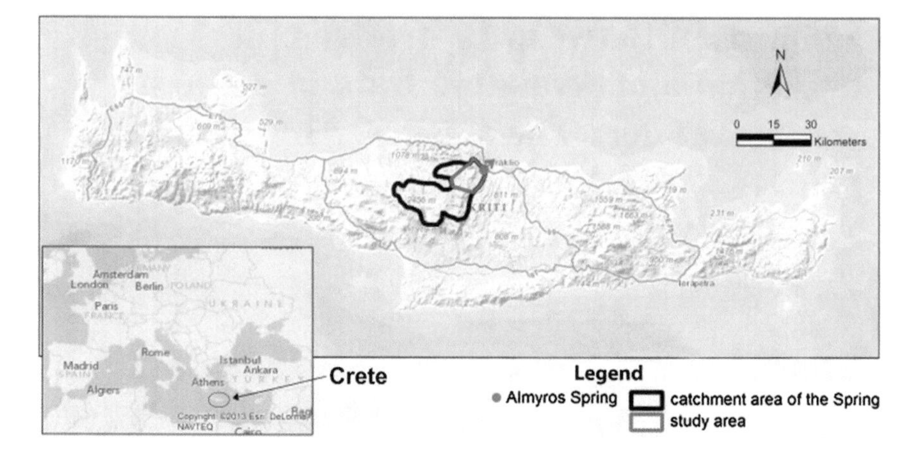

Fig. 1 Location of the Almyros spring, Heraklion city, Crete Island, Greece

Because of the spring's importance for water supply of the city of Heraklion, since 1964 many hydrogeological investigations have been carried out with only few practical results as far as the exploitation of its fresh water resources is concerned (Arfib et al. 2007; Alexakis and Tsakiris 2010).

In order to determine alternative methods for extracting the freshwater from the karst formation, it is necessary to firstly understand how the spring operates in relation to the structure of the karst aquifer, and secondly, to analyze the dynamic interaction between the freshwater of the spring and the sea. In this paper, the first issue is investigated.

For the simulation of groundwater flow in saturated porous media, the literature has to show a range of options and software packages. In this work, the "MOD-FLOW" (Mc Donald and Harbaugh 1988; Harbaugh et al. 2000) simulation code was applied in order to test the applicability and versatility of this numerical model to simulate the groundwater flow. In order for a groundwater model to be accurate, reliable, and robust, it requires a tremendous amount of information and a high level of understanding of the aquifer functioning. The model requires to compiling detailed information on the geological formations, groundwater flow directions, groundwater recharge, hydrogeological parameters, extraction or injection of flow rates from wells, and the groundwater quality characteristics.

2 Hydrogeological Settings

The hydrological basin of Almyros is underlain by pre-Neogene and Neogene geological units (Fig. 2). The geological basement of the area is the semi-autochthon "permeable" unit of "Platenkalk" (Triassic—Eocene), which consists of carbonate rocks and dominates on the greatest part of the basin. It is extended to the southern part of the region.

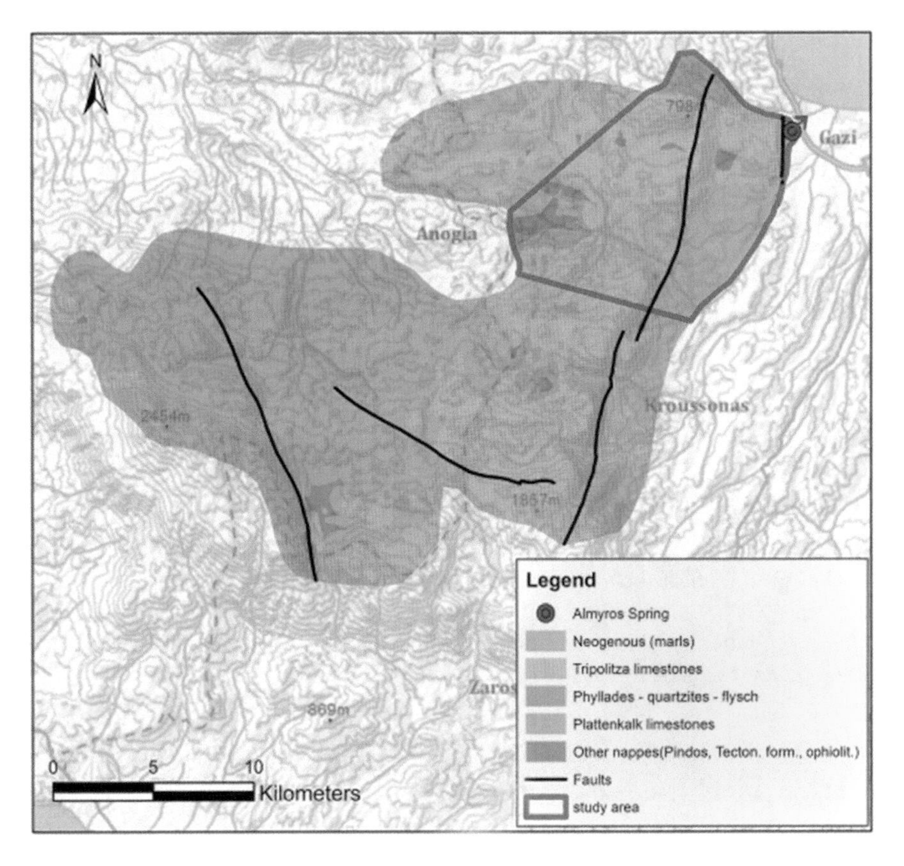

Fig. 2 General map of the catchment area: structural geology, elevations and cities. (Data source: institute for Mediterranean studies, Crete)

The "phyllitic-quartzitic" zone is overthrusted onto the previously discussed unit. It consists of impermeable, alternating pelitic schists and quartzites and extends over the northern part of the region. The "Tripolis" geotectonic zone overlays the "phyllitic-quarzitic" series and is composed of faulted and karstified thick-bedded carbonate rocks. All these series form the mountains on the west of the plain of Heraklion. They are separated from this plain by a great north-south fault, which affects the area's physiography. This fault has been partially overlain by the Neogene impermeable deposits, which consist mainly of Pleistocene age horizontal or subhorizontal beds of yellowish marls.

The spring is located in the faulted zone at the contact between the pre-Neogene and Neogene rocks. The coastal aquifer of Almyros develops in the karstified and folded pre-Neogene carbonate rocks, and the homonymous brackish spring is one of the main discharge points. Brackish water occurs as a result of fresh and seawater mixing conducted in the karstic conduits. Precipitation within the catchment area (800 mm/yr) plays a critical role in the Almyros spring regime as it is closely related to the amount of discharged water and its quality.

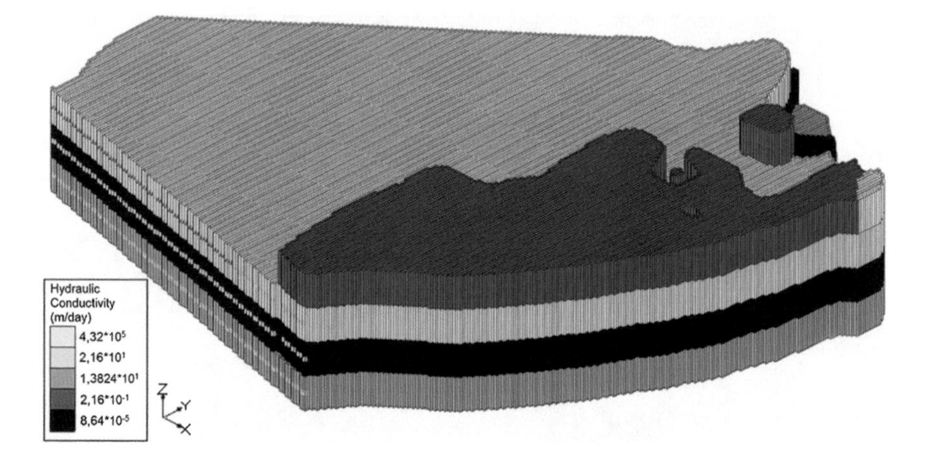

Fig. 3 Spatial distribution of hydraulic conductivity (m/day) in the numerical model

The simulation of the karstic aquifer was limited within its area of low altitude (Fig. 3), as it is the only one that can be practically useable for drilling wells or infiltration galleries. The extent of this region reaches 72 km^2, and the main flow development has a Southwest-Northeast direction.

3 Data: Field Measurements

Existing data and analyses of previous studies were considered for the compilation of the conceptual and mathematical models (Zampetakis 2001). For the mathematical modeling, the following data were used: monthly measurements of rainfall for 13 years period from 1987 to 2000 (Department of Land Reclamation, Crete Region), 20 lithological sections (IGME), lithological sections of 15 boreholes and the results of the respective pumping tests (IGME), monthly level measurements in 15 wells covering a total period of 13 years from 1987 to 2000 (IGME, DEYAH), and monthly flow measurements of Almyros spring covering a total period of 13 years from 1987 to 2000 (IGME, DEYAH). Apart from the time series of meteorological parameters that show little gaps, all other time series show continuity and cover the full period stated above.

4 Design of the Numerical Model

The numerical model consists of four horizontal layers, occupying a total area of 72 km^2. It was discretized into 197 rows and 182 columns, consisting of 36.181 cells with dimensions of 60 $\times$ 60 m. Based on hydrogeological data, the 4 layers are: L1: Neogene; L2: Limestone "Tripolis"; L3: Phyllitic-Quarzitic and L4: Platenkalk.

For the definition of the lithological sections, the hydrogeological and topographical maps of the area as well as the results of previous research work in this basin were used (Zampetakis 2001). In order to better simulate the main evolution mechanisms that control groundwater flow, the numerical mesh has been refined in some areas of interest, i.e., the margins of the study area, wells, and the spring, so that boundary conditions can be set more precisely.

4.1 Boundary Conditions

The analysis of piezometric data indicates that the principal direction of groundwater flow is from southwest to northeast. It was suggested that under ideal conditions the boundaries of the modeling domain coincide with the natural hydrogeological boundaries (Anderson and Woesner 1992; Panagopoulos 1996). The north and east boundary areas have general conditions of steady flow and zero flux. The south and west regions of the model receive water through underground transfusions of the karst system. To simulate this mechanism, the transient flow was defined with the use of the module "wells of MODFLOW". The simulation of pumping from wells was also made by using the same package.

The MODFLOW package drain was used to calculate the flow directed outside the simulation area. Thus, a source is discharged at a rate that depends on the value of the conductance coefficient and the resulting hydraulic gradient. The values were selected ranging from 4×10^3 m^2/d to 19×10^3 m^2/d.

4.2 Distribution of Hydraulic Parameters

For the compilation of the mathematical model, the definition of hydraulic conductivity, porosity, specific storage, and specific yield are required. The hydraulic parameter values that were used as input data in the designed model have been obtained from an older study (Zampetakis 2001), which was based on the pumping test analyses and on the geological and lithological characteristics of the system.

Hydraulic conductivity values range from 8.64×10^{-5} to 21.6 m/day, except for the region around the Almyros spring where values are much bigger, as indicatively illustrated in Fig. 3. Porosity values range from 1 to 15 %, specific yield values from 0.1 to 10 %, and specific storage values from 3.3×10^{-7} to 1×10^{-4} m^{-1}.

The calculation of water recharge by direct infiltration of rainfall and underground lateral transfusions was done with rainfall values of 13 years period in monthly time steps (meteorological stations of Anogeia, Kroussonas, Foinikia, and YEB). For each formation a different rate of active infiltration was calculated mainly based on lithological characteristics: 40–55 % for karst, 10–25 % for silt, and 5–8 % for other formations. The model adjustment was based on its ability to reproduce the measured values of spring water supply and the variation of the aquifer piezometric heads in selected positions.

4.3 Calibration-Verification of the Model

The simulation period was defined from 1 November 1987 to 1 November 2000. Water budget and water level data were provided by the Land Reclamation Department, Crete Region. Each month was considered as one calibration period, so that the total number of calibration periods was 156. The time step was taken equal to 1 day.

The criteria for calibrating the model were the evolution of piezometric heads at selected observation points, the water flow rate at the Almyros spring and its monthly variation. The numerical model was initially calibrated under steady flow conditions and subsequently under transient flow conditions. The following criteria were considered during the calibration process: (a) the water balance and (b) the differences between measured and computed piezometric heads at selected observation points. Solving the system of equations of groundwater flow was performed using the PCG2-MODFLOW package and the convergence criterion was used at the level difference between successive approximate solutions of 0.01 m.

The period 1987–1988 was selected in order to define the steady state flow conditions, because the year before the wells started working was characterized by steady average rainfall conditions with no extreme stresses imposed on the system. A total of 10 observation points were used. A maximum of 2 m deviation between simulated and measured heads was assumed acceptable at this stage of calibration.

For the calibration in transient flow conditions, the selected period was 1987–1994, because during this period a satisfactory volume of integrated data could be found. Hydrographs in 10 observation points were used for this calibration. A maximum of 5 m deviation between simulated and measured heads was assumed acceptable at this stage of calibration.

The verification of the model was done for a period of 6 years (1994–2000).

5 Results

The water balance produced by the model in steady state conditions agrees well with the reference water balance estimated using the available field data (Zampetakis 2001). For the steady state condition, the graph showing the measured versus the computed piezometric heads was resulted without significant deviations (Fig. 4).

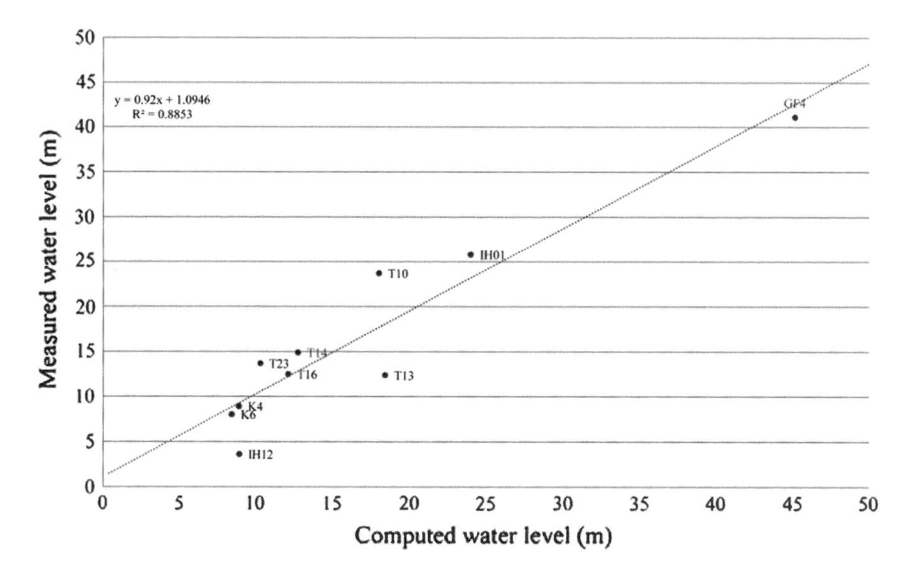

Fig. 4 Measured and computed groundwater levels at some observation points

The model has the ability to simulate with sufficient precision the variation of the piezometric surface without systematic error deviations. The Fig. 4 shows some relative weakness for an accurate simulation of the extreme values and abrupt changes, which occur frequently in a karst system.

The study of hydrographs at selected monitoring points suggests that the results obtained from the calibrated model in transient state conditions matches relatively well the field data, as illustrated in the spring flow hydrographs shown in Fig. 5.

The ModFlow cannot directly represent karstic formations, faults, tectonic forms, and sources. The source could not be designed as a point, as the absence of Source Package processing compels the use of Drain Package, which requires dimensions to be inserted. Thus, Almyros spring, was designed as a drain area and not as a discharge position. The inability of the program to simulate discontinuities, can create big differences from field measurements. These errors were corrected, where possible, through "local settings".

The differences between measured and simulated values of spring outflow are probably due to specific characteristics of the hydrodynamic function of the karst system and the limited capabilities of an exact simulation of these using a Darcian type of flow through porous media. The extreme minimum values of field measurements can be attributed to the influence of pumping. The extreme maximum values of spring outflow are due to abrupt changes of rainfall that probably mobilize additional karstic fractures that are not taken into account by the model.

The average hydrodynamic functioning of the karst source is simulated satisfactorily, without very significant deviations from the measured field values (Fig. 5). Apart from particular operating mechanisms of the karstic spring, which

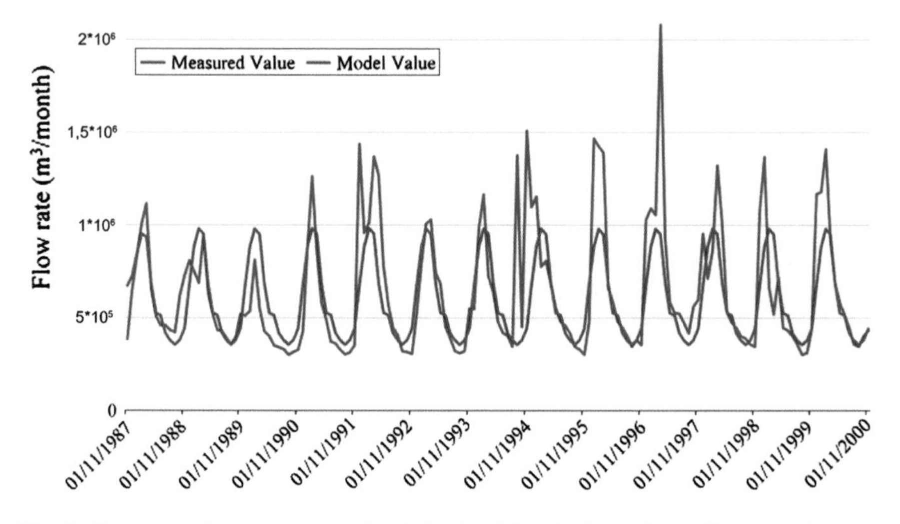

Fig. 5 Comparison between measured and simulated flow hydrographs at Almyros spring

cannot be described accurately by a Darcian mathematical code, part of the observed differences may also be attributed to measurement errors.

Sensitivity analysis that was performed with variation of parameters in the range of ± 50 % from setting values, showed significant sensitivity to changes received from rainfall and underground lateral transfusion variations. No significant sensitivity was shown to the change of the water volume pumped, or the change in storage rates. By accepting the ability of the model to simulate the main function and evolution mechanisms of the karst system of the Almyros aquifer, alternative scenarios for water resources management in the region will be investigated in the near future, under average, extreme wet, and extreme dry periods.

6 Conclusions

The application of a Darcian type numerical modeling of karstic systems, such as the Almyros aquifer, should be considered with much of precaution. Although generally satisfactory results can be achieved, high outflow peaks are not simulated accurately. These occur after abrupt changes of rainfall, that may activate additional karstic faults that are not taken into account by the model. The quality of the model output and its ability for simulating the main hydrodynamic mechanisms of a karstic aquifer is generally acceptable and could be used for analyzing alternative scenarios of groundwater resources management. The collection of additional data with better spatial resolution and longer time duration will possibly enhance the capability of the model. We may conclude that the application of the

Darcy's law approximation for simulating the mean hydrodynamics of karst aquifers is possible, provided that a clear knowledge of their geometry and their physical characteristics are given in form of multi annual field data. Also the understanding of limitations of this kind of models for an accurate description of local characteristics of the flow, such as the spring's outflow peaks and the sea water intrusion, is important for the use of the model in practical applications of groundwater resources management.

References

Alexakis D, Tsakiris G (2010) Drought impacts on karstic spring annual water potential. Application on Almyros (Heraklion, Crete) brackish spring. Desalin Water Treat 16:1–9

Anderson M, Woessner W (1992) Applied groundwater modeling. Simulation of flow and advective transport. Academic Press, London, p 381

Arfib B, de Marsily G, Ganoulis J (2007) Locating the zone of saline intrusion in a coastal karst aquifer using spring flow data. Groundwater 45(1):28–35

Harbaugh AW, Banta ER, Hill MC, McDonald MG (2000) MODFLOW-2000, the US Geological Survey modular groundwater model-user guide to modularization concepts and the groundwater flow process. US Geological Survey Open-File Report: 00–92

Mc Donald M, Harbaugh A (1988) A modular three-dimensional finite-difference ground-water flow model. In: Techniques water-resources investigations, book 6, chapter A1, USGS, Washington, p 750

Panagopoulos A (1996) A methodology for groundwater resources management of a typical alluvial aquifer system in Greece. Ph.D. thesis, University of Birmingham, Birmingham, p 251

Zampetakis G (2001) Monitoring balance of groundwater and surface waters of Crete, Brackish Aquifers, Volume III, Almyros Heraklion spring report. Institute of Geology and Mineral Exploration (IGME), Regional Office of Crete, Rethymno

Assessing Freshwater Resources in Coastal Karstic Aquifer Using a Lumped Model: The Port-Miou Brackish Spring (SE France)

B. Arfib and J.-B. Charlier

Abstract Freshwater resources in coastal aquifers are restricted by seawater intrusion. Studying brackish spring can be an appropriate approach to assess this saline intrusion and to elaborate a conceptual model of the karst aquifer. The aim of this study is to model the hydrogeological flows and the salinity of a brackish spring using a lumped numerical model (Rainfall-Discharge-Salinity), and to quantify the freshwater discharge available. The model is based on a classical karst model composed of connected reservoirs representing the main storage elements of the karst aquifer, which can be deduced from the analysis of discharge and salinity recorded time series. The model was successfully applied on the Port-Miou spring (400 km^2), in SE France, which is one of the main submarine springs around the Mediterranean Sea. Four reservoirs were used to model spring discharge and salinity: a SOIL reservoir feeding a DEEP brackish reservoir impacted by seawater intrusion, and two FAST and SLOW reservoirs representing the shallower fresh-water resource. We showed that the spring water is always brackish more or less diluted by freshwater during flood events. These results improved the conceptual hydrogeological model of the Port-Miou spring and showed the effectiveness of lumped models to simulate discharge and salinity in coastal karst aquifers.

1 Introduction

Coastal karstic aquifers may represent important groundwater resources. Exploitation of such aquifer type requires firstly an evaluation of available freshwater. Secondly, withdrawal must be done with a special attention to the saline intrusion processes. Classical scheme of seawater intrusion can be viewed as a more or less

B. Arfib (✉)
Aix-Marseille University, CEREGE, Case 67, 3 Place Victor Hugo, 13003 Marseille, France
e-mail: arfib@cerege.fr

J.-B. Charlier
BRGM, 1039 Rue de Pinville, 34000 Montpellier, France

B. Andreo et al. (eds.), *Hydrogeological and Environmental Investigations in Karst Systems*, Environmental Earth Sciences 1,
DOI 10.1007/978-3-642-17435-3_35

thick transition zone between fresh and salt water. But this simplified view cannot explain the occurrence of large brackish submarine groundwater discharge (SGD) to the ocean through karstic springs. Saline intrusion distribution in karst aquifer appears more complex compared to porous aquifer, due to the heterogeneity of karst media. Many authors (e.g., Stringfield and Legrand 1971; Arfib et al. 2007; Fleury et al. 2007) have pointed out that submarine karstic springs are connected to preferential flow below the sea level within karst conduits.

The salinity of such springs may be highly variable in time, and is usually related to discharge variations. Thus, in order to assess freshwater resources, a modeling approach is needed, which allows both discharge and salinity to be simulated. Conceptual or reservoir models (frequently called Rainfall-Discharge models) can be used to test hypotheses in order to understand the main hydrogeological processes (e.g., Hartmann et al. 2012). Mixing models can be combined with Reservoir models to simulate discharge as well as chemical variations of spring water (Charlier et al. 2012).

In this setting, the aim of this study is (1) to develop a conceptual model of saline intrusion within a coastal karstic aquifer, and (2) to assess freshwater resources in the aquifer, modeling both discharge and salinity with a reservoirs model. The study site is the Port-Miou spring in south France, which is a large submarine spring having a mean annual discharge of around 7 $m^3 \cdot s^{-1}$.

2 Study Site

The Port-Miou aquifer is located in south-east of France (Fig. 1), along the Mediterranean Sea, in a carbonate environment (Jurassic and Cretaceous limestone, dolostone, and mixed siliciclastic-carbonate rocks). Two main submarine springs, Port-Miou and Bestouan, drain the aquifer to the sea in the bay of Cassis. Port-Miou is one of the main coastal karst springs of Europe (Tulipano et al. 2005; Custodio 2010). The recharge area extends on 400 km^2, mainly over natural landscape made of hills and karst plateau and polje (Cavalera 2007). The climate is Mediterranean given the mean annual rainfall ranging between 500 and 1,000 mm with storm events of high intensities (up to 100 mm per day), and drought during the summer season.

The spring outlet has been explored by cave-divers. They discovered a huge saturated karst conduit developed more than 2 km inland (Fig. 1), with a diameter reaching 20 m. The exploration ended at 223 m deep below the sea level, in a karstic shaft (Meniscus 2013). One kilometer eastward, the submarine spring of Bestouan gives access to a 3 km long saturated karst conduit.

In the 1970, a submarine underground dam has been built in the saturated karst conduit of Port-Miou at 500 m upstream the spring. This dam stopped the seawater intrusion in the subhorizontal conduit, and allows measurements of groundwater salinity coming from the aquifer before the mixing zone with seawater at the outlet. This dam is now an in situ Laboratory to study the SGD and its impact on

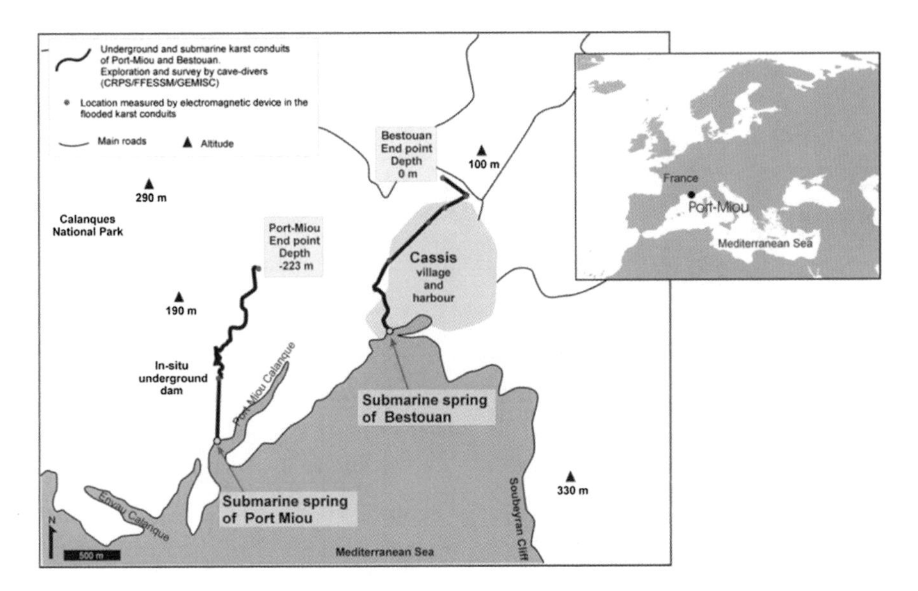

Fig. 1 Location of the Port-Miou case study in France

the sea. Daily rainfall and Potential Evapo-Transpiration (PET) were recorded by METEO France, and spring discharge and salinity were measured in the main conduit upstream the dam.

3 Conceptual Model of the Coastal Karst Aquifer

Figure 3 presents 18 months of daily data, showing that the spring is brackish with salinity values ranging from 2 to 12 kg/m^3. During the six main rainfall events, salinity decreases when discharge increases, showing that both parameters are anti-correlated. Even during highest flood events, groundwater remains brackish (2 kg/m^3), polluted by seawater intrusion in depth.

The time series analysis gives insights on the hydrogeological functioning of the aquifer, detailed herein and illustrated in Fig. 2a:

- The groundwater salinity comes from a deeper reservoir, as salinity dilution occurred systematically during flood events. From salinity evolution during recession periods, we may hypothesize that an almost constant salinity inflow occurred with a value of around 14 kg/m^3.
- The aquifer is a typical karst aquifer, with large change in discharge during flood events. The mean response time of a few days shows that a fast infiltration occurs.

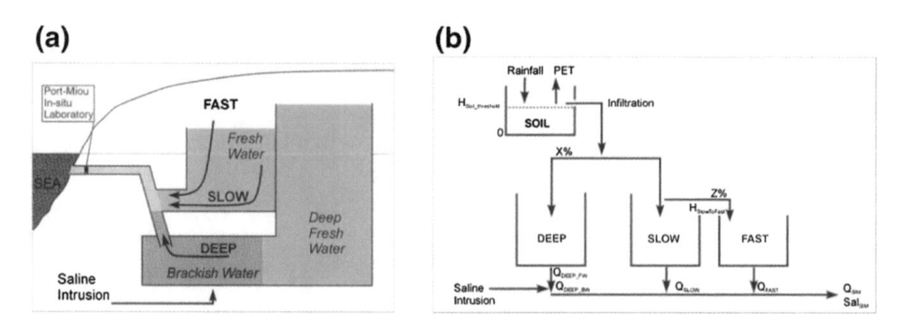

Fig. 2 Conceptual model of the Port-Miou coastal karst aquifer (**a**), and Structure of the corresponding rainfall-discharge-salinity model (**b**)

- The groundwater storage appears high according to the long recession curve, showing that a slower component is drained by the aquifer during low flow periods.

Consequently, our conceptual model shows that saline water due to seawater intrusion is mixed in depth with freshwater in a deep brackish water reservoir. A continuous discharge from this deep brackish reservoir is mixed with freshwaters coming from a shallower reservoir, having typical karst response (fast and slow components) to rainfall events.

The conceptual model (Fig. 2a) summarizes water flows from each compartment involved and transfer processes that link them. In order to verify the main hypothesis about the hydrogeological processes highlighted earlier, a modeling approach is presented in the next section. The experimental step was conducted at the aquifer scale, and thus a lumped approach was chosen for the modeling step. The modeling constraints were: (i) to equally represent discharge and salinity fluctuations; (ii) to integrate the three DEEP, SLOW, and FAST reservoirs; and (iii) to accurately simulate the water and mass budget.

4 Modeling Approach

4.1 Model Structure

Structure and parameters of the conceptual model are presented in Fig. 2b. This model is developed from a modeling approach recently described and tested with good performances on rainfall-discharge-solute data by Charlier et al. (2012). Our model is based on four connected reservoirs. A SOIL reservoir that recharges the aquifer is partitioned into:

- a DEEP reservoir of brackish water, with low variations of discharge. This reservoir represents the storage function of the aquifer, with a long residence time. Saline intrusion affects this reservoir.
- a SLOW reservoir of fresh water, to represent the storage and the long recession after rainfall events.
- a FAST reservoir of freshwater, to represent the fast infiltration and rapid transfer through the aquifer during high-water event.

Discharge Q $[L^3 \cdot T^{-1}]$ at the outlet of each reservoir is calculated as a function of the reservoir stock H using a linear relation:

$$Q = H_{out} * \text{Area with } H_{out} = H * kr \quad (1)$$

where kr $[T^{-1}]$ is a constant characterizing the recession curve of the reservoir. Once Q are calculated for each reservoir, then salinity is calculated using the following mass balance equations:

$$Q_{Spring} * \ Sal_{Spring} = Q_{BW} * \ Sal_{BW} + Q_{FW} * \ Sal_{FW} \quad (2)$$

$$Q_{Spring} = Q_{BW} + Q_{FW} \quad (3)$$

where Sal is the salinity $(M \cdot L^{-3})$, with subscript Spring for the spring, subscript BW for brackish water reservoir, and subscript FW for fresh water reservoir.

In the DEEP reservoir, the model assumes that the salinity is constant, equal to 14 kg/m^3 (Sal_{BW}). To account for the seawater discharge feeding this reservoir, deep brackish water discharge (Q_{DeepBW}) is calculated from the discharge of freshwater (Q_{DeepFW}) and salinity as follows:

$$Q_{DeepBW} = Q_{DeepFW} * (Sal_{SW}/(Sal_{SW} - Sal_{BW})) \quad (4)$$

where Sal_{SW} is the salinity of the seawater (i.e. 38 kg/m^3). The percentage of recharge flowing from SOIL to DEEP reservoir is given by the X parameter. SLOW and FAST partitioning given by the Y parameter varies from Y1 to Y2 according to water level in the SLOW reservoir using a threshold parameter ($H_{Slow/Fast}$).

4.2 Parameterization and Calibration Strategy

The model inputs are the rainfall and PET, and outputs are simulated discharge and salinity, which were compared to the observed ones to test model performances. The model needs a total of nine parameters that may be fixed or optimized. From the results of previous works using recession hydrograph analysis, kr_{DEEP}, kr_{SLOW}, and kr_{FAST} and the constant salinity of the DEEP brackish reservoir (Sal_{BW}) were fitted. Finally, five parameters needed to be optimized: $H_{SOIL\_threshold}$, $H_{SLOW/FAST}$, X, Y1, Y2.

Table 1 Calibration values of the model parameters: calibration 1 from Dec. 2010 to Oct. 2011, and calibration 2 from Oct. 2011 to Jun. 2012

Parameter	Signification	Calibration 1	Calibration 2
$H_{SOIL\_threshold}$	Threshold value to activate recharge from the soil	34 mm	34 mm
Kr_{DEEP}	Recession coefficient for the reservoir DEEP	0.0001 day^{-1}	0.0001 day^{-1}
Kr_{SLOW}	Recession coefficient for the reservoir SLOW	0.008 day^{-1}	0.008 day^{-1}
Kr_{FAST}	Recession coefficient for the reservoir FAST	0.16 day^{-1}	0.19 day^{-1}
$H_{SLOW/FAST}$	Threshold value in SLOW reservoir	55 mm	55 mm
X	Percentage of recharge of the DEEP reservoir	0.50	0.54
Y1	Percentage of recharge of the FAST reservoir (when $H_{Slow/Fast}$ is not exceeded)	0.25	0.27
Y2	Percentage of recharge of the FAST reservoir (when $H_{Slow/Fast}$ exceeded)	0.55	0.64
Sal_{BW}	Salinity of the DEEP reservoir	14 kg/m^3	14 kg/m^3

The quality of the simulation for the spring discharge Q and salinity Sal is evaluated by the Nash-Sutcliffe (NS) efficiency criterion. In order to optimize simulations on low and high values (recessions and flood peak as well as dilutions and brackish water inflows), we compute NS on $\sqrt{Q}$ and $\sqrt{Sal}$, respectively. Calibration procedure takes also into account the water budget over the calibrated period, optimizing simulations for relative error inferior to 5 % on volume and mass transport. Calibration was carried out manually by a trial–and-error procedure.

4.3 Application on the Study Site

The parameterization results of the calibration are given in Table 1. To avoid the question of the short duration time series, leading to difficult assessment of model performances, a split-sample test (Klemeš 1986) was conducted. This test considers that both subperiod of the whole time series (Dec. 2010 to Oct. 2011, and Oct. 2011 to Jun. 2012) should be used in turn for calibration and validation. Modeling results showed good accuracy on discharge and salinity simulation for calibration and an acceptable performance for validation: NS $\sqrt{Q}$ was 0.80 and 0.69, and NS $\sqrt{Sal}$ was 0.79 and 0.62, respectively. The fairly results in validation may come from the flood event distribution. In fact, only one large flood event in November 2011 occured, leading to split the time series in a first subperiod with small events, and a second one with a large event. The range of discharge and salinity is thus different for both subperiods, given more difficulties to optimize model parameters. The water budget is equilibrated in the calibration as well as the validation procedure (Error on volume of 2.5 and 13.6 %, and Error on mass of 1.1

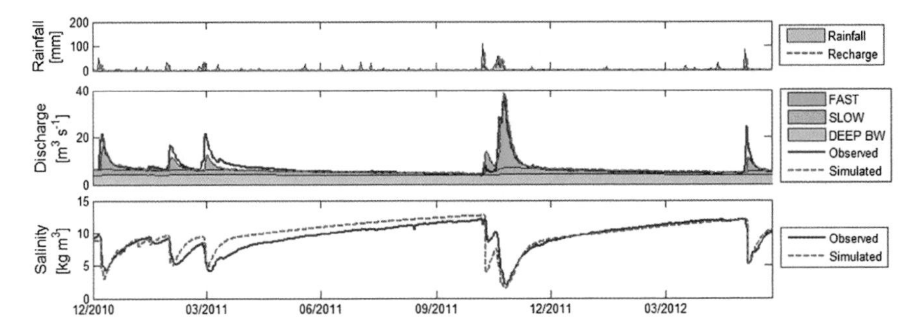

Fig. 3 Simulated and observed data at the Port-Miou spring; calibration was done from Oct. 2011 to Jun. 2012, and validation from Dec. 2010 to Oct. 2011

and 0.9 %, for calibration and validation, respectively). Example of simulated recharge, discharge, and salinity are shown in Fig. 3. The general shape of the simulated discharge and salinity plots is well reproduced.

5 Discussion and Conclusion

The aim of our study was to develop a conceptual model of saline intrusion within the coastal karstic aquifer of Port-Miou, and to assess freshwater resources in the aquifer using a lumped model. One original aspect of our model is the use of the salinity for the discharge calibration, and conversely. This gives more robustness in the modeling approach, because all parameters influence both discharge and salinity. For instance, Kr_{DEEP} and Kr_{SLOW} govern discharge recession as well as salinity increase during low water levels. This is mainly the main point of this model, which is relatively parcimonious (nine parameters), when it is used to simulate two independent time series. Advantages of such reservoir model are a simple implementation and an easy adaptation to various case-studies.

We can deduce from the conceptual and numerical models that groundwater flow is spread in several compartments in the aquifer, with their own hydrogeological behavior. These compartments, viewed as reservoirs for modeling purpose, are all connected together to the main karst conduit of the Port-Miou spring.

In the deep part of the aquifer, the freshwater is mixed with saltwater. About 50 % of the effective rainfall (parameter X) supplies this deep reservoir (corresponding to 2.5 m^3/s for the studied period). This aquifer has a strong inertial behavior ($k = 10^{-4}$ d^{-1}), which could be coherent with a large storage, with smooth and low changes in hydraulic gradient (there is no karstic-type functioning). The mixing zone should be so wide and the transit time so long that the aquifer provides some brackish water with an almost constant salinity and discharge. This brackish water is drained by the karst conduit of Port-Miou several kilometers upstream the spring, and at least 223 meters below the sea level. This reservoir seems not to be affected by high rainfall events. This behavior is very different from the one observed in another main coastal Mediterranean karst, at the Almyros spring of Heraklio (Crete, Greece). In this latter case, Arfib and Marsily (2004) showed that the discharge of seawater to the main karst conduit connected to the outlet changes during high-water events. Furthermore, this discharge can be stopped or reversed (and consequently the spring water became fresh). The Port-Miou model shows that changes in water level in the aquifer, even during high discharge events, can be insufficient to decrease the saline intrusion in order to obtain drinking water.

The Fast and Slow reservoirs can be viewed as the two shallower components of the freshwater flowing through an aquifer not affected by saline intrusion. The Fast reservoir represents the flow through conduits, with a very low storage, but a high transmissivity. This Fast freshwater is the main component of flood events when spring discharge is more or less higher than 8 m^3/s (Fig. 3). The Slow reservoir simulates the baseflow of the freshwater aquifer. This freshwater aquifer dilutes the brackish water coming from the deep reservoir in the main karst conduit connected to the spring. For the studied period, the fast and slow flows represent on average 1.2 and 1.5 m^3/s of freshwater at the spring, respectively.

From our results, we can conclude that groundwater is available as freshwater resource in two specific compartments: (1) in the shallower karst aquifer not affected by saline intrusion, and (2) in the deep aquifer, upstream the saline intrusion zone. Further work is still needed to precisely localize the connection zone between both compartments.

Coastal karstic springs are characterized by salinity changes, usually with a decrease in salinity when the discharge increases. In the Mediterranean context, it has been shown that springs can be fresh during high discharge, like in the Almyros of Heraklio case study (Arfib et al. 2007), with the deep saline intrusion stopped due to groundwater height increase. On the contrary, the Port-Miou case-study shows that the saline intrusion can be almost constant in time, and that the salinity variations are mainly due to dilution of the deep brackish water by a shallower fresh aquifer. Finally, this paper shows that a lumped model is an effective tool to characterize the mechanism of salinity changes.

Acknowledgments This paper is part of the KarstEAU project funded by the "Agence de l'Eau Rhône-Méditerranée-Corse", "Conseil Général du Var", "Conseil Général des Bouches-du-Rhône" and "Région Provence Alpes Côte d'Azur". We thank N. Mauger (MS Student) who helped in this project. We also thank "Météo France" for the rainfall/ETP data, and the "Conservatoire du Littoral" for the access to the Port-Miou in situ underground laboratory.

References

Arfib B, de Marsily G (2004) Modeling the salinity of an inland coastal brackish karstic spring with a conduit-matrix model. Water Resour Res 40(11):1–10

Arfib B, de Marsily G, Ganoulis J (2007) Locating the zone of saline intrusion in a coastal karst aquifer using springflow data. Ground Water 45(1):28–35

Cavalera Thomas (2007) Study of the hydrogeological functioning and the catchment area of the submarine spring of Port-Miou (South-East France). PhD. thesis. University of Provence-Marseille. 403p.[in French] http://tel.archives-ouvertes.fr/tel-00789232

Charlier JB, Bertrand C, Mudry J (2012) Conceptual hydrogeological model of flow and transport of dissolved organic carbon in a small Jura karst system. J. Hydrol. 460–461:52–64

Custodio E (2010) Coastal aquifers of Europe: an overview. Hydrogeol J 18:269–280

Fleury P, Bakalowicz M, de Marsily G (2007) Submarine springs and coastal karst aquifers: a review. J Hydrol 339:79–92

Hartmann A, Lange J, Vivó Aguado À, Mizyed N, Smiatek G, Kunstmann H (2012) A multi-model approach for improved simulations of future water availability at a large Eastern Mediterranean karst spring. J Hydrol 468–469:130–138

Klemeš V (1986) Operational testing of hydrological simulation models. Hydrol Sci J 31(1):13–24

Meniscus X. (2013) Spring of Port Miou: -223 m on 7 May 2012. Spelunca n°132. 23-27

Stringfield VT, Legrand HE (1971) Effects of karst features on circulation of water in carbonate rocks in coastal areas. J Hydrol 14:139–157

Tulipano L, Fidelibus MD, Panagopoulos A (eds) (2005) COST action 621 Groundwater management of coastal karstic aquifers, p 367 (COST Office)

Groundwater Flow Modeling in a Karst Area, Blau Valley, Germany

C. Neukum, J. Song, H.J. Köhler, S. Hennings and R. Azzam

Abstract Karst aquifers typically have complex flow patterns as a result of the depositional heterogeneities and large conduits from dissolution features. Various field measurements were carried out to build a hydrogeological conceptual model of the Jurassic aquifer close to Blaubeuren (Germany), including drilling data, well tests, and geophysical surveys. These data were assembled to simulate the groundwater flow at a site where a possible new pumped-storage plant is in planning approval. The favorable layout of the lower reservoir is designed without sealing to the connected karst aquifer underneath. Approving this construction concept, various measures and strategies have to be formulated by using numerical modeling to limit possible adverse impacts on the construction site and its environment, considering that the Blau Valley requires high level of protection. Based on the conceptual model, the aquifer system is represented by a three-dimensional finite element model using the FEFLOW numerical code. The model is calibrated for steady-state and transient conditions by matching computed and measured piezometric levels (November 2012—March 2013) from over 30 observation wells to estimate the best-fitted spatial distribution of both hydraulic conductivity and storage coefficient in the aquifer. The model is used to analyze quantitatively the flow regime, the groundwater mass balance, and the aquifer hydraulic properties of this karst area. The results of the study provide necessary information regarding the hydraulic behavior of the aquifer in order to plan the construction phase and the subsequent operation of the pumped-storage plant.

Keywords Groundwater modeling · Karst aquifer · Pumped-storage plant · Germany

C. Neukum (✉) · J. Song · R. Azzam
Department of Engineering Geology and Hydrogeology, RWTH Aachen University, Lochnerstr. 4-20, 52064 Aachen, Germany
e-mail: neukum@lih.rwth-aachen.de

H.J. Köhler · S. Hennings
Dr. Köhler & Dr. Pommerening GmbH, Am Katzenbach 2, 31177 Harsum, Germany

B. Andreo et al. (eds.), *Hydrogeological and Environmental Investigations in Karst Systems*, Environmental Earth Sciences 1,
DOI 10.1007/978-3-642-17435-3_36

1 Introduction

The excavation pit of a limestone quarry in southern Germany was proposed as a location for construction of a pumped-storage plant (PSP) to even out frequent changes between electricity shortages and surpluses. This can result in a series of ecological problems due to the planned construction scenario, which includes no sealing of the basin floor and flanks. Therefore, investigation and understanding of the groundwater dynamics, which is caused by the construction and operation of the pumped storage plant, is very important for protection of the local ecological environment.

Karst aquifers are generally highly heterogeneous. They are dominated by secondary (fracture) and tertiary (conduit) porosity and may exhibit hierarchical permeability structure or flow paths (Scanlon et al. 2003). Groundwater flow predictions in karst aquifers require the use of models that represent a sufficiently large number of potentially important physical features to ensure an appropriate characterization of system behavior. In this work, an equivalent porous media distributed model was built to simulate the local groundwater flow conditions through the faulted and fractured limestone in the area of Blau Valley. Automated inverse method was applied to optimize the model parameters to achieve the best-fit to available observations. This model was used thereafter to analyze quantitatively, the flow regime, the groundwater mass balance, and the aquifer hydraulic properties, so that the present and alternative construction scenarios of PSP in the future works can be tested and the associated impact on the environment of the Blau Valley.

2 The Study Area

The study area is located near the city of Blaubeuren, Germany, with an area of about 8 km^2 (Fig. 1a). The outcropping rocks in the study area can be divided into three main groups viz. Quaternary sediments, white limestone of Jura (upper and lower massive limestone), and Tertiary sediments of lower freshwater molasses. The regional geological conditions in the Blau Valley are characterized by fluvial deposits and weathering products of hard rocks, which date back to the Quaternary period and lies directly above the carbonate rocks of Malm. In the quarry area, the major stratigraphic units are the distinctly fissured and mild to moderate karstified white limestone of Jura. The white Jura continues downward and toward southeast direction with a slight inclination of 1° to 2° under the sediments of lower fresh water molasses.

The main geometric-structural and hydrogeological characteristics of the aquifer system were reconstructed on the basis of the general geologic reconstruction using well log data from 47 boreholes. This system consists of two main aquifers, which have fundamentally different hydrogeological characteristics. The Quaternary

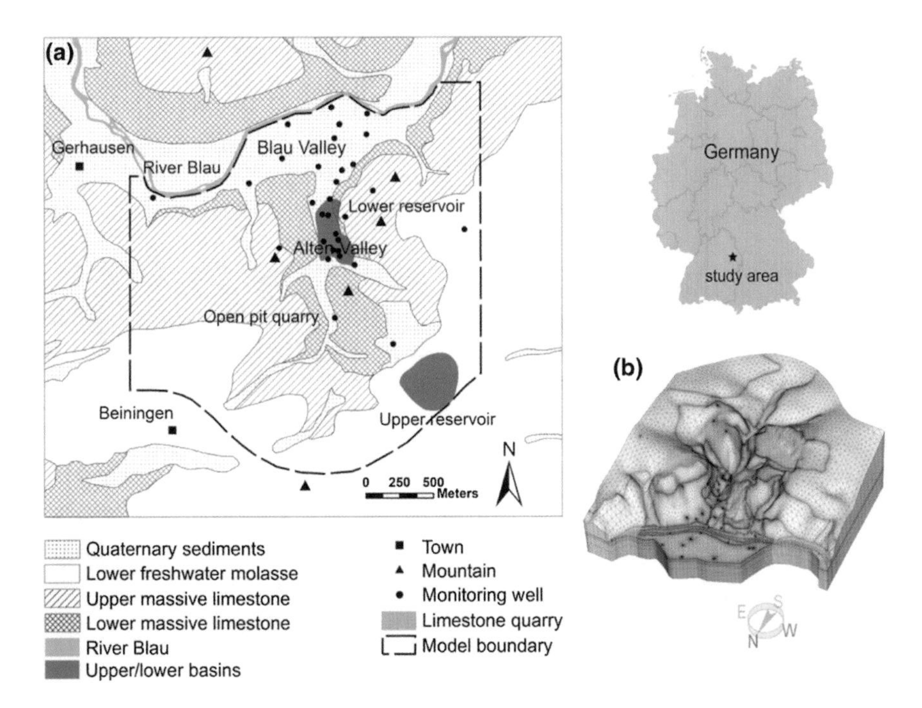

Fig. 1 **a** Location of study area in southern Germany with geological formations; **b** Computation grid for the three-dimensional model

sediments in the Blau Valley are made up of several gravely and silt layers, which combine to form a single-layer porous aquifer. This porous aquifer reaches a maximum depth of 27 m in the Valley at an elevation of about 475 m above the mean sea level (m asl). Beneath the porous aquifer, that is the regional spread karst aquifer in the entire area. The karst aquifer is up to 50 m thick and shows highly variable water pathways in vertical and lateral extent based on the lithological facies and formation conditions. The karst aquifer in the entire area is appropriately assumed to be an equivalent porous media using "Equivalent continuum approach", because both the groundwater flow within the rock matrix and through karst formations can be averaged into a bulk conductivity of the model's cell in such scale (Ford and Williams 2007). Collected data show that the karst aquifer is unconfined in the outcrop area outside the Blau Valley, but confined in the Blau Valley area due to the presence of partially low permeable valley deposits. These two aquifers are partially separated by a low permeability alluvial clay layer in the valley area, so the hydraulic connection between the two units is ensured by an indirect leakage. However, at the borders of the valley the alluvial barrier layer is often missing, where the porous aquifer is mainly supplied by lateral inflow from karst aquifer.

The aquifer system is primarily recharged by direct infiltration of precipitation. The mean annual precipitation observed at Blaubeuren weather station is 793 mm, with major rains occurring in spring and fall. The second major source of recharge is lateral groundwater flow from southern directions. Groundwater flows from south to north in the unconfined section of the aquifer and generally northeast in the confined section to discharge at River Blau, which is the major creek in this area and flows through the Blau Velley approximately in E-W direction.

3 Numerical Modeling

The numerical groundwater model is introduced by using FEFLOW (finite element subsurface flow system) working under both steady-state and transient conditions. The model was discretized into eight layers with 0.5 million nodes and 0.8 million elements to represent all geological units. The element sizes vary from 50 m near the edges of the model to 0.5 m at the observation points (Fig. 1b). In the Blau Valley, the top six layers represents the porous aquifer, while at other places, the top seven layers represent the karst aquifer.

3.1 Boundary Conditions

The boundary conditions assigned to the numerical model derive directly from the conceptual reconstruction of the aquifer system (Fig. 2a). The River Blau was defined as the northern edge of the model. The Cauchy boundary condition was assigned on this edge and constrained by the river level measurements, i.e., the water level of the river equals the groundwater level. The western edge of the modeled area acts as no-flow boundary. The Dirichlet boundary condition was applied on the southern edge with a constant head value of 540 m asl and on the eastern edge with linear decreased values from 540 to 520 m asl in S–N direction. A no-flow boundary condition was applied to the bottom of the model, and spatially distributed recharge was defined at the top model surface. Pumping wells were assigned to the interior to the model to represent water withdrawals.

3.2 Calibration

Basically, two steps were followed in modeling the aquifer: a steady state model was developed to calibrate the spatial distribution of hydraulic conductivity and in-/out-transfer coefficients to match the measured piezometric surface, and a transient model was run for a 3 months period (November 2012—March 2013) by using daily recharge to optimize the distribution of storage coefficient such that groundwater level differences between recharge events were matched. Both steady state and transient models were calibrated using inversion implemented with Parallel PEST (Doherty 2004) and the pilot points method (Doherty 2003).

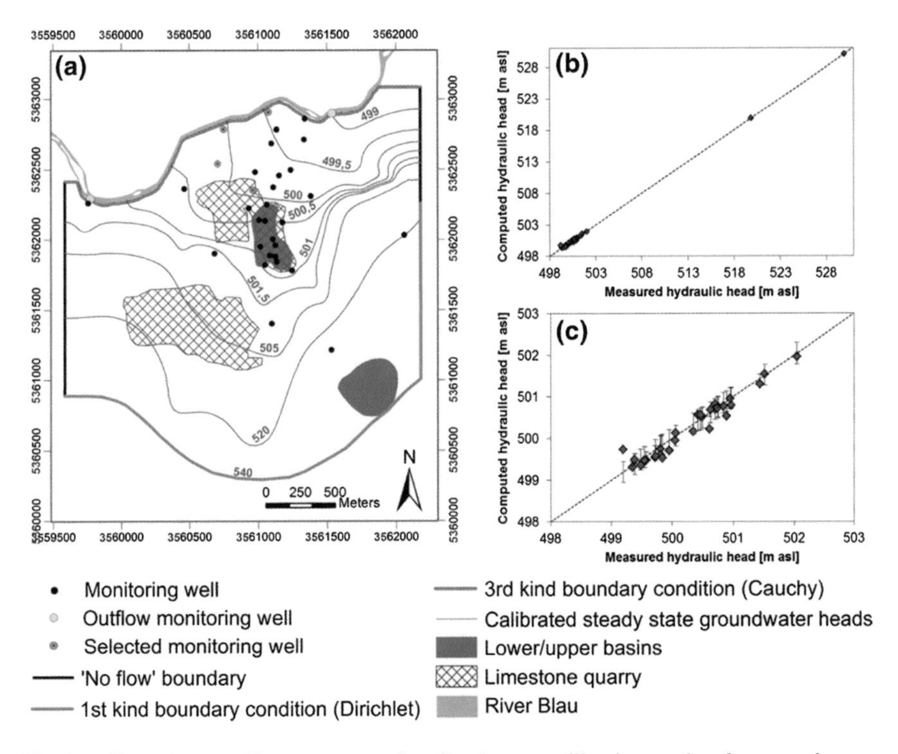

Fig. 2 **a** Groundwater table contour map; **b**, **c** Steady state calibration results of measured versus observed heads in different scale

Measured water levels on 24 Nov. 2012 were used to evaluate the steady state model calibration because the recharge was close to stable condition on this day, after a dry period of several weeks, and the number of synoptic water level measurements (33) was greatest. Spatial distribution of recharge zones were defined under the condition of land uses and topography. The distributed recharge rate on 24 Nov. 2012 was then assigned to the steady state model. Under the consideration of the anisotropy, 3,654 pilot points distributed in different zones with different spacing were created to calibrate the conductivity of the entire model. By using the regularization method of SVD (Singular Value Decomposition)-Assist, PEST identified then 140 super-parameters representing combination of the 3,654 parameters that are uniquely identifiable given the observation dataset, to optimize the distribution of parameter values. The calibrated steady state groundwater heads and a contour view of the computed heads are plotted in Fig. 2a, b, c), which show the scatter plot of simulated versus measured heads after the calibration in different scale. The allowed absolute difference was defined of 0.1 m in the Blau Valley, and in the limestone area should be smaller than 0.25 m. The average difference of all observation points after calibration is about 0.13 m. As shown in the scatter plots in Fig. 2c, the computed values at four observation wells exceed the defined value

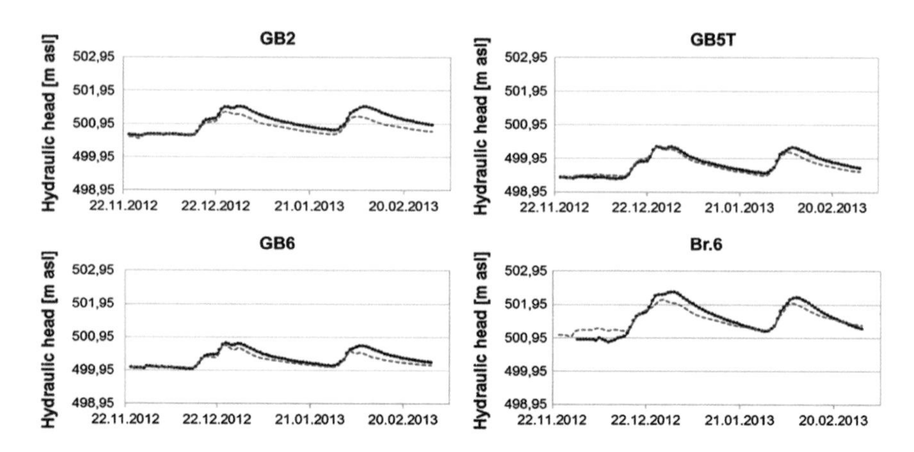

Fig. 3 Comparison between computed (*red*) and measured (*black*) groundwater level at four selected observation wells from 24 Nov. 2012 to 01 Mar. 2013

range, which is possibly caused by the locally confined situation nearby these three points.

Simulated heads and the calibrated distribution of hydraulic conductivity from the steady state model were used as input for the 3 months' period transient model. The model was calibrated with the observed data from 36 wells for transient conditions. The aim of transient-state calibration was to simulate the variations of water table, which were recorded at the monitoring wells, as accurately as possible. The transient simulation started on 24 Nov. 2012 and ended on 01 Mar. 2013, lasting a total of 98 days. The time varying water level of the River Blau (transfer at northern boundary) and inflows from the surface (distributed recharge on the top layer) were assigned to the transient model. The results of the calibration under transient conditions are shown in the plots of Fig. 3, where results are compared with measurements at four selected observation points from three different model areas, i.e., at the River Blau, in the Blau Valley, and in the quarry area. The computed groundwater levels are generally in agreement with the collected values from the observation wells. The Nash-Sutcliffe model efficiency coefficient at GB2, GB5T, GB6, and Br.6 is about 0.59, 0.90, 0.79, and 0.82, respectively. Nash-Suttcliffe efficiencies can range from $-\infty$ to 1. An efficiency of 1 corresponds to a perfect match between model and observations. Values of >0.5 are generally acceptable for fitting measured to simulated groundwater heads.

3.3 Water Budget

The steady state model has a positive balance of about 155 m^3/day (about 1.5 % of the total rate). This shows a satisfactory equilibrium between inflows and outflows. The meteoric recharge on 24 Nov. 2012 is of about 855 m^3. From southern sector, the groundwater flows into the modeled area for a total of about 1,770 m^3/day and then flows toward the River Blau. On the northern edge the transfer effect between the River Blau and aquifer is significant. A total of about 10,622 m^3/day of groundwater flows into the River Blau and the porous aquifer in the Blau Valley receives 8,155 m^3/day of river water from Blau. The model under the transient condition shows a positive balance of about 9,893 m^3/day due to the strong meteoric infiltration (about 10,700 m^3/day) during the simulation reference period.

4 Conclusions and Future Works

Modeling of karstic systems as equivalent porous media is suitable in simulating regional groundwater flow using regional aquifer parameters. The presented model can be used to estimate various measures and strategies to limit possible adverse impacts on the construction site and its environment, considering that the Blau Valley requires high level of protection. This model is still not completely reliable because of the incomplete knowledge of the aquifer system. Future research will improve the numerical groundwater model through further piezometric and hydrometric surveys. The hydraulic and dispersive properties of the aquifer system will be clearer by further pumping tests. The presented model represents the basis for future rock-mechanical simulations of the lower basin. The ascertainable hydraulic states obtained from the model would then be coupled with a rock-mechanical model to investigate their effects on mechanical properties of rock mass, especially on the transient states and stress changes in rock mass during operation of the pumped storage plant.

References

Doherty J (2003) Ground water model calibration using pilot points and regularization. Ground Water 41(2):170–177

Doherty J (2004) PEST model-independent parameter estimation user manual, 5th edn. Watermark Numerical Computing, Brisbane

Ford D, Williams P (2007) Karst hydrogeology and geomorphology. Wiley, Chichester pp 562

Scanlon BR, Mace RE et al (2003) Can we simulate regional groundwater flow in a karst system using equivalent porous media models? Case study, Barton Springs Edwards aquifer, USA. J Hydrol 276:137–158

Multi-scale Assessment of Hydrodynamic Properties in a Karst Aquifer (Lez, France)

A. Dausse, H. Jourde and V. Léonardi

Abstract The present study focuses on the hydrodynamic characterization of the Lez karst aquifer (Southern France) on the basis of hydraulic field tests performed at different scales of space and under distinct hydrological conditions. Depending on the water level conditions, the organization of the flow paths linked to the geological structure of the reservoir changes and a compartmentalization of the system due the hierarchization of hydraulic connections to the main flow paths was assessed. For the same parameter characterized at borehole scale and at regional scale, a difference of $10–10^5$ has been quantified. This quantification of hydrodynamic parameters provides important constraints on multiscale modeling and the characterization of main flow paths in such a karst system.

1 Introduction

Groundwater in karst aquifers constitutes the main available water resource in many areas, especially in Mediterranean regions where environmental pressures increase the problems of water scarcity (Iglesias et al. 2007). Understanding groundwater flows and transport processes in such a context is therefore of prime importance to reach a sustainable management of groundwater resources (Bakalowicz 2004). Because the high level of heterogeneity, characterizing groundwater flows in karst aquifers at different scales of space and time remains a challenge.

In this study, the monitoring of groundwater flow is performed at several scales of time and space, within a single Mediterranean karstic carbonate aquifer, the Lez karst aquifer, located in the South of France. The method is based on the analysis

A. Dausse (✉) · H. Jourde · V. Léonardi
Hydrosciences Montpellier UMR 5569 (CNRS, IRD, UM1, UM2), Université Montpellier, 2 CC MSE, 34095 Montpellier Cedex 5, Montpellier, France
e-mail: amelie.dausse@um2.fr

B. Andreo et al. (eds.), *Hydrogeological and Environmental Investigations in Karst Systems*, Environmental Earth Sciences 1,
DOI 10.1007/978-3-642-17435-3_37

of different hydrodynamic tests (pumping tests, slug, and injection tests) at regional scale, experimental field site scale, and borehole scale. The aim of this paper is to explain how to allocate the hydrodynamic parameters as a function of the spatial and temporal scale of observation to improve the understanding of flow dynamics and to estimate the storage capacity of the aquifer.

2 Study Sites and Methodology

2.1 *Regional Scale*

The Lez aquifer is a fractured and karstic carbonate reservoir located in the South of France (Fig. 1a). The aquifer is composed of a 400 m thick formation of Jurassic and Beriasien limestones and barred in some areas by Lower Cretaceous marls and marly limestones. Important tectonic features, like the NE-SW Corconnes fault network across the aquifer, brings the Jurassic and Beriasien karstic formation (West) in contact with the impermeable formations (East). It gives rise to one main outlet located 22 km from the Mediterranean coast and at the elevation of 65 m above sea level, the Lez spring, and several seasonal overflowing springs. The Lez spring is a Vauclusian spring that drains an aquifer of around 380 km^2, with a highly variable natural flow range from 0 to 8 $m^3.s^{-1}$ during flood events. The main of karst conduit upstream of the Lez spring is pumped for water supply since 1981 with sequential pumping at a flow rate average about 1.1 $m^3.s^{-1}$. These large pumping rates impact the hydrodynamics of the karst aquifer at the regional scale. A relatively dense groundwater monitoring network (Fig. 1a) permits the spatial distribution of the hydrodynamic properties of the aquifer to be assessed from the hydrodynamic response to pumping at the Lez spring.

2.2 *Experimental Field Site and Borehole Scale*

The Terrieu experimental field site is located 4.6 km from the Lez spring (Fig. 1b). The site consists of 22 uncased boreholes, with an average depth of 60 m, laid out in a 1500 m^2 area (Fig. 1b). The boreholes are quite near to each other (average distance about 5 m). They intersect fractured and karstified Upper Jurrasic and Lower Cretaceous limestone locally forming an unconfined aquifer. All the boreholes reached the saturated zone and piezometric levels vary from according to the season.

A four day pumping test with a discharge rate of 50 $m^3.h^{-1}$ was performed in P0 when the groundwater level was relatively high (66.5 m asl and overflow at the Lez spring). Hydrodynamic monitoring was performed in P0 and in fifteen

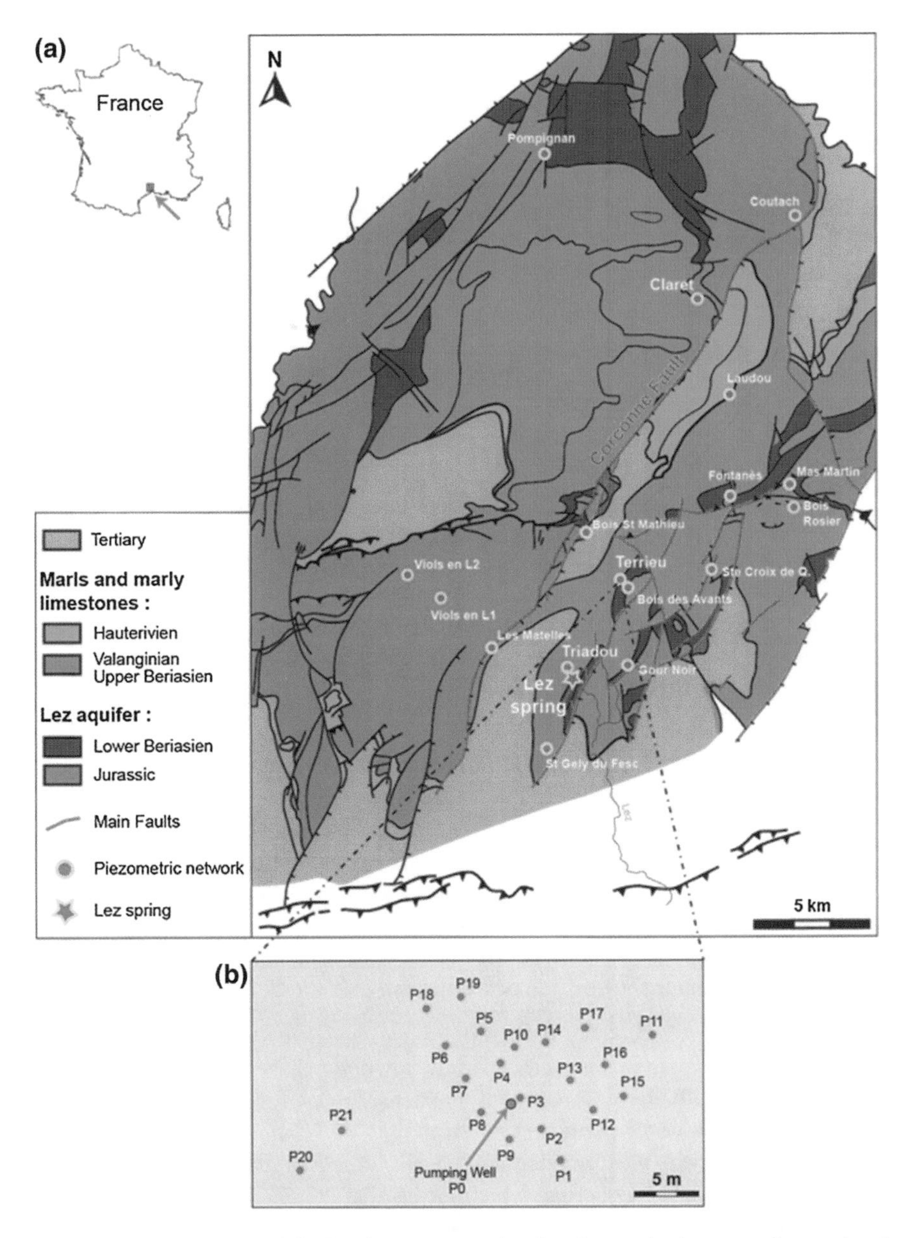

Fig. 1 **a** Geological map of the Lez karst system, showing the monitoring network at regional scale; **b** Monitoring network at the scale of the Terrieu experimental field site

observation boreholes. In borehole P15, a straddle packer isolated a karst conduit to identify the specific response of this main flow path. Straddle-packer tests (slug and injections tests) were performed in four boreholes of the Terrieu experimental field site to determine the vertical distribution of transmissivity at borehole scale.

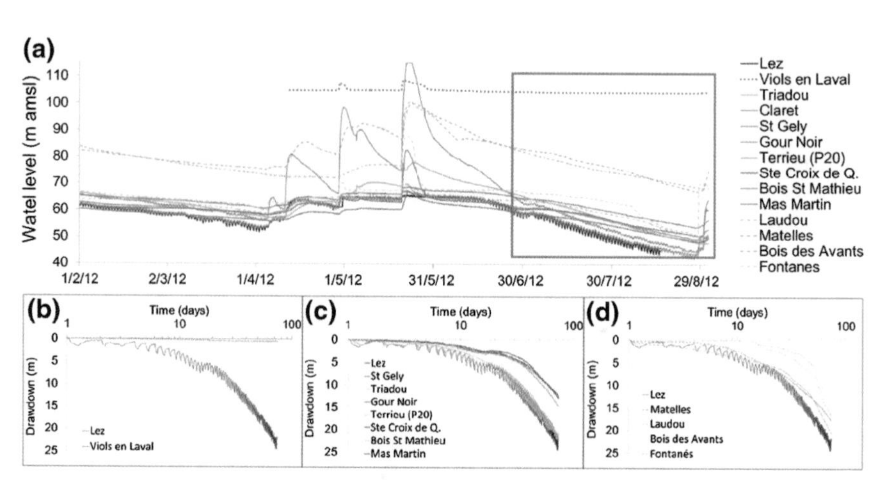

Fig. 2 Regional scale hydrodynamic response to pumping at the Lez spring. **a** Hydrodynamic behavior of Lez system; **b** borehole without drawdown due to the Lez pumping; **c** boreholes with similar Lez drawdowns responses; **d** boreholes with complex drawdowns responses

3 Spatial Scaling of Hydrodynamic Properties

3.1 At the Regional Scale (Fig. 2)

During summer, very little rain and large pumping rates are the cause for a cessation of flow at the springs; when this happens, we can assume that the decrease in water level is mainly due to pumping at the Lez spring. This hypothesis is consistent with piezometric data gathered before the construction of the Lez pumping station that showed a stabilization of groundwater level during summer and very low discharge at the spring (Fabris 1970).

The comparison of drawdowns during the low water period shows different behaviors according to the position of wells relative to the Corconnes fault network:

(i) west of the Corconnes fault (Pompignan and Viols-le-Fort area), the stability of water levels with no drawdown suggests that there is no connection with the compartment impacted by pumping at the Lez spring during low water periods (Fig. 2b);

(ii) east and near to the Corconnes fault, wells present similar drawdown evolution to those at the Lez spring, which means that there is a good hydraulic connectivity with the main flow path network. Also, drawdowns at Terrieu site, Claret Brissac, and Triadou wells exhibit periods of drawdown and recovery due to daily variation of pumping rate at the Lez spring, which suggests direct connection to the main flow path in a confined environment (Fig. 2c).

(iii) other wells show drawdowns dissimilar from drawdown at the Lez spring, which may be interpreted as a lower hydraulic connectivity with the spring (Fig. 2d).

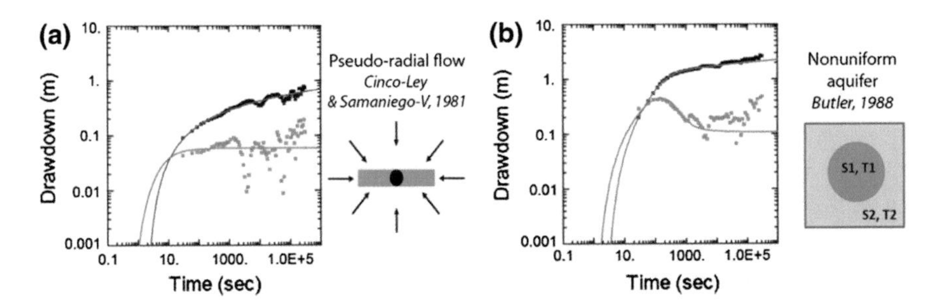

Fig. 3 Hydrodynamic responses at local scale to pumping in P0. **a** Pseudo-radial flow response for boreholes crossing the main flow path; **b** nonuniform aquifer response

The hydrodynamic response to the pumping at the Lez spring is mainly linked to the position of wells relative to the NE-SW Corconne fault. These faults systems (in red in Fig. 1) organize the main flow paths with a function of hydraulic conductor for east zone and a function of barrier fault at west zone.

Seasonal drawdown from June to August 2012 was analyzed like a long-term pumping test in an equivalent porous medium. The transmissivity values assessed with the Theis model (1935) varied between 1.4×10^{-2} and 6.7×10^{-2} $m^2.s^{-1}$, which means that the order of magnitude of the transmissivity was obtained assuming an equivalent porous medium. In addition, the order of magnitude of the storage varied between 1.7×10^{-4} (conduit storage) and 5.7×10^{-2} (diffuse storage).

3.2 At Experimental Field Site Scale (Fig. 3)

During the pumping test, the Terrieu site boreholes presented two kind of drawdown responses (Fig. 3a, b). Some boreholes and the isolated conduit in P15 with packer system showed homogeneous hydrodynamic properties with pseudo-radial flow response (graph a). These boreholes are located on the main flow path and main directions of fracturing identified at local scale by Jazayeri et al. (2011). The other boreholes showed heterogeneous responses (graph b), which can be interpreted assuming a nonuniform aquifer with the model of Butler (1988). These boreholes with heterogeneous pressure responses are crossing fissured matrix with indirect connection to the main flow path.

The interpreted hydrodynamic properties showed that the order of magnitude of transmissivity is the same for all these boreholes (1.1×10^{-2} to 1.9×10^{-2} $m^2.s^{-1}$), which is consistent with the order of magnitude of the overall transmissivity previously assessed at the regional scale. The storage varied between 3.0×10^{-5} (diffuse storage) and 4.7×10^{-2} (conduit storage). Diffusivity values (T/S ratios) are in agreement with the geological properties of the site, i.e., $D > 5$ $m^2.s^{-1}$ for boreholes intersecting conduits, $0.2 < D < 0.5$ for fractured zones, and $D < 0.07$ $m^2 s^{-1}$ for fissured matrix zones.

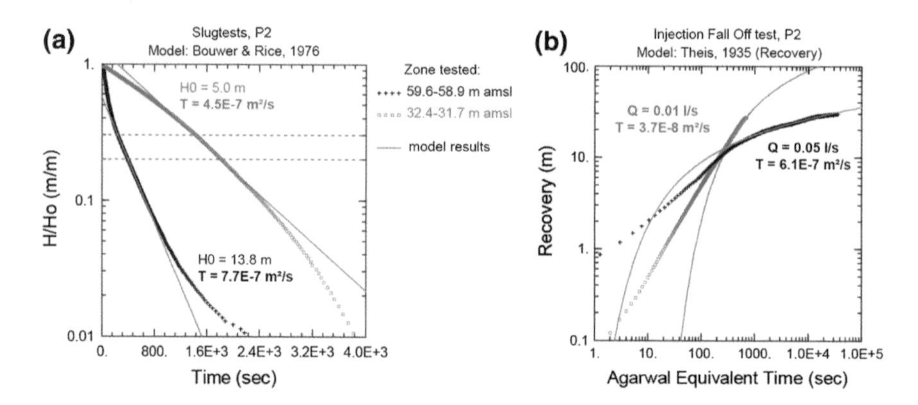

Fig. 4 Estimation of transmissivity from **a** slug test and **b** injection data analysis

3.3 At Borehole Scale (Fig. 4)

In the Terrieu experimental field site, the tested section of the boreholes correspond to horizons where fractures and conduits were detected on the basis of temperature and conductivity logs and video logs. Transmissivity values obtained varied from 4.9×10^{-7} and 8.9×10^{-4} $m^2.s^{-1}$ for slug tests with Bouwer and Rice (1976) interpretation and from 3.7×10^{-8} and 1.3×10^{-5} $m^2.s^{-1}$ for injection tests with Theis (1935) recovery interpretation. Note that these transmissivities do not reach high values due to the limited injection rate of the testing equipment that did not allow testing of highly productive zones. Because of the specific setup of these packer tests, no storage values could be determined. Figure 4 presents slug tests (a) and injection tests recovery (b) interpretations at two fractured zones at P2 borehole.

Most of the time, there is one order of magnitude of difference between transmissivity estimated from slug test and from injection test for a same zone. Slug tests are extremely sensitive to altered, near-well conditions (Butler and Healey 1997). Negative skin, due for example to open fracture intersecting borehole, can produce slug-test estimated transmissivity that may be orders of magnitude higher than the average transmissivity of the formation in the vicinity of the borehole.

4 Temporal Scaling of Hydrodynamic Properties

At the regional scale, piezometric data show that the most diffusive boreholes present daily drawdowns due to the effects of pumping at the Lez spring.

The piezometric data from April to July 2012 are plotted in Fig. 5a for the Triadou borehole. The pumping effect, for the same rate (Fig. 5b), is different

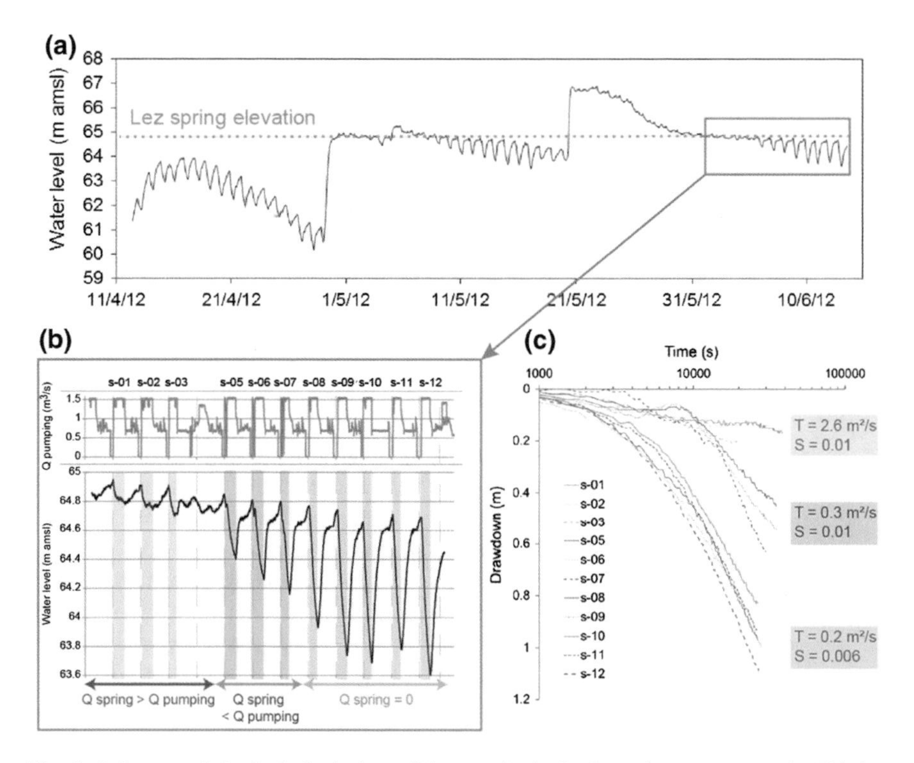

Fig. 5 Influence of the hydrological condition on the hydrodynamic response at the Triadou borehole. **a** Hydrodynamic behavior of the Triadou borehole from April to May 2012; **b** Lez pumping effects depending on hydrological situation; **c** Comparison of drawdowns effects

depending on the period on the hydrological cycle. The general pattern is that drawdowns are low when there is flow at the Lez spring and are high when there is no flow. Change of water pressure presents two behaviors: first the increasing of drawdown due to decrease of water level and natural spring rate, secondly a stabilization of drawdown independently of water level decrease.

The effect of pumping was analyzed by comparing the hydraulic parameters (transmissivity T and storage S) obtained assuming an equivalent homogeneous medium with the Theis model interpretation.

We observed that (Fig. 5c):

(i) when the flow rate at the Lez spring is higher than the pumping rate, the transmissivity and storage aquifer are high. The average values of transmissivity are greater than 1 $m^2.s^{-1}$ and storage is about 1.1×10^{-2};

(ii) during transition period (Q spring < Q pumping), average transmissivity decrease about one order of magnitude ($T = 0.3\ m^2.s^{-1}$) and the order of magnitude of storativity remains the same ($S = 1.7 \times 10^{-2}$);

(iii) when there is no flow at the Lez spring, both transmissivity and storage values decrease.

The values determined for different hydrologic conditions showed that natural flow in the karst increases intrinsic karst hydrodynamic properties. As already shown by Jazayeri et al. (2011) on the Terrieu experimental field site, the water table level influences the connectivity of some observation wells to the high permeability flow path network. As a consequence, the hydraulic parameters are impacted and the transmissivity value decreases. To characterize the reservoir at regional scale, it is more appropriate to focus on the hydrodynamic properties analysis when there is no flow at the spring.

5 Conclusion

This study shows the importance of the water level and the structure of karst aquifer on groundwater flow organization. At regional scale, both the hydrological conditions (i.e., high or low water level) and geological compartmentalization that impact the hydraulic connectivity, control the hydrodynamic properties. At the experimental field site scale, a pumping test highlighted the heterogeneous flow pattern that can be linked to the position of boreholes and the main geological features. At the borehole scale, hydraulic tests revealed a high range of hydrodynamic properties (transmissivity from 10^{-8} to 10^{-3} $m^2.s^{-1}$) depending on the investigated part of the aquifer (matrix, fracture, or drain). Understanding hydrodynamic properties distribution at different time and space scales provided new insights into the organization and functioning of the reservoir. This analysis is the first step toward multiscale modeling using conceptual models to represent the characteristics of the main flow paths.

References

Bakalowicz M (2004) Karst groundwater: a challenge for new resources. Hydrogeol J 13:148–160
Bouwer H, Rice RC (1976) A slug test method for determining hydraulic conductivity of unconfined aquifers with completely or partially penetrating wells. Water Resour Res 12(3):423–428
Butler JJ (1988) Pumping tests in nonuniform aquifers—The radially symmetric case. J Hydrol 101:15–30
Butler JJ, Healey JM (1997) Relationship between pumping-test ans slug test parameters: scale effect or artifact? Groundwater 36(2):305–313
Cinco-Ley H, Samaniego-V F (1981) Transient pressure analysis: finite conductivity fracture case versus damaged fracture case. SPE 10179:5–7
Fabris (1970) Contribution à l'étude de la nappe karstique de la source du Lez. PhD Thesis, University of Montpellier, France
Iglesias A, Garrote L, Flores F, Moneo M (2007) Challenges to manage the risk of water scarcity and climate change in the Mediterranean. Water Resour Manage 21:775–788
Jazayeri Noushabadi MR, Jourde H, Massonnat G (2011) Influence of the observation scale on permeability estimation at local and regional scales through well tests in a fractured and karstic aquifer (Lez aquifer, Southern France). J Hydrol 403:321–336
Theis CV (1935) The relation between the lowering of the piezometric surface and the rate and duration of discharge of a well using groundwater storage. Am Geophys Union Trans 16:519–524

KARSTMOD: A Generic Modular Reservoir Model Dedicated to Spring Discharge Modeling and Hydrodynamic Analysis in Karst

H. Jourde, N. Mazzilli, N. Lecoq, B. Arfib and D. Bertin

Abstract On the basis of the characterization of the different karst subsystems (Soil/Epikarst—Unsaturated Zone—Saturated Zone) and mathematical models developed on specific sites, we propose an adjustable modeling platform of karst for both the simulation of spring discharge at outlets and the analysis of the hydrodynamics of the compartments considered in the model. This platform was developed within the framework of the KARST observatory network initiative from the INSU/CNRS, which aims to strengthen knowledge-sharing and promote cross-disciplinary research on karst systems at the national scale.

1 Introduction

Karst basins result from several polyphased digging and erosion processes that affect heterogeneous and fractured carbonated rocks. In such terrains, dissolution and collapsing lead to a network of heterogeneous and anisotropic underground voids, particularly difficult to apprehend. From a numerical modeling point of

H. Jourde
Université Montpellier 2, UMR 5569 Hydrosciences, Montpellier, France

N. Mazzilli (✉)
UAPV, UMR1114 EMMAH, 84914 Avignon, France
e-mail: naomi.mazzilli@univ-avignon.fr

N. Mazzilli
INRA, UMR1114 EMMAH, 84914 Avignon, France

N. Lecoq
Université de Rouen, UMR 6143 M2C, Rouen, France

B. Arfib
Aix-Marseille Université, CEREGE, Marseille, France

D. Bertin
Géonosis, Saint Jean Du Pin, France

B. Andreo et al. (eds.), *Hydrogeological and Environmental Investigations in Karst Systems*, Environmental Earth Sciences 1,
DOI 10.1007/978-3-642-17435-3_38

view, water flowing within conduits can reach velocities of several meters per second, while water flowing through fractures or the fissured matrix travels much more slowly. In the light of this discrepancy, karst aquifers exhibit nonlinear hydrodynamic behaviors that include threshold effects, which are naturally difficult to quantify and forecast, but must be accounted for in models.

Groundwater flow modeling is recognised as a major tool for analyzing hydrological processes and for water resource management. It is also a challenging task in karst environments where knowledge of the system geometry may prove difficult to gather. Global compartment models are based on physically sound structures and equations that are selected by the modeler as being representative of the main processes at stake, together with semiempirical ones (Abbott and Refsgaard 1996). Because they require little assumptions on the system geometry and perform well in the prediction of time series data, such models are popular for karst water resource management (Fleury et al. 2009; Hartmann et al. 2013).

The structure of compartment models is a trade-off between adaptability (the model must be able to represent a large variety of hydrological conditions) and parsimony (parameters must be identifiable and overfitting must be prevented) (Perrin et al. 2001). Good modeling practice includes careful sensitivity analysis and uncertainty assessment (Refsgaard et al. 2007).

KARSTMOD is an adjustable modeling platform for both the simulation of spring discharge at karst outlets and the analysis of the hydrodynamics of the compartments considered in the model. This platform is developed within the framework of the KARST observatory network initiative from the INSU/CNRS, which aims to strengthen knowledge-sharing and promote cross-disciplinary research on karst systems at the national scale. The objective of KARSTMOD is to provide a publicly available platform for global, compartment modeling, which provides (i) numerically stable solutions (ii) sensitivity analyses tools and (iii) uncertainty assessment tools.

2 Model

2.1 Structure

The proposed model structure is based on the well-established conceptual model of karst (vertical variability of the hydrodynamic properties, related to the Soil/Epikarst, Unsaturated Zone—Saturated Zone subsystems). The variability of the hydrodynamic properties within a given subsystem (e.g., Matrix-conduit interactions) is also accounted for. The proposed model structure is deliberately parsimonious, with the purpose to avoid over-parameterization.

In its most complete form, the platform offers three compartments organized on a two-level structure (Fig. 1): (i) compartment E (higher level) and (ii) compartments M and C (lower level). In its most simplified form, the structure is reduced

Fig. 1 Model structure

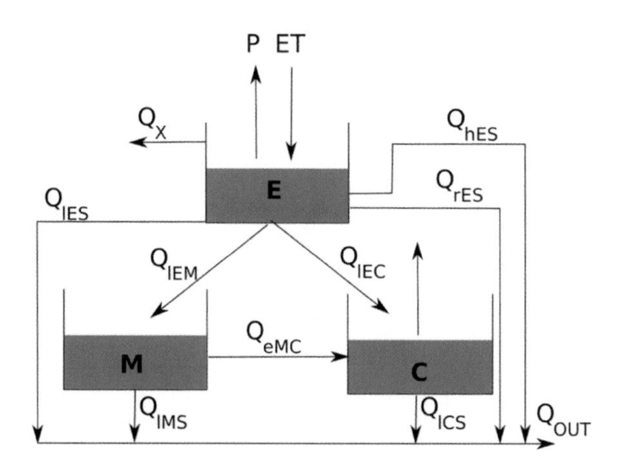

to the only compartment E, which cannot be deactivated. All compartments may be defined as either having a minimum water level or with no lower boundary.

Further developments of the KARSTMOD platform may include some modifications to the current model structure. In particular, a third level may be introduced in order to account for longwave hydrodynamic behavior related to the deep saturated zone of karst, especially for vauclusian karst system that are widely present around the Mediterranean basin.

2.2 Governing Equations

The model has three balance equations:

$$\frac{dE}{dt} = P - ET - Q_X - Q_{lES} - Q_{lEM} - Q_{lEC} - Q_{rES} - Q_{hES}$$

$$\frac{dM}{dt} = Q_{lEM} - Q_{eMC} - Q_{lMS}$$

$$\frac{dC}{dt} = Q_{lEC} + Q_{eMC} - Q_{lMC} - Q_{pump}$$

where E, M, C are the water levels in the compartments E, M, and C, respectively, P is the precipitation rate, ET is the evapotranspiration rate, Q_{PUMP} is the discharge outtaken from compartment C, Q_X is the discharge flowing out of the model from compartment E and Q_{lES}, Q_{lEM}, Q_{lEC}, Q_{lMS}, Q_{lCS}, Q_{rES}, Q_{eMC} are internal fluxes. The general notation of the internal fluxes is of the form Q_{xAB} where A stands for the compartment from which the flux is taken, and B stands for the compartment which receives the flux. Internal fluxes may obey either a linear discharge-water level relation (Q_{rES}), or be activated when the water level reaches a given threshold (Q_{lES}, Q_{lEM}, Q_{lEC}, Q_{lMS}, Q_{lCS}). Such threshold effects allow reproducing the nonlinear response of karst to precipitation events. The flux Q_{hES} introduces an hysteretic discharge-water level relation. The flux Q_{eMC} from reservoir M to reservoir C is proportional to the difference in water level in both reservoirs. Q_{eMC} may take either positive or negative values. The model discharge at the outlet Q_{out} is defined as $Q_{out} = A \times (Q_{rES} + Q_{lES} + Q_{yES} + Q_{lMS} + Q_{lCS})$ where A is the recharge area of the catchment.

3 Software Implementation

KARSTMOD is a free Jar-packaged software. The minimum requirement to run KARSTMOD is JRE (Java Runtime Environment) 1.6. The KARSTMOD.JAR file can be executed by simply double-clicking the JAR file, on operating systems Microsoft Windows, Mac, and Linux. The jar archive and the user manual can be downloaded from the SO Karst website (www.sokarst.org).

3.1 Workflow

In the first step, the user defines the model structure and fluxes through the graphical interface.

Then, the platform can be used in either "simple run" or "calibration" mode. In the simple run mode, all model parameters are set by the user, allowing a trial-and-error calibration and empirical sensitivity analysis of the model. This mode can be especially useful for educational purpose. The automated calibration mode is based on a sobol, pseudo-random sampling of the parameter space (Sobol 1985).

The platform also provides various graphical outputs: (i) measured and simulated discharge time series, (ii) cumulated measured and simulated volumes, (iii) dotty plots of the calibration criteria against the parameter value, which allow a preliminary analysis of the model sensitivity, (iv) internal model discharge and water levels time series, which brings insights into the model functioning and can be used for comparison with field discharge or water table data.

3.2 Graphical User Interface

The application window can be decomposed into seven areas (see Fig. 1):

1. The *model area* displays the model structure. Inactive compartments and fluxes are grayed out. All elements can be activated or deactivated by pressing the space bar or by clicking on the element except the upper compartment, which cannot be deactivated.

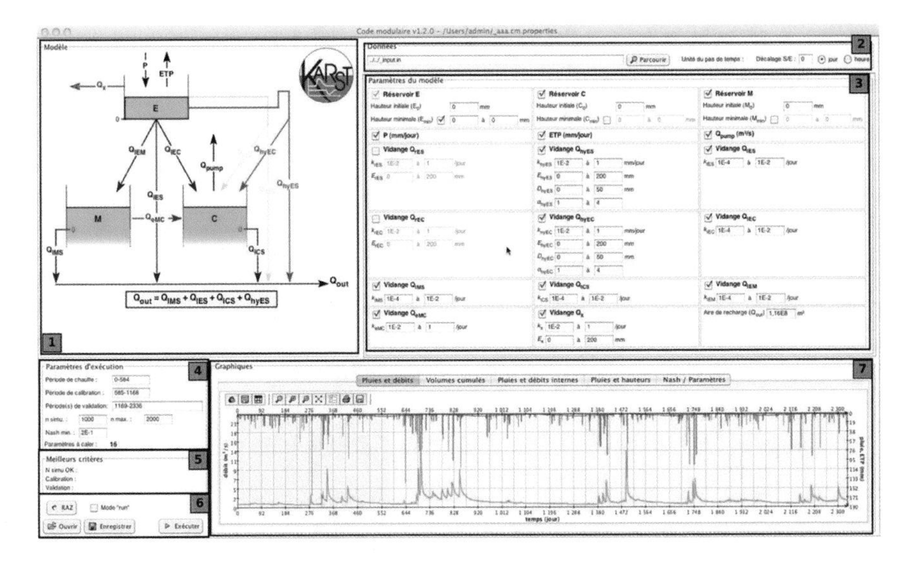

Fig. 2 Interface overview: *1* Model area, *2* Input area, *3* Data area, *4* Execution parameters area, *5* Results area, *6* Control area, *7* Graphics area

2. The *input area*, where input files for rainfall, evapotranspiration, measured discharge, and pumped discharge are specified.
3. The *data area*, where the parameter set (simple model run) or the parameter bounds (model calibration) are defined.
4. The *execution parameters area*. The warm-up, calibration, validation periods and the parameters for the model calibration are defined here.
5. The *results area*: displays the highest value of the objective function for the parameter set tested (calibration mode), or the value of the objective function for the only parameter set tested (simple run mode).
6. The *control area* allows the user to save the current model structure and parameter sets or to open former realizations, and to run the model.
7. The *graphics area* provides several automated plots (Fig. 2).

3.3 Forthcoming Functionalities

In addition to the automatic parameterization, the Monte carlo procedure included in the platform allows adding sensitivity analysis and probabilistic prediction tools. The next version of the software will include the following functionalities:

- additional objective functions and the possibility to run a multi-objective calibration,
- calculation of variance-based Sobol sensitivity indexes,
- prediction bounds assessment.

Once tested the relevance of the model to reproduce the temporal variability of flows at different sites, a last step will be to integrate some physical and chemical indicators (electrical conductivity or turbidity) to better constrain the model and simulate their dynamics.

4 Summary

The platform can be regarded as a useful tool for analyzing the hydrological processes for karst systems, and for assessing the impact of management policies, for example when pumping is performed directly in the karst network. Thanks to its friendly interface, no programming skills are required to run the modeling platform. KARSTMOD will therefore prove especially useful for teaching and occasional users.

From a research point of view, both the various graphical outputs (measured and simulated discharge, cumulated measured and simulated volumes, dotty plots of the calibration criteria against the parameter value, and internal model discharge and water levels), and the forthcoming functionalities (multi-objective calibration run, variance-based sensitivity indexes, prediction bounds assessment) may give new insights into the physical functioning of karst within each compartment.

Acknowledgments This work was funded by the network of research sites on KARST (www.sokarst.org), one observatory network of INSU (CNRS).

References

Abbott MB Refsgaard JC (1996) Distributed hydrological modelling. In: Abbott MB, Refsgaard JC (ed) Kluwer Academic Publishers, Dordrecht

Fleury P, Ladouche B, Conroux Y, Jourde H, Dörfliger N (2009) Modelling the hydrologic functions of a karst aquifer under active water management—The Lez spring. J Hydrol 365(3–4):235–243. doi:10.1016/j.jhydrol.2008.11.037

Hartmann A, Weiler M, Wagener T, Lange J, Kralik M, Humer F, Mizyed N, Rimmer A, Barberá JA, Andreo B, Butscher C, Huggenberger P (2013) Process-based karst modelling to relate hydrodynamic and hydrochemical characteristics to system properties. Hydrol Earth Syst Sci 17:3305–3321. doi:10.5194/hess-17-3305-2013

Perrin C, Michel C, Andreassian V (2001) Does a large number of parameters enhance model performance? Comparative assessment of common catchment model structures on 429 catchments. J Hydrol 242(3–4):275–301. doi:10.1016/S0022-1694(00)00393-0

Refsgaard JC, van der Sluijsb JP, Højberg AL, Vanrolleghemc PA (2007) Uncertainty in the environmental modelling process—A framework and guidance. Environ Model Softw 22(11):1543–1556. doi:10.1016/j.envsoft.2007.02.004

Sobol IM (1985) Points which uniformly fill a multidimensional cube. Math Cybern 2, Znanie, Moscow (in Russian)

Relating Land Surface Information and Model Parameters for a Karst System in Southern Spain

A. Hartmann, M. Mudarra, A. Marín, B. Andreo and T. Wagener

Abstract GIS-based methods are often used to assess the spatial distribution of mean annual recharge rates of karstic aquifers, but they typically do not provide temporal information about the dynamics of recharge. Numerical models are able to assess the temporal dynamics of recharge but they often provide only a single time series of recharge without any information on spatial distributions of recharge. In this study, we compare a process-based numerical karst model—in which the spatial variability of karst properties is considered statistically using analytical distribution functions—with an independently applied GIS-based recharge estimation method. We find that both methods produce similar spatial distributions of recharge rates. We further demonstrate that similarity between the two methods can only be achieved if the numerical model is calibrated with discharge and hydrochemical data. Using this similarity, we explore the value of the relations between the spatial input information of the GIS-based method and the analytical distributions of the process-based karst model for combined application of the two methods at sites without calibration data.

1 Introduction

GIS-based methods that use spatial information of geology, soil types, vegetation, mean annual precipitation, etc. are often used to derive time averaged spatial distribution of karst recharge (Andreo et al. 2008). Despite their great value for assessing recharge rates in space, GIS-based methods do not provide information about how recharge might change over time. On the other hand, numerical models can be used to obtain time series of recharge. But such models often provide only a

A. Hartmann (✉) · T. Wagener
Department of Civil Engineering, University of Bristol, Bristol, UK
e-mail: aj.hartmann@bristol.ac.uk

M. Mudarra · A. Marín · B. Andreo
Department of Geology and Centre of Hydrogeology, University of Málaga (CEHIUMA), 29071 Málaga, Spain

B. Andreo et al. (eds.), *Hydrogeological and Environmental Investigations in Karst Systems*, Environmental Earth Sciences 1,
DOI 10.1007/978-3-642-17435-3_39

single-time series of recharge rates without any spatial information, or, if they do provide spatial information, are highly parameterized without adequate observations for constraining their parameters. In this study, we use spatial mean annual recharge rates obtained using a GIS-based recharge estimation method (APLIS, Andreo et al. 2008) and compare it with the temporal information of recharge dynamics provided by a lumped process-based karst simulation model (VarKarst, Hartmann et al. 2013a), which considers the spatial variability of karst properties through analytical distribution functions. Within this comparison, we explore the value of hydrochemical information for the calibration of the VarKarst model. We further evaluate the usefulness of the APLIS input parameters to assess the shape of the VarKarst analytical distribution that controls the recharge dynamics and hence their usefulness in providing a priori estimates of the VarKarst distribution.

2 Study Site

The Villanueva del Rosario system is located approximately 30 km north of the city of Malaga, Southern Spain (Fig. 1). Rainfall mainly occurs in autumn and winter, with a mean annual precipitation of 760 mm (1968/69–2009/10). The Villanueva del Rosario system consists of 400–450 m thick Jurassic dolostones and limestones, which are enveloped by Upper Triassic clays and evaporite rocks (mainly gypsum) at the bottom, and by Lower Cretaceous-Paleogene marly-limestones and marls at the top. Discharge occurs mainly through Villanueva del Rosario spring (260 L/s annual mean discharge rate) situated at 770 m ASL, on the northern border of the carbonate outcrops (Fig. 1). Additional background information about the hydrogeological characteristics of the Villanueva del Rosario system have been described in other papers (Mudarra et al. 2014).

3 Methodology

3.1 Estimation of Recharge

The GIS-based approach to estimate the average spatial distribution of recharge is the APLIS method (Andreo et al. 2008). It was developed to estimate the mean annual recharge within carbonate aquifers, expressed as a percentage of precipitation. This is done by using average annual precipitation, its spatial distribution, and several physical variables (Durán et al. 2004): altitude (*A*), slope (*S*), lithology (*L*), infiltration landforms (*I*) and soil type (*S*):

$$\overline{R_i} = \frac{A_i + P_i + 3L_i + 2I_i + S_i}{0.9} \cdot F_{h,i} \tag{1}$$

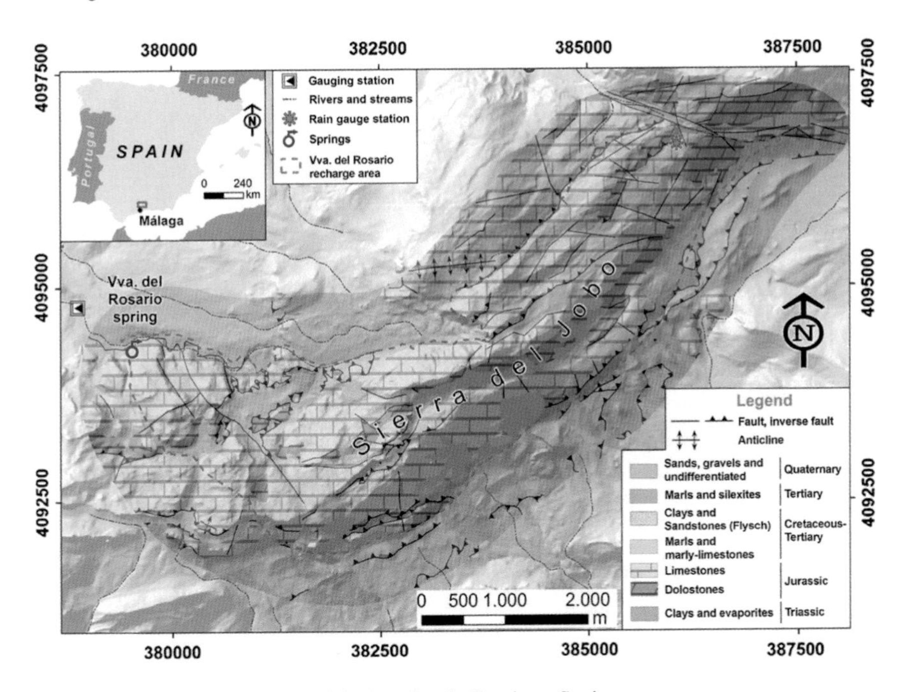

Fig. 1 Map of the study area and its location in Southern Spain

where $\overline{R_i}$ [-] is the mean annual recharge rate at a location i, and $F_{h,i}$ acts as a correction factor that adapts to values between 0.1 and 1 depending on the permeability of the aquifer permeability (Marín 2009). The spatially variable input parameters (A, P, L, I, S) are usually found by remote sensing and field surveys and transformed into rankings between 1 and 10. The APLIS method was developed from studies carried out in eight carbonate aquifers in southern Spain, which are representative of a wide range of climatic and geologic characteristics, and was applied successfully at various locations in Southern Spain (Andreo et al. 2008) and other worldwide sites.

The process-based VarKarst model is used to assess the temporal evolution of recharge (Fig. 2). The model was previously developed for another karst system in Southern Spain (Hartmann et al. 2013a) and it was shown that it is capable to be applied to various settings in the Mediterranean, Middle Europe and the Middle East when hydrochemical data is available for its calibration (Hartmann et al. 2013b). The VarKarst model includes spatial variability using Pareto functions that are applied to a set of N model compartments:

$$V_{S,i} = V_{\max,\mathrm{S}} \cdot \left(\frac{i}{N}\right)^{a} \tag{2}$$

where $V_{S,i}$ [mm], $i = 1 \ldots N$, is the distribution of soil storage capacities among the model compartments that is controlled by the maximum soil storage capacities

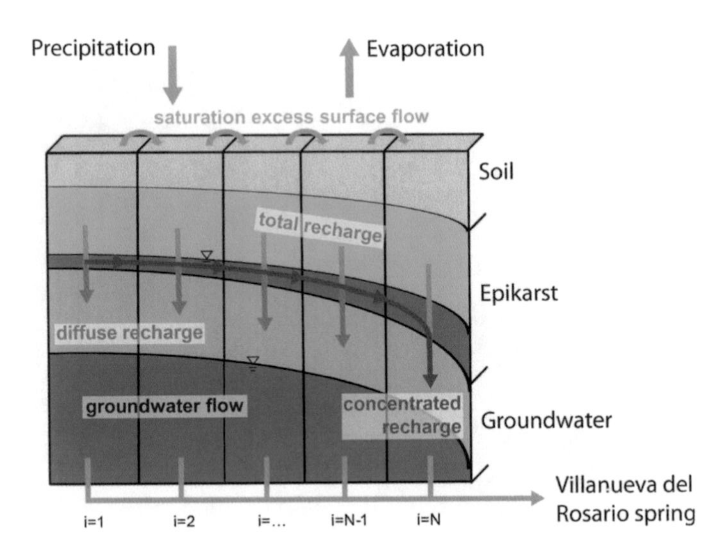

Fig. 2 Schematic structure of the VarKarst model (Hartmann et al. 2013a; modified)

$V_{max,S}$ [mm] and the distribution coefficient a [-]. The same distribution used express the variability of the epikarst storage capacities and the epikarst storage coefficients, making it a major control for the simulated recharge dynamics. Storage and flow dynamics of the VarKarst model follow the linear reservoir concept, i.e. discharge at time t is linearly dependent on the volume stored at time t (for more elaborations see Hartmann et al. 2013a). For calibration, a 3-year record (hydrological years 2006/07–2008/09) of precipitation, temperature, spring discharge and Cl, NO_3 and SO_4 concentrations (precipitation and discharge) are available. To explore the contribution of hydrochemical data, calibration with discharge only, and calibration with both discharge and hydrochemistry is performed.

3.2 Comparison of the Two Methods

To compare the two methods, we establish cumulative recharge distributions of the APLIS average recharges and of the time averaged recharge rates of the N VarKarst model compartments. To explore similarities between the APLIS input parameters and the VarKarst analytical distribution (Eq. 2), we establish cumulative distributions of both. The VarKarst distribution has to be re-scaled since the APLIS input parameters are transformed into rankings

$$X_i = \min(X) + [\max(X) - \min(X)] \cdot \left(\frac{i}{N}\right)^a \tag{3}$$

where X_i, $i = 1 \ldots N$, is the distribution of estimate of the APLIS input parameter X (A, P, L, I, or S) using the VarKarst analytical distribution function. The parameters of lumped models have to be estimated from a priori information (such as soil types) when no calibration data is available. The comparison of the hereby obtained cumulative distributions will show which of the APLIS input parameters may be used as an a priori proxy for the shape of the VarKarst distribution function.

4 Results and Discussion

Figure 3 shows the results of the APLIS method and the VarKarst model. While APLIS indicated that there are areas in the study site that reach up to 80 % of average recharge, the VarKarst model indicates that during wet conditions (e.g. 2008/09) recharge can reach temporally up to 100 % before it falls down to zero between the rainfall events. The comparison of the cumulative distributions of mean annual recharge rates from the APLIS method with the mean annual recharge rates from the N VarKarst model compartments (Fig. 4) shows that there is a strong coincidence for exceedance probabilities >0.7. This is only true though when discharge and hydrochemistry are used for calibration; the cumulative distributions differ much more when only discharge is used.

We find linear correlations of $r^2 = 0.35$–0.89 when we compare the cumulative distributions of the APLIS input variables with the analytical distribution that controls the recharge processes in the VarKarst model (Eq. 3, Fig. 5). An improved agreement could be obtained by reversing the VarKarst analytical distribution for the distributions of the rankings of P (slopes) and I (infiltration landforms). That way, we find high correlations for the rankings of altitude (A, $r^2 = 0.69$), the slopes (P, reversed, $r^2 = 0.94$), the lithology (L, $r^2 = 0.80$) and the infiltration landforms (I, reversed, $r^2 = 0.69$). There is no obvious relation with soils (S). However, the ranking of the APLIS input variables results in step-like cumulative distributions, which makes the interpretation of the obtained r^2 difficult.

Given that a 5×5 m^2 digital elevation model is available, we can derive cumulative distributions for the altitudes and slopes of the study area without using the APLIS rankings; hence avoiding the step-like shape of their distributions (Fig. 6). That way, the agreement with the VarKarst analytical distribution increases to $r^2 = 0.89$ for the distribution of altitudes, but it decreases to $r^2 = 0.47$ for the slopes. Hence, the distribution of altitudes appears to be the most reliable proxy for the shape of the VarKarst distribution function—at least at the study site. The correlation between the altitude distribution and the shape of the analytical function in VarKarst is most probably due to the increase of carbonate rock outcrops and a decrease of soil depths with higher altitudes (Mudarra 2012). More

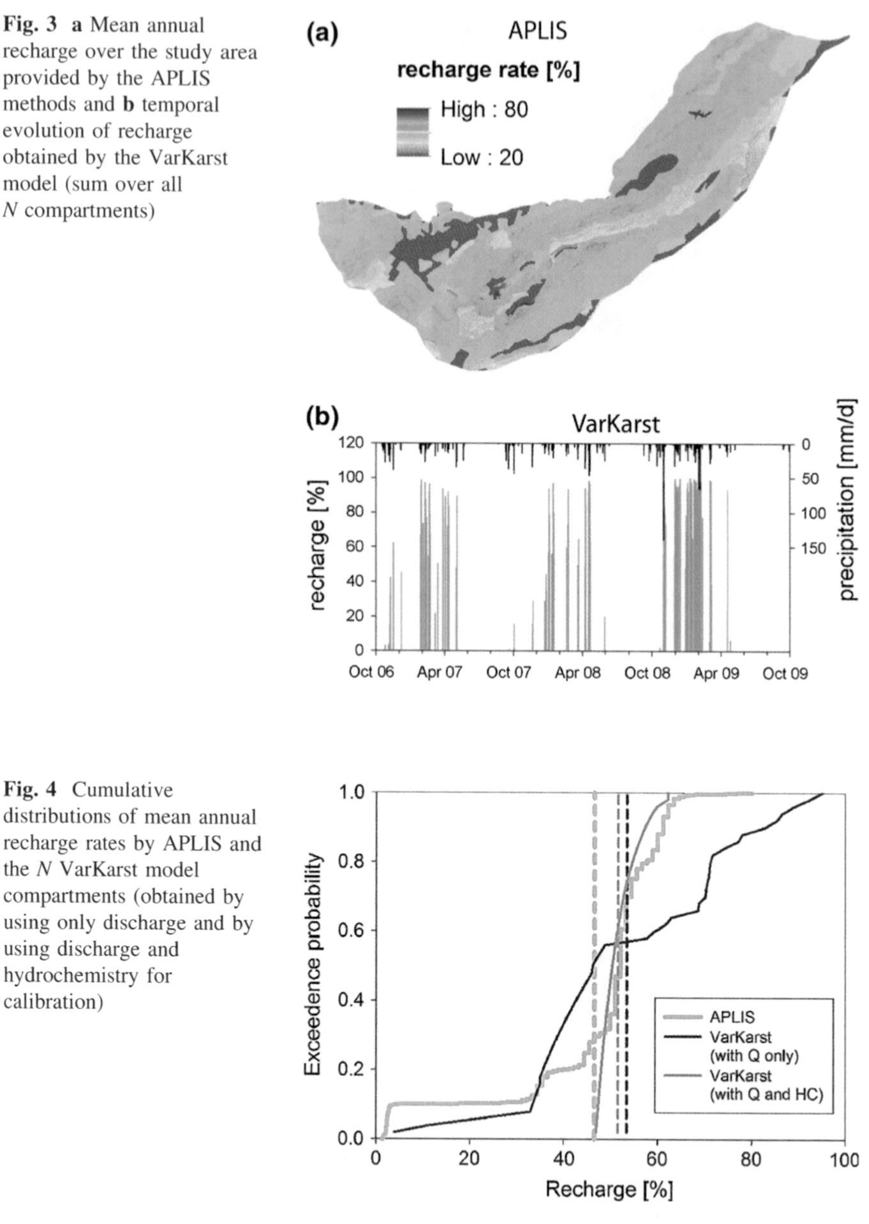

Fig. 3 **a** Mean annual recharge over the study area provided by the APLIS methods and **b** temporal evolution of recharge obtained by the VarKarst model (sum over all *N* compartments)

Fig. 4 Cumulative distributions of mean annual recharge rates by APLIS and the *N* VarKarst model compartments (obtained by using only discharge and by using discharge and hydrochemistry for calibration)

detailed observations and analysis than possible here are necessary to understand the relations to the other input variables of APLIS distribution (slopes, lithology and infiltration landforms).

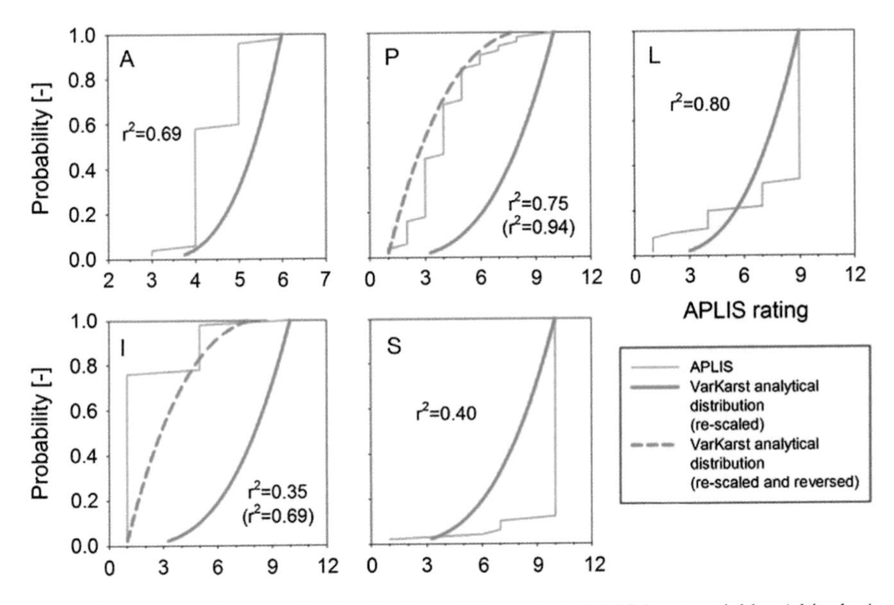

Fig. 5 Comparison of cumulative distributions of the ranked APLIS input variables (altitude *A*, slopes *P*, lithodology *L*, inftration landforms *I*, soil *S*) and the shape of the VarKarst analytical distribution function; r^2 of reversed distributions are indicated in *brackets*

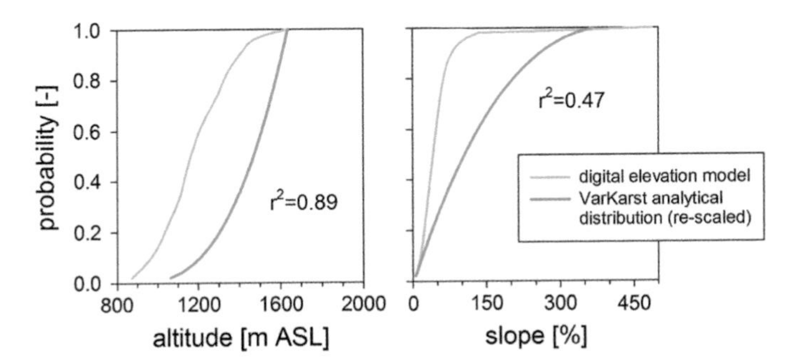

Fig. 6 Comparison of cumulative distributions of altitudes and slopes derived from a 5 × 5 m^2 digital elevation model and the VarKarst analytical distribution function (slopes are compared to the reversed distribution)

5 Conclusions

This study shows that two different methods to estimate recharge in karst areas produce similar distributions of mean annual recharge rates, even though they are based on independent information. The APLIS method is based on field surveys and remote sensing data. It provides spatial estimates of annual recharge rates.

The VarKarst model, on the other hand, is based on calibration through observed time series of spring discharge and hydrochemistry. It provides information about temporal evolution of recharge and spatial variability is only considered implicitly using an analytical distribution function. We show that a good agreement between the two methods can be achieved if the VarKarst model is calibrated with both discharge and hydrochemical observations. Furthermore, we show that among the APLIS input parameters, altitude distribution is most correlated with the analytical distribution that controls the dynamics of recharge in the VarKarst model. In similar regions, the altitude distribution might therefore be used as proxy to assess the shape of the VarKarst distribution function when no calibration data is available. Further and more detailed exploration of its relation to the other APLIS input variables in future studies may even allow a combined application of GIS-based methods and numerical models to fulfil the urgent need for large-scale karst water resources estimation (Hartmann et al. 2014).

Acknowledgments This work is a contribution of the Research Group RNM-308 of Junta de Andalucía to the projects IGP 598 of UNESCO and CGL2012-32590 of DGICYT. Furthermore, this work was supported by a fellowship within the Postdoc Programme of the German Academic Exchange Service (DAAD).

References

Andreo B, Vías J, Durán JJ, Jiménez P, López-Geta J, Carrasco F (2008) Methodology for groundwater recharge assessment in carbonate aquifers: application to pilot sites in southern Spain. Hydrogeol J 16:911–925

Durán JJ, Andreo B, Vías J, López-Geta J, Carrasco F, Jiménez P (2004) Classification of carbonate aquifers in the Betic Cordillera in accordance with recharge rates (in Spanish). Boletín Geológico y Minero 115:199–210

Hartmann A, Barberá JA, Lange J, Andreo B, Weiler M (2013a) Progress in the hydrologic simulation of time variant recharge areas of karst systems—exemplified at a karst spring in Southern Spain. Adv Water Resour 54:149–160

Hartmann A, Weiler M, Wagener T, Lange J, Kralik M, Humer F et al (2013b) Process-based karst modelling to relate hydrodynamic and hydrochemical characteristics to system properties. Hydrol Earth Syst Sci 17:3305–3321

Hartmann A, Goldscheider N, Wagener T, Lange J, Weiler M (2014) Karst water resources in a changing world: review of hydrological modeling approaches. Rev Geophys. doi:10.1002/2013rg000443

Marín AI (2009) The application of GIS to evaluation of resources and vulnerability to contamination of carbonated aquifer. Test site Alta Cadena (Málaga province) Thesis. Type, University of Málaga (Spain)

Mudarra M (2012) Importancia relativa de la zona no saturada y zona saturada en el funcionamiento hidrogeológico de los acuíferos carbonáticos. Caso de la Alta Cadena, sierra de Enmedio y área de Los Tajos (provincia de Málaga) Thesis. Type, University of Malaga, Malaga, Spain

Mudarra M, Andreo B, Marín AI, Vadillo I, Barberá JA (2014) Combined use of natural and artificial tracers to determine the hydrogeological functioning of a karst aquifer: the Villanueva del Rosario system (Andalusia, southern Spain). Hydrogeol J. doi:10.1007/s10040-014-1117-1

Neural Networks Model as Transparent Box: Toward Extraction of Proxies to Better Assess Karst/River Interactions (*Coulazou* Catchment, South of France)

L. Kong-A-Siou, H. Jourde and A. Johannet

Abstract Karst catchments frequently exhibit complex exchanges between surface and subterranean flow. While the swing between surface flood and underground flood is complex, the ability to predict such behavior would be of great interest for flood forecasting and water recharge assessment. To this end an innovative methodology is proposed to visualize internal variables of a neural network model. It proves to be efficient to extract internal variables highly correlated to measured signals previously identified as proxy of the karst-river exchanges. The study focuses on a small Mediterranean catchment where karst/river interactions control the dynamic and genesis of surface floods. But the methodology is generic and can be applied to any catchment provided the availability of a sufficient database.

1 Introduction

Karst aquifers represent important underground resources for water supplies. Nevertheless, such systems are currently underexploited because of their heterogeneity and complexity, which make work fields and physical measurements expensive, and frequently not representative of the whole aquifer. Due to complex processes combining geodynamic, eustatism, and dissolution processes, karst hydrosystems are heterogeneous in space. Such a complexity leads to various hydrological behaviors at different timescales, involving processes in the unsaturated or saturated zones. Another complexity comes from the interaction between

L. Kong-A-Siou · H. Jourde
HydroSciences Montpellier, Université Montpellier II, Place E. Bataillon, 34095 Montpellier Cedex 5, France

A. Johannet (✉)
Ecole Des Mines d'Alès, LGEI, 6 Avenue de Clavières, 30319 Alès Cedex, France
e-mail: anne.johannet@mines-ales.fr

B. Andreo et al. (eds.), *Hydrogeological and Environmental Investigations in Karst Systems*, Environmental Earth Sciences 1,
DOI 10.1007/978-3-642-17435-3_40

river and the aquifer as karst groundwater may contribute to runoff (Jourde et al. 2007; Bailly-Comte et al. 2008, 2012). The karst aquifer can also attenuate surface floods or act as a natural flood control dam (Bailly-Comte et al. 2009; Jourde et al. 2014). Two different behaviors can thus be identified in such basins that can move from one state to another in a sole flood event. In order to study these behaviors, systemic approach is chosen considering rainfall and runoff time series as input–output signals of a complex karst/river system. In this context, neural networks are used as transparent box (dedicated architecture containing several subnetworks, Johannet et al. 2008), in order to extract hidden variables from the model, which should give indications about karst-river exchanges. To this end, apart from the introduction, the second part of the paper presents neural network modeling in the particular case of a transparent box model. In the third part an overview of the *Coulazou* catchment is given. The fourth part presents results and discussion.

2 Neural Networks as Transparent Box

Artificial neural networks (ANN) are nonlinear machine learning model. In hydrology, the multilayer perceptron is commonly used due to its property of "Universal approximation" (any nonlinear function can be implemented by this architecture) and parsimony (relatively to other statistical models). Multilayer perceptron and the model selection methodology are presented in (Kong-A-Siou et al. this book). For an extended description of ANN, reader should refer to (Dreyfus 2005).

The model used is this study is a composition of several multilayer perceptron (see Sect. 3.2). In this section, specific information on gray box and transparent box modeling are given.

2.1 Graybox Modeling Using Neural Networks

ANNs are mainly considered as black-box models. However, recent works focused on the ability to interpret these models in terms of physical meaning. (Johannet et al. 2008) and (Jain and Kumar 2009) both demonstrated the feasibility of building a neural network able to deliver physically interpretable information. The proposed methodology, based on (Johannet et al. 2008), is composed of three steps: (i) the representation of the hydrosystem behavior as a block-diagram, (ii) the design of the neural model following the block-diagram, each box being represented by a multilayer perceptron, or a linear neuron, and (iii) the training of the model. This "knowledge-based box" permits the introduction of physical constraints in the NN model. It aims at constraining the model to give physically interpretable information. Two kinds of interpretations can stem from the model: hidden information, as potential evapotranspiration for example, that can be extracted from the output of a specific hidden neuron (Johannet et al. 2008); time transfers and contribution of various area of the basin can also be obtained from the parameters of the model (Kong-A-Siou et al. 2013).

2.2 Extraction of Hidden Information

As shown by Johannet et al. (2008), hidden information can be extracted from a neural network with a constrained architecture. Usually, all variables of the model (see universal approximator in Kong-A-Siou et al. this book) are connected to all neurons of the hidden layer. In a constrained architecture, variables are only connected to several sub networks in order to represent several physical phenomena. In this sense one can say that the architecture is constrained by the high level postulated knowledge about the system functioning. In the cited study, the idea was to build a subneural network (universal identificator) to identify the evapotranspiration (ET) signal from an a priori evapotranspiration signal. This subnetwork provides a "simulated" ET signal which is taken into account in the main NN model, a rainfall-runoff model. The a priori ET signal is a simple Gaussian curve with a maximum during summer and a minimum in winter.

The "simulated" ET was obtained by the subnetwork "output" which is actually a hidden neuron of the main NN model. This hidden neuron output was called "hidden ETP". The shape of this hidden ETP has a maximum in autumn and this shape was reproduced with different sizing of the NN model (number of hidden neurons) which proves the robustness of the method. The constrained model permitting the extraction of "hidden ETP" was called "transparent box model."

The present study aims at using a transparent box model to estimate the water level in the karst for two points identified as proxies of karst-river exchanges.

3 Case Study: The *Coulazou* River Karst Catchment

3.1 Catchment Presentation

The ephemeral *Coulazou* river, located near *Montpellier* (Southern France) crosses the *Aumelas Causse* karst system. The catchment comprises two parts: upstream, a non-karst area of about 20 km^2 with very few infiltrations; downstream, a karst area of about 40 km^2 with multiple karst-river interactions.

Huge rainfalls (cumulated rainfall could reach 200 mm in 1 day as in September 2005) can generate flash floods on the impervious upstream watershed. Downstream, over 15 swallow holes have been described along the *Coulazou* river crossing the karst, which result in direct stream-aquifer interactions. Outlets of the

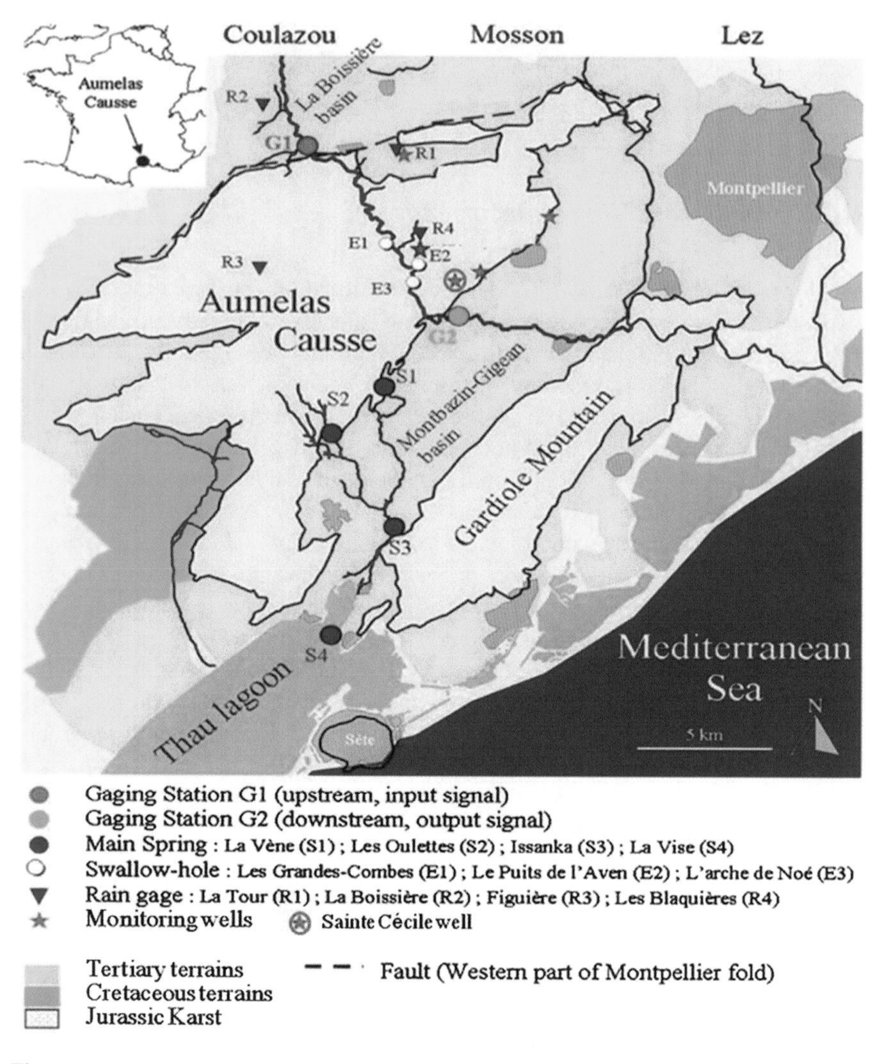

Fig. 1 Map of the catchment (after Bailly-Comte et al. 2008)

karst aquifer are the *Vène* Spring in high-flow condition, *Issanka* spring used for water supply, and the *Vise* submarine spring. A monitoring network has been settled (Jourde et al. 2007), including: (i) two gauging stations in the *Coulazou* river, located upstream and downstream the karst, (ii) four rain gauges distributed over the catchment and, (iii) water level, temperature and electrical conductivity measurements in wells and caves (Fig. 1). Discharge at the *Vène* spring is recorded at 5 or 10 min time step: these data allowed classifying various types of karst/river interactions (Bailly-Comte et al. 2009). Initial water level measured in the karst aquifer (*Sainte-Cécile* well) gives information about the hydraulic connection

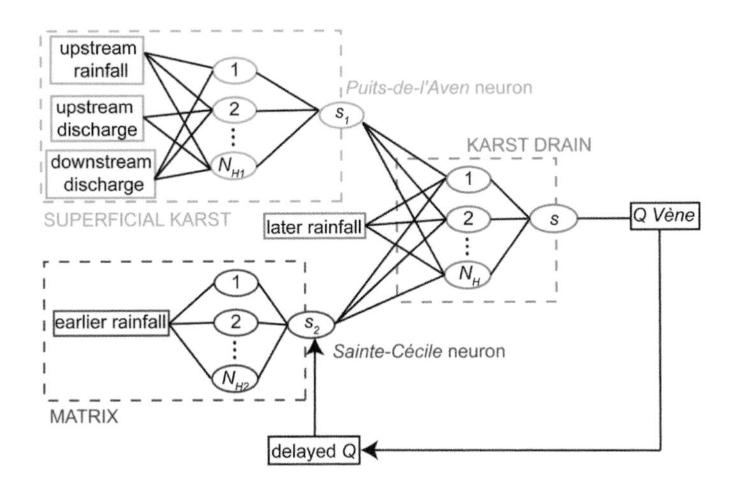

Fig. 2 Transparent box model for proxies extraction

between the river and the aquifer (perched stream or connected stream). The flow direction (gaining or losing stream) during flood can be inferred from temperature and electrical conductivity measured in a cave located in the riverbed (*Puits-de-l'Aven* Cave). These proxies of karst-river exchanges will be used in the work presented herein and compared to internal variables of the model.

3.2 Transparent Box Model

According to the previous studies (Jourde et al. 2007; Bailly-Comte et al. 2008, 2009), and as suggested by (Johannet et al. 1994), the karst system behavior is drawn following three boxes: (i) the matrix, (ii) the superficial karst watershed, and (iii) the underground karst drainage network (conduits). In the neural network model, each box corresponds to a universal approximator (see Kong-A-Siou et al. this book, and Fig. 2). Starting from this architecture, the presented work aims at "opening" the black box to see if the outputs of these hidden neurons could be compared to actual physical variables as suggested by (Johannet et al. 2008). To this end, hidden neurons S_1 and S_2 are compared to the water level in the *Puits-de-l'Aven* Cave and the *Sainte-Cécile* Well, respectively. It is pointed out that the NN model is trained to correctly reproduce the *Vène* discharge using input data: rainfall over the upstream watershed, upstream discharge, downstream discharge, earlier rainfall, and later rainfall. Water level measurement of *Puits-de-l'Aven* and *Sainte-Cecile* are not introduced in the model.

The available database includes 16 flood events, split as follows: 12 events in the training set, two events in the stop set and two events in the test set. Thanks to

Table 1 Maximal correlation (C_c) and corresponding delay (D_c) between measured water levels and hidden neurons output (S_1 and S_2—see Fig. 2) for *Puits-de-l'Aven* and *Sainte-Cécile*

	Coulazou upstream	*Sainte-Cécile*		*Puits-de-l'Aven*	
Flood	Maximal water height (cm)	C_c	D_c	C_c	D_c
1	128	0.73	0		
3	253	0.36	−8 h	−0.60	−1h15
3bis	74	0.65	−13h15		
4 (stop set)[1]	85	0.75	−30 min	−0.86	0
5 (test set)[1]	94	0.79	0	−0.80	0
6	181	0.63	0	−0.41	0
7	86	0.75	2h30	−0.95	−30 min
8	98	0.46	1h30	−0.88	0
10	69	0.55	1 h	−0.38	0
11	143	0.51	−11 h	−0.82	−2h30
12	60	0.51	−1h45	−0.92	0
16	128	0.66	−4h15		
17	44	0.71	−3h30		
19 (stop set)[1]	97	0.57	3 h	−0.53	−2h45
21	67	0.60	1h45	<0.2	
22 (test set)[1]	87	0.65	2 h	−0.72	−2h15

[1] See Sect. 3.2. Training set includes floods without indication

the utilization of regularization methods, the selection of variables can be roughly done; only the initialization and the number of hidden neurons N_H, N_{H1}, N_{H2} are accurately selected. In this study a 15 min time step is used.

4 Results and Discussion

Given the model presented in the previous section, the outputs of hidden neurons S_1 and S_2 were compared with the water levels in *Puits-de-l'Aven* cave and *Sainte-Cécile* well. It must be pointed out that the neuron output is normalized in [−1, +1]; the comparison focuses then on the relative variations. To this end, the cross-correlographs of neuron outputs versus water levels were calculated for each flood: the maximal cross-correlation C_c and the corresponding delay D_c are provided in Table 1. The C_c values prove good to very good correlations between the neuron outputs and the measured water level in *Sainte-Cécile* and *Puits-de-l'Aven*, including the "test" floods. An important delay is generally observed for *Sainte-Cécile* reflecting the poor synchronization of the model.

Figure 3 presents the limnigraphs for the test set floods. For each limnigraph, the beginning of the flood is poorly represented. This is probably due to the choice

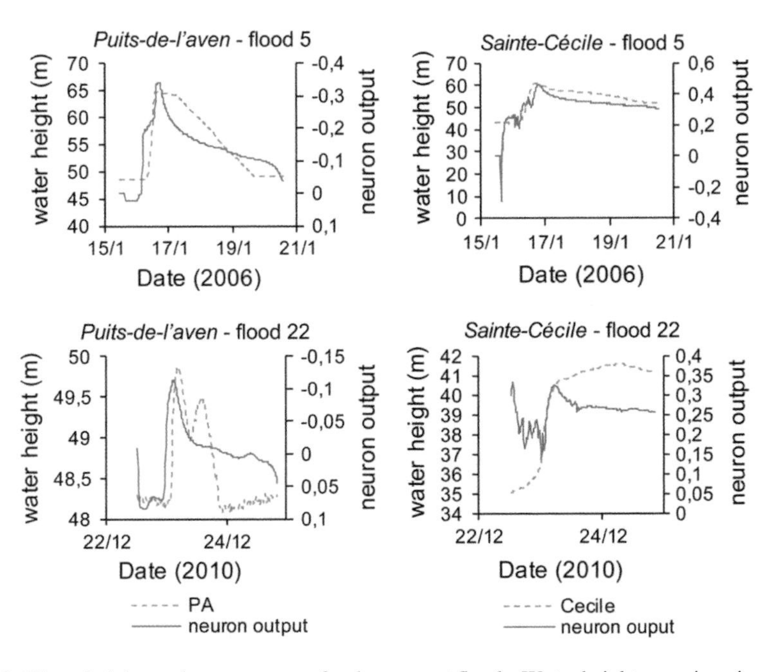

Fig. 3 Water heights and output neuron for the two test floods. Water heights are given in meters NGF (French ordnance datum)

of initial condition at the beginning of each flood, which is set at the value of the end of the previous flood (floods are concatenate in a unique continuous time series, even if it is not the case in actual world). However, thanks to training, variations of the water heights are very well represented for the flood 5 in both cases. For flood 22 in *Puits-de-l'Aven*, a delay appears during the rising up, according to the delay D_c (Table 1). These results confirm the efficiency of the NN model architecture to represent hidden information. Results from this preliminary study should be improved by analyzing each flood separately in order to better choose initial conditions and to improve constrains. For example the extreme values of the neurons (*Puits-de-l'Aven, Sainte-Cécile)* should be constrained using the water level extreme values. In addition, several architectures may be tried depending on the hypothesis that can be postulated on the behavior of the hydrosystem and its relation with the surface river.

5 Conclusion

In the context of karst catchments swinging between surface flood and underground flood, an innovative method, based on neural networks, was exerted to visualize physically interpretable signals inside the neural model. The signals visualization is obtained by constraining the neural network architecture using a block-diagram representation of the postulated behavior of the system. With the example of the *Coulazou* hydrosystem (South-eastern France), this method proved that a neural network model can deliver hidden, physically interpreted signals. This method will promise interesting developments to design virtual and efficient proxies of complex behaviors.

References

Bailly-Comte V, Jourde H, Roesch A, Pistre S (2008) Mediterranean flash flood transfer through karstic area. Environ Geol 54:605–614

Bailly-Comte V, Jourde H, Pistre S (2009) Conceptualization and classification of groundwater-surface water hydrodynamic interaction in karst watersheds: case of the karst watershed of the Coulazou River (Southern France). J Hydrol 376:456–462

Bailly-Comte V, Borrell V, Jourde H, Pistre S (2012) A conceptual semidistributed model of the Coulazou River as a tool for assessing surface water–karst groundwater interactions during flood in Mediterranean ephemeral rivers. Water Resour Res 48(W09534):14

Dreyfus G (2005) Neural networks methodology and applications, 1st edn. Springer, Berlin

Jain A, Kumar S (2009) Dissection of trained neural network hydrologic models for knowledge extraction. Water Resour Res 45

Johannet A, Mangin A, D'Hulst D (1994) Subterranean water infiltration modelling by neural networks : use of water source flow. Int Conf Artif Neural Netw ICANN 94(2):1033–1036

Johannet A, Vayssade B, Bertin D (2008) Neural networks: from black box towards transparent box application to evapotranspiration modeling. Int J Comput Intell 4

Jourde H, Roesch A, Guinot V, Bailly-Comte V (2007) Dynamics and contribution of karst groundwater to surface flow during Mediterranean flood. Environ Geol 51(5):725–730

Jourde H, Lafare A, Mazzilli N, Belaud G, Neppel L, Dörfliger N, Cernesson F (2014) Flash flood mitigation as a positive consequence of anthropogenic forcing on the groundwater resource in a karst catchment. Environ Earth Sci 71(2):573–583

Kong-A-Siou L, Cros K, Johannet A, Borrell-Estupina V, Pistre S (2013) KnoX method, or knowledge extraction from neural network model. Case study on the Lez karst aquifer (southern France). J Hydrol 507:19–32

Neural Networks for Karst Spring Management. Case of the *Lez* Spring (Southern France)

L. Kong-A-Siou, V. Borrell-Estupina, A. Johannet and S. Pistre

Abstract Karst hydrosystems constitute important water resource but their recharge and emptying process are poorly known and quantified. Water resource management is thus difficult. Nevertheless, it is a major issue when rainfall is not uniformly distributed during the year, as in Mediterranean climate. This study proposes a method based on neural networks permitting to simulate karst emptying as a function of the pumping volume during the dry period. Applied to the *Lez* karst system, the model provides excellent simulations of the water level at the main outlet of the system by using mean pumping discharge and zero rainfall hypothesis during dry period. An arbitrary extreme scenario is also provided by introducing a mean pumping volume.

1 Introduction

Karst aquifers represent 10 % of the earth's surface and provide drinking water for about 25 % of the global population (Ford and Williams 2007). However, karsts are known for their structural heterogeneity, which implies complex recharge and emptying. In addition, the Mediterranean specific climate causes both important rainfalls during the autumn and dryness during summer. Karst water level forecasting during exploitation is thus a major issue for karst spring management, especially for regions where rainfall are nonuniformly distributed during the year. Considering the complexity of the water height prediction in karst aquifers, the systemic approach is chosen. Following this approach, pumping discharge and

L. Kong-A-Siou · A. Johannet (✉)
Ecole Des Mines d'Alès, 6 Avenue de Clavières, 30319 Alès Cedex, France
e-mail: anne.johannet@mines-ales.fr

L. Kong-A-Siou · V. Borrell-Estupina · S. Pistre
Hydrosciences Montpellier, Université Montpellier II, Place E. Bataillon, 34095 Montpellier Cedex 5, France

B. Andreo et al. (eds.), *Hydrogeological and Environmental Investigations in Karst Systems*, Environmental Earth Sciences 1,
DOI 10.1007/978-3-642-17435-3_41

rainfall are considered as the input signals of a complex system; and water level in the karst outlet as the output of this system. Neural networks are nonlinear machine learning models, able to approximate nonlinear relations without a priori information about the system. This work proposes thus using neural networks to simulate water level drawdown in a karst aquifer during the exploitation of the main outlet of this hydrosystem. Apart from the introduction, the second part of the paper presents neural network models and the methodology used to design the karst water height model. In the third part, the case study on the *Lez* Spring, a Mediterranean exploited karst spring is presented along with its specific problematics in particular testing *scenarii* of pumped flow rate. Finally, the fourth part provides results and discussion.

2 Neural Networks for Water Level Prediction

Artificial neural networks (ANN) have been increasingly applied in the field of hydrology over the past couple of decades (Maier and Dandy 2000; Filho and Dos Santos 2006; Corzo and Solomatine 2007; Toukourou et al. 2011). In the field of hydrology, their great interest is that they do not necessitate a priori knowledge about the system they represent and are able to implement nonlinearity. In the following, a brief presentation of neural network modeling applied to hydrology and karst modeling is proposed. For more details on neural networks, the interested reader is referred to (Dreyfus 2005).

2.1 Nonlinear Modeling

As demonstrated by (Hornik et al. 1989), the multilayer perceptron, a specific artificial neural network is capable of "Universal Approximation." This property means that any nonlinear function can be implemented by this architecture: the solution exists. The multilayer perceptron is a neural network (feed-forward or recurrent) having one hidden layer of N_c neurons and one linear output neuron (Fig. 1). Multilayer perceptron is also parsimonious in comparison regarding other statistical models (Barron 1993).

The model proposed is intended, at a discrete time kT (T is the sampling period, $k \in N^{+}$) or more simply at discrete time k, to simulate water level in karst conduit. The neural model was designed based on (Nerrand et al. 1993):

$$P(k) = g_{\mathrm{NN}}\left[P(k-1), P(k-2), \ldots, P(k-d+1), r_i(k), r_i(k-1), \ldots, r_i(k-w+1), Q_p(k)\right]$$

where g_{NN} is the nonlinear function implemented by the neural network, Q_p is the pumped discharge, w and d are the widths of sliding temporal windows conveying,

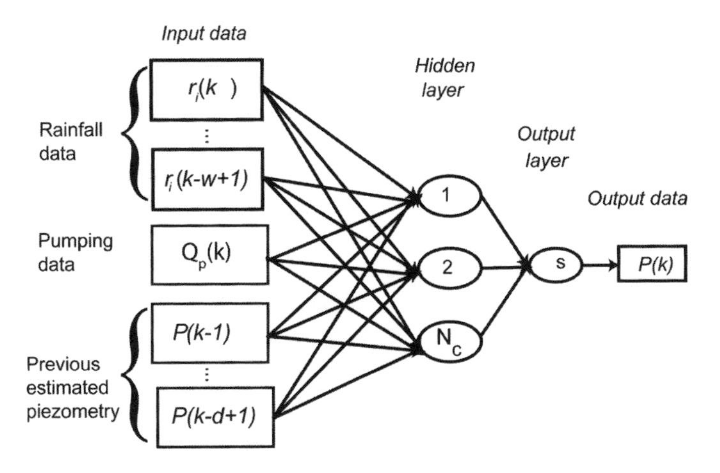

Fig. 1 Multilayer perceptron as universal approximator for rainfall water level modeling

respectively, the rainfall and the previous water-level information. Such a model is dynamic, given its dependence on previous output values $P(k\text{-}1)\ldots\ P(k\text{-}d+1)$. Calibration step (called *training* for neural networks modeling) aims at minimizing the squared error cost function by the Levenberg-Marquardt algorithm (Hagan and Menhaj 1994). Thanks to the training procedure the function g_{NN} is calculated without any analytical expression of the implemented process.

2.2 Model Selection

A major issue of machine learning model is overfitting due to the bias-variance dilemma. In order to avoid overfitting and keep generalization capabilities of the model, regularization methods must be used. As studied on (Kong-A-Siou et al. 2011b) early stopping (Sjöberg et al. 1995) combined with cross-validation (Stone 1974) gives satisfactory results. Early stopping is used to stop the training before a complete convergence and cross-validation method is used to choose the appropriate level of complexity (adjusted by varying w, d, and N_c). For each w, d, N_c values combination, a cross-validation score is calculated as described in (Kong-A-Siou et al. 2011a). Neural networks are very sensitive to the parameters initialization. Totally, 50 randomly chosen initializations are thus tried. The cross-validation score S is calculated for each initialization of each investigated architecture (combination of w, d, and N_c values). The ten initialization which provide the best score S are selected and the median M_{10} of these score is calculated. The architecture corresponding to the best M_{10} is selected. With this architecture, the initialization corresponding to the best cross-validation score is chosen. In order to quantify rigorously the quality of the model, quality criteria are calculated on test sets, set apart from training and stopping datasets.

3 Case Study: The *Lez* Hydrosystem

3.1 Presentation of the Lez Hydrosystem

The *Lez* karst system is located in the south of France near the *Montpellier* conurbation (400,000 inhabitants). Its main outlet, the *Lez* spring is tapped for fresh water supply.

The *Lez* spring hydrogeological basin is estimated to be 380 km^2 area (Thiery et al. 1983). The recharge area is estimated at 130 km^2, including main aquifer outcrops and secondary aquifers (Fig. 2). Exchanges between secondary aquifers and the main *Lez* aquifer are poorly known. In addition, the *Lez* spring is not the sole outlet of the main aquifer. Several secondary temporary outlets exist, particularly along the faults (Fig. 2). These faults split the watershed into compartments. Specifically, the *Corconne* fault complicates the exchanges between the eastern and the western part of the aquifer as it puts in contact impervious formations with karstified formations. In addition, it can also works as drain (Bérard 1983), leading the water to the *Lez* spring as shown by tracer experiments.

The main outlet exploitation consists of four boreholes in the main karst drain (Fig. 3). A Public Utility Decree (PUD) allows pumping a maximum of 1,700 l/s, with a maximal drawdown of 30 m below the spring altitude. The drawdown is measured in a well near the pumping borehole (Fig. 3). In order to maintain a minimal flow in the *Lez* river, the PUD imposes a release discharge of 160 l/s. The Mediterranean climate directly affects the spring exploitation. It has dry summers and rainy autumn with intense rainfall events: precipitation can reach as high as 300 mm in 3 days as demonstrated in 2002.

As a consequence, the aquifer is highly exploited during the summer, with an important drawdown and the *Lez* spring depletion. Due to the DUP, pumping is controlled during the summer in order to never exceed a drawdown of 30 m; for this reason, a kind of control of the drawdown is performed during summer, which imposes a constant groundwater level (plateau) inside the spring. At the beginning of the autumn, the drawdown can be completely canceled with a sole rainy event.

For these reasons, it is of major importance for the spring management to study the relation between pumping and drawdown during summer, and predict the drawdown as a function of the exploitation of the spring. As the karst is not homogeneous, this relation is nonlinear and NN modeling thus appears as an interesting tool for this purpose.

3.2 Neural Network for the Lez Water Level Prediction

The universal approximator presented in Sect. 2.1 is used. For the forecasting of the ground water level at time k, inputs include rainfall at time (k to k-w + 1), pumping volume at time k, estimated previous water level at time k-1 to k-d and

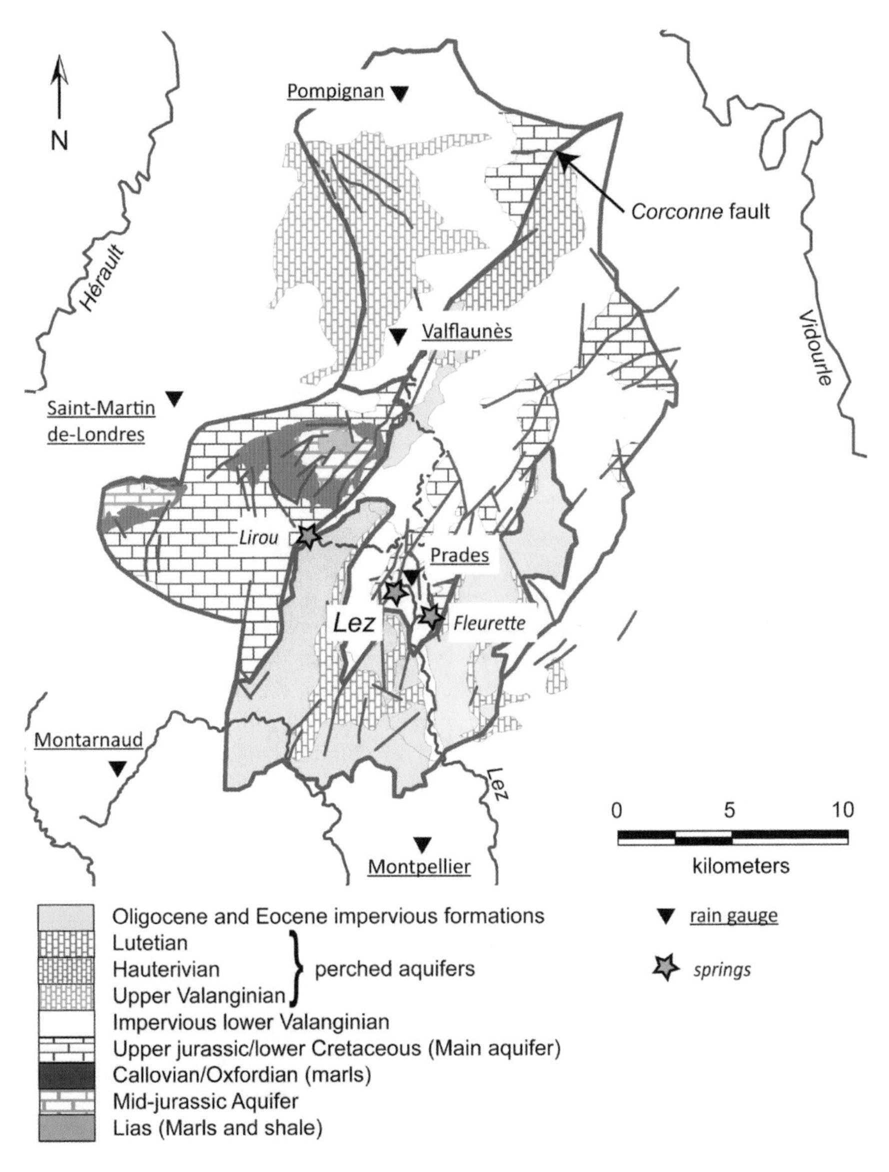

Fig. 2 *Lez* hydrosystem map (from Kong-A-Siou et al. 2011b)

potential evapotranspiration (*PE*) estimated with a Gaussian signal at time k (Fig. 1). Windows width w and d and number of hidden neurons N_c are selected using cross-validation as explained in Sect. 2.2 and in (Kong-A-Siou et al. 2011b). The data available span 8 years, from 1997 to 2004. Since the water level undergoes the greatest variation during summer, calendar years are considered, with the test year being 2003 and the stop set 2001.

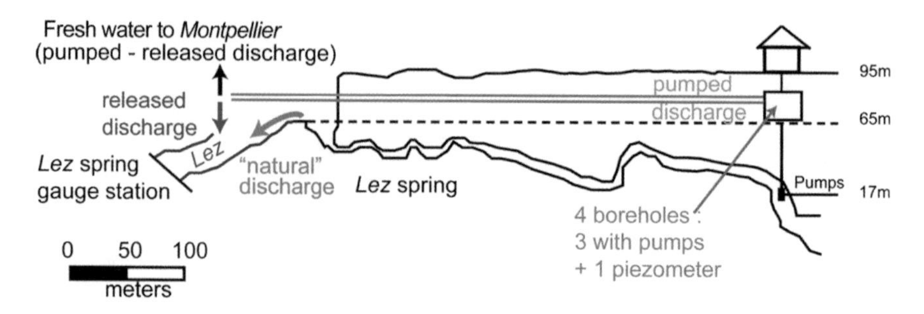

Fig. 3 *Lez* spring management

Table 1 Selected neural network complexity and variables

w (rainfall width)	d (previous discharge)	PE	Q_p	N_c
12	1	1	1	3

The designed model is thus the following (Table 1).

In order to assess the model efficiency, a test set must be chosen different from the training set. To this end, we propose to choose the driest summer of the database (year 2003) in order to evaluate the ability of the model to generalize to extreme configuration (summer 2003 is considered as the most important drought of the period). The more severe hypothesis: zero rainfall during the dry period (1st June–31 August), is applied and allows, at the end of May, to provide forecasts up to the end of August (up to 3 months lead time). Three various pumping *scenarii* were evaluated during this period: (i) actual measured pumping volume, (ii) mean pumping volume, (iii) mean pumping volume times two (Fig. 4). The last is not a realistic hypothesis as the maximal pumped volume is decided by Public Utility Decree but it is still interesting to assess the model generalization capability. However, a pumping trial should actually be necessary to assess the quality of this simulation.

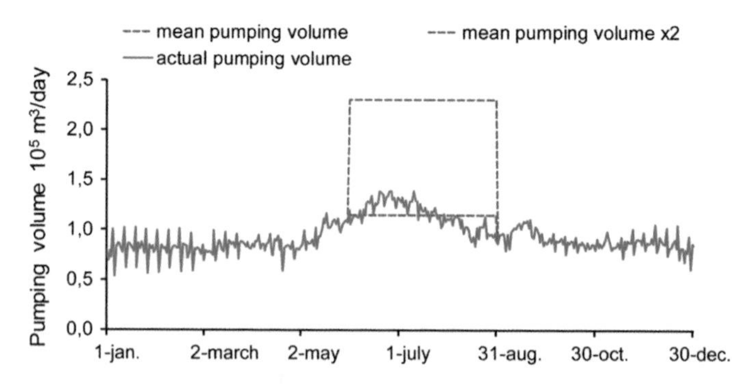

Fig. 4 Actual pumping signal and hypothesis for the test summer 2003

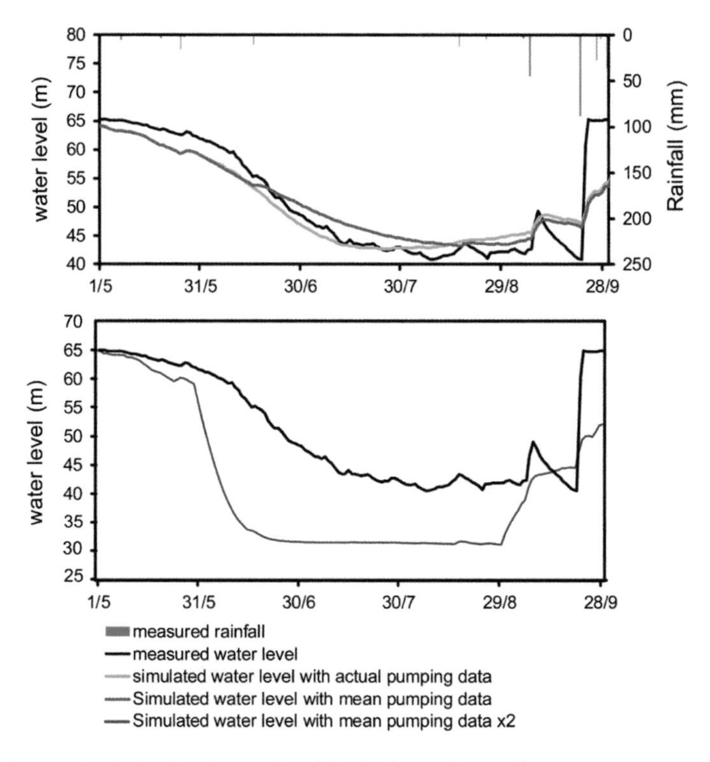

Fig. 5 Measured and simulated water level in the *Lez* spring well

4 Results and Discussion

Measured and simulated ground water levels of the three experiments previously presented are shown in Fig. 5. For the first experiment, corresponding to the actual pumped volume, the simulation is excellent, with a Nash criterion of 0.85. The drawdown and the slope are correctly simulated. However, the rising up, due to the huge rainfall event of 22th September is not correct. It is easy to understand remembering that the model does not receive actual rainfall but zero rainfall. The second experiment, including mean pumped volume provides similar results, with a 0.83 Nash criterion. This good quality simulation suggests that detailed information of the pumped volume is not necessary to obtain a correct simulation of the drawdown. Let us note that for all simulations, the maximal estimated drawdown is up than 90 % of the observed drawdown. In considering these excellent simulations, an arbitrary extreme pumping scenario is proposed corresponding to the double of the actual mean pumping volume. The simulation suggests that the maximal drawdown should be about 150 % of the actual maximal drawdown. Of course, this is hypothetical. The drawdown needs to be verified by pumping trial. Nevertheless, it is interesting to note that the groundwater level draws a plateau as it is in the actual pumping scenario. This behavior is not difficult to reproduce for a neural model because the neurons of the hidden layer implement a derivable "threshold" function (sigmoid function). Moreover, it can be noticed that the plateau estimated by the model depends nonlinearly on the pumping scenario.

5 Conclusion

In the context of karst water resources management, a model based on neural networks was designed in order to simulate the groundwater drawdown at the main outlet of a large karst hydrosystem under exploitation. This systemic approach was chosen as it does not necessitate a priori information about the karst system. Three rainfall scenarii were evaluated in the worse configuration: with zero rainfall during 3 months. The model shows excellent forecasts on the more severe drought of the database revealing one more time that neural networks are able to generalize to dataset outside of their training examples. Such a model appears thus as very promising tool to improve karst water resource management.

References

Barron AR (1993) Universal approximation bounds for superpositions of a sigmoidal function. IEEE Trans Inf Theory 39:930–944

Bérard P (1983) Alimentation en eau de la ville de Montpellier. Captage de la source de Lez étude des relations entre la source et son réservoir aquifère. Rapport n°2 : Détermination des unités hydrogéologiques. BRGM Montpellier

Corzo G, Solomatine D (2007) Knowledge-based modularization and global optimization of artificial neural network models in hydrological forecasting. Neural Networks 20(4):528–536

Dreyfus G (2005) Neural Networks Methodology and Applications, 1st edn. Springer, Berlin

Ford D, Williams P (2007) Karst Hydrogeology and Geomorphology. Wiley, Chichester

Filho AJ, dos Santos CC (2006) Modeling a densely urbanized watershed with an artificial neural network, weather radar and telemetric data. J Hydrol 317(1–2):34–48

Hagan MT, Menhaj MB (1994) Training feedforward networks with the Marquardt algorithm. Neural Networks 5(6):989–993

Hornik K, Stinchcombe M, White H (1989) Multilayer feedforward networks are universal approximator. Neural Networks 5(2):359–366

Kong-A-Siou L, Johannet A, Borrell Estupina V, Pistre S (2011a) Complexity selection of a neural network model for karst flood forecasting: the case of the Lez Basin (southern France). J Hydrol 403:367–380

Kong-A-Siou L, Johannet A, Borrell Estupina V, Pistre S (2011b) Optimization of the generalization capability for rainfall–runoff modeling by neural networks: the case of the Lez aquifer (southern France). Environ Earth Sci 65:2365–2375

Maier HR, Dandy GC (2000) Neural networks for the prediction and forecasting of water resources variables: a review of modelling issues and applications. Environ Model Softw 15:101–124

Nerrand O, Roussel-Ragot P, Personnaz L, Dreyfus G, Marcos S (1993) Neural networks and nonlinear adaptive filtering: unifying concepts and new algorithms. Neural Comput 5:165–199

Sjöberg J, Zhang Q, Ljung L, Benveniste A, Deylon B, Glorennec PY, Hjalmarsson H, Juditsky A (1995) Nonlinear black-box modeling in system identification: a unified overview. Automatica 31(12):1691–1724

Stone M (1974) Cross-validatory choice and assessment of statistical predictions. J Roy Stat Soc: Ser B (Methodol) 36(2):111–147

Thiery D, Bérard P, Camus A (1983) Captage de la source du Lez. Etude de relation entre la source et son réservoir aquifère - rapport n°1 : recueil des données et établissement d'un modèle de cohérence. BRGM Montpellier

Toukourou MS, Johannet A, Dreyfus G, Ayral PA (2011) Rainfall-runoff modeling of flash floods in the absence of rainfall forecasts: the case of Cevenol flash floods. J Appl Intell 35:178–189

Nonlinear System Engineering Techniques Applied to the Fuenmayor Karst Spring, Huesca (Spain)

J.A. Cuchí, D. Chinarro and J.L. Villarroel

Abstract Fuenmayor is a modest karst spring that tapes a small limestone aquifer near Huesca, Spain. A previous paper proposes a transfer function between effective rainfall and discharge concluding that Fuenmayor has an acceptable linear response. However, the linear model does not estimate adequately the response to some events where the nonlinearities are evidenced. To deal with the nonlinear characteristics of Fuenmayor, it is proposed a black-box model based on the Hammerstein-Wiener block-oriented structure. It is composed by a linear dynamic system surrounded by two static nonlinearities at its input and output. Seven different configurations of blocks are presented. Their efficiency has been evaluated by the Nash–Sutcliffe model efficiency coefficient. A good result is obtained with a configuration where the linear block is a second-order transfer function, with a zero and seven unit delays. The first nonlinear block is a piecewise polynomial and the second block has been suppressed. The running test draws out a maximum Nash-Sutcliffe efficiency of $E = 0.9383$.

1 Introduction

Mathematical models are powerful tools for analyse and simulate systems. Linear systems are usually modelled by the so-called Transfer Function (TF) that relates the input and the output of the system. Sampled systems are time continuous

J.A. Cuchí (✉) · J.L. Villarroel
Aragón Institute of Centre of Engineering Research (I3A), University of Zaragoza, Mariano Esquillor Street, 50071 Zaragoza, Spain
e-mail: cuchi@unizar.es

J.L. Villarroel
e-mail: jlvilla@unizar.es

D. Chinarro
San Jorge University, 50071 Zaragoza, Spain
e-mail: dchinarro@usj.es

B. Andreo et al. (eds.), *Hydrogeological and Environmental Investigations in Karst Systems*, Environmental Earth Sciences 1,
DOI 10.1007/978-3-642-17435-3_42

systems with their inputs and outputs sampled where the Discrete TF is used. DTF is defined as the ratio of the Z transform of the output and the Z transform of the input. If the system is linear, this function is rational with numerator and denominator polynomials of the complex variable z:

$$G(z) = \frac{Y(z)}{X(z)} = \frac{b_0 + \ldots + b_m z^{-m}}{a_0 + \ldots + a_n z^{-n}}$$

Thus, the behaviour of the system is described as a set of parameters a_i, b_i. There are two main approaches to obtain the model TF of a system: (1) Physical modelling and (2) Identification where a black-box model is derived from the observation of the inputs and outputs without knowledge of the internal structure or physical laws. Parametric identification can be viewed as the process of establishing the parameters of a TF. When the system is not linear, for black-box approaches, a block-oriented model is frequently used. Consists of a cascade connection of static (memoryless) nonlinear and linear blocks. One of most used structure is the Hammerstein-Wiener (HW) model where a linear block model is surrounded by two nonlinear blocks. The TF linear block represents the dynamic component of the model. The two nonlinear blocks emulate the static nonlinearities in the system, (Eskinat et al. 1991). They are implemented using several nonlinear estimators as deadzone, saturation, piecewise, sigmoidnet or wavenet proposed by Ljung (2007) . The blocks models are popular because they are easier to implement than other classical heavy-duty, nonlinear models. HW model admits two simplifications. The Hammerstein model has a nonlinear component followed by a linear component. In contrast, a Wiener model is the opposite.

To evaluate the quality of different models, there are several efficiency coefficient methods. For the present work, the Nash-Sutcliffe efficiency coefficient (NSEC) has been used, defined by Nash and Sutcliffe (1970) as:

$$E = 1 - \frac{\sum_{k=1}^{N} (y(\mathrm{kT}) - \hat{y}(\mathrm{kT}))^2}{\sum_{k=1}^{N} (y(\mathrm{kT}) - \bar{y})^2}$$

The coefficient E can range from $-\infty$ to 1. An efficiency value $E = 1$ corresponds to a perfect match of model output to the measured data. An efficiency value of 0 ($E = 0$) indicates that the model is as accurate as the mean of the observed data, whereas an efficiency less than zero ($E < 0$) occurs when the observed mean is a better predictor than the model.

The Identification Techniques for Linear System were applied for the karst aquifer of Fuenmayor (FMY) by Cuchí et al. (2014). Using a black-box Output Error (OE) model (Ljung and Glad 1994; Ljung 1999), a suitable TF with two poles, one zero and seven delays was defined to estimate the behaviour of the spring. The NSEC offers a value of 0.816, concluding that FMY has an acceptable linear response to the rainfall. However, it is evident that the spring has a no total lineal behaviour as show Chinarro et al. (2010, 2012). For this reason, the objective of the paper is to apply nonlinear identification techniques of System Engineering to FMY in order to analyse the nonlinearities of the karst system.

2 Description of the Test Site

Fuenmayor spring (740 m asl) is the outlet of a small mountain karst aquifer in the Guara Formation (Eocene limestone) and hosted by the modest San Julián overthrust described by Millán (1996).

Average total rainfall is near 800 mm under a continental Mediterranean regime with Spring and Fall frontal rains and sparse but violent Summer thunderstorms. The mean temperature is 12°C. The aquifer characteristics were initially presented by Pascual and Trilla (1974), Chinarro et al. (2012) and Cuchí et al. (2014). The recharge area (Fig. 1) is currently estimated on 9.8 Km^2 (Oliván 2013) covered by very shallow soils, bare limestone with hohlkarren and light rillenkarren, and gelifraction scree. Maximum storage of the soil has been estimated around 30 mm by Villarroel and Cuchí (2004) and Oliván (2013). Rainfall and discharge were measured by an automatic monitoring station located at the spring from June 2002 to June 2005. Data were collected hourly in order to control the known fast response of the spring, following Ogata (1987) as sampling should be performed eight or ten times during the rise time transient response. Hourly evapotranspiration to calculate the effective rainfall has been estimated by the Eagleman method. Relative humidity data are obtained from the Monflorite airport, 16 km south. The dataset has been analyzed to try different HW models with Matlab applications which offer the appropriate nonlinear estimators.

3 Results and Discussion

Considering FMY as a linear system and applying parametric identification from Ljung and Glad (1994) and Ljung (1999), a TF has been obtained between discharge and effective rainfall:

$$\frac{0.01488(1 + 0.9980z^{-1})}{1 - 1.988z^{-1} + 0.9876z^{-2}} z^{-7}$$

It is a second-order function with seven delays (7 h) similar to the lag of response observed by the spring keepers. Poles (roots of denominator) and zeros (roots of numerator) are complex quantities that determine the dynamic behaviour of the aquifer system. In this case, the poles are $z_1 = 0.9993$ and $z_2 = 0.9883$ corresponding to recession coefficients $a_1 = 0.0168$ days^{-1} and $a_2 = 0.2825$ days^{-1}.

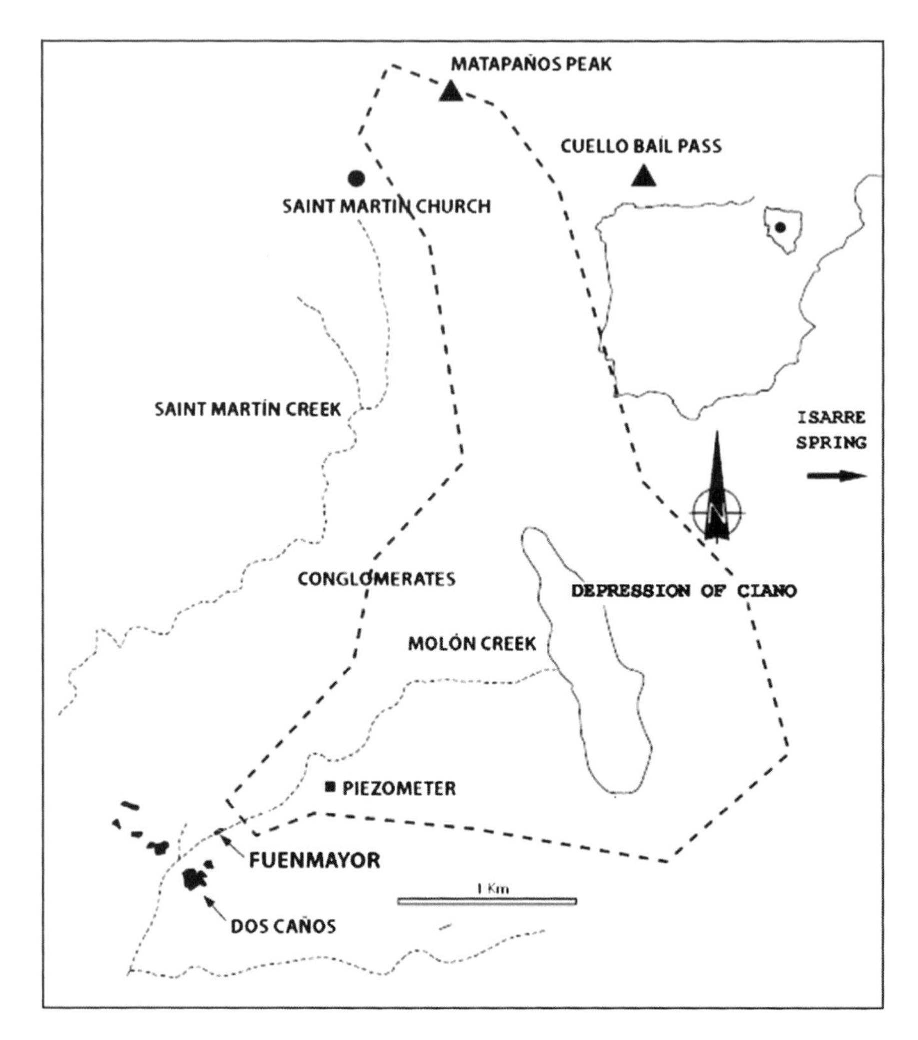

Fig. 1 Fuenmayor karst spring and estimated limits of recharge area

The first value is similar to the last depletion coefficient calculated by Trilla and Pascual (1974) using the Maillet method. Recession coefficients are in the exponential of the temporal expressions of the hydrograms and are related to transfer function poles by the equation $z = e^{-\mathrm{aT}}$ (Ogata 1987).

The best obtained TF for FMY has NSCE of 0.84. The value is slightly different from the presented by Cuchí et al. (2014) and Chinarro (2014), since evapotranspiration has been recalculated prior to perform the soil balance. The obtained NSCE is a relative high fit, that points out that the aquifer has an acceptable linear response. However, an analysis of the coherence function graph (Fig. 2) shows that the response is high at low frequencies but declines after an approximate value of 0.07 days^{-1}, corroborated by wavelet coherence presented by Chinarro (2014).

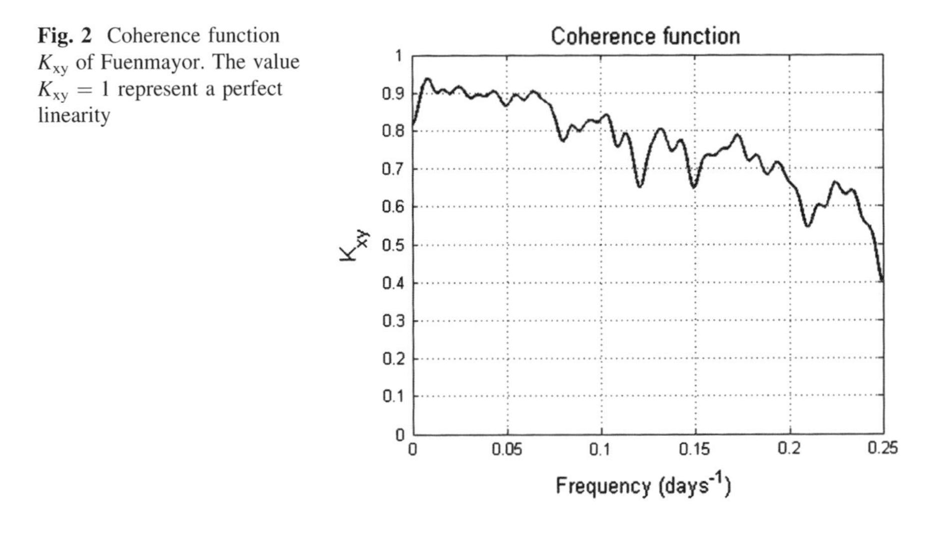

Fig. 2 Coherence function K_{xy} of Fuenmayor. The value $K_{xy} = 1$ represent a perfect linearity

That means that singular episodes of rainfall of high intensity and short duration, as summer thunderstorms, that have high frequency components shows a nonlinear behaviour. On the contrary, long duration rains of moderate variations of intensity produce a response in the system well simulated by linear models.

In order to analyse the nonlinear response of FMY, several black-box models based on the HW block-oriented structure have been tested. The central block of every model is a linear TF. The used software changes slightly the adjustment coefficients, the gain, in every case, but the same number of delays, poles and zeros is retained. The quality of the model is estimated by the NSEC. Several nonlinear simulations with estimators as dead zone, saturation and piecewise were tested. Only seven models have been presented in Table 1. Sigmoidnet and wavenet functions without a hydrogeological equivalent for karst behaviour were discarded.

A step-by-step methodology has been followed in order to establish the sensitivity of the system behaviour to small changes in the nonlinearities. As a first task, the identification work has been done on the input nonlinear block, using piecewise polynomials, dead zone and saturation equations (models 1–5). A second phase is focused on the output nonlinear block (model 6). The last phase includes both nonlinear blocks (model 7).

For the first phase, the addition of an input nonlinear block, the use of a piecewise improves the NSCE. Single piecewise models (1, 2, and 3) increase the fit value until 0.90. The maximum value obtained is 0.9383 for a simulation with 50 breakpoints (model 4). The case should be considered as a mathematical artefact as it is evident the difficulty to find a geological situation for that scenario. The models with a low number of breakpoints, as models 1 and 2, show a change of behaviour around a rainfall value of 10–13 mm hr^{-1} (Fig. 3). The models show a decreasing response after a breakpoint suggesting a change in the recharge or in fate of flow of underground water. Two behaviours of the system can be defined. The discharge response for rainfalls of low intensity, lower than 10 mm^{-1}, may be

Table 1 Different Hammerstein-Wiener block-oriented structure black-box models for the rainfall-discharge relationship at Fuenmayor aquifer

Model	Input non linearity	Linear block	Output non linearity	NSEC
0	None	$\frac{0.01488(1+0.9980z^{-1})}{1-1.988z^{-1}+0.9876z^{-2}}z^{-7}$	None	0.8400
1	Piecewise 1breakpoint	$\frac{1+0.9981z^{-1}}{1-1.988z^{-1}+0.9882z^{-2}}z^{-7}$	None	0.9023
2	Piecewise 2breakpoint	$\frac{1+0.9981z^{-1}}{1-1.988z^{-1}+0.9881z^{-2}}z^{-7}$	None	0.9040
3	Piecewise 3breakpoint	$\frac{1+0.9981z^{-1}}{1-1.988z^{-1}+0.9882z^{-2}}z^{-7}$	None	0.9031
4	Piecewise 50breakpoint	$\frac{1+0.9973z^{-1}}{1-1.986z^{-1}+0.9864z^{-2}}z^{-7}$	None	0.9383
5	Saturation in 2.38 mm/h	$\frac{0.02016(1+0.9985z^{-1})}{1-1.989+0.989z^{-2}}z^{-7}$	None	0.8945
6	None	$\frac{1+0.9978z^{-1}}{1-1.988z^{-1}+0.9883z^{-2}}z^{-7}$	Piecewise 10 breakpoint	0.8576
7	Piecewise 1 breakpoint	$\frac{1+0.9982z^{-1}}{1-1.989z^{-1}+0.9891z^{-2}}z^{-7}$	Piecewise 10 breakpoint	0.9098

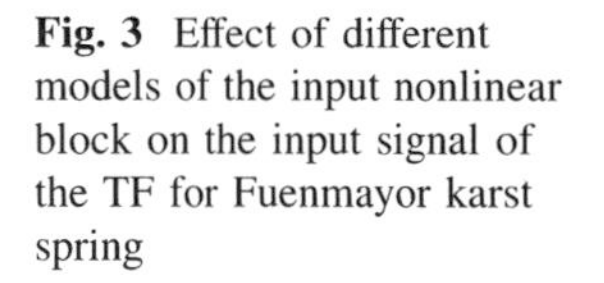

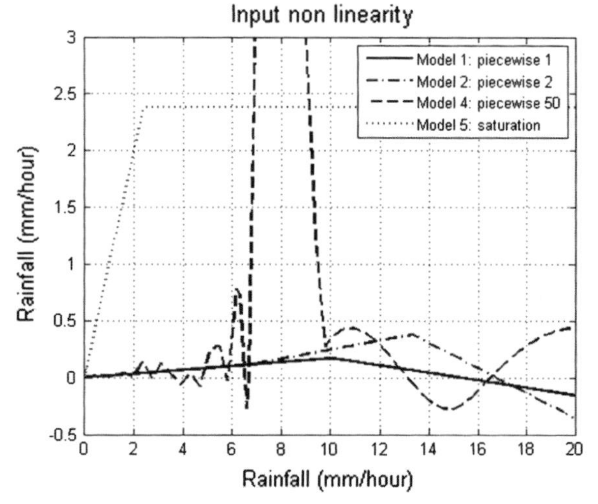

Fig. 3 Effect of different models of the input nonlinear block on the input signal of the TF for Fuenmayor karst spring

well simulated by linear models. After a threshold of rain intensity, the response of the system decreases as rain intensity increase. The effect may suggest an alternative transfer of water entered in the recharge area of the system. FMY has no know tropleins. The neighbour Dos Caños spring has a different chemistry and temperature that FMY. However, a hydraulic connection was detected with the aquifer taped by the neighbour outlet area of Dos Caños spring and associated minor springs, during a pumping test (Villarroel and Cuchí 2004). Other possibility is the loss of rainwater by surface or underground flow to the Saint Martin creek, feasible by topographic reasons. A third scenario is an underground transfer of water to Isarre spring (30T X: 723089 Y: 4680005 Z: 805) located approximately 1,100 m east of the estimated perimeter of FMY aquifer. The results of a simulation with a saturation input block is presented as model 5, with a FIT of 0.8945. The results, presented in Fig. 3, shows also a stabilisation of the response

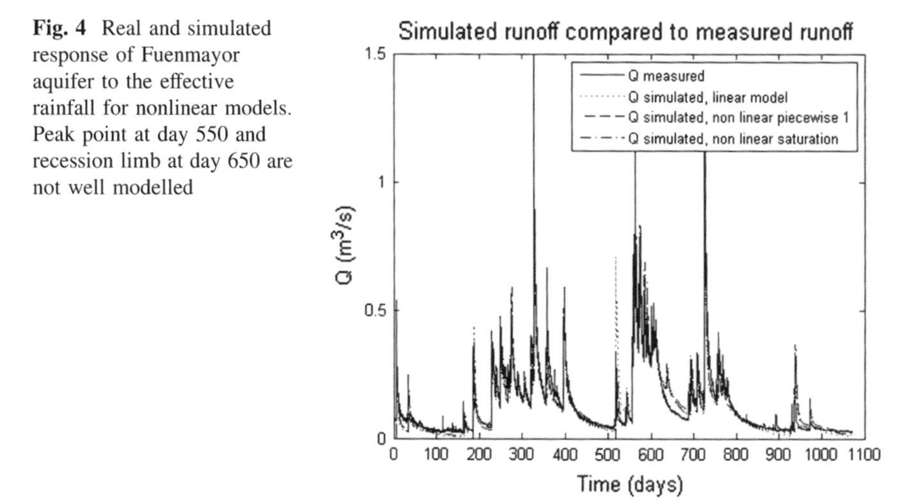

Fig. 4 Real and simulated response of Fuenmayor aquifer to the effective rainfall for nonlinear models. Peak point at day 550 and recession limb at day 650 are not well modelled

after events with low rainfall intensity. However, the optimal threshold for the tested family is obtained for a value of 2.38 mm^{-1}, indeed a very low value. The use of a dead zone does not produce any improvement of NSCE from the linear TF model and is not presented. FMY is a permanent spring and the use of the effective rainfall as the input data compensates the effect of the epikarst zone filtering the low-intensity rainfall episodes. But it is possible than this model may be useful for tropplein springs.

During the second phase, it was observed that the effect of a single output nonlinear block produced lesser effect than the input nonlinear block. The model number 6 of Table 1 has increased its NSCE in only 0.01875 points over that obtained by the simple linear model. Figure 3 shows that the output nonlinearity is, in practice, a near straight line. The low effect of the output nonlinear block can be also assessed by comparing models 1 and 7 of Table 1. The increase of NSCE is very low, from 0.9023 to 0.9098, adding to model 1 one output piecewise nonlinear block. But the addition of an input block to model 6 shows a clear fit increase. A view on the graph of the transformation of the response signal, the output block, after the lineal TF for models 6 and 7 shows a near linear behaviour almost indifferent to the number of pieces of the polynomial.

Such above discussions indicate that the behaviour of FMY spring can be modelled with a Hammerstein-Wiener block oriented structure with good results. However, some parts of the real hydrograph are not well simulated as it is showed in Fig. 4. The most obvious difference is the discrepancy at the recession limb around the day 650. The peak near day 500 is also overestimated. One possibility may be due to the effect of irregular precipitations, locally known as “andalocios”, over the recharge area. That is a typical meteorological pattern on late Spring, with rainfall controlled by a topographic effect of the Pyrenees front.

4 Conclusions

The behaviour of the relationship rainfall-discharge for Fuenmayor Spring can be represented by a nonlinear Hammerstein technique, obtaining an increase of the Nash-Sutcliffe efficiency coefficient over that obtained with only a linear Transfer Function. The best results are obtained with a piecewise nonlinear input block. In other hand, the addition of a nonlinear output block is irrelevant. However, geologic sense should be applied to avoid mathematical artefacts. Evidently more work is needed.

Acknowledgments The work was supported by the researching group GTE (Grupo de Tecnologías en Entornos Hostiles) of the Univ. of Zaragoza. Authors are grateful to the Engineering Department of Huesca city; the valuable assistance of R. Fortuño, the spring keeper and the access permits from the authorities of the Natural Park of Sierra and Canyons of Guara.

References

Chinarro D, Cuchí JA, Villarroel JL (2010) Application of wavelet correlation analysis to the karst spring of Fuenmayor. San Julián de Banzo, Huesca, Spain. In: Andreo B et al (eds) Advances in Research in Karst Media, vol 1. Springer, Berlin, pp 75–81

Chinarro D, Villarroel JL, Cuchí JA (2012) Wavelet analysis of Fuenmayor karst spring, San Julián de Banzo, Huesca Spain. Environ Earth Sci 65(8):2231–2243

Chinarro D (2014) System Engineering applied to Fuenmayor karst aquifer (San Julián de Banzo, Huesca) and Collins glacier (King George Island, Antarctica). Springer. Serie Springer Thesis, Ph.D thesis, U. Zaragoza.

Cuchí JA, Chinarro D, Villarroel JL (2014) Linear system techniques applied to the Fuenmayor karst spring, Huesca (Spain). Environ Earth Sci 79:1041–1060

Eskinat E, Johnson SH, Luyben WL (1991) Use of Hammerstein models in identification of nonlinear systems. AIChE J 37:255–268

Ljung L (1999) System identification. In: Theory for the user, 2nd edn. Prentice Hall, Englewood Cliffs

Ljung L, Glad T (1994) Modeling of Dynamic Systems. Prentice Hall, Englewood Cliffs

Ljung L (2007) System identification toolbox for use with Mathlab. Version 7. The MathWorks, Inc. Natick

Millán H (1996) Estructure and cinematic of the front of the south Pyrenean overthrust at the external Sierras of Aragón. Ph.D. thesis, U. Zaragoza. p 396

Ogata K (1987) Discrete-Time Control Systems. Prentice-Hall International Editions, Englewood Cliffs

Oliván C (2013) Delimitation, recharge evaluation and hydrodynamics of the aquifer drained by the karst spring of Fuenmayor (Prepyrenees of Aragon). Ph.D. thesis, U. Zaragoza. p 196

Trilla J, Pascual I (1974) Analysis of hydrographs of a karst spring (Fuenmayor, Huesca). Agua 87:20–28

Villarroel JL, Cuchí JA (2004) Cualitative response from May 2002 to April 2003 of the karst spring of Fuenmayor (San Julián de Banzo, Huesca) to rainfall and atmosphere temperature. Bol Geol Min 115(2):237–246

Controlling Factors of Wormhole Growth in Karst Aquifers

Y. Cabeza, J.J. Hidalgo and J. Carrera

Abstract Flow and water discharge in karst aquifers are controlled by the conduit network. Therefore, understanding karst conduit formation is important to conjecture the aquifer topology, i.e., conduit density and size, and to predict the aquifer dynamics. Conduits are generated by preferential pathways to flow known as wormholes that grow competing with each other. The success of a wormhole is determined by its ability to drive water away from its neighbors. Once a wormhole forms, water tends to flow along this preferential path thus reducing the availability of water for the enlargement of less developed wormholes. Wormhole growth is then controlled by the flow rate, the dissolution mechanisms and the heterogeneity of the hydraulic conductivity field. In this work, we propose two conceptual models to describe the geometry of the wormhole capture zone and its effect on the surrounding wormholes. First, we consider a cross-section intersecting the wormhole longitudinally. Second, we consider a radial model centered in the wormhole. These models are representative of field (fracture) and laboratory (tube), respectively. We perform a series of steady state simulations to obtain the dependence of the capture zone on the wormhole's geometry. This naturally leads to a relation between the wormhole's geometry and the density of wormholes because only one wormhole grows within a capture zone.

Keywords Karst · Wormhole · Capture area

Y. Cabeza (✉) · J.J. Hidalgo · J. Carrera
IDAEA-CSIC, Barcelona, Spain
e-mail: yoar.c.d@gmail.com

Y. Cabeza
Technical University of Catalonia (UPC), Barcelona, Spain

B. Andreo et al. (eds.), *Hydrogeological and Environmental Investigations in Karst Systems*, Environmental Earth Sciences 1,
DOI 10.1007/978-3-642-17435-3_43

1 Introduction

Karst aquifers are highly heterogeneous media that form when the carbonate rock is dissolved by aggressive water. These aquifers are triple-porosity systems, in which flow is controlled by the topology of the conduit network (White 2002; Gabrovšek and Dreybrodt 2001). Characterizing this conduit network is a very difficult task but understanding how it forms and behaves may help to infer its topology and properties.

The initial stage of a conduit network is an unstable dissolution front. Experiments (Detwiler et al. 2003) and simulations (Cheung and Rajaram 2002; Szymczak and Ladd 2004) have shown that the dissolution front can derive into channels originated by differences in the hydraulic conductivity field. The dynamics of the dissolution fronts are controlled by the feedback between reaction, porosity changes, and concentration of flow (Szymczak and Ladd 2011). The instabilities within the dissolution front grow competing with each other for the availability of water (Hoefner and Fogler 1988; Robert H. Nilson and Stewart K. Griffiths 1990). This growth generates more permeable pathways known as wormholes. When, for any reason, one of these wormholes grows larger that its neighbors, it becomes a preferential flow-path that concentrates the flow reducing the disposable water for less developed wormholes (Szymczak and Ladd 2006).

The effect of flow focus by a developed wormhole has a direct effect on the density of wormholes that successfully grow in a system. Prediction of wormhole density is a topic of interest not only for karst cave formation, but also in fields like CO_2 sequestration or for the oil industry, where artificial increment of porosity by acidification is a common practice. (Fredd and Fogler 1998; Cohen et al. 2008)

The aim of this work is to analyze how a developed wormhole affects the flow. We assume that the separation between two successful wormholes is proportional to the wormhole capture area. We consider two different conceptual models (cross-section and radial) and performed sequences of steady state simulations varying the wormhole length to find an expression that relates the wormholes capture area with its geometry.

2 Wormhole Growing in a Fracture

We consider a planar fracture in a rectangular domain in which a wormhole of length L_w is formed (Fig. 1). Using the symmetry of the system, we model a square of length L/L_w with the wormhole on the left side. Boundary conditions prescribe head at the top and at the bottom of the domain, so that the flow is parallel to the wormhole. The wormhole is simulated as a Dirichlet boundary with head equal to the head at the top. This condition is equivalent to consider the wormhole as an infinitely permeable zone. Using L_w as the length scale, the dimensionless governing equation and boundary conditions are:

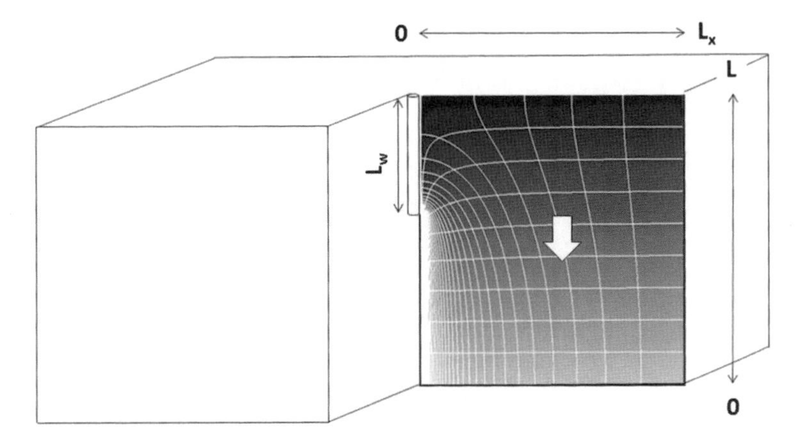

Fig. 1 Geometry for the cross-section model. A plane within a fracture where a wormhole is growing on its *left* side is represented. The *gray* area is the model section, *white lines* are potentiometric surface and *stream lines*. *White arrow* represents direction of flow

$$\nabla^2 h = 0$$

$$h = H_1 \text{ at } y = L$$

$$h = H_2 \text{ at } y = 0$$

$$h = H_1 \text{ at } x = 0 \text{ and } L \geq y \geq L_w$$

where h is head, L is domain length and H_1 and H_2 are prescribed head at the top and at the bottom, respectively.

The dimensionless flow rate within the wormhole is expressed as:

$$Q_w = \frac{Q_w}{qL_w} \int_{\frac{L}{L_w}-1}^{\frac{L}{L_w}} q \cdot n \, \mathrm{dy} = f\left(\frac{L}{L_w}, \frac{L_x}{L_w}\right)$$

where Q_w is the flow within the wormhole and q is Darcy's flow.

We perform a series of steady state models varying the wormhole's length on each simulation. Figure 2 shows that the ratio of flow rate within the wormhole and total flow rate (Q_T) varies linearly with the length of the wormhole. In the initial stages, when the domain can be considered infinite, this relation flow rate/ wormhole length keeps a 1:1 proportion.

The wormhole's capture area (CA) in a fracture plane can be calculated as:

$$\mathrm{CA} = \frac{Q_w}{Q_T} L$$

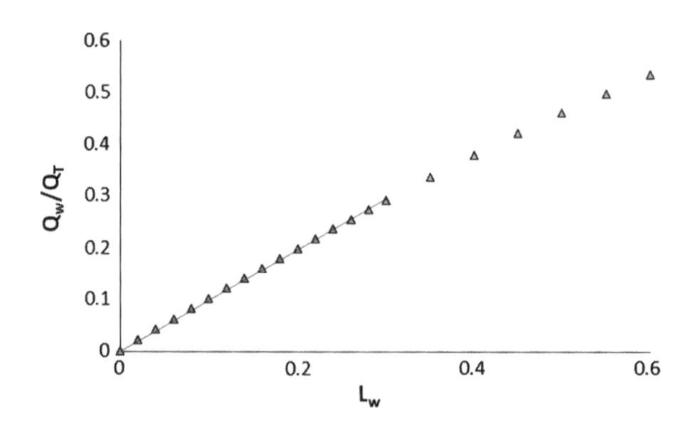

Fig. 2 Relation between flow rate within the wormhole/total flow rate and wormhole length for the fracture simulations. The relation is $y = 0.9856x$, with $R^2 = 0.9994$

So, in this case, the wormhole's capture area equals to the relation between flow rate within the wormhole and total flow rate.

3 Wormhole Growing in a Cylindrical Core

We simulate a cylindrical core of radius R and height L where a single wormhole is growing at its center (Fig. 3). In this case, the wormhole is characterized by its length L_w and radius r_w.

Using L_w and r_w as z and r scales, respectively, the dimensionless governing equation and boundary conditions are written as:

$$\left(\frac{1}{r}\frac{\partial}{\partial r}r\frac{\partial h}{\partial r}\right) + \left(\frac{r_w}{L_w}\right)^2\left(\frac{\partial^2 h}{\partial z^2}\right) = 0$$

$$h = H_1 \text{ at } z = L \text{ and } r_w < r < R$$

$$h = H_1 \text{ at } L_w - L < z < L \text{ and } r = r_w$$

$$h = H_2 \text{ at } z = 0 \text{ and } r_w < r < R$$

$$\nabla h \cdot \mathbf{n} = 0 \text{ at } r = R \text{ and } 0 < z < L$$

In this case, the flow rate within the wormhole is:

$$Q_w = \int_{L-L_w}^{L} 2\pi r_w K \frac{\partial h}{\partial r}\,\mathrm{dz}$$

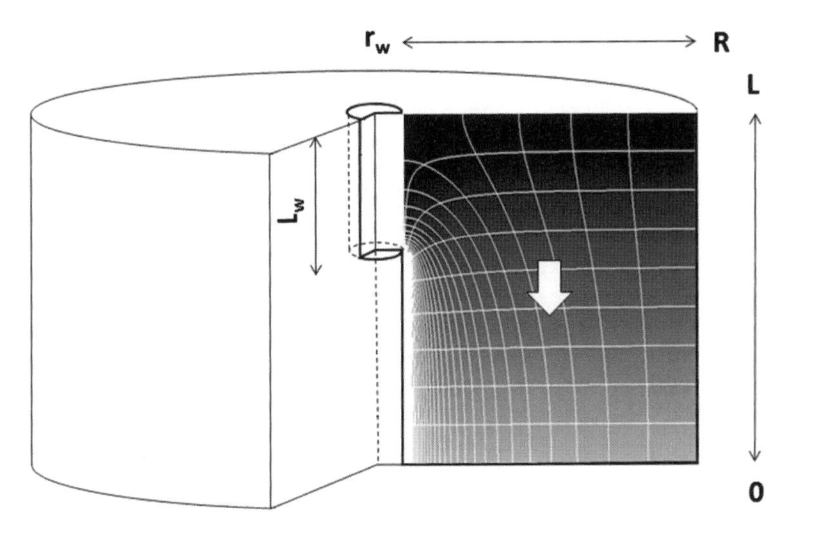

Fig. 3 Geometry of the radial model. The figure represents a core where a wormhole is growing at its center. The *gray* area is the model section. *White lines* are potentiometric surface and *stream lines*. *White arrow* represents direction of flow

The dimensional analysis of the governing equations suggests that the ratio between the flow that the wormhole takes from its surroundings and the total flow rate is proportional to the square of its length and a function of the relation between its radius and its length. That is,

$$\frac{Q_W}{Q_T} \propto 2\pi L_w^2 f\left(\frac{r_W}{L_W}\right)$$

where $f(r_w/L_w)$ is equal to:

$$f\left(\frac{r_W}{L_W}\right) = \frac{1}{\ln\left(\frac{L_W}{r_W}\right)}$$

We perform steady state simulations for different values of r_w (10^{-4}, 5.10^{-4}, 10^{-3}, 5.10^{-3}, 10^{-2}, 5.10^{-2}, and 10^{-1}) varying the wormhole length. Figure 4 confirms the validity of the assumption for the solution of $f(r_w/L_w)$ In absence of boundary effects, the simulated Q_w/Q_T is linearly proportional to the proposed expression with a coefficient of 0.2.

The capture area in this case is:

$$\mathrm{CA} = \frac{Q_W}{Q_T} = \frac{\frac{2\pi}{5} L_W^2}{ln\left(\frac{L_W}{r_W}\right)}$$

This expression does not work if $r_w > L_w$ because $\ln(L_w/r_w)$ becomes negative.

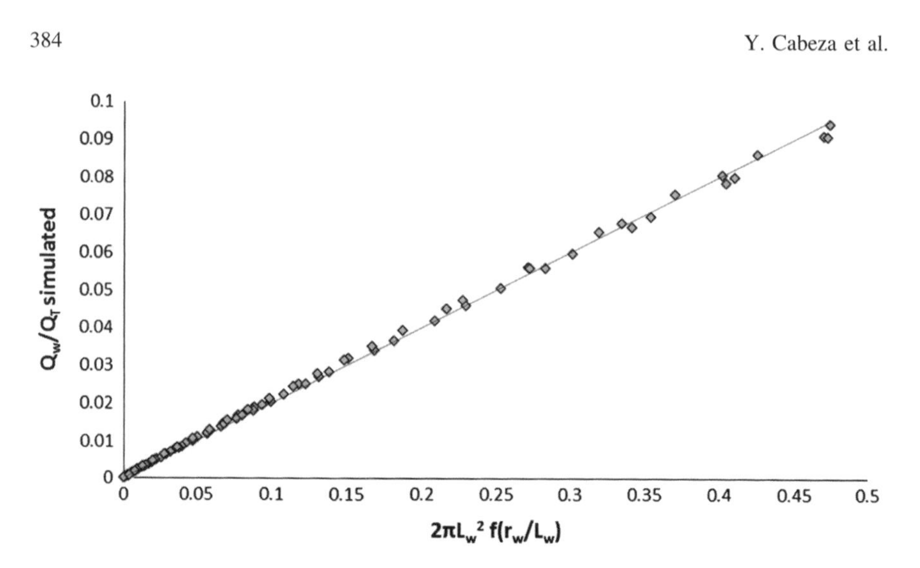

Fig. 4 Relation between flow rate within the wormhole/total flow rate and the hypothesis for the capture area in a cylindrical core. The results displayed correspond to wormholes with radius between 0.0001 and 0.01. The relation is $y = 0.2x$ and $R^2 = 0.9983$

4 Conclusions

In order to characterize the effect that the concentration of flow due to a preferential pathway (or wormhole) produces on the flow field of a given system, a sequence of steady state simulations have been performed, under two different conceptual models (cross-section and radial).

Results show that it is possible to characterize the flow that a wormhole takes from its surroundings by a simple analytical expression. For a fracture, the capture area is proportional to L_w and for a cylindrical core it is proportional to L_w^2 and $\ln(r_w/L_w)$. The proposed expression for radial flow models works in any case where $L_w > r_w$. By combining these results with the growth rate of a wormhole in a given system, it would be possible to characterize completely the wormhole behavior.

However, additional work is necessary for combining these results for the flow behavior with the chemical processes that occur in a karst system, like dissolution, and the corresponding wormhole growth rate.

References

Cheung W, Rajaram H (2002) Dissolution finger growth in variable aperture fractures: role of the tip-region flow field. Geophys Res Lett 22:2075

Cohen CE, Ding D, Quintard M, Bazin B (2008) From pore scale to wellbore scale: Impact of geometry on wormhole growth in carbonate acidization. Chem Eng Sci 63:3088–3099

Detwiler RL, Glass RJ, Bourcier WL (2003) Experimental observations of fracture dissolution: The role of Péclet number in evolving aperture variability. Geophys. Res. Lett 30, 1648

Fredd CN, Fogler HS (1998) Influence of transport and reaction on wormhole formation in porous media. AIChE J 44:9

Gabrovšek F, Dreybrodt W (2001) A model of the early evolution of karst aquifers in limestone in the dimensions of length and depth. J Hydrol 240:206–224

Hoefner ML, Fogler HS (1988) Pore evolution and channel formation during flow and reaction in porous media. AIChE J 34(1):45–54

Nilson RH, Griffiths SK (1990) Wormhole growth in soluble porous materials. Phys Rev Lett 65:13

Szymczak P, Ladd AJC (2004) Microscopic simulations of fracture dissolution. Geophys Res Lett 31:23606

Szymczak P, Ladd AJC (2006) A network model of channel competition in fracture dissolution. Geophys Res Lett 33:L05401

Szymczak P, Ladd AJC (2011) Instabilities in the dissolution of a porous matrix. Geophys Res Lett 38:L07403

White WB (2002) Karst hydrology: recent development and open questions. Eng Geol 65(2–3):85–105

An Example of Karst Catchment Delineation for Prioritizing the Protection of an Intact Natural Area

V. Ristic Vakanjac, Z. Stevanovic, M. Aleksandra, B. Vakanjac and C.I. Marina

Abstract The paper presents the method and results of an assessment and delineation of the catchment area of the karstic spring "Perućac," which is the main drainage area of Tara Mt. in western Serbia. It is the northeastern, inner zone of the classic Dinaric karst characterized by a dominant extension of highly karstified Triassic limestones, a poorly developed hydrographic network with many ponors, and dense forests. Tara Mt. is one of five national parks in Serbia. The goal of the project was to identify a high-priority protection zone within the preserved natural area. Based on data on the discharge regime of the Perućac Spring, and by applying the multiple nonlinear correlation method, the size of the direct catchment area was assumed to be 79.3 km^2, which was 5–18 % larger than that estimated by previous surveys.

1 Introduction

Tara National Park, established in 1981, is one of five national parks in Serbia. It encompasses the largest part of Tara Mt. with an average altitude between 800 and 1,200 m a.s.l. Tara NP has an area of 19.175 ha with a protected buffer zone

V. Ristic Vakanjac (✉) · Z. Stevanovic · C.I. Marina
Faculty of Mining and Geology, Belgrade, Serbia
e-mail: vesna.ristic@rgf.bg.ac.rs

Z. Stevanovic
e-mail: zstev_2000@yahoo.co.uk

C.I. Marina
e-mail: marinacokorilo@googlemail.com

M. Aleksandra
Natural History Museum, Belgrade, Serbia
e-mail: amaran@nhmbeo.rs

B. Vakanjac
Faculty for Applied Ecology Futura, Belgrade, Serbia
e-mail: borivac@gmail.com

B. Andreo et al. (eds.), *Hydrogeological and Environmental Investigations in Karst Systems*, Environmental Earth Sciences 1,
DOI 10.1007/978-3-642-17435-3_44

around it of 37.584 ha (Radović 2007) (Fig. 1). Together with the adjacent protected areas (Zaovine and Nature Park Šargan-Mokra Gora) Tara NP represents the center of the future Drina Biosphere Reserve.

Due to the specific geologic, geomorphologic, hydrologic, pedologic, and climatic features, Tara Mt. is among the most important centers of Balkan and European ecosystems and species diversity. It is characterized by varied communities of old deciduous and mixed coniferous forests that represent a unique example of well-preserved forests in SE Europe with numerous endemic and relict plant and animal species. About one-third of the total flora and fauna of Serbia can be found within Tara NP, including representatives of almost 1,000 plant, 135 bird, and 53 mammal species (Maran 2012). The "queen of all endemic species in Europe" is the Serbian spruce (*Picea omorika*) discovered in 1876 by the famous Serbian botanist Josif Pančić. Preserved forests are habitats for viable populations of diverse fauna including the endemic Pančić' grasshopper, several endangered species such as the brown bear, otter, European pine marten and chamois. Within the canyons of Tara Mt., there are many traces of prehistory and the ancient Roman and Byzantine cultures.

The study area is situated at the border between Serbia and Bosnia and Herzegovina, which refers from a geotectonic point of view to the Inner Dinarides of western Serbia. For the most part, it consists of karstified Triassic limestones, featuring a poorly developed hydrographic network with a large number of ephemeral and sinking streams. Given that the altitude is above 1,000 m, precipitation and the thickness of the snow cover in winter are significant contributors to the rates of infiltration and aquifer recharge. Most of the terrain belongs to the Drina River basin, which forms a large portion of the natural border between these two countries. Apart from the Drina and its tributaries, the reservoir of the Bajina Bašta Hydroelectric Power Plant at Perućac and the pumped-storage reservoir of the Bajina Bašta II HPP at Zaovine on Tara Mt. are of great economic significance.

The goal of the research was to assess the catchment area of Perućac Spring, the most important and largest karstic spring on Tara Mt., for the purposes of prioritizing protection and implementing stringent preventative measures in the national park.

2 Geology and Hydrogeology of Study Area

Tara Mt. has a long geological history, from Paleozoic to Quaternary. The most prevalent rocks are Paleozoic shales, Permian-Triassic sandstones with intrusions of igneous rocks and Mesozoic carbonate complex (the Triassic and Upper Cretaceous limestones and dolomites) that constitute about 30 % of the Tara Mt. In addition, Jurassic ophiolites, Oligocene-Miocene lacustrine sediments and Quaternary fluvial deposits have significant distribution in the wider area.

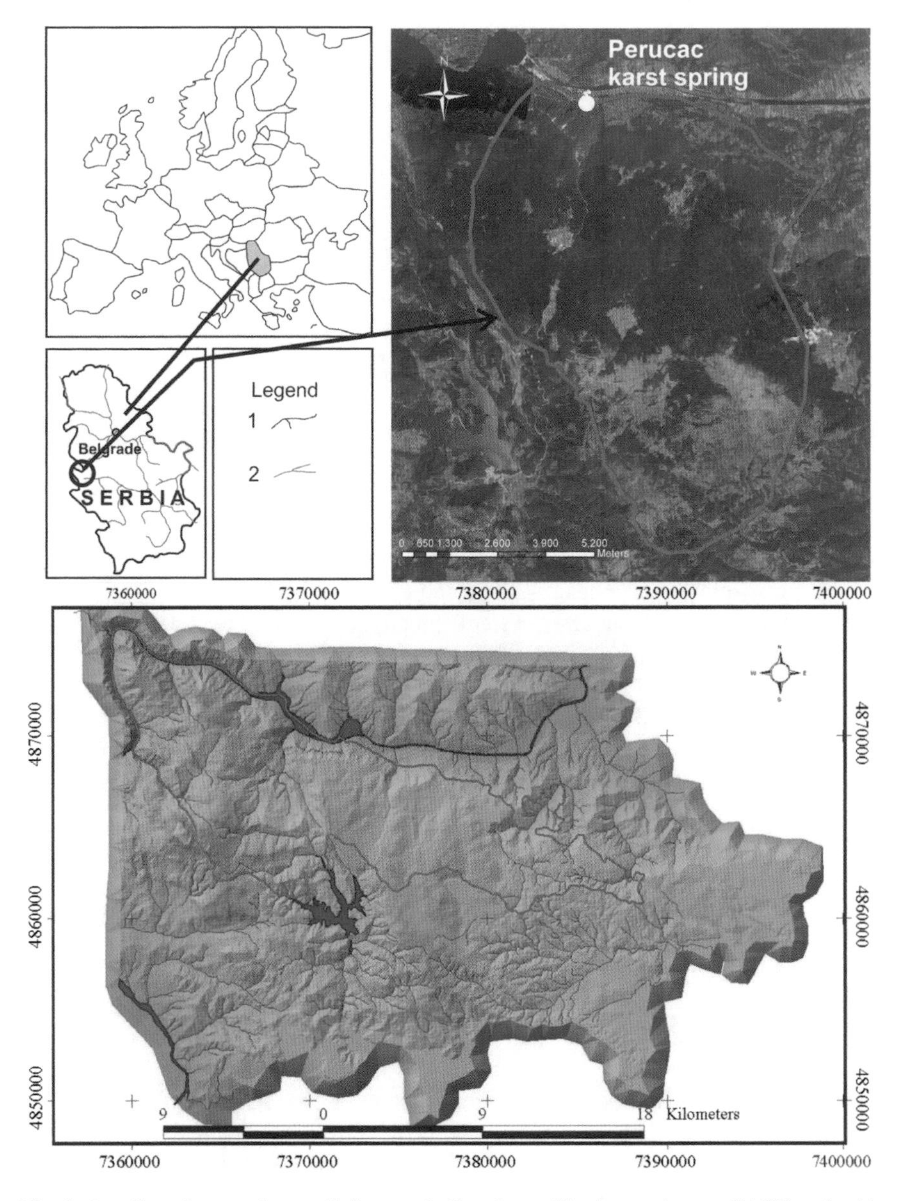

Fig. 1 Locality of research area (*left corner*), Google satellite image (source SASPlanet) with catchment area of kasrt spring Perucac (*right corner*) and border of Tara NP on DEM (after Radović et al. 2005) (*down*)

In the 1960s, Energoproject of Belgrade undertook comprehensive surveys in connection with a feasibility study and later with the construction of the Bajina Bašta Dam, one of the largest hydropower projects in Serbia. Krešić (1984) provided data on the Perućac Spring discharge regime, and Milanović (2005)

Fig. 2 Perućac spring

discussed the potential impact of the underground reservoir at this spring. Jemcov (2008) assessed the water budget and predicted the effects of groundwater extraction, while Jemcov et al. (2008) and Živanović (2011) studied the vulnerability of the Tara Mt. aquifer system. Finally, Jemcov et al. (2010) evaluated several options for tapping the Perućac Spring and proposed a procedure for selecting the best technical solution.

Massive Middle and Upper Triassic limestones constitute the main formation in the study area. Fissured rocks are generally semi-permeable or permeable, acting as hydrogeological barriers, while rocks featuring karst porosity, where faults serve as main groundwater pathways, are highly permeable.

The karst aquifer is recharged by infiltration of precipitation via ponors and a series of fractures and fissures. Average annual precipitation totals are high, often reaching 1,000 mm of the water column, and the precipitation regime is relatively uniform during the year (i.e., long precipitation-free periods are very rare). In most of the terrain, the presence of tall vegetation prevents direct infiltration and has an adverse impact on the karst aquifer's recharge: it reduces the infiltration rate. Given that some 80 % of the land area is densely forested, the effect of evapotranspiration is pronounced. The hydrographic network is developed in areas made up of serpentinites and peridotites, which intensify runoff at interfaces with the carbonate rocks of Tara Mt., where the water sinks quickly and recharges the aquifer (allogenic recharge).

The Perućac Spring discharges at the foot of Tara Mt. at an elevation of 234 m a.s.l (Fig. 2). It is situated 1.7 km downstream from the Bajina Bašta HPP on the Drina River. The karst aquifer is formed in Middle and Upper Triassic limestones, and discharged at the point of contact between these rocks and Triassic sandstones (Scythian stage) and the Paleozoic Drina Complex. The current zone of emergence of the Perućac Spring is also the lowest point of aquifer discharge and the lowest level of karstification, since no siphonal circulation has been recorded in the upland or in the spring floor. From the point of emergence, the spring forms a

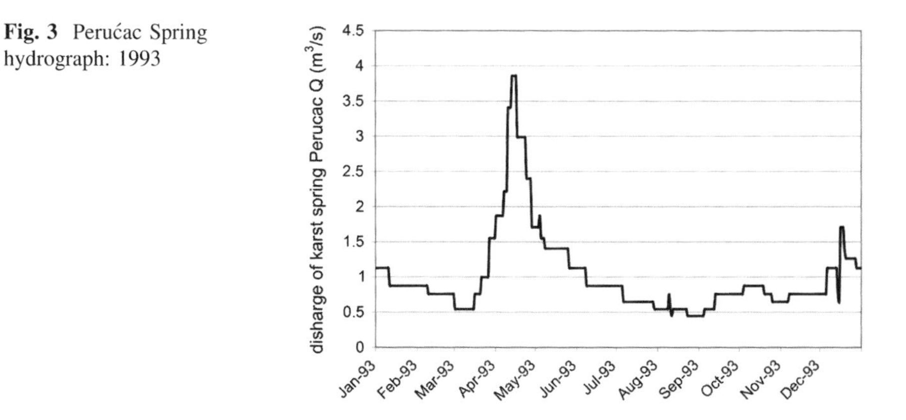

Fig. 3 Perućac Spring hydrograph: 1993

stream that empties into the Drina. The stream is 365 m long and has for that reason been named the Godina (Year) River.

The discharge rate of the Perućac Spring was initially monitored in the 1960s, before the dam was erected, then again in 1987 and 1988 by the Jaroslav Černi Institute, and from November 1991 to 2 July 1992 by Energoproject-Hydroengineering. Since then, the discharge has been recorded by the Bajina Bašta HPP personnel. The discharge regime of the Perućac Spring was assessed using the available dataset spanning 7 years (1992–1998). Typical spring discharge hydrograph is presented in Fig. 3. This figure shows only one peak, caused by simultaneous spring rain and snowmelt.

During 1956–1958, recorded discharges were in interval 1.34–6.1 m^3/s with an average of around 4 m^3/s (Krešić 1984). The annual average discharge rates of the Perućac Spring in 1993–1998 were in the interval from 1.004 (1993) to 1.579 m^3/s (1995). The absolute maximum was recorded on 13 and 14 April 1994 (9.815 m^3/s), and the absolute minimum on 30 and 31 October 1995 (0.446 m^3/s).

3 Method

The nearest meteorological station at Zlatibor Mt. (alt. 1,029 m) was selected to assess the effect of precipitation on the discharge rate of the Perućac Spring. A coefficient of correlation of 0.74 was derived from correlating annual values, while the analysis of the effect of monthly precipitation totals on spring discharge rates yielded 0.31. A cross-correlogram with a time step of $t = 1$ day, for a time period of 7 years, was produced for daily analysis (Fig. 4). In general, the residence time between the catchment area and the rain gauge station on Zlatibor Mt. is 2 days. Additionally, there are two more peaks: one after 8 days and the other after 26 days. The peaks are a result of the release of water from the snow cover accumulated during the winter months.

Catchment size and evapotranspiration needed to be defined for water budgeting purposes. A model was developed for cases where hydrogeological research had not

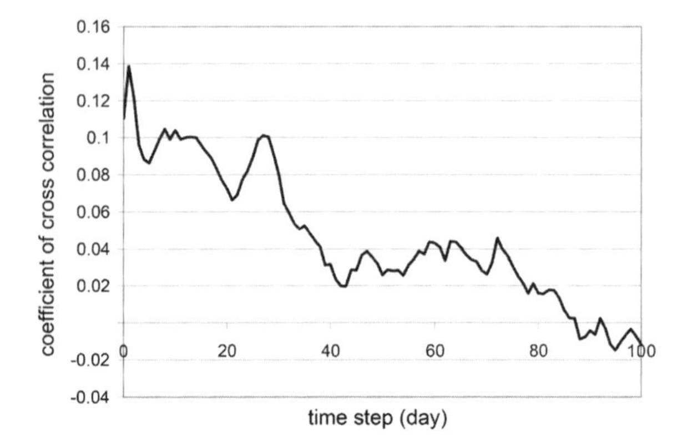

Fig. 4 Cross-correlogram

been sufficiently extensive to determine the catchment size and where the karst discharge dataset was not long enough. The model is comprised of two levels:

- Level 1: Completing the time series of mean monthly discharges by means of multiple nonlinear correlation (MNC), and
- Level 2: Delineation of the catchment area and karst aquifer water budgeting.

The task at Level 1 was to establish causal relationships based on available long-term monthly hydrometeorological data from the extended area, as well as available (usually short-term) karst spring data. The essence of the model is to establish linear correlation functions between standardized variables of the time series of mean monthly discharges and the climate parameters relevant to the karst spring discharge regime (Prohaska et al. 1977, 1979, 1995; Ristić 2007). This method was used to simulate the time series of mean monthly discharges of the Perućac Spring for the analytical period 1960–2009.

The basic equation for assessing a karst aquifer water budget with a monthly time step is:

$$P_{ij} = h_{ij} + E_{ij} + \left(V_{ij} - V_{i,j-1}\right) = h_{ij} + E_{ij} \pm \varDelta_{ij} \tag{1}$$

where: P_{ij}—monthly precipitation totals in the karst watershed [mm], h_{ij}—mean monthly karst spring discharge layer [mm], E_{ij}—monthly sums of actual (real) evapotranspiration in the karst watershed [mm], V_{ij}—water volume of the considered karst aquifer in the *j*th month [mm], $\varDelta_{ij}$—change in water reserves of the karst in the *j*th month [mm].

Given available data, the water budget equation (Eq. 1) featured two unknown quantities: E_{ij} and $\varDelta_{ij}$. A new boundary condition was therefore introduced to assess the "real (actual) evapotranspiration" E_{ij} in the karst. The analytical values of daily sums of potential evapotranspiration derived applying a modified Thornthwaite method (Thornthwaite and Mather 1957; Ristić 2007; Stevanović et al. 2010; Ristić et al. 2013), were used to assess the first approximation of monthly sums of real evapotranspiration from the karst based on known mean daily air temperatures and real monthly sums of insolation.

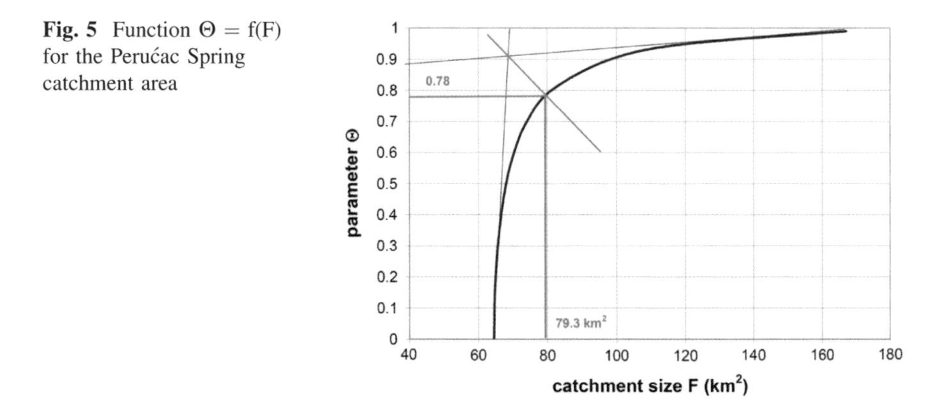

Fig. 5 Function $\Theta = f(F)$ for the Perućac Spring catchment area

A procedure that gradually approaches the solution to the water budget equation (Eq. 1) was followed to estimate daily sums of "real" evapotranspiration, assuming that:

- the initial volume of stored karst groundwater was equal to the volume at the end of the pre-defined analytical period, or $V_0 \approx V_K$ where 0 and K are the beginning and end of the selected analytical period, respectively;
- the distribution of daily sums of actual (real) evapotranspiration was nonlinear, such that for rainy days the derived daily sums of potential evapotranspiration were considered to be daily sums of real (actual) evapotranspiration, or: $E_{ik} = \mathrm{PET}_{ik}$ while for all the other later days the actual (real) daily sums of evapotranspiration decreased according to the following formula:

$$E_{j(k+\tau)} = \Theta^{2\tau} \cdot \mathrm{PET}_{ik} \quad (2)$$

  where: E_{ik}—actual daily sum of evapotranspiration [mm], and τ—1, 2, 3,..., m—time step in days;
- the catchment was delineated for different values of Θ ($\Theta = 0, 0.1, 0.2,\ldots, 0.9$), applying the rule $V_0 \approx V_K$.
- The function $\Theta = f(F)$ was constructed. The average of the centerline tangents of the function $\Theta = f(F)$, and the function itself represented the real catchment size, in the specific case of the Perućac Spring (Fig. 5).

4 Results and Discussion

Apart from the catchment size, the outputs from the model included daily and monthly real evapotranspiration levels for a given Θ. The long-term real evapotranspiration was 445.6 mm. The maximum annual real evapotranspiration was 575.7 mm, and the minimum was 310.0 mm. Monthly water volumes of the karst

Table 1 Perućac Spring karst aquifer: summary of water budget for the period 1960–2009

Spring	F	P	E	h	Q_{av}	q	W	φ
	km^2	mm	mm	mm	m^3/s	l/s/km^2	10^6 m^3	
Perućac	79.3	987.4	445.6	547.5	1.376	17.35	43.39	0.554

aquifer were computed using the generated input time series of monthly precipitation totals, real evapotranspiration and runoff layers for the catchment area of Perućac Spring.

Based on the values derived above, Table 1 shows the results for the basic elements of the water budget equation for the study area. The values pertain to a long-term average reduced to the period 1960–2009. The following parameters are shown in Table 1: catchment size F (km^2), average annual precipitation P (mm), average annual evapotranspiration E (mm), runoff layer h (mm) as $h = P - E$, annual change in volume ΔV (mm) as $\Delta V = h - (P - E)$, long-term average discharge rate Q_{av} (m^3/s), long-term average runoff modulus q (l/s/km^2), volume of water W (10^6 m^3) discharged during an average hydrological year, and long-term average runoff coefficient φ.

Previous assessments of the size of the Perućac Spring catchment area (F) varied from 65 to 75 km^2 (Krešić 1984; Milanović 2005; Jemcov et al. 2010; Živanović 2011). The size of 79.3 km^2 derived following the above procedure showed a maximum variation of 18 %, i.e., a larger catchment than previously estimated on the basis of field surveys. With regard to the groundwater budget in an average hydrological year, the Perućac Spring was found to have drained some 45 % of all precipitation, which was more realistic and less than estimated by some previous studies. For instance, due to large volume of drained waters as resulted from measurements in period 1956–1958 it has been assumed that effective infiltration might be even 80 % for the estimated catchment of 75 km^2 and still with "catchment deficit." The average discharges for 1956–1958 were around 4 times larger than those for 1993–1998, which led to such a conclusion. Therefore, the installation of permanent monitoring stations on ephemeral and small streams, which provide allogenic recharge to the karst aquifer through numerous ponors, is strongly recommended and needs to be consistent with the previously proposed protection approach for ponors as restricted areas (first level of protection according to national legislation).

5 Conclusions

The karstic Perućac Spring is the main drain of Tara Mt. and the largest spring in the Dinaric karst of western Serbia. The fact that its potential is still underutilized despite abundant reserves and that the drainage regime is relatively stable qualify this spring for inclusion on the list of the most prospective alternative sources for public water supply in Serbia. By applying method of multiple nonlinear correlation (MNC) the surface of Perućac Spring basin is calculated on 79.3 km^2 which is for larger than previously estimated for 5–18 %, (depending on information source). With this value an approximated effective infiltration is 45 % of total precipitation in an average hydrology year. The Perućac Spring and its catchment area need to be given priority when protection measures for Tara National Park are considered. The spring's catchment occupies some 45 % of preserved natural area, such that the delineation of the catchment area, supported by other aquifer surveys and vulnerability assessments, and the implementation of stringent preventative measures within the zone are required.

References

Jemcov I (2008) Karst aquifer water balance and solutions optimization of their water tapping, on the examples from Serbia, PhD thesis, Faculty of Minig and Geology, University of Belgrade, p 396

Jemcov I, Milanovic S, Milanovic P (2010) Decision support procedure for constructing karst underground reservoir—a case study on Perucac karst spring (Western Serbia). In: Andreo B, Carrasco F, Durán JJ, LaMoreaux JW (eds) Advances in research in Karst media. Springer, Heidelberg, pp 415–421

Jemcov I, Zivanovic V, Colic S, Milanovic S (2008) Vulnerability evaluation of karst massif Tara groundwater—supporting of sustainable management of National Park. In: Proceedings of the karst and speleology, Book IX, SANU, pp 65–80, Belgrade (printed in Serbian)

Kresic N (1984) Hydrogeology of karstic terrains of Drina River basin upstream of Bajina Bašta on the territory of SR Serbia (in Serbian). MS thesis, Fac. Min. & Geol. University of Belgrade

Maran A (2012) Geoconservation of the Cretaceous marine geosites from Serbia: Boljevac and Mokra Gora. Archive of the Bucharest, University, p 210 + 4 annexes (unpublished doctoral thesis)

Milanovic S (2005) Perućac underground spring and storage. In: Milanović P, Stevanović Z, Radulović M (Eds) Excursion Guide of the International Conference KARST 2005, Belgrade-Kotor, pp 30–31

Prohaska S, Petkovic T, Simonovic S (1977) Nonlinear mathematical model for extension and interrupts filling of hydro- meteorological data. In: Proceedings of Jaroslav Cerni Institute for the Development of Water Resources, No 587, Belgrade (printed in Serbian), pp pp 25–34

Prohaska S, Petkovic T, Simonovic S (1979) Mathematical model for spatial transfer and interpolation of hydro-meteorological data. In: Proceedings of Jaroslav Cerni Institute for the Development of Water Resources, vol 64, Belgrade (printed in Serbian)

Prohaska S, Ristic V, Srna P, Marcetic I (1995) The use of mathematical VMC model in defining Karst spring flows over the years. In: XV Congress of the Carpatho-Balkan Geological Association, 4/3, str., Atina, pp 915–919

Radovic D (2007) Evolving GIS at Tara National Park (Serbia). Bocconea 21:183–191

Radović D, Stevanovic V, Markovic D, Jovanovic S, Dzukic G, Radovic I (2005) Implementation of GIS technologies in assessment and protection of natural value of Tara National Park. Arch Biol Sci 57(3):193–204, Belgrade

Ristic V (2007) Simulation model developed for calculation of karst springs daily discharges, Faculty of Minig and Geology, University of Belgrade, PhD thesis, p 325

Ristic Vakanjac V, Prohaska S, Polomcic D, Blagojevic B, Vakanjac B (2013) Karst aquifer average catchment area assessment through monthly water balance wquation with limited metodological data set: application to Grza spring in Eastern Serbia. Acta Carstologica Slovenia 42(1):109–119

Stevanovic Z, Milanovic S, Ristic Vakanjac V (2010) Supportive methods for assessing effective porosity and regulating karst aquifes. Acta Carstologica Slovenia 39(2):313–329

Thornthwaite CW, Mather JR (1957) Instructions and tables for computing potential evapotranspiration and the water balance. Publication in climatology 10, Drexel Institute of Technology, Centerton

Zivanovic V (2011) Pollution vulnerability assessment of groundwater—examples of karst in Serbia. MS thesis, Faculty of Mining and Geology, University of Belgrade, p 212

Assessment of Groundwater Vulnerability in Croatian Karstic Aquifer in Jadro and Žrnovnica Springs Catchment Area

J. Loborec, S. Kapelj, D. Dogančić and A.P. Siročić

Abstract The groundwater vulnerability assessment is based on the evaluation of physical, chemical, and biological properties of the environment which can provide a certain degree of protection to the groundwater from contamination. In this paper are presented results of application four different methods (SINTACS, EPIK, PI, and COP) for groundwater vulnerability assessment in karstic aquifer in Croatia. The main objectives of the study were to apply the methods on test site where, so far, groundwater vulnerability was never assessed, then to modify the methods in order to improve adaptation to the research field and available data. After comparing vulnerability maps obtained by all four presented methods, it was determined that the most appropriate method is COP + K, which was additionally modified. Although presented methodology is not new, it can be used as a background for land-use planning, because it identifies parts of the catchment area that are, due to its natural features, more vulnerable to human impact. It can be also applied as an additional tool in groundwater protection for delineation of protection zones, and provides very useful data in various fields of water management, especially for karst area in Croatia.

Keywords Groundwater • Intrinsic vulnerability • Karst aquifer • Jadro and žrnovnica

J. Loborec (✉) · S. Kapelj · D. Dogančić · A.P. Siročić
Faculty of Geotechnical Engineering, University of Zagreb, Hallerova aleja 7, 42000 Varaždin, Croatia
e-mail: jloborec@gfv.hr

S. Kapelj
e-mail: sanja.kapelj@zg.t-com.hr

D. Dogančić
e-mail: ddogan@gfv.hr

A.P. Siročić
e-mail: anita.pticek.sirocic@gfv.hr

B. Andreo et al. (eds.), *Hydrogeological and Environmental Investigations in Karst Systems*, Environmental Earth Sciences 1,
DOI 10.1007/978-3-642-17435-3_45

1 Introduction

Available supplies of high quality water are becoming major issue of today and future development in the world. Republic of Croatia is well-known as a country with a wealth of high-quality water resources and that is why great attention must be given to maintain this wealth. Almost half of the country's territory is covered by karstified carbonate rocks that contain large amount of high-quality groundwater. In these areas, karstified carbonate rocks are typically covered by a thin, irregularly distributed soil layer and are intersected by a multitude of interconnected fractures facilitating the rapid infiltration of surface water. Groundwater flows through the conduits and fissures with high velocity and with a relatively short retention time. That allows the quick and far-reaching spread of potential contamination from the surface (Goldscheider 2005). For that reason, karst aquifer systems are extremely vulnerable and easily threatened by human activities.

The vulnerability of groundwater to contamination is estimated since the 1970s , when Margat (1968) introduced this term. Vrba and Zaporozec (1994) said that vulnerability is relative, unquantifiable, and immeasurable characteristic of aquifer system, an idea that some areas are more vulnerable to groundwater contamination than others due to its intrinsic features. Through the years, scientists have developed and presented numerous methods for vulnerability assessment. Vulnerability assessment is a highly subjective process and depends largely on applied methods. It is not uncommon that the application of different methods on the same test field leads to significant differences in results (vulnerability map). Therefore, at the beginning of the twenty-first century, as part of the COST 620 project "Vulnerability and risk mapping for the protection of carbonate (karst) aquifers," scientists have proposed a framework approach that would balance the vulnerability assessment process and that would make the results in individual regions comparable. It is called "European approach" and it proposes three basic parameters that influence aquifer recharge from the ground surface to the water table: overlaying layers (O), concentration of flow (C), and precipitation regime (P) (Daly et al. 2002; Zwahlen 2004). "European approach" does not provide exact methodology and rating of each factor, but only rough guidelines that each country can adapt to their specific conditions and available data.

Main aim of this paper is to present results of application of several different methods for assessing intrinsic groundwater vulnerability: SINTACS (Civita and De Maio 2000), EPIK (Doerflinger et al. 1999), PI (Goldscheider 2005), and COP (Vías et al. 2006) to a deep karstic aquifer in Dalmatia, Croatia. SINTACS, EPIK, and PI methods are developed before "European approach" and the COP method was created in accordance with its recommendations. Each of these methods was modified and tested based on the typical characteristics (parameters) of the test site.

2 Characteristics of the Study Area

The catchment area of Jadro and Žrnovnica springs with its approximate area of 560 km^2 is one of larger aquifer systems in Dinaric type of karst. It is located in the middle of Dalmatia, in the wider hinterland of the city of Split. It is part of the Cetina river basin. The springs are located in the foothills of the Mosor and Kozjak Mts. in the contact zone between permeable carbonate sedimentary rocks and the coastal flysch belt. Jadro is situated at an altitude of 35 m a.s.l., while Žrnovnica lies at the altitude of 90 m a.s.l. According to the available hydrological data, discharge of the Jadro spring ranges between 3.60 and approximately 78 m^3/s. Žrnovnica joins several smaller springs upwelling in the broader area. The minimum measured discharge is up to 250 l/s, while the maximum discharge is up to 19.1 m^3/s). The Jadro spring is used for the water supply of Split, surrounding settlements and the towns of Solin, Kaštela, and Trogir (up to 300,000 people). Žrnovnica is used for the water supply of the village of Žrnovnica and irrigation of agricultural areas. The catchment area is predominantly composed of carbonate rocks, limestone and dolomite, and flysch sediments, while other rocks occur more rarely. From geological point of view, Carbonate rocks early Mesozoic (Cretaceous) built over 75 % of the study area. Tertiary and Quaternary rocks and sediments build something less than 18 % of the area, while others (Permian, Triassic, and Jurassic) occupy only about 7 % of the catchment surface. On the hydrogeological map, five main categories of rocks according to their permeability are singled out (Fig. 1). Detailed description hydrogeological characteristics of observed rocks can be found in interpretation of the Basic Hydrogeological Map of the Republic of Croatia, Split and Primošten sheets (Fritz and Kapelj 1998). A vast of different karst features can be found in the catchment area. Particularly interesting features are the sinkholes, which is allocated over 10,000 individual objects, as well as large number of swallow holes, pits, and caves. They are used in the vulnerability assessment as an important indicator of karstification since they indicate on developed epikarst zone and zone of potential rapid infiltration water from surface into the underground (Ford and Wiliams 2007; Kapelj et al. 2004).

Data from earlier conducted researches show an increasing trend in the quantity of a number of contaminants in the spring water due to economic expansion in the catchment area which was until recently poorly populated and developed. Lack of appropriate protection of drinking water sources has also contributed to this problem. Therefore, last few years was conducted a comprehensive research of Jadro and Žrnovnica springs catchment area (Kapelj et al. 2006, 2008, 2009, 2012). Results of the Study were analysed by GIS tools and used as input data for mapping groundwater intrinsic vulnerability.

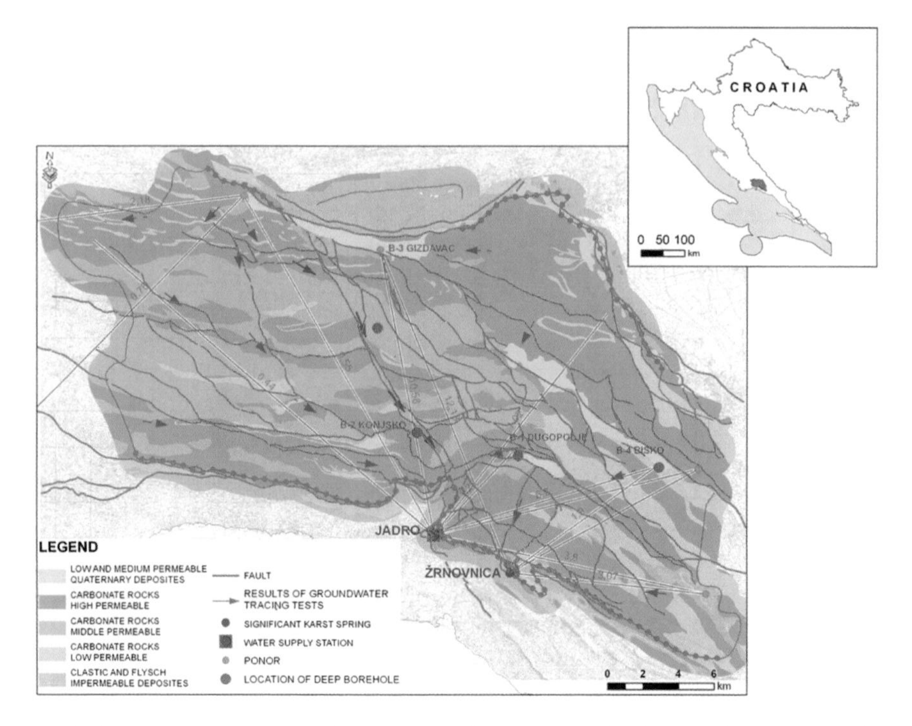

Fig. 1 Hydrogeological map of the catchment (*Inset* map of Croatia with the location of the study area)

3 Vulnerability Mapping

3.1 SINTACS Method

The SINTACS method of vulnerability assessment is based on evaluation of seven parameters: depth to groundwater (S), impact of effective infiltration (I), capacity of attenuation in the unsaturated zone (N), capacity of attenuation in the soil/sediment cover (T), hydrogeological properties of the aquifer (A), hydrogeologic conductivity range of the aquifer (C), and hydrogeological role of the terrain slope (S). Each parameter is in the range of values between 1 and 10, where higher value indicates greater aquifer vulnerability. In addition to the standard method, the impact of the sinkholes on the vulnerability assessment was tested (Kapelj et al. 2013; Loborec 2013). By the standard procedure, the value of C parameter (hydrogeologic conductivity range of the aquifer) is estimated based on the coefficient of hydraulic conductivity (K), which is directly dependent on the type of rock that builds aquifer. Since a large portion of the catchment area is build of permeable carbonate rocks, the zoning of parts with different coefficient of permeability (due to different degree of fracturing and karstification) is very difficult, especially if there is no enough data from field research. Therefore is in the analysis included an

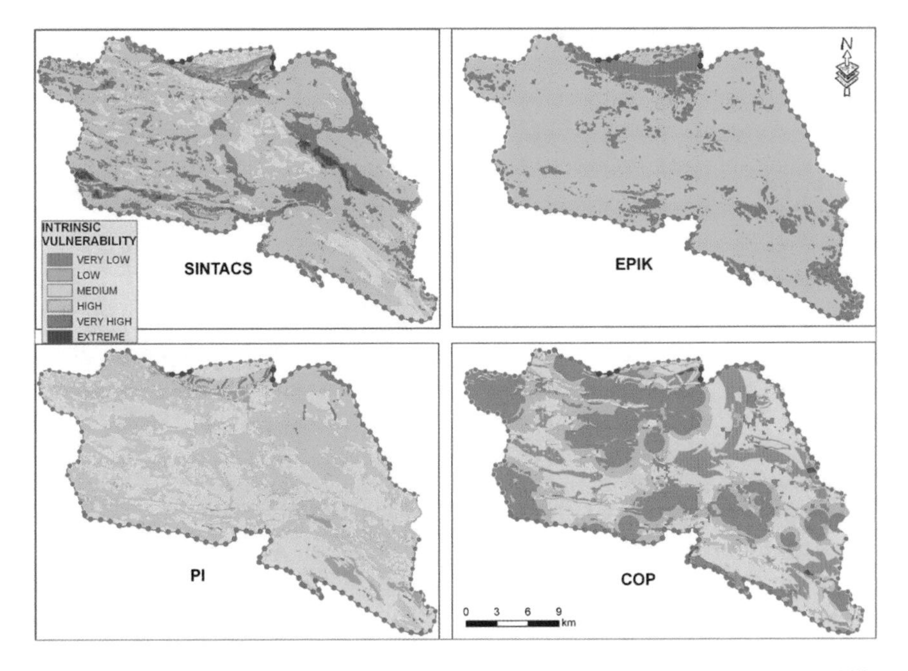

Fig. 2 Resulting vulnerability maps obtained by application of four different vulnerability methods (SINTACS, EPIK, PI, and COP)

additional indicator of karstification and thus the permeability of the terrain. By analyzing the spatial distribution of sinkholes was obtained the map of sinkhole density as number of sinkholes in the area 250 × 250 m, which is combined with a map of the distribution the parameter C obtained by the standard method. So, obtained modified parameter C is included in the calculation of the vulnerability index and finally for the vulnerability mapping by SINTACS method (Fig. 2).

3.2 EPIK Method

EPIK method was developed exclusive for karst aquifers and represents a certain basis of the European approach. It assesses vulnerability by evaluating four main attributes: epikarst (E), protective cover (E), infiltration conditions (I), and karst network development (K). At the end, each factor is multiplied with weighting factor (in range from 1 to 3) in order to emphasize impact of certain attributes in the calculation of protective factors and thus the final vulnerability map. When using this method on Jadro and Žrnovnica, catchment area assessment of E factor was problematic. According to recommendations, individual class attribute E is defined based on the fracturing of terrain, or by the presence of karst geomorphologic forms. As already mentioned, on the catchment area is present a multitude of geomorphological forms, and therefore on the large part of catchment is estimated very high class of vulnerability. Therefore, in this method is also used sinkhole density map, where certain classes of density were used to evaluation factor E. Because of that very high class of vulnerability has significantly decreased (Fig. 2) what is more realistic situation.

3.3 PI Method

This method was developed within the COST 620 project. Although it seems like it is the simplest method because it evaluate only two parameters, function of protective cover (P) and infiltration conditions (I), it is outlined in detail and includes more subfactors than the previous methods, specially parameter P, for which is necessary to determine seven separate subfactors. Vulnerability classes are obtained through the protective factor $\pi = P * I$. Here was also taken into account the density of sinkholes when assessing subfactor F (fracturing) in the total protective function value, but there was no significant impact of this modification. When defining I parameter, according to the guidelines, on quite large part of catchment as dominant flow process was obtained surface runoff. Since the soil data used in this assessment are derived from Hydropedological map scale 1:300,000, results are taken with reserve and recommendation is to conduct detailed studies of soil characteristics in this area (Fig. 2).

3.4 COP Method

COP method is the youngest of all applied methods and it is fully coherent (methodological and terminological) with the guidelines of the "European approach" (assessing three main parameters as mentioned in Chap. 1). Assessment of parameter O (protective function of overlaying layers) is similar to PI method, but it is better adapted to the conditions in our area as well as available data (especially as far as the impact of fracturing and karstification and soil data). Parameter C distinguish swallow hole recharge area with the impact zone 5,000 m, which is in this case too much, due to the rapid sinking of water into the underground (Kapelj et al. 2005, 2007), but has significant impact on final vulnerability map, as seen in in Fig. 2.

4 Results and Discussion

By comparing the results of the intrinsic vulnerability assessment (vulnerability maps), obtained by four methods presented, it is evident that there are similarities but also significant differences between the results.

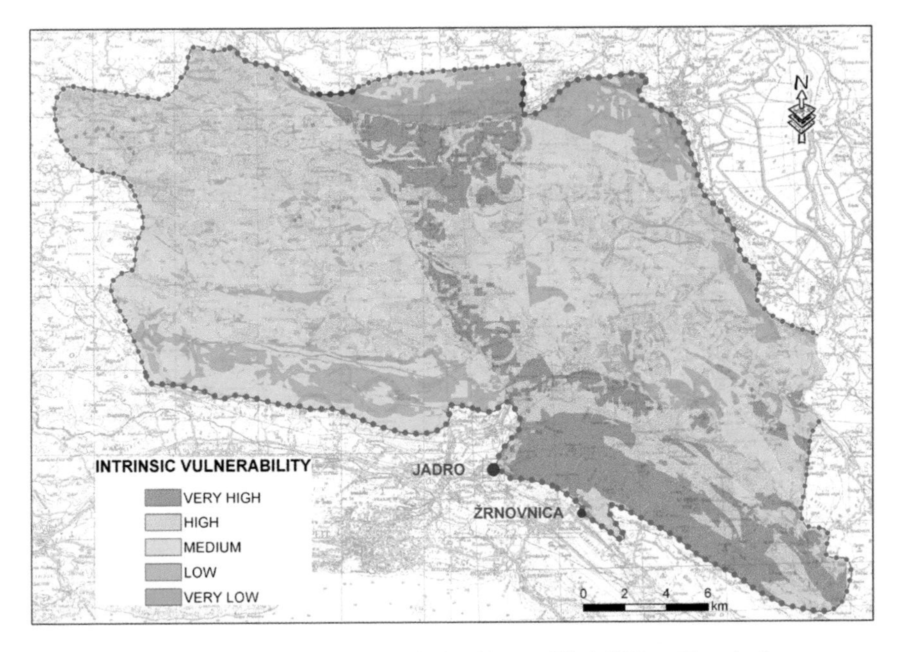

Fig. 3 Source intrinsic vulnerability map obtained by modified COP + K method

In SINTACS method, it is difficult to define with sufficient precision individual parameters, as in karst there are no detailed measurements of certain features (such as depth to groundwater, infiltration, thickness, and characteristics of soil, hydraulic conductivity). Interpolation of these features at a very heterogeneous and anisotropic field, such as the Dinaric karst, is unfortunately only an assumption, which reduces the reliability of the method. Also, it is important to note that SINTACS method does not take into account the swallow holes as particularly vulnerable places, and as the most vulnerable parts, allocates lowland areas with small slope angle that collect water from the higher parts and conduct underground. Disadvantage of EPIK method in this case is generalization and subjectivity in application without detailed guidelines. The method is applicable in areas with very limited database for which you want to define a global view of vulnerability without some significant details. Also, overestimate the protective role of soil based on the too little data. PI and COP methods, although resemble have given significantly different results. The main disadvantage of PI method is use of soil data (eFC for P determination and saturated hydraulic conductivity for I determination), while COP method deals with texture and thickness of soil layers which are more reliable and accessible data. Another difference is that COP method separately estimates the impact of precipitation, although it was not been taken into account dilution with increasing precipitation, based on some measurements. PI method give special attention to the catchment area of sinking streams (they are no sinking streams in this area), and overestimate the occurrence of surface runoff. While COP method overestimates the zone of influence of swallow holes (which is easy to deal with, allows easier adjustment occurrence of surface runoff by spatial analyses of slopes and hydrographic tools.

It is important to emphasize that in this case are about mapping intrinsic vulnerability of water sources (karst springs), so it is necessary to consider the characteristics of groundwater flow in the saturated zone of the aquifer toward the springs (the so-called K factor in the "European approach"). That has been done by defining K factor by method proposed by Ravbar and Goldscheider (2007) and Andreo et al. (2009), with a few modifications, which at the end, has been confirmed as the most suitable for the application in Jadro and Žrnovnica springs catchment area (Fig. 3).

5 Conclusions

Vulnerability maps can be used as a very useful additional tool in water protection and land-use planning. For that reason, it is important to determine the most appropriate method that could be applied to all areas with similar geological, hydrogeological, and hydrological characteristics, and so the results would be uniform and comparable. The applied methods of karst groundwater vulnerability assessment have shown very different results. Based on the comparisons, it was concluded that the COP + K method is the most appropriate to the conditions of the study area. It uses the available data, which are arranged in very suitable parameter estimation, emphasizes the most vulnerable areas and with respect to several modifications gives the most acceptable results, as confirmed by comparison with the results of some field research and matching with the proposed zones of protection of these sources. In order to vulnerability mapping become even more reliable method and also with the aim of exclusion of subjective interpretation, it is necessary to make additional efforts in development of validation procedures.

References

Andreo B, Ravbar N, Vías JM (2009) Source vulnerability mapping in carbonate (karst) aquifer by extension of the COP method: application to pilot sites. Hydrogeol J 17:749–758

Civita M, De Maio M (2000) SINTACS R5, a new parametric system for the assessment and automating mapping of groundwater vulnerability to contamination. Pitagora Editor, Bologna, p 226

Daly D, Dassargues A, Drew D, Dunne S, Goldscheider N, Neale S, Popescu IC, Zwahlen F (2002) Main concepts of the European approach for (karst) groundwater vulnerability assessment and mapping. Hydrogeol J 10(2):340–345

Doerfliger N, Jeannin PY, Zwahlen F (1999) Water vulnerabiliy assessment in karst environments: a new method of defining protection areas using a multi-attribute approach and GIS tools (EPIK method). Environ Geol 39(2):165–176

Ford DC, Wiliams PW (2007) Karst hydrogeology and geomorphology. John Wiley & Sons, UK 562

Fritz F, Kapelj J (1998) Osnovna hidrogeološka karta Republike Hrvatske M 1:100 000, listovi Split i Primošten. [Basic Hydrogeological Map of the Republic of Croatia 1:100 000, sheets Split and Primošten—in Croatian] Institut za geološka istraživanja, Zagreb

Goldscheider N (2005) Karst graundwater vulnerability mapping: application of a new method in the Swabian Alb, Germany. Hydrogeol J 13:555–564

Kapelj J, Kapelj S, Singer D (2004) Spatial distribution of dolinas and its significance for groundwater protection in karst terrain. In: Ortega A (ed) Grounwater flow understanding from lokal to regional scales XXXIII Congress IAH & 7th Congress ALHSUD, La Octava Casa, Zacatecas

Kapelj S, Kapelj J, Biondić R, Singer D, Picer M (2005) The approach of the groundwater vulnerability assessment in the area of the Zadar town with regard to PCB and other contaminants. In: Book of proceedings, Third Croatian geological congress, Opatija, pp 197–198

Kapelj S, Kapelj J, Biondić R, Biondić B, Kovač I, Tušar B, Prelogović E, Marjanac T, Andrić M, Kovačić D, Strelec S, Gazdek M (2006) Studija upravljanja vodama sliva Jadra i Žrnovnice—Prva faza studijsko istraživačkih radova. [Water managenet study of the Jadro and Žrnovnica springs recharge area, first phase—in Croatian] EVV:1/2005. Hrvatske vode, Split, unpublished

Kapelj S, Kapelj J, Singer D, Obelić B, Horvatinčić N, Babinka S, Suckow A, Brianso H L (2007) Risk assessment of groundwater in the area of transboundary karst aquifers between the Plitvice Lakes and Una River catchment. In: Nakić Z (ed) Second international conference on water in protected areas, Kopriva—graf, Zagreb, pp 86–90

Kapelj S, Kapelj J, Marjanac T, Prelogović E, Cvetko-Tešović B, Biondić B, Ivanković T, Jukić D, Denić-Jukić V (2008) Studija upravljanja vodama sliva Jadra i Žrnovnice—Druga faza studijsko istraživačkih radova. [Water managenet study of the Jadro and Žrnovnica springs recharge area, second phase – in Croatian] EVV:9/2007. Hrvatske vode, Split, unpublished

Kapelj S, Kapelj J, Dogančić D, Loborec J, Ivanković T, Cvetko-Tešović B, Milanović D (2009) Studija upravljanja vodama sliva Jadra i Žrnovnice—Treća faza studijsko istraživačkih radova. [Water managenet study of the Jadro and Žrnovnica springs recharge area, third phase—in Croatian] EVV:21/2008. Hrvatske vode, Split, unpublished

Kapelj, S., Kapelj, J., Dogančić, D. & Loborec, J. (2012): Studija upravljanja vodama sliva Jadra i Žrnovnice—Četvrta faza studijsko istraživačkih radova [Water managenet study of the Jadro and Žrnovnica springs recharge area, fourth phase—in Croatian] EVV:21/2008. Hrvatske vode, Split, unpublished

Kapelj S, Loborec J, Kapelj J (2013) Assessment of aquifer intrinsic vulnerability by the SINTACS method. Geol Croat 66(2):119–128

Loborec J (2013) Procjena rizika od onečišćenja podzemnih voda u kršu na području sliva izvora Jadra I Žrnovnice. [Risk assessment of contamination of groundwater in karst in the recharge area of the Jadro and Žrnovnica springs—in Croatian] PhD Thesis Faculti of mining, geology and Petroleum engineering, University of Zagreb

Margat J (1968) Vulnérabilitié des nappes d'eau souterraine à la pollution. BRGM Publication 68 SGL 198 HYD, Orléans

Ravbar N, Goldscheider N (2007) Proposed methodology of vulnerability and contamination risk mapping for the protection of karst aquifers in Slovenia. Acta Carsolog 36(3):397–411

Vías JM, Andreo B, Perles MJ, Carrasco F, Vadillo I, Jiménez P (2006) Proposed method for groundwater vulnerability mapping in carbonate (karstic) aquifers: the COP method. Hydrogeol J 14:912–925

Vrba J, Zaporozec A (eds) (1994) Guidebook on Mapping groundwater vulnerability. International Contribution to Hydrogeology (IAH), Hannover, p 131

Zwahlen F (2004) (ed) Vulnerability and risk mapping for the protection of carbonate (karst) aquifers. European Commission, Directorate—General for Research, European research area: structural aspects—COST, p 620, 279

Extension of DRASTIC Approach for Dynamic Vulnerability Assessment in Fissured Area: Application to the Angad Aquifer (Morocco)

M. Amharref, R. Bouchnan and A.-S. Bernoussi

Abstract In this study, we consider the dynamical aspect of groundwater vulnerability in fissured medium. This aquifer is considered as intermediary between porous and karst media. For these two media, there are several vulnerability assessment methods as DRASTIC for porous media and EPIK for karst ones. For fissured media, we used in this study the F-DRASTIC method adapted by introducing a new parameter, F, reflecting the fissures effects on the vulnerability. The dynamical aspect of vulnerability is proofed through two vulnerability maps realized, for the Angad region, Morocco, for two different years. These years are chosen according to their significant level of regional climate variations. These maps were made using a geographic information system, GIS. The results analysis has shown, first, the dynamical aspect of vulnerability, but also the effect of the fissure parameter on vulnerability and its variations.

Keywords Groundwater pollution · Fissured aquifer · Fracturing index · Vulnerability dynamic

1 Introduction

Groundwater vulnerability to pollution has firstly been defined, in relation to the unsaturated zone characteristics (Margat 1968). Later, some authors have proposed the integration of the saturated zone characteristics to reflect assimilative groundwater phenomena (Vrba and Zaporozec 1994; Gogu and Dassargues 2000;

M. Amharref (✉) · R. Bouchnan · A.-S. Bernoussi
Equipe: Géoinformation et Aménagement Du Territoire, Faculté Des Sciences et Techniques, Bp. 416, Agadir, Morocco
e-mail: amharrefm@yahoo.fr

R. Bouchnan
e-mail: bouchnanrachid@gmail.com

A.-S. Bernoussi
e-mail: a.bernoussi@fstt.ac.ma

B. Andreo et al. (eds.), *Hydrogeological and Environmental Investigations in Karst Systems*, Environmental Earth Sciences 1,
DOI 10.1007/978-3-642-17435-3_46

Aller et al. 1987; Maxe and Johansson 1998). Other authors introduce specific characteristics to pollutants, allowing the distinction between intrinsic and specific vulnerability (Gogu and Dassargues 2000). More than a hundred methods for assessing the vulnerability of groundwater pollution have been developed around the world. Vulnerability, defined as the degree to which a system undergoes harm due to the exposure to stress factors, is dynamic (Turner et al. 2003). However, few studies have been conducted to quantify changes in vulnerability under varying environmental conditions (Luers et al. 2003). In this work, we show, through an application on Angad aquifer, that the vulnerability is dynamic. We used the F- DRASTIC method (Bouchnan et al. 2013) for two different years.

2 Study Area

The studied area (393 km^2), is a part of the Angad aquifer. It is located in north-eastern Morocco near the Algerian-Moroccan border. It is bounded on the north by the chain of Beni Snassen. In the West, It is bounded by Megrez and Harraza reliefs and Bou Houria plain. In the South, it is limited by Jebel Hamra. To the East, it is in the hydrogeological continuity with Marnia plain in Algeria (Fig. 1).

This aquifer is constituted by a subsided basin framed by a graben system with subverticales faults (Lahrach 1999). The facies of this aquifer are characterized by the presence of a fissured basaltic within sedimentary rocks. The role of hydraulic relay played by these fissured rocks between the deep groundwater of Jebel Hamra and the Angad groundwater was confirmed (Mortier et al. 1967). These rocks are the best aquifers with permeability ranging between 10^{-4} and 10^{-3} m/s (Quang Trac et Simonot 1971).

This area is characterized by a semiarid continental climate with dry summer and temperate winter. The annual rainfall varies from year to year. Indeed, annual changes during the period 1967–2004 shows that amplitudes can reach 300 mm between two successive years (Fig. 2).

The application of the standardized precipitation index on the period 1935–2003 shows a drought condition before 1956 and normal-to-wet phase until 1970, followed by a drought period (Sebbar et al. 2011).

The groundwater level variations, depending on rainfall variations, showed a response with a delay of 6 to 11 months (30 control points). This observation allows to conclude that this aquifer recharge is not only done directly by the rain that fell on this plain (Quang Trac et Simonot 1971).

3 F-DRASTIC Method: Principle

The F-DRASTIC method, adapted for fractured aquifers, is based on the addition of a fracturing index, Fr, as eighth parameter. The vulnerability index is calculated according to Eq. (1):

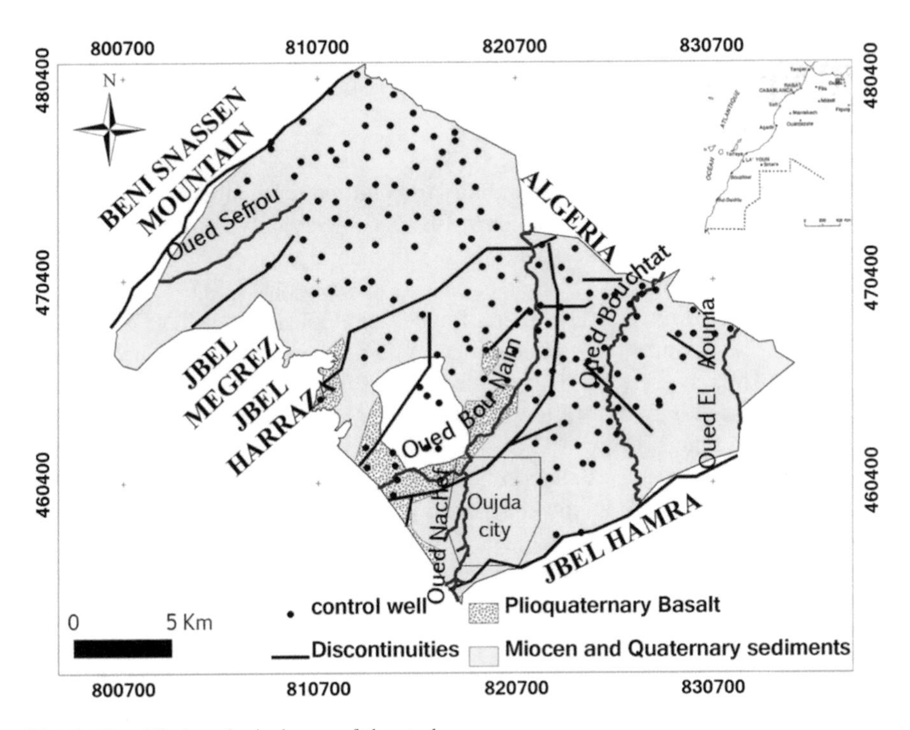

Fig. 1 Simplified geological map of the study area

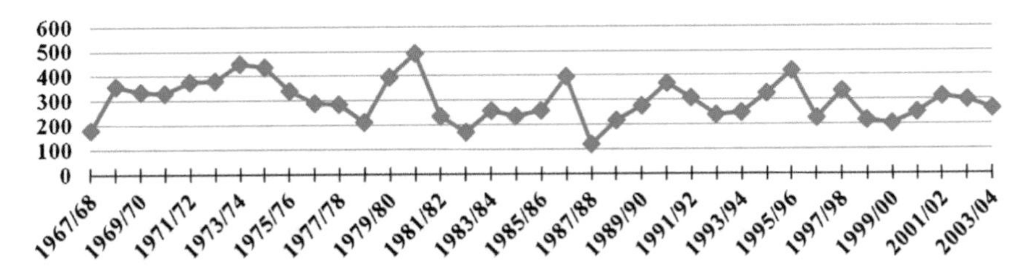

Fig. 2 Annual variation in rainfall (ABHM-Oujda, Hamri Station—No. 5921)

$$IV = ID + (Fr \times 5) \tag{1}$$

IV: vulnerability index; ID: DRASTIC index; Fr: fracturing index; 5: weight.

The fracturing index rates vary between 0, for nonfractured zones, and 10 for highly fractured zones.

The fracturing index "F" is evaluated according to the geometrical factors of the discontinuities and fissured rocks in the unsaturated zone. It is based on the relation between the thickness and depth of fissured rocks in the unsaturated zone, the arrangement of the fissures, the discontinuities density and the vertical hydraulic conductivity of the vadose zone (Bouchnan et al. 2013).

4 Dynamic Vulnerability: Application on Angad Aquifer

4.1 Problem Statement

The mapping of groundwater vulnerability to pollution risks, initially conceived as a means of representing a general condition, can convey only a limited amount of static, numerical information on the factors in natural protection of groundwater against potential pollution (Margat et Suais-Parascandola 1987). However, this idea was discussed by several authors. Ravbar and Goldscheider have considered the strong groundwater table fluctuations in response to precipitation events or snowmelt. They suggest the introduction of a subfactor for the frequency and duration of the activity of swallow holes (Ravbar and Goldscheider 2006, 2007).

In this reflection, F-DRASTIC parameters are variables at different timescales. The problem is to know if the variation of these parameters has an effect on vulnerability scores. Indeed, the changes in parameters related to the lithology of saturated and vadose zones, the fracturing index and hydraulic conductivity depend on groundwater level fluctuations and vertical heterogeneity of aquifers. Therefore, vulnerability defined on the basis of these parameters should be considered as a dynamic character.

4.2 Static Parameters Mapping

The soil map (S) was established on the basis of the study of nature, thickness, and porosity of the soil (DPA-Oujda 1989). This zone is characterized by the dominance of carbonate and isohumic soils occupying 40 and 36 % of this area. The fersiallitic and poor soils cover 20 %. The urban area, occupying 4 % is approached as the nonaggregated clay described in the DRASTIC method.

The topography map (T) was generated based on the DTM (Digital Terrain Model) of the Angad plain. The slope of this area varies between 0 and 24.3 %. The low slopes of less than 5° occupy the major part of this zone with 94.4 %.

4.3 Dynamic Parameters Mapping

For the five time-dependent parameters, we studied the variations of each parameter between the two selected years: 1995 and 2004.

The recharge (R) results from: effective infiltration, river infiltration, irrigation returns and losses of sewerage and drinking water systems. Among these elements, the effective infiltration is highly variable from one year to another.

The variation in other elements between these 2 years is expected to be very low. The standard classification of recharge is similar for both years. It does not

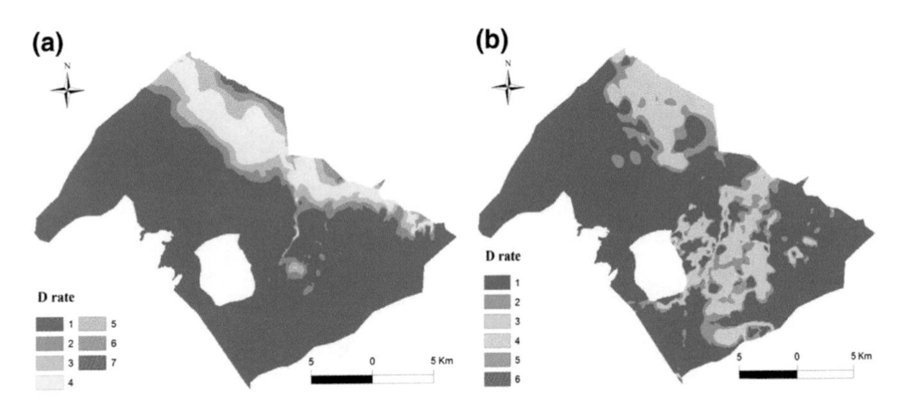

Fig. 3 Depth rates maps **a** for 1995 and **b** for 2004

allow the visualization of changes in the recharge values. The high recharge values are localized in Bou Naim river and urban area.

Changes in the two parameters related to Infiltration (I) and Aquifer (A) depend on annual fluctuations in groundwater level and vertical facies heterogeneity. The examination of stratigraphic drilling logs shows that the dominant facies of the saturated and unsaturated zones for both years are invariable.

The depth maps (D) of the 2 years show an average variation of groundwater level around 5.7 m, between 1995 and 2004 (Fig. 3). The shallow depths are generally located in the area bounded by the Bouchtat and Bou Naim rivers in northeast of Oujda city and in the northeast of this plain.

The variations in depth rates concern 37.8 % of this area between these 2 years. These variations, with different magnitudes, mainly concern the northeast of the plain and the zone bounded by Bou Naim and Bouchtat rivers in the northeast of Oujda city.

For the hydraulic conductivity (C), the corresponding annual values are derived according to the relation (2) of the equivalent hydraulic conductivity in stratified soil to horizontal flow (Fig. 4).

$$K_e = \Sigma K_i \times L_i / \Sigma L_i \tag{2}$$

Ke: equivalent hydraulic conductivity; Ki and Li: hydraulic conductivity and thickness of the layer i.

Changes in permeability between the 2 years concern primarily, the south and central part of the study area. They cover 21.6 % of this area.

The fracturing index (F) depend on the discontinuities density and the fissured rocks depth, that are static, and the fissured rocks thickness in the vadose zone, which is dynamic depending on groundwater level fluctuations (Fig. 5). The variations concern 9.6 % of the study area. They cover the central part around the nonaquifer zone characterized by the presence of fissured basalt.

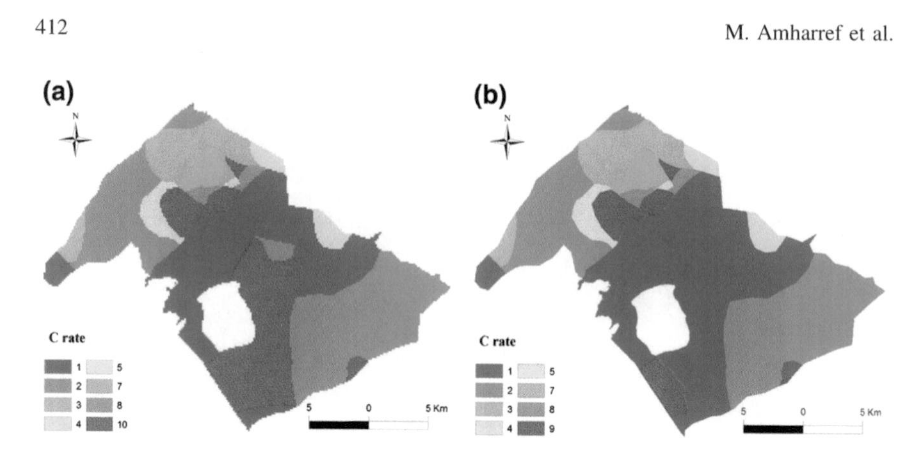

Fig. 4 Hydraulic conductivity rates maps **a** for 1995 and **b** for 2004

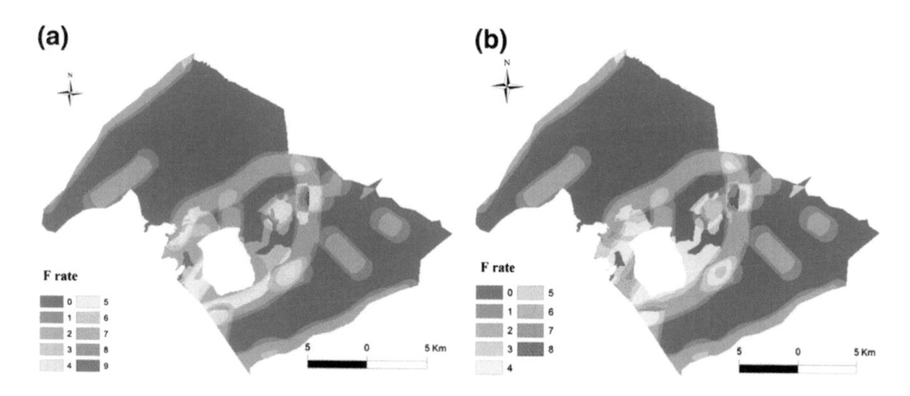

Fig. 5 Fracturing index maps **a** for 1995 and **b** for 2004

4.4 F-DRASTIC Vulnerability Maps and Results Interpretations

The F-DRASTIC vulnerability maps for 1995 and 2004 years show a great variation of the spatial distribution of the vulnerability degrees (Fig. 6). These variations concern 55.6 % of the studied area for the vulnerability index and 29.7 % for the vulnerability degree. This observation leads the dynamic aspect of vulnerability.

The map removal sensitivity analysis was performed on the vulnerability variations according to the dynamic parameters variations between the two study years.

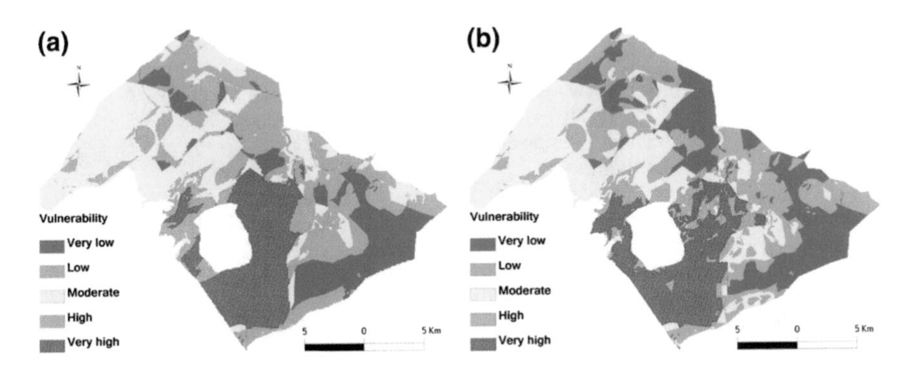

Fig. 6 F-DRASTIC vulnerability maps; **a** for 1995 **b** for 2004

5 Conclusion

In this study, we have shown, through the application of the F-DRASTIC adapted method on the Angad aquifer, that the aquifer vulnerability is dynamic. It varies in time according to changes of some dynamic parameters. Indeed, the two obtained maps for two different years are different: an important variation of the spatial distribution of the vulnerability degrees is observed. The vulnerability variations, in the considered area, are very sensitive to the variation in hydrogeological conditions, related to the depth, hydraulic conductivity, and the fracturing index.

Therefore, the vulnerability maps allow only average state estimation of aquifers, and cannot be considered as guidance tools for the land use planning. To overcome this problem, risk maps have to be considered.

Acknowledgments We thank the Moulouya River Basin Agency, the National Meteorological Station of Angad, and the Agriculture Delegation of Oujda (Morocco) for their fruitful collaboration.

References

Aller L, Bennet T, Lehr JH (1987) DRASTIC: a standardized system for evaluating groundwater pollution potential using hydrogeological settings. USEPA/600/2-87-036, Oklahoma

Bouchnan R, Amharref M, Samed Bernoussi A, Chanigui M (2013) Approche F-DRASTIC pour l'évaluation de la vulnérabilité des aquifères fracturés à la pollution: application sur l'aquifère d'Angad (Maroc). In : Proceedings of the 4th international Congress "Water, Waste & Environment (EDE4)", Morocco, Agadir

DPA-Oujda (1989) Etablissement d'une étude pédologique de reconnaissance au 1/100.000 ème en vue de la mise en valeur agricole dans la zone du couloir Taourirt-Oujda, Rapport Général. Ministére de l'Agriculture et de la Reforme Agraire, Maroc, pp 163

Gogu RC, Dassargues A (2000) Current trends and future challenges in groundwater vulnerability assessment using overlay and index methods. Environ Geol 39:549–559

Lahrach A (1999). Caractérisation du réservoir Liasique profond du Maroc oriental et étude hydrogéologique, modélisation et pollution de la nappe phréatique des Angads. Thèse de doctorat, Sciences de la Terre, Hydrogéologie-Géothermie, Université Sidi Mohammed Ben Abdellah - Faculté des Sciences et Techniques. Maroc, Fès-Saiss

Luers AL, Lobell DB, Sklar LS et al (2003) A method for quantifying vulnerability, applied to the agricultural system of the Yaqui Valley. Mexico Glob Environ Chang 13:255–267

Margat J (1968) Vulnérabilité des nappes d'eau souterraine à la pollution. Base de la cartographie. Rapport BRGM 68 SGC 198 Hyd, BRGM, Orléans, France

Margat J et Suais-Parascandola MF (1987) Mapping the vulnerability of groundwater to pollution. Some lessens from experience in France. In: Duijvenbooden W, Waegeningh HG (eds) Vulnerability of soil and groundwater to pollutants: international conference Noordwijk aan Zee, The Netherlands, March 30-April 3, 1987. Proceedings and information/TNO committee on hydrological research. The Hague: TNO, 1987

Maxe L, Johansson PO (1998) Assessing groundwater vulnerability using travel time and specific surface area indicators. Hydol J 6:441–449

Mortier F, Quang Trac N, Sadek M (1967). Hydrogeology of the volcanic rock formations of North-East Morocco in Hydrology of fractured rocks. In: Proceedings of the Dubrovnik Symposium, 1965. IASH-UNESCO. Ceuterick, Louvain. Belgium, pp 327–333

Quang Trac N, Simonot M (1971) Le couloir Taourirt-Oujda in Ressource en eau du Maroc. Tom 1:279–287

Ravbar N, Goldscheider N (2006). Integrating temporal hydrologic variations into karst groundwater vulnerability mapping—examples from Slovenia. In: Goldscheider N (ed) Proceedings of the 8th conference on limestone hydrogeology. Université de Franche-Comté Besançon, Université de Franche-Comté, Neuchâtel, Switzerland, 21–23 Sept 2006, pp 229–233

Ravbar N, Goldscheider N (2007) Proposed methodology of vulnerability and contamination risk mapping for the protection of karst aquifers in Slovenia. Acta Carsologica 36(3):461–475

Sebbar A, Badri W, Fougrach H et al (2011) Etude de la variabilité du régime pluviométrique au Maroc septentrional (1935–2004). Secheresse 22:139–148. doi:10.1684/sec.2011.0313

Turner BL, Kasperson RE, Matson PA et al (2003) A framework for vulnerability analysis in sustainability science. Proc Natl Acad Sci USA 100:8074–8079

Vrba J, Zaporozec A (1994) Guidebook on mapping groundwater vulnerability, vol 16. IAH, Hannover

Validation of Vulnerability Assessment Using Time Series Analysis—the Case of the Korentan Spring, SW Slovenia

G. Kovačič and N. Ravbar

Abstract In the shallow Orehek karst aquifer in southwest Slovenia, the use of hydrological data analysis for the purpose of validating the assessment of water source vulnerability was tested. The appropriate criteria for the aquifer and groundwater flow characterization were identified and major drawbacks highlighted. Results of water budget calculations were used to determine the extent of the catchment of the Korentan spring, which is the main outflow from the aquifer. The vulnerability assessment was verified by autocorrelation and cross-correlation analyses of available daily hydrological time series data. The small variability of the Korentan spring water temperature and electric conductivity time series points to the dominance of autogenic recharge and that sinking streams in the catchment contribute to the spring to a minor degree. The analysis indicated relatively small storage capacity of the aquifer and its high degree of karstification and rapid groundwater flow. The results justify the small proportion of highly vulnerable areas and lower vulnerability for marginal parts of the aquifer. Time series analyses proved to be time- and cost-effective, but have limited applicability for vulnerability validation purposes, as they do not provide direct and clear spatially resolved information on the vulnerability of the catchment.

G. Kovačič (✉)
Faculty of Humanities, University of Primorska, Titov trg 5,
6000 Koper, Slovenia
e-mail: gregor.kovacic@fhs.upr.si

G. Kovačič
Science and Research Centre of Koper, University of Primorska,
Garibaldijeva 1, 6000 Koper, Slovenia

N. Ravbar
Karst Research Institute ZRC SAZU, Titov trg 2, 6230 Postojna, Slovenia
e-mail: natasa.ravbar@zrc-sazu.si

N. Ravbar
Urban Planning Institute of the Republic of Slovenia, Trnovski pristan 2,
1127 Ljubljana, Slovenia

B. Andreo et al. (eds.), *Hydrogeological and Environmental Investigations in Karst Systems*, Environmental Earth Sciences 1,
DOI 10.1007/978-3-642-17435-3_47

1 Introduction

The assessment and mapping of groundwater vulnerability to contamination are increasingly used for water sources protection and management, especially in karst environments (Vrba and Zaporozec 1994; Zwahlen 2004; Foster et al. 2013). Therefore, many assessment methods have been developed so far that mainly differ in the selection and evaluation of different parameters. Their application and a comparison of results showed that the methods are liable to subjectivity and unreliability in the assessment procedure (Gogu et al. 2003; Neukum and Hötzl 2007; Ravbar and Goldscheider 2009). The results are often influenced by various factors such as the availability and quality of data and its interpretation.

To avoid bias in research and to ensure a reliable interpretation of vulnerability indices, the validation of vulnerability maps should be done by default. Until now, validation unfortunately has not become standard practice and there is no commonly accepted methodology. Tracing techniques using artificial tracers are the most straightforward and most commonly used validation method (Goldscheider et al. 2001; Perrin et al. 2004; Andreo et al. 2006; Ravbar and Goldscheider 2007; Neukum et al. 2008). However, cost-effective artificial tracing can most often only be applied in limited areas or over a small surface.

The present study tested the usefulness of simultaneous auto- and cross-correlation analysis of daily hydrological data sets for vulnerability validation purposes. The appropriate criteria were identified to evaluate the resulting spring vulnerability map. The case study was done for the Orehek karst aquifer in southwest Slovenia.

2 Description of the Test Site

The Orehek aquifer is a well-defined shallow karst aquifer that covers an area of around 9 km^2. The aquifer consists of an anticline of Cretaceous and Palaeocene limestone (Fig. 1) that is partially thrust over in the southwest by the Eocene flysch that surrounds it (Gospodarič et al. 1970; Petrič and Šebela 2004; Kovačič and Petrič 2007). The carbonate rocks are well karstified. In the north, the surface is covered with a thin layer of rendzina soil, and in the southern part with brown carbonate soil of various depths. The area is influenced by a subcontinental climate with an average annual precipitation of about 1,670 mm and an average annual runoff of about 900 mm (1971–2000).

The aquifer is recharged by several small intermittent sinking streams on the south (e.g., Čermelice, Orehovške ponikve). Two intermittent springs, Poliček and Mrzla jama, and the permanent Korentan spring drain the aquifer on its north and northeast margin. Tracer test results proved the connection of the southern part of the aquifer and the Orehovške ponikve sinking stream with the Poliček spring during high water periods. During low water periods, the underground waters of the southern part of the aquifer flow toward the Pivka River, situated further to the east (not shown in Fig. 1) (Gospodarič et al. 1970). The second tracer test done during a

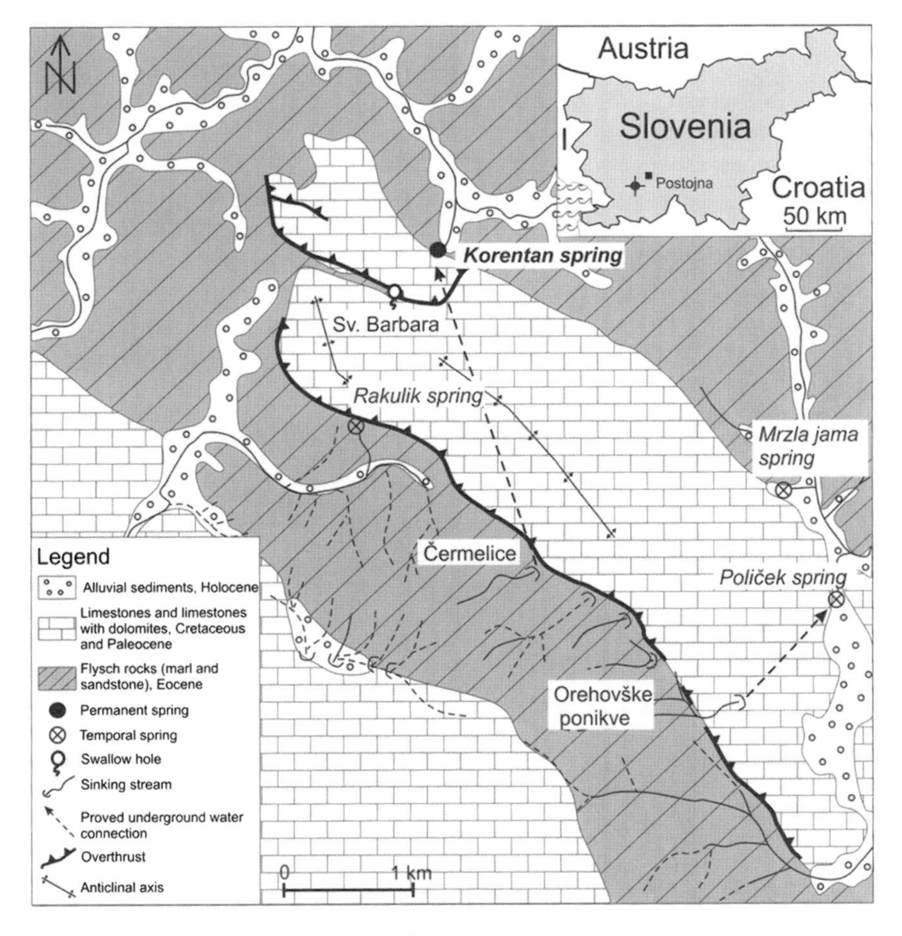

Fig. 1 Hydrogeological setting of the studied area

medium water period proved the connection of the Čermelice sinking stream with the Korentan spring. The mean linear groundwater flow velocity was estimated to be 25 m/h and the tracer recovery at 71 % (Schulte 1994).

The Korentan spring, the main outflow from the Orehek aquifer, is a typical karst spring characterized by rapid, strong responses to precipitation events. The spring's discharge ranges from a few l/s to about 3 m^3/s with an average of 0.2 m^3/s.

3 Methodology

The analysis is based on the existing intrinsic vulnerability map using the "Slovene Approach" (Ravbar and Goldscheider 2007). The evaluation procedure and results are presented in Ravbar et al. (2013).

A water budget calculation for the Orehek aquifer was carried out to determine more precisely the catchment area of the Korentan spring. Mean annual precipitation and evapotranspiration data (1971–2000) provided by the Environment Agency of Slovenia (EARS 2013) was used.

An autocorrelation analysis of spring discharge (Q), temperature (T), and electric conductivity (EC) daily time series and a cross-correlation analysis of daily precipitation P–Q, P–EC, and P–T data were performed to determine the behavior of the aquifer, its reaction to precipitation events, and its storage capabilities.

For performing the autocorrelation and cross-correlation analyses, the daily time series (Q, EC, and T) of the Korentan spring in the 2004 hydrological year (September 23, 2003, to August 31, 2004; duration 344 days) were used. Daily precipitation data from the Postojna meteorological station for the same period (EARS 2013) was used as relevant for the area studied.

Using the results of the time series analysis, the adequacy of the vulnerability class distribution of the resulting vulnerability map was examined and verified.

4 Results and Discussion

On the vulnerability map (Fig. 2), highly vulnerable areas occupy 1.6 % of the studied area and are attributed to caves, excavation sites, roads in the spring's catchment area, and to the sinking Čermelice stream and swallow holes (e.g. Sv. Barbara) and their surroundings. Moderate vulnerability is assigned to 54.2 % of the studied area, i.e., to the major part of the catchment and to barren areas, caves, sinking streams, and swallow holes within the aquifer but outside the Korentan catchment area. Areas of low vulnerability extend over 44.2 % of the area and cover dolines in the Korentan catchment area and majority of the Orehek aquifer outside the Korentan catchment area.

The vulnerability classes can be converted into protection zones. Areas of high vulnerability can be classified as Protection Zone I, areas of moderate vulnerability as Protection Zone II (inner protection zone), and areas of low vulnerability as Protection Zone III (outer protection zone).

The validation of obtained results should follow the main concepts of vulnerability (Zwahlen 2004) and be done independently of the map making processes. The following criteria are the most relevant in validating vulnerability indices: (i) travel time of a possible contaminant; (ii) its peak concentration, and (iii) duration of contamination. These indices can be most clearly determined by performing artificial tracer tests.

This approach has three minor drawbacks: (i) tracer type and mass, injection points and mode are subject to human choice, control, and error; (ii) tracer detection depends on its detection limit (which affects the calculation of travel time and duration of appearance); (iii) cost-conscious artificial tracing is most often hampered by a limited number of injection and sampling points.

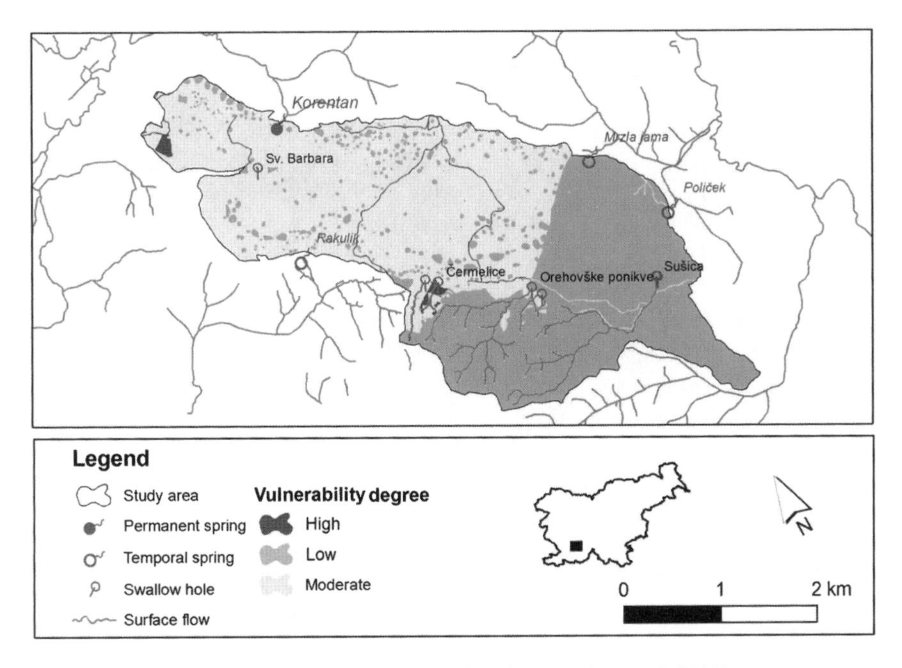

Fig. 2 Korentan spring vulnerability map (modified from Ravbar et al. 2013)

Statistical analysis can be used as an alternative or a complementary technique to support the credibility of vulnerability maps. The present study used a time series analysis of daily hydrological data sets.

The analysis of a spring's hydrological data provides a better understanding of a karst system's response to recharge and indirectly provides information regarding the structure and functioning of karst aquifers (Mangin 1984; Box et al. 1994). The autocorrelation function quantifies the memory effect of the system, which gives indirect information on the storage capacity and the degree of karstification of the system. Generally, a low memory effect is often related to a small storage capacity and high karstification where rapid infiltration and fast flow in conduits are the dominant conditions (Larocque et al. 1998; Panagopoulos and Lambrakis 2006). Such types of aquifers are characterized by short travel times, which also minimize the degradation and dilution of a possible contaminant.

Cross-correlation functions imply the transformation of input signals (precipitation or concentrated infiltration via ponors) to output signals (karst spring, well) and indicate the degree of karstification of a karst system. The delay, which is the time lag between lag 0 and the lag of the maximum value of the cross-correlation coefficient ($r_{xy}(k)$), gives an estimation of the pressure pulse transfer times through the aquifer (Panagopoulos and Lambrakis 2006).

Low memory effect and short transit times indicate high groundwater vulnerability. Higher storage capacities and longer transit times mean a lower degree of vulnerability as the possible contaminant arrival is delayed and its concentration significantly reduced or the contaminant does not arrive at all.

In this study, the water budget analysis for the 2004 hydrological year corresponds very well to the average values of the 1971–2000 reference period. The results suggest that the catchment area of the Korentan spring encompasses approximately 6 km^2. This estimate supports the tracer test results that the southern part of the aquifer is not drained by or may only marginally contribute to the Korentan spring. The lower degrees of vulnerability assigned to this part of the aquifer are thus justified.

The results of daily time series analysis typically show very low memory effect for the P (1 day) and for the Korentan spring Q (3 days; Fig. 3). The storage capacity of the Orehek aquifer is therefore small, meaning that the spring reacts instantly to precipitation events in the catchment, which is characteristic of small and well-karstified karst systems (Kovačič 2010). The fast reaction of the spring's discharge to precipitation events is also indicated by the shape of the P–Q cross-correlation function (Fig. 4; $r_{xy}(0) = 0.70$), where the reaction of the spring's discharge to precipitation events becomes statistically insignificant after only 2 days ($r_{xy} = 0.17$). Similar results, high values of $r_{xy}(0)$ ranging from 0.54 to 0.74 and their rapid drop below 0, were determined for several small and well-karstified springs in the nearby Unica River basin (Kovačič 2010) and also for the well-karstified aquifer of the Vipava karst spring ($r_{xy}(0) = 0.8$) north of the studied area (Jemcov and Petrič 2009).

The small storage capacity of the Orehek aquifer indicates the rapid infiltration of precipitation, short retention time in a relatively shallow vadose zone, and the rapid flow of infiltrated water through fissures and conduits toward the Korentan spring. Generally, this indicates that the Orehek aquifer is shallow and well karstified, which is important for the estimation of the flow within the saturated zone. Importantly, the results of the daily P–Q cross-correlation analysis are consistent with the Korentan spring vulnerability map that generally indicates the moderate vulnerability of the Orehek aquifer.

According to Larocque et al. (1998) and Kovačič (2010), in certain hydrological conditions and in the case of binary karst systems, cross-correlation analyses of EC time series between swallow holes and springs could provide valuable information on the hydrogeological functioning of karst systems equal to that from tracer tests. This means that the time variability of dissolved particle concentration in water could be used for calculating apparent flow velocities of water between swallow holes and springs. The results could be used to verify the vulnerability indices of sinking rivers and swallow holes.

Unfortunately, such an analysis for the studied aquifer was not performed because the EC time series of the sinking streams were not available. However, some conclusions can be deduced on the basis of the Korentan spring EC and T data variability. The small variations of these values (range T = 2.7° C, range EC = 67 μS/cm) and the high memory effect of EC (53 days) and T (49 days) show that the autogenic recharge is of much greater importance for the spring than the concentrated recharge from the Čermelice sinking stream (Figs. 1 and 3). This

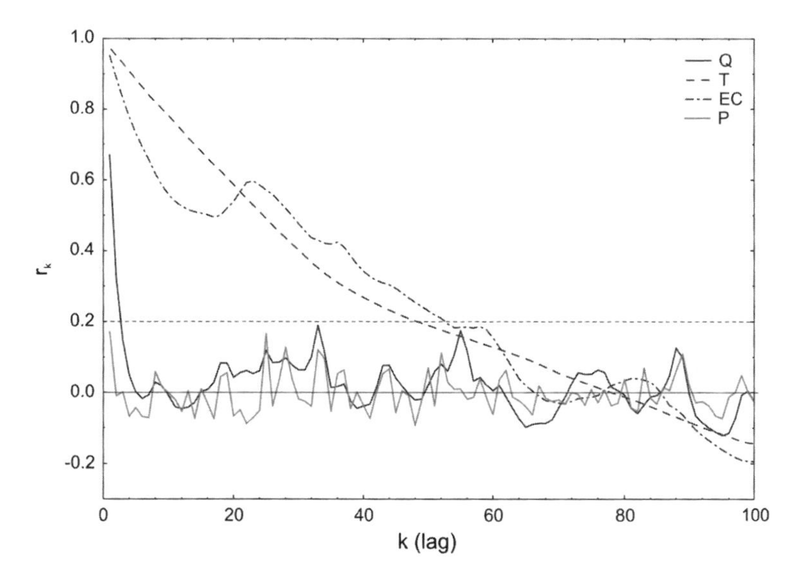

Fig. 3 Autocorrelation functions of daily discharge (Q), temperature (T), and electrical conductivity (EC) time series of the Korentan spring and precipitation (P) time series from the Postojna meteorological station in the 2004 hydrological year

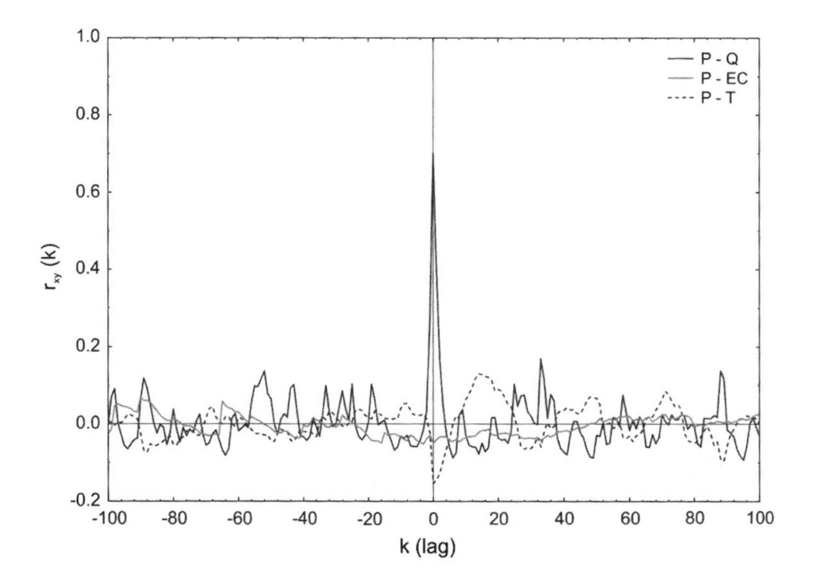

Fig. 4 Cross-correlation functions of daily precipitation (P) from the Postojna meteorological station as input and discharge (Q), temperature (T), and electrical conductivity (EC) of the Korentan spring as output in the 2004 hydrological year

explains the relatively very small portion (1.6 %) of highly vulnerable areas that are mainly attributed to permanent sinking streams and swallow holes. The low variability of T and EC values of the Korentan spring and also the results of P–EC and P–T cross-correlation analyses (values of r_{xy} functions around 0 indicate that the correlations are not statistically significant) show that the contribution of the sinking streams to its total recharge is minor (Fig. 4). For a comparison, the high variability of T and EC values (range T = 21.8° C, range EC = 370 μS/cm) of the nearby sinking Pivka River is pronounced in the Unica spring (range T = 10.5° C, range EC = 190 μS/cm), indicating a direct and rapid connection from the swallow hole to the spring reflecting the important contribution of the Pivka River to the Unica spring recharge (Kovačič 2010). The variability of T and EC values of the Unica spring is lower due to the important recharge from the other part of its catchment where autogenic recharge dominates.

5 Conclusion

The study performed showed the usefulness of simultaneous auto- and cross-correlation analyses of daily hydrological data sets for the purpose of validating the assessment of water source vulnerability. The following analyses proved to be especially useful for the aquifer and groundwater flow characterization: (i) autocorrelation analyses of the spring's Q, T, and EC; (ii) P–Q cross-correlation analysis; (iii) Q–Q, EC–EC, and T–T cross-correlation analyses between swallow holes and springs (not performed in this study, but in Kovačič 2010); and (iv) water budget calculation. These analyses reveal the degree of the aquifer's storage capacity, its karstification degree and functioning, and the reaction of the spring to precipitation events.

Time series analyses proved to be mainly cost- and time-effective. However, their applicability is limited because the results do not provide sufficient and straightforward information on the hydrogeological functioning of the karst system, nor clear information on the degree of vulnerability of the selected surface areas. As such, time series analyses should not be a stand-alone method for verifying vulnerability results but could be a complementary technique to support the validation of source vulnerability maps.

In the present study, the lower vulnerability of the outer part of the Korentan spring corresponds to the results of the autocorrelation analysis of EC and T time series of the spring that show sinking streams having very little influence on the spring's water and that the Korentan spring is mainly recharged by the diffuse infiltration of precipitation water. On the other hand, the moderate vulnerability of the major part of the Orehek aquifer corresponds to the low memory effect of the Korentan spring. The validation demonstrated that the vulnerability assessment is, at least for the studied test site, accurate.

Acknowledgments This manuscript is a contribution to UNESCO/IUGS IGCP 513 Project "Global Study of Karst Aquifers and Water Resources." Special thanks to Dr. Metka Petrič for the permission to use the Korentan spring data.

References

Andreo B, Goldscheider N, Vadillo I, Vías JM, Neukum C, Sinreich M, Jiménez P, Brechenmacher J, Carrasco F, Hötzl H, Perles JM, Zwahlen F (2006) Karst groundwater protection: First application of a Pan-European Approach to vulnerability, hazard and risk mapping in the Sierra de Líbar (Southern Spain). Sci Total Environ 357:54–73

Box GEP, Jenkins GM, Reinsel C (1994) Time series analysis: forecasting and control, 3rd edn. Prentice Hall, New Jersey

Environment Agency of the Republic of Slovenia (EARS) (2013) Mean monthly precipitation and evapotranspiration data in the reference period 1971–2000. The Archive of the Environment Agency of the Republic of Slovenia, Ljubljana

Foster S, Hirata R, Andreo B (2013) The aquifer pollution vulnerability concept: aid or impediment in promoting groundwater protection? Hydrogeol J 21:1389–1392

Gogu RC, Hallet V, Dassargues A (2003) Comparison of aquifer vulnerability assessment techniques. Application to the Neblon river basin (Belgium). Environ Geol 44(8):881–892

Goldscheider N, Hötzl H, Fries W, Jordan P (2001) Validation of a vulnerability map (EPIK) with tracer tests. In: Mudry J, Zwahlen F (eds) 7th Conference on Limestone hydrology and fissured media, Besançon, France, University de Franche-Comté, Besançon, 20th–22nd Sept 2001, pp 167–170

Gospodarič R, Habe F, Habič P (1970) Orehovški kras in izvir Korentana (The karst of Orehek and the source of the Korentan). Acta carsologica 5:95–108

Jemcov I, Petrič M (2009) Measured precipitation versus effective infiltration and their influence on the assessment of karst system based on results of the time series analysis. J Hydrol 379:304–314

Kovačič G (2010) Hydrogeological study of the Malenščica karst spring (SW Slovenia) by means of a time series analysis. Acta Carsologica 39(2):201–215

Kovačič G, Petrič M (2007) Karst aquifer intrinsic vulnerability mapping in the Orehek area (SW Slovenia) using the EPIK method. In: Witkowski AJ (Ed) Groundwater vulnerability assessment and mapping: selected papers from the groundwater vulnerability assessment and mapping international conference, Ustroń, Poland, 2004, Taylor and Francis, London

Larocque M, Mangin A, Razack M, Banton O (1998) Contribution of correlation and spectral analyses to the regional study of a large karst aquifer (Charente, France). J Hydrol 205:217–231

Mangin A (1984) Pour une meilleure connaissance des systèmes hydrologiques à partir des analyses corrélatoire et spectrale. J Hydrol 67:25–43

Neukum C, Hötzl H (2007) Standardization of vulnerability maps. Environ Geol 51:689–694

Neukum C, Hötzl H, Himmelsbach T (2008) Validation of vulnerability mapping methods by field investigations and numerical modelling. Hydrogeol J 16(4):641–658

Panagopoulos G, Lambrakis N (2006) The contribution of time series analysis to the study of the hydrodynamic characteristics of the karst systems: application on two typical karst aquifers of Greece (Trifilia, Almyros Crete). J Hydrol 329:368–376

Perrin J, Pochon A, Jeannin PY, Zwahlen F (2004) Vulnerability assessment in karstic areas: validation by field experiments. Environ Geol 46:237–245

Petrič M, Šebela S (2004) Vulnerability mapping in the recharge area of the Korentan spring. Slovenia Acta Carsologica 33(2):151–168

Ravbar N, Goldscheider N (2007) Proposed methodology of vulnerability and contamination risk mapping for the protection of karst aquifers in Slovenia. Acta Carsologica 36(3):461–475

Ravbar N, Goldscheider N (2009) Comparative application of four methods of groundwater vulnerability mapping in a Slovene karst catchment. Hydrogeol J 17:725–733

Ravbar N, Kovačič G, Marin AI (2013) Abandoned water resources as potential sources of drinking water—a proposal for management of the Korentan karst spring near Postojna. Acta Geographica Slovenica 53(2):295–316

Schulte U (1994) Geologische und hydrogeologische untersuchungen Im Karst Von Orehek (Slowenien). Diplomarbeit, Universität Karlsruhe, Deutschland

Vrba J, Zaporozec A (1994) Guidebook on mapping groundwater vulnerability, vol 16. IAH, International Contributions to Hydrogeology, Heise

Zwahlen F (ed) (2004) Vulnerability and risk mapping for the protection of carbonate (karst) aquifers. Final report of COST Action 620. European Commission, Directorate-General XII Science, Research and Development, Brussels

Safeguard Zones and Activities Permitted Cartography: Application in Carbonate Aquifers of Southern of Spain

A. Jiménez-Madrid, F. Carrasco and C. Martínez

Abstract Carbonate aquifers constitute a water reserve of essential importance for human supply. Accordingly, suitable protection measures should be established in order to meet the requirements of the Water Framework Directive (WFD). The objective of this paper is to discuss a methodology for defining protection zones for carbonate bodies containing groundwater intended for human consumption, these zones being determined by mapping the activities permitted within them. This approach constitutes an effective tool for land-use planning, enabling the appropriate location of human activities to ensure no adverse effects are produced on the quality of drinking water, in accordance with WFD criteria. The results obtained in a karst groundwater body in southern Spain show the percentage of land that must be protected to preserve the quality of water intended for human consumption, and its territorial distribution, demonstrating the value of this method in land-use planning.

1 Introduction

Groundwater constitutes a basic resource in Europe, as is evident in countries such as Austria, Germany, Italy and Denmark, where more than 70 % of the population's water supply comes from groundwater (Martinez Navarrete et al. 2008). In

A. Jiménez-Madrid (✉)
CRN Consultores, Ríos Rosas, 19, E-D, 3°A, 28003 Madrid, Spain
e-mail: ajimenez@crnconsultores.com

F. Carrasco
Department of Geology and Centre of Hydrogeology, University of Málaga (CEHIUMA), 29071 Málaga, Spain
e-mail: fcarrasco@uma.es

C. Martínez
Geological Survey of Spain, Madrid, Spain
e-mail: c.martinez@igme.es

B. Andreo et al. (eds.), *Hydrogeological and Environmental Investigations in Karst Systems*, Environmental Earth Sciences 1,
DOI 10.1007/978-3-642-17435-3_48

Europe, carbonate aquifers are especially important in terms of the supply of water for human consumption, as they extend beneath 35 % of the surface area and their resources make up 50 % of the total water available for supply in Europe (COST 65 1995).

The protection of groundwater was highlighted as a priority objective among European environmental policies with the entry into force in 2000 of the Water Framework Directive, WFD (EU 2000), and more specifically in 2006 with the Groundwater Directive on the protection of groundwater against pollution and deterioration (EU 2006).

According to the WFD, the existence of groundwater in sufficient quality and quantity for human consumption is an absolute necessity, and therefore appropriate protection systems for this resource must be created and implemented. Under Article 7.3 of the WFD, the definition of groundwater protection zones to meet the requirements of water for human consumption is an optional measure, but one that is strongly recommended, as it enables security measures to be centralised to limit the deterioration of groundwater quality and thus reduce the need for purification (Martínez-Navarrete et al. 2011).

Furthermore, as observed by Gómez Orea (2002), any model of water management will be insufficient if it fails to include predictions and measures for balanced, sustainable territorial development. Water is essential to land use, and so such a model must take into account the question of water management.

To achieve the goals set by the WFD and to integrate groundwater protection into land-use management, this paper considers the development of a method for mapping permitted activities, as a land-use instrument for direct application with which appropriate decisions can be taken regarding the introduction of new activities, the modification of existing ones or the location of new groundwater extraction points.

2 Study Area

The Sierra de Cañete (Subbetic zone) is located in the Western Mediterranean, in southern Spain (Fig. 1). This mountain range has a surface area of approximately 55 km^2 and receives a mean annual precipitation of over 1,000 mm.

The materials outcropping here are Jurassic limestones and dolomites that have been made permeable through fissures and karstification. The karst is poorly developed. The most representative exokarstic landform is that of limestone pavement, in Penibetic materials. There are very few areas of preferential infiltration (such as dolines and uvalas), and only occasionally small and shallow sinkholes are present.

Due to tectonic phenomena, the Sierra de Cañete is divided into several aquifers (compartments) (Fig. 1). These aquifers are recharged by rainwater infiltration. Groundwater flow occurs through fissures and conduits towards discharge springs that are located at different topographic heights according to aquifer divisions; a moderate karstic regime is apparent (Junta de Andalucía 2002).

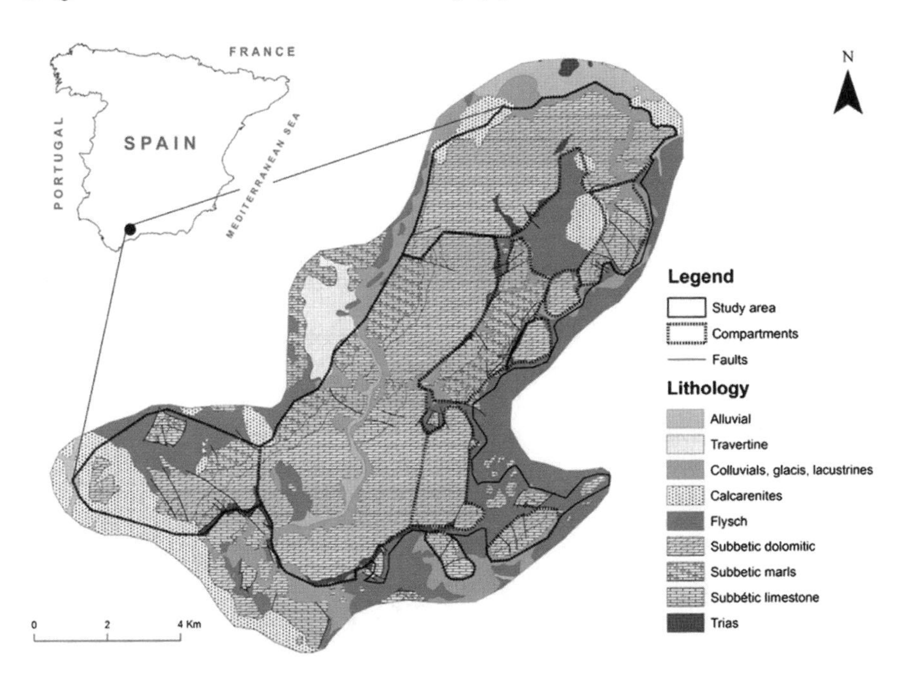

Fig. 1 Geographic and hydrogeologic setting of Sierra de Cañete

3 Methodology

According to the WFD (Article 7.3), groundwater protection zones (GPZ) are areas to be established, optionally, within which appropriate protective measures are applied to limit the deterioration of the quality of groundwater used for human consumption and to reduce the amount of purification required. This option is strongly recommended in view of the very large dimensions of groundwater bodies in many EU Member States.

Jiménez-Madrid et al. (2011) presented the GPZ method for determining such safeguard zones as a means of protection for carbonate groundwater bodies used for drinking water supply. In this method, the first step is to establish the risk of groundwater contamination (RI index) by combining the characterisation of external pressures (IP index) and the evaluation of the intrinsic vulnerability to contamination (DRISTPI index). The second step is to identify existing supply points and their inputs, i.e. the areas within the territory that provide the water used for human consumption. This goal is independent of other measures that may be required in other areas to comply with WFD requirements. Finally, the existing security perimeters are identified in the protection zones obtained.

Based in COST 620 European approach, IP index is proposed here to characterise the risk of each identified pressure, the IP factor (intensity of pressures). Its definition is based on the following premises: definition of the IP factor, removal of the Rf factor and substitution of the Qn factor. Once the existing pressures have been inventoried, they are classified according to five categories of pressure intensity by assigning a value ranging from 1 to 5, which are then integrated into the methodology for delineating safeguard zones. Overlapping pressures are characterised by the sum of their respective IP values (Jiménez-Madrid et al. 2011).

The DRISTPI index is a new method for assessing the intrinsic vulnerability in all types of aquifers. The aim is to develop a single method based on the original DRASTIC method characterised by a plurality of applications and functionality. As a fundamental requirement, there are two scenarios that need to be considered: scenario 1, which relates to materials that are highly karstified, and scenario 2, which extends over the rest of the area where there is a lower degree of karstification. It also considers the removal of parameters, the modification of ranges and weights and the incorporation of a new ''preferential infiltration'' parameter, which is specific for each scenario (Jiménez-Madrid et al. 2013).

As a part of this work and based on indications set out in COST Action 620, the Risk Index (RI) is defined to assess the risk of groundwater contamination, which is determined as the product of the pressure index and the vulnerability assessment index (Jiménez-Madrid et al. 2011). To integrate these two factors and achieve the proposed RI index, the results obtained have to be reclassified into an index that ranges between 1 (minimum) and 5 (maximum). The product of these two indices gives rise to five categories of risk: very low risk, low risk, moderate risk, high risk and very high risk (Table 2).

Once the zoning safeguards are established, the activities permitted are mapped, as a dynamic instrument to be applied directly in the study area in order to determine where pressures can be located such that no negative impact is produced on the quality of groundwater intended for human consumption.

In the present study, the mapping of permitted activities was conducted taking into account the risk of contamination with respect to the entire water body included in the protection zones. To do so, the intensity of the pressures was characterised according to the values of the IP index summarised in Table 1, reclassified to values from 1 (very low) to 5 (very high), following the criteria of Jiménez-Madrid (2012). Vulnerability to contamination as an intrinsic property of the medium is evaluated on a scale from 1 to 5 by the DRISTPI method (Jiménez-Madrid et al. 2011).

By combining these parameters, the risk of contamination can be determined by a double-entry matrix (Table 2) as follows:

$$\text{Existing IP} \times \text{DRISTPI} = \text{Existing RI}$$

As can be seen in Table 2 five kinds of groundwater contamination risk are proposed. The mapping of permitted activities is based on the following algorithm:

$$(\text{Existing IP} + \text{Allowed IP}) \times \text{DRISTPI} \leq \text{moderate RI}$$

Table 1 Pressures and their intensity. IP index (Jiménez-Madrid et al. 2011)

Pressure	Pressure intensity	
	Minimum	Maximum
Urban activities	35 waste water treatment plant	75 recycling centre
Industrial activities	25 quarry	100 nuclear waste
Agricultural activities	15 gardens	60 irrigation with wastewater
Livestock-farming activities	25 grazing	45 discharge of slurry

Table 2 Evaluation of contamination risk. RI index (Jiménez-Madrid et al. 2011)

RI INDEX		VULNERABILITY (DRISTPI INDEX)					CLASSES OF RISK
		1	2	3	4	5	
PRESSURE (IP index)	1	1	2	3	4	5	VERY LOW
	2	2	4	6	8	10	LOW
	3	3	6	9	12	15	MODERATE
	4	4	8	12	16	20	HIGH
	5	5	10	15	20	25	VERY HIGH

The aim is to allow in each zone only new activities whose pressure intensity (Table 1 when added to that of existing activities does not exceed the mean value (12) of the RI index, corresponding to a moderate risk of contamination.

By integrating all available information using spatial analysis tools from a geographic information system, and implementing the necessary algorithms, a map of permitted activities can be generated and used to manage the territory within each cell of a pre-defined space in order to protect the groundwater supplies used for human consumption and to perform appropriate land-use planning. Six classes of permitted activities within the safeguard zones can be distinguished:

- Class 0: Negative and 0 values. No further activities permitted.
- Class 1: Activities permitted provided the resulting IP index, when added to that of existing pressures, does not exceed 19.
- Class 2: Activities permitted provided the resulting IP index, when added to that of existing pressures, does not exceed 39.
- Class 3: Activities permitted provided the resulting IP index, when added to that of existing pressures, does not exceed 59.
- Class 4: Activities permitted provided the resulting IP index, when added to that of existing pressures, does not exceed 79.
- Class 5: All activities permitted provided the resulting IP index, when added to that of existing pressures, does not exceed 100.

The mapping of permitted activities was obtained from the results of the IP, DRISTPI and RI indices proposed by Jiménez-Madrid et al. (2011), although in fact this could be achieved using any of the existing indices and methodologies currently employed to assess the risks presented by environmental pressures and the intrinsic vulnerability to contamination, provided that their results are reclassified to values from 1 (very low) to 5 (very high).

To determine the IP index from the existing pressures and the activities permitted in a given area, the index for each concept must be added; in addition, the number of existing activities must be taken into consideration, as the incorporation of a particular activity in isolation, with the corresponding IP index, will not have the same effect as if ten activities of the same type were already present.

4 Results

Protection of drinking water in the study area, the Sierra de Cañete, requires 81.3 % of the area to be defined as GPZ, with different degrees of restriction. Figure 2 (Jiménez-Madrid et al. 2011) shows the mapping of permitted activities obtained for these safeguard zones, and the respective percentages of the total study area. As can be seen, the activities permitted are mainly those with an IP index classed as moderate or less (accounting for 45.9 % of the surface area).

As can be seen, in 2.1 % of the area no further activity can take place. This class 0 is located mainly in quarries, where there is preferential flow and current activities are contaminating. The zones where only activities with a very low pressure intensity (Class 1) are allowed account for 2.9 % of the surface area. These zones correspond to riverbeds, where infiltration phenomena occur, and to the northwestern part of the study area, where the thickness of the unsaturated zone is not very great and the aquifer is more vulnerable to a possible contamination event.

23.2 % of the territory is classified as Class 2, which allows the introduction of activities considered to present low or very low pressure intensity. This class of permitted activities is located in the carbonate outcrop where grazing currently takes place, and includes the karrenfields of Sierra del Padastro and Sierra del Padrastrillo. Activities with an IP classed as moderate or less IP (Class 3) are allowed in 45.9 % of the total study area, which constitutes the remaining limestone materials, where the presence of environmental pressure is at present nonexistent.

Class 4 of permitted activities, i.e. those with an IP not exceeding 79, can take place in 7.2 % of the territory, where the material outcropping is of low permeability and less vulnerable to contamination. Class 5, which allows the execution of activities with a pressure intensity exceeding 79, is not represented in the Sierra de Cañete.

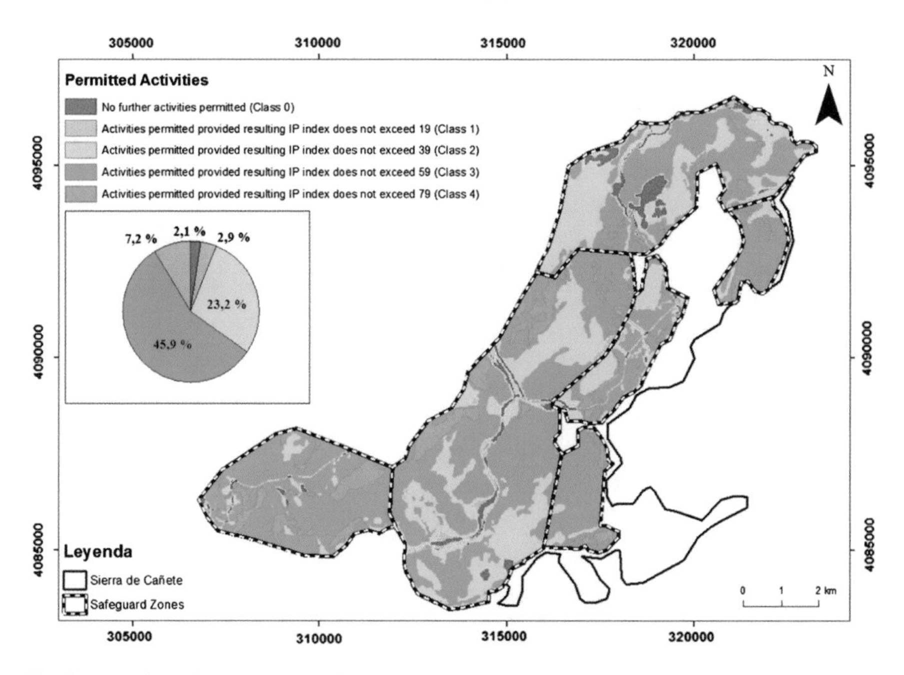

Fig. 2 Mapping of permitted activities in Sierra de Cañete (Malaga, Spain)

5 Conclusions

To fulfil their purpose, such protection zones must be incorporated into land-use planning. To this end, we propose a methodology to obtain a map of permitted activities. Such a map can be used by planners to select, in a flexible way, areas where pressure-creating activities may be allowed, and to determine the characteristics of such activities such that the WFD objective of preserving water used for human consumption continues to be met.

The mapping of permitted activities is based on the application of a double-entry matrix to assess the risk of groundwater contamination. This procedure involves calculating an index of intensity of environmental pressures (IP) in combination with an index of intrinsic vulnerability to pollution (DRISTPI) to characterise the risk of contamination.

Application of the proposed algorithms, using spatial analysis tools associated with a geographic information system, yielded five classes of permitted activities with varying degrees of restrictions, together with a class 0 corresponding to areas where no new activities may take place. Each of these five classes is illustrated by a list of permitted activities, based on the corresponding IP index.

The methodology proposed for mapping permitted activities was applied in the Sierra de Cañete. Preliminary results reflect a predominance of Class 2 territory, where activities with an IP index of moderate or less can be introduced, while only

5 % of the territory (Class 0 and Class 1) restricts new activities to those presenting very low pressure intensity, or allows no new activities at all.

Zoning to protect groundwater for human consumption, together with the use of permitted activities mapping as a preventive strategy, constitutes a useful approach to land-use planning, although this method requires further research to determine the economic impact that might be provoked by its implementation, especially as concerns changes in land use which in turn may have a social impact and produce territorial imbalances. These are factors that need to be taken into consideration and evaluated.

References

COST 65 (1995) Hydrogeological aspects of groundwater protection in karstic areas. Final report (COST action 65), European Commission, Brussels, Luxembourg

Gómez Orea D (ed) (2002) Ordenación territorial. Mundi Prensa y Agrícola Española, Madrid

Jiménez-Madrid A, Carrasco F, Martínez-Navarrete C (2011) Protection of groundwater intended for human consumption: a proposed methodology for defining safeguard zones. Environ Earth Sci 65(8):2391–2406

Jiménez-Madrid A, Carrasco F, Martínez C, Gogu RC (2013) DRISTPI, a new groundwater vulnerability mapping method for use in karstic and non-karstic aquifers. Q J Eng Geol Hydrogeol. doi:10.1144/qjegh2012-038

Jiménez-Madrid A (2012) Estudio metodológico para el establecimiento de zonas de salvaguarda en masas de agua en acuíferos carbonatados utilizadas para consumo humano. Aplicación de la directiva marco del agua. Serie: Tesis Doctorales, N° 20. Depósito Legal: MA-9814–2012. Publicaciones del Instituto Geológico y Minero de España. ISBN: 978-84-7840-872-6

Junta de Andalucía (2002) Estudio hidrogeológica de la Sierra de Cañete (Málaga). Consejería de Obras Públicas y Transportes, Sevilla

Martinez Navarrete C, Grima Olmedo J, Durán JJ, Gomez Gomez JD, Luque Espinar JA, De la Orden Gomez JA (2008) Groundwater protection in mediterranean countries after the European water framework directive. Environ Geol 54:537–549

Martínez-Navarrete C, Jiménez-Madrid A, Sánchez-Navarro I, Carrasco F, Moreno-Merino L (2011) Conceptual framework for protecting groundwater quality. Int J Water Resour Dev 27(1):227–243

European Union (2000) Water framework directive. Directive 2000/60/EC of the European parliament and of the council establishing a framework for the community action in the field of water policy. OJ L 327, 22.12.2000

European Union (2006) Groundwater directive 2006/118/EC of the European parliament and of the council of 12 December 2006 on the protection of groundwater against pollution and deterioration. OJ L 372, 27.12.2006

Assessment of Groundwater Contamination in Yucatan Peninsula (Mexico) by Geostatistical Analysis

R. Alcaraz, E. Graniel, A.F. Castro and I. Vadillo

Abstract The objective of this study was to assess the pollution of groundwater in Yucatan aquifer. The study area is centered on the city of Merida, the main city of the Yucatán state and its surroundings. This aquifer supplies groundwater to urban, agricultural, and industrial activities. Due to its high degree of karstification, it is a very vulnerable area because water and any substance within infiltrates rapidly and could affect the quality of groundwater. Therefore, a groundwater sampling campaign was done to study different pollutants, such as NO_3^-, Cl^-, fecal coliforms, and organophosphorus compounds. Data were treated with geostatistical tools and different maps were obtained by kriging. The main sources of pollution are located close to the city of Merida, in addition to isolated points of rural or industrial areas. Pollution plume moves toward North or Northwest area. NO_3^-, Cl^-, organophosphorus compounds, and fecal coliforms are an issue of concern for groundwater quality.

R. Alcaraz (✉) · A.F. Castro · I. Vadillo
Group of Hydrogeology, Faculty of Science, University of Malaga, 29071 Malaga, Spain
e-mail: ruben_alcaraz_navarro@hotmail.com

A.F. Castro
e-mail: antoniof.castro@juntadeandalucia.es

I. Vadillo
e-mail: Vadillo@uma.es

E. Graniel
Group of Hydraulics and Hydrology, Faculty of Engineering, University Autónom of Yucatán, 150 Cordemex, Mérida, Mexico
e-mail: graniel@uady.mx

A.F. Castro
Department of Wastes and Soil, Environmental Office of Junta de Andalucía, 29007 Málaga, Spain

B. Andreo et al. (eds.), *Hydrogeological and Environmental Investigations in Karst Systems*, Environmental Earth Sciences 1,
DOI 10.1007/978-3-642-17435-3_49

1 Introduction

Several cities in Yucatan Peninsula, such as Merida, Valladolid, or smaller urban areas, obtain its water supply from groundwater. The higher permeability due to karstification causes a significant potential for pollution (Alcocer and Escobar 1999). The objective of the study was the evaluation of the pollution in the area close to the city of Merida and between Merida and the sea border. In order to achieve an efficient management of groundwater supplies, it was necessary an evaluation of the polluting agent that was present; this evaluation allows us to detect the pollution focus that are affecting directly or indirectly to the aquifer. Nitrate, chloride, fecal coliforms, and organophosphorus compounds were analyzed for this project.

1.1 Site Description, Previous Studies, and Specific Objectives

The study is centered on Merida (Fig. 1), which is the biggest city in the Yucatan Peninsula. Different sources of contamination exist at the study area, as sewage systems, agriculture, and industries. The main industrial activity is located toward North. In the area predominates subhumid tropical climate with a rainfall close to 1,200 mm, a temperature of 26 °C and an ETP of 2,100 mm.

Geologically, the subsoil is formed by a sequence of calcareous sediments of marine origin from Tertiary age (Butterlin and Bonet 1960; Bonet and Butterlin 1962). The oldest rocks are located on the South and they are emerging in Sierrita de Ticul, these rocks are dolomitized silicified or recrystallized.

Groundwater goes from areas with higher rainfall located in the South toward the coast, where aquifer discharges to the sea (Graniel et al. 2003, 2009). The high hydraulic conductivity is a combination of primary rock permeability, fracturing, and the karstic conduits. The dissolution of carbonate rock occurs according to the content of calcium carbonate and the acidic water in karst process, where the heterogeneity of the system increases continuously promoting the storage and movement of groundwater (Villasuso et al. 2011).

Geostatistical methods are currently the most widely used for the treatment of spatial data, because of the quality of the results obtained and its operation (Chica-Olmo and Luque-Espinar 2003). They are based on the *Theory of Regionalized Variables* based on the spatial variation of regionalized variable, i.e., explicitly consider the spatial variability of the phenomenon studied through intrinsic random function called semivariogram (or most commonly, variogram). The variographic or structural analysis includes the estimation and interpretation of the experimental variogram, their adjustment to a theoretical model, and model validation. This requires combining knowledge of the variable studied and geostatistical abilities. The kriging is the geostatistical estimation method that provides the *most likely value* of the variable in a nonexperimental point. It is a linear,

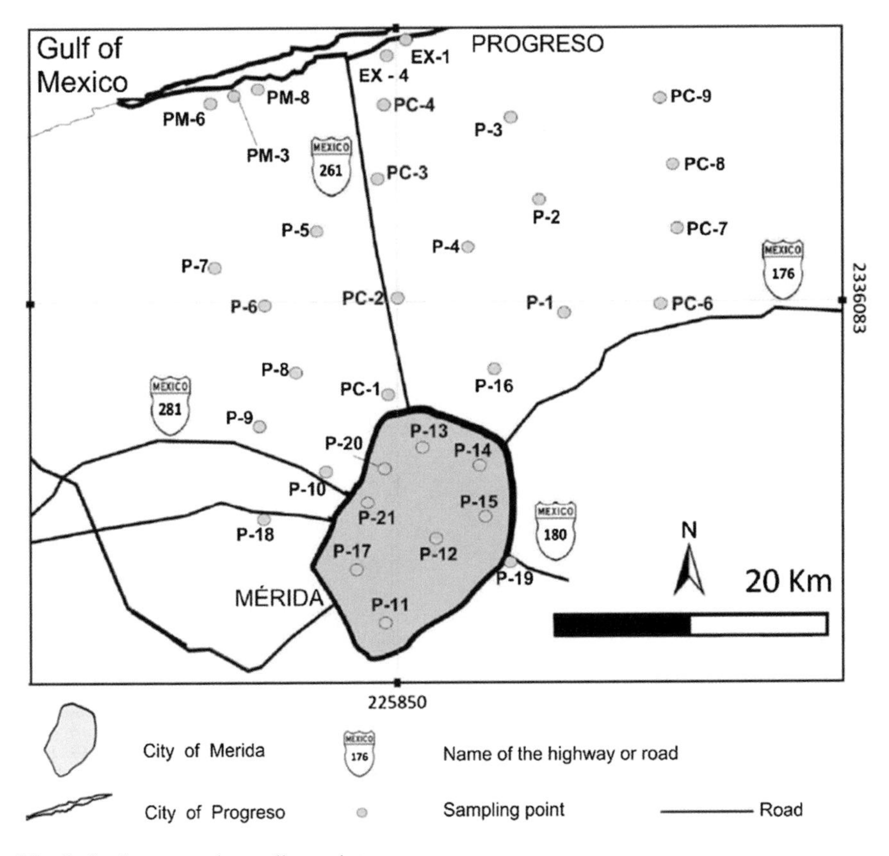

Fig. 1 Study area and sampling points

unbiased, and optimal estimator. There is a variety of methods each adapted to the various situations that occur in spatial estimation problems. In this article, the experimental semivariograms and fitted models (Fig. 2) and the corresponding mappings of the four variables listed (Figs. 3 and 4) were prepared using ordinary kriging, allowing it to be accompanied by a corresponding map of standard deviations. The support block used was 1 km^2.

2 Results and Discussion

2.1 Nitrates and Chloride

The analysis of nitrate shows a very stable variogram (Fig. 2), spherical type, and relatively low range. There is a correlation between data for distances up to 6 km. The highest concentration in NO_3^- is in the center of Merida (Fig. 3), this

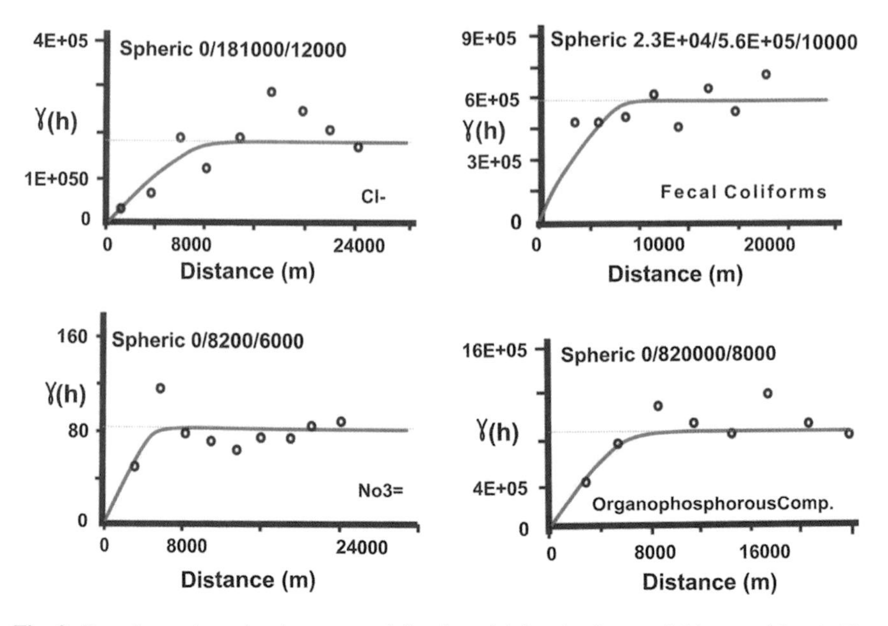

Fig. 2 Experimental semivariograms and fitted model for the four variables considered. The *gray line* indicates the variance

concentration is related with urban activities. It is observed a high concentration which moves in the same way as groundwater. The city with its urban activities produces this compound and the movement of groundwater displaces it, making up the geometrical form showed in Fig. 3. In addition to the concentration in the city, other points exists where the presence of NO_3^- is higher than in other places into the area, this situation is because there are some agricultural activities together with industrial activities.

Cl^- semivariogram reveals a spatial continuity without nugget effect and ranges up to 12 km (Fig. 2). Chlorides show a high spatial continuity, up to 12 km. Highest values coincide with rural areas between Merida and the North coast, where the values reach 2,000 mg/L.

2.2 *Fecal Coliforms and Organophosphorus Compounds*

The semivariogram for fecal coliforms (FC) is similar to the rest of variables range (10 km), but is the only case with a random component (low nugget effect). FC are concentrated in urban areas, such as Merida and in smaller villages (Fig. 4). This situation causes that groundwater has a high concentration of FC, being one of the most important pollutants. However, their distribution does not match those of NO_3^-.

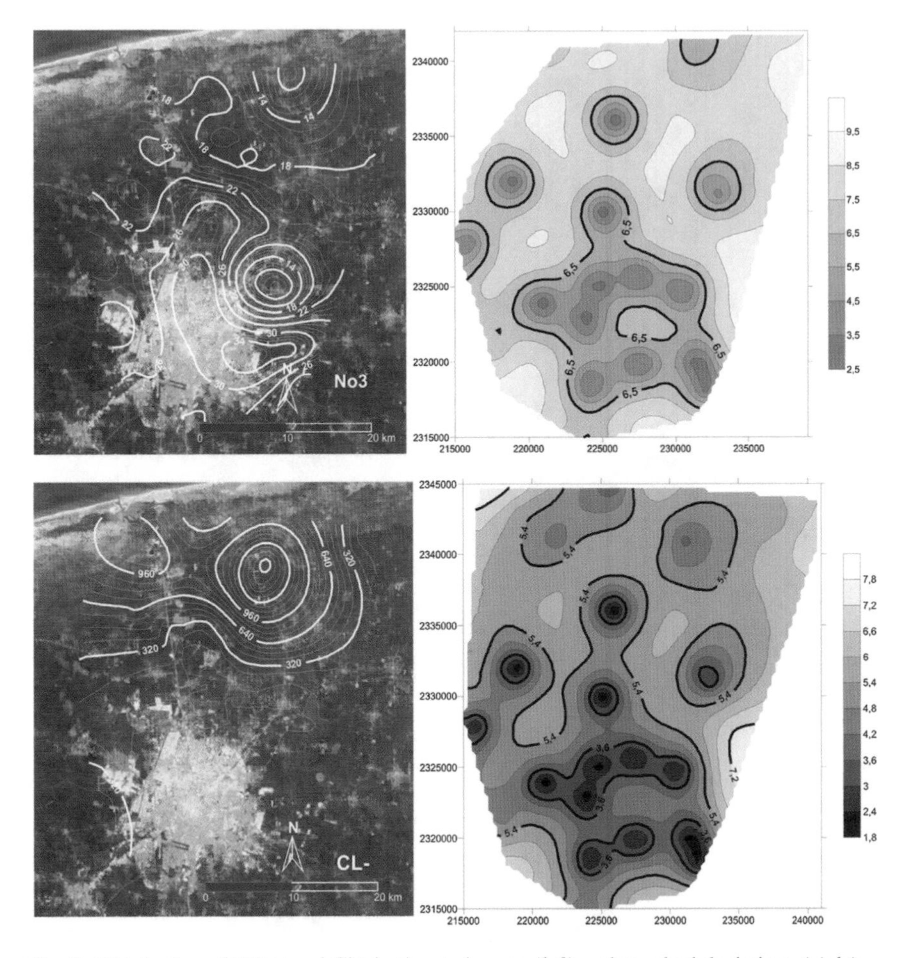

Fig. 3 Distribution of NO_3^- and Cl^- in the study area (*left*) and standard deviations (*right*)

Organophosphorus compounds are used for different functions like manufacturing of nitrocellulose and cellulose acetate, like ematicides (Thionazin) and agrochemicals (forate) or like insecticides (dimethoate, disulfoton, parathion, and Methlyparathion). Figure 2 shows that the semivariogram reaches 8 km scope, i.e., data are correlated for the same or lower values than this range distance, with no appreciable random component.

Map in Fig. 4 revealed a high concentration of organophosphorus compounds in the city of Merida but with some hot spots toward the north in several rural areas. In several samples, have been surpassed the maximum limits established by current regulations: NOM-127-SSA1-1994.

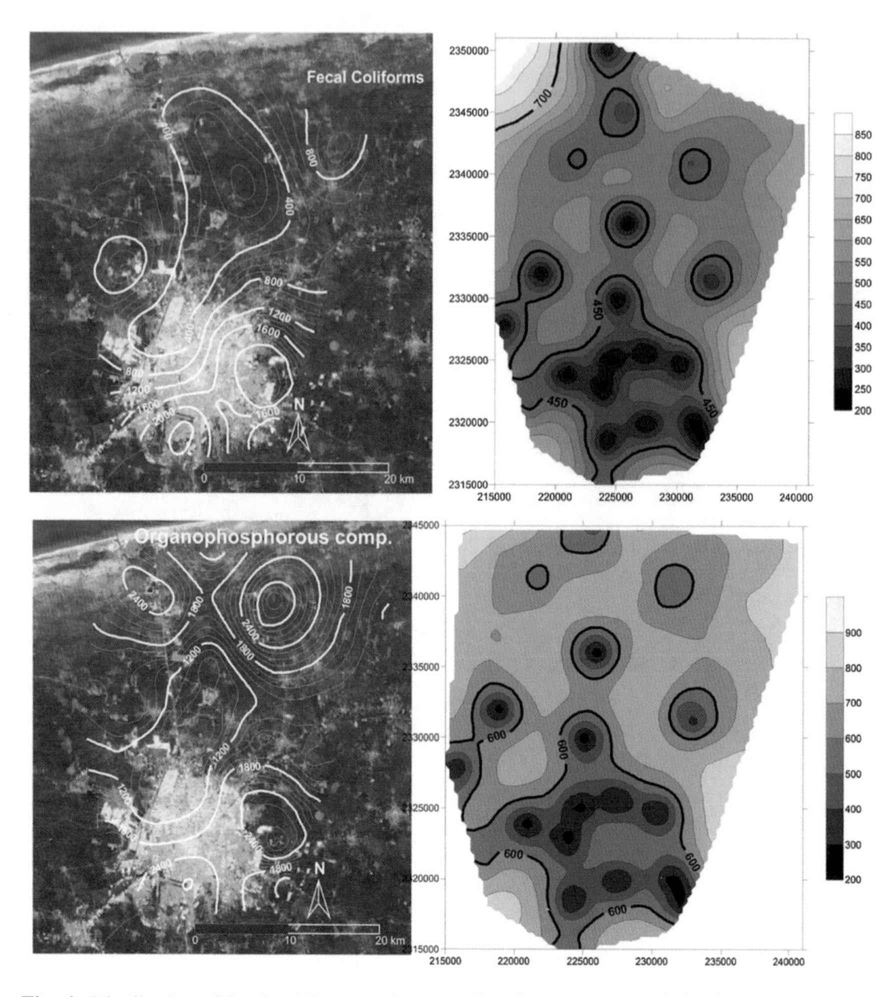

Fig. 4 Distribution of fecal coliforms and organophosphorus compounds in the study area (*left*) and standard deviations (*right*)

3 Conclusion

The main problem of Yucatan aquifer is not the quantity of water resources but the quality. The karst terrain makes it very vulnerable to pollution. The high concentration of nitrates is centered in Merida. There are also some high concentration points along the study area, sometimes due to livestock and agricultural activities in the area, which is not very high. An analysis using geostatistical techniques (ordinary kriging) allowed a spatial estimation of the studied variables.

Chloride concentration is higher in the north of the study area. Contamination of this parameter is due to the activity of the surrounding urban activities (sewage systems) and marine intrusion that occurs from the Gulf of Mexico to the aquifer.

Fecal coliforms are undoubtedly one of the most important and troubling pollutants in the area, the higher concentration is within the city of Merida and the villages of the place.

For some of the organophosphorus compounds, the maximum limits established by current regulations (NOM-127-SSA1-1994) are overpass.

The main pollution sources are the ineffective sewage structure, and even the lack of it, both from urban activities. Merida and Progreso cities and some small towns and villages in the area jointly with industrial activities and livestock activities are the main sources of pollution.

Acknowledgments The first author (Rubén Alcaraz) acknowledges the funding from project 148167 "Evaluación de la calidad del agua en el Acuífero kárstico de Yucatán"of the Mexican Council of Science and Technology (CONACYT-CONAGUA) and the studentship offered by Yucatan University. Rubén Alcaraz also acknowledges to the Master of Water Resources and Environment of the University of Málaga (Spain) the opportunity to participate in this project.

References

Alcocer J, Escobar E (1999) Contaminación del agua subterránea en la península de yucatán. Technical Report of Merida Autonomous University, México

Butterlin J, Bonet F (1960) Información básica para la interpretación geohidrológica de la península de yucatán. Secretaría de Recursos Hidráulicos, México D.F

Bonet F, Butterlin J (1962) Stratigraphy of the northern part of the Yucatan Peninsula. New Orleans Geological Society, New Orleans, pp 52–57

Chica-Olmo M, Luque-Espinar JA (2003) Creación de mapas de calidad de aguas subterráneas mediante métodos de Krigeaje. Boletín geológico y minero 114(3):299–310

Graniel E, Carrillo J, Cardona A (2003) Dispersividad de solutos en el karst de Yucatán. Ing Rev Acad 7(003):49–56

Graniel E, Pacheco A, Coronado V (2009) Origen de los sulfatos en el agua subterránea del Sur de la Sierrita de Ticul, Yucatán. Ing Rev Acad FI-UADY 13(1):49–58. ISSN:1665-529X

Villasuso M, Sánchez I, Canul C, Casares R, Baldazo G, Souza J, Poot P, Pech C (2011) Hydrogeology and conceptual model of the karstic coastal aquifer in northern Yucatan State, Mexico. Trop Subtrop Agroecosyst 13:243–260

An Emblematic Case of Pollution of Wells and Karst Springs Supplying the City of Ragusa (South-Eastern Sicily)

R. Ruggieri

Abstract This paper deals with the occurrence of pollution in two important karst springs, supplying the aqueduct of the city of Ragusa, which happened in September 2010 and it is still taking place. Both springs show higher values of ammonia and the presence of salmonella, elements ascribable to wastewater of animal origin. According to this, investigations identified a number of farms present within the springs-protected areas as likely to have caused the pollution. These were imposed by ordinances to build adequate storage tanks for the animal wastewater. Paradoxically, the construction of these tanks led to a further worsening of the state of pollution, as the latter from episodic, linked to rainfall, became continuous due to the overflowing of wastewater from the tanks never emptied, as it was ascertained. A geological and geochemical study, preparatory to the execution of tracer tests, conducted by the Water Department of Genio Civile of Ragusa and ARPA (Regional Agency for the Environmental Protection), allowed a hydrogeological characterization of the recharge area and the definition of the hydrologic regime of the springs, that in this case, resulted as interconnected. Follow-up tests with fluorescent tracers, carried out on a few farms, were then interrupted because of the opposition of one of the owners. From that moment on, everything stops as for the research of the origin of the pollutant, while at the same time the situation gets worse, both in terms of environment, for the devastating effect on the ecosystem of the Ciaramite stream due to the spill in the riverbed of the polluted water springs, and for the resulting pollution of two municipal drinking water wells placed at the confluence of the Ciaramite stream with the Irminio river. The lack of further drinkable water determined the starting of the crisis of the city water distribution system having to turn to a supply of chance with tank trucks and shifts that created situations of considerable discomfort to major part of the citizens. At present, after 3 years from the start of the polluting event, despite the ordinances issued by the

R. Ruggieri (✉)
Assessorato Infrastrutture e Mobilità, Sicily Region, Hyblean Center
for Speleo-Hydrogeological Researches, Ragusa, Italy
e-mail: info@cirs-ragusa.org

B. Andreo et al. (eds.), *Hydrogeological and Environmental Investigations in Karst Systems*, Environmental Earth Sciences 1,
DOI 10.1007/978-3-642-17435-3_50

City Hall toward a number of livestock farms who do not comply with the collection of waste and its disposal, as a result we assist to the loss of a spring and the ecological degradation of the Ciaramite stream valley.

1 Introduction

In September 2010, technicians of the Ragusa Municipality Water Service detected the presence of elements related to a possible water pollution of the Oro and Misericordia springs, such as turbidity with yellowish coloration. They warned both the Water Genio Civile of Ragusa Regional Office, which is responsible for matters concerning the licenses and authorizations of public waters, and the ARPA Office (Environmental Protection Agency), which is responsible for the quality control of groundwater. Following the first sampling and analyses that actually detected the presence of pollutants beyond the limits permitted by the Drinking Water Regulations (in particular NH_4), the Municipality excluded the water of the springs from the city aqueduct, while trying to identify and resolve the cause of that pollution. Meanwhile, the judicial authority of Ragusa began an investigation with the consulting of both the Genio Civile di Ragusa and the ARPA.

The surveys and inspections carried out in the recharging area of the springs, detected the presence of 12 potential sources of pollution from livestock farms which were not complying with the law as regard to the storage of organic manure and sewage collection, which were both due to the herd of cattle belonging to those farms. The notification of such violations of the law leads to the issue of complaints and ordinances that required farms to put in place statutory requirements regarding the collection and processing of cattle manure. Therefore, as a result of these dispositions, the farms in question in March/April 2011 start the adjustment work, consisting in the construction of tanks for the collection of sewage. At the same time, a judicial/administrative inquiry began and hydrogeological investigations were initiated by the Water Section of the Sicily Region—Dipartimento Regionale Tecnico, and the water monitoring was conducted for the qualitative aspects by the ARPA and the ASP (Public Hygiene Laboratory), for the chemical and microbiological aspects, respectively.

The above surveys are aimed at the hydrogeological characterization of the springs regime and recharging area, at their constant monitoring of chemical and microbiological testing, and the establishment of a program of tracer tests aimed at identifying the causes of pollution.

2 Geographic and Geological Context

The area in question is located in south-eastern Sicily known as the Hyblean Mountains (Fig. 1). This is an extensive plateau with the highest elevation of 1,000 m.a.s.l, in its northern sector (Mount Lauro carapace), sloping toward the

Fig. 1 The Hyblean limestone plateau in the south-eastern part of Sicily

east and south to the sea level, while to the west with a staircase pattern is connected with the Vittoria—Gela plain. From the geostructural point of view, the plateau constitutes a foreland sector, which was slightly deformed during the Sicilian-Maghrebian chain orogeny. The south-western sector of the Hyblean Plateu, where the area of investigation is located, is dominated by carbonates of the Ragusa Formation (Rigo and Barbieri 1959; Grasso et al. 2000).

3 Hydrogeological and Karst Features

Over the plateau, the principal aquifers are associated with the Ragusa Formation carbonates. Several partly confined aquifers occur within the Irminio Member at depths of 100–150 m and are separated by variable thickness of marly intercalations. A deeper and better confined aquifer within the formation is provided by the Leonardo Member. The vulnerability of the carbonates aquifers is high when they form the exposed karst surface. Karst dissolution along with the fracture systems of the Ragusa Formation carbonates started at the end of the Miocene and has continued to develop in exposed areas throughout the Pliocene and Quaternary (Ruggieri and Grasso 2000).

4 General Piezometric Framework and Regime of Springs

The survey of phreatimetric data has allowed the reconstruction of the piezometric groundwater flow of the entire province of Ragusa. In this way, the recharge areas of the hydrogeological south-central complex, bounded by two major watersheds, were defined, as well as the three main hydrostructures (Fig. 2). In such a basin, two hydrogeological substructures are identified and in one of which, the Petraro-Biddemi-Irminio hydrokarst system, falls the area of recharge of the polluted springs.

In Fig. 3 are highlighted some distinctive features of the two springs related to the variability and complexity of the circulation systems in fractured-karst limestone. In particular, the diagram shows the discharges of the Oro spring characterized by a cyclical decline with peak flows and values close to zero, whereas the discharge curve related to the Misericordia spring shows a rapid decrease typical of a water flow concentrated in karst conduits. The above springs supplying the city of Ragusa since 1996 are formally preserved by the establishment of protected areas.

5 Program of Investigations and Tracing Tests

Following some meetings, between the Ragusa Municipality authorities and the regional technical authorities, a program of investigations was defined concerning the hydrogeological and geochemical characterization of the area in exam. Both investigations were finalized to the proposal and the performing of chemical tracer tests for detecting the causes of the pollution of the springs. The chemical monitoring, which began in September 2010, showed the presence of high values of NH_4, K, Na, and Cl, whereas the microbiological analysis showed the presence of salmonella bacteria. In particular, the time series analysis pointed out that the pollution of Oro and Misericordia springs, appearing for the first time in September 2010, the levels of NH_4 in late 2010 decreased, and disappeared until April 2011; then pollution appeared again in May 2011, being persistent from that moment on especially in the Oro spring (Fig. 4). The pollution of the springs from May 2011, unlike the precedent period where the rains had triggered it, conveying widely scattered wastewater into groundwater, especially on the ground without any protection. The rainfall at the end of 2011, had, however, produced only a dilution effect of the levels of contamination, which then, in periods of low rainfall, returned to high values. It should be emphasized the contemporary state between the reappearance of pollution in the springs and the period of realization of the manure and sewage collection tanks that were made by the farmers by following the orders issued by the Ragusa Municipality. With the high concentrations of the pollutants, prevailed the assumption of a constant loss from an underground storage tank, which was continuously fed by wastewater due to cattle

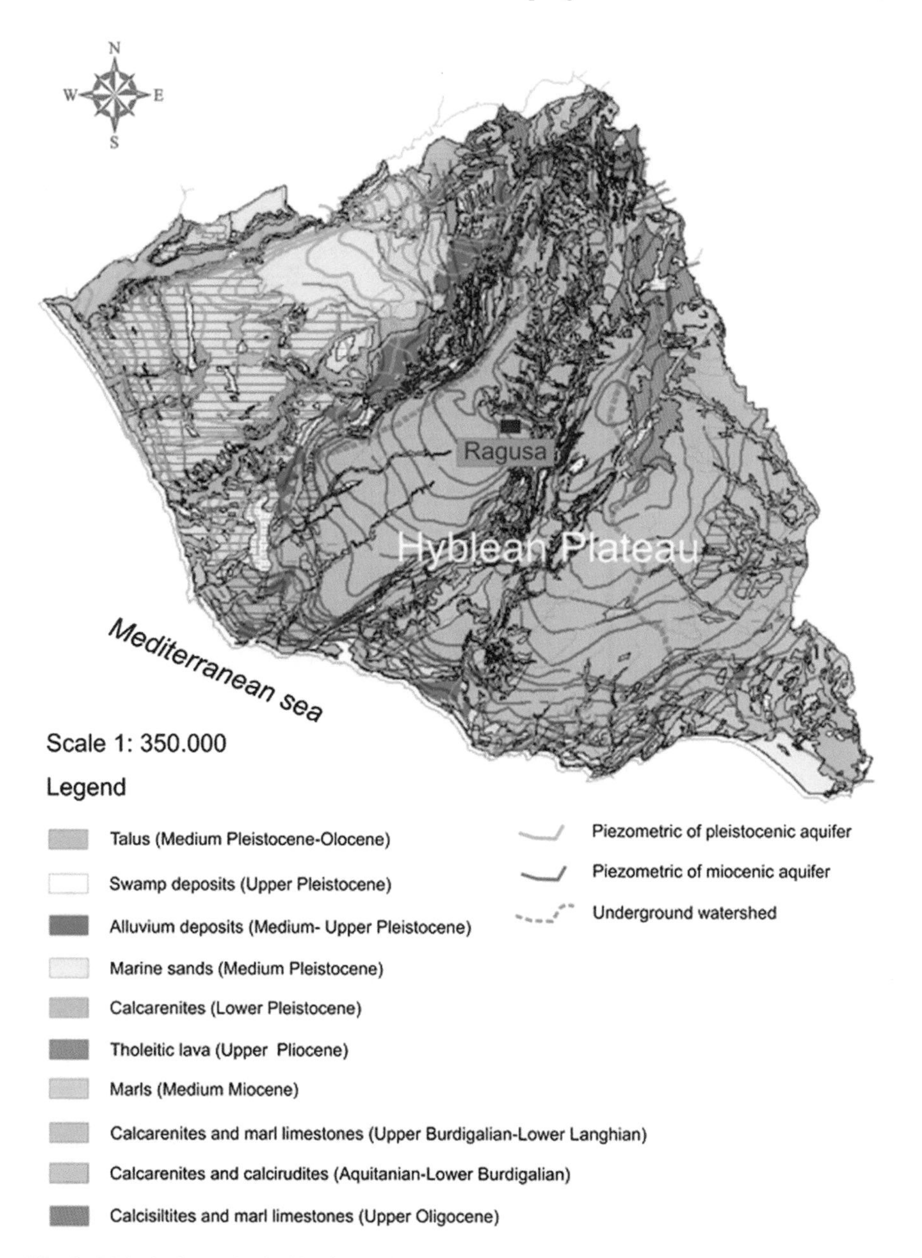

Fig. 2 Main hydrogeological basins and watersheds (hatched *red lines*) of the central-southern sector of the Hyblean plateau, with: *green lines* showing groundwater piezometric levels of the quaternaty sand-calcarenitic aquifer, *blue lines* groundwater piezometric levels of the miocenic limestone aquifer (Ruggieri 2005). The legend reports the main lithological terrains outcropping in the basin

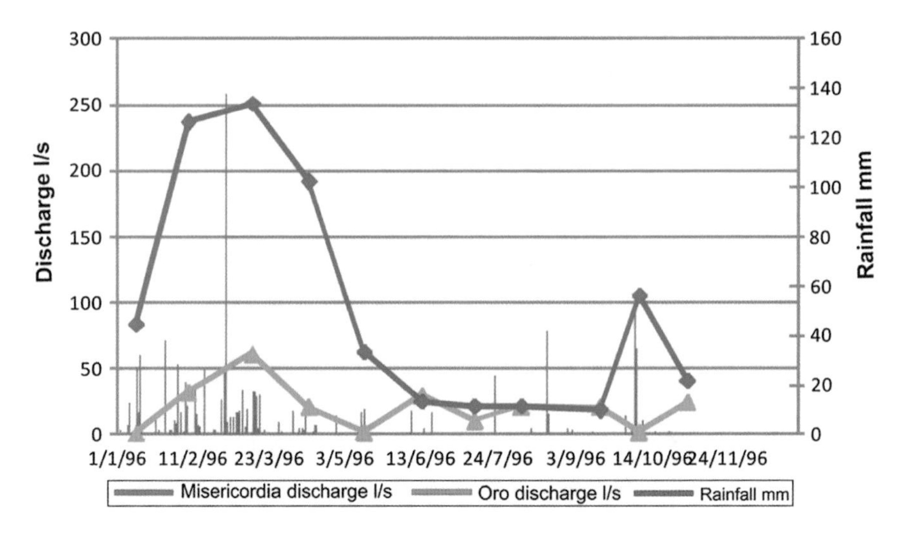

Fig. 3 Diagram of comparison oro-misericordia springs discharge/rainfall

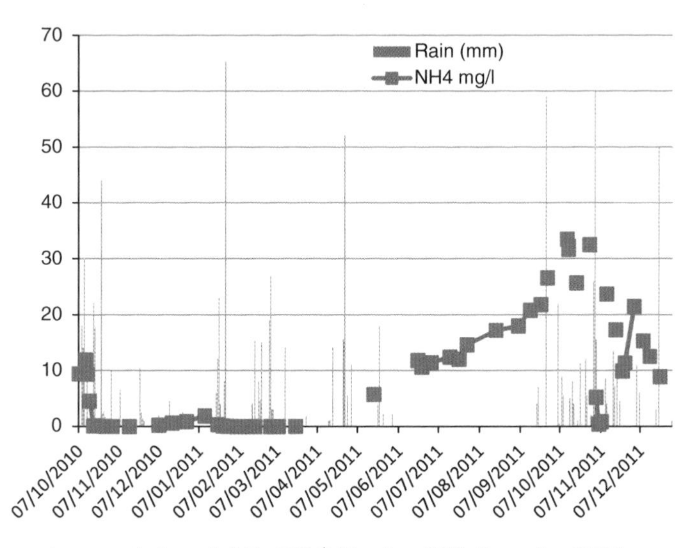

Fig. 4 Oro spring: correlation rainfall—NH_4^+ (October 2010–December 2011)

breeding. This loss would create a persistent stream of liquid manure, which seeping through the fractured ground, would be conveyed quickly to the karst conduits of the aquifers drained by the springs.

Based on the results from chemical and hydrogeological investigations, the area within which to research the causes that led to the pollution of the Oro and Misericordia springs was delimited. In that area, the points of potential contamination were identified which were constituted by cattle wastewater container tanks built by the farmers. Once these points of potential wastewater leakage were identified, the Genio Civile and Arpa proposed to the Municipal of Ragusa to carry out tracing tests (USEPA 1988, 2003; Field 2003; Benischke et al. 2007; Behrens et al. 2001; Goldscheider et al. 2008) using safe nontoxic artificial dye tracer (in order to identify which of these points was the cause of the propagation of the pollution toward the springs.

The proposed and performed tests consisted of the injecting of an artificial dye tracer (sodium fluorescein and/or Tinopal CBS-X) at a point absorber (wastewater tanks) and the following daily monitoring of water emerging from the Oro and Misericordia springs using activated charcoal bags, in order to verify the possible detection of the tracer.

On March 26, 2012, the tracer test started with the injection of a given quantity of sodium fluorescein in the hole trench containing the wastewater tank of the named Farm 1. From that the following points were made:

1. Fluorescein injected between 11 and 11.30 am on March 26 was detected with the charcoal bag taken after 24 h of placing in the Oro spring;
2. In the Misericordia spring, the fluorescence was detected with the third charcoal bag, between 48 and 72 h, remaining constant in intensity also in the subsequent samples;

As the presence of fluorescein ended at the springs, and after checking the initial condition of chemical concentration was checked with blank samples, on 23 May 2012 the second test with the injection of Tinopal CBS-X began, this time into the wastewater tank of named Farm 2, distant from the Oro and Misericordia springs 1,030 and 1,090 m, respectively. In the days following the injection of the tracer, there had been no evidence of its presence in the waters flowing from the springs.

6 Results and Conclusions

Since the possibility to carry out other tests at the moment have been denied by the farmers, and pending that the competent authorities, with new and more specific orders, could command the execution of new tests, at the present, based on the investigations so far carried out, it is possible formulate the following considerations:

1. The diagram of comparison rainfall-discharges shows that, although the springs have different but contiguous recharge areas, they belong to the same karst system and interconnect each other as the underground watershed vertically move, in relation to the rainfall seasonal regime. This fact allows the interchange of waters and the consequent diffusion of the pollution from one spring to the other.
2. Tests with dye tracer (sodium fluorescein) injected and flushed into the tank excavation of the named Farm 1, had given positive results with the response of the tracer within 24 h in the Oro spring and within 72 h in the Misericordia spring since the injection.

3. Although the second test with Tinopal CBS-X tracer, injected into the tank of Farm 1, had not showed positive evidences, since the impossibility to carry out further tests (in drilled well and/or outside the tank), one cannot exclude, incontrovertibly, that also this farm cannot, however, cause the pollution of the spring karst aquifer.

It follows that, regardless on how more or less normally the farms carry out the treatment of sewage and manure of livestock, the presence of the latter is definitely not compatible with the legislation and regulatory requirements governing the areas of protection of drinking waters. For which reason, the Municipality of Ragusa must assume responsibility for deciding whether to tolerate the presence of these farms and therefore give up to the waters of the springs, with a simultaneous increase of costs due to an increased consumption of electric energy to replace the missing waters from drilled wells, quantified it in about €170,000 per year, or find alternative solutions, such as the removal and treatment of wastewater stored in the tanks, at their own expense, at the municipal sewage treatment plant or, alternatively, with the planning and construction of a biogas conversion plant.

References

Behrens H, Beims U, Dieter H, Dietze G, Eikmann T, Grummt T, Hanisch H, Henseling H, Käss W, Kerndorff H, Leibundgut C, Müller-Wegener U, Könnefahrt I, Scharenberg B, Schleyer R, Schloz W, Tilkes F (2001) Toxicological and ecotoxicological assessment of water tracers. Hydrol J 9(3):321–325

Benischke R, Goldscheider N, Smart CC (2007) Tracer techniques. In: Goldscheider N, Drew D (eds) Methods in Karst Hydrogeology. International Contributions to Hydrogeology. Taylor & Francis, London, UK, pp 147–170

Field Malcom S (2003) A review of some trace-test design equations for tracer-mass estimation and sample-collection frequency. Environ Geol 43:867–881

Grasso M, Pedley HM, Maniscalco R, Ruggieri R (2000) Geological context and explanatory notes of the "Carta Geologica del settore centro-meridionale dell'Altopiano Ibleo". Mem Soc Geol Italy 55:45–52 (I tav. f.t.1–92)

Goldscheider N, Meiman J, Pronk M, Smart C (2008) Tracer tests in karst hydrogeology and speleology. Int J Speleol 37(1):27–40

Rigo M, Barbieri R (1959) Stratigrafia pratica applicata in sicilia. Boll Serv Geol d'It 80(2-3): 351–441

Ruggieri R, Grasso M (2000) Caratteristiche stratigrafiche e strutturali dell'altipiano ibleo e sue implicazioni sulla morfogenesi carsica. Speleol Iblea 8:19–35

Ruggieri R (2005) Il sistema idrogeologico del settore centro-meridionale Ibleo (sicilia sud-orientale). Speleol Iblea 11:17–37

United States Environmental Protection Agency (1988) Application of dye-tracing techniques for determining solute-transport characteristics of ground water in karst terranes

U.S. Environmental Protection Agency (EPA) (2003) Tracer-test planning using the efficient hydrology tracer-test design /EHTD program. National Center for Environmental Assessment, Washington DC (EPA/600/R-03/034)

Attenuation of Bacteriological Contaminants in Karstic Siphons and Relative Barrier Purifiers: Case Examples from Carpathian Karst in Serbia

L. Vasić, Z. Stevanović, S. Milanović and B. Petrović

Abstract Karstic groundwater, because of its unique hydrological characteristics, is extremely sensitive to contamination by pathogens. For this reason more attention has recently been paid to the relationship between pathogens and the hydrogeological and geological characteristics of karst aquifer. This paper presents causes of contamination of three large sources located in the karst aquifers in the Carpathian-Balkanides in eastern Serbia. The bacteriological analyses and their correlation with physical and chemical characteristics in seasonal intervals provide an insight into the functioning of studied karst aquifers. It has been confirmed that ascending springs which drain deeper siphonal systems or the presence of adjacent porous aquifers as an additional purifier barrier mitigate bacterial waves and have much better water quality than gravity springs with an unstable discharge regime. For the latter, typical fast draining does not allow the activation of the attenuation capacity of the aquifer system.

Keywords Karst · Microbiology · Groundwater circulation · Attenuation

1 Introduction

Karst groundwater is the main resource for water supply for around 25 % of the world population (Ford and Williams 2007) but compared to other geological formations water from karst aquifers is the most vulnerable to pollution. In general, the natural quality of karst water can be described as excellent, and in Serbia where the percentage of karstic waters used is around 15 % of total drinking

L. Vasić (✉) · Z. Stevanović · S. Milanović · B. Petrović
Centre for Karst Hydrogeology, Department of Hydrogeology, Faculty of Mining and Geology, University of Belgrade, Djusina 7, Belgrade, Serbia
e-mail: lilydhg@rgf.bg.ac.rs

B. Andreo et al. (eds.), *Hydrogeological and Environmental Investigations in Karst Systems*, Environmental Earth Sciences 1,
DOI 10.1007/978-3-642-17435-3_51

waters, the majority of the water utilities employ only elementary chlorination as a water treatment. However, due to high vulnerability if a chemical or bacteriological contaminant is present, a high risk of pollution exists. Connection between karst water circulation and bacteriological contamination was studied by many authors (Kelly et al. 2009; Davraz and Varol 2012; Sinreich et al. 2014). Bacteria could reach karst groundwater in many ways. Pathogenic and faecal bacteria can enter the underworld by precipitated and infiltrated water that washed away the pollution caused by agricultural or livestock activities, by seepage from septic tanks, because of decomposition and decaying of suspended sediments, and in many other ways (Bitton and Gerba 1983; Thorn and Coxon 1992).

In most cases, the bacteriological quality of karst water resources in the area of the Carpathian-Balkanides holds persistently to the outflow regime and thus increased turbidity, which occurs during the period of high water, is often used as an indicator of pollution (Stevanović et al. 1998). The springs where the water drains from the aquifer in the form of small reservoirs or puddles open the possibility of bacterial growth due to lengthy exposure of the water to light and higher temperatures. The main objective of the paper is to define how the type of underground karst water circulation can affect the transport of microbial contamination or influence the autopurification process. Depending on the conditions, the shortest life of *E. Coli* is 70–210 days, which proves how old a pollution of this type is. Experimental determinations of the life cycle of different microorganisms in limestones are very important. At low temperatures (4–8 °C) life cycles vary from 40 days for *Salmonella* (at contamination rates to 10^5 microbes/l of water) to 120 days for *enterococci* (at contamination rates to 10^8 microbes/l of water) (Gavich et al. 1985; Stevanović and Dragišić 1996).

2 Case Examples of Bacterial Pollution of Some Karstic Springs in the Carpathian Karst

In the period preceding the 1960s several hydric epidemic cases occurred in rural Serbia due to the contamination of locally used spring water. Then after, sanitary conditions significantly improved and control of such sources strengthened. Bacterial contamination was observed at several karstic springs which drain unconfined aquifers and were characterized by gravitational or siphonal circulation, but the level of bacteriological impurities was regularly lower for siphonal spring water.

Water from Krupac and Kavak Springs are tapped for water supply of the city of Pirot in SE Serbia. Former is a classic example of a gravitational spring with large annual discharge fluctuations, while late is characterized by a steady outflow regime and ascending circulation in the discharge zone. Due to the large dimension of the karst channels in the Krupac spring and the insufficient time the water spends in the ground, natural purification of infiltrated water cannot occur; hence most of the "impurities" remain (Milanović and Vasić 2011). In contrast to the Krupac spring, the bacteria that indicate anthropogenic pollution at the ascending Kavak spring (streptococci or coliform bacteria) were not observed: Good groundwater quality is the result of a deeper siphonal circulation, and prolonged aquifer—water contact increases attenuation capacity resulting in a stable chemical and microbiological composition (Milanović and Vasić 2011). An additional positive factor for the quality of Kavak water is that the spring discharges not directly from the limestones but by passing thin overlying beds of Pliocene-Quaternary sand and clayey sands (Fig. 1). It is thus recent sediments that provide a kind of purifying barrier for the karstic waters.

A comparative diagram of some qualitative and quantitative characteristic seasonal changes in gravity Krupac and ascending Kavak Springs is shown on Fig. 2. As a result of rapid water exchange, aerobic mesophylls are present in all samples taken from the gravity spring, while in the ascending spring water aerobic mesophyll bacteria were found in only one sample (30/1 ml). Similarly, faecal streptococci were observed in many samples from gravity spring, while in the water of ascending spring these bacteria were found only twice, probably as the result of intense rainfall and nearby infiltration.

An example of accidental bacteriological contamination of spring water is gravity Mirovo Spring in the central part of the Carpathian arch of eastern Serbia, tapped for water supply to Boljevac and nearby villages. Repeated analyses of karst water indicated its favourable physical and chemical properties. However, the pollution of spring water was indicated in August 1982 by intestinal infections caused by water-borne bacteria and virus Echo 11 among over a thousand inhabitants of Boljevac and surrounding communities. The bacterial contamination has been repaired by elementary chlorination, which has, though, only a mitigating effect without eliminating the source of the contamination (Stevanović and Dragišić 1996).

To identify the source of the pollution, the extensive survey including simultaneous hydrology measurements was conducted in wider catchment. The small village some 3.5 km upstream from Mirovo Spring was identified as a source of pollution.

Without a sewerage system, domestic waste water is discharged directly into small stream. Downstream of the village, the stream runs over highly karstified Mesozoic limestone, partly percolating over fractures and masked swallow holes identified as major pollution points. Therefore, fast infiltration of contaminated surface waters, gravity underground flow, absence of attenuation capacities, and temporary insufficient chlorination resulted with this hydric epidemic.

Another example of the transport of pollution from a remote part of the catchment area has been recorded in the waters of Nemanja Spring tapped to supply water to the town of Ćuprija. Surface streams in the upper part of the basin flow through several villages without any sewage system being installed. In the limestone section of the profile beyond the spring, the greatest part of these waters infiltrates the area certified by simultaneous hydrometry. The Nemanja spring has several levels of discharge, and various conditions of circulation resulted in the appearance of different bacteria (Stevanović et al. 1998).

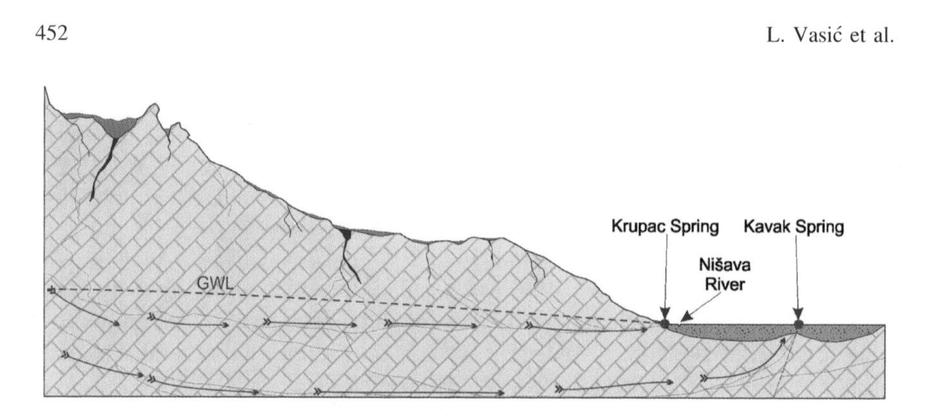

Fig. 1 Gravitational (Krupac spring) and siphonal (Kavak spring) groundwater circulation

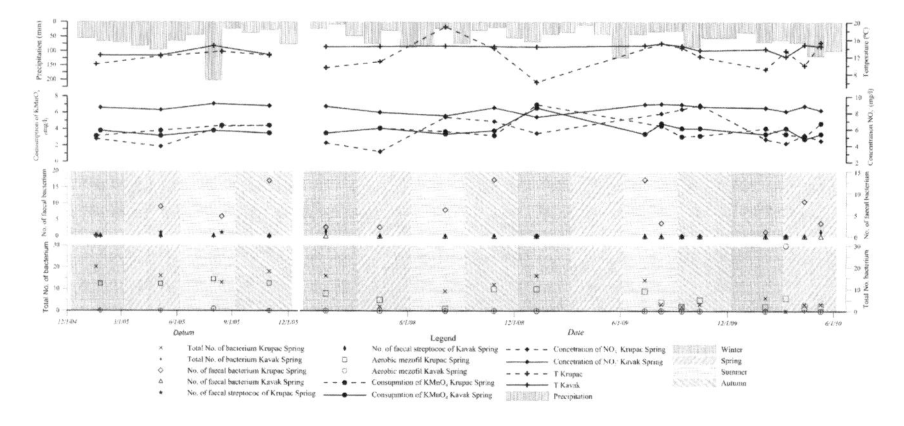

Fig. 2 Comparative diagram of some qualitative and quantitative characteristic seasonal changes in Krupac and Kavak springs (based on analysis of labs of medical centre pirot)

A consistent analysis of bacterial purity of raw water from karst springs from 1995 to 1997 (a hundred control analyses yearly on average) indicated that only 15–20 % of the samples were fit for use. The registered contaminations were in range of 20–280 coliform bacteria (*E. coli* dominant) per 1 ml of water was identified (Stevanović et al. 1998).

Interrelationships among the bacterial contamination, groundwater circulation, discharge and precipitations have been analysed (Stevanović et al. 1998). Rainfalls affected the discharge of the springs at higher elevations (no. 2 on Fig. 3) after 24–30 h on average (and after around 48–60 h at the springs at lower elevations (no. 1 on Fig. 3). The conducted tracing test confirmed a velocity of around 1 km/day (0.012 m/s). This indicates direct and rapid propagation to the higher gravity drainage system, and activation of attenuation capacities, however limited, of the karst aquifer in deeper reservoirs and siphons.

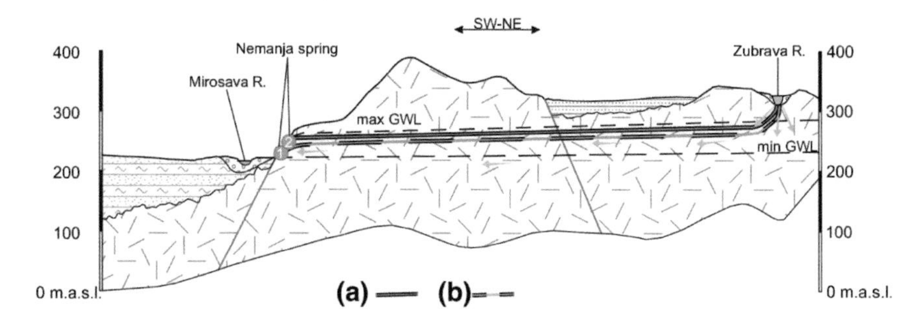

Fig. 3 Scheme of pollutant transport in karst aquifer. **a** high intensity; **b** lower intensity (from Stevanović et al. 1998)

3 Discussion and Conclusion

The design of the conceptual model for defining the dual circulation of karst groundwater and its impact on the accidental bacterial contamination was validated by analysing the discharge mechanism of these three sources and several other karst springs in the Carpathian karst. The basic parameters taken into account are the way of pollutants, kind of their appearance, the transport of the pollution, the existence of relative barrier purifiers and the manner of appearance of pollution at the discharge point. In order to establish a general "input–output" model the following layers and their relation databases should be defined and established:

- Hydrogeological database—defining the type of circulation in certain karst channels, the flow rate of groundwater, retention time in the rocks as well as changes in the physical and chemical parameters over time. Furthermore, it is necessary to define groundwater budget elements and seasonal behaviour of aquifers;
- Topography and morphology database which directly influences the rate of infiltration and runoff;
- Geological database—collection of maps of lithological features and tectonic structures which provides a specific set of data for the creation of the model;
- Database of human, agricultural and livestock activities in different seasons and their visualization on the specific layer.

In order to understand the specifics of the circulation of groundwater in the karst of the Carpathian-Balkanides and bacteriological contamination, a basic conceptual hypothesis was created (Milanović et al. 2013):

- The type of the recharge of the karstic aquifers;
- The diameter and length of karst channels through which groundwater circulates, and the capacity and type of karst springs;
- The velocity of the groundwater flow i.e. the time that bacteria spend in the aquifer;

- The sort of aerobic or anaerobic conditions in the karst underground, depending on the structure and development of the karst aquifers.

This conceptual model is therefore based on two fundamental principles: *the time* that water spends in the underworld and *the distance* that groundwater traverses from the point of infiltration to the water intake structure i.e. karst spring (Vasić et al. 2013).

Karst systems generally have two clearly significant points of impact of pollution on groundwater quality. The problem is actually a deficiency in the purification processes of infiltrated water: firstly—the zone of infiltration through swallow holes (ponors), percolation through the vadose zone or systems of fractures to the fully saturated zone, or groundwater level; secondly—a drainage area, or an area where at a very small distance the mixing of groundwater and surface water occurs in the immediate vicinity of the spring.

The layout of the influence of the surface agents on the karst groundwater area in the case of gravitational and siphoned circulation is shown on Fig. 4.

According to data from described case examples, it can generally be concluded that the circulation of groundwater in karst and its influence on bacterial contamination depend on flow conditions of the two main types and of course many sub-varieties as well:

1. Gravitational circulation of groundwater in the saturation zone due to rapid water exchange, short residence time and the characteristics of aerobic conditions, in terms of prevailing gravity flow (Krupac Spring, Mirovsko Spring, Nemanja Upper Spring).
2. Siphonal circulation of groundwater due to fracture porosity, where the water circulates through the channels and cracks of cm/mm in size, where the circulation is very slow, and the water exchange is slow. In this case anaerobic conditions may also be created, with longer retention of water in the underground (in studied ascending springs Kavak and Nemanja Lower Spring).

Of the three analyzed sources two have a dual discharge mechanism, an upper and lower discharge zone. In both cases the reaction of the lower-positioned zone is slow drainage and better water quality; siphons, therefore, can be indicated as a relative barrier for bacterial contaminants.

As seen from the Kavak spring case the attenuation capacity can be additionally stimulated by the presence of an overlying bed which acts as a purifying barrier. Figure 5 shows the two pressurized channels within the karstic aquifer: (a) a channel partially filled with sediments which may reduce the level of bacteria, and (b) a drainage point and adjacent porous aquifer which may totally purify contaminated water. That of course depends on the potentiometric pressure in the pipe which influences velocity, permeability of adjacent porous media and residence time in both the main karstic and the adjacent aquifer (Stevanović in press).

The slow and partly siphonal circulation of water at some of the springs suggests the activation of the self-purifying properties and reduction of the contamination concentration. Also, one of the key elements in the analysis of the

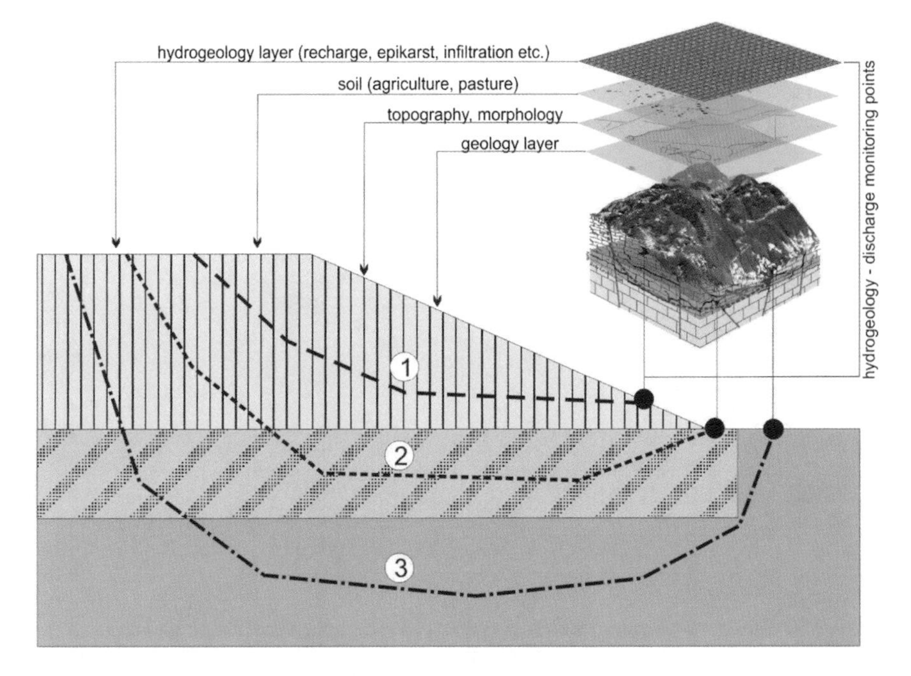

Fig. 4 Conceptual model of draining and circulation of the karstic springs, *1* gravitational circulation/rapid water exchange/large cavities; *2* siphonal circulation/rapid water exchange/drainage channels of moderate size; *3* siphonal circulation/slow water exchange/small dimension drainage channels, fissures and voids

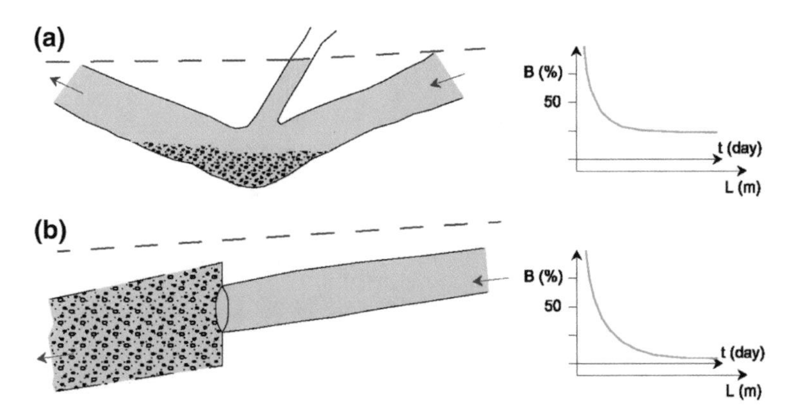

Fig. 5 The pressurized karstic drainage siphons: **a** partially filled with sediments, and **b** with adjacent porous aquifer. The diagrams show a reduction of bacteria (*B*) in the percentage based on residence time (*t*) and the length of the channel (*L*) i.e. distance between infiltration and discharge points (after Stevanović in press)

occurrence of bacterial contamination in karst springs is the zone of discharge and water sampling locations. In fact, neglect of the sampling locations can often give a misleading picture of the bacteriological water quality of some karst springs. Since karst springs often drain to the small lake depression (Vauclusian type) where water exchange is very slow, the influence of local factors that may bring pollution (bacteriological) is significant, while the same water at a certain depth in the karst channels often has less registered bacteria than on the surface or none at all. Accordingly, only water directly from the karst channels showing accurate microbiological characteristics should be sampled.

References

Bitton G, Gerba CP (eds) (1983) Groundwater pollution microbiology. Wiley, New York

Davraz A, Varol S (2012) Microbiological risk assessment and sanitary inspection survey of Tefenni (Burdur/Turkey) region. Environ Earth Sci. doi:10.1007/s12665-011-1332-1

Ford DC, Williams P (2007) Karst hydrogeology and geomorphology. Wiley, New York

Gavich IK (ed) (1985) Metody ohrany podzemnyh vod od zagrjaznenija i istošćenija. Nedra, Moscow

Kelly WR, Panno SV, Hackley KC et al (2009) Bacteria contamination of groundwater in a mixed land-use karst region. Water Qual Expo Health. doi:10.1007/s12403-009-0006-7

Milanović S, Vasić LJ (2011) Hidrogeološka osnova zaštite podzemnih voda u karstu na primeru. Beljanice Vodoprivreda 252–254:165–173

Milanović S, Stevanović Z, Vasić LJ et al (2013) 3D modelling and monitoring of karst system as a base for its evaluation and utilization: a case study from eastern Serbia. Environ Earth Sci. doi:10.1007/s12665-013-2591-9

Sinreich M, Pronk M, Kozel R (2014) Microbiological monitoring and classification of karst springs Environ Earth Sci. doi:10.1007/s12665-013-2508-7

Stevanović Z, Dragišić V (1996) Some cases of accidental karst water pollution in the Serbian carpathians. Theor Appl Karstol 8:137–144

Stevanović Z, Jemcov I, Dokmanović P et al (1998) An example of bacteriological contamination of a captured spring. In: Proceedings of IAH/AIH conference on "gambling with groundwater", Las Vegas, pp 173–178

Stevanović Z (ed) Characterization of karst aquifer. In: Characterization and engineering of karst aquifer. Springer, Heidelberg (in press)

Thorn RH, Coxon CE (1992) Hydrogeological aspects of bacterial contamination of some western Ireland karstic limestone aquifer. Environ Geol 20:65–72

Vasić LJ, Milanović S, Petrović B et al (2013) Uticaj cirkulacije podzemnih voda u karstu na pojavu bakteriološkog zagađenja. Vodoprivreda 264–266:219–229

Impact of a Tunnel on a Karst Aquifer: Application on the Brunnmühle Springs (Bernese Jura, Switzerland)

A. Malard, P.-Y. Jeannin and D. Rickerl

Abstract Tunnel drilling in karst regions often leads to major disturbances in the hydrogeological functioning of aquifers and flow-systems. Numerous examples are documented in Switzerland and induced significant costs, which were not or rarely anticipated (e.g.: Flims, Jeannin et al. 2009). The Ligerztunnel is one of these example. The tunnel was built a few hundreds of meters upstream from the Brunnmühle spring, which contributes to the drinking water supply of communities of Twann and Ligerz. During the construction, a major karst conduit with a huge discharge rate was intersected in a side exploration tunnel. Overflowing water was diverted into the Twannbach canyon. In the main section, smaller conduits were found and drained outside by pipe leading water close to the Brunnmühle spring. Actually, authorities want to add a safety tunnel parallel to the main tunnel. In this view, SISKA is in charge of evaluating the hydrological disturbances on the spring regime. The paper presents the approach applied to assess the potential effect of the drilling of a new tunnel near to a group of karst springs and pumping wells. The approach combines available spatial information and a hydraulic model. The KARSYS approach is first applied on this system in order to set up a 3D geological and hydrogeological model of the karst aquifer and the related systems. The spatial distribution of karst conduits within the massif is assessed based on a speleogenetical and inception horizons model (KarstALEA method). Inferring from these models, a karst conduits network is generated. The hydraulic model of the downstream part of the conduits network, which concerns the close vicinity of the safety tunnel project, is precisely calibrated using head and discharge data. Flow in this conduits network is then simulated using SWMM 5.0 in order to reproduce the hydrological responses of the different outlets (permanent springs, drainage devices, overflow springs, etc.).

A. Malard · P.-Y. Jeannin · D. Rickerl (✉)
Swiss Institute for Speleology and Karst Studies, Rue de La Serre 68,
2301 La Chaux-de-Fonds, Switzerland
e-mail: info@isska.ch

B. Andreo et al. (eds.), *Hydrogeological and Environmental Investigations in Karst Systems*, Environmental Earth Sciences 1,
DOI 10.1007/978-3-642-17435-3_52

1 Introduction

Communities of Twann and Ligerz (Bernese Jura, CH) obtain water from the Brunnmühle spring, which is a karst spring emerging from Malm karst aquifer. In 1980s the construction of the Ligerztunnel, which crosses the karst aquifer 100 m upstream from the spring induced serious hydrological disturbances to the karst hydraulics. Active karst conduits have been intersected in the tunnel and required the drilling of a pipe to discharge karst waters out of the tunnel, close to the spring (Bollinger and Kellerhals 2007). This considerably modified the dynamics of the karst aquifer: the seasonal regime of the spring as well as the functioning of the overflow springs located 40 m above the permanent outlet. Nowadays the construction of a safety gallery parallel to the tunnel raises the question of new hydrological disturbances on the spring. The communities already considered to replace the actual supply device by two wells drilled further upstream in the same aquifer. The question to be addressed is therefore to assess the potential disturbances of the new gallery on the spring and wells.

As the study is still in course, only the steps of the proposed approach are presented in the following note. This approach is an extension of the KARSYS approach (Jeannin et al. 2013). The evaluation is based on five main steps:

i. Build a geological and hydrogeological 3D model of the site in order to characterize the underground drainage zone of the system and its catchment area over the surface using the KARSYS approach;
ii. Establish a speleogenetic model of the site to assess the potential organization of the karst conduits network. At the same time, the analysis of existing caves and karst features will be used to identify the inception horizons;
iii. Design a hydraulic model of the system using the hydrological karst features (permanent and overflows springs, catchment areas), the artificial drainage devices (tunnel, drainage pipes) and the results of the speleogenetic and inception models;
iv. Reproduce the discharge regime of the Brunnmühle permanent springs and the overflow outlets;
v. Test various scenarios of hydrological disturbances on the spring regime caused by the tunnel construction. As far as possible, disturbances will be extended to pumping wells, which are being equipped as the future drinking supply for the communities;

2 Context

The site is located at the foothill in the Eastern part of the Jura Mountain (canton of Bern) along Lake Biel (see Figs. 1 and 2). The geological context is composed of South-East dipping pile of Jurassic and Cretaceous limestone, underlain by Oxfordian Marls (aquiclude). The tunnel develops approximately along the strike

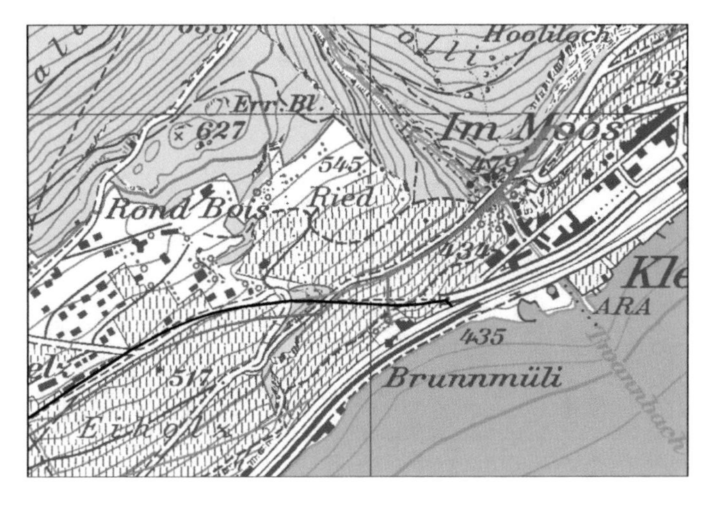

Fig. 1 Site location *Black line* existing main tunnels; *Green line* existing exploration tunnel; *Red line* security gallery (project), *Orange lines* future railway and road tunnels. The Brunnmühle spring (*blue*) emerges at 433 m a.s.l close to Lake Biel. Other springs are visible in the Twannbach canyon (Im Moos)

of bedding planes and the Brunnmühle spring emerges at the base of the massif, close to the elevation of the lake Biel, but at the top of the limestone series. Several other springs emerge in the Twannbach canyon. Due to the existing exploration tunnel, which intersected a major karst conduit, the natural hydrological regime of the karst system was very probably modified. This must be considered in the perspective of a hydraulic simulation and impact assessment.

As sketched on Fig. 2 the Brunnmühle spring emerges at the contact between the Portlandian limestone and the Purbeckian marls. It is mainly fed by the Malm aquifer (upper Jurassic limestone, from Sequanian to late Portlandian, ~600 m of thickness) and partially dammed by the Purbeckian marls. All tunnels are located within the Malm aquifer, close to the groundwater table.

Holiloch and Gisheren are emissive caves, which become active during high-flow events. They act as overflow springs of the system, especially during the spring melt period (https://www.google.ch/maps/@47.093352,7.151663,3a,75y,331.3h,77.88t/data=!3m4!1e1!3m2!1snPY6RkNP2cjiAKW9Pb2xEA!2e0?hl=fr). Their entrances are located 44 m above the elevation of the Brunnmühle spring indicating that the hydraulic head considerably rises during high-flow conditions.

3 Approach

The first step of the KARSYS approach (Jeannin et al. 2013) consists in building a geological 3D model of the site depicting the geometry of the Malm aquifer. Figure 3 depicts the geological structure of the site. The existing and projected

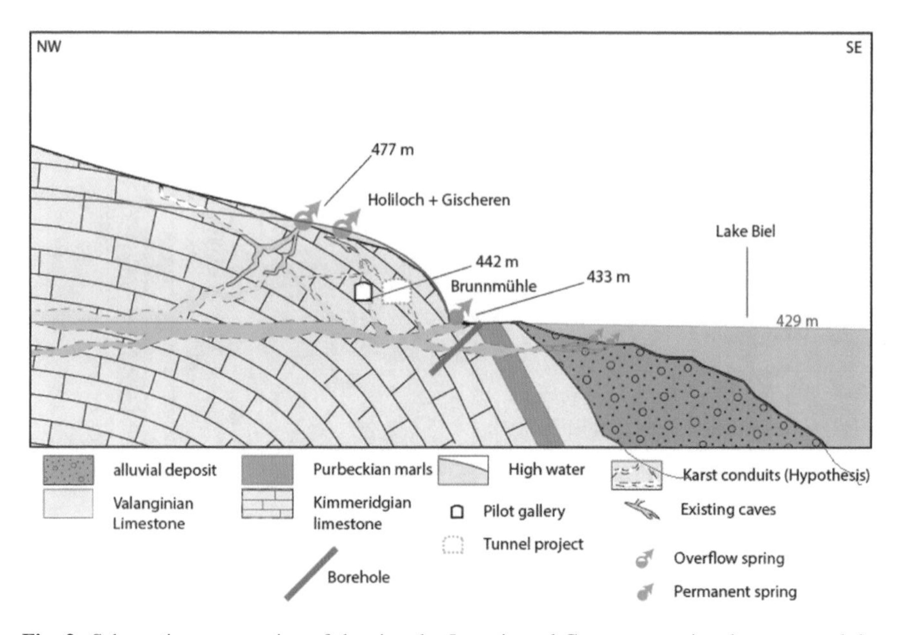

Fig. 2 Schematic cross-section of the site, the Jurassic and Cretaceous units plunge toward the South-East. The Brunnmühle spring emerges at the contact between the portlandian limestone and the purbeckian marls

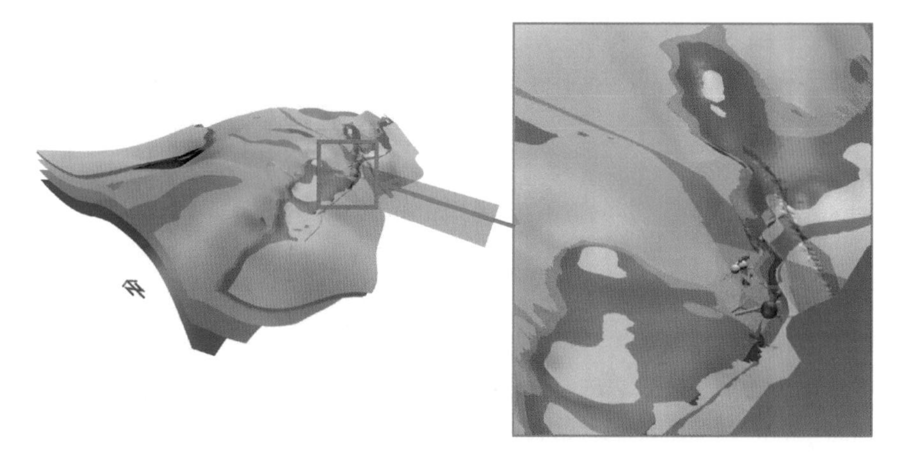

Fig. 3 Perspective view of the geological 3D model of the site (8km along the east-western axe and 6km along the north-southern axe); the tunnel (*green pipe*) develops along the bedding planes below the Purbeckian marls (*red layer*)

tunnels (green pipes) all develops within the Malm limestone, i.e., Purbeckian Marls (red layer on Fig. 3).

Hydrological features (mainly springs, but also boreholes and caves indicating the position of the water table) are added to the model in order to build a hydrogeological 3D model of the site. The model shows that the Brunnmühle

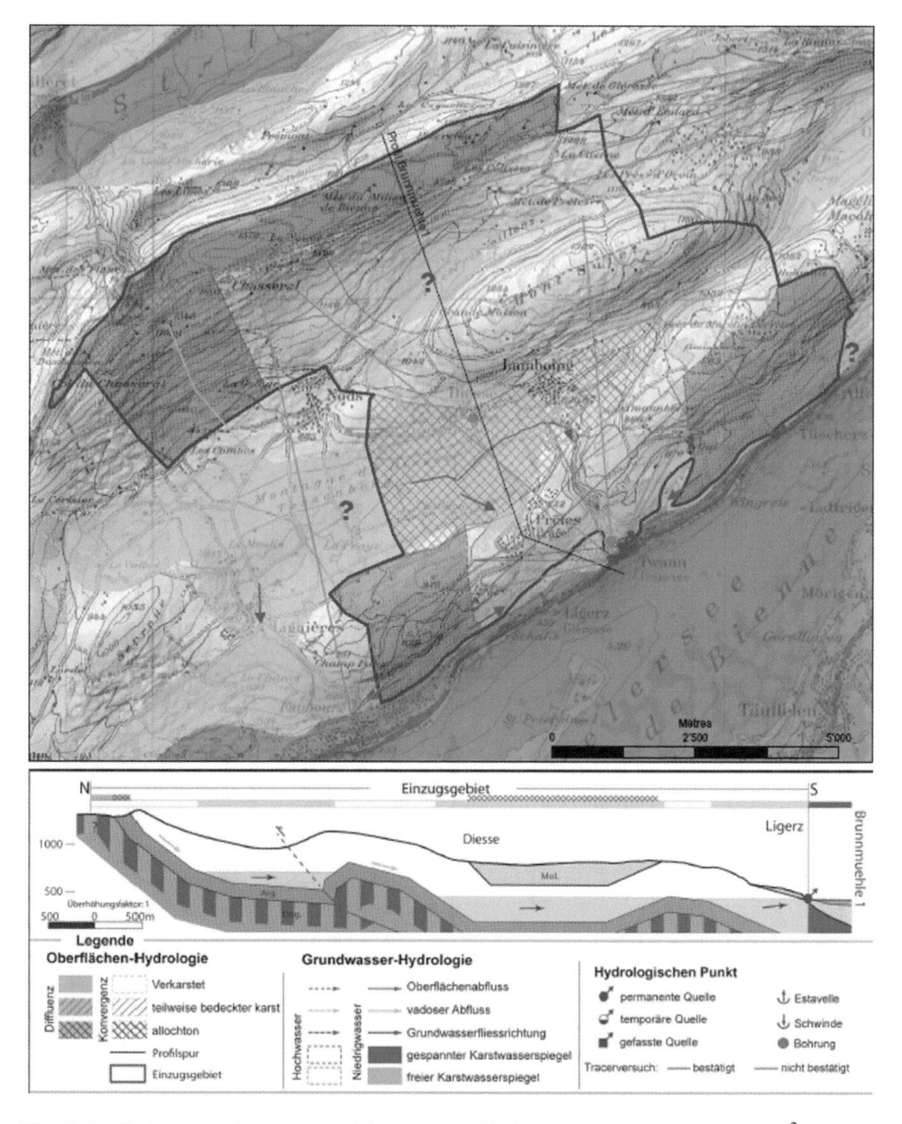

Fig. 4 Preliminary catchment area of the Brunnmühle karst system; the area is 66 km^2 wide but presents divergent parts with the adjacent karst systems

spring emerges at the lowest outcrop of Malm limestone, confirming the damming effect of the Purbekian marls (red layer). Knowing the geometry of the aquifer and assuming an almost horizontal water table at low water conditions, the extension of groundwater bodies, the underground drainage zone, and the respective catchment area feeding the system can be delineated (Fig. 4).

The second step of the approach is to set up a speleogenetical model of the site in order to hypothesize the organization of the karst conduits within the massif.

The density and the geometry of existing caves and karst evidences in this area will be analyzed and compared with the geological model in order to assess the main horizons of karstification (called "inception horizon", Filipponi 2009) along some specific stratigraphic limits or due to tectonic disturbances. Such model makes it possible to assess the supposed organization of the karst network. This approach is known as the KarstALEA method (Filipponi et al. 2012).

The aim of the third step of the approach is to generate a pipe network according to the 3D hydrogeological and speleogenetical models, and to make a first guess of the flow parameters of the generated pipes. A series of hydraulics and speleogenetical principles are used to generate a conduit network linking the catchment area according to its topography and infiltration characteristics to the main outlets (springs) of the system. The generator provides files directly compatible for SWMM 5.0 (Rossman 2004), which is a robust pipe-flow modeling tool.

For the fourth step, the model will be calibrated using first existing head and discharge data acquired in the neighboring of the spring. The hydraulic characteristics and the topology of the pipes in the vicinity of the spring will be calibrated using a series of various hydraulic situations (e.g., overflow conditions of the respective springs, measured heads, and discharge rates). This network will be included into the pipe network generated at catchment scale. This model will be calibrated using a recharge model assessing infiltration on various parts of the catchment area.

The fifth step will include the simulation of the intersection of various karst conduits (pipes) by the projected tunnels. The first simulations show that the effect can range from almost insignificant to very considerable depending on the topological position of the intersected karst conduit.

4 Discussion

One main uncertainty of the proposed approach is related to the lack of discharge rate measurements (short or incomplete hydrographs) of the different outlets. This is especially true for artificial tunnels, where discharge rates have not been measured correctly until now (April 2014).

Another difficulty is that no data are available for all springs before the construction of the first tunnel, making a comparison with the natural situation almost impossible.

5 Conclusion

The KARSYS approach was then applied to the Brunnmühle karst system in order to define the underground drainage pattern and the catchment area of the spring. In a second step a speleogenetical model was established in order to design the organization of the karst conduits within the massif. Then, these parameters are being injected within a hydraulic model—SWMM 5.0® to simulate the regime of the Brunnmühle spring and its overflow springs in the Twannbach canyon. According to the speleogenetical model, several scenarios of potential disturbances will be tested to assess the effects on the springs' regime, as well as on the future borehole.

As a preliminary discussion, it appears that simulating such systems may be quite complicated as data and measurements do not exist for all outlets. These data and measurements may also be obsolete as artificial galleries (or drainage) probably modified the natural functioning of the aquifer.

As the study is still in progress, results could not be discussed yet. Only the approach was presented here. Results will be presented during the conference and compared to the initial approach.

References

Bollinger D, Kellerhals P (2007) Umfahrungstunnel Twann (A5): Druckversuche in einem aktiven Karst. Bull angew Geol 12/2:49–61

Filipponi M (2009) Spatial Analysis of Karst Conduit Networks and Determination of Parameters Controlling the Speleogenesis along Preferential Lithostratigraphic Horizons. Ecole polytechnique fédérale de Lausanne (EPFL), Suisse, PhD dissertation. 305 p

Filipponi M, Schmassmann S, Jeannin PY, Parriaux A (2012) KarstALEA: Wegleitung zur Prognose von karstspezifischen Gefahren im Untertagbau—Forschungsprojekt FGU 2009/003 des Bundesamt für Strassen ASTRA, unpubl. rep. Schweizerischer Verband der Strassen- und Verkehrsfachleute VSS, Zürich, Schweiz

Jeannin PY, Eichenberger U, Sinreich M, Vouillamoz J, Malard A et al (2013) KARSYS: a pragmatic approach to karst hydrogeological system conceptualisation. Assessment of groundwater reserves and resources in Switzerland. Environ Earth Sci 69(3):999–1013

Jeannin PY, Häuselmann P, Weber E, Wildberger A (2009) Impact assessment of a tunnel on two karst springs, flims, Switzerland. In: Proceedings of the 15th international congress of speleology—Kerrville, vol 3. Texas, United States of America, p 1537 19-26 July 2009 (contributed Papers)

Rossman LA (2004) Storm Water Management Model. User's manual version 5.0, EPA, unpubl. rep. U.S. Environmental Protection Agency, Cincinnati, OH

Hydrogeological Risks of Mining in Mountainous Karstic Terrain: Lessons Learned in the Peruvian Andes

D. Evans

Abstract Managing mine waste is one of the greatest challenges facing mining companies globally. Over the next 20 years, more than 10 billion of tons of waste will be generated in Peru based on the projected mining rates from existing mines and new projects. Most of Peru's metal ore deposits are located at high elevations within a narrow, tectonized carbonate-rock belt extending over 2,000 km in length. Many of the limestone formations are karstified and characterized by high recharge and percolation rates, well-developed subsurface drainage and complex flow patterns. Mine tailings and waste rock facilities are preferably located close to the mine and processing operations and are often underlain, at least partially, by karstic limestone. Mining companies prefer to place mine waste in unlined basins when it can be shown that "natural hydraulic containment" will limit seepage to environmentally acceptable levels. Such a demonstration relies on detailed geological and hydrogeological investigations and the development of sound conceptual groundwater flow models. The depth of epigenetic karstification is greatly influenced by carbonate purity, bedding dip, faulting intensity, and oxidation of sulfide minerals (ARD). In addition, hypogenic karstification is common, particularly in areas with recent volcanism and hydrothermal activity. Detailed three-dimensional geological and numerical seepage models are used to predict seepage and environmental risk although simulating the effect of the channel network is often challenging.

1 Introduction

The Andes are the result of crustal deformation due to subduction of oceanic crust (or Nazca plate) beneath the South American continental plate. Carbonate and other sedimentary rocks, which were originally horizontal and below sea level, are

D. Evans (✉)
FloSolutions S.A.C, Lima, Peru
e-mail: devans@flosolutions.com

B. Andreo et al. (eds.), *Hydrogeological and Environmental Investigations in Karst Systems*, Environmental Earth Sciences 1,
DOI 10.1007/978-3-642-17435-3_53

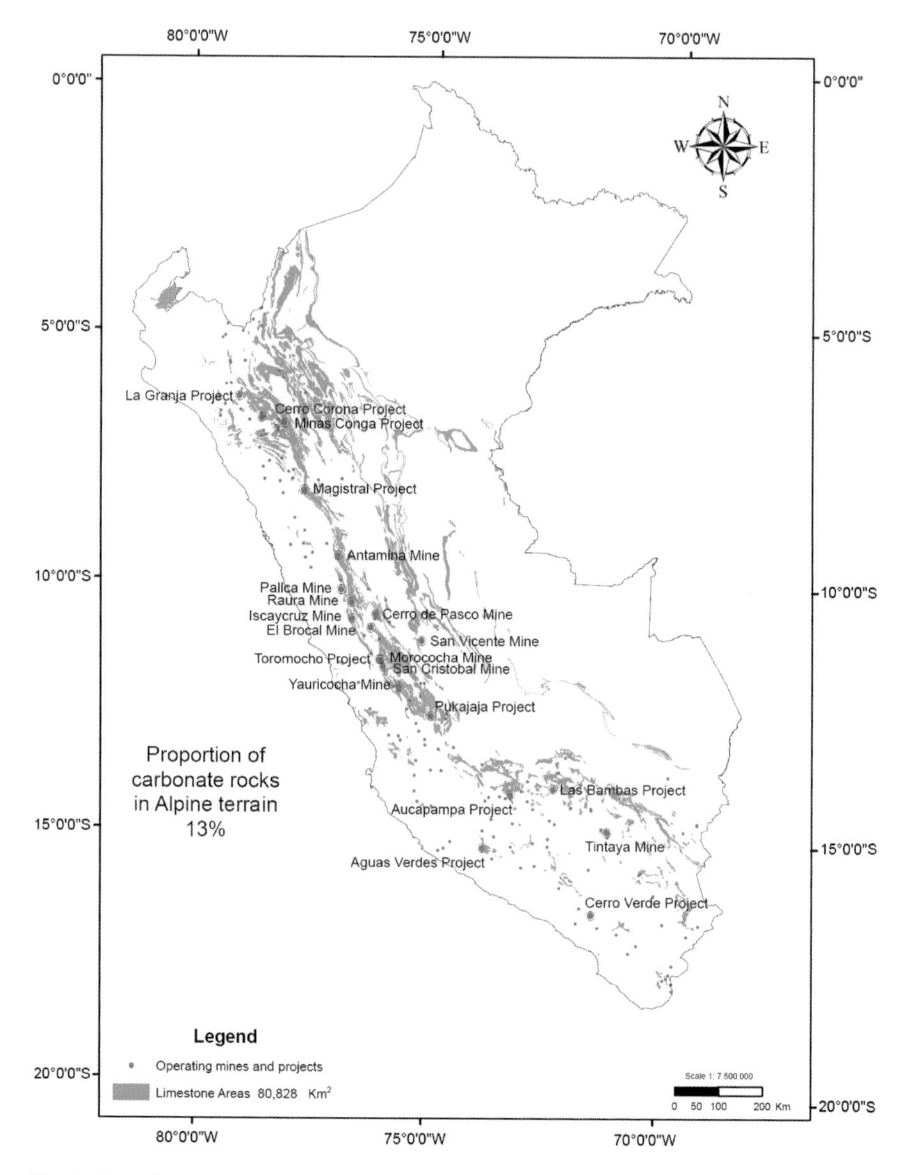

Fig. 1 Map of Peru showing areas of carbonate terrain (*green areas*) and all operating mines and mining projects

now folded, faulted, and sometimes completely overturned. Carbonate rocks are brittle and subject to fracturing by these tectonic forces. These uplifted, tightly folded, and faulted carbonate terrains are host to over half of the base and precious metal deposits occurring as porphyry, skarn, vein, and replacement-type deposits (Fig. 1). Most of the mines are located between elevations 3,500 and 5,000 m a.s.l where precipitation rates are typically between 1,000 and 1,500 mm/year. Most of the carbonate formations are karstified and characterized by high recharge and percolation rates, well-developed subsurface drainage, and complex flow patterns.

Within northern Peru, karstifiable carbonate rocks include the Cretaceous aged Cajamarca, Yumagual, Mujarrún, and Chulec Formations. Within central Peru, the Cretaceous aged Jumasha Formation and the Jurassic aged Condorsinga/ Aramachay Formations (Pucará Group) are the most karstifiable. Karstic limestone formations in southern Peru include the Jurassic aged Gramadal and Soccosani Formations and the Cretaceous aged Arcurquina and Ferrobamba Formations.

Karst areas typically represent unfavorable conditions for the hydraulic containment of mine waste facilities, as there is high risk of seepage losses. The waste facilities associated with mines are often sited within the same karstic carbonate terrains, mainly due to economic, property, or geologic constraints, putting local and potentially regional karstic aquifers at risk of contamination. Local communities rely heavily on seeps, springs, and rivers associated with these karstic aquifers for potable and irrigation water. Protecting karst aquifers located within new mining areas can be challenging due to the hydrogeologic complexity of karst systems and mine waste development restrictions imposed on mine planners. Environmental issues related to developing mine waste storage areas in karstic terrain include: (1) surface subsidence caused by sinkhole development and potential loss of tailings or mine water; (2) contamination of karst aquifers and springs; and (3) piping of tailings into karstic conduits. In some cases, lining the facilities is not feasible due to steep, irregular terrain, and potential for liner deformation.

The potential for deep, well-developed karst and deep phreatic levels depends on several factors including CaCO3 purity, bedding thickness, bedding inclination, degree of faulting and fracturing, topography and hydrologic base level conditions, climate, and acidity of the source water (Goldscheider and Drew 2007). The intensity of the faulting and folding has a pronounced effect on the karstification depth, depth of groundwater circulation, and degree of regional flow. Steeply dipping strata, found within Peru's thrust-fold belts (Fig. 2), helps direct flow along bedding planes promoting deep dissolution. Deep karst shafts form in steeply inclined carbonate strata.

Karst features include small-scale surface dissolution features such as karren (photo 1a); karren fields (b); deep solution channels along fault zones (c); grikes (vertical or subvertical fissures developed by solution along a joint) (d); karst shafts or "simas" (e); caves (f), sinkholes or "dolines" (g), springs (h), poljes (i). The unconfined compressive strength of the limestones in Peru is typically high, which is necessary for the development of large scale dissolution features.

2 Risk of Seepage from Mine Waste Facilities

Leaching of metals and other soluble constituents occur as water percolates through the mine waste. The fate of this seepage depends on the hydraulic containment of the waste facility, the attenuation of the seepage along the flow path,

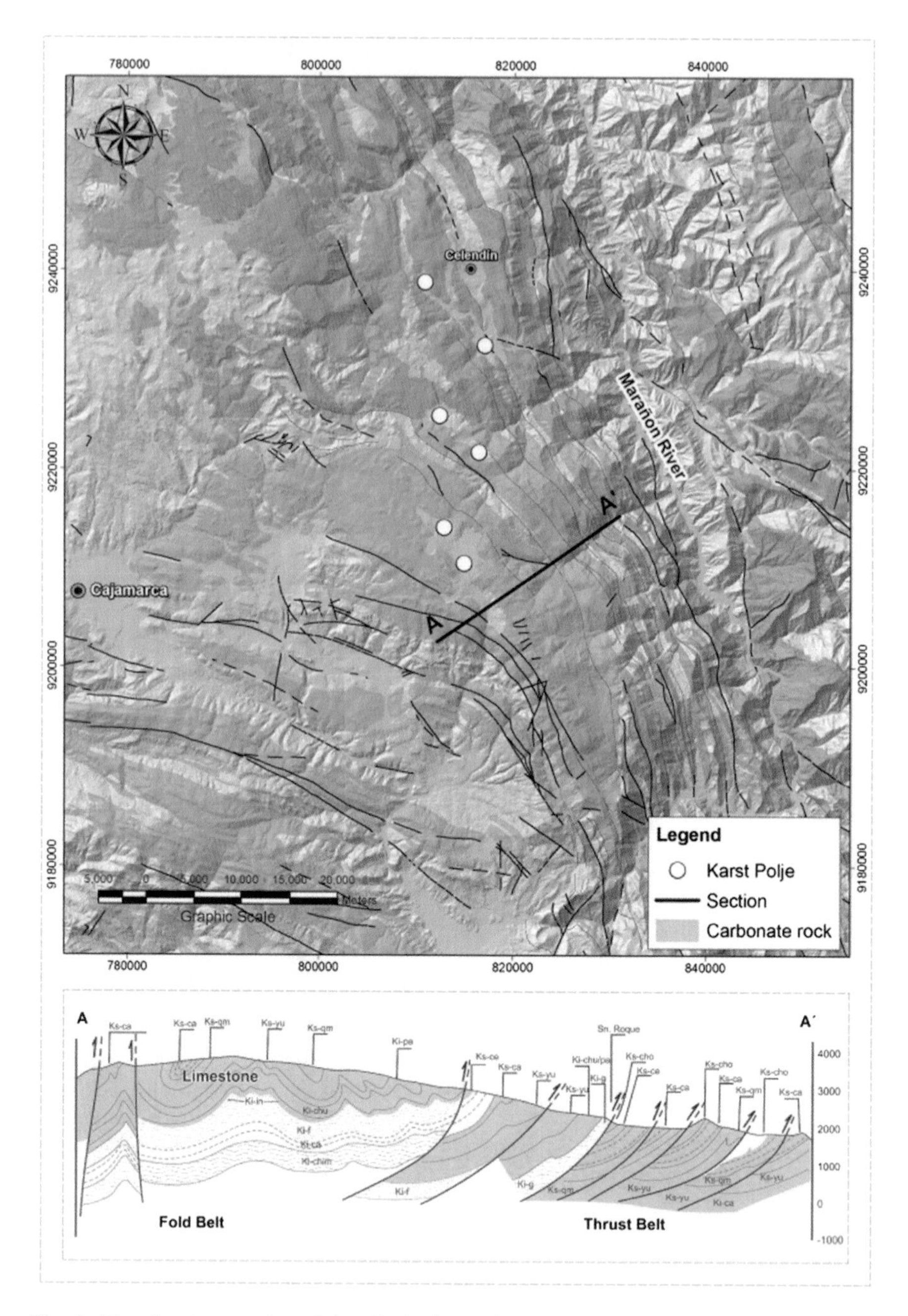

Fig. 2 Map showing a major mining district in northern Peru with intense folding and faulting. Deep karstification and regional flow is common in this district

and the ability to capture the seepage before discharging to sensitive receivers. The environmental risk of related to waste rock facilities is considerably less than tailings facilities since they are typically well-drained and, by comparison to tailings facilities, do not significantly change the original phreatic levels associated with the foundation. Hydraulic containment issues can arise, however, due to dipping strata, faulting, and karstic foundations, which can shift the groundwater divide from the topographic divide beneath the waste dump. Tailings facilities, on the other hand, are commonly operated under saturated conditions with the phreatic level equal to the pond.

Unlined tailings facilities are at greater risk of seepage losses compared to waste rock piles due to the greater hydraulic potential, particularly if karstic limestone lies beneath or adjacent to the facility. Seepage from an unlined tailings facility will cause water table mounding to occur in the foundation and ridgeline areas. Problems arise when the phreatic levels in the waste impoundment exceed the piezometric levels in the surrounding ridges allowing seepage to escape laterally to aquifers with lower hydraulic potential. Foundation seepage from waste impoundments can be recovered at the downstream toe of the tailings dam or waste rock pile through the use of seepage collection and pump-back systems; however, if seepage is lost from the foundation to karstic systems the potential for recovery is difficult (and commonly impossible) due to the hydrogeologic complexity of the karst.

3 Hydraulic Conductivity and Hydraulic Containment Relationships

Distinct ground water flow regimes develop within tectonized karstic carbonate rocks, which are directly related to the hydraulic conductivity and drainability of the rock mass. Fault zones have an extremely important role in the groundwater movement in the Peruvian Andes. The anisotropy caused by fault zones for example can greatly affect the flow direction, concentration of contaminated seepage, and therefore the impact on water resources (in particular karst springs). Assessing the structure and properties of fractured rock and karst aquifers presents severe practical problems because of their anisotropic and severely heterogeneous nature. Yet estimation of the hydrogeologic properties of the formations is necessary to assess the hydrogeological suitability of proposed mine waste sites.

Hydraulic conductivity (K) data from over 2,000 hydraulic conductivity (mainly Lugeon) tests in limestone at several mining projects throughout Peru is presented in Fig. 3. The plot shows K versus depth below ground surface and the geomean for each 50 m depth increment. The tests were conducted in valley floors (where karst is commonly concentrated), valley slopes, and ridgeline areas surrounding the proposed and existing mine waste sites. Falling head tests were conducted in very low K zones and constant head tests were conducted in the very

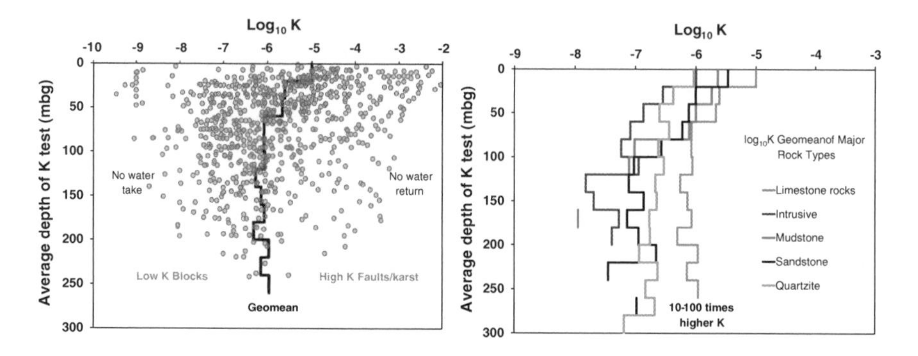

Fig. 3 Hydraulic conductivity (K) data from over 2,000 hydraulic conductivity tests in limestone at several mining projects throughout Peru (*right* figure). The *left* figure shows the geomean hydraulic conductivity values of the limestone with depth compared to the geomean hydraulic conducitivty of other non-karstic rock types

high K zones with no return of injection water. Discrete packed-off intervals typically ranged between 5 and 10 m. It should be noted that the actual bulk hydraulic conductivity of the rock mass could be higher than the packer testing results due to: (1) scale effect, and (2) borehole skin effects (i.e., permeability reduction of the fracture zone caused by drilling).

The hydraulic-conductivity measurements ranged over eight orders of magnitude, from the lower detection limit of the equipment at 1×10^{-10} m/s to 1×10^{-2} m/s. Drilling programs have shown that extreme hydraulic conductivity changes can occur over very short distances with K values as high as 10^{-1} m/s in karstified fault zones immediately adjacent to wall rock exhibiting K values less than 10^{-8} m/s. According to White (2007) the practical boundary between fracture permeability and conduit permeability occurs at an aperture of about one centimeter. Turbulent flow starts at these fracture apertures, which correspond to hydraulic conductivities roughly around 10^{-2} m/sec. The matrix hydraulic conductivities of most compacted limestones are in the range of 10^{-11}–10^{-9} and for most practical purposes can be ignored; therefore, carbonate rocks with hydraulic conductivity values higher than 10^{-9} m/s could be considered fractured. Despite the large number of tests, the data is considered mostly representative of the fractured rock mass rather than the karstic system due to the difficulty in encountering karstic features. Comparative studies of hydraulic properties measured in different karst aquifers have shown that, regardless of the range of porosity measured in the aquifer matrix, conduits typically account for less than 1 % of the porosity of the aquifer, but more than 95 % of the permeability (Worthington et al. 2000). It is extremely important to note that if *no karstic conduits are found in drilling programs it does not preclude their existence since the potential for hitting karstic conduits within the phreatic zone is extremely low.*

Figure 3b shows the geomean hydraulic conductivity versus depth of the limestone compared to clastic sedimentary and intrusive rocks. The plot shows a reduction in hydraulic conductivity with depth for all of the clastic and intrusive rocks with the limestone geomean hydraulic conductivity remaining consistently high to depths of 300 m. The reduced depth decay of hydraulic conductivity for the limestones is attributed to fracture dissolution. With lower rates of decrease in K with depth for the soluble carbonate rocks, the depth of groundwater circulation increases, discharge to local systems decreases with more water entering regional groundwater system. This is significant for hydraulic containment of mine waste facilities with the potential for regional movement of seepage to regional surface water drainages.

The depth of the phreatic surface and degree of hydraulic containment surrounding a new mine waste facility depends largely on the hydraulic conductivity and drainability of the rock mass. Distinct hydrogeological regimes are present in a tectonized, karstic limestone: (1) the epikarst layer, which is commonly perched due to high permeability contrasts with the underlying, less fractured/karstic limestone; (2) porous matrix and microfractures forming the blocks between fracture/fault zones; (3) high permeability solution fractures; (4) fault zones, which can have complex fracturing and karstification; and (5) principal karstic channels (Fig. 4). Ground water flow velocities in the channels can be hundreds of meters per day, whereas flow within the matrix or slightly fractured blocks can be as low as a few meters per year. The water within the fractures, conduits, and the rock matrix in karstic limestone is often poorly connected and significant differences in head can exist between a relatively unfractured limestone and an open conduit connected to a karst spring. A deep karst drain at the intersection of a bedding-parallel fault and vertical fault (as shown) could have the potential for shifting the groundwater divide significantly away from the topographic divide, potentially beneath the mine waste areas.

The potential for finding these deep karstic drains or the most permeable zone of a fault by drilling and testing is extremely low. Factors of safety should be applied to these water levels to account for the potential for more transmissive fracture systems. The factor of safety is highly site-specific and should take into consideration the purity of the carbonate rock, bedding thickness, degree of folding and faulting, stratal dip, and potential for confining layers.

4 Hydraulic Containment Concepts

A common misconception with mine developers is that hydraulic containment can be achieved if the *phreatic level in the surrounding ridges is higher than the highest phreatic level in the waste impoundment* (Evans and Goldscheider 2007). This is the premise for the "hydraulic trap" concept, which is also used in other types of waste depositories. In reality, the potential for hydraulic containment will be determined by the piezometric distribution between the waste impoundment, the containing ridges, and receiving valleys. It is therefore important during the feasibility stage to install multi-level piezometers within the ridge and valley areas

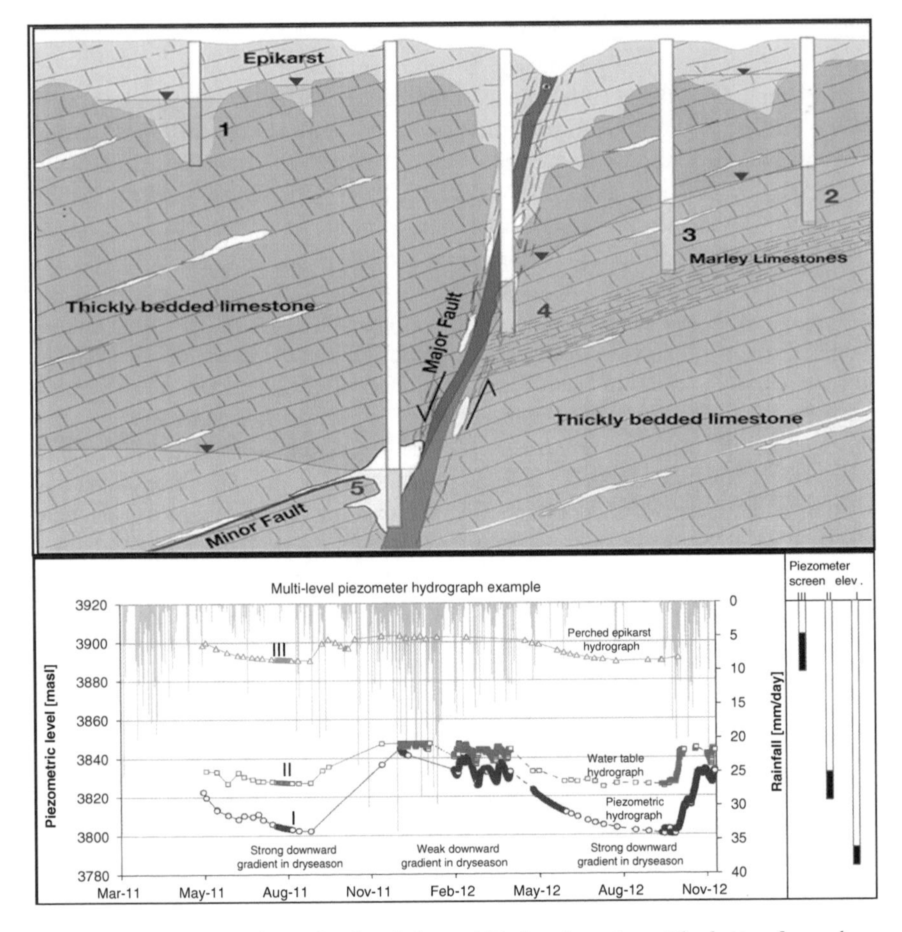

Fig. 4 Concept of groundwater level variations within karstic systems. The *bottom figure* shows a multi-level piezometer hydrographs associated with a perched aquifer in the epikarst compared to hydrographs in the saturated zone

to define vertical hydraulic gradients and piezometric distributions between the recharge and discharge areas.

Natural hydraulic containment may not exist if the waste facility is surrounded by deep valleys with strong hydraulic gradients between the waste facility and these valleys. If the foundation of the valley is permeable, hydrostatic conditions within the valley may prevail with hydraulic head at the base of the final tailings similar to that of the tailings pond. The dip of the carbonate strata and the hydraulic gradients between the mine waste facility and adjacent groundwater discharge areas is extremely important. Four hydrogeological scenarios are presented in Fig. 5 showing the potential for uncontrolled seepage losses to adjacent watersheds from waste areas developed adjacent to karstic terrain. The risk of seepage losses increases substantially in when the dip of the adjacent limestone

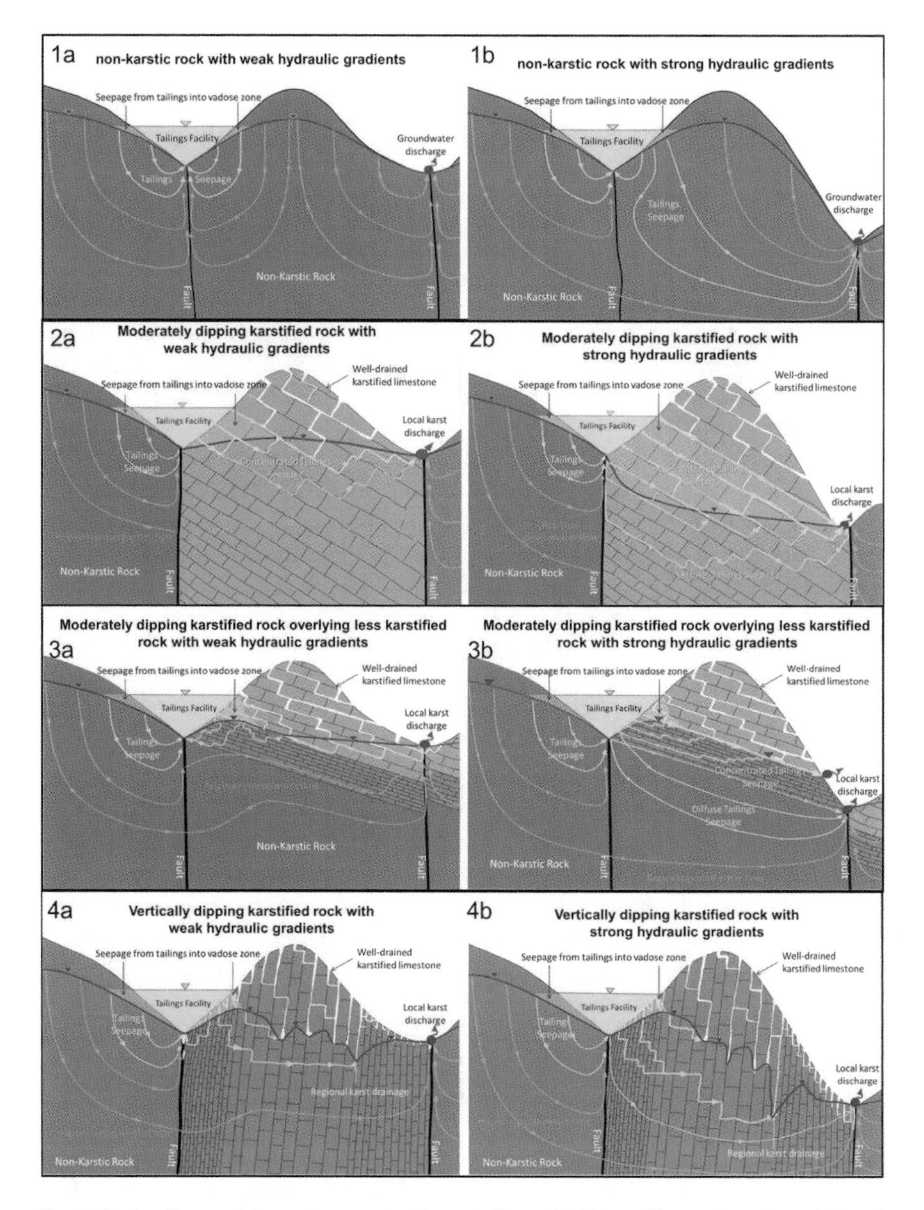

Fig. 5 Hydraulic containment concepts for a tailings facility with non-karstic and karstic foundations with varying degrees of hydraulic gradient and stratal dip

strata and hydraulic gradients increase. The near vertical strata allows for deeper dissolution along bedding planes and contacts, creating the greatest potential for deep karst and deep karst drainage, and ground water contamination at a regional scale.

Numerical ground water flow modeling of the waste facilities will help predict the phreatic and piezometric levels based on expected recharge during operation and after closure; however, determining the lowest possible phreatic level in a karstic limestone ridge surrounding a proposed waste facility can be challenging.

5 Modeling of Tectonized Karstic Systems

Tectonized karstic groundwater systems are generally considered difficult to model mathematically because of their extreme hydrogeological complexity and non-unique nature. Some of the challenges limiting the capabilities of modeling of these systems include: (1) the geometry and hydraulic properties of the faults and karstic network is typically poorly understood, so it is often inferred; (2) the hydraulic properties can vary by orders of magnitude over short distances; and (3) the hydraulic interconnectivity of the fault zones and karstic conduits are typically poorly understood. Nevertheless, groundwater flow models are still the best tools for quantitatively assessing the risks associated with developing mine waste facilities in karstic and fractured rock systems. Although the model may not be able to predict what *will* happen in reality due to the challenges listed above, it can be used to predict what *can* and *cannot* happen given a variety of scenarios.

Hydrogeological models are a simplification of reality and cannot incorporate each and every known geological feature, which might control the groundwater movement. Since oversimplification of groundwater flow models can lead to significant predictive errors, incorporation of faults within numerical models is often warranted when assessing hydraulic containment and risk of impact (Dufour et al. 2012). Since karstification is commonly associated with faults, inclusion of highly transmissive anisotropic faults within the model is a more common approach to account for the karstic channeling. If the karstic conduit system is reasonably well-known then inclusion of this in the model is considered.

6 Conclusions

Managing mine waste in karstic mountainous terrain is one of the greatest challenges facing mining companies in Peru. A thorough understanding of the karst hydrogeology is a precondition for environmental permitting of new waste facilities and delineation of ground water protection zones. This understanding includes determination of the depth of the karst and the ultimate related water levels, delineation of the catchments of the karst springs, and general ground water flow patterns. Since karstic features and their effects on the ground water flow system are not directly observable, building a strong conceptual model of potential flow patterns within the vadose and phreatic zones is essential. Site selection studies need to be performed during the pre-feasibility stage of mine development to determine the best locations of new mine waste facilities, and to reduce the risk of impacts to water resources. Finding the karstic conduits responsible for draining a carbonate aquifer can be a bit like finding a needle in a haystack—even with the most sophisticated geophysical programs and other techniques. Care must be taken when interpreting piezometric levels collected from piezometers, which do not intersect the karstic system, and conservative factors of safety should be applied. Because of the problems associated with drilling in complex terrains, dye-tracer and speleology studies have proved to be very cost-effective in helping define the ground water catchments surrounding the proposed mine waste facilities (Evans et al. 2005).

References

Dufour R, Evans D, Cho J, Renard P, Mariethoz G, Carpentier A (2012) Balancing complexity and simplicity in numerical groundwater flow models for mining projects in mountainous settings. In: Proceedings from FEFLOW users conference. Berlin, Germany

Evans D, Letient H, Aley T (2005) Aquifer vulnerability mapping in Karstic Terrain—Antamina Mine, Peru. In: Proceedings from the society of mining engineers conference, 2005 Salt Lake City, USA

Evans D, Goldscheider N (2007) Aquifer vulnerability and risks associated with developing mine waste facilities in Karstic Alpine Terrain, Peru. In: Proceedings from XXVIII convención minera, Arequipa, Perú

Goldscheider N, Drew D (eds) (2007) Methods in karst hydrogeology. Int Contrib Hydrogeol 26:276. (London: Taylor & Francis)

White WB (2007) A brief history of karst hydrogeology: contributions of the NSS. J Cave Karst Stud 69(1):13–26

Worthington SRH, Ford DC, Beddows PA (2000) Porosity and permeability enhancement in unconfined carbonate aquifers as a result of solution. In: Klimchouk AB, Ford DC, Palmer AN, Dreybrodt W (eds) Speleogenesis, Evolution of karst aquifers Huntsville: National Speleological Society, pp 463–472

Karstic Hydrogeology of the Uchucchacua Underground Mine (Perú) and Its Interaction with Surface Waters

D. Apaza-Idme, A. Pulido-Bosch and F. Sánchez-Martos

Abstract Karstic landforms abound in the vicinity of the Uchucchacua mine (4,500–5,000 m a.s.l.). These help infiltration and the rapid inflow of water, which penetrates even the deepest working levels of the mine. These inflows affect mining operations and the mine has to be drained. In wet months, high volumes of mine water must be pumped, and these volumes are drastically reduced in dry months. Mine water is collected and drains by gravity along the Patón Tunnel. The mine dewatering has created drawdown cones, so modifying natural groundwater flow patterns. Surface water and groundwater are interconnected because of the karst, and this influences the volume of mine water that must be pumped from the mine. Direct infiltration occurs via dolines and sinkholes, and along the main interconnected open faults and mineral veins, leading to very high infiltration rates overall. The current study aims to evaluate the impact of meteoric water on mining operations.

Keywords Mountain karst · Karstic aquifer · Underground mines · Mine drainage

1 Introduction

The Uchucchacua mine is situated in the Central Highlands of Peru, in Oyón province (Lima district) at an elevation of between 4,500 and 5,000 m a.s.l., close to the watershed of the Huaura basin. Precipitation mainly occurs between December and March, with an annual mean of 1160 mm. The mean annual temperature is 4 °C.

D. Apaza-Idme (✉)
Hidroandes Consultores S.A.C, Lima, Perú
e-mail: dapaza@hidroandes.com.pe

A. Pulido-Bosch · F. Sánchez-Martos
G I. Recursos Hídricos y Geología Ambiental, University of Almería, Almería, Spain

B. Andreo et al. (eds.), *Hydrogeological and Environmental Investigations in Karst Systems*, Environmental Earth Sciences 1,
DOI 10.1007/978-3-642-17435-3_54

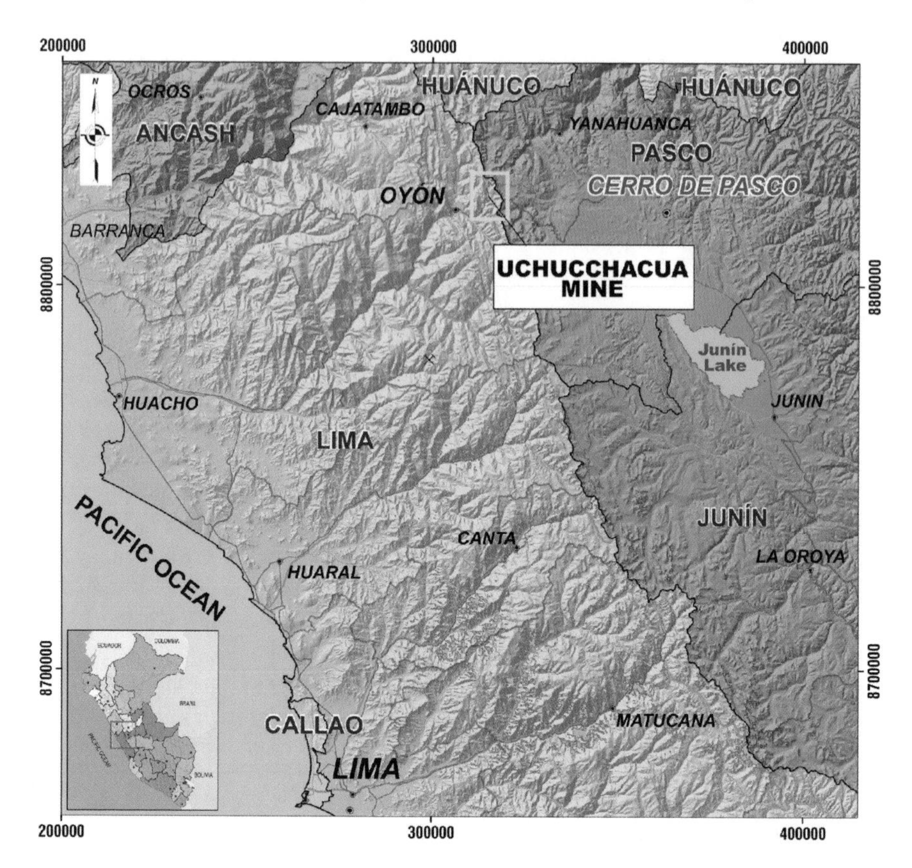

Fig. 1 General geographic location

Underground mining operations have continued since colonial times in the three sectors of Carmen, Socorro and Huantajalla. The mineral deposits are epigenetic, arising from the infill of fractures (veins), flow channels and from metasomatic replacement from hydrothermal solution. Numerous old mine workings of this silver deposit still exist in the Nazareno, Mercedes, Huantajalla and Casualidad areas of the Central Highlands. In the early 1960s, the Buenaventura Mines Company began prospecting and, in 1975, they installed an industrial plant. Today, this has a treatment capacity in excess of 4,000 tonnes/day.

The current study aims to assess how surface water affects the underground mine workings, and determine the extent of meteoric water flow into the mine (Apaza 2013), since this inflow creates the need for intensive pumping in the mine during wet periods. However, the need for pumping is drastically reduced during dry months (Fig. 1).

2 Materials and Methods

The geology of the area indicates an extensive outcrop of Upper Cretaceous calcareous rocks, known as the Jumasha Formation, which is 1,460 m thick. This underlies the marl of the Calendín Formation and overlies the sandstone of the Goyllarisquizga Formation, which outcrops in the southwestern part of the study area.

Limestone is the main aquifer in the area, which is formed by three large units. The first is the Lower Jumasha formation, up to 570 m thick, composed of alternating marly limestone and nodular limestone with chert. The second unit—the Middle Jumasha formation—has an estimated thickness of 485 m and comprises pale grey limestone alternating with nodular limestone with some marly horizons. The Upper Jumasha formation, with an estimated thickness of 405 m, exhibits a more diverse lithology of fine-grained limestone, with carbonaceous schist at the base and beige marly limestone at the top (Jacay 2005). The area suffered intense glacial activity in the Pleistocene period, creating U-shaped valleys above 3,800 m. a.s.l., which were subsequently infilled by lateral and frontal moraines as the glaciers receded. Towards the north, there are extensive moraine trails (INGEMMET 1996) (Figs. 2 and 3).

Tectonic activity favoured the intrusion of dacite, riolite and mineral-rich hydrothermal solutions that formed the veins of the mineral deposit exploited for its silver sulphosalt minerals (proustite and pyrargyrite). Zinc (blende) is extracted as a subproduct, and there is also a wide range of other minerals.

The hydrogeology of the area is dominated by karstic limestone; there is a small proportion of fractured limestone, which are frequent in the high-altitude zones. The most important aquifer corresponds to the limestone of the Jumasha Formation; of these, the Middle Jumasha limestone are the most soluble and show greater karstification (Evans et al. 2005). Here, there is greater percolation of rainwater, with rapid flow towards the saturated zone. From the saturated zone, water flows laterally and intercepts the deepest levels of the underground mine.

The hydrogeological mapping allowed identification of the hydrogeological units based on their lithology. An inventory of springs and *bofedales* (spring-fed wetlands in the *altiplano*), infiltrations into the mining galleries, as well as flow in boreholes, discharges from anomalous structures during the drilling of hydrogeological boreholes and excavation of soil pits have been done. In addition, two hydrochemical sampling surveys were done (wet season and dry season) over a network of river and lagoon sites. Three groundwater surveys were completed, including flow gauging of springs and mine water; piezometric level measurements have also been done.

The investigation involved drilling of thirteen boreholes in February 2014, with different deeps, between 40.2 and 150 m. These were subject to hydraulic tests, including Lefranc, Lugeon and slug-tests at 8 m intervals. A log of the borehole lithology was also made, describing discontinuities and karstified reaches, and measurement of water levels.

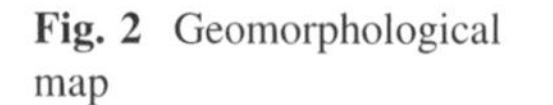

Fig. 2 Geomorphological map

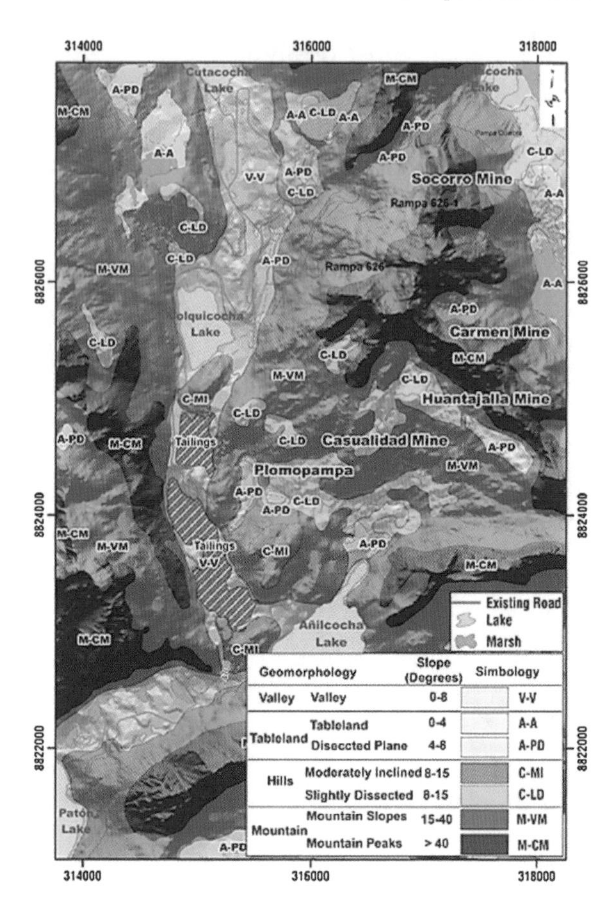

3 Results and Discussion

The data of hydraulic tests were grouped into three units: moraine sediments, massive limestone and slightly karstified fractured limestone. Values were classified according to rock type and size of grain (in the case of unconsolidated sediments). The data are shown in Table 1.

Recharge occurs by infiltration of rainwater and snowmelt. The greatest recharge is produced over areas of bare limestone rock and through the deposits of granular sediments that retain runoff water. The dolines with sinkholes are the greatest recharge points. The mean recharge rate is estimated to be between 40 and 70 % of the total precipitation. This is a moderate–high rate, highly conditioned by the anisotropy of limestone.

The hydrogeological catchment exceeds the hydrological one and receives recharge from outside its watershed. The piezometry indicates a concentration of flows towards the south-west, which coincides with the pattern of surface drainage. In dry months, there is no significant groundwater discharge. In wet months, by

Fig. 3 Geological map

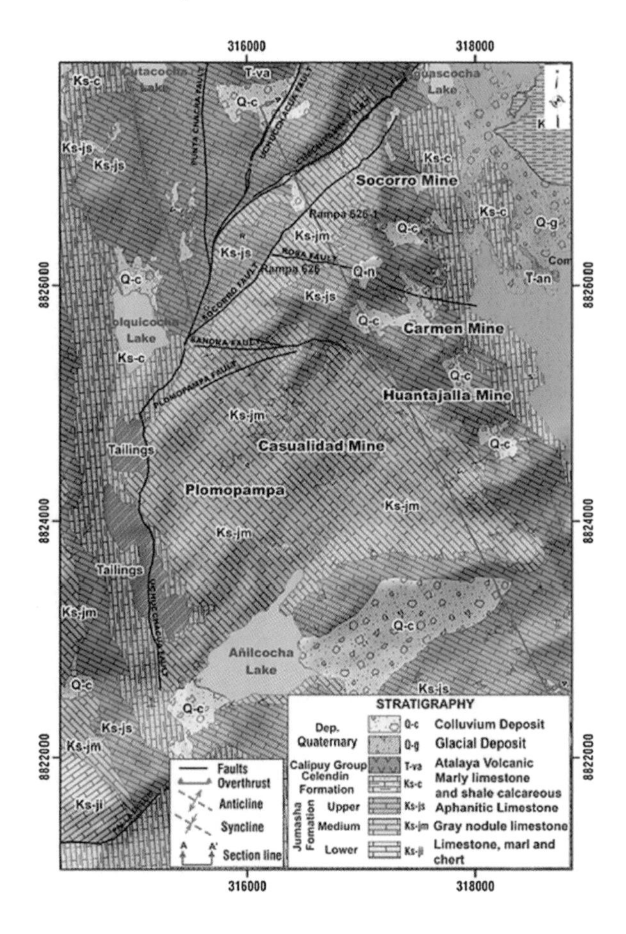

Table 1 Summary of permeability of hydrogeological units

Hydrogeological unit	Hydraulic conductivity (m/d)				Degree of permeability	Classification
	Min.	Med.	Max.	Geom.		
Moraine	0.07	0.10	0.13	0.07	Low	Aquitard
Massive limestone	0.0001	0.010	0.03	0.01	Very low	Aquiclude
Fractured limestone	0.10	3.10	18.60	0.70	Moderate–High	Aquifer

Min: Minimum; Med: Mean, Max: Maximun and Geom.: Geometric Mean

contrast, springs rise in the low points of the basin, reaching peak flows of 693 l/s, located in fractures areas, zones of lithological contact and fluvioglacial sediments. It is calculated that the largest discharges occur towards Añilcocha Lake, which is the lowest hydrological point (Fig. 4).

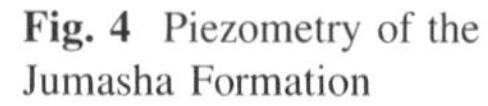

Fig. 4 Piezometry of the Jumasha Formation

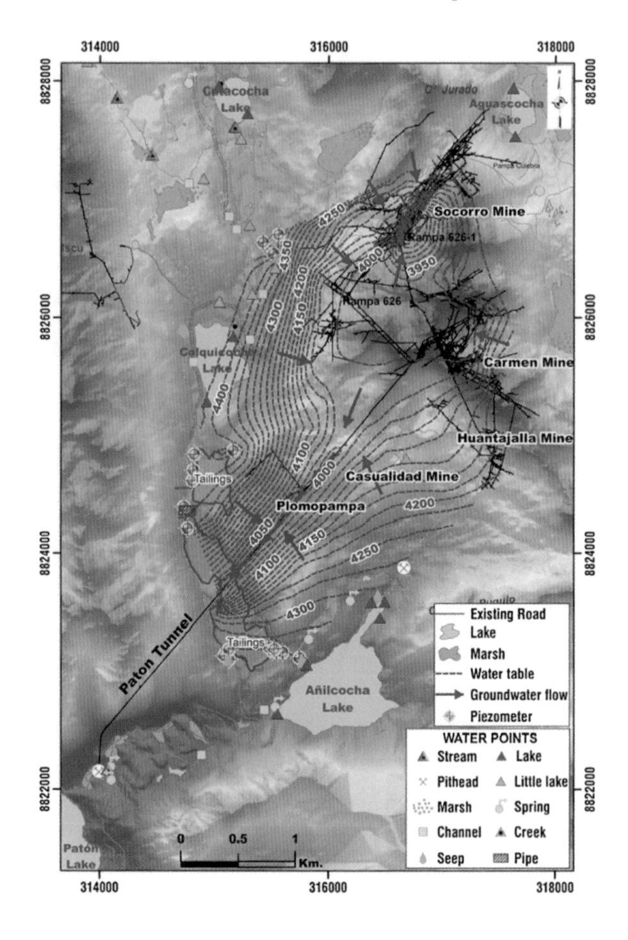

Mining operations have modified the natural flow regime creating drawdown cones, which cause mine drainage to converge towards the Patón tunnel, which serves as the drain for the mine. Underground mining in the Uchucchacua mine occurs in three zones: Socorro, Carmen and Huantajalla, situated 470 m below the ground surface. Ore extraction occurs between 4,550 and 4,120 m. a.s.l., with an even deeper level between 4,120 and 3,990 m a.s.l. The mine is dewatered using a system of submersible pumps. The water from higher and lower levels is collected at the 3,990 m level and diverted into a drainage tunnel that takes it to the Patón lagoon (Fig. 5).

At the time of writing, seasonal pumped flows of mine water oscillate between 141 l/s in the dry months and 1,504 l/s in wet months (Fig. 6). The discharges vary through the mine, with the greatest flows coming from levels 3,990 and 4,180 m. There was a marked increase in inflow between April 2013 and February 2014. Mine drainage occurs via two main drainage systems (Patón and Huantajalla tunnels extending to the ground surface), which discharge between 2,022 l/s in wet months and 146 l/s in dry months.

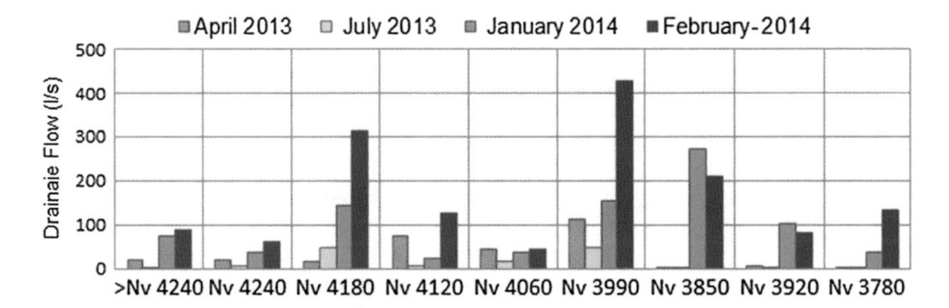

Fig. 5 Evolution of mean monthly pumped flows

Fig. 6 Three-monthly evolution of pumped water in the mine

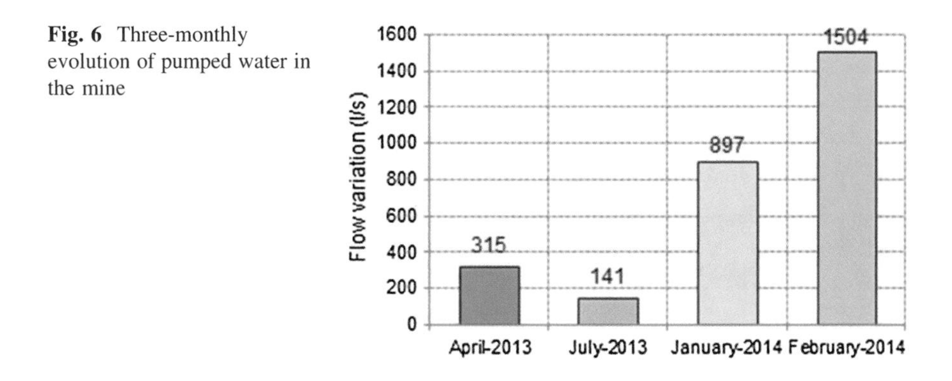

Interconnection between surface water and groundwater clearly encourages inflow into the deeper levels of the mine, especially at the 3,780 m a.s.l. mine level. As indicated above, infiltration occurs through dolines, sinkholes and the interconnected open faults and main mineral veins. These favour a high rate of infiltration—in the range of 40–70 %—due to the large percolation capacity that derives from the pronounced karstification of the aquifer (Fig. 7).

The dominant water type of the groundwater in this area is calcium sulphate. TDS of these waters varies from the minima observed in the springs (208–434 mg/l) to the peak TDS observed inside the mine (521–1,680 mg/l). There is also calcium bicarbonate waters, which exhibit the lower TDS values (124–198 mg/l). Water inside the mine interacts with water in the mineralised zone, which is slightly acid. As a consequence of mixing with percolation water, it becomes alkaline, with a low metalloid content.

The interconnection that exists between surface water and groundwater clearly influences the volume of water reaching the deepest part of the mine, especially at the 3,780 m.a.s.l. level. This influence is demonstrated by the chloride content. Groundwater springs and resurgences have very low chloride content (0.6–2.0 mg/l), while water samples from inside the mine has higher values (1.2–13.0 mg/l). Nevertheless, in February 2014 (corresponding to a wet period), chloride values fell drastically as a result of mixing with inflows derived from rainwater. At this time,

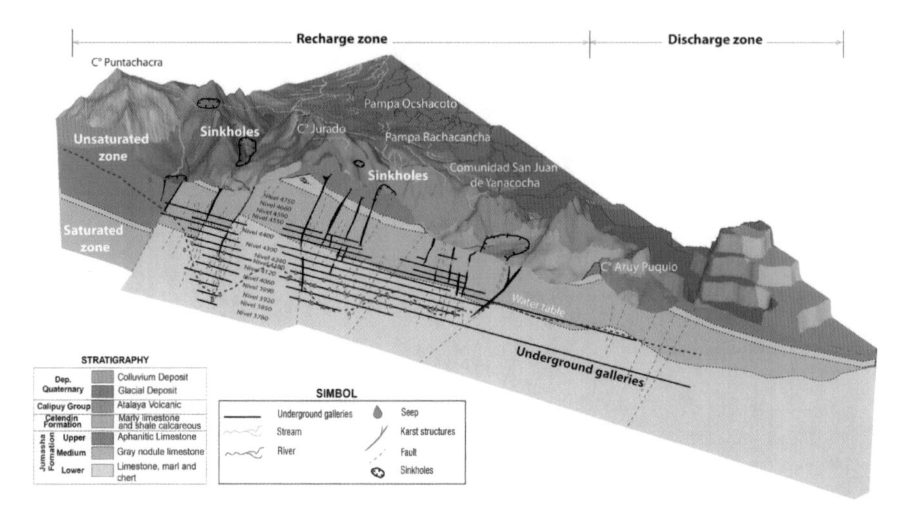

Fig. 7 Conceptual hydrogeological model of the Uchucchacua mine

groundwater samples from the upper levels of the aquifer fluctuated between 0.1 and 1.0 mg/l, while concentrations in water collected inside the mine were between1.1 and 10.3 mg/l.

4 Final Considerations

The Uchucchacua mine is situated in the karstic limestone of the Jumasha Formation, whose numerous karstic landforms allow rapid percolation and flow into the deeper working levels of the mine. These flows affect mining activities. The dewatering operations have modified the natural flow regime, creating drawdown cones and diverting mine water into the Patón tunnel, which acts as a drain for the mine.

Presently, mean monthly flows of pumped mine water are between 141 l/s (dry months) and 1,504 l/s (wet months). The discharge is not homogeneous through the mine, with the greatest flows coming from the mine levels at 3,780 and 4,180 m a.s.l. There was a marked increase in flow between April 2013 and February 2014. The more saline mine waters have a calcium sulphate type, with TDS of 521–1,680 mg/l, while less saline waters have a calcium bicarbonate type and TDS of 124–198 mg/l.

This study highlights the interconnection that exists between surface waters and groundwater in the study area and shows how surface flows influence the volume of water that has to be drained from the deeper levels of the mine, especially from the 3,780 m level. Direct infiltration occurs through dolines, sinkholes, interconnected open faults and mineral veins, with a high rates of recharge.

The conceptual model proposed will be refined on the basis of new hydrochemical data, data from newly-installed piezometers, including water level measurements and monitoring of groundwater and mine water discharges. The results of these studies will be used to propose management actions aimed at reducing the rate of percolation into the karst.

Acknowledgments We grateful for the helpful corrections and suggestions made by anonymous reviewers, which improved the article. Similarly to the Compañía de Minas Buenaventura S.A.A, for permission to use the information from mine.

References

Apaza D (2013) Evaluación Hidrogeológica de Las Labores Subterráneas—Mina Uchucchacua, Fase I. Informe inédito Compañía Minas Buenaventura S.A.A. Hidroandes Consultores S.A.C, San Isidro

INGEMMET (1996) Boletín Geología del Cuadrángulo Oyón, Hoja 22-J

Jacay J (2005) Análisis de la sedimentación del sistema cretáceo de los Andes del Perú Central. Revista del Instituto de Investigación FIGMMG 8(15):49–59

Evans D, Letient H, Aley T (2005) Aquifer vulnerability mapping in karstic terrain Antamina mine. Perú Infominer 3(6):1–13

Caves and Mining in Brazil: The Dilemma of Cave Preservation Within a Mining Context

A.S. Auler and L.B. Piló

Abstract The exploitation of mineral reserves in Brazil, especially limestone and iron ore, is currently restricted due to the existence of caves. The vast majority of caves documented in the country over the last 4 years (approximately 3,000) have been identified through environmental studies conducted for mining operations. To determine whether a cave should be protected or not, a series of criteria were formally established by recent (2008/2009) federal laws. Four classes of cave relevance were formally designated, based primarily on geological and biospeleological criteria. *Maximum Relevance* caves must be protected, together with a 250 m buffer zone. *High Relevance* caves may be removed, provided that two other high relevance caves, preferably within the same geological unit and containing similar characteristics, are permanently protected. However, the acquisition of areas containing caves, especially within iron ore regions, has become extremely difficult due to the high price of iron ore. *Medium Relevance* caves may be subject to removal, but speleological compensation must be applied. *Low Relevance* caves may be mined with no need for environmental compensation. Although these laws occasionally permit cave destruction, their ambiguous specifications and numerous criteria produce a highly restrictive scenario in which approximately 85 % of all caves are categorized as Maximum or High Relevance. The situation is further exacerbated by the very low minimum length of 5 m for any void to be classified a cave, producing a high number of caves regardless of lithology. Conducting a full cave environmental study, besides being financially costly, takes approximately 1.5 years to complete, primarily due to the requirement to perform two biospeleological sampling events during dry and wet seasons. The protection of several caves within mining areas has significantly decreased access to exploitable reserves, causing caves to remain under severe economic pressure. While Brazilian law emphasizes cave preservation, it provides no specific provisions for the protection of other karst features or karst aquifers.

A.S. Auler (✉) · L.B. Piló
Instituto Do Carste, Rua Brasópolis 139, Belo Horizonte, MG 30150-170, Brazil
e-mail: aauler@gmail.com

B. Andreo et al. (eds.), *Hydrogeological and Environmental Investigations in Karst Systems*, Environmental Earth Sciences 1,
DOI 10.1007/978-3-642-17435-3_55

1 Introduction

The rapid growth of mining activities in Brazil due to high prices on metal commodities in the international market (such as iron ore), coupled with internal market growth for construction and agriculture (cement and lime) has threatened caves and karst features. Over the last decade, new cave-related laws have called for substantial changes in the ways cave studies must be performed in areas marked for mining operations. Prior to detailed studies, any void over 5 m in horizontal and vertical length must be protected, together with a 250 m buffer zone surrounding the cave. The preservation or removal of caves depends on a detailed, 1–1.5 year-long study to determine the degree of cave significance. Cave significance studies require the analysis of a large number of criteria, mostly geological and biological. With a sudden surge of cave studies beginning in approximately 2008, there are now cave significance studies for over 1,000 caves, accounting for roughly 10 % of all caves registered in the national cave database (CECAV 2014).

Of the four possible outcomes of a cave significance study (Maximum Relevance, High Relevance, Medium Relevance and Low Relevance), only Maximum Relevance provides complete protection for a cave. However, depending on the criteria, even this degree of protection may not be permanent because additional studies may provide new data that may decrease the relevance status. While a High Relevance cave may be mined away, it requires two caves of the same rock type with similar characteristics to be preserved, which poses additional difficulties. This compensation arrangement makes areas that contain relevant caves extremely valuable to mining companies. Medium Relevance caves may be quarried away, pending cave- and karst-related environmental compensation that may not involve cave preservation. Low Relevance caves are not subject to protection or environmental compensation.

Cave-relevance studies performed so far have, for the vast majority of cases, classified caves of Maximum and High Relevance. Medium and Low Relevance caves have been largely absent. These highly restrictive criteria impose serious limitations on mining activity expansion given the very low minimum length of 5 m being classified a cave, thus resulting in a large number of caves, in addition to the requirement for a 250 m buffer zone and degree of relevance (Maximum or High). These criteria currently present some of the most serious restraints to mining development in Brazil and have had severe economic and social impacts on mining-dependent areas.

This challenging dilemma has precipitated intense pressure on the government to adopt a less restrictive legal framework. However, the increase in the number of cave studies has led to a complete reshaping of speleology in Brazil, with an ever-increasing number of professional speleologists. Although caves are in the spotlight, protected by a complex legal apparatus, there is no mention of other karst features, creating an unjustified bias towards the preservation of caves. This paper provides an overview of the environmental studies created under Brazil's recent cave laws and of the ever-growing challenge of preserving significant caves without impairing economic development.

2 Legal Background

Prior to 1988, Brazil imposed no specific laws regarding caves. During the elaboration of Brazil's new constitution, a lobby of cavers succeeded in including caves as property of the union, together with lakes, rivers, sea and natural resources, including those of the subsoil (Brasil 1988). This new approach was the starting point from which a large number of legal initiatives took effect. Being both classified as property of the union, caves and mines became equally dependent on federal-level government decisions.

Two years later, IBAMA (The Brazilian Institute of Environment and Renewable Natural Resources) stated through its Directive 887 that caves should be included in environmental studies and could not be impacted (IBAMA 1990). The Directive proposed that caves be used for technical or scientific purposes only, including tourism and educational and cultural activities. This same Directive also made compulsory the need for permits when sampling or performing scientific studies in caves. Directive 887 also introduced the need to protect a 250 m buffer zone perpendicular to the limit (walls) of the mapped cave. In the same year, Federal Decree 99556 re-emphasized the need to protect all caves, stating that caves belong to Brazilian heritage and should be integrally protected (Brasil 1990). Thus, at this early stage, cave law was extremely restrictive because no cave, regardless of its size or importance, could be impacted. If all caves were to be protected, it would result in the shutdown of numerous mining projects. However, despite the law, in many situations speleological studies were not performed, which led to the loss of many caves.

In 1997, the National Center for Research and Conservation of Caves (CECAV) was created as a specialized department of IBAMA. With offices in cave-dense areas, CECAV activities included, at this early stage, the assessment of environmental studies related to caves in mining areas. The professional speleologists of CECAV provided much-needed expertise and represented a conservative force towards limiting cave impact by mining projects. Currently, CECAV is no longer involved in environmental licensing, and has instead shifted its focus towards cave research and management in federally protected areas.

A major legal change occurred in 2004, when Normative Resolution 347 from CONAMA (National Environmental Commission) (CONAMA 2004) recognized that caves may vary in environmental importance. For the first time, it was legally accepted that studies on ecological, scientific, cultural and scenic aspects should be performed to determine the significance of a cave. The same resolution instituted the creation of CANIE, a national cave database managed by CECAV.

The two most recent laws, Federal Decree 6640 (Brasil 2008) and Normative Instruction 2 (MMA 2009) provided rubrics outlining four classes of cave significance and the criteria involved in their determination. Maximum relevance caves must possess at least one of the following: (i) a unique or rare genesis, (ii) notable dimensions in length, area or volume, (iii) unique speleothems, (iv) geographical isolation, (v) essential shelter for threatened species, (vi) essential habitat for

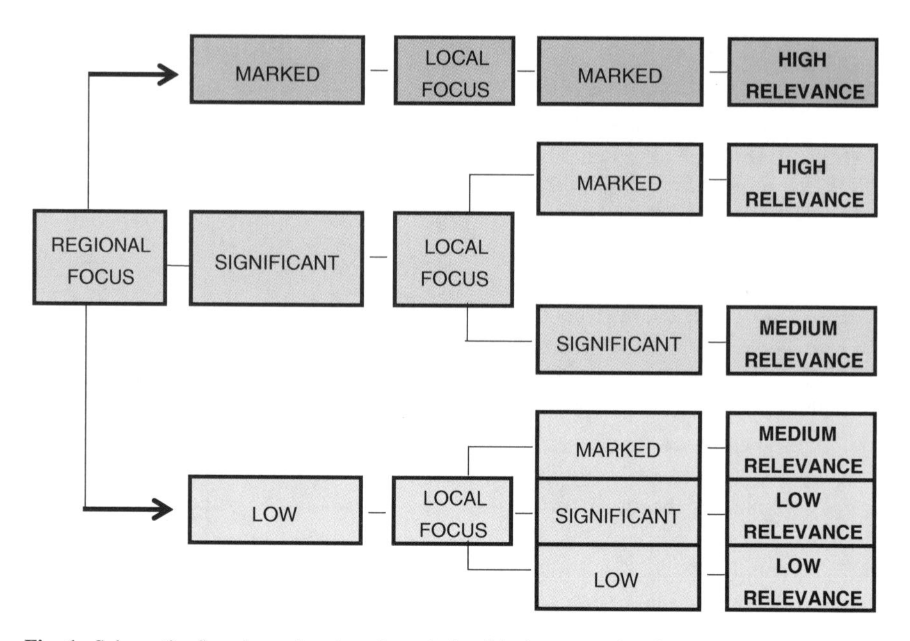

Fig. 1 Schematic flowchart showing the relationship between the degrees of relevance and importance of criteria (marked, significant and low) at both local and regional scales of analysis (MMA 2009)

endemic or relict troglobites, (vii) habitat for a rare troglobite, (viii) unique ecological interactions, (ix) significant paleoenvironmental importance and (x) significant historic, cultural and religious value.

Identifying a High Relevance cave is more complex, relying on comparisons between several items across caves both at a local and regional scale (Fig. 1). The local scale is defined as a continuous geomorphological feature (ridge, plateau, etc.) containing caves. The regional scale is the speleological unit, including a larger karst or cave area displaying physiographic homogeneity. A high relevance cave should be of high importance at both the local and regional scales, or of medium importance at the regional scale but of high importance at the local scale. A minimum of 22 criteria must be taken into consideration, and the presence of only one is enough to grant a cave High relevance status (Table 1).

Medium and Low Relevance caves are those which do not meet the Maximum or High relevance criteria. Due to the extensive number of criteria listed, caves are rarely classified as Medium or Low Relevance.

The legal apparatus, besides being complex, contains concepts and procedures that are either dubious or incorrect because the method was not tested prior to publication. The definition of a cave, for example, refers to a natural underground cavity accessible to humans, regardless of its dimensions and with or without a natural entrance (Brasil 2008). For instance, small passages accessible only to cave fauna and entrance-less caves detected by geophysics fall into a grey zone in which it is not possible to ascertain whether they should be considered or not. The

Table 1 Criteria that must be matched to classify a cave of high relevance (MMA 2009)

Criteria
I—Type locality
II—Established populations of species with important ecological role
III—New taxons
IV—High species richness
V—High relative abundance of species
VI—Peculiar fauna composition
VII—Troglobites that are not rare, endemic or relict
VIII—Troglomorphic species
IX—Obligatory trogloxenes
X—Population of exceptional size
XI—Rare species
XII—High length when compared to other caves at regional scale
XIII—High area when compared to other caves at regional scale
XIV—High volume when compared to other caves at regional scale
XV—Significant presence of rare speleogenetic features
XVI—Perene lake or underground drainage, with marked influence over any of the criteria in this table
XVII—Diversity of chemical deposition with many types of speleothem and varied processes of deposition
XVIII—Notable configuration of speleothems
XIX—High influence of the cave over the karst system
XX—Interrelation of the cave with a maximum relevance cave
XXI—Aesthetic and scenic values of national and international significance
XXII—Systematic public visitation with a regional or national scope

These criteria refer to marked importance at both regional and local scales (Fig. 1). If none of these are met, additional criteria (significant importance at the local scale) must be applied

subjective wording used in some definitions, such as "notable" or "unique" among many others, produce differences in interpretation depending on the experience of the consultant; therefore, studies conducted by different teams can produce different relevance diagnoses.

Due to the highly restrictive nature of the legal framework, most mining projects with identified caves are facing losses in exploitable reserves. The economic impact of caves has led to a growing lobby of mining industry professionals advocating for less rigid laws to release the mines at the expense of imperiling caves.

3 A Review of Cave Significance Studies in Brazil

The promulgation of Decree 6640 in late 2008 prompted a marked increase in cave-related consulting work. Not only mining operations but any new development, regardless of area or cave potential, was required to commission a

Fig. 2 GPS-equipped cave-prospecting teams of 2–3 members walk alongside one another, creating 50 m-wide parallel lines that represent the walking route. Line spacing depends on cave potential and visibility (which varies with rock type, vegetation and relief). Ideally, spacing equals the distance of visibility/2. Areas between lines may be checked if necessary. Areas containing high-potential features such as dolines and rock outcrops must be inspected in more detail. Orange circles denote a 250 m buffer zone of identified caves

speleological assessment. The requirement applied to projects including reservoirs, electricity lines, urban developments, roads and railroads. Some of these new developments, such as transmission lines through the Amazon or very large reservoirs, can span hundreds of thousands of hectares, requiring a workforce of speleologists not readily available in Brazil. This situation created a demand for cavers that substantially altered the nature of caving activities in Brazil, which was previously dominated by non-professionals.

The first step in any speleological assessment involves cave identification, which is carried out by one or several 2–3-person teams walking 20–150 m alongside one another depending on cave potential (based on rock type) and visibility (which is usually controlled by relief, and vegetation especially) (Fig. 2). Considering the 5 m cave length limit, even low potential rocks may yield caves. Limestone areas may result in high cave densities on the order of 0.35 caves/hectare. Iron ore areas present lower densities, averaging approximately 0.15 caves/hectare. The largest cave assessment project ever performed in Brazil involved the characterization of a limestone area containing 750 caves.

Cave dimension is one of the most critical factors in determining cave significance. Length, depth, area and volume must be measured with a high degree of accuracy. A BCRA 5D grade cave survey (Day 2002) is the standard format for cave mapping, producing a high-precision cave map that determines cave length through the conservative definitions presented by Chabert and Watson (1981). Cave depth and area are easily obtained using cave surveying software and CAD drawing packages. Cave volume is a more complex parameter, and its precision depends on the number of cross-sections present, which are averaged by area according to the formula $V = (h_1 + h_2 + \cdots h_n) / n$ (where v is the cave volume, *h* is the average height of cross-sections and *n* is the number of cross-sections). Average height is calculated by dividing the cross-section area by the distance between the cross-section extremes. Survey work tends to progress at an average of two to three caves (with approximate dimensions of 20 m) per day for a mapping team of three people.

Geological assessment requires the analysis of several parameters related to the cave geomorphological context (whether associated with dolines, scarps, valleys, etc.), rock type, rock structure (including structural measurements), hydrological qualitative assessment, speleothem description, clastic sediment characterization and a brief nondestructive paleontological assessment. An experienced, two-person team is able to perform geological assessments of 2–3 caves a day.

Currently, laws require a seasonal approach to cave fauna characterization. Wet (austral summer) and dry (winter) sampling events are required. Because most caves are small, the number of external, non-cave obligate organisms tends to be high and the typical species-accumulation curve never stabilizes. Using the map, cave floor area is divided into a series of 10 m^2 quadrants with a 30-min sampling effort for each quadrant, resulting in average of approximately 110 specimens per cave. A three-member-strong biospeleological team can successfully sample 2–3 caves a day. The number of species collected tends to be extremely high, with some of the largest projects exceeding 30,000 specimens. Screening and identification tends to be time-consuming and external expert advice is needed to obtain the species level. The identification of potential troglobites requires a careful assessment by specialists. Over 100 new species have been sampled during cave environmental studies, only a small fraction of which have been scientifically described.

The total effort required for a complete cave assessment study is significant, considering laboratory and report-production time and involves a team of no less than 25 specialized personnel over a period of approximately 1.5 years. In a sample of 386 caves within iron rocks studied by Carste Consultores Associados, 17.6 % were of maximum relevance, 68.1 % were of High relevance and 12.7 % were of Medium Relevance. Only 1.6 % of the total sample was classified of Low Relevance. Considering both limestone and iron ore, the most frequent criteria that confers Maximum Relevance to caves is habitat for a rare troglobite (69 %) followed by notable dimensions in length, area or volume (14 %) for a total sample of 797 caves. For High Relevance caves, high relative abundance of species is the commonest parameter, occurring in 32 % of all caves considering the same sample.

4 Impact on Mining Projects

As was previously mentioned, the density of caves within iron ore and limestone tends to be high; thus, as a rule, with a minimum length of 5 m for cave identification, any area with rock outcrops will inevitably contain caves. With an obligatory buffer zone of 250 m (which results in a minimum protected area of 20 ha), cave existence can impose considerable restrictions on mining development.

Iron ore caves present an especially challenging situation due to their peculiar genesis. The original bedrock (Banded Iron Formation—BIF) comprises alternating layers of iron and silica. As silica tends to be geochemically more mobile, the removal of silica creates high-grade ore and the initial rock porosity that eventually evolves into caves. Thus, there is a well-marked relationship between the occurrence of iron ore caves and high-grade ore. Thus, cave preservation will likely restrict mining development. However, full exploitation of iron resources will lead to a loss of caves. This dilemma is not without serious economic and social consequences, as significant decreases in exploitable area (sometimes in excess of 50 % of the original planned mine) and delays from speleological studies lead to severe financial drawbacks.

Iron ore prices have increased dramatically over the last decade, due mostly to the expansion of the Chinese market. Prices have soared from USD $13.82 per dry metric ton in December 2003 to USD $135.79 in December 2013 (Index Mundi 2014), a staggering rise of 883 %, prompting the expansion of existing mines and the development of new ones.

In a simple modeling exercise, a single cave located within an iron ore area in Brazil (such as Carajás in northern Brazil) together with its 250 m buffer zone will represent a major loss in ore reserves (Fig. 3). Considering current ore prices and an initial protection buffer of 20 ha for a 200 m deep stepped bench, the total weight would exceed 280 billion tons, totaling in excess of USD $38 billion for Carajás high-grade ore. Depending on the site, this can be regarded as minimum because some deposits are in excess of 300 m in depth (Trendall et al. 1998). Although the same considerations are valid for other rock types, the total financial loss will be much smaller owing to less-valuable mineralization.

Since 2008, over 1,000 caves have had full environmental studies completed for various mining projects. As of December 2013, only approximately 30 caves have been granted approval for removal. Considering the vast majority of caves have been classified as high relevance, this delay is due mostly to difficulties revolving around environmental compensation. The requirement to choose two caves of equal relevance and characteristics for each high relevance cave to be removed has proved a daunting task, and in iron ore areas especially. Due to the relationship between caves and high-grade ore, there is increasing demand and limited availability for these highly priced areas. Furthermore, an additional environmental assessment is required for caves to be used for compensation. The impossibility of applying the "2 per 1" rule has prompted the release of the Directive 30 (ICMBIO 2012), which opens new possibilities, including the provision of financial support

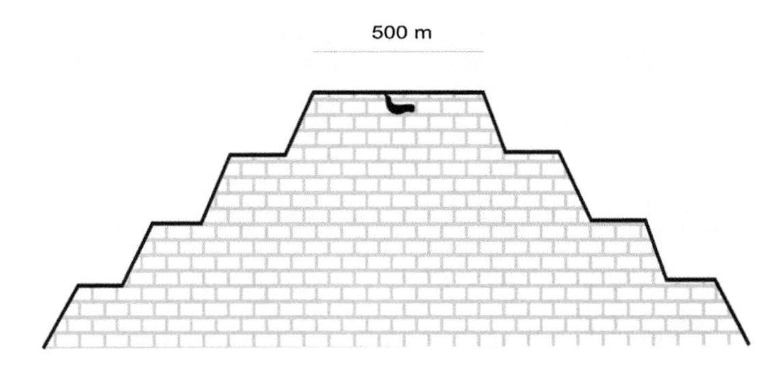

Fig. 3 In a simple simulation, a cave, with its 250 m radius of protection, plus the vertical extent of the ore body (in excess of 300 m in the Carajás deposit) may represent financial losses on the order of more than $USD 20 billion

for the creation or maintenance of conservation units with caves. These new possibilities are now being explored by several mining companies, although due to highly bureaucratic processes not a single approval has been granted.

5 Final Remarks

The need to distinguish between significant and non-significant caves to allow for mining expansion has profoundly altered the cave and mining situation in Brazil. The presence of caves is now considered a major restriction to mine development, which has led to a focus of cave assessment studies and especially those concerned with biospeleology. The rather restrictive nature of the law has also resulted in increased political pressure to speed up the environmental study process, which has in turn solicited counter-pressure from environmentalist groups.

This new impetus towards cave studies has resulted in a spectacular increase in the volume of cave data. The number of new caves identified surpasses 1,000 per year, a very large amount considering that by 2008, only approximately 7,000 caves had been registered in the Brazilian cave database. New geological and biological findings have resulted in the discovery of a great number of new, cave-adapted species and minerals, and many of these findings represent proprietary data that are yet to be published.

With the restrictive nature of the law and the expansion and operation of several mines at peril, government involvement is ever-increasing. Economic and social impacts due to cave preservation are considerable and are thus prompting increasing pressure on cave specialists.

None of the existing laws take the karst system into consideration. Although their relationship to caves is recognized, karst features remain unprotected. This paradoxical situation results in a cave-focused approach that fails to address karst as an integrated system, leading to potential fragmentation and losses in both geo- and biodiversity.

Acknowledgments This study is based on work developed at Carste Consultores Associados Ltda. We acknowledge assistance from Cristiano Marques, Gustavo Perroni and Geraldo V. Santos in data processing and interpretation.

References

Brasil (1988) Constituição da República Federativa do Brasil de 1988. Presidência da República. http://www.planalto.gov.br/ccivil_03/constituicao/constituicao.htm. Acessed 27 Jan 2014

Brasil (1990) Decreto Federal 99.556. Presidência da República. http://www.planalto.gov.br/ccivil_03/decreto/1990-1994/D99556.htm. Acessed 27 Jan 2014

Brasil (2008) Decreto Federal 6.640. Presidência da República. http://www.planalto.gov.br/ccivil_03/_Ato2007-2010/2008/Decreto/D6640.htm. Acessed 27 Jan 2014

CECAV (2014) Cadastro Nacional de Informações Espeleológicas—CANIE. ICMBIO. http://www.icmbio.gov.br/cecav/canie.html. Acessed 27 Jan 2014

Chabert C, Watson R (1981) Mapping and measuring caves: a conceptual analysis. NSS Bulletin 43:3–11

CONAMA (2004) Resolução N. 347. Diário Oficial 176:54–55

Day A (2002) Cave surveying. British Cave Research Association, Buxton

IBAMA (1990) Portaria N° 887 de 15 de junho de 1990. Diário Oficial 117:11844

ICMBIO (2012) Instrução Normativa N. 30. Instituto Chico Mendes de Biodiversidade. http://www.icmbio.gov.br/cecav/downloads/legislacao.html. Acessed 27 Jan 2014

Index Mundi (2014) Iron ore monthly prices. Index Mundi. http://www.indexmundi.com/commodities/?commodity=iron-ore&months=120. Acessed 27 Jan 2014

MMA (2009). Instrução Normativa N. 2. Ministério do Meio Ambiente. http://www.icmbio.gov.br/cecav/images/download/IN%2002_MMA_criterios_210809.pdf. Acessed 27 Jan 2014

Trendall AF, Basei MAS, de Laeter JR, Nelson DR (1998) SHRIMP zircon U-Pb constraints on the age of the Carajás formation, Grão Pará group. J S Am Earth Sci 11:265–277

Definition of Microclimatic Conditions in a Karst Cavity: Rull Cave (Alicante, Spain)

C. Pla, J.J. Galiana-Merino, J. Cuevas-González, J.M. Andreu, J.C. Cañaveras, S. Cuezva, A. Fernández-Cortés, E. García-Antón, S. Sánchez-Moral and D. Benavente

Abstract Rull Cave (Alicante, SE Spain) is a shallow karstic cavity located in metre-thick beds of Miocene limestone conglomerate and overlain by soil with little profile development. A microenvironmental monitoring system was installed in order to record the exchange between the cave and the external atmosphere. Data were collected every 15 min over a period of 14 months (from 22/11/2012 to 13/01/2014). Both radon and CO_2 concentration values changed over the course of the annual cycle and were strongly controlled by the difference in air temperature between the exterior and the cave atmosphere. Wavelet transform was applied to the data to determine the influence of visitors on the environmental parameters controlling the cave's microclimate.

Keywords Karst · Gas exchange · Microclimatic measures · Rull cave

C. Pla (✉) · J. Cuevas-González · J.M. Andreu · J.C. Cañaveras · S. Cuezva · D. Benavente
Dpto. Ciencias de la Tierra y del Medio Ambiente, Universidad de Alicante, Campus San Vicente del Raspeig S/N, San Vicente del Raspeig, 03690 Alicante, Spain
e-mail: c.pla@ua.es

C. Pla · J.C. Cañaveras · D. Benavente
Laboratorio de Petrología Aplicada, Unidad Asociada CSIC-Universidad de Alicante, Campus San Vicente del Raspeig S/N, San Vicente del Raspeig, 03690 Alicante, Spain

J.J. Galiana-Merino
Dpto. Física, Ingeniería de Sistemas y Teoría de la Señal, Universidad de Alicante, Campus San Vicente del Raspeig S/N, San Vicente del Raspeig, 03690 Alicante, Spain

S. Cuezva · A. Fernández-Cortés · E. García-Antón · S. Sánchez-Moral
Museo Nacional de Ciencias Naturales (CSIC), José Gutiérrez Abascal, 2, 28006 Madrid, Spain

B. Andreo et al. (eds.), *Hydrogeological and Environmental Investigations in Karst Systems*, Environmental Earth Sciences 1,
DOI 10.1007/978-3-642-17435-3_56

1 Introduction

Underground cavities have been widely studied in order to assess the behaviours, patterns and trends of current microclimatic factors (Fernández-Cortés et al. 2009, 2011 among others). Fluctuations in temperature, relative humidity or barometric pressure enable determination of annual cycles, which can then be studied in detail to understand underground systems. These fluctuations govern the exchange of gases such as CO_2 and ^{222}Rn through the soil membrane between underground and environmental atmospheres, and determine the outgassing and isolation stages of a subterranean atmosphere (Cuezva et al. 2011). The dynamics of these processes can be studied by applying new techniques such as wavelet analysis (Galiana-Merino et al. 2014) to time series.

The aim of this study is to understand the relationship between measured microclimatic parameters and the behaviour of trace gases (CO_2 and ^{222}Rn) by analysing variations inside and outside Rull Cave, and to quantify the influence of visitors. Wavelet analysis is applied to data in order to determine the relationships prevailing between factors. A microenvironmental system was installed inside the cave to obtain sufficient information about climate parameters and their influence on the natural dynamics of the underground system in relation to outside environmental conditions.

2 Field Site Description

Rull Cave (38° 48′ 40″ N; 0° 10′ 38″ W) is located in the north eastern sector of Alicante province, on the Spanish Mediterranean coast (30 km from the coast line). The cave is located in the Vall d'Ebo basin, which originated during the late Miocene in response to compressive dynamics affecting underlying Cretaceous limestones. This basin was filled by massive series of limestone conglomerates deposited on top of massive clayey marls (middle Miocene). Rull Cave is located in these series of massive conglomerates characterised by considerable textural and petrophysical complexity (de Carvalho et al. 2013). The overlying soil has little profile development with few horizons. Both the fracture system and the mineral composition of the host conglomerates determine the infiltration process of meteoric waters and the exchange of gases. The cavity presents an almost rounded shape, and comprises a total area of 1535 m^2. The ceiling reaches a maximum height of 20 m in the central chamber, and the relative thickness of the overlying host rock varies from 9.3 to 22.3 m depending on the point selected. The study area is characterised by a thermo-Mediterranean sub-humid climate (Rivas-Martínez 1984). Inside the cave, calcite speleothems such as stalactites, columns, curtains or crusts are common, and fallen blocks of different sizes are present due to old ceiling collapses. Currently, the cavity has a single entrance shut by a door located at the top part of the cave, at the highest level. The door has a considerable area (approximately 3 m^2). Rull Cave is a tourist cavity, which has been equipped to receive visitors (during the period studied, Rull Cave received an average of 40 people a day), and spotlights and concrete corridors have been installed to allow people to move around easily when visiting the cave.

3 Methodology

A microenvironmental monitoring system was installed to record microclimatic data on cave air. The system consisted of one complete station with a data logger (dataTaker DT50, Grant Instruments Ltd., Cambridge, UK) and a set of sensors. Data were recorded at fifteen minute intervals. Air temperature and relative humidity were measured by a HygroClip S3 sensor (Pt100 1/10 DIN temperature sensor and a Rotronic humidity sensor). Atmospheric pressure was measured with a silicon capacitive sensor (Vaisala BAROCAP). A Ventostat 8002 sensor (Telaire, USA) was used to measure CO_2 concentrations, while radon concentrations were determined with a Radim 5 Monitor. Outside the cave, a weather station with an independent data logger (HOBO Onset, USA) recorded air temperature, relative humidity, rainfall and wind (speed and direction) values every 15 min. An analysis of the recorded measurements was carried out to determine the prevailing patterns in Rull Cave. In addition, analysis using the software package, Environmental Wavelet Tool (Galiana-Merino et al. 2014), which facilitates wavelet analysis and filtering, cross-correlation, entropy of curves or spline analyses was applied to the obtained data.

4 Prevailing Microclimatic Conditions

The microclimatic records obtained over a period of 14 months (November 2012–January 2014) are shown in Fig. 1. Rull Cave presented high thermo-hygrometric stability throughout the annual cycle. Recorded values of inside temperatures revealed an average value of 15.89 °C. Temperature oscillations were ± 1.14 °C. Relative humidity inside the cave was always higher than 95 %; thus, the cave remained under a state of saturation. The average CO_2 concentration was 1927 ppm, although values in excess of 4000 ppm were recorded when the outdoor temperature was higher than the cave air temperature. Radon concentration showed a similar behaviour; while minimum values were detected from November to March, higher values were reached in summer. The average ^{222}Rn concentration inside Rull Cave was 1520 Bq m^{-3}, although values reached nearly 4000 Bq m^{-3}.

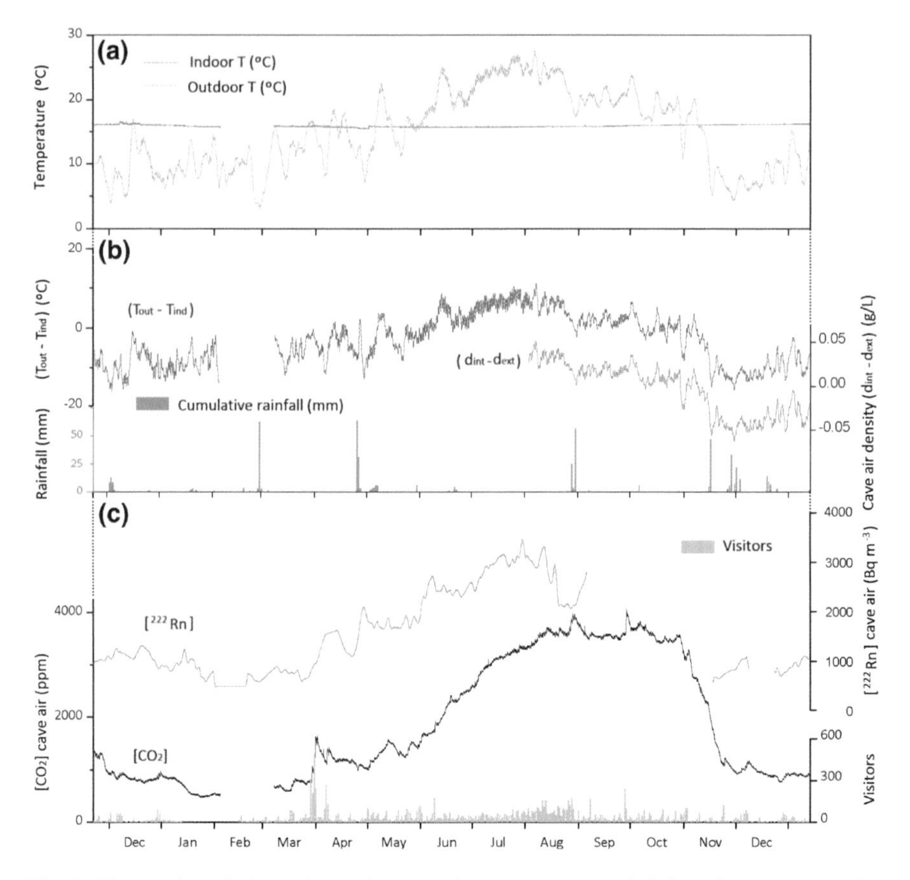

Fig. 1 Time series of the main environmental parameters recorded in rull cave (November 2012–January 2014). **a** Evolution of indoor and outdoor air temperature. **b** Rainfall (absolute daily values), air temperature difference ($T_{out}-T_{ind}$) and air density difference ($d_{int}-d_{ext}$). **c** Carbon dioxide and radon content of the cave air. Total daily visitors

5 Discussion

Cave ventilation was strongly dependent on relationships between indoor and outdoor temperatures (Fig. 1). Density differences between the cave and the outside air were primarily responsible for the isolation or outgassing state of the cavity (García-Antón et al. 2014). Furthermore, relative humidity, barometric pressure, cave geometry and the influence of visitors were also significant factors. In winter, ventilation was predominantly driven by air density differences between the cave and the external atmosphere, and air renewal processes maintained radon and CO_2 levels below the average annual concentration. In contrast, from May to November, temperatures inside the cave remained below the external temperature; the cave functioned as a trap for cold air and as a net CO_2 reservoir. In general, the ^{222}Rn baseline

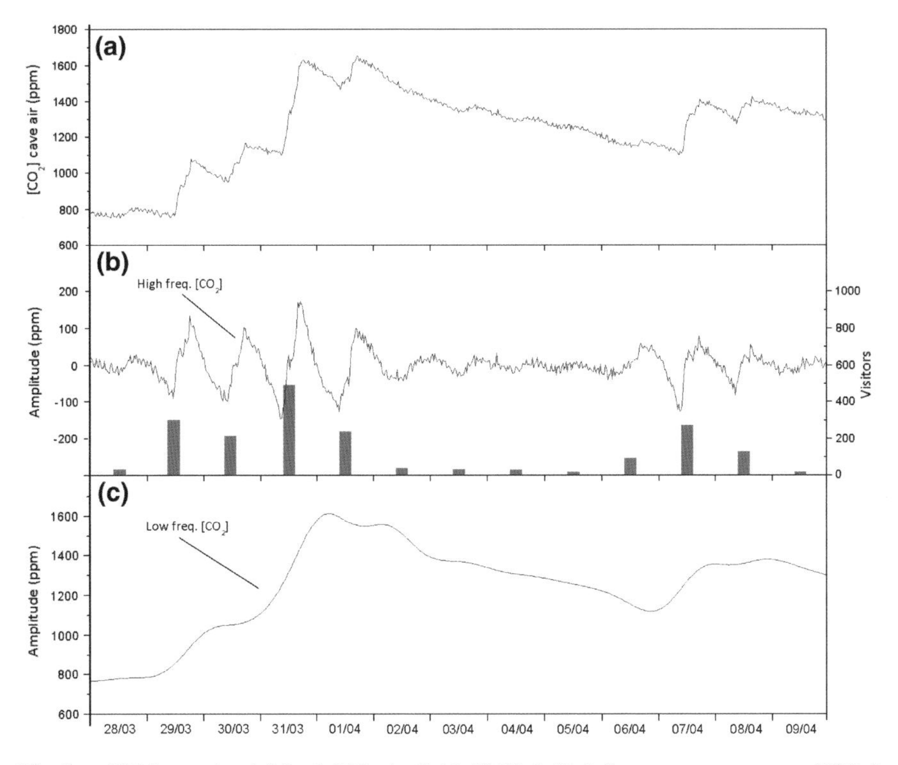

Fig. 2 **a** [CO_2] raw signal (March 28th–April 4th 2013). **b** High frequency component of [CO_2] raw signal. **c** Low frequency component of [CO_2] raw signal. Filtered signals (**b** and **c**) are obtained by applying DWT using 5 and Daubechies 10 as scale and mother wavelet filter, respectively

followed the same pattern as that observed for CO_2. Throughout the entire cycle, temperature difference and CO_2 concentration were totally dependent (Fig. 1).

As Rull Cave is a tourist cavity, visitors exert an important influence on the microclimatic conditions inside the cave. In order to enhance detection of this influence, the raw CO_2 signal was filtered by applying Discrete Wavelet Transform (DWT). The presence of visitors directly affected short-term CO_2 concentrations, subsequently buffered by natural dynamics. When applying DWT, high frequency events (such as visitors) can be clearly distinguished from the signal's natural trend (corresponding to low frequency events). Thus, the influence of visitors can be eliminated and the continuous trend of the signal highlighted. From March 28th to April 9th 2013, 1,928 people visited the cavity, and the CO_2 signal was studied over this period (Fig. 2). Those days with maximum visitors could be clearly identified in the high frequency component of the CO_2 concentration. From 29/03 to 1/04, more than 200 people a day visited the cave. The high frequency signal extracted from measured CO_2 concentrations revealed, for instance, an increase of nearly 200 ppm in comparison with the natural trend of raw records on March 31st, when 492 people entered the cave. The disturbance in CO_2 concentrations caused by visitors was clearly enhanced when both components of the original signal were separated. The number of visitors was directly proportional to the increase in CO_2 concentration. After some days (29/03-01/04) of continuous visits, CO_2 pattern tends to decrease those days with minor visitors. This decrease is stopped when a consecutive period of 3 days (06/04-08/04) with a considerable amount of visitors (more than 90 people) enter into the cave.

6 Conclusions

The seasonal pattern of trace gases in Rull Cave is related to cave ventilation and gas exchange with outside air, mainly controlled by meteorological factors. The main parameters governing the cave's microclimate are differences between outdoor and indoor air temperatures and densities, among others. Furthermore, in tourist caves such as Rull, the presence of visitors exerts a direct, short-term influence on CO_2 concentrations. Wavelet analysis has proved a useful tool for analysing and quantifying the disturbance caused by people. The advantages of wavelet analyses applied to microclimatic time series has recently been demonstrated. By implementing wavelet transform, a signal can be filtered and divided into different frequency components. Thus, the continuous trend of the signal can be clearly distinguished from high frequency events, providing an extremely useful insight into the relationships between microclimatic parameters and trace gas concentrations in underground systems.

Acknowledgments This research was financed by the Spanish Ministry of Science and Innovation (CGL2011-25162). A pre-doctoral research fellowship was awarded to C. Pla for this project.

References

Cuezva S, Fernández-Cortés A, Benavente D, Serrano-Ortiz P, Kowalski AS, Sánchez-Moral S (2011) Short-term CO_2(g) exchange between a shallow karstic cavity and the external atmosphere during summer: Role of the surface soil layer. Atmos Environ 45:1418–1427

de Carvalho L, Pla C, Galvañ S, Cuevas-González J, Andreu JM, Cañaveras JC, Benavente D (2013) Caracterización Petrográfica y Petrofísica de la Roca Encajante de la Cueva del Rull (Vall d'Ebo, Alicante). Macla 17:39–40

Fernández-Cortés A, Sánchez-Moral S, Cuezva S, Cañaveras JC, Abella R (2009) Annual and transient signatures of gas exchange and transport in the Castañar de Ibor cave (Spain). Int J Speleol 38(2):153–162

Fernández-Cortés A, Sánchez-Moral S, Cuezva S, Benavente D, Abella R (2011) Characterization of trace gases' fluctuations on a 'low energy' cave (Castañar de Íbor, Spain) using techniques of entropy of curves. Int J Climatol 31(1):127–143

Galiana-Merino JJ, Pla C, Fernández-Cortés A, Cuezva S, Ortiz J, Benavente D (2014) Environmental wavelet tool: continuous and discrete wavelet analysis and filtering for environmental time series. Comput Phys Commun (in press)

García-Anton E, Cuezva S, Fernández-Cortés A, Benavente D, Sánchez-Moral S (2014) Main drivers of diffusive and advective processes of CO_2-gas exchange between a shallow vadose zone and the atmosphere. Int J Greenh Gas Control 21:113–129
Rivas-Martínez S (1984) Pisos bioclimáticos de España. Lazaroa 5:33–43

Natural Ventilation of Karstic Caves: New Data on the Nerja Cave (Malaga, S of Spain)

C. Liñán and Y. del Rosal

Abstract In the Nerja Cave, there is a natural convective airflow which follows a seasonal model common in caves known as "chimney" characterized by, at least, two entrances at different altitudes. To explain this model, contradictory to the known entrances of the Nerja Cave, located at the same height, some research has been done in a surrounding cavity, known as the Pintada Cave. The obtained results confirm the existence of a physical connection between the Nerja Cave and the Pintada Cave, inaccessible to humans, and describe a very simplified, general model of airflow circulation between them, which allows for the removal of anthropogenic impact in the Nerja Cave during the most visited season in the year.

1 Introduction

The Nerja Cave, a good of cultural interest, in the category of Archeological Place and an internationally recognized heritage sight of Geological Relevance, is one of the most important tourist caves in Spain, with about 485,541 visitors annually for the period 1988–2013. The cavity, with a horizontal development and a volume of 300,000 m^3 (Fig. 1), has three entrances, which are located at 158, 161, and 162 m above sea level (SEM 1985). About a third of the cave, the Tourist Galleries, is

C. Liñán (✉) · Y. del Rosal
Nerja Cave Foundation, Research Institute, Carretera de Maro s/n, 29787 Malaga
Nerja, Spain
e-mail: cbaena@cuevadenerja.es; crilinbae@uma.es

Y. del Rosal
e-mail: yolanda@cuevadenerja.es

C. Liñán
Centre of Hydrogeology of University of Malaga, 29071 Malaga, Spain

C. Liñán
Faculty of Science, Department of Geology, University of Malaga, 29071 Malaga, Spain

B. Andreo et al. (eds.), *Hydrogeological and Environmental Investigations in Karst Systems*, Environmental Earth Sciences 1,
DOI 10.1007/978-3-642-17435-3_57

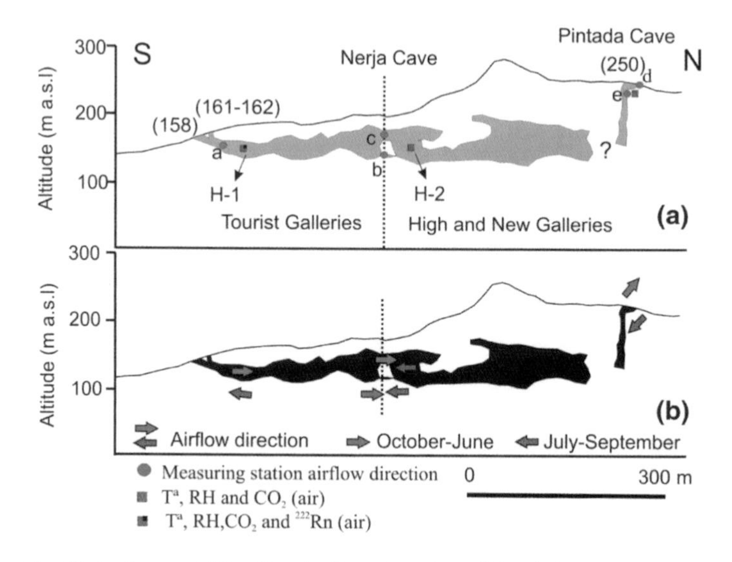

Fig. 1 **a** The Nerja Cave and The Pintada Cave: location of the airflows measurement stations (**a–e**) and of the sensors. **b** Direction of airflows measured in the Nerja Cave and Pintada Cave during 2013

open to tourists while the other part, the High and New Galleries, is only visited by researchers and reduced groups of tourist. The cave has a microclimatic station comprising of various sensors that measure, with hourly intervals, temperature, relative humidity, and air concentration of ^{222}Rn and CO_2, among other parameters (Carrasco et al. 2001; Liñán et al. 2009). Furthermore, a weather station measures the environmental parameters outside the cave (Liñán et al. 2007).

The Pintada Cave is a small cavity near the Nerja Cave (Fig. 1) which has been explored for some time by the possibility that the two caves were connected (GEMA 1976). The speleological explorations were unsuccessful so, in 1979, workers began to drill a well in the Pintada Cave, with the aim of building an artificial connection with the Nerja Cave. The well, which reached 75–80 m deep, was finalized in 1982 without reaching the intended aim.

2 Natural Ventilation in the Nerja Cave

In the Nerja Cave, there is a convective airflow due to the difference between the outside and inside air density (Cañete 1997). The ventilation rate of the cave is maximum in winter, 3.21 m^3/sec, and minimum in summer, 0.04 m^3/sec (Dueñas et al. 1999).

Since 2008, radon concentration in the cave air is measured by a sensor RADIM 5WP (Rosal et al. 2010). From November to April, the outside air is cooler and denser than the cave air (Fig. 2) so the air easily enters inside the cave, displacing the indoor air, and decreasing the concentration of ^{222}Rn in the cavity air to values of the order of 80 Bq/m^3. From May to October, the outside air is warmer and less dense than the cave air, so it reduces the outside air inlet and the concentration of ^{222}Rn in the air of the cave is higher, to reach the daily average values over the 400 Bq/m^3 from July to September.

In some sectors of the Nerja Cave, airflow can be discerned with a certain intensity (Fig. 1a), its direction can be crudely measured, by dusting silty material. The measurements show that from October to June, the airflow direction is predominantly from the Tourist Galleries to the deeper galleries of the cave (Fig. 1b). By contrast, in July, August, and September, the airflow direction goes from the deeper galleries to the Tourist Galleries. This circulation model is similar to the one observed in cavities known as "chimney" (Choppy 1982; Buecher 1999), characterized by, at least, two entrances at different altitudes. During the winter, the colder and denser outside air enters the cavity through the entrance at bottom, its temperature gradually increases and decreases its density and exits the cavity through the entrance located at a higher level. In the summer, the indoor air is colder and denser than the outside air, so the air comes out of the cave through the entrance located at the bottom while it is replaced by the warmer outside air, which enters the cave through the entrance located at a higher altitude.

As the three known Nerja Cave entrances are on a similar altitude (between 158 and 162 m.a.s.l.), the existence of a possible connection between the Nerja Cave and the Pintada Cave (250 m.a.s.l.) was reconsidered although it was not practicable. To confirm this, in April 2013 air flow measures in the Pintada Cave began. Shortly afterward, in June 2013, a sensor VAISALA GM70 equipped with probes

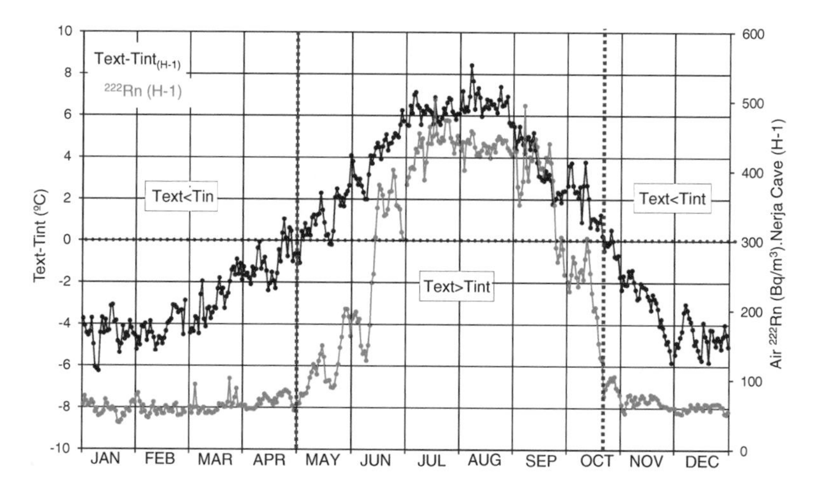

Fig. 2 Temporal evolution of the temperature differences *Text-Tint* versus radon air concentrations measured in the Nerja Cave (Tourist Galleries). Daily average data for the period 2008–2013. *Text* external temperature; *Tint* internal temperature

for the measurements of air temperature, relative humidity, and air CO_2 concentration was installed in the Pintada Cave. This was done in order to identify the arrival of air from the Nerja Cave, *marked* by the antropic impact. At the same time, in the Nerja Cave measurements of airflow and environmental parameters continued.

3 Results and Discussion

The results obtained from the 114 airflow measurements made in 2013 (Fig. 1a) show that from the middle of October to June, there is an airflow incoming in the Nerja Cave and outgoing through the Pintada Cave (Fig. 1b). From July to September, the direction of airflows is reversed: the outside air enters through the Pintada Cave and heads to the Nerja Cave. These results are consistent with the existence of a possible connection between both cavities.

Furthermore, the environmental parameters analysis shows that since the beginning of the data series up to July 3rd, the CO_2 air concentrations of the Pintada Cave are between 250 and 2,000 ppm, with a mean value 478 ppm (Fig. 3). Periodically CO_2 increases are also detected in the air of the Pintada Cave. Since July 5th, CO_2 increases disappear almost completely in the data series of the Pintada Cave and more uniform concentrations of CO_2 in the air appear between 250 and 580 ppm, with a mean value of 381 ppm. Until July 3rd, CO_2 increases observed in the Pintada Cave would be associated with the air from the Nerja Cave, as a result of the CO_2 increase produced by the visitors.

On July 5th, the airflow inversion has already occurred. Thus, the external air, with values of atmospheric CO_2, comes into the Pintada Cave, through the New and High Galleries, arrives at the Tourist Galleries (where the CO_2 values are increased by human impact) and finally goes outside the Nerja Cave. The CO_2 measured in the air of the High Galleries in the Nerja Cave (Fig. 3, H-2) confirms this point as it shows from July a progressive decrease until near to 480–500 ppm as a result of dwindling supply of anthropogenic CO_2. Before July 5th (from 13rd to 18th June), a short episode with low CO_2 concentrations in the Pintada Cave is detected, and high concentrations of ^{222}Rn in the Nerja Cave which corresponds to a timely reversal of general ventilation, in response to a temporary rise in temperature differences external–internal (Fig. 4).

This general air circulation continues until October 9th, when higher values of CO_2 are registered in the Pintada Cave (Fig. 5). The decrease in temperature differences external–internal produces a new reversal of air circulation and an increase in the ventilation of the karstic network. Before that (from September 28th to September 30th) a point reversal airflow is determined due to the occurrence of high concentrations of CO_2 in the Pintada Cave as well as in the High Galleries in the Nerja Cave.

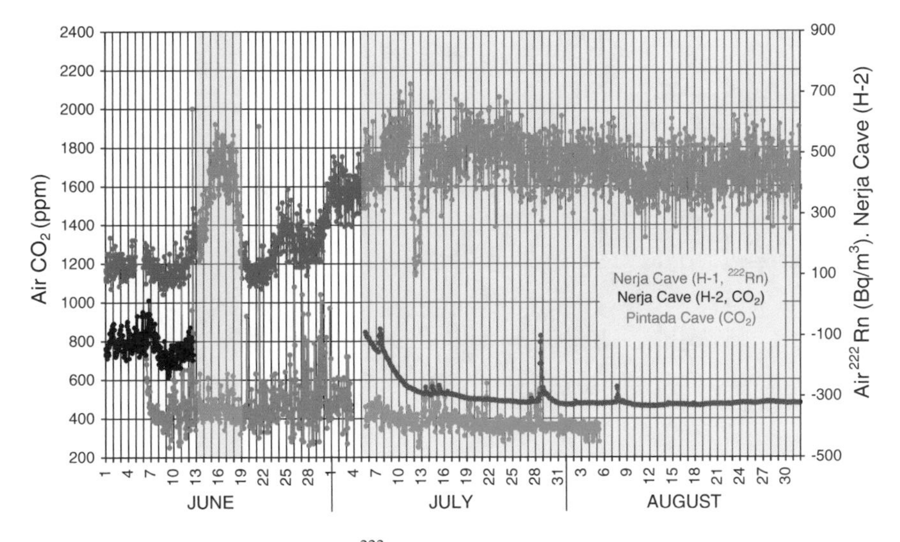

Fig. 3 Temporal evolution of the air ^{222}Rn and air CO_2 concentration in the Nerja Cave and air CO_2 concentration in the Pintada Cave, from June to August 2013

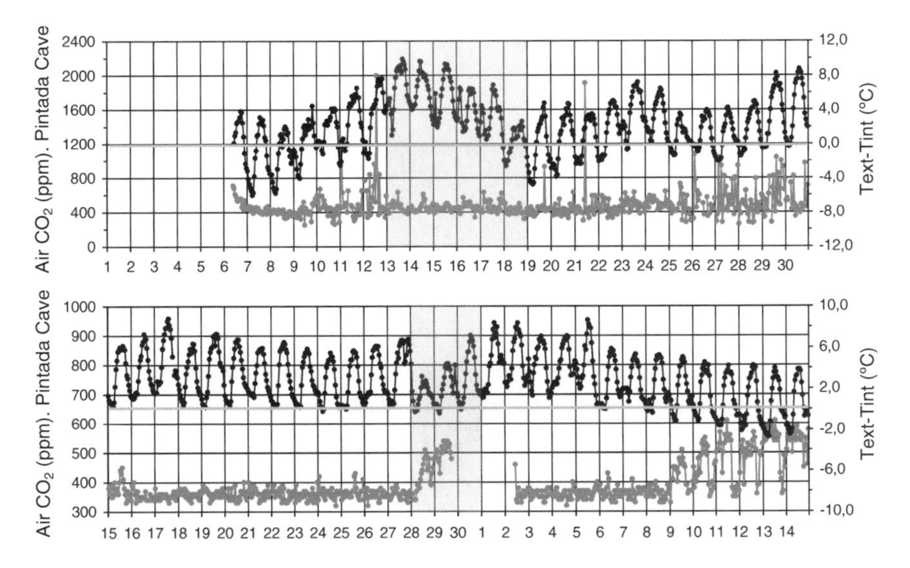

Fig. 4 Temporal evolution of the air CO_2 concentration in the Pintada Cave (*red*) and the temperature difference Text-Tint (*black*) from June 1st to 30th (*top*) and September 15th to October 14th, 2013 (*below*)

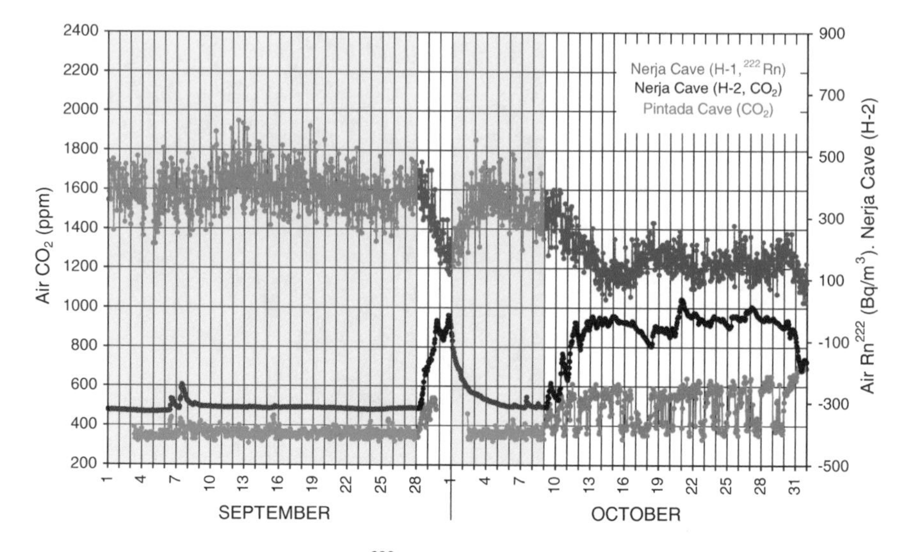

Fig. 5 Temporal evolution of the air ^{222}Rn and air CO_2 concentration in the Nerja Cave (H-1 and H-2 Halls, respectively) and air CO_2 concentration in the Pintada Cave, during September and October 2013

4 Conclusions

The results confirm the existence of a physical connection between the Nerja Cave and the Pintada Cave, not accessible to humans, and establish a very simplified general model of airflow circulation. From October to June, outside air enters the Tourist Galleries of the Nerja Cave, which is enriched in CO_2 as a result of the contribution of visitors. This anthropogenic CO_2 goes to the High and New Galleries of the Nerja Cave and finally arrives at the Pintada Cave and exits. From July to September, the airflow direction is reversed: the outside air, with atmospheric CO_2 values, enters the Pintada Cave, through the High and New Galleries of the Nerja Cave, arrives at the Tourist Galleries (where it is enriched with anthropogenic CO_2), and finally exits. This airflow model contributes to the elimination of anthropogenic impact in the Nerja Cave during the summer, when the cave receives the highest number of visitors. The existence of a new access to the karstic network of the Nerja Cave (the Pintada Cave) is a topic that must necessarily be considered for appropriate management and conservation.

Acknowledgments We thank Jose Manuel Cabezas Cabello for his collaboration in the translation of this paper.

References

Buecher R (1999) Microclimate study of Kartchner Caverns. Arizona. J Cave Karst Stud 61(2):108–120

Cañete S (1997) Concentraciones de Radón e intercambio de aire en la Cueva de Nerja. MD thesis, Universidad de Málaga

Carrasco F, Vadillo I, Liñán C, Andreo B, Durán JJ (2001) Control of environmental parameters for management and conservation of Nerja Cave (Málaga, Spain). Acta Carsologica 31(1):105–122

Choppy J (1982) Dynamique de l'air. Phenomenes karstiques (1). Spéléo Club de Paris and Club Alpin Français. Paris, France

Del Rosal Y, Garrido A, Montesino A, Liñán C (2010) Estudios del radón en la Cueva de Nerja (Málaga). In: Durán JJ, Carrasco F (eds) Cuevas: Patrimonio, Naturaleza, Cultura Y Turismo. ACTE, Madrid

Dueñas C, Fernández MC, Cañete S, Carretero J, Liger E (1999) ^{222}Rn concentrations, natural flow rate and the radiation exposure levels in the Nerja Cave. Atmos Environ 33:501–510

GEMA (1976) Exploración en la Cueva de Nerja. Jábega 13:60–68

Liñán C, Simón MD, del Rosal Y, Garrido A (2007) Estudio preliminar del clima en el entorno de la Cueva de Nerja (Andalucía, provincia de Málaga). In: Durán JJ, Robledo P, Vázquez J (eds) Cuevas Turísticas: Aportación al Desarrollo Sostenible. IGME, Madrid

Liñán C, Carrasco F, Calaforra, JM, del Rosal Y, Garrido A, Vadillo I (2009) Control de parámetros ambientales en las Galerías Altas y Nuevas de la Cueva de Nerja (Málaga). Resultados preliminares. In: Durán JJ, López–Martínez J (eds) Cuevas turísticas, cuevas vivas. ACTE, Madrid

SEM (1985) La Cueva de Nerja. Patronato de la Cueva de Nerja, Málaga

Environmental Study of Cave Waters: A Case Study in Las Herrerías Cave (Llanes, Spain)

M. Meléndez, M. Jiménez-Sánchez, I. Vadillo, H. Stoll, M.J. Domínguez-Cuesta, D. Ballesteros, E. Martos, L. Rodríguez-Rodríguez and J. García-Sansegundo

Abstract Las Herrerías Cave (Llanes, North Spain) is a relevant cave due to its cave palaeolithic paintings. It was declared of Good Object of Cultural Interest by the Asturias Regional Government. It is located in a karst Carboniferous aquifer which surface involves farming, livestock and tourist use, as well as a discrete mining activity and an old quarry converted into a waste area above the karst massif. A hydrogeological research was carried out from 2007 to 2010 focusing on the impact of land use on the cave. The aim of this work is to highlight the importance of using hydrochemistry monitoring to determinate the impact of the

M. Meléndez (✉)
Geological Survey of Spain, C/Matemático Pedrayes 25, 33005 Oviedo, Spain
e-mail: m.melendez@igme.es

M. Jiménez-Sánchez · H. Stoll · M.J. Domínguez-Cuesta · D. Ballesteros · E. Martos · L. Rodríguez-Rodríguez · J. García-Sansegundo
Department of Geology, University of Oviedo, c/Arias de Velasco s/n, 33005 Oviedo, Spain
e-mail: mjimenez@geol.uniovi.es

H. Stoll
e-mail: hstoll@geo.umass.edu

M.J. Domínguez-Cuesta
e-mail: mjdominguez@geol.uniovi.es

D. Ballesteros
e-mail: ballesteros@geol.uniovi.es

E. Martos
e-mail: arikieva@gmail.com

L. Rodríguez-Rodríguez
e-mail: laurarr@geol.uniovi.es

J. García-Sansegundo
e-mail: j.g.sansegundo@geol.uniovi.es

I. Vadillo
Faculty of Sciences, Department of Geology, University of Málaga, Campus de Teatinos sin, 29071 Malaga, Spain
e-mail: Vadillo@uma.es

B. Andreo et al. (eds.), *Hydrogeological and Environmental Investigations in Karst Systems*, Environmental Earth Sciences 1,
DOI 10.1007/978-3-642-17435-3_58

land use on cave waters. Several sampling campaigns of rainfall, the waters inside and outside the cave and El Boláu spring were carried out. A sampling device was installed inside the cave to collect discrete samples of dripwater each 48 h in 1.5 l bottles. The array contains 24 bottles and was deployed for 48 days intervals. Results evidence that calcium bicarbonate is the dominant chemical component in all the sampled water. Some of them, collected under the waste area, show high values in NO_3^-, Ba^{2+}, TOC, K^+, SO_4^{2-}, Silicium, Fe, Si, $PO_4^{3-}$and Sr^{2+}, being indicative of an adverse effect on the quality of drip water.

1 Introduction

The characterization and evaluation of the hydrochemistry of karst waters is essential for a better understanding of the human impact derived from land use. The study of cave waters chemistry is useful to define the potential source of cave contamination and some authors have considered it as the best way to get a success protection (Ravbar and Goldscheider 2007). Soil and vadose zone of karst aquifers act as protective covers to groundwater (Mudry et al. 2003), but are also the way in which contaminants travel towards underground waters.

Las Herrerías Cave (Llanes, North Spain) shows prehistoric paintings declared as a Good Object of Cultural Interest by the Asturias Regional Government. Karst waters (including dripwaters, flowing streams and springs) are sensitive indicators of land use impact in the cave (Pulido-Bosch et al. 1997), which is the physical environment of the paintings. The environmental study of the cave has involved geomorphological, hydrogeological and karst vulnerability studies with the delineation of protection areas (Jiménez-Sánchez et al. 2008a, b; Domínguez-Cuesta et al. 2010; Meléndez et al. 2011; Marín et al. 2012). Moreover, a detailed hydrochemical study was developed, including dripwater monitoring and several sampling water campaigns. The aim of this work is to highlight the importance of hydrochemical studies to establish land use impacts on cave waters, using Las Herrerías Cave as a case study.

2 Setting

Las Herrerías Cave (43° 23.99′ N, 4° 45.96′ W) is located at the foot of the Cuera Range, just 3 km from the Cantabrian Coast (Fig. 1) in the North of Spain. The cave includes 748 m of horizontal conduits developed between 38 and 61 m a.s.l. (Domínguez-Cuesta et al. 2010). The cavity is under a small hill located between two rivers that flow to the North. The climate of the area corresponds to the oceanic domain, with mean rainfall values reaching 1,000 mm/year and mean annual temperature reaching 10 °C (González Taboada and Anadón Álvarez 2011). Inside

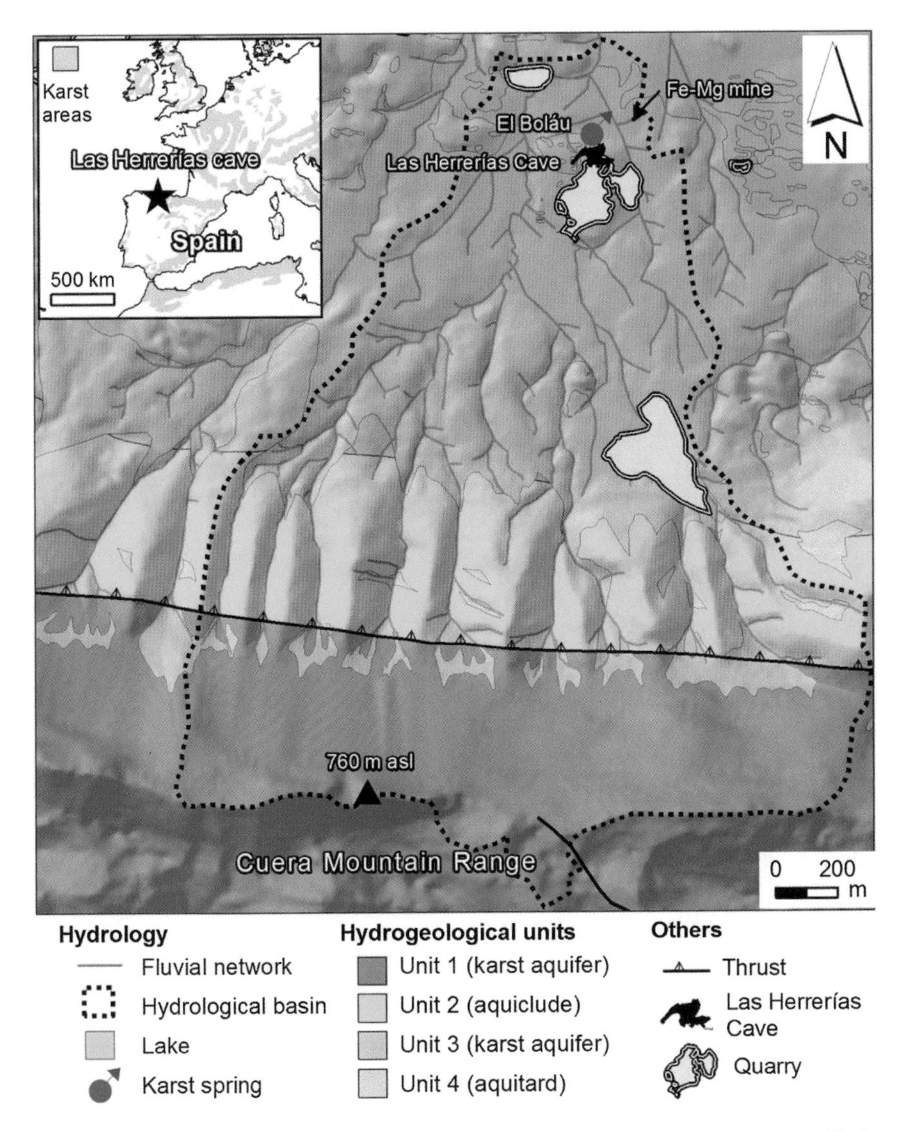

Fig. 1 Setting of the study area with the main hydrogeological units (modified from Meléndez et al. 2011)

the cave, the temperature parameter range is between 12.25 and 12.05 °C, with only 0.2 °C variation, clearly influenced by the external temperature (Jiménez-Sánchez et al. 2008a, b).

From a geological point of view, the cave surroundings are formed by ordovician quartzite, devonian sandstone and carboniferous limestone modified by a sub-vertical south-directed E-W trending thrust (Meléndez et al. 2011). The cave is sited in a karst massif developed in the carboniferous limestone and covered by quaternary deposits up to 2 m thick. These deposits are mainly originated by the combination of karst, torrential and gravitational processes. The area presents well-developed soil, scrublands and small forest modified by scattered farming, stock breeding and minor tourist use, as well as an old Fe–Mg mine and some quarries. One of these quarries located over the study cave is developed and nowadays used to store waste building material that could be a potential source of cave water contamination, showing from 0 to 4 m thickness.

Four hydrogeological units have been defined (Meléndez et al. 2011): unit 1 is formed by karstified carboniferous limestone that represents the south karst aquifer; unit 2 includes ordovician to devonian quartzite and sandstone and represents an aquiclude; unit 3 corresponds to the karstified carboniferous limestone of the north karst aquifer, where the study cave is placed; and unit 4 involves quaternary detrital deposits that are interpreted as aquitards. The main inputs of the north karst aquifer (unit 1) correspond to the vertical infiltration of water drained by fluvial streams coming from unit 4. The underground waters flow north to El Boláu karst spring (37 m a.s.l.), with 3–30 l/s discharge.

3 Methodology

The methodology of work includes water sampling in Las Herrerías Cave and its surroundings, cave dripwater monitoring, chemical analysis and Principal Components Analysis (PCA). Water sampling was done to characterize the hydrochemistry of the cave waters. The sampling points include water rainfall, seven cave dripwaters, two cave streams and El Boláu spring. From February 2007 to January 2008 a battery-powered dripwater collector was installed in a place called *Sargazos in Pinturas hall* in the northeast area of the cave. The device was designed to collect dripwater during each 48 h in discrete 1.5 l bottles. The array contains 24 bottles and it was deployed for 48 days intervals (Fig. 2). Chemical analysis was done in the Laboratory of Hydrogeology of the University of Málaga, measuring the ions by Ionic Chromatography (IC) and the Total Organic Carbon (TOC) by a TOC analyzer.

4 Results and Discussion

The Sargazos dripwater monitoring shows a predominance of bicarbonate and calcium components related to the lithology of the bedrock. However, values of nitrate and sulphate are higher than could be expected in a system with a carbonate lithology. Also, NO_2^- is present in this area unlike in other areas of the cave (Table 1).

The temporal evolution of the Sargazos dripwater evidences several behaviour patterns of the principal ions (Fig. 3). In this way, waters sampled from June to

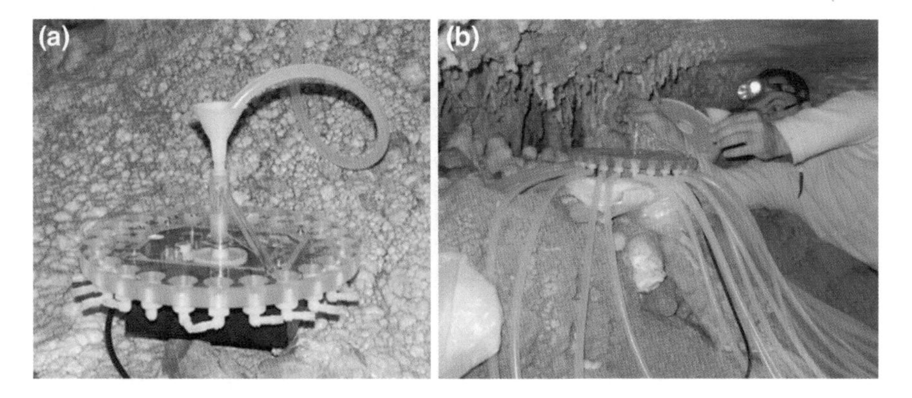

Fig. 2 **a** Details of the device during its assembly. **b** View of the device with the sample collector installed; each *white* tube collects the water of the dripstone during 48 h

Table 1 Average composition of hydrochemistry parameters of Sargazos dripwater

DOC	HCO^{3-}	SO_4^{2-}	Cl^-	NO_3^-	NO_2^-	F^-	Ca^{2+}	Mg^{2+}	Na^+	K^+	Si	Sr^{2+}	Ba^{2+}	Fe	Mn	P
4.06	199.8	22.7	15.3	5.8	0.51	0.04	111	2.9	11.1	0.4	2.2	84	7	31	4	43

DOC is the Dissolved Organic Carbon. Units are expressed in mg/l, except Sr^{2+}, Ba^{2+}, Fe, Mn and P that are depicted in μg/l

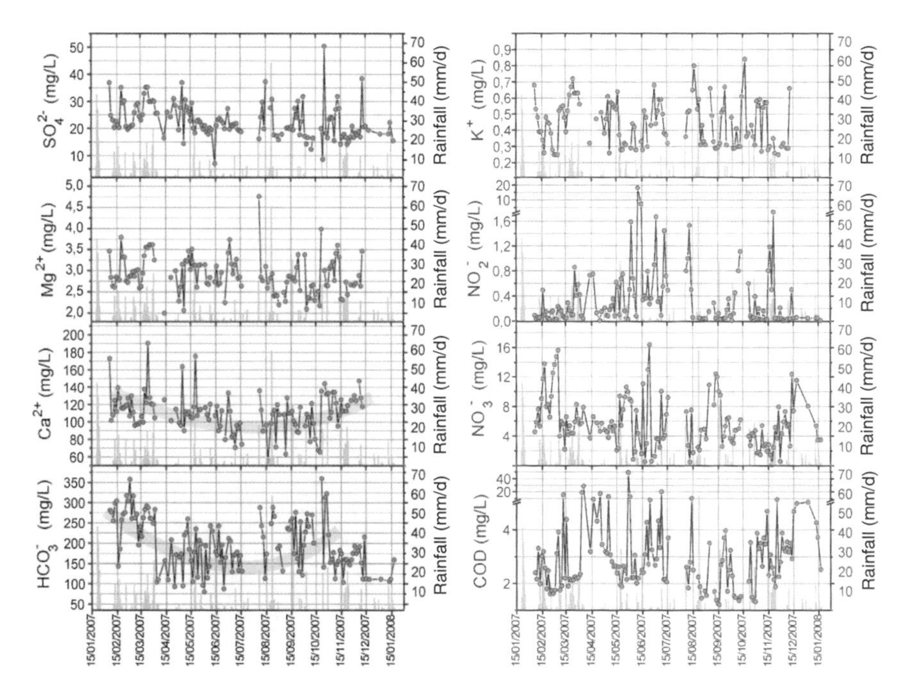

Fig. 3 Temporal evolution of the hydrochemical parameters in Sargazos dripwater from February 2007 to January 2008

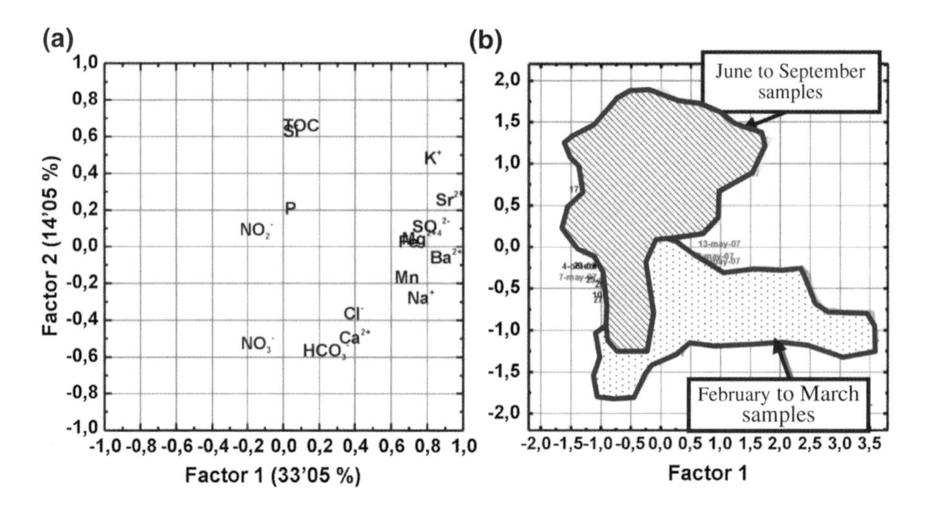

Fig. 4 PCA representation 16 variables and 108 samples from Sargazos dripwater. a Variables and b Sampled

September present values of Ca^{2+} a little lower than during the rest of the year. The behaviour of Na^{+} and Cl^{-} is similar, although the lower values were measured from July to October SO_4^{2-} values from August to December are, as well, little lower than the first months of the year. Finally, the Mg^{2+} variation is not considered to be significant.

The PCA provides very valuable information about the origin of the cave waters. It is used to identify how correlated the variables are. Besides, samples are represented by hydrochemical affinity so it can be recognized as hydrochemical or temporal evolution patterns. For the Sargazos dripwater two main components explain 47 % of the total variance (Fig. 4). The first one can explain 33 % and includes eight variables (Na^{+}, K^{+}, Mg^{2+}, Sr^{2+}, Ba^{2+}, Fe, Mn and SO_4^{2-}), being interpreted as corresponding to the soil influence. The second explains 14 % of the total variance and is governed by five variables (DOC, Ca^{2+}, Si, HCO_3^{-} and NO_3^{-}) and includes samples influenced by rainfall. So, two origins of the cave waters are defined.

The results obtained from the rest of the sampling of waters from the Las Herrerías Cave and its surroundings show that principal ions are bicarbonate and calcium, and the second anion in importance is chloride or sulphate. It is noted the high values of SO_4^{2-}, NO_3^{-} and Cl^{-} in the east of the cave. Piper Diagram (Fig. 5) shows the groups of water types. The groups situated in the east show an increase in NO_3^{-} and Cl^{-}.

In this case, the PCA treatment data generated four components that explain 83 % of the total variance. The first one can explain 46 % and is contributed by Na^{+}, Mg^{2+}, Sr^{2+}, Ba^{2+}, Cl^{-}, SO_4^{2-} and NO_3^{-}. The second component explains 16 % and is conditioned by Na^{+}, K^{+} and NO_3^{-}.

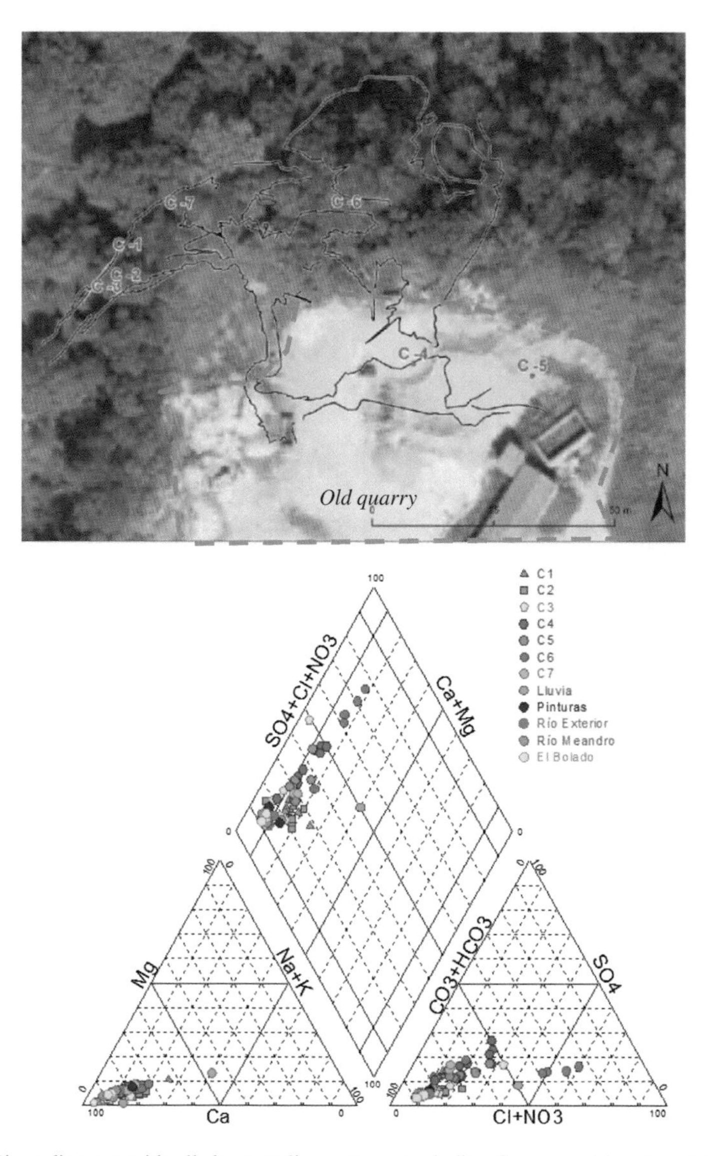

Fig. 5 Piper diagram with all the sampling waters, excluding Sargazos dripwater, July 2007–August 2008. At the top is shown the position of cave surface as an aerial photograph. An old quarry, converted in a waste area, can be observed in the South

In the graph of the samples shown in Fig. 6, it can be seen that factor 1 groups all the components associated to meteoric origin (Na^+), to soil (Sr^{2+}, Ba^{2+}, Cl^-, SO_4^{2-} and NO_3^-) or to an anthropic influence (Cl^-, SO_4^{2-} y NO_3^-). In the sample graph (Fig. 6b), samples C4, C5 and C6, located to the left of the diagram correspond to the waters from the east of the cave and under the old quarry converted in a waste area or surroundings. These samples show higher concentration of Cl^-, SO_4^{2-} and NO_3^-.

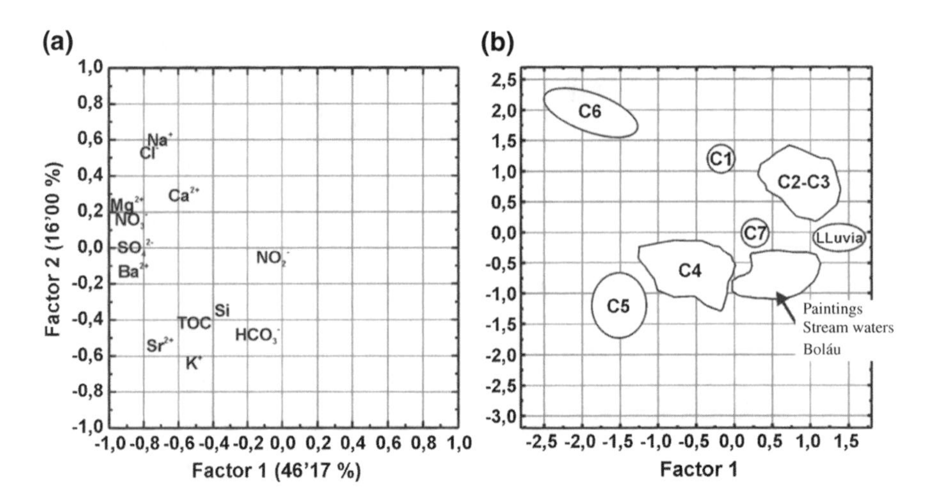

Fig. 6 PCA representation for all the sampling waters, excluding Sargazos dripwater. **a** Variables and **b** Sampled

One aspect of this research was to discern the possible affection of land use on the natural water quality of the cave, especially the impact of the old quarry converted in a waste area. The continuous monitoring of the dripwaters with reference C has revealed a negative impact of this plant in the area under the old quarry. This impact is evident in most C4, C5 and C6 drips water points which have higher concentration on the following parameters: nitrate, barium, dissolved organic carbon, iron, potassium, phosphorus, silicon, sulphate and strontium. Among the highlights are shown high concentrations of nitrate in the samples near the quarry, where values reach up to 10 times higher than in the drip points located in the northwest sector of the cavity. The highest value of nitrate has been measured in C6 drips, which is situated in Fig. 6b. at the area of Na^{+}–Cl^{-} and also presents a high content in SO_4^{2-}, so an anthropic influence can be deduced.

5 Conclusions

As previously shown in other studies, in Las Herrerías Cave, hydrochemistry involves fundamental tools to study human impact on natural dripwaters in karst systems. The installation of a continuous sampling device also allows knowing the temporal variations of chemical parameters. The PCA constitutes a useful technique to discern sample groups and how the variables are correlated, providing valuable information about the origin of the cave waters. Also, the representation of the samples allow establishing hydrochemical affinity so it can recognize hydrochemical or temporal evolution patterns.

The research developed in Las Herrerías Cave waters has resulted in the detection of anthropic influence, especially in the surrounding area of an old quarry used to store waste. This is valuable to improve the previously defined protection areas and to establish the measures to be taken to preserve properly prehistoric paintings.

It has been shown that samples under an old quarry have high concentrations of nitrate, barium, dissolved organic carbon, iron, potassium, phosphorus, silicon, sulphate and strontium. It is worth remarking that these results clearly show the sensitivity of the karst system to contamination, and so, the importance of special protection. The research findings demonstrated the importance of such studies for sustainable use of soil to protect palaeolithic paintings in karst caves.

Acknowledgments This research was funded by the Asturias Government (CN-06-177 contract).

References

Domínguez-Cuesta M, Jiménez-Sánchez M, Rodríguez-Rodríguez L, Ballesteros D, Meléndez M, Martos E, García-Sansegundo J (2010) Uso de la Geomorfología y el SIG para caracterizar el impacto de las actividades mineras en zonas kársticas. In: Berrezueta Alvarado E, Domínguez-Cuesta MJ (eds) Técnicas Aplicadas a La Caracterización Y Aprovechamiento de Recursos Geológico-Mineros: Volumen I: Descripciones Metodológicas. Red Minería XXI, CYTED, Instituto Geológico y Minero de España, Oviedo, Spain, pp 80–90

González Taboada F, Anadón Álvarez R (2011) Análisis de escenarios de cambio climático en Asturias. Oficina para la Sostenibilidad, el Cambio Climático y la Participación, Gobierno del Principado de Asturias, Oviedo, Spain

Jiménez-Sánchez M, Domínguez Cuesta MJ, García-Sansegundo J, Stoll H, Vadillo I, Rodríguez-Rodríguez L, Aranburu A (2008a) Estudio preliminar de la geomorfología de la cueva de Herrerías y su entorno (Llanes, Asturias, Noroeste de España). In: Trabajos de Geomorfología En España, 2006–2008. X Reunión Nacional de Geomorfología. Cádiz, Spain, pp 45–48

Jiménez-Sánchez M, Stoll H, Vadillo I, López-Chicano M, Domínguez-Cuesta MJ, Martín-Rosales W, Meléndez-Asensio M (2008b) Groundwater contamination in caves: four case studies in Spain. Int J Speleol 37(1):53–66, 14

Marín AI, Andreo B, Jiménez-Sánchez M, Domínguez-Cuesta MJ, Meléndez-Asensio M (2012) Delineating protection areas for caves using contamination vulnerability mapping techniques: the case of Herrerías Cave, Asturias. Spain J Cave Karst Stud 74:103–115

Meléndez M, Jiménez-Sánchez E, Martos E, Domínguez-Cuesta M, Rodríguez-Rodríguez L, Ballesteros D (2011) Aplicación de ensayos de trazadores a la caracterización hidrogeológica del acuífero kárstico de la Cueva de Las Herrerías (Llanes, España). In: Berrrezueta Alvarado E, Domínguez-Cuesta M (eds) Técnicas Aplicadas a La Caracterización Y Aprovechamiento de Recursos Geológico-Mineros. Volumen II: Procesos Experimentales. Red Minería XXI, CYTED, Instituto Geológico y Minero de España, Oviedo, Spain, pp 109–123

Mudry J, Coxon C, Kilroy G, Kapelj S, Surbeck H, Vadillo I (2003) Specific vulnerability. Contaminants in carbonate karst groundwater (inorganic contaminants). Vulnerability and risk mapping for the protection of carbonate aquifers. Action COST 620:36–43

Pulido-Bosch A, Martín-Rosales W, López-Chicano M, Vallejos A (1997) Human impacts in a touristic cave (Aracena, Spain). Environ Geol 31(3/4):142–149

Ravbar N, Goldscheider N (2007) Proposed methodology of vulnerability and contamination risk mapping for the protection of karst aquifers in Slovenia. Acta Carsologica 36(3):397–411

Climate-Driven Changes on Storage and Sink of Carbon Dioxide in Subsurface Atmosphere of Karst Terrains

A. Fernández-Cortés, S. Cuezva, E. García-Antón, M. Alvárez-Gallego, D. Benavente, J.M. Calaforra and S. Sánchez-Moral

Abstract A comprehensive environmental monitoring programme has been recently launched in Ojo Guareña cave system (Burgos, Spain), one of the longest caves in Europe, aimed to assess the magnitude of the spatiotemporal changes of CO_2 (g), on daily and synoptic timescales in the cave–soil–atmosphere profile. CO_2 concentration of cave air is usually close to atmospheric background but huge daily oscillations of CO_2 levels, ranging 680–1,900 ppm/day on average, have registered during periods when exterior air temperature oscillates every day around cave air temperature. These daily variations of CO_2 content are hidden once the air temperature outside is continuously below cave temperature and a prevailing advective-renewal of cave air is established, so that daily-averaged concentrations of CO_2 reach minimum values close to 500 ppm. The spatiotemporal pattern of CO_2(g) provides evidence that the amounts of carbon that might be sequestered and then emitted (CO_2) from subsurface air located in the uppermost part of the vadose zone could be noticeable at local or regional scale by considering long subterranean systems as Ojo Guareña karst.

A. Fernández-Cortés (✉) · S. Cuezva · E. García-Antón · M. Alvárez-Gallego
S. Sánchez-Moral
Museo Nacional de Ciencias Naturales (MNCN-CSIC), José Gutiérrez Abascal, 2, 28006 Madrid, Spain
e-mail: acortes@mncn.csic.es

A. Fernández-Cortés
Geomnia Natural Resources SLNE, Madrid, Spain

S. Cuezva · D. Benavente
Laboratorio de Petrología Aplicada, Universidad de Alicante, Campus San Vicente Del Raspeig s/n, 03690 Alicante, San Vicente Del Raspeig, Spain

J.M. Calaforra
Departamento Biología y Geología, Recursos Hídricos y Geología Ambiental, Universidad de Almería, Ctra. Sacramento s/n, 04120 Almería, La Cañada de San Urbano, Spain

B. Andreo et al. (eds.), *Hydrogeological and Environmental Investigations in Karst Systems*, Environmental Earth Sciences 1,
DOI 10.1007/978-3-642-17435-3_59

1 Introduction

The vadose zone of karst terrains may contain large amounts of gases occupying the air-filled, cracks and voids of soil, bedrock or unconsolidated sediment (Benavente et al. 2010; Serrano-Ortiz et al. 2010; Bourges et al. 2012). Recent assessments of subterranean cavities have allowed prompting that they act as sinks/reservoirs or net emitters/sources of CO_2 depending on atmospheric conditions (Fernández-Cortés et al. 2011; Cuezva et al. 2011; García-Antón et al. 2014).

Here, we contribute with newfangled data and understanding of $CO_2(g)$ variations obtained at small scale in a detailed field study of a shallow atmosphere belonging to a longer subterranean system (Ojo Guareña karst, Spain). This study combined continuous multi-parameter monitoring of the cave-soil-atmosphere (main climatic data and gas composition) with geochemical tracing using $\delta^{13}C$–CO_2, in order to assess the climate-driven fluctuations of $CO_2(g)$ into subsurface and its spatiotemporal pattern of exchange with soil and atmosphere.

2 Study Site

Ojo Guareña cave system, one of the longest caves in Europe, is distributed over five overlapping levels comprised of passages up to 10 m high and 20 m wide with three main entrances and several other minor cavities (Grupo Espeleológico Edelweiss 1986). Over 100 km of galleries, developed in Cretaceous limestone, have been surveyed (Puch 1998). No previous study on the microclimate of this cave has been published. The present study has been conducted in a sector some 3 km in length, with three subsectors for continuous monitoring: Edelweiss, Gours Hojas and Museo Cera and a network of 16 air- sampling points (Fig. 1).

3 Methods

A specific monitoring, sampling and analysis programme was designed and implemented to obtain data of $CO_2(g)$ contents and physical parameters of cave air, either via continuous logging by installed network of sensors and spatial-distributed and spot air sampling by seasonal short surveys.

Hourly data of the main cave-air parameters were registered at three cave locations named: Edelweiss, Gours Hojas and Museo Cera, besides the meteorological conditions at exterior located in Dolencias doline (see Fig. 1). To date, a complete suite of time series have been registered from August 2013 to February 2014. Each monitoring station consists of a HOBO Pro v2 datalogger with built-in temperature and relative humidity sensors and a HOBO-U21 datalogger (Onset Computer Corporation, Bourne, MA, USA). A set of special sensors were connected

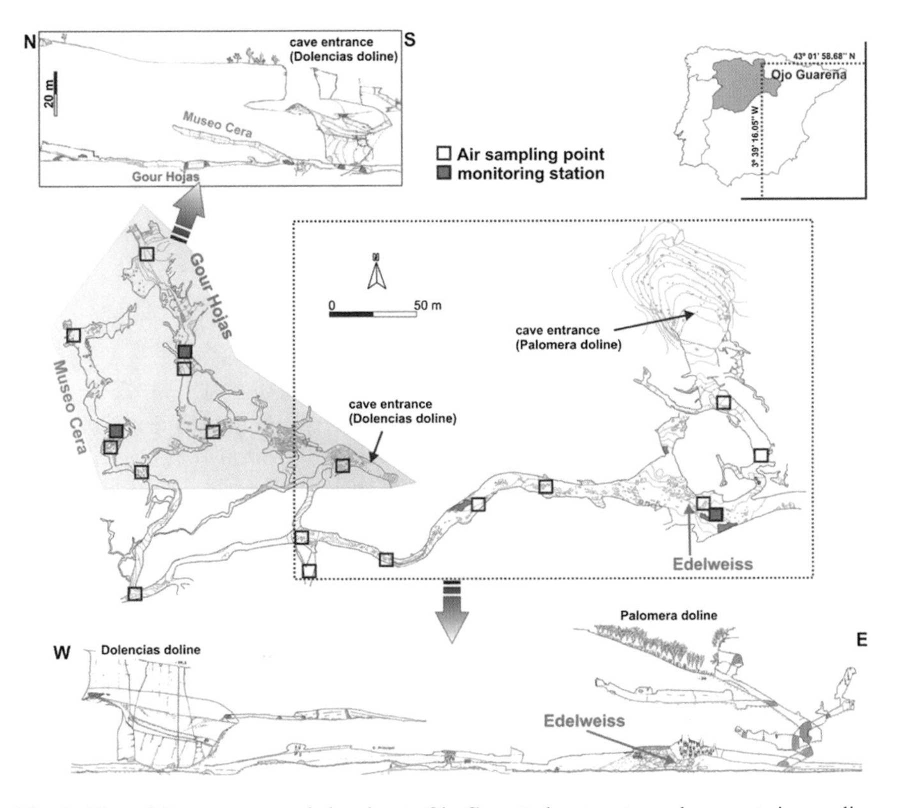

Fig. 1 Map of the cave sectors belonging to Ojo Guareña karst system where spot air sampling and continuous monitoring have been conducted. Detailed cross-sections of galleries are displayed in relation to surface geomorphology and two of the main entrances to the subterranean system (Dolencias and Palomera dolines). Cross-sections modified from Grupo Espeleológico Edelweiss (1986)

to this datalogger for the narrow range of measurements expected, including: air temperature, relative humidity and barometric pressure. The outdoor weather station (HOBO U22) also includes a tipping-bucket rain gauge for rainfall measurements. Carbon dioxide concentration of cave air was registered with a K-33 ELG autonomous monitor (CO2meter, Ormond Beach, FL, USA) based on non-dispersive infrared technology and sampling method by gas diffusion. The ^{222}Rn concentration of cave air was continuously registered (sampling interval of 30 min) using a Radim 5WP radon monitor (GT-Analytic KG, Innsbruck, Austria).

Spot air-samplings were bimonthly conducted in a predefined network of points spatially distributed inside cave, soil and exterior, during four short surveys (<3 h long). Soil air was extracted using a micro-diaphragm gas pump (KNF Neuberger, Freiburg, Germany) of 3.1 l/min at atmospheric pressure. Cave and exterior air were sampled using a low flow pump and saved in 1 L Tedlar bags with lock valves. Bag samples were analysed within 48 h of sampling for CO_2 concentrations and isotopic signal $\delta^{13}CO_2$ by a laser-based analyser (Picarro G2101-i, California, USA) that employs cavity ring-down spectroscopy (Crosson 2008). Precisions of 200 and 10 ppb are guaranteed by manufacturer for $^{12}CO_2$ and $^{13}CO_2$, respectively, over a wide operating range, with a resulting precision less than 0.3 ‰ for $\delta^{13}C$ (CO_2) after 5 min of analysis.

4 Results and Discussion

Carbon dioxide within the subterranean atmosphere of Ojo Guareña shows a stair-step pattern with strong shifts depending on the temperature relationship between cave air and atmosphere. Thus, high summer and low autumn-winter levels have been registered at each cave location during the study period (Fig. 2). CO_2 concentration of cave air usually reaches the atmospheric background (~400 ppm) at some point during the day, especially in those locations near to the main cave entrances and exposed to high ventilation rates (Edelweiss monitoring station, Fig. 2). Advection is the mechanism that controls the CO_2 variations into cave atmosphere and it varies seasonally and throughout the day, to the extent that the climate-driven forces do due to changes of meteorological conditions. The advective flux injects the outer air to the cave air and extracts subterranean air. This process is triggered either by local temperature differences (thermal buoyancy by different densities in warm and cold air), by intense wind or by a combination of both. The advective influx of outer air during colder months (when cave air temperature is above temperature at exterior) also entails a prevailing fall of air temperature in those locations near to cave entrances (−2.25 °C in Gours Hojas and −4 °C in Edelweiss site) and, even, a depletion of radon content of air at the most remote and isolated cave locations, e.g. radon content of air from Museo Cera reaches a background levels around 500 Bq/m^3 (Fig. 2). The dripping water samples analysed in Museo Cera have a partial presure of CO_2 (pCO_2) ranging $10^{-2.35}$–$10^{-1.80}$ bar, which is considerably higher than pCO_2 of cave air during winter ($10^{-3.12}$ bar on average). Consequently, an increase in drip rate due to previous intense and frequent rainfall events seems to provoke some slight increments in CO_2 concentration by drip outgassing, even during the prevailing air renewal of Museo Cera (e.g. end of November and January).

The logging equipments have also registered huge daily oscillations of CO_2 levels ranging 680–1,900 ppm/day on average (from August to early October), in accordance with two opposite patterns controlled by outside air temperature: (1) CO_2 depletion (−47 to −166 ppm/h, on average) joined to sharp falls of air temperature (−0.01 to −0.06 °C/h on average, with maximums ranging −0.15 to −0.57 °C/h) due to a nocturnal and early morning inlet of outside cold air and (2) an increase in CO_2 concentration (+58 to +182 ppm/h, on average) coeval to a recovery of cave air temperature during daytime. These daily variations of CO_2 content are hidden once the air temperature outside is continuously below cave temperature and a prevailing advective-renewal of cave air is established, so that

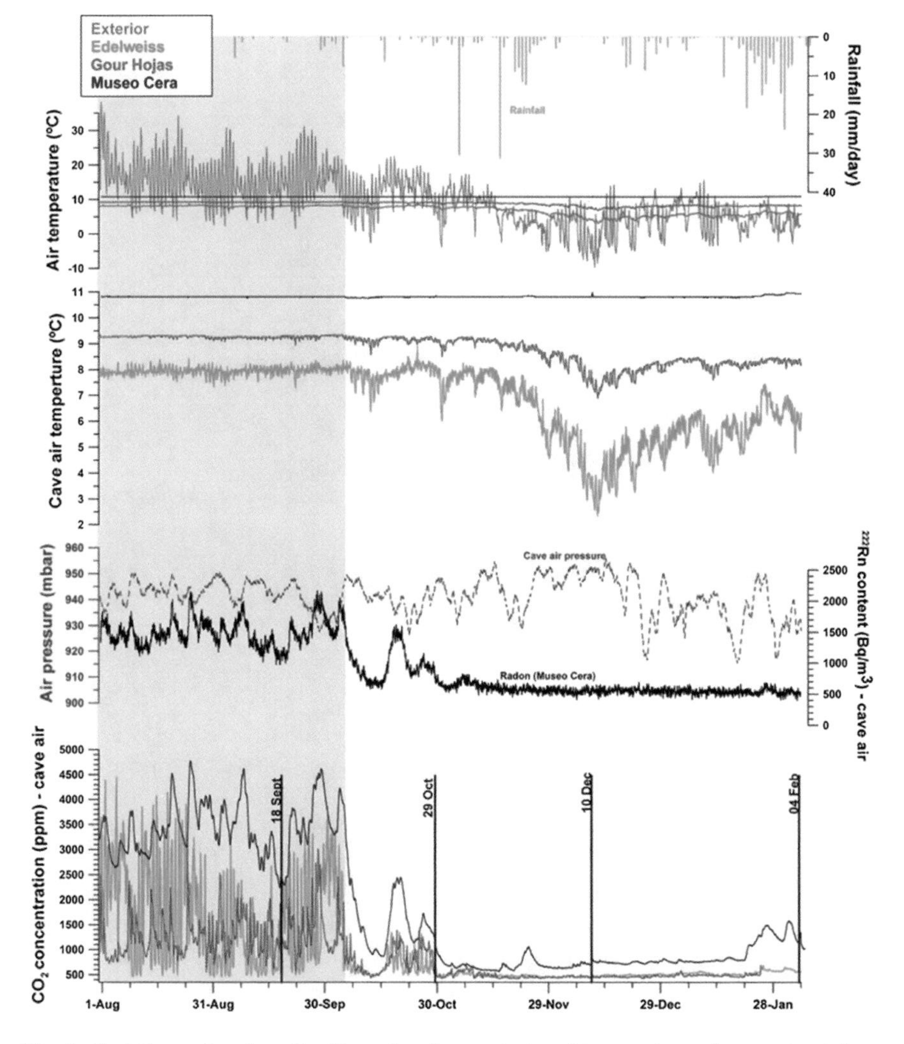

Fig. 2 Variations of carbon dioxide and radon content and temperature of cave air at three locations: Edelweiss, Gours Hojas and Museo Cera (see Fig. 1) in relation to weather changes (temperature and rainfall) and variation of cave air pressure. Vertical *black lines* indicate the dates when spot air-samplings were conducted in a predefined network of points spatially-distributed inside cave (see Fig. 1)

daily-averaged concentrations of CO_2 and ^{222}Rn reach minimum values close to 500 ppm and 700 Bq/m^3, respectively.

The seasonal and daily fluctuations of gases in function of temperature oscillation depend on the morphology of each cave sector and its location relative to exokarst. Thus, the circadian pulses of gases in function of external temperature are smoothed in the inner cave locations (e.g. Museo Cera) where, conversely, variations of CO_2 and ^{222}Rn are correlated with medium-term changes of air

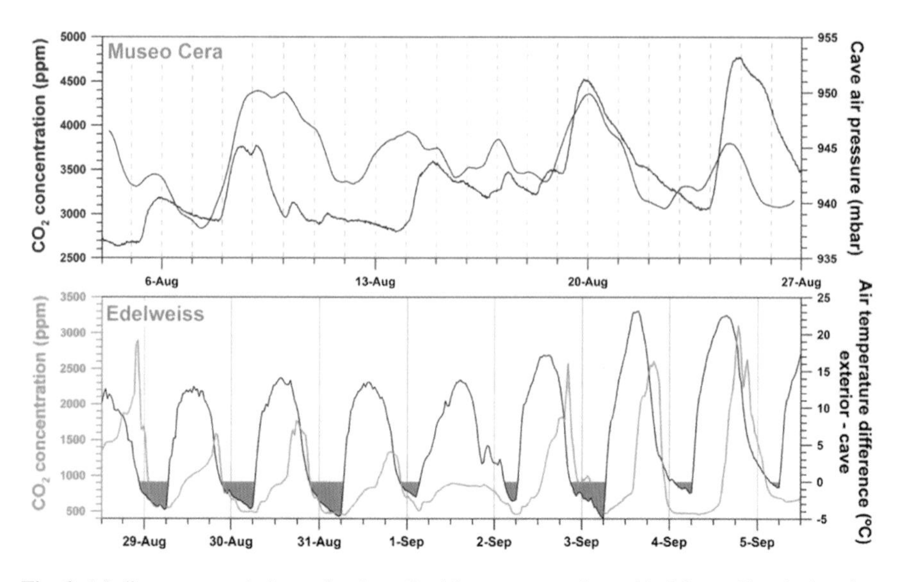

Fig. 3 Medium-term variations of carbon dioxide contents registered in Museo Cera in function of cave air pressure changes, and daily pulses of CO_2 levels in Edelweiss primarily controlled by air temperature differences (exterior-cave)

pressure (Fig. 3). On the contrary, the cave sectors near to entrance as Edelweiss (see Fig. 1) reach daily the minimum values when outside air temperature falls below cave air temperature, primarily during night hours (Fig. 3). These daily decreases in gas concentrations are even triggered before exterior temperature is below cave temperature, so it suggests that colder and CO_2-depleted air masses from other cave locations can also be displaced to Edelweiss gallery.

During periods with strong oscillations of CO_2 levels at daily scale (Fig. 3 for Edelweiss gallery) the sharp falls in CO_2 content are followed by CO_2 increments with similar magnitude during daylight hours. The increase in CO_2 concentration is coeval to a rise of air temperature difference between exterior and cave, but CO_2 maximums are reached with some hours delayed with respect to maximum air temperature difference. In the absence of assessing the GHGs' contents and the isotopic signal $\delta^{13}CO_2$ by continuous real-time monitoring with CRDS techniques during several daily cycles, we hypothesise two possibilities as mechanism responsible of the daily recovery of CO_2 content of cave air after a preceding nocturnal ventilation by advective mechanisms: (1) an intense diffusive flux of soil-derived CO_2 to the cave; (2) an intense CO_2 flux coming from deeper levels of the karst system.

CO_2 concentration of cave air is broadly the result of mixing the background atmospheric CO_2 with soil-produced CO_2, which fits a Keeling model with a mean

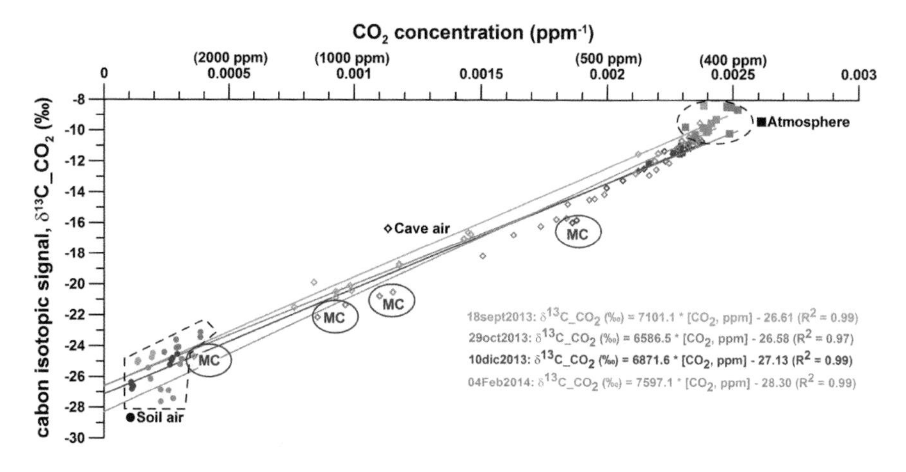

Fig. 4 Time evolution of Keeling models for CO_2 (isotope ratios are plotted against the inverse of CO_2 concentration) with three-component mixture: background atmosphere (*solid squares*), cave air (*open rhombus*) and soil air (*solid circles*). MC denotes air samples for Museo Cera

y-intercept value of −27.1 ‰ (Fig. 4). The CO_2-soil recharge by diffusion prevails during end of summer but the cave air becomes well-mixed with atmosphere during colder months (December–February). Air from more isolated galleries (e.g. Museo Cera) always has a higher CO_2 content and $\delta^{13}C[CO_2]$ below −20 ‰ with respect to other cave locations and considering the same air-sampling survey, which points to a gas inlet with a mainly edaphic origin. Conversely, lower CO_2 contents and a $\delta^{13}C[CO_2]$ that tend to local atmospheric background (~9 ‰), denote high air renewal of cave atmosphere.

Air samplings were conducted at times of day with high-ventilation rates, so it has enabled to draw up a spatial distribution of the main cave sectors for storage or sinks of carbon dioxide (Fig. 5). A prime example of isolated sector is the Museo Cera that is topographically higher than the rest of the cave sectors (Fig. 1). The air thermal stratification creates a motionless trap of warm and less dense air and it contributes to the CO_2 entrapment in the Museo Cera, hindering the air exchange with outer atmosphere. Microenvironmental data show that it operates as a motionless atmosphere that practically does not take part in the daily aerodynamic process between cave and the external atmosphere, however, it is controlled by: (1) 'breathing pulses' due to air pressure unbalances (cave-exterior) that involves periodic but not daily changes in direction of airflow and (2) seasonal (winter/summer) renewal of cave advective processes. The diffusive flux of soil-derived CO_2 prevails under these environmental conditions, so its isotopic signal $\delta^{13}C[CO_2]$ reaches here the minimum values compared to other cave locations (Fig. 5).

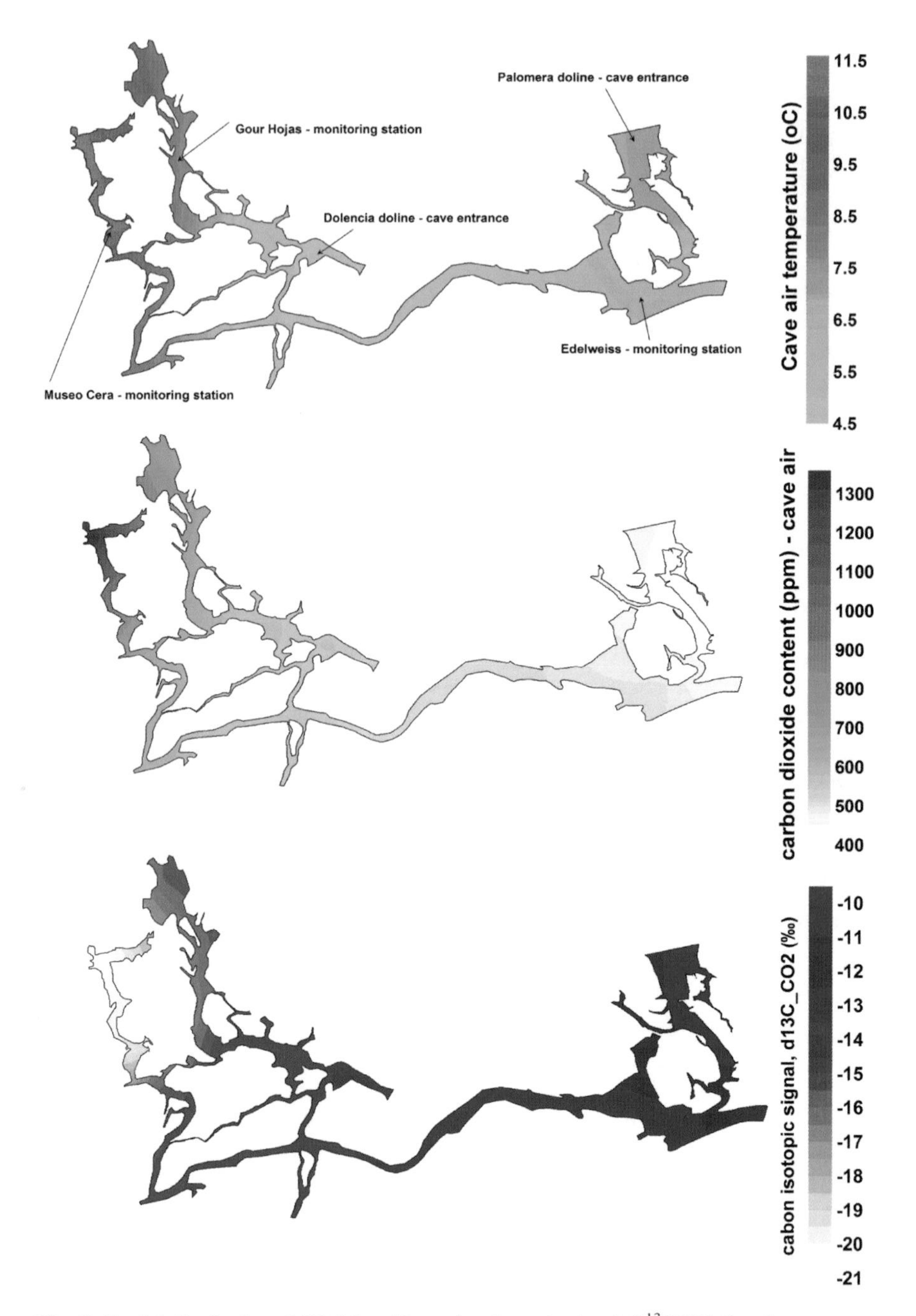

Fig. 5 Spatial distribution of CO_2(g) and its carbon isotopic signal ($\delta^{13}C[CO_2]$) and temperature of air from main cave sectors belonging to Ojo Guareña karst system (see Fig. 1). Data for each contour map correspond to mean values from four bimonthly spot air samplings (see Fig. 2), showing the isolated areas and those with a prevailing air exchange with exterior

5 Conclusion

Most subsurface air located in the uppermost part of the vadose zone from karst is very far from being saturated with organic carbon. The results of this monitoring study show that the amounts of carbon that might further be sequestered (CO_2) are with a potentially significant impact of the local or regional atmospheric $CO_2(g)$ budget. The spatiotemporal pattern of carbon dioxide described here provides evidence that atmospheric air which is inhaled into dynamically ventilated caves can then return to lower troposphere as CO_2-rich, cave air, even at short-term scale (daily).

Acknowledgments This research has been funded by the Fundación Patrimonio Natural belonging to Regional Government of Castilla y León. We thank the cave managers for their funding support and collaboration throughout the entire investigation. Authors were funded by several pre- and postdoctoral grants from the Spanish Ministry of Economy and Competitiveness (A.F–C: programme Torres Quevedo, co-financed European Social Funds; E.G-A: CSIC JAE-Predoctoral grant; S.C: programme Juan de la Cierva, and M.A-G: MEC FPI-Predoctoral). This research was supported by the Spanish Ministry of Science and Innovation: project CGL2010-17108, and in collaboration with project CGL2011-25162.

References

Benavente J, Vadillo I, Carrasco F et al (2010) Air carbon dioxide contents in the vadose zone of a Mediterranean karst. Vadose Zone J 9:126–136

Bourges F, Genthon P, Genty D et al (2012) Comment on carbon uptake by karsts in the Houzhai Basin, southwest China by Junhua Yan et al. J Geophys Res Biogeosci 117:G03006

Crosson ER (2008) A cavity ring-down analyzer for measuring atmospheric levels of methane, carbon dioxide, and water vapor. Appl Phys B-Lasers Opt 92:403–408

Cuezva S, Fernandez-Cortes A, Benavente D et al (2011) Short-term $CO_2(g)$ exchange between a shallow karstic cavity and the external atmosphere during summer: role of the surface soil layer. Atmos Environ 45:1418–1427

Fernandez-Cortes A, Sanchez-Moral S, Cuezva S et al (2011) Characterization of trace gases' fluctuations on a low energy cave (Castañar de Ibor, Spain) using techniques of entropy of curves. Int J Climatol 31:127–143

Garcia-Anton E, Cuezva S, Fernandez-Cortes A et al (2014) Main drivers of diffusive and advective processes of CO_2-gas exchange between a shallow vadose zone and the atmosphere. Int J Greenhouse Gas Control 21:113–129

Grupo Espeleológico Edelweiss (1986) Complejo kárstico de Ojo Guareña. In: Diputación Provincial de Burgos (ed) Kaite, vol 4–5. Diputación Provincial de Burgos, Burgos, Spain

Puch C (1998) Grandes Cuevas Y Simas de España. Espeleo Club de Gracia, Barcelona

Serrano-Ortiz P, Roland M, Sanchez-Moral S et al (2010) Hidden, abiotic CO_2 flows and gaseous reservoirs in the terrestrial carbon cycle: review and perspectives. Agric For Meteorol 150(3):321–329

A Field Analog of CO_2-Closed Conditions in a Karstified Carbonate Aquifer (Nerja Cave Experimental Site, South Spain)

J. Benavente, I. Vadillo, C. Liñán, F. Carrasco and A. Soler

Abstract We present new data that illustrate the hydrochemical evolution of groundwater along a flow line in the Triassic marbles around the Nerja Cave, South Spain. Water dissolves calcite and dolomite, and then $CaSO_4$. The environment is locally rich in CO_2 (up to near 60,000 ppmv) and consequently the water increases significantly its content in Ca^{2+}, Mg^{2+}, HCO_3^- and SO_4^{2-} along the flow, with EC values between 500 and 900 μS/cm. The pH values are typically in the 7–8 range, and the equilibrium PCO_2 of the water varies between $10^{-1.5}$ and $10^{-2.5}$ atm. In the considered flow line there is a relatively deep borehole (S2: 380 m; 280 m saturated) that shows pH values around 10 and equilibrium PCO_2 of 10^{-6} atm, with EC values generally in the 150–200 μS/cm range. Most of its solutes derive from rainwater concentration, together with the dissolution of carbonate minerals in a system closed to CO_2. For this reason we consider S2 to be

J. Benavente (✉)
Water Research Institute, University of Granada, c/Ramón y Cajal, n° 4, 18071 Granada, Spain
e-mail: jbenaven@ugr.es

I. Vadillo · C. Liñán · F. Carrasco
Department of Geology and Centre of Hydrogeology, University of Málaga (CEHIUMA), 29071 Málaga, Spain
e-mail: vadillo@uma.es

C. Liñán
Nerja Cave Foundation, Carretera de Maro s/n, 29787 Málaga, Nerja, Spain
e-mail: cbaena@cuevanerja.com; crilinbae@uma.es

F. Carrasco
e-mail: fcarrasco@uma.es

A. Soler
Faculty of Geology, Department of Crystallography, Mineralogy and Mineral Deposits, University of Barcelona, Martí i Franqués, 08028 Barcelona, Spain
e-mail: albertsolergil@ub.edu

B. Andreo et al. (eds.), *Hydrogeological and Environmental Investigations in Karst Systems*, Environmental Earth Sciences 1,
DOI 10.1007/978-3-642-17435-3_60

a field analog of such conditions. The nearly stagnant water of this well also shows evidence of sulphate reduction. Unlike its solute contents, isotopically (δ^2H and $\delta^{18}O$) the water of S2 does not show any modification with respect to the other points along the flow line.

1 Introduction and Objectives

When stagnant or quasi-stagnant conditions develop along groundwater flow, it can be expected to find water with very low renewal rate. In addition, it can become isolated from significant CO_2 sources coming from the soil horizons, thus leading to relatively anomalous compositions compared to those arising from the more frequent conditions found in CO_2-open systems. In this study, by way of repeated hydrochemical sampling, we analyze the composition of the groundwater in what can be conceived as a man-made field analog of the conditions which determine the nearly complete isolation of that water from CO_2 sources. Our analog is a relatively deep (380 m; 280 m saturated), fully cased borehole (S2) drilled during 1999–2001 in a highly karstified Triassic calcite-dolomite aquifer, which also shows some thin interbedded layer of schists (Las Alberquillas aquifer). Our study has the additional interest that the air in the vadose zone around the S2 borehole has high concentrations of CO_2 (up to 60,000 ppmv: Benavente et al. 2010). In the saturated zone of the aquifer, near the studied borehole, there is a pumping well (CW: 150 m depth) used for the supply of the cave facilities. This well was built in 1998 to substitute an adjacent one (165 m, built in 1984), uninstalled at present, which can be used for sampling purposes. Another monitored point is the most significant spring in the area: the Maro spring: MS, with outflow peaks of > 1 m^3/s.

Associated to the MS discharge are deposits of travertines. The fourth point considered in our study is an uninstalled well (EW) situated approximately 1 km upstream S2 in the general flow direction. These points belong to a monitoring network (the Nerja Cave Experimental Site, NCES), which also includes a number of shallow boreholes designed to monitor the vadose zone. In Fig. 1 we represent the general location of the study area, and the situation of the NCES and the four sampling points considered. The chemical and isotopic data of samples from these points illustrate the way the groundwater flow modifies its composition in the vicinity of the NCES. In this work we focus on the differences that the water in the S2 presents compared with other nearby points of the saturated zone, and we explain such differences in light of its isolation from the groundwater flow, and particularly due to the local depletion of CO_2.

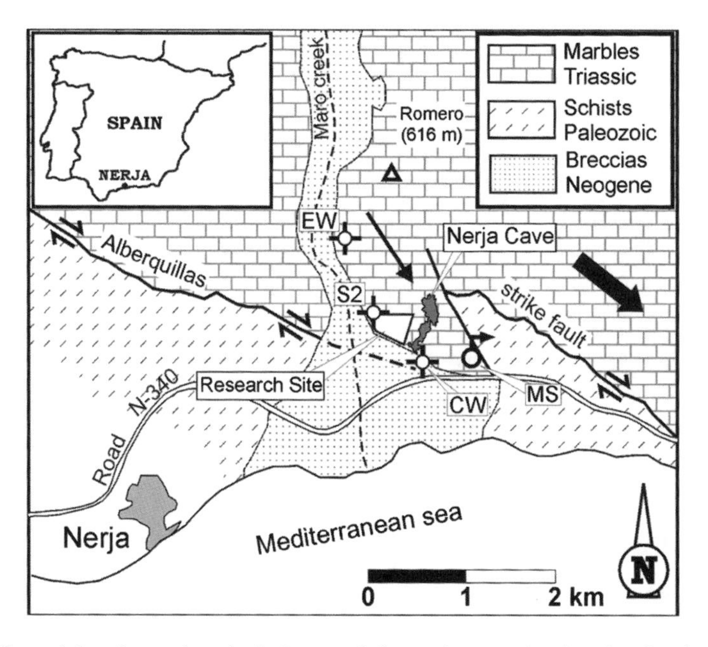

Fig. 1 General location and geological map of the study area showing the situation of the sampling points. Arrows indicate the local and regional flow directions

2 Hydrogeological Setting

The area belongs to the Alpujarride Complex of the Betic Cordillera. In the study area the Triassic carbonate formation dips gently toward the south, and its thickness can be more than 400 m. To the north it becomes folded and fractured. This formation is underlain by a thick sequence of schists, mostly Palaeozoic. To the south of the Nerja Cave, outcrops of the metapelitic materials contact those of carbonate formation by means of the so-called Las Alberquillas fault (Fig. 1), which has proven to be active in recent geologic times. The net outcome of its activity since late Miocene has been the relative elevation of the northern block, in which the cave is developed (Guerra et al. 2004).

The study area has a slightly dry, mild, Mediterranean climate. The general hydrodinamical and hydrochemical characteristics of Las Alberquillas aquifer in the vicinity of the Nerja Cave were stated in the work of Andreo and Carrasco (1993). A regional flow circulation from NW (mountain areas) to SE (Mediterranean Sea) was proposed. Nevertheless, piezometric evidences of lateral disconnection between aquifer blocks, as well as transmissivity measurements indicating values in the 10–1,000 m^2/d range suggested that local flow systems were likely to occur (Andreo and Carrasco 1993). One of these systems develops around the Nerja Cave, draining toward the points MS and CW. The 50 m difference in the water level altitude is a consequence of the heterogeneities mentioned before. Groundwater level is some

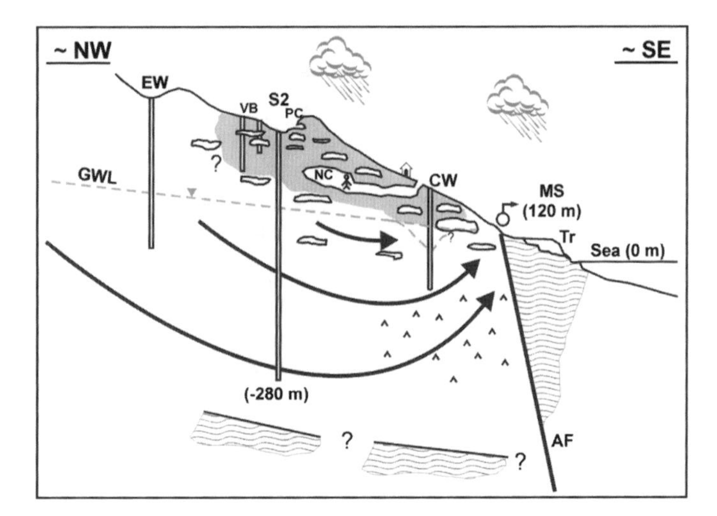

Fig. 2 Hydrogeological sketch (not to scale) of a cross section along the main flow in the aquifer in the study area. Some flow lines and altitudinal data are included. *GWL* groundwater level. *VB* vadose boreholes. *Tr* travertines. *NC* Nerja Cave. *PC* Pintada Cave. *Curved lines* represent schists. *Chevron symbols* represent $CaSO_4$ dissolution. *Shaded* sectors represent the recorded CO_2-rich vadose environment. See text and Fig. 1 for references of the sampling points (EW, S2, CW and MS)

meters below the Nerja Cave floor. The aquifer crops out in a rugged mountainous area, with peaks of more than 1,500 m a.s.l. Its recharge comes mainly from the precipitation over the permeable outcrops. The main hydrochemical signature of this local system, pointed out in previous works (Andreo and Carrasco 1993; Liñán et al. 2000), is the important increase in calcium and sulphate contents of the groundwater drained by these two points and the relatively high equilibrium PCO_2 values in CW water: from 8,500 (late summer) to 19,500 ppmv (springtime). Figure 2 depicts schematically a hydrogeological cross-section which intends to coincide with the flow direction and includes the points considered in this study. As can be seen in the figure, there are a number of cavities, the most important of them open to tourist visits is the Nerja Cave (NC). Other cavities have been recognized both above and below the NC (as, for instance, Pintada Cave: PC in Fig. 2); some others have been crossed by the vadose boreholes (VB) and finally others have been identified by geophysical surveys (Vadillo et al. 2012). Most of these caves are developed mainly in the horizontal direction, although it is logical to imagine that vertical discontinuities can interconnect them and favor the general vadose ventilation which typically follows different patterns in summer than in winter (Benavente et al. 2011). Even artificial conducts can play the same role, as is the case of a shaft drilled between PC and NC.

3 Methods

Hydrochemical sampling has been done at different dates from 2005 to 2013. Samples in MS and CW (when pumping) were taken directly in the outflow. Sampling in EW, S2 and CW (when not in operation) was carried out with a 0.5 L HDPE bailer. This device allows to sample at different depths (100, 200 and 300 m in S2). Electrical conductivity (EC), temperature (T), redox potential (ORP), pH and dissolved oxygen (DO) measurements were performed in the field. Major ions were analyzed by ionic chromatography. An elemental analyzer provided the inorganic carbon (HCO_3^-) and total organic carbon (NPOC). Isotopic measurements of δ^2H and $\delta^{18}O$ were obtained by laser spectroscopy and expressed in terms of delta (‰) units relative to the Vienna Standard Mean Ocean Water (V-SMOW). The precisions are ±0.5 ‰ (δ^2H) and ±0.1 ‰ ($\delta^{18}O$). Analytical results providing an ionic-balance error of more than 5 % have not been taken into account in further interpretation. The code PHREEQC (Parkhurst and Appelo 1999) was used in speciation calculations as well as in hydrochemical modeling.

Rainwater samples are taken after precipitation events from a rain gauge collector at the NCES. This task is done by the cave's scientific staff and provides hundreds of samples. Measurements of pH, EC and major constituents were carried out within 24 h of sampling. Both types of samples can thus stay some hours in the collector devices before their analysis. So, for the rainwater samples an equilibration process with the carbonate dust in atmosphere is likely to occur.

A permeability ("slug") test was carried out in S2 by way of introducing 15 L of water, then measuring the decay or recovery of the water level back toward the static condition as a function of time. The interpretation was done by the Hvorslev (1951) method. The result of the "slug" test in S2 indicates a value of hydraulic conductivity of the order of 10^{-9} m/s.

4 Results

Table 1 shows the analytical results obtained in point S2, and the average values of other points of interest (CW, EW and MS) and rain samples. Figure 3 is a Piper plot of the different analysis in the previous table. Figure 4 is a binary plot of the contents of the two stable isotopes of the water molecule analyzed. The altitude of the water level in S2 is near 60 m a.s.l. with few variations along the sampling campaigns. Water from S2 has generally pH values in the 10–11 range. Temperature is variable depending on the dates, whereas in winter it increases slightly with depth: from 19.4 to 19.6 °C, in summer the maximum values (22–24 °C) are found just below the water level. In most of the samples the EC ranges between 150 and 200 µS/cm. The maximum DO value is 3 mg/L. TOC is usually in the 2–10 mg/L range, although one sample shows an exceptionally high value of 26 mg/L. Calculated equilibrium PCO_2 values are near $10^{-6.5}$ atm. Saturation

Table 1 Dates and depth of sampling, and analytical results of the samples of the studied points

	SAMPLE	DATE	EC	T	pH	DO	TOC	Ca^{2+}	Mg^{2+}	Na^{+}	K^{+}	HCO_3^-	Cl^-	SO_4^{2-}	NO_3^-	F^-	δ^2H	$\delta^{18}O$
BOREHOLE S2	100 m	Jul-2005	178	22,6	10,20	1,3	-					37,30	35,82	3,07	1,91	0,65	-	-
	100 m	Feb-2006	212	19,6	10,15	2,1	9,59	6,08	1,06	27,23	15,16	35,90	43,32	1,63	0,14	0,74	-	-
	150 m	Feb-2006	168	19,5	10,34	1,7	2,79	5,65	0,94	18,03	8,59	36,60	25,01	0,36	0,09	0,54	-	-
	200 m	Feb-2006	165	19,5	10,19	2,1	3,41	7,64	2,87	15,28	7,63	46,10	23,12	0,22	0,07	0,47	-	-
	250 m	Feb-2006	200	19,6	9,63	2,1	1,82	8,84	9,55	14,82	7,52	81,00	22,38	0,18	0,07	0,38	-	-
	300 m	Feb-2006	170	19,6	10,24	1,7	2,59	5,16	1,03	17,06	10,71	37,10	25,40	0,22	0,07	0,57	-	-
	100 m	May-2007	167	21,6	10,44	1,9	26,24	5,55	1,20	18,59	15,85	13,22	-	-	-	-	-	-
	110 m	May-2007	163	21,4	10,53	1,7	6,96	5,27	0,54	17,48	10,11	11,73	-	-	-	-	-	-
	150 m	May-2007	166	22,6	10,52	1,6	8,33	5,45	0,70	17,47	9,32	11,66	-	-	-	-	-	-
	100 m	Jun-2011	181	24,1	10,34	2,9	-	5,86	0,76	24,00	10,85	-	24,26	4,99	0,76	0,45	-38,57	-6,13
	200 m	Jun-2011	182	23,0	10,44	3,0	-	5,39	0,19	25,16	11,08	-	23,77	4,94	0,81	0,45	-39,24	-6,41
	100 m	Jan-2012	184	19,4	10,60	1,8	-	5,53	0,36	24,64	10,64	-	24,56	5,14	0,81	0,42	-38,08	-6,76
	150 m	Jan-2012	182	19,6	10,55	1,4	-	7,36	0,36	24,34	10,99	-	24,05	5,09	0,70	0,43	-38,07	-6,75
	100 m	Jun-2013	173	23,0	10,28	2,1	2,67	9,87	0,77	19,32	18,33	55,40	38,74	5,15	1,00	0,41	-40,68	-6,75
	150 m	Jun-2013	172	22,3	10,32	1,0	2,73	9,79	0,77	20,29	8,58	52,00	25,92	5,06	0,76	0,41	-40,55	-6,85
	100 m	Oct-2013	210	21,5	10,14	1,3	5,28	15,71	2,50	29,19	8,02	26,75	36,95	6,86	0,85	0,56	-39,35	-6,66
		Max	212	24,1	10,60	3,0	26,24	15,71	9,55	29,19	18,33	81,00	43,32	6,86	1,91	0,74	-38,07	-6,13
		Min	163	19,4	9,63	1,0	1,82	5,16	0,19	14,82	7,52	11,66	22,38	0,18	0,07	0,38	-40,68	-6,85
RAIN WATER		m	49	-	7,80	-	-	8,90	2,35	4,60	2,80	23,60	10,70	12,00	1,80	-	-30,90	-5,16
ESPARTO WELL		m	553	20,2	7,17	6,1	1,49	20,49	17,67	6,62	2,38	224,38	27,13	17,17	3,38	0,11	-34,18	-5,89
CAVE WELL		m	900	21,8	7,23	6,3	0,94	95,68	48,44	22,04	9,46	232,15	41,00	159,30	12,55	0,20	-38,56	-6,64
MARO SPRING		m	722	19,3	7,62	8,9	0,36	113,90	30,09	11,30	2,40	204,72	20,18	200,23	1,44	0,50	-44,22	-7,65

All the contents in mg/L, except EC (μS/cm), T (°C) and stable isotopes (‰). (*m* mean value, *max* maximum, *min* minimum)

indices (SI) with respect to calcite, dolomite and gypsum are near 1.0, 2.5 and −3.0, respectively. The hydrofacies correspond mostly to the Na–Cl and Na–HCO_3 types (Fig. 3). Nitrate contents are generally less than 1 mg/L.

Groundwater sampled in EW, CW and MS shows average EC values in the 500–900 μS/cm range, and pH between 7.1 and 7.7. Average temperature ranges between 19 and 22 °C. DO is in the 6–9 mg/L range, and TOC is below 1.5 mg/L. Calculated equilibrium PCO_2 average values are $10^{-2.4}$ atm (MS), $10^{-1.8}$ (CW) and $10^{-1.6}$ (EW). Approximate average values for the SI in MS are 0.2 (calcite), 1.4 (dolomite) and −1.3 (gypsum); in CW are 0.1 (calcite), 1.2 (dolomite) and −1.4 (gypsum); in EW are −0.5 (calcite), 0.3 (dolomite) and −2.8 (gypsum). The hydrofacies of EW is Mg–HCO_3, for MS it is Ca–SO_4, and in the case of CW it has mixed features: Ca, Mg–SO_4–HCO_3. Average nitrate contents are > 1 mg/L, with a maximum of near 12 mg/L in CW.

The isotopic composition (Fig. 4) in S2 shows differences in function in the dates and depth of samples. Overall, there is no significant deviation with respect to the position of the average CW water in the diagram. The other two groundwater plots are clearly out of the S2 composition, indicating lower (MS) and higher (EW) contents. The point representing average local precipitation has highest contents. In the points of the June 2011 campaign, it is particularly noticeable the difference in contents with the sampling depth. The same can be identified in the June 2013 campaign (Fig. 4).

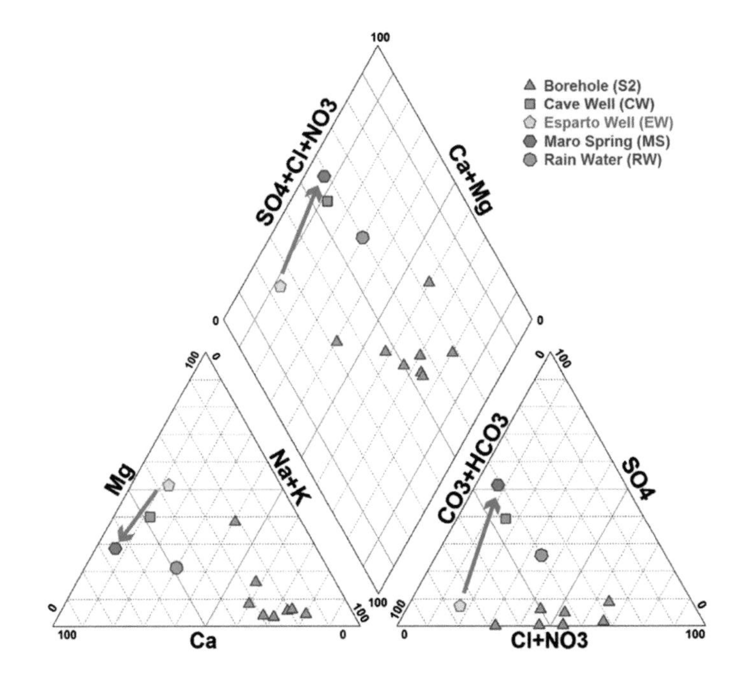

Fig. 3 Piper plot of the samples of Table 1. The arrow shows the main flow path

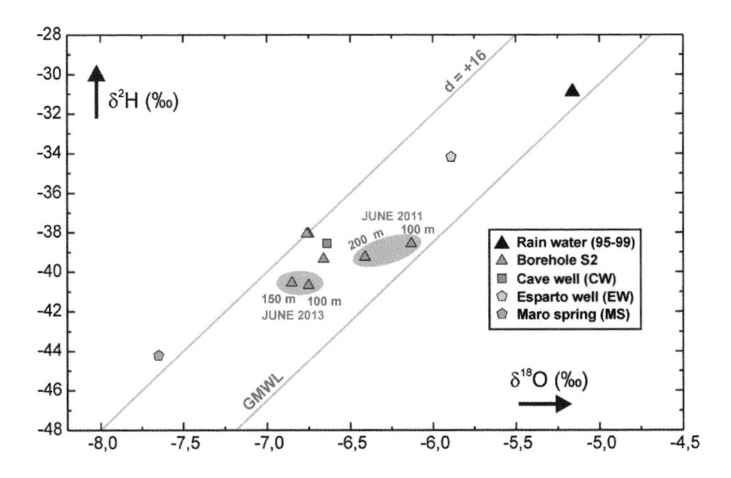

Fig. 4 Plot of the isotopic compositions of samples of Table 1

5 Discussion

In correspondence with Andreo and Carrasco (1993), Table 1 indicates similar results in the case of MS. CW composition in our study indicates higher values of EC and T, and particularly SO_4^{2-} and Ca^{2+}. Chloride and nitrate remain quite similar, and HCO_3 decreases. These changes may be induced by the extraction regime in the new well, with the consequence of getting water coming from deeper flows with higher contents in dissolved gypsum (Fig. 2). This mineral phase was confirmed by $\delta^{34}S$ values (15, 3 ‰).

MS chemical composition is similar to CW, although more diluted due to the effect of local recharge inputs favored by a karstic network. Due to this, it is colder with high DO and pH average values. Karstic influence in the hydrodinamical and hydrochemical data in this spring was deduced in a previous study (Liñán et al. 2000). In fact, Moral and Benavente (2010) suggested that the spring could be linked with a cave system (Fig. 2). The hydrochemical evolution from EW to MS involves: (1) the dissolution of calcite and dolomite along the saturated flow or by the effect of local recharge in the sector of high CO_2 vadose contents (Benavente et al. 2010) and (2) the dissolution of $CaSO_4$ (Fig. 3) which happens at the level of CW and MS points. It is worth to point out that neither gypsum nor anhydrite has been identified in outcrops of the study area, but these minerals are common in materials with similar degree of metamorphism in geologically similar nearby areas. Due to its hydrogeological character, in MS one could expect influences of both deep flow lines as well as near surface ones. Although water mixing happens, its isotopic composition (Fig. 4) reveals recharge at higher altitudes than the other points. On the contrary, the EW isotopic composition indicates a lower altitude of recharge (Fig. 2). Nitrates and, specially, TOC are regarded as good tracers for local recharge in the area (Batiot et al. 2003). Its average contents in EW, CW and MS corroborate the relatively higher influence of this recharge in the first point. Nitrate in CW is anomalously high (>10 mg/L), and is interpreted as a particular incidence of recycled water percolating from the garden sector near the entrance to the Nerja Cave, a circumstance which has been identified in previous studies (Liñán et al. 2008).

Although during the drilling operations it was found a number of karstification evidences in the saturated zone, and it was able to identify inputs of water at different levels of the column, the water sampled in S2 can be considered in the present conditions nearly stagnant, provided the very low value of permeability found on it. This is the consequence of being completely cased with unscreened tube after its drilling. In spite of its situation in the flow line joining roughly EW with CW and MS (Fig. 2), its composition does not reflect at all the general hydrochemical processes mentioned above. Water inside S2 has Cl^-, Na^+ and K^+ contents which are likely to be derived from simple concentration of precipitation water. The comparison with P composition indicates that some sulphate has suffered reduction, which is in accordance with the low DO values and the decrease in HCO_3^-. TOC values are, in some of the S2 samples, anomalously high (>5 mg/L: Table 1), even taking into account the relatively high values found in previous surveys (from 1 to 5 mg/L: Batiot el al. 2003). It is suggested that oxidation of this OC favors sulphate reduction. It is particularly illustrative the nearly depletion in CO_2 that the very low PCO_2 value and high pH value suggests. It is likely that water in S2 has evolved toward equilibrium in calcite and dolomite in a CO_2-closed system.

The isotopic contents of the S2 samples are, however, fully coherent with its supposed recharge sources. They are quite similar to the data in the nearby CW. Furthermore, it is possible to identify in the S2 isotopic contents the effects of a certain stratification which is congruent with the basic flow net depicted in Fig. 2. This circumstance can be inherited from the moment of its drilling, ended in 2001, and not subjected to further modifications. Chemically, however, the water has experienced the significant modifications commented above.

6 Conclusion

In Nerja research site (South Spain) located over a karstified Mediterranean carbonate environment (calcite and dolomite Triassic marbles) it was found an analog to groundwater conditions in a CO_2-closed system. The analog is a borehole (S2) with an unscreened casing in all its length (380 m). Water column inside S2 reaches nearly 200 m. Sampling in S2 shows pH values around 10 and PCO_2 of 10^{-6} atm. With EC values of less than 200 µS/cm, most of its solute contents seem to derive from the concentration of rainwater, together with the dissolution of carbonate minerals in a system closed to CO_2 and sulphate reduction process. The latter can be linked with the relatively high TOC values. These hydrochemical characteristics contrast with those that have been identified in the groundwater flow system in which the point S2 is included. The general trend is the dissolution of calcite and dolomite in an environment that, for the local recharge through the vadose zone near the cave, is marked by equilibrium with very high PCO_2 values. Near the main natural outflow of the system-Maro spring—the main hydrochemical process is the dissolution of $CaSO_4$.

The water isotopic data obtained for the S2 are coherent with the flow model proposed for the study area. In this case, the data for what can be considered nearly stagnant water are similar to a nearby pumping well, although their hydrochemical characteristics are completely different.

Acknowledgments This research was funded by the Cueva de Nerja Foundation and done in the framework of Andalusia's Research Groups RNM-126 and RNM-308. We also thank the Geological Survey of the Spanish Ministry of Environment for drilling the boreholes. We acknowledge the anonymous reviewers for all changes and comments that improved the text.

References

Andreo B, Carrasco F (1993) Estudio hidrogeológico del entorno de la Cueva de Nerja. Trabajos Cueva Nerja 3:163–187

Batiot C, Liñan C, Andreo B, Emblanch C, Carrasco F, Blavoux B (2003) Use of TOC as tracer of diff use infiltration in a dolomitic karstic system: The Nerja Cave (Andalusia, Southern Spain). Geophys Res Lett 30:2179–2183

Benavente J, Vadillo I, Carrasco F, Soler A, Liñán C, Moral F (2010) Air CO2 contents in the vadose zone of a Mediterranean karst. Vadose Zone J 29:647–659
Benavente J, Vadillo I, Liñán C, Carrasco F, Soler A (2011) Hydrochemical ventilation effects in a karstic show cave and in its vadose environment, Nerja, Southern Spain. Carbonates Evaporites 26:11–17
Guerra A, Serrano F, Ramallo D (2004) Geomorphic and sedimentary Plio-Pleistocene evolution of the Nerja area (Northern Alboran basin, Spain). Geomorphology 60:89–105
Hvorslev MJ (1951) Time lag and soil permeability in groundwater observations. US Army Corps of Engineers Waterway Experimentation Station Bulletin 36
Liñán C, Andreo B, Carrasco F (2000) Caracterización hidrodinámica e hidroquímica del manantial de Maro (Sierra Almijara, provincia de Málaga) (Hydrodinamic and hydrochemistry of Maro spring). Geogaceta 27:95–98
Liñán C, Carrasco F, Vadillo I, Garrido A (2008) Estudios hidrogeológicos en la Cueva de Nerja (Málaga) (Hydrogeological studies in the Nerja Cave). Publs. del Instit. Geol. y Min. de España, serie Hidrogeología y Aguas Subterráneas 25:673–684
Moral F, Benavente J (2010) Importancia del transporte de CO2 a través del aire de la zona saturada de los sistemas kársticos. Algunos ejemplos de la Cordillera Bética. En: Durán JJ, y Carrasco F (eds) Cuevas: Patrimonio, Naturaleza, Cultura y Turismo, Madrid. Asociación española de Cuevas Turísticas, 169–182
Parkhurst DL, Appelo CAJ (1999) User's guide to PHREEQC (version 2): a computer program for speciation, batch-reaction, one-dimensional transport, and inverse geochemical calculations. Water-Resour Investi Rep 99:4259
Vadillo I, Benavente J, Neukum C, Grützner C, Carrasco F, Azzam R, Liñán C, Reicherter K (2012) Surface geophysics and borehole inspection as an aid to characterizing karst voids and vadose ventilation patterns (Nerja research site, S. Spain). J Appl Geophys 82:153–162

Terrestrial Laser Scanning for 3D Cave Reconstruction: Support for Geomorphological Analyses and Geoheritage Enjoyment and Use

A. Marsico, M. Infante, V. Iurilli and D. Capolongo

Abstract Recent developments in laser scanning techniques and digital modelling provide powerful tools for the knowledge, management and preservation of the underground. A Terrestrial Laser Scanner (TLS) survey was performed to create a 3D virtual model of the Santa Croce cave (Apulia, Southern Italy); this is well known for its Palaeolithic and Neolithic finds (a Neanderthal thigh-bone among all) and used as a showcave both for its natural and historic heritage. The survey included chimneys, passages and also the surface over the cave to acquire a model of the entire system. Data were acquired by processing the resulting points cloud to pursue three main purposes: karst hazard management, education and geoheritage preservation. Thanks to the virtual model, the opportunity to visit the site through a virtual tour, also showing hardly accessible details, is an exciting way to discover the underground environment. Therefore, interactive virtual image can be utilised to promote the site as an important tool to disseminate knowledge and to increase interest in the Earth sciences in society at large.

A. Marsico (✉) · M. Infante · V. Iurilli · D. Capolongo
Dipartimento di scienze della Terra e Geoambientali, Università degli Studi di Bari "Aldo Moro", via E. Orabona, 4, 70125 Bari, Italy
e-mail: antonella.marsico@uniba.it

M. Infante
e-mail: marcolaser@libero.it

V. Iurilli
e-mail: vincenzo.iurilli@uniba.it

D. Capolongo
e-mail: domenico.capolongo@uniba.it

B. Andreo et al. (eds.), *Hydrogeological and Environmental Investigations in Karst Systems*, Environmental Earth Sciences 1,
DOI 10.1007/978-3-642-17435-3_61

1 Introduction

Dry valleys and caves are among the most common features of geodiversity in karst plateaus, although they are often intensely modified by man (Gray 2004). Caves in particular, as a result of human presence for long times, have become part of the ancestral culture and are important elements of the place and community identity. The "Regional Register of Caves" reveals an unexpectedly large number of caves subjected to guided tours that total a few thousands visitors per year. This form of enjoyment leads to the need for a better knowledge of the environmental system for multiple purposes, such as: (1) a scientifically based information sharing, (2) preservation of the sensitive elements of the system, educating visitors to the protection and (3) awareness of the possible hazard factors, on the surface and in the subsurface.

The Santa Croce cave is a showcave both for its natural and historic heritage: the cave sites in an interesting geomorphological location, a wide "lama" (local name for dry valleys) with steep sides, and Palaeolithic and Neolithic findings were discovered inside it. Thus, the site is under some type of preservation, such as that of Superintendence for Archaeological Heritage of Apulia. From the hazard point of view the cave and the surroundings fall within the "geomorphological hazards areas" as mapped by the River Basin Authority of Apulia; it is also recorded in the Regional Register of the Geosites. In order to provide an accurate 3D model of the cave, better than the already available speleological survey (Francescangeli et al. 1995), the terrestrial laser scanner (TLS) has been used.

The TLS technique is becoming a frequently used research tool to investigate topics in geomorphological (Lim et al. 2010; Infante et al. 2012) and sedimentological/structural fields (Franceschi et al. 2009) and it has also been applied in studies concerning archaeological investigations in caves (Rüther et al. 2009; Lerma et al. 2010; Puchol et al. 2013). A unique 3D model allows to collect many exhaustive multidisciplinary information and provides data for a lot of applications. The rapid data collection rate significantly reduces potentially hazardous field survey time. The high density of points permits a shape to be tightly fitted, improving the precision of its geometric characteristics (Hiremagalur et al. 2007). It allows an alternative form of access to the inaccessible part of the site and to monitor the status of a potential degradation. Moreover, the virtual reality technology can be easily applied for site and tourism planning and management, marketing (Guttentag 2010), geoheritage-oriented education and research, providing a digital reference and record for future generations (Rüther et al. 2009).

2 Geological Setting

The study cave is located in the northern part of the Murge plateau, in the Apulia region (Southern Italy), where the Mesozoic limestone of *Calcare di Bari* Fm. outcrops (Fig. 1). In the Plio-Pleistocene age, tectonics shaped the plateau in steps,

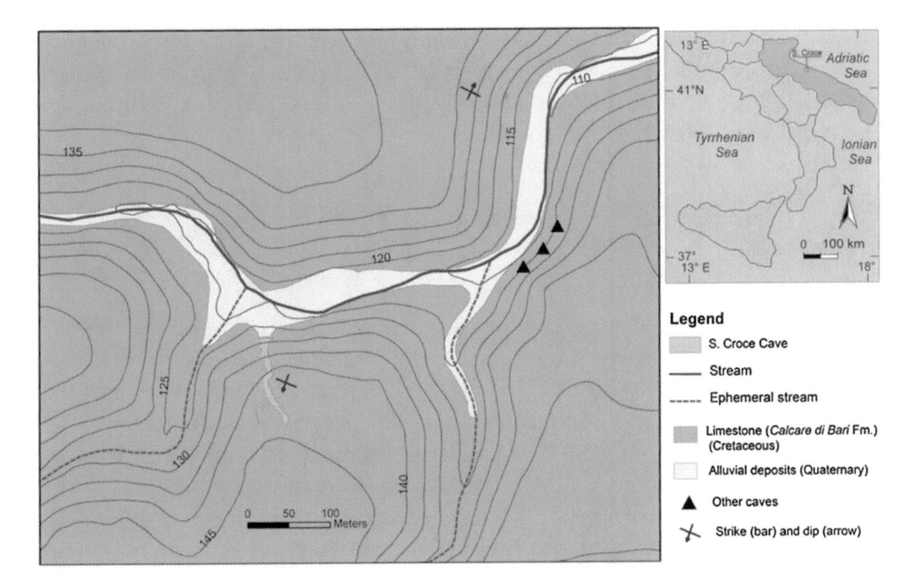

Fig. 1 Santa Croce cave location and geological setting

with elongation parallel to the coastline, whose edges were deeply cut by dry valleys here called "lama". "Santa Croce" is the name of an ancient rural settlement, as much as of the valley and of the nearby cave.

This "lama" crosses the graben of lower Murge (Pieri et al. 1997) cutting the Cretaceous carbonate sequence; close to the cave it is entrenched in the *Palorbitolina lenticularis* Limestone member (Luperto Sinni and Masse 1992) a micritic, hazel brown or yellowish limestone of the Lower Aptian age. The lithofacies is characterized by microfossiliferous calcarenite and calcisiltite rocks with intercalated stromatolitic limestone. Dolomitization process occurs in some layers of this unit (Francescangeli et al. 1995). North of the valley the carbonate sequence shows an antiform structure (Martinis 1961) with a WNW-ESE striking axis, so that the area of the caves lies on its southern flank.

The Santa Croce cave system consists of a main tunnel, one hundred or so metres long and six metres wide, and several less easily accessible passages that are in part linked together or clogged up by recent (colluvial) or ancient (*terra rossa*) soils. Recent surveys suggested a paleokarst origin of the cave since the most is fossilised by the *terra rossa* middle Pleistocene deposits (Del Gaudio et al. 2002; Iurilli 2010). Inside the cave, joints and bedding planes are well visible on the exposed bedrock, because of the lack of calcite speleothems. Dissolution produced both overlapping metric domes, that sometime stretch into chimneys, and alveoli, probably due to moisture condensation, as it appears to be the present-day process. The cave has an irregular floor with elevations ranging from 123 to 115 m above sea level. The cave filling is made up of a sequence of a few metre thick Quaternary sediments overlying the limestone basement. Two archaeological survey tests were carried out during the 1950s, when many findings of the

Mousterian lithic industries (Middle Palaeolithic) and a *Homo sapiens neanderthalensis* thigh-bone were found inside the cave. In the 1990s, others archaeological surveys were carried out and a Neolithic mat (6,500-years old) made of interlaced vegetable fibres was found.

3 Methods

Time of flight measurement technology is based on the principle of sending out a laser pulse and observing the time taken for the pulse to reflect from the object and return to the instrument sensor. The distance range is combined with high-resolution angular encoder measurements to provide the three-dimensional location of the point (Hiremagalur et al. 2007). The result obtained is a point cloud and it can be used for several purpose. A high-speed, high-accuracy terrestrial laser scanner Leica HDS 3000 has been used (Leica Geosystems 2006).

Survey process was developed inside and over the cave. Because of the cave shape and the presence of domes and chimneys at the roof, scanner was set up at 37 strategically positioned points; the ground surface on top was detected positioning the TLS in 25 locations. In fact, multiple scans at different locations are required to capture the entire survey site, due to the scanner line-of-flight being obscured by structures in one or more of the scans location. Scans resolutions range between 10 and 30 mm both in horizontal and in vertical spacing in the cave, while the scans on surface have 80–100 mm spacing to obtain the raw ground morphology. Each scan had to contain the highest resolution control points (targets) of previous scan thus providing overlaps to join scans into one single point cloud of the whole cave. To perform survey of whole area, workflow started from the inner side of the cave proceeding towards the entrance and then getting scans of countryside above the cave.

The data processing, by the Cyclone© software, generates 3D point cloud composed of millions of points according to the spatial resolution previously arranged; then each scan was cleaned of unwanted objects, i.e. some devices that could alter the digital surface rendering. The widespread vegetation in the ground surface was removed by filter tools based on the user-defined ground identification. The union of the scans was achieved by joining overlapping targets. Processing the dataset generates two main results: point clouds and closed surfaces. The first is considered as groups of sparse objects and measurements and cross sections are really similar to real shape. With the latter it is possible to obtain continuous shape to gain volumes and other measures, but shapes do not always correspond to real forms. The 3D model easily defines the rock thickness between the cave roof and the land surface: several topographic profiles describe the whole system better than the previous survey. It can be used for several purposes such as site management, as well the geological and geotechnical survey and the enhancement and promotion of the geoheritage.

Starting from the 3D model, a user defined 3D animation of the cave can be created: moving on the point cloud, different cave features can be pointed out according to the aim the 3D model was created for, to give a full visual knowledge of the site.

4 Results and Discussion

S. Croce cave is easy to visit, but the survey process faced difficulties due to the rock surface geometry and shapes. However, the point cloud preserved with good accuracy much of rock details (Fig. 2) such as fractures, karstic shapes, sedimentary patterns and fractures, particularly those enlarged by dissolution. A further benefit is the possibility to insert virtual tools, such as lines and planes, for better shape description. The cave system can be inscribed in a 90 m × 18 m rectangle; the main gallery is described by two main pipes of 35 and 47 m connected by a 16 m long curved elbow. On a horizontal plane the point cloud highlights the "S" shape whose width reaches 5.4 m (Fig. 3). Despite the model giving an incomplete description of the roughness of the roof, since irregular forms hide surfaces in the backside, it is possible to evaluate rock thickness and geometry anyway. Overall, 18 chimneys were identified: their average length is about 6 m and all of them show a roughly conical shape with smooth surfaces; highest features are often hidden from bottom field of view. However, the virtual model shows the relationships between the cave voids and the outer land surface (Fig. 2): it is possible to define spots where the top of the cave is close to the external surface.

The secondary narrow tube was discovered thanks to the partial emptying of the red earth deposit. It is about 1 m wide, its average height is less than 2 m and is about 50 m long. It consists of a sinuous connection to another segment parallel to main gallery (Fig. 3). It must be pointed out that both ends of this latter are still completely filled.

Cross-sections are the best tools to easily read and describe geometrical patterns. In 1995 a cave survey was performed, according to the standard speleological method, interpolating few data and drawing several cross-sections in order to deliver a cave plant showing horizontal path and vertical geometry. Comparing the actual survey with the old one (Fig. 3b), the former is far apart closer to the real. To appreciate TLS good results, sections derived from digital model were matched with drawings from the old survey methods, and the improvement is very clear due to the better accuracy. Handmade profiles allowed accurate measurements only where easily accessible, and only on chosen sections: moreover, they were affected by the operator point of view that generated deformed shapes. Handmade tracing had to be very smooth, whereas digital model preserves the detail chosen during data acquiring. On the 3D cave model the archaeological area is clearly visible (Figs. 3, 4): this is in the lowest zone of the floor, and extends about 14 m in length. 3D model

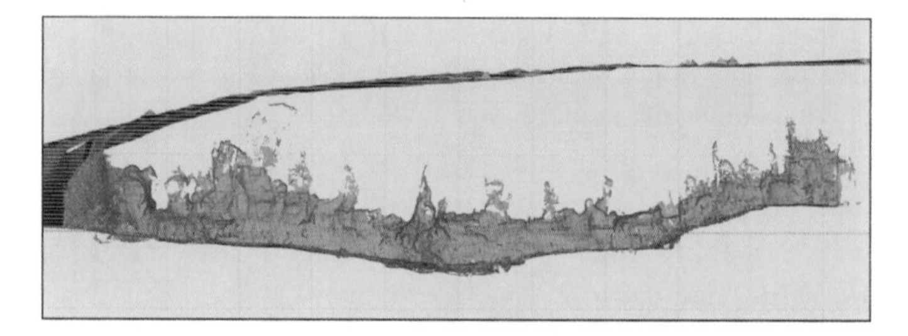

Fig. 2 Longitudinal section through the whole system showing cave morphology (*grey*) and hillside shape: chimneys and domes are well visible at the cave roof. The side of the *yellow square* is equal to 10 m

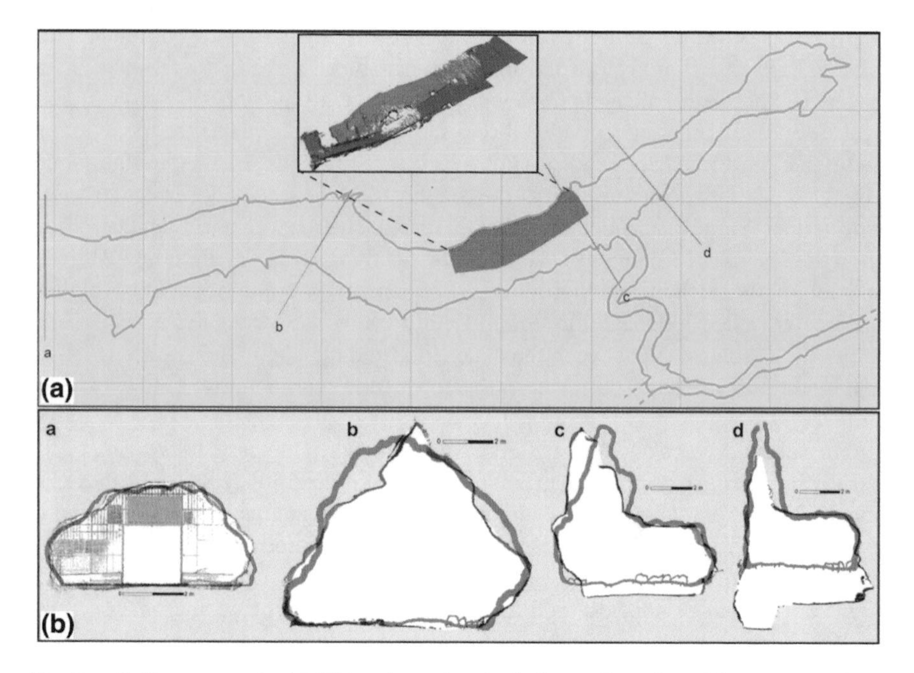

Fig. 3 **a** S. Croce cave plan highliting the archaeological area (in *red*), and four cross-sections: the side of the *yellow square* is equal to 10 m; **b** comparison between the 1995 (*red outline*) and the actual survey sections: in sections *c* and *d* difference in the floor level is due to cave emptying by recent excavations

can be a virtual platform where all objects collected during archaeological digs can be added in accurate spatial position, and can be a tool for teaching purpose.

The survey allows to highlight some limestone blocks, in the ceiling as much as on cave walls, bounded by joints and suspected of potential instability. This led to start geological hazard studies by researchers in geomechanics. The 3D model shows its usefulness in different ways: (1) it is possible to extract faithful representations of the cave walls and ceilings, very useful for graphical annotations *on*

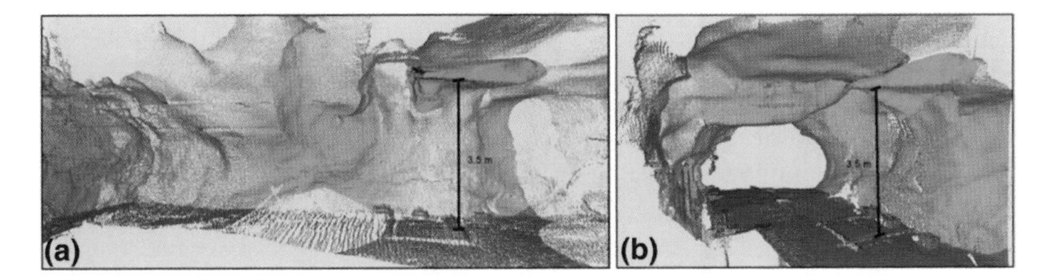

Fig. 4 Sketches showing the archaeological dig (*the red surface*) and one of the rock blocks (*green*) bounded by joints (about 3.5 m above the cave floor); **a** longitudinal section; **b** cross section in perspective view

site; (2) off site, once the rock blocks are known, it is possible to assess their volumes as well as any other geometric parameter that may be useful to measure. One such block, located near the archaeological area, is shown in Fig. 4.

The laser scanning survey has its main aim in helping knowledge about geotechnical features; however, digital model should play a good role in divulging knowledge about the cave. To achieve as better as possible this role, many features can be highlighted by visual effects: digital surfaces can be draped in a lot of colour patters, e.g. by texture mapping, showing point cloud with natural hue scheme or by merging photographs with 3D shapes. Elevation maps, giving multiple colour patterns to describe height range, are also useful. Furthermore, 3D animation can be created to make a virtual journey inside the cave: a bird's eye view can highlight more remarkable features or it simply represents an exciting way of access to the cave.

5 Conclusions

The TLS approach is an improvement in the cave survey techniques used until a short while ago to get reliable data without other on-site measurements. In fact, due to local difficulties, it is often easier to measure dip and strike off the joints in the model than on-site, decreasing working time, ecological disturbance and operators number and increasing safety. Thanks to the 3D output, both professional and common users can focus on details just observed on site. Cave surveys show the limits due to site accessibility with the device. There has been a lack of data in the highest and most irregular ceiling parts, since the instrument always worked on walking level, even if some higher scan points were necessary. Data voids are the main limits of this method, so often users should interpolate data with the help of proper techniques.

Since the site has both geological and archaeological relevance, a set of documents can be issued for tourist information or teaching purposes. Moreover, 3D animation can stand in for the real cave visit when the site is inaccessible due to weather conditions or archaeological diggings.

However, TLS improved the cave knowledge and contributes to site management and enhancement. Unlimited cross sections, cave dimensions and joints measures are foundations for further studies, while pictures and 3D animation can improve cave website and other documents to spread knowledge to no-experts, such as tourists, pupils and local administrators.

References

Del Gaudio V, Iurilli V, Lillo A, Pagliarulo P (2002) Osservazioni geofisiche, geologiche e mineralogiche su una cavità carsica di interesse archeologico (Grotta di Santa Croce, Bisceglie-Ba). Atti Terzo Convegno di Speleologia Regionale, 6-7 Dicembre 2002 Castellana Grotte, Grotte e Dintorni, 231–240

Francescangeli R, Iurilli V, Lorusso D, Reina A, Ruina G (1995) La grotta di Santa Croce in agro di Bisceglie (BA). Indagini geologiche e speleologiche (Report). Bisceglie (Italy)

Franceschi M, Teza G, Preto N, Pesci A, Galgaro A, Girardi S (2009) Discrimination between marls and limestones using intensity data from terrestrial laser scanner. ISPRS J Photogram Remote Sens 64:522–528

Gray M (2004) Geodiversity: valuing and conserving abiotic nature. John Wiley and Sons, Chichester

Guttentag DA (2010) Virtual reality: applications and implications for tourism. Tour Manag 3:637–651

Hiremagalur J, Yen KS, Akin K, Bui T, Lasky TA (2007) Creating standards and specifications for the use of laser scanning in caltrans projects. AHMCT Research Report UDC-ARR-07-06-30-01

Infante M, Marsico A, Pennetta L (2012) Some results of coastal defences monitoring by ground laser scanning technology. Environ Earth Sci 67(8):2449–2458

Iurilli V (2010) Le forme carsiche. In *Il patrimonio geologico della Puglia*, Suppl. a Geologia dell'Ambiente/2010, ISBN: 978-88-906716-4-7; 61–74. http://sigeapuglia.blogspot.it/

Leica Geosystems (2006) http://leica-geosystems.com/hds

Lerma JL, Navarro S, Cabrelles M, Villaverde V (2010) Terrestrial laser scanning and close range photogrammetry for 3D archaeological documentation: the Upper Palaeolithic Cave of Parpalló as a case study. J Archaeol Sci 37:499–507

Lim M, Rosser NJ, Allison RJ, Petley DN (2010) Erosional processes in the hard rock coastal cliffs at Staithes, North Yorkshire. Geomorphology 114:12–21

Luperto Sinni E, Masse JP (1992) Biostratigrafia dell'Aptiano in facies di piattaforma carbonatica delle Murge baresi (Puglia, Italia meridionale). Riv. It. Paleont. Strat. 98(4):403–424

Martinis B (1961) Sulla tettonica delle Murge nord-occidentali. Acc. Naz. dei Lincei, Rend. Cl. Sc. Fis. Mat. e Nat., 31

Pieri P, Festa V, Moretti M, Tropeano M (1997) Quaternary tectonic activity of the Murge area (Apulia foreland, Southern Italy). Annali di Geofisica XL/5:1395–1404

Puchol OG, McClure SB, Senabre JB, Villa FC, Porcelli V (2013) Increasing contextual information by merging existing archaeological data with state of the art laser scanning in the prehistoric funerary deposit of Pastora cave, eastern Spain. J Archaeol Sci 40(3):1593–1601. doi:10.1016/j.jas.2012.10.015

Rüther H, Chazan M, Schroeder R, Neeser R, Held C, Walker SJ, Matmon A, Horwitz LK (2009) Laser scanning for conservation and research of African cultural heritage sites: the case study of Wonderwerk Cave, South Africa. J Archaeol Sci 36:1847–1856

A Laser Technique for Capturing Cross Sections in Dry and Underwater Caves

A. Schiller and S. Pfeiler

Abstract For the acquisition of 3D-geometry data of groundwater conduits in the coastal Karst plain of Yucatan near the town of Tulum a novel laser scanning device was developed and applied. The method is derived from similar industrial systems and for the first time adapted to the specific measurement conditions in underwater cave systems. The device projects a laser line over the whole cave perimeter at a certain position. This line represents the intersection of a plane with the caves wall. Through proper design and calibration of an imaging system it is possible to derive the true scale geometry of the perimeter by special image processing techniques. Through caption of regularly spaced images it is possible to reconstruct the true scale 3D-shape of a tunnel if position and attitude data is incorporated. The method provides easily operable acquisition of the 3D-data of caves in clear water with superior resolution and speed and facilitates the measurement in underwater tunnels as well as in dry tunnels significantly. The data gathered represent crucial input to the study of state, dynamics and genesis of a complex karst water regime.

1 Introduction

The Ox Bel'Ha Karst conduit system is located at the south-east coast of the Yucatan peninsula in the region of the growing town of Tulum, Mexico (Fig. 1). In the subsurface, and below the city, the whole area is nerved by a vast network of underwater caves and conduits developed in nearly horizontally layered limestone. The uppermost layer of the karst aquifer represents practically the only freshwater resource of the region. Below the freshwater layer there is saltwater intruding from the sea and reaching deep regions. The freshwater is endangered by rapid urban

A. Schiller (✉) · S. Pfeiler
Geological Survey of Austria, Wien, Austria
e-mail: arnulf.schiller@geologie.ac.at

B. Andreo et al. (eds.), *Hydrogeological and Environmental Investigations in Karst Systems*, Environmental Earth Sciences 1,
DOI 10.1007/978-3-642-17435-3_62

development and insufficient wastewater management. In this context, sustainable water management as well as protection of the barrier reef and the nearby Sian Ka'an biosphere reserve (green marked in Fig. 1) require a better understanding of the water resource and its potential (Gondwe 2010). To achieve this, collaborations of local NGO, exploration divers, the University of Neuchâtel and the Geological Survey of Austria are in progress within a representative part of the conduit system (Fig. 2) with the objective to acquire crucial input data for hydrological modelling by means of standard and innovative methods (Vuilleumiere 2011) .

2 Principle

The method presented herein is derived from similar laser scanning methods as applied for measuring tasks in industrial processes (e.g. Kannala et al. 2008; Matsui et al. 2009), adapted to the special measurement conditions in (underwater) caves. The device (Figs. 3, 7) consists in principle of a camera with fish eye lens and a laser head projecting a laser line over the whole perimeter of a tunnel. Both components of the device are connected through a rigid bar (made of aluminium) preserving a defined geometry of the system as shown in Fig. 3.

The laser line is imaged by the camera. Taking into account the geometry of the set-up as well as the distortion of the imaging system the true scale shape of the laser line can be directly calculated from the image of the laser line. The whole set-up is designed so that that this calculation can be accomplished with a relative simple mapping function that converts the image coordinates (px, py) of a point on the laser line into (x, z)-coordinates on a cross-section plane through the tunnel (Fig. 4). Since the optical distortion of the imaging system is axial-symmetric referred to as optical axis, the mapping function is axial-symmetric as well. This calculation can be done for every point along the imaged laser line. Herewith the problem is comfortably solved in the ideal case. The real case emerges more complex due to geometric errors in the instruments design as well as light refraction at the water/dome/air interfaces. This can be considered by careful calibration of the system.

2.1 Positioning

In the presented test survey the cross section positions are related to the 'stick line'. The stick line represents the geometry of a cave system as a series of connected straight lines, similar to sticks (Figs. 2, 5). In reality it is a cord attached in the cave onto rocks and other suitable features. The stick line is measured by exploration divers with compass, depth meter and scale tape in dead reckoning technique. The offset of a laser scanned cross-section to the stick line is visually well-defined by the intersection of the laser plane with the cord as indicated by a bright dot. The position along the line is defined by equally spaced intervals.

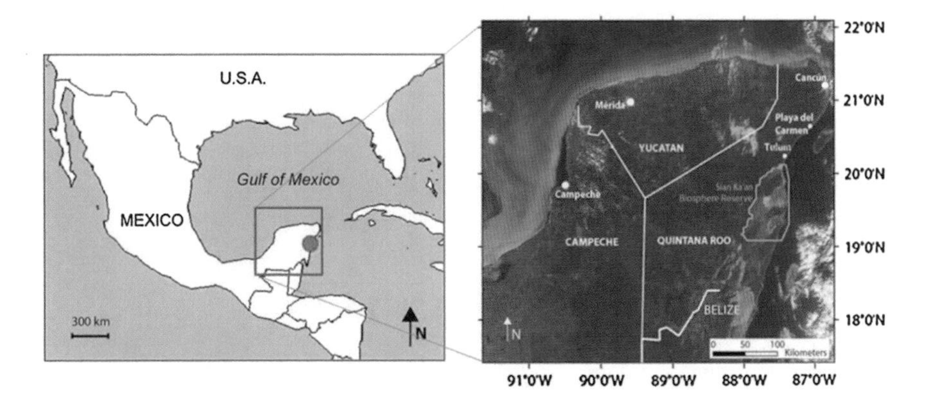

Fig. 1 Location of the survey area (*red point*, *red circle*)

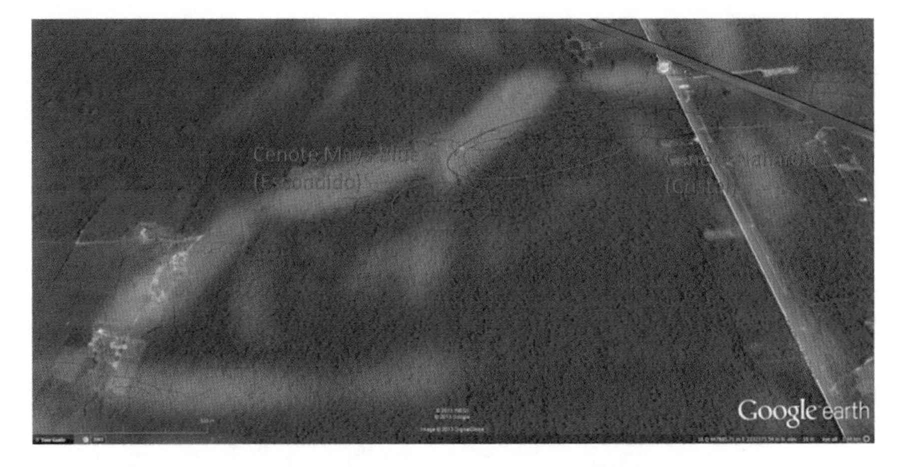

Fig. 2 Part of the cave system chosen for field tests. *Blue lines* survey data (*stick lines*). *Red ellipses* studied parts

2.2 Processing

The first processing step is digitizing the image data. This can happen automatically by standard image processing techniques (Fisher and Naidu 1991). However, manual editing in advance gives the opportunity of interpreting gaps in the laser line (shadows or side tunnels or light absorbed, Fig. 5). The second step is transforming the data from pixel coordinates into real-world coordinates. In principle this is done with a lookup table as soon as the mapping function is known from calibration. The third processing step is the compilation of successive cross sections from successive shots to any 3D representation of the scanned tunnel by incorporating attitude and position information (Figs. 5, 6), error estimations are given in Table 1.

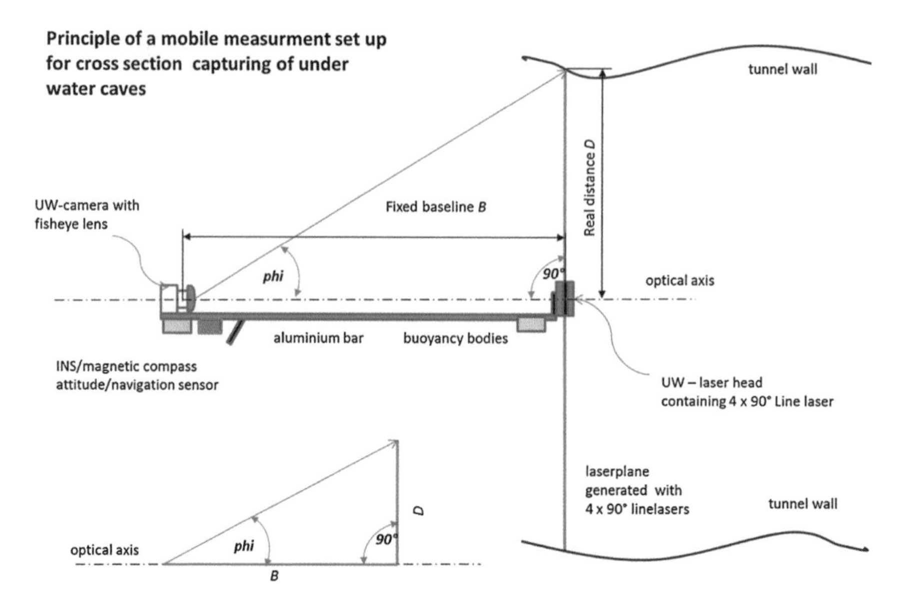

Fig. 3 Basic principle of the scanning method

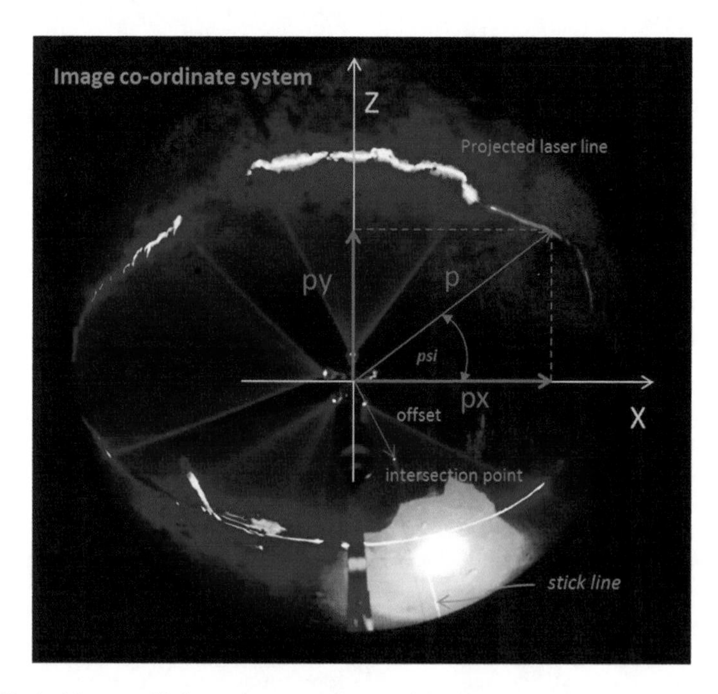

Fig. 4 Typical image with laser plane coordinates of the image system

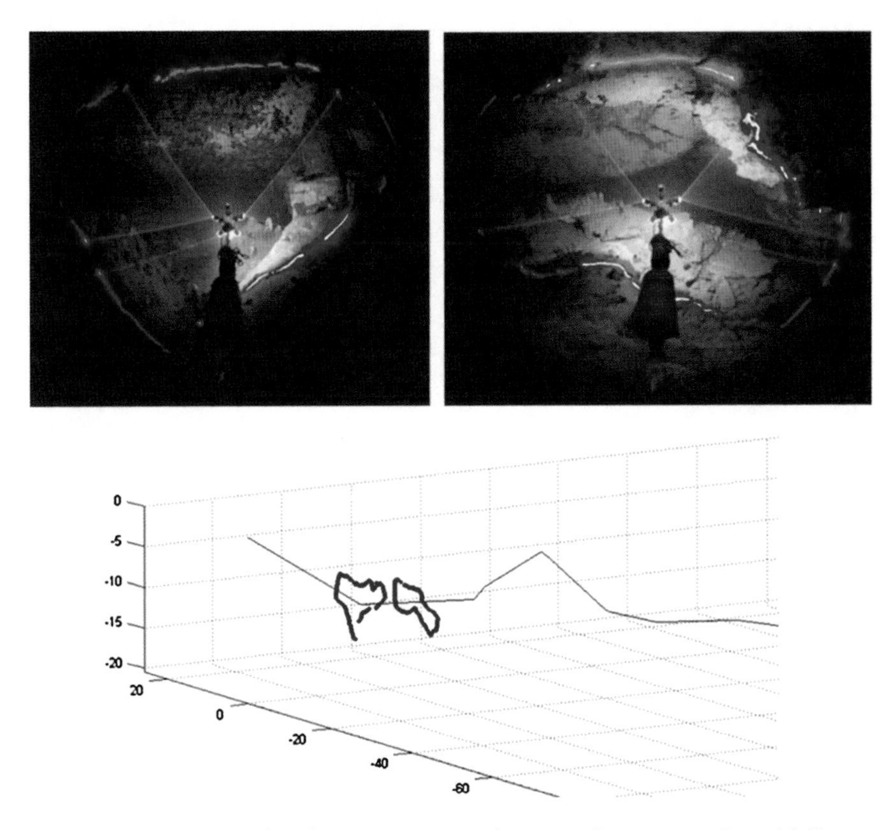

Fig. 5 *Top* Two cross section shots. *Bottom* mapped cross sections connected to stick line

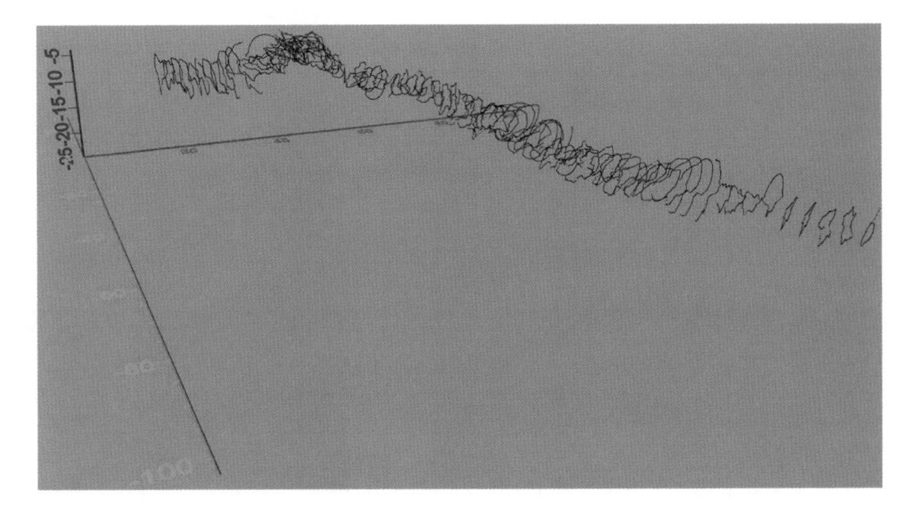

Fig. 6 Ox Bel Ha system, Cenote Maya *blue*, A-tunnel, 77 shots

Table 1 Magnitude of errors

Error source	Magnitude
Calibration	1 mm to 5 cm at 15 m center distance
Digitizing	1 mm to 5 cm at 15 m center distance
Attitude + positioning by diver	cm in cross section, decimeters along the stick line
Stick line	Decimeters to meters depending on length and complexity of survey

Table 2 Field test scans

Cenote maya blue (Escondido)	Branch	Number of shots	Diver
First test	A-tunnel	125	Robbie Schmittner
Second test	A-tunnel	77	Richie Schmittner
Third test	Dead zone	71	Richie Schmittner

3 Field Test

A prototype system was designed in the second half of 2012 and prepared for a first field test in Tulum in March 2013 (Fig. 7). The tunnels scanned are located in the Ox Bel Ha system and accessible through Cenote Maya Blue (Fig. 2, Table 2). Five line lasers (four blue and one red) have been chosen and installed with batteries and tilt switch into an underwater light housing each. The imaging system consists of a DSLR camera with 4.5 mm circular fisheye lens. At each shot the device was levelled horizontally and adjusted parallel to the line. In course of the under water tests about 270 cross sections have been captured in two tunnels during three dives of approximately 50 min each. With an average separation of approximately 3 m this gives some 800 m of scanned tunnel.

4 Results

The tests showed that the device is well balanced and easily operable under water. After digitizing and mapping the data was visualized in cross sections combined with stick line data as shown in Fig. 6. The first prototype worked well and delivered processible data. In order to minimize possible errors (Table 2) specific advancements are in progress and will be tested during the next survey 2014. Main improvements are a new laser head for achieving a laser line without gaps and the

Fig. 7 Richie Schmittner starting the second dive in A-tunnel

integration of attitude data for processing cross sections captured with arbitrary orientation of the device. With this handy device important geometric parameters can be quickly captured in underwater as well as in dry caves. The method gives several thousand perimeter points with one shot, i.e. in 0.2 s—so the acquisition speed, resolution and information density are superior to other methods. The geometry data is crucial for estimating the hydraulic head losses and fluxes in the conduits and to constrain the groundwater flow model developed by the collaboration partner (Vuilleumiere 2011). Furthermore, this extensive high resolution geometric data provides the opportunity for analysis regarding hierarchy and development of karst conduit networks.

Acknowledgments We thank for the great support by Amigos de Sian Ka'an, Robert and Richard Schmittner (Xibalba Diving Center), Bil Phillips (Speleotech), Simon Richards, the Austrian Science Fund which finances the Austrian part of the project Xibalba (I994-N29).

References

Fisher RB, Naidu DK (1991) A comparison of algorithms for subpixel peak detection. In: Proceedings of the 1991 British Machine Vision Association conference (BMVAC 1991), pp 217–225

Gondwe BRN (2010) Exploration, modelling and management of groundwater-dependent ecosystems in karst—the Sian Ka'an case study, Yucatan, Mexico. Ph.D. thesis, Technical University of Denmark, Kongens Lyngby, 86p

Kannala J, Brandt SS, Heikkilä J (2008) Measuring and modelling sewer pipes from video. Mach Vis Appl 19(2):73–83

Matsui K, Yamashita A, Kaneko T (2009) 3-D shape reconstruction of pipe with omni-directional Laser and omni-directional camera. In: Proceedings of the 3rd international conference of Asian society for precision engineering and nanotechnology (ASPEN2009), 1A2-15, pp 1–5
Vuilleumier C (2011) Stochastic modeling of the karstic system of the region of Tulum (Quintana Roo, Mexico), MSc thesis, University of Neuchâtel (Switzerland), 37p

Climate Variability During the Middle-Late Pleistocene Based on Stalagmite from Órganos Cave (Sierra de Camorra, Southern Spain)

C. Jiménez de Cisneros, E. Caballero, B. Andreo and J.J. Durán

Abstract Paleoclimatic reconstruction in southern Spain has been investigated in this study using stable isotope analyses from Órganos Cave (Southern Spain). A combination of $\delta^{18}O$ and $\delta^{13}C$ analyses with uranium-series dating and δD values of the water extracted from stalagmite fluid inclusions makes it possible to obtain an isotopic record during 340–370 ky BP (MIS10). The profile shows the climatic evolution in the area and we can see a colder early stage with small fluctuations that change to warmer conditions at the end, coinciding with the start of the interglacial stage MIS9.

Keywords Speleothems · Paleoclimate · Pleistocene · Stable isotopes · Fluid inclusions

1 Introduction

Variations in $\delta^{13}C$ and $\delta^{18}O$ values of calcite speleothem have been widely recognized as a powerful tool to reconstruct the paleoclimate/environment of terrestrial areas (e.g. Banner et al. 1996; Bar-Matthews et al. 2000; McDermott 2004; Fairchild et al. 2006; McDermott et al. 2006; Spötl and Mangini 2007). During the late twentieth century, the main focus of research was on $\delta^{18}O$ records in

C. Jiménez de Cisneros (✉) · E. Caballero
Instituto Andaluz de Ciencias de la Tierra, CSIC-UGR, Avenida de Las Palmeras 4, 18100 Granada, Armilla, Spain
e-mail: concepcion.cisneros@iact.ugr-csic.es

B. Andreo
Department of Geology and Centre of Hydrogeology, University of Málaga (CEHIUMA), 29071 Málaga, Spain
e-mail: andreo@uma.es

J.J. Durán
Geological Survey of Spain, Instituto Geológico y Minero de España (IGME), C/Ríos Rosas, 23, 28003 Madrid, Spain

B. Andreo et al. (eds.), *Hydrogeological and Environmental Investigations in Karst Systems*, Environmental Earth Sciences 1,
DOI 10.1007/978-3-642-17435-3_63

speleothems as paleotemperature indicators. Support for a paleoclimatic record based on growth frequencies of speleothems can be obtained from the dated deposits themselves, by variations in ^{13}C and ^{18}O of calcite along the growth axes of deposits that formed in isotopic equilibrium with the groundwater. The precise dates that can be obtained on these records using the U-Th technique potentially enable high chronostratigraphic resolution (Edwards et al. 1987; Musgrove et al. 2001; Marshall et al. 2009).

Fluid inclusions, representing fossil seepage water, are formed in speleothems when small volumes of drip water become trapped within the precipitating calcite. The oxygen isotope ratios of speleothem calcite and hydrogen isotope ratios of speleothem fluid inclusion are water-isotope tracers, which enable the calculation of paleotemperatures (Mathews et al. 2000; Dennis et al. 2001; Genty et al. 2002; Verheyden et al. 2005; Vonhof et al. 2006; Jiménez de Cisneros et al. 2011; Jiménez de Cisneros and Caballero 2013).

In this paper we present a paleoclimate record of the western Mediterranean. The stalagmite from southern Spain (Málaga province) is dated by ^{230}Th/^{234}U, ranging in age from 340 to 370 ky BP. Continental records during this period are scarce not only in the western Mediterranean (Jiménez de Cisneros et al. 2003; Durán et al. 2004; Muñoz-García et al. 2007; Dominguez-Villar et al. 2008; Moreno et al. 2010; Bartolomé et al. 2012; Jiménez de Cisneros and Caballero 2013; Durán et al. 2013) but also worldwide, which enhances the interest of this work.

2 Location and Geological Setting

Órganos or Mollina cave is located in Malaga province (Southern Spain, Fig. 1) at 650 m above sea level and was formed in Lower Jurassic dolomites of the External Zone of the Betic Cordillera. The unsaturated zone above the cave is several tens of metres thick and comprises fissured but poorly karstified dolomites and limestones with diffuse flow behaviour. Stalagmite sample was collected in the inner part of the cave, which has only one entrance. Hence dripwaters would not have been subject to evaporation or rapid degassing, which is important for stable isotope variations in the calcite. It is a stalagmite EM-MO over 56 cm long formed by laminated calcite. Colour is predominantly white with some alternating black bands. Thin-section analysis indicates suitability for preserving a quasi-continuous climate record in the whole of the sample. Examination of thin-sections reveals that apparent texture/colour changes are due to the alternating density of fluid inclusion and minor clays or organics. Crystal fabric is predominantly large and columnar in nature favouring deposition in equilibrium with its corresponding drip water (Kendall and Broughton 1978).

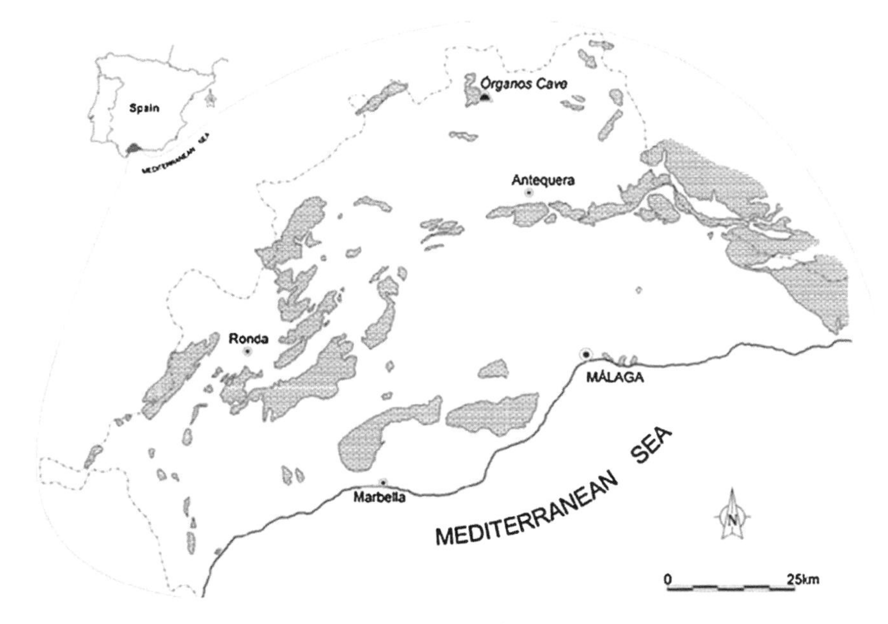

Fig. 1 Location map, showing the situation of the Órganos Cave, southern Spain

3 Analytical Methods

The mineralogical composition of the samples was determined by X-ray diffraction, using a Panalytical X'PERT pro diffractometer. The stalagmite was cut along its growth axis after retrieval from the cave in order to expose its growth laminae and to permit checks for a secondary alteration.

The sampling for this isotopic study was carried out using a dental-drill. Samples for stable isotope analyses of less than 10 mg were drilled. The drill was cleaned with HCl and water in between samples, in order to avoid sample contamination. Samples for stable isotope analysis were typically taken at 2–5 mm intervals along the growth axis of the stalagmite (Fig. 2). In total, 65 samples were analysed, including 12 supplementary samples to check the isotopic equilibrium with the Hendy Test (Hendy 1971). CO_2 was extracted from calcite at 70° by reaction with H_3PO_4 (McCrea 1950) using a VG Micromass preparation line. Data were corrected for fractionation using the carbonate-phosphoric acid fractionation factors for calcite. Results are reported as ‰ versus the PDB standard for carbonates. Paleotemperatures were calculated using the Sharp (2007) equation. Analytical precision was 0.1‰ for $\delta^{18}O$ and 0.05‰ for $\delta^{13}C$ based on replicate measurements of an internal carbonate standard. The isotopic composition of percolating water from which the flowstone calcite precipitated can be determined by analysing water released from fluid inclusions trapped within the carbonate matrix (Jiménez de Cisneros et al. 2011).

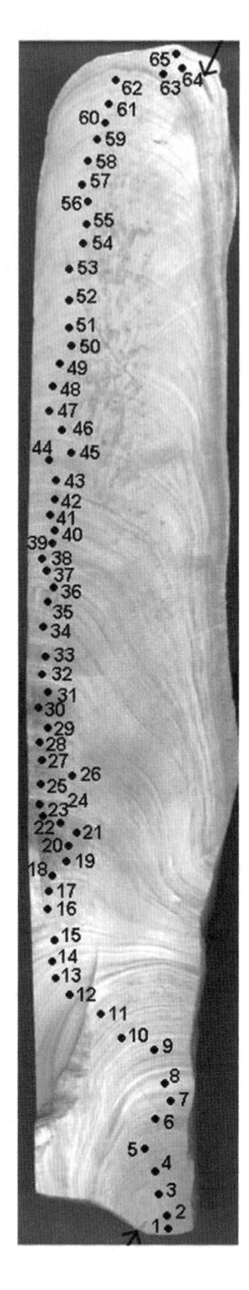

Fig. 2 Stalagmite EM-MO from Órganos Cave formed during MIS10. Small black circles show the position of the samples used for stable isotope analyses

U-series analyses were carried out using multi-collector inductively coupled plasma mass spectrometry (MC-ICPMS). The small calcite samples required (100–500 mg) mean that thin growth phases and hiatuses can be easily resolved.

High precision (typically ~0.5 %) allows growth phase length to be more precisely constrained. U-Th isotopic measurements were undertaken on a ThermoFinnigan Neptune MC-ICPMS, (Bristol Isotope Group facilities, University of Bristol, UK).

4 Results

Results from the XRD analysis show that the studied material consists of calcite. Most of the calcite is homogeneous, made of clear palisade crystals with fine growth laminations. Speleothem mineralogy and fluid inclusions are interpreted to be primary, which permits to conclude that fluid inclusion isotope composition is undisturbed since the time of formation.

The stalagmite was tested with the Hendy test for isotopic equilibrium. Samples taken along a single growth layer should show no signs of kinetic isotopic fractionation. The test results show a variation of less than 0.4‰ in $\delta^{18}O$ values and no sequential enrichment in the $\delta^{13}C/\delta^{18}O$ plot. This indicates that speleothem growth is in isotopic equilibrium with the drip water and that the calcite isotopic values do not reflect kinetic effects.

Figure 3 shows the variation of $\delta^{18}O$ and $\delta^{13}C$ from the base to the top of the stalagmite versus time. The $\delta^{18}O$ values vary from −4.30 ‰ to −5.95 ‰ and the $\delta^{13}C$ vary from −3.63 ‰ to −6.03 ‰. We have extracted inclusion water using the method by Jiménez de Cisneros et al. (2011). Water released from inclusions was analysed and the mean values were −4.7‰ $\delta^{18}O_w$ and −29.2‰ δD. Analytical uncertainties are 1.5‰ for δD and 0.5‰ for $\delta^{18}O$.

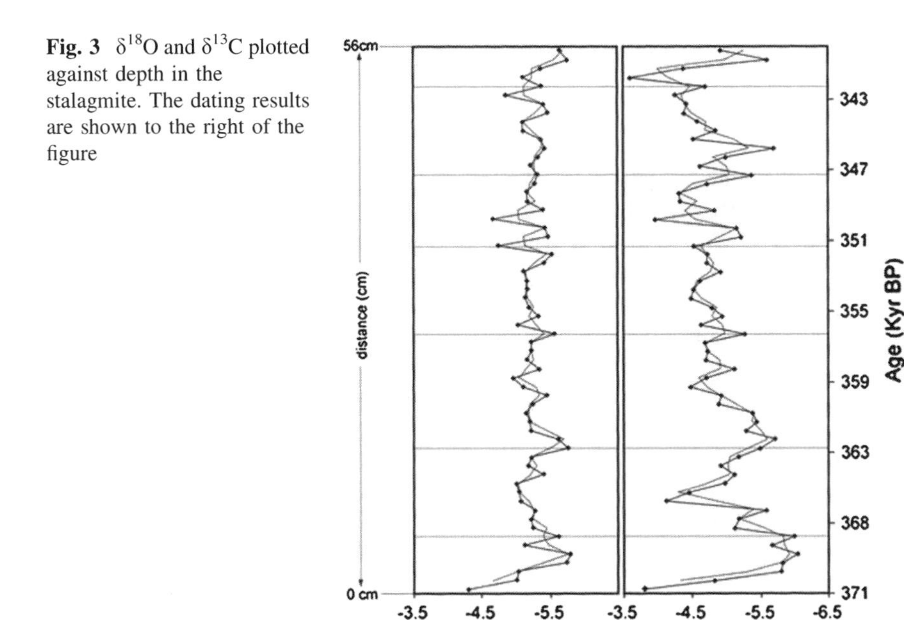

Fig. 3 $\delta^{18}O$ and $\delta^{13}C$ plotted against depth in the stalagmite. The dating results are shown to the right of the figure

The calculated ages from Uranium series $^{230}Th/^{234}U$ are 340–370 ky from the base to the top of the stalagmite. Precipitation rates were calculated by measuring the distance between points of known age. If a uniform stalagmite growth rate is considered, this means an accretion rate of about 1.8 cm/1,000 years. This rate is interpreted to represent consistent flow rates of meteoric water through the local country rocks into the cave.

5 Discussion

Calcite of stalagmites mainly originates from the seepage of rainwater. Although temperature is the major controlling variable for $\delta^{18}O$ carbonate values in speleothems, it should be stated that there are two competing temperature effects controlling the speleothem $\delta^{18}O$ values. These are an increase in the $\delta^{18}O$ (less negative) values of meteoric water related to warming and a decrease in the $\delta^{18}O$ (more negative) speleothem data corresponding to increased temperatures within the cave.

The $\delta^{18}O$ and $\delta^{13}C$ values for stalagmite when combined with the $^{230}Th/^{234}U$ ages, help to reconstruct a continuous paleoclimate record over the 340,000–370,000 years ago, being the interval during which this stalagmite was formed. This period is characterized by values of precipitation, temperature and vegetation cover very favorable in this area for the development of speleothems (Durán 1996; Durán et al. 2004). Samples precipitated under thermodynamic equilibrium conditions are expected to display the following criteria: a) no correlation between $\delta^{18}O$ and $\delta^{13}C$ along a single growth layer and b) constant $\delta^{18}O$ and variable $\delta^{13}C$ values along a single growth layer. The stalagmite samples yielded positive results in the Hendy test (Hendy 1971), suggesting that calcite precipitation occurred under conditions of (or near to) the isotopic equilibrium. $\delta^{18}O$ values remained almost constant in each growth layer, showing only minor changes (under 0.4 in all cases). Figure 3 shows the carbon and oxygen stable isotope records obtained along the growth axes of stalagmite. These plots depict no significant correlation between carbon and oxygen values, a finding that also supports isotopic equilibrium during precipitation.

An alternative test of thermodynamic equilibrium is by comparing the $\delta^{18}O$ values of fluid inclusions with those of the surrounding calcitic matrix. Assuming that the inclusions and the surrounding calcite are coeval, and that the oxygen isotope composition of trapped fluids did not vary over time, published water/calcite equilibrium calibrations can be used to reconstruct cave paleotemperatures, and to assess whether these temperature predictions are physically realistic. To obtain the absolute temperature variation for a given $\delta^{18}O$ variation an independent estimate of $\delta^{18}O_w$ should be done, using fluid inclusion studies. The mean values of paleotemperature obtained were 17–18 °C and in this case we used the $\delta^{18}O_w = -4.7‰$ mean value of water extracted in the samples (Sharp 2007). The oxygen isotope record shows an overall increasing trend from the eldest growth

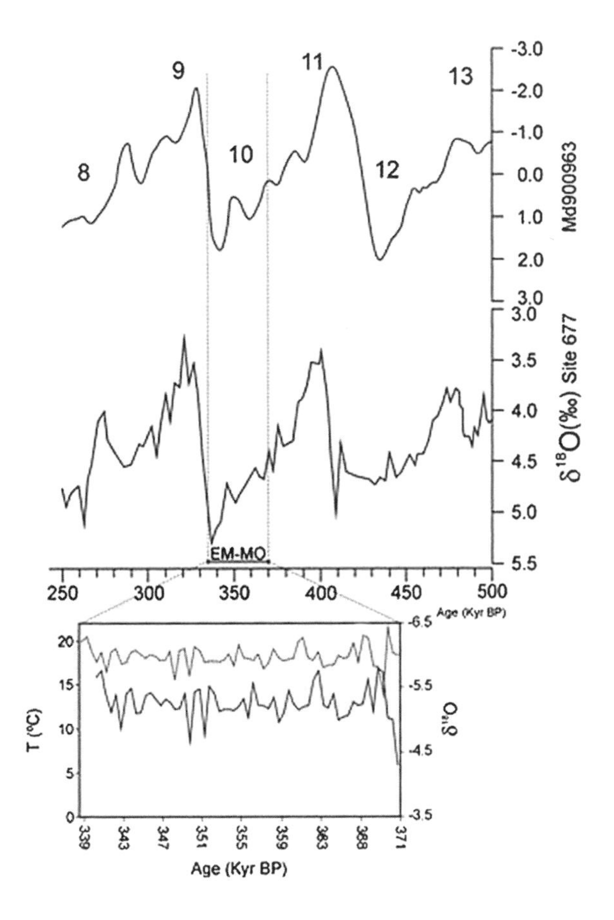

Fig. 4 The growth period of the stalagmite EM-MO matches the glacial period (MIS10) in the global sea level curve from ODP Site 677 (Shackleton et al. 1990) and MD 99003963 (Bassinot et al. 1994). δ^{18}O record obtained in the stalagmite from Órganos Cave

period (17–20 °C) which may be interpreted as representing the Middle–Late Pleistocene late warming (340,000–370,000 years ago).

A temperature-time profile using δ^{18}O from the speleothem as proxies for temperature is plotted on Fig. 4 together with those obtained from site 677 (Shackleton et al. 1990) and MD99003963 (Bassinot et al. 1994). The speleothem record contains several very sharp changes in δ^{18}O being the sign of rapid cooling and warning trends. In order to interpret δ^{18}O fluctuations in a paleoclimatic sense, the relationship between δ^{18}O and temperature must be assessed. The relatively enriched oxygen isotope ratios in the oldest part of the stalagmite (end of MIS10) indicate a colder period followed by a slightly warmer one (start of MIS9). Our paleoclimate records are generally concordant with data obtained in isotopic studies of deep marine sediments, confirming that we are recording global climatic events (Williams et al. 1988; Dotsika et al. 2010; Moreno et al. 2013).

6 Conclusions

This study presents terrestrial Middle–Late Pleistocene paleoclimate results from a stalagmite from southern Spain (Málaga). The stalagmite is dated by $^{230}Th/^{234}U$, ranging in age from 340 to 370 ky BP.

$\delta^{18}O$ values for individual speleothem layers are consistent with equilibrium deposition of individual speleothem layer. Paleoclimatic information was gained by dating periods of stalagmite growth and measuring $\delta^{18}O$ and δD of speleothem calcite and fluid inclusions respectively.

The isotopic record of stalagmite shows a deposition under colder conditions (MIS10) followed by a warmer period (MIS9).

Acknowledgments This work was completed thanks to the support of the Research Projects CGL2007-61876/BTE and CGL2013-45230-R and it is a contribution in the frame of Associated Partnership "Geoquímica Avanzada" CSIC-UMA. Thanks to an anonymous reviewer for useful comments on the manuscript. Thanks to Angel Caballero at the drawing office.

References

Banner JL, Musgrove M, Asmerom Y, Edwards RL, Hoff JA (1996) High-resolution temporal record of Holocene ground-water chemistry; tracing links between climate and hydrology. Geology 24:1049–1053

Bar-Matthews M, Ayalon A, Kaufman A (2000) Timing and hydrological conditions of Sapropel events in the Eastern Mediterranean as evident from speleothems. Soreq Cave, Israel. Chem Geol 169:145–156

Bartolomé M, Moreno A, Sancho C, Hellstrom J, Belmonte A (2012) Cambios climáticos cortos en el Pirineo central durante el final del Pleistoceno superior y Holoceno a partir del registro estalagmítico de la cueva de Seso (Huesca). Geogaceta 51:59–62

Bassinot FC, Labeyrie LD, Vincent E, Quidelleur X, Shackleton NJ, Lancelot Y (1994) The astronomical theory of climate and age of the Brunes-Matuyama magnetic reversal. Earth Planet Sci Lett 126:91–108

Dennis PF, Rowe PJ, Atkinson TC (2001) The recovery and isotopic measurement of water from fluid inclusions in speleothems. Geochim Cosmochim Acta 65(6):871–884

Domínguez-Villar D, Wang X, Cheng H, Martín-Chivelet J, Edwards RL (2008) A high-resolution late Holocene speleothem record from Kaite Cave, Northern Spain: $\delta^{18}O$ variability and possible causes. Quatern Int 187:40–51

Dotsika E, Psomiadis D, Zanchetta G, Spyropoulos N, Leone G, Tzavidopoulos I, Poutoukis D (2010) Pleistocene palaeoclimatic evolution from Agios Georgios Cave speleothem (Kilkis, N. Greece). Bull Geol Soc Greece 43 (2):886–895

Durán JJ (1996) Los sistemas kársticos de la provincia de Málaga y su evolución: contribución al conocimiento paleoclimático del Cuaternario en el Mediterráneo occidental. Ph.D. Thesis University of Madrid, Spain

Durán JJ, Barea J, López-Martínez J, Rivas A, Robledo P (2004) Panorámica del karst en España, In: Andreo, B., y Durán JJ (eds) Investigaciones En Sistemas Kársticos Españoles. Publicaciones del Instituto Geológico y Minero de España. pp 15–25

Durán JJ, Pardo-Igúzquiza E, Robledo PA, López-Martínez J (2013) Ciclicidad en espeleotemas: ¿qué señales climáticas registran? Boletín Geológico y Minero 124(2):307–321

Edwards RL, Chen JH, Wasserburg GJ (1987) 238U-234U-230Th-232Th systematics and the precise measurement of time over the past 500,000 years. Earth Planet Sci Lett 81:175–192

Fairchild IJ, Smith CL, Baker A, Fuller L, Spötl C, Mattey D, McDermott M, E.I.M.F. (2006) Modification and preservation of environmental signals in speleothems. Earth-Sci Rev 75:105–153

Genty D, Plagnes V, Causse Ch, Cattani O, Stievenard M, Falourd S, Blamart D, Ouahdi R, Van-Exter S (2002) Fossil water in large stalagmite voids as a tool for paleoprecipitation stable isotope composition reconstitution and paleotemperature calculation. Chem Geol 184:83–95

Hendy CH (1971) The isotopic geochemistry of speleothems. I. The calculation of the effects of differents modes of formation on the isotopic composition of speleothems and their applicability as paleoclimatic indicators. Geochim Cosmochim Acta 35:801–824

Jiménez de Cisneros C, Caballero E, Durán JJ, Vera JA, Juliá R (2003) A record of Pleistocene climate from stalactite, Nerja Cave, South Spain. Palaeogeo Palaeoclim Palaeoecol 189:1–10

Jiménez de Cisneros C, Caballero E, Vera JA, Andreo B (2011) An optimized thermal extraction system for preparation of water from fluid inclusions in speleothems. Geologica Acta 9(2):149–158. doi:10.1344/105.000001646

Jiménez de Cisneros C, Caballero E (2013) Paleoclimate reconstruction during MIS5a based on a speleothem from Nerja Cave, Málaga, South Spain. Nat Sci doi:10.4236/ns.2013

Kendall AC, Broughton PL (1978) Origin of fabrics in speleothems composed of columnar calcite crystals. J Sediment Petrol 48(2):519–538

Marshall D, Ghaleb B, Countess R, Gabities J (2009) Preliminary paleoclimate reconstruction based on a 12500 year old speleothem from Vancouver Island, Canada: stable isotopes and U-Th disequilibrium dating. Quatern Sci Rev 28:2507–2513

Matthews A, Ayalon A, Bar-Matthews M (2000) D/H ratios of fluid inclusions of Soreq cave (Israel) speleothems as a guide to the Eastern Mediterranean Meteoric Line relationships in the last 120 ky. Chem Geol 166:183–191

McCrea JM (1950) On the isotopic chemistry of carbonates and a paleotemperature scale. J Chem Phys 18:849–857

McDermott F (2004) Palaeo-climate reconstruction from stable isotope variations in speleothems: a review. Quatern Sci Rev 23:901–918

McDermott F, Schwarcz HP, Rowe PJ (2006) 6. Isotopes in speleothems. In: Leng MJ (ed) Isotopes in palaeoenvironmental research. Springer, Dordrecht, pp 185–226

Moreno A, Stoll H, Jiménez-Sánchez M, Cacho I, Valero-Garcés B, Ito E, Edwards LR (2010) A speleothem record of rapid climatic shifts during last glacial period from Northern Iberian Peninsula. Global Planet Change 71:218–231

Moreno A, Belmonte A, Bartolomé M, Sancho C, Oliva B, Stoll H, Edwards IR, Cheng H, Hellstrom J (2013) Formación de espeleotemas en el noreste peninsular y su relación con las condiciones climáticas durante los últimos ciclos glaciares. Cuadernos de Investigación Geográfica, Universidad de La Rioja 39 (1):25–47

Muñoz-García MB, Martín-Chivelet J, Rossi C, Ford DC, Schwarcz HP (2007) Chronology of termination II and the Last Interglacial Period in North Spain based on stable isotope records osf stalagmites from Cueva del Cobre (Palencia). J Iberian Geol 33(1):17–30

Musgrove M, Banner JL, Mack LE, Combs DM, James EW, Cheng H, Edwards RL (2001) Geochronology of late Pleistocene to Holocene speleothems from central Texas: implications for regional paleoclimate. Geol Soc Am Bull 113(12):1532–1543

Shackleton NJ, Berger A, Peltier WR (1990) An alternative asatronomical calibration of the lower Pleistocene timescale based on ODP Site 677. Trans R Soc Edinb Earth Sci 81:251–261

Sharp Z (2007) Principles of stable isotope geochemistry. Pearson Prentice Hall, Upper Saddle River

Spölt C, Mangini A (2007) Speleothems and paleoglaciers. Earth Planet Sci Lett 254:323–331

Verheyden S, Genty D, Cattani O, Van Breukelen M (2005) Characterization of fluid inclusions in speleothems: heating experiments and isotopic hydrogen composition of the inclusion water. Geophys Res Abstr 7:02772

Vonhof HB, van Breukelen MR, Postma O, Rowe PJ, Atkinson TC, Kroon D (2006) A continuous-flow crushing device for on-line δ^2H analysis of fluid inclusion water in speleothems. Rapid Commun Mass Spectrom 20:2553–2558

Williams DI, Thunell RC, Tappa E, Rio D, Raffi I (1988) Chronology of the Pleistocene oxygen isotope record: 0–1.88 my BP. Palaeogeo Palaeoclim Palaeoecol 125:51–73

Trace Elements in Speleothems as Indicators of Past Climate and Karst Hydrochemistry: A Case Study from Kaite Cave (N Spain)

J.A. Cruz, J. Martín-Chivelet, A. Marín-Roldán, M.J. Turrero, R.L. Edwards, A.I. Ortega and J.O. Cáceres

Abstract A stalagmite that grew during the Holocene (between 4.9 and 0.9 ka BP) in Kaite Cave (Ojo Guareña Karst Complex, Burgos, N Spain) has been analyzed by Laser-Induced Breakdown Spectroscopy (LIBS) with the aim of reconstructing secular variations in the hydrochemistry of the karst system, in turn related to changes in the environment outside the cave. LIBS analyses yield significant changes in Mg/Ca and Sr/Ca intensity ratios through the stalagmite, which reveal consistent trends and patterns at decadal to centennial scales. The origin of the observed changes in Mg/Ca and Sr/Ca ratios is discussed in the framework of the cave system and the regional climatic variability, particularly the changes in precipitation.

J.A. Cruz (✉) · J. Martín-Chivelet
Departamento de Estratigrafía, Facultad de Ciencias Geológicas, Universidad Complutense de Madrid, 28040 Madrid, Spain
e-mail: jcruzmartinez@ucm.es

J.A. Cruz · J. Martín-Chivelet
Instituto de Geociencias (CSIC, UCM), C/José Antonio Nováis 12, 28040 Madrid, Spain

A. Marín-Roldán · J.O. Cáceres
Departamento de Química Analítica, Facultad de Ciencias Químicas, Universidad Complutense de Madrid, 28040 Madrid, Spain

M.J. Turrero
Departamento de Medioambiente, Ciemat, Avda. Complutense 22, 28040 Madrid, Spain

R.L. Edwards
Department of Geology and Geophysics, University of Minnesota, 310 Pillsbury Drive SE, Minneapolis, MN 55455, USA

A.I. Ortega
Centro Nacional de Investigación sobre la Evolución Humana CENIEH, Paseo Sierra de Atapuerca S/N, 09002 Burgos, Spain

B. Andreo et al. (eds.), *Hydrogeological and Environmental Investigations in Karst Systems*, Environmental Earth Sciences 1,
DOI 10.1007/978-3-642-17435-3_64

1 Introduction

Although most stalagmites commonly consist of nearly pure calcium carbonate, different elements (such as Mg, Sr, and Ba) dissolved in the percolating waters can be incorporated as trace elements to the carbonate precipitates. The concentration of trace elements in a speleothem depends in fact on multiple factors determined by the cave environment, the host rock, the karstic percolating system, and the surface conditions and processes (e.g. Verheyden et al. 2000; Sinclair et al. 2012; Tremaine and Froelich 2013). As stalagmites usually grow during long periods of time, they can record changes in trace element concentration (as well as in other physical and chemical properties) which can be interpreted in terms of past changes in karst hydrochemistry and climate.

The use of speleothem trace-elements as high-resolution proxies of past karstic or climatic conditions requires analytical techniques capable of detecting the small compositional changes in microsamples which represent time intervals as short as a season or a year. Given that speleothems yield time series of hundreds or thousands of years, the analytical techniques should be practical and relatively inexpensive; i.e., capable of giving thousands of data from samples without spending years in lab work or an excessive economical cost.

In the task of searching for new high-resolution methods for the reconstruction of past climate changes, the *Laser Induced Breakdown Spectroscopy* (LIBS) provides an enormous potential, as the technique gives enough analytical sensitivity to the most abundant trace elements from micro samples of $CaCO_3$, and is remarkably rapid and affordable (Vadillo et al. 1998; Fortes et al. 2012). The LIBS represents an analytical method capable of detection point-to-point differences in atomic composition of elements that can be of relevance for paleoclimatic analyses. This technique supposes some advantages as high speed of analysis, little or no sample preparation, and high sensitivity.

In this study, a Holocene stalagmite retrieved from Kaite Cave (Ojo Guareña Karst Complex, Burgos Province, Northern Spain) has been analyzed by means of the LIBS technique with the aim of: (1) testing the potential of the technique for analyzing speleothems of a cave which hosts broader paleoclimate and cave-monitoring projects, (2) obtaining a first time series of changes in Mg/Ca and Sr/Ca intensity ratios through a stalagmite that records four millennia of nearly continuous growth, and (3) interpreting these geochemical data in the frame of the hydrochemistry of the karst system, in turn related to changes in environment climate outside the cave.

2 Studied Material: Stalagmite BUDA-100

Buda-100, the stalagmite chosen for this work, is the object of a broader paleoclimatological project which includes microstratigraphy, geochronology, and stable isotopes analyses. Only the features of interest for this study will be herein briefly described.

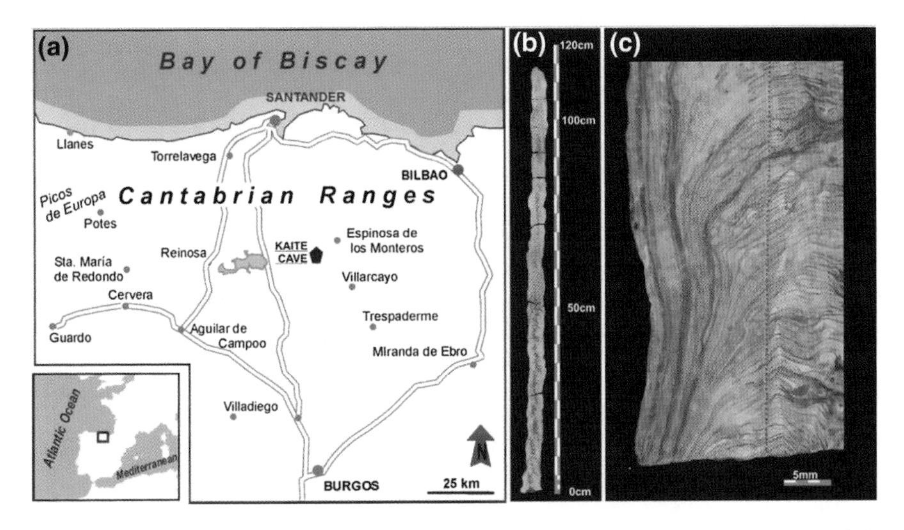

Fig. 1 **a** Location map of the Kaite cave in northern Spain. **b** Longitudinal sections of the Holocene stalagmite used in this study. **c** Thick section of the stalagmite with the microcraters produced by LIBS analyses on the sample

Buda-100 was fallen and broken in several pieces when collected in the Buda Hall of the Kaite Cave (890 m.a.s.l., Burgos, Province, N Spain) (Fig. 1a–c). Kaite is a small ($\sim$350 m length) and relatively shallow (12–20 m below the surface) isolated cave located over the six known karstic levels that form the main system of the Ojo Guareña Karst Complex. The host rocks of the cave are Upper Cretaceous limestones which show partial dolomitization.

The stalagmite is cylindrical and very elongated (112 cm long $\times$ 4–6 cm thick), and exclusively consists of calcite (with columnar and dendritic fabrics). Remarkably, its internal stratigraphy shows a pervasive pattern defined by submillimetric annual lamination.

The age of the stalagmite, based on 13 ^{230}Th radiometric dates yielded by ICP-MS in the Minnesota Isotope Laboratory, ranges from 4.9 to 0.9 ka BP. Nearly continuous growth has been demonstrated by integrating absolute ages and annual lamina counting. However, three minor hiatuses, located, respectively, around 4.55, 3.40, and 1.65 ky BP punctuate that record.

3 Methodology

LIBS measurements were obtained using a Q-switched Nd:YAG laser (Quantel, Brio model) operating at 1,064 nm, with a pulse duration of 4 ns. Samples were placed over an X–Y–Z manual micro-metric positionator with a 0.5 mm stage of travel at every coordinate to ensure that each laser pulse impinged on the speleothem. The laser beam was focused on the surface of the speleothem with a

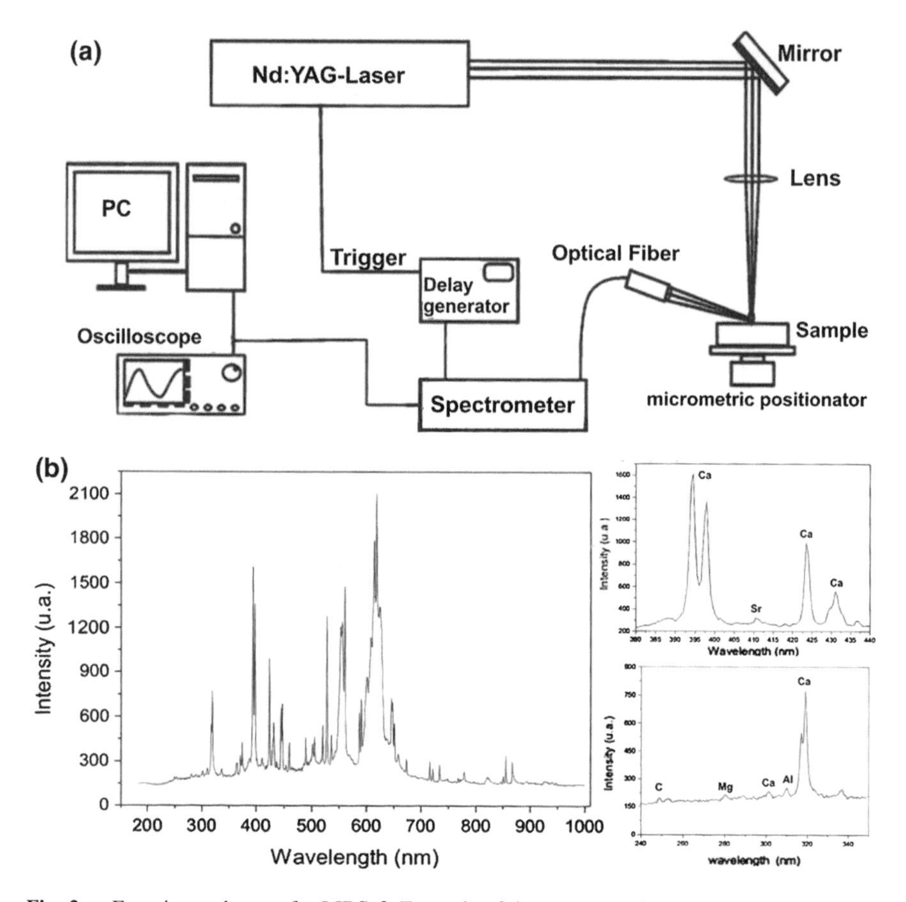

Fig. 2 **a** Experimental setup for LIBS. **b** Example of the spectrum obtained by LIBS, and more detailed emission lines at *right*

100 mm focal-distance lens. This large working distance allowed easy sample manipulation and plasma light collection while the focusing provided by the lens enabled extremely precise placement of the beam on the speleothem. Emission from the plasma was collected, and then focused into an optical fiber, coupled to a spectrometer. The spectrometer system was a user-configured miniature single-fiber system EPP2000 StellarNet (Tampa, FL, U.S.A.) with a charged coupled device detector (CCD) with a spectral resolution of ±0.5 nm. The wavelength range used was from 200 to 1,000 nm. The detector integration time was set to 100 ms (Fig. 2).

Data were obtained by averaging 20 laser shots at each sample position after 3 laser shots used for cleaning the surface of the speleothem section while the lateral resolution was established at 500 μm.

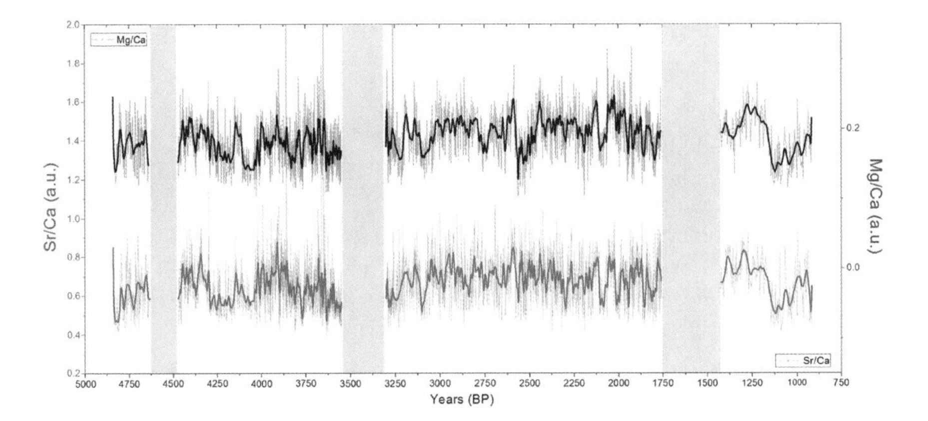

Fig. 3 Time series for Mg/Ca and Sr/Ca intensity ratios in Buda100, where *grey areas* represent hiatus

4 Results

We have performed 2,400 analyses through the longitudinal section of Buda-100 to construct a time series for changes in Mg and Sr content through the stalagmite being expressed as variations in the Mg/Ca and Sr/Ca intensity ratios. These series are plotted in Fig. 3, where running averages are also included as a basic filter to discern the main patterns of variation. No clear long-term trends are observed in any of the series, but a consistent variability can be recognized in both cases. Both series show clearly different patterns, as revealed by the low correlation values obtained when considering the whole data. However, several time intervals of decades to centuries show much higher correlation values indicating a notable covariance (Fig. 3).

One important part of the results deals with the evaluation of the technique for the analysis of sample like Buda-100, so this study could serve as a reference for further analyses on other stalagmites from Kaite and other caves. In this task, we have developed two control tests as a way to check the reliability and accuracy of LIBS, as well as the representativeness of the obtained data and series. First, we have checked for the uniformity of the results along a single (annual) growth layer. The test was based on 12 successive analyses and revealed a remarkably lateral continuity in both Mg/Ca and Sr/Ca intensity ratios. On the other hand, we performed a transect of 140 analyses parallel to the main profile (and separated 3 mm from it), which showed good replicability of the series.

5 Interpretation and Discussion

Mg and Sr contents in the calcite of stalagmites are given by the distribution coefficient of each element: $K_{Tr} = (Tr/Ca)_{calcite}/(Tr/Ca)_{solution}$.

In the case of Mg, the distribution coefficient (K_{Mg}) is modulated by temperature, an aspect which derived in considering that trace element as a possible speleothem paleothermometer in pioneer works (e.g., Gascoyne 1983). However, according to the experimental work by Huang and Fairchild (2001) about the temperature-dependence of K_{Mg} under laboratory conditions, the temperature changes in the Kaite cave estimated for the late Holocene (Martín-Chivelet et al. 2011), would be too small to generate significant changes in the Mg/Ca ratios through time in the stalagmite Buda-100.

In the case of the Sr, the distribution coefficient (K_{Sr}) does not show significant dependence on temperature (Verheyden et al. 2000), but on other parameters, particularly the calcite precipitation rate (e.g., Tesoriero and Pankow 1996; Fairchild and Treble 2009), especially for growth speeds over 0.5 mm/yr (Gabitov and Watson 2006). For Buda-100, growth rates are in that order or larger, and our data suggest that this parameter had strong control over the geochemical respond in the speleothem.

Thus, if those distribution coefficients are assumed to be approximately constant through time, variations in Mg/Ca and Sr/Ca ratios through the stalagmite need to be examined in terms of changes in the composition of dripwaters. In the case of Mg and for Kaite cave, the changes could be related to various processes:

(a) Differential rates of dissolution of calcite and dolomite in the bedrock, which is favored in the case of Kaite cave by the mixed dolomitic–calcitic nature of the Cretaceous carbonates. Because the Mg content of dolomite, this process will affect near exclusively to this element: the aqueous leachates should have lower Mg/Ca relative to the bedrock because of preferential calcite dissolution, but that ratio can rise by increasing water residence times, an aspect that took place during the drier time intervals when the seepage flow decreases. Similar cases were described by Fairchild et al. (2000). On the other hand, if the Sr content in the dolomite is different than in the calcite, changes in the incongruent dissolution of the host rock can determine differences in the Sr concentration of the resulting waters. Frequently, the Sr content of the dolomite is significantly lower than that of the calcite (e.g., Kretz 1982). When this occurs, an inverse relationship should be expected between Mg/Ca and Sr/Ca in drip waters and also in the resulting speleothems (e.g., Roberts et al. 1998). Finally, it should be mentioned that in a recent paper, Sinclair (2011) has shown that Mg and Sr can be also preferentially released from calcite during incongruent calcite dissolution (ICP), inducing a variability in Sr/Ca and Mg/Ca of drip waters that need to be further investigated in our study case.

(b) Prior calcite precipitation (PCP), which is favoured by the climatic conditions of the Kaite area (Turrero et al. 2008). When percolating, the waters can pass from a dissolution regime to a precipitation regime as they get into areas with lower pCO_2 than that which they have previously equilibrated. The resultant degassing leads to calcite precipitation and, consequently also to water enrichment in Mg and Sr relative to Ca. PCP is enhanced during dry climatic periods (e.g., Fairchild and Treble 2009), and thus, the higher values in both trace elements of the dripwaters and of the speleothems should reflect climatic conditions of low rainfall.

(c) Existence of additional, non-limestone sources of Sr: It can be, in the case of Kaite, related to the presence of argillaceous material in the limestone or, more probably, in the soil above the cave. Sr can thus be supplied both from carbonates and clays, and changes in the Sr/Ca of percolating waters be explained in terms of differences in these sources, which can also have a climatic origin (which need to be more deeply investigated).

For Buda-100, Sr/Ca and Mg/Ca show different patterns through the stalagmite. Noticeably, the two records yield variable degrees of correlation depending on the time intervals. For some periods of several decades to several centuries, a significant positive covariation between the two ratios can be clearly observed (Fig. 3), whereas for other intervals the correlation is less evident. Kaite cave monitoring during the last 10 years reveals consistent correlation between Sr/Ca and Mg/Ca at interannual scales (Turrero et al. 2014 this volume). Mg/Ca ratios seems to be more sensitive to changes in rainfall than Sr/Ca, although both parameters present some correlation at seasonal scales.

Verheyden et al. (2000), in one of the few papers that consider longer time periods in a stalagmite, show that the degree of covariation between Mg/Ca and Sr/Ca can change significantly through time, which is interpreted mainly as the result of changes in residence time. In our case, different behavior on the water–rock interactions through time could be modulating the different relationships between Sr/Ca and Mg/Ca, although these interactions need to be further investigated. Current monitoring in Kaite Cave, and further research in Sr isotopes and other paleoclimate proxies such as stable isotopes will contribute to better define the changes in the concentration of trace element series of the speleothems, and thus their use as paleohydrological and paleoclimate proxies.

6 Conclusions

The LIBS is a promising analytical technique for getting long series of geochemical data form microstratigraphic series such as those contained in carbonate speleothems. The technique provides enough analytical sensitivity to the most abundant trace elements from micro samples of $CaCO_3$, and is remarkably fast and affordable. In this work, the LIBS has been used to obtain a high-resolution record of Mg/Ca and Sr/Ca ratios through an Holocene stalagmite from Kaite Cave (Ojo Guareña Karst Complex, Burgos, N Spain) in northern Spain. The origin of the observed changes in Mg/Ca and Sr/Ca ratios is discussed in the framework of the hydrochemistry of the karst system as well as the regional climatic variability, particularly the changes in regional rainfall.

Acknowledgments Contribution to project CGL2010-21499-BTE and researches groups "Paleoclimatology and Global Change" and "Laser Induced Breakdown Spectroscopy (LIBS)" from the UCM (Spain). We thank the facilities and permissions given by the Junta de Castilla y León (Spain) for accessing and working in the Ojo Guareña Natural Monument. The collaboration of the Grupo Espeleológico Edelweiss (Exma. Dip. Prov. Burgos) is also greatly acknowledged. Thanks are extended to S. Moncayo, M.B. Muñoz García, and S. Manzoor for their help and advice in this work.

References

Fairchild IJ, Borsato A, Tooth AF et al (2000) Controls on trace element (Sr–Mg) compositions of carbonate cave waters: implications for speleothem climatic records. Chem Geol 166:255–269

Fairchild IJ, Treble PC (2009) Trace elements in speleothems as recorders of environmental change. Quat Sci Rev 28:449–468

Fortes J, Vadillo I, Stoll H et al (2012) Spatial distribution of paleoclimatic proxies in stalagmite slabs using laser-induced breakdown spectroscopy. J Anal At Spectrom 27:868–873

Gabitov RI, Watson EB (2006) Partitioning of strontium between calcite and fluid. Geochem Geophys Geosyst 7:Q11004. http://dx.doi.org/10.1029/2005GC001216

Gascoyne M (1983) Trace element partition coefficients in the calcite -water system and their palaeoclimatic significance in cave studies. J Hydrol 61:213–222

Huang Y, Fairchild IJ (2001) Partitioning of Sr^{2+} and Mg^{2+} into calcite under karst analogue experimental conditions. Geochim Cosmochim Ac 65:47–62

Kretz R (1982) A model for the distribution of trace elements between calcite y dolomite. Geochim Cosmochim Ac 46:1979–1981

Martín-Chivelet J, Muñoz-García MB, Edwards RL et al (2011) Land surface temperature changes in Northern Iberia since 4000 yr BP, based on δ13C of speleothems. Global Planet Change 77:1–12

Roberts MS, Smart PL, Baker A (1998) Annual trace element variations in a Holocene speleothem. Earth Planet Sc Lett 154:237–246

Sinclair DJ (2011) Two mathematical models of Mg and Sr partitioning into solution during incongruent calcite dissolution: implications for dripwater and speleothem studies. Chem Geol 283:119–133

Sinclair DJ, Banner JL, Taylor FW et al (2012) Magnesium and Strontium systematics in tropical speleothems from Western Pacific. Chem Geol 295:1–17

Tesoriero AJ, Pankow JF (1996) Solid solution partitioning of Sr^{2+}, Ba^{2+} and Cd^{2+} into calcite. Geochim Cosmochim Ac 60:1053–1063

Turrero MJ, Garralón A, Martín-Chivelet J et al (2008) Seasonal and interannual changes in Ca and Mg of dripping waters in Kaite Cave (Spain). Geochim et Cosmochim Acta 71(15–1):A962

Turrero MJ, Garralón A, Sánchez L et al (2014) Variations in trace elements of drip waters in Kaite Cave (N Spain): significance in terms of present and past processes in the karst system. In this volume

Tremaine DM, Froelich PN (2013) Speleothem trace element signatures: a hydrologic geochemical study of modern cave dripwaters and farmed calcite. Geochim Cosmochim Ac 121:522–545

Vadillo JM, Vadillo I, Carrasco F et al (1998) Spatial distribution profiles of magnesium and strontium in speleothems using laser-induced breakdown spectrometry. Fresenius' J Anal Chem 361:119–123

Verheyden S, Keppens E, Fairchild IJ et al (2000) Mg, Sr and Sr isotope geochemistry of a Belgian Holocene speleothem: implications for paleoclimate reconstructions. Chem Geol 169:131–144

Variations in Trace Elements of Drip Waters in Kaite Cave (N Spain): Significance in Terms of Present and Past Processes in the Karst System

M.J. Turrero, A. Garralón, L. Sánchez, A.I. Ortega, J. Martín-Chivelet, P. Gómez and A. Escribano

Abstract Drip-water chemistry in karstic caves can vary at seasonal to inter-annual scales in response to climatic factors such as temperature, rainfall, and seasonality, which determine changes in the hydrological and hydrochemical processes of the percolating waters in their paths from the atmosphere to the cave. In this paper the characterization of stalagmite forming drip-waters based on long-term (years) time-series data is presented as a key task for understanding the geochemical behavior of a specific system, the Kaite Cave (N Spain). The work focuses on the relationships between rainfall, drip rates, drip-water calcium concentration, and drip-water trace elements amount (e.g., Mg and Sr); as indicators of hydrologic processes defining the karst system and controlling speleothem growth and composition patterns.

1 Introduction

Carbonate speleothems can be excellent indicators of past climate variability, and for this reason, they have been extensively studied during the last two decades. In this task, the analysis of variations in trace elements concentration through individual stalagmites has a high potential for gaining knowledge on the

M.J. Turrero (✉) · A. Garralón · L. Sánchez · P. Gómez · A. Escribano
Departamento de Medioambiente, CIEMAT, Avda. Complutense 22, 28040 Madrid, Spain
e-mail: mj.turrero@ciemat.es

A.I. Ortega
CENIEH, Paseo Sierra de Atapuerca S/N, 09002 Burgos, Spain

J. Martín-Chivelet
Departamento de Estratigrafía, Facultad de Ciencias Geológicas, Universidad Complutense de Madrid, 28040 Madrid, Spain

J. Martín-Chivelet
Instituto de Geociencias (CSIC, UCM), C/José Antonio Nováis 12, 28040 Madrid, Spain

B. Andreo et al. (eds.), *Hydrogeological and Environmental Investigations in Karst Systems*, Environmental Earth Sciences 1,
DOI 10.1007/978-3-642-17435-3_65

environmental changes occurring outside the cave during the speleothem growth (e.g., Roberts et al. 1998; Fairchild et al. 2000, 2001; Hellstrom and McCulloch 2000; Verheyden et al. 2000; Huang et al. 2001; Treble et al. 2003; Johnson et al. 2006; Cruz et al. 2007; Sinclair et al. 2012; Tremaine and Froelich 2013). However, this potential appears in many cases masked by the difficulties in unequivocally interpreting those geochemical records, which commonly depend on multiple, usually interdependent, environmental or climatic factors at a specific site. In this situation, long-term research programs in the specific cave sites where paleoclimate studies are being performed, involving environmental, hydrochemical, and geochemical monitoring, must be accomplished for reducing uncertainties and strengthening paleoclimate reconstructions based on speleothem trace elements.

Against that perspective, this work presents a long-term (years) time-series chemical record of drip water obtained from monitoring Kaite Cave (N Burgos, Spain). The study is part of a project for characterizing the climate changes in Northern Spain from speleothem records (e.g., Martín-Chivelet et al. 2006, 2011; Domínguez-Villar et al. 2008; Cruz et al. 2014). The paper focuses on spatial and temporal variations in the calcium concentration and trace elements record (Mg and Sr) of drip waters collected from different points at Kaite cave and its relevance to interpret the paleoclimate signals recorded in the speleothems of this area.

2 Site Location and Monitoring Program

Kaite is the cave where the study was performed (Fig. 1). Located in the northern part of the Burgos Province (N Spain), it is an isolated and relatively shallow cave (12–20 m below the surface) which belongs to the Ojo Guareña Karst Complex (Martín-Merino 1986). Kaite in included in the Ojo Guareña Natural Monument and has an additional protection because its archaeological remains, being closed to general public.

The sampling and monitoring program commenced on 2003 at Las Velas Hall and is still active. The frequency of visits to collect data and samples is four times in a year, coinciding with seasonal changes during the year. The shortage of visits allows maintaining the karstic system undisturbed.

The selected site for water sampling is located at the end of a gallery, Las Velas Hall, separated 340 m from the main entrance and communicated with the main gallery by a very narrow passage (Fig. 1). The seepage water in Las Velas Hall is frequent, with permanent dripping during the year, and the speleothems are abundant, some of which are growing at the present time. The monitoring site (see sampling site in Fig. 1) is characterized by a stable cave climate: the temperature is 10.40 ± 0.04 °C (Fig. 1) and reflects the mean annual temperature outside the cave, the relative humidity exceeds always 99 % and there are not significant air currents (Turrero et al. 2004, 2007).

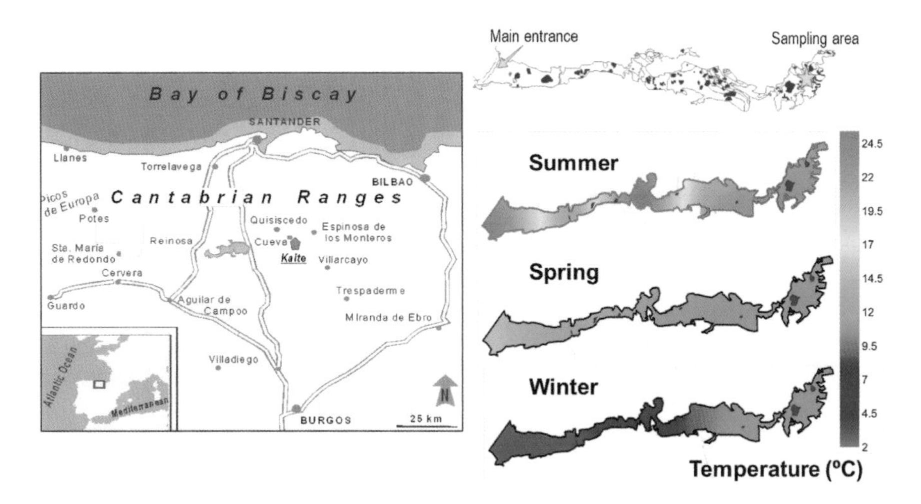

Fig. 1 Location of the Kaite cave (*left*), and plan of the cave with the main entrance and the sampling site indicated (*right*). Below the plan are the temperature data (average of several years and different seasons; autumn temperature is very similar to the spring temperature, so it is omitted in the plot) evidencing the isolation of the sampling area in terms of temperature, which is quite stable along the year

There are three main sources of data on drip water rate and chemistry at Las Velas Hall, named KT-WL, KT-WM and KT-WR. All of them have a continuous but inhomogeneous drip water discharge during the whole year (Fig. 2), with a definite seasonal effect.

The total volume of dripping water collected seasonally in the sites is consistent with an advective flow through the surrounding rock, although a dual infiltration process exists: a slow flow (KT-WL), with an averaged seasonal and annual drip rate of $\sim$0.02 mL/min and a moderate to rapid flow (KT-WR), ranging from 0.02 to 2.25 mL/min (mean value 0.65 mL/min). Between them, an intermediate flow (KT-WM), with a mean flow of 0.07 mL/min, has also been monitored. Thus, in spite of the proximity between the points flow rates indicate spatial variability in the infiltration system.

All water samples (aliquots representative of the total amount of the seasonal sampling for each point) are filtrated through Millipore filters with pore size 0.45 μm for complete chemical characterization. Preservation previous to analysis is undertaken according to the constituents under analysis. Trace elements were determined after bringing the samples to pH < 1.5 with ultrapure nitric acid. Non-acidified samples were used to determine major cation and anion concentrations. The trace elements were determined by inductively coupled plasma-optical emission spectrometry (ICP-OES). Data quality is assessed by replicate samples and sample blanks.

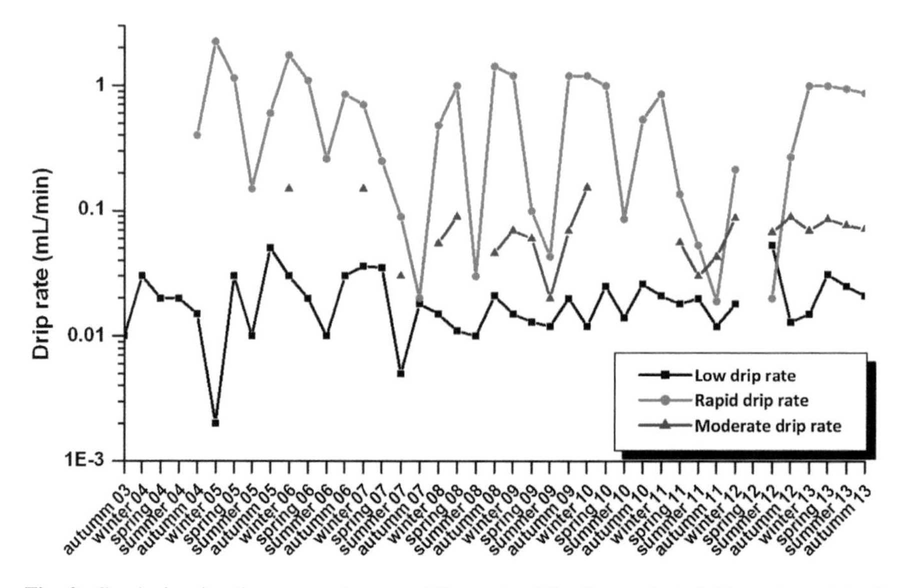

Fig. 2 Graph showing the averaged-seasonal flow rate of the three selected drip water points at Las Velas Hall. A continuous flow occurs in the three sites, although the differences between them indicate spatial variability in the infiltration system

3 Results and Discussion

(a) *Calcium concentration in drip-waters and rainfall*

Rainfall in the area and calcium concentration in the drip water has been taken as parameters to illustrate variations between the sites monitored at Las Velas Hall (Fig. 3). Calcium concentration varies in the three drip waters in a similar way, being low and quite constant in the slow drip water point, and high and variable in the rapid and moderate drip water points, although all of them catch the rainfall trend. For the discussion, the seasonal and inter-annual variability of calcium concentration is considered separately.

Seasonal behavior The variations in rainfall and calcium concentration with time show similar patterns: as rainfall increases, e.g. during rainy seasons, so does calcium concentration in drip water at the three considered sites (Fig. 3 left). Therefore the variation in the calcium concentration in the drip water catches what is occurring outside the cave and should be explained in hydrologic terms. In this sense, lower rainfall means less water available for recharging the system, while more rainfall means more water available; after rainy periods water percolates through cracks filling totally the open spaces in the epikarst and dissolving the rock, so increasing calcium concentration in water. However, the remarkable difference in calcium concentration between the slow and the fast drip points suggests that during drier periods other parameters such as residence times and

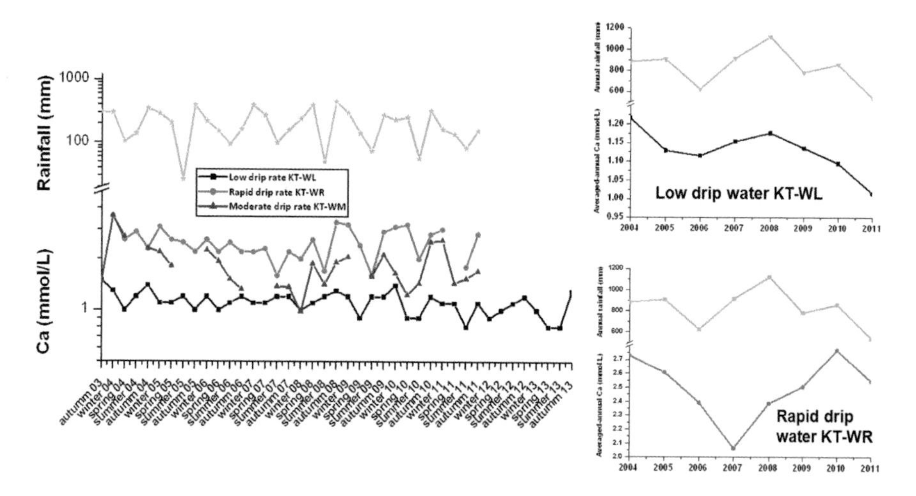

Fig. 3 Graph showing the averaged-seasonal (*left*) and annual (*right*) variation in Ca concentration in the drip water and the seasonal and annual rainfall during monitoring time

flow paths through the epikarst and the unsaturated zone should be considered to explain drip water chemistry.

Inter-annual behavior Direct correlation between annual rainfall and the (averaged) annual calcium concentration is observed in the monitored points (Fig. 3, right), but more markedly in the slow drip rate site. Two main minima on calcium concentration can be observed in the considered interval (2006 and 2011), which coincide with rainfall minima attributable to the two most recent drought periods in N Spain. On the contrary, the rainy hydrological years of 2006/2007, 2007/2008, and 2008/2009 are recorded in the cave drip waters by the highest concentrations of calcium.

Changes between dripping points Despite the described correlations between Ca and rainfall at seasonal and inter-annual scales, significant differences are observed between the records of the dripping points. As these are located very close in the same cave hall, the discussion on the geochemical processes is assumed to deal with flow paths and residence time.

(b) *Mg/Ca and Sr/Ca and epikarst processes*

In order to recognize possible hydrologic changes in the epikarst (e.g., residence time), Mg/Ca and Sr/Ca molar ratios are analyzed for discussion since relationships between Ca, and Mg and Sr in waters have been proved as excellent indicators to infer the nature of reactions in the epikarst and the unsaturated zone (e.g. Huang and Fairchild 2001; Treble et al. 2003; Cruz et al. 2007; Tremaine and Froelich 2013).

Figure 4 shows the fluctuating molar ratios of Mg/Ca, Sr/Ca with time, obtained after 10 years of monitoring in the three described drip water points in Las Velas Hall. It should be noted that Sr/Mg ratio is quite constant through time (Sr/Mg_{KT-}

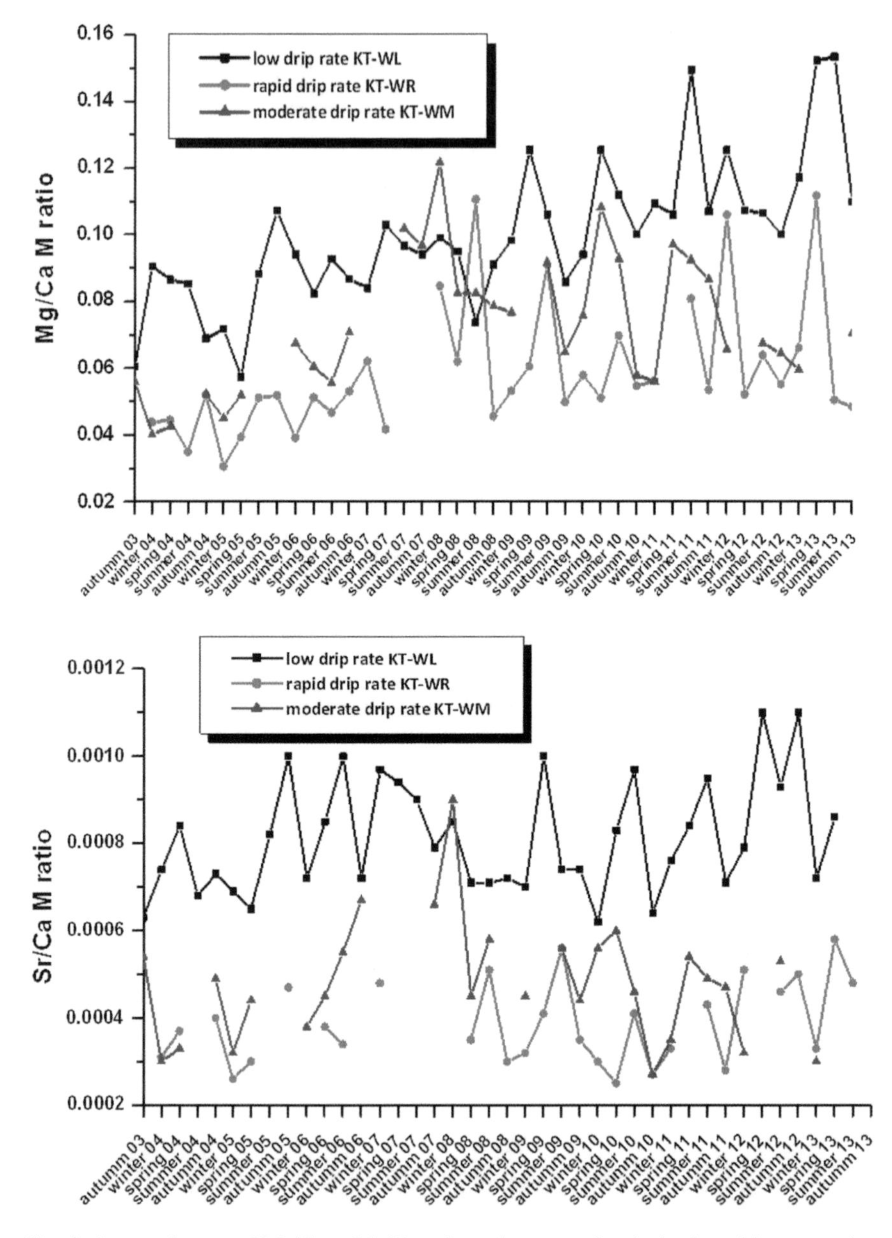

Fig. 4 Averaged-seasonal Mg/Ca and Sr/Ca molar ratios versus time in the three drip water points

$_{WL}$ = 0.009 ± 0.001 and Sr/Mg$_{KT\text{-}WR}$ = 0.007 ± 0.001) which, after Tremaine and Froelich (2013), should allow to interpret Mg and Sr variations in terms of dry/wet periods. On the other hand, the source of the magnesium is the dissolution of low-magnesium calcites of the overburden; calcium does not correlate with magnesium ($r = 0.12$ in KT-WL and $r = 0.14$ in KT-WR), reflecting that drip water should not be affected by a dilution process when rainfall increases (Tooth and Fairchild 2003; Baldini et al. 2006), so attention should be paid to dissolution–precipitation processes in the epikarst and the vadose zone.

There is good correlation between calcium and Mg/Ca both in KT-WL ($r = 0.71$) and KT-WR ($r = 0.74$), with a progressive increase in the molar ratio since the beginning of the monitoring up till now (Fig. 4). This trend is also present but much more attenuated in the case of the Sr/Ca molar ratio.

Considering the net decrease in calcium concentration recorded for the same period in both KT-WL and KT-WR (Fig. 3), and the sharp increase in the Mg/Ca ratio (less clear in Sr/Ca ratio) in KT-WL related to KT-WR, a process affecting preferentially to the slow drip water point and much less markedly to the rapid drip water point should be considered. Prior calcite precipitation at the epikarst and/or unsaturated zone is invoked as the process causing progressive calcium consumption and therefore the inter-annual decreasing in the calcium concentration and the increasing in the Mg/Ca molar ratio in KT-WL This process could also be occurring at KT-WR but at less extension, since its response to rainy periods is much more rapid and therefore it is assumed that the flow path is also more rapid than at KT-WL and the residence time of the water feeding the drip point is lower. That means a significant storage may occur in the matrix, and in other parts of the system, like the epikarst, at KT-WL, driving slower fluxes and giving time to the above described processes.

Prior calcite precipitation is considered a long-term (years) water–rock interaction process as a consequence of a prolonged dryness situation as the reflected in this work (Fig. 3). At seasonal scale (Fig. 4) during summers there is a higher evapotranspiration and the recharge is lower than the rest of the year for which the effect of prior calcite precipitation should increase, being the general inter-annual trend superimposed to that seasonal trend.

4 Conclusions

A multi-year cave monitoring program performed in Kaite has revealed key features to calibrate paleoclimate signals recorded in the speleothems. The relationship between calcium concentration and rainfall indicates that variations in this parameter should be interpreted mostly in hydrologic terms. In response to the rainfall decrease, the inter-annual monitoring shows a progressive decrease in calcium concentration and an increase in the Mg/Ca ratio, and less evident in Sr/Ca ratio. The rapid drip water point has lower and less variable Mg/Ca and Sr/Ca ratios than the slow drip water point, as the result of lower residence times indicative of rapid flow paths in the system.

This work enhances the importance, within cave monitoring programs, of tracking different dripping points during timescales long enough to accurately understand the geochemical processes occurring in the karst systems, and thus to unequivocally interpret signals preserved in speleothems.

Acknowledgments Contribution to project CGL2010-21499-BTE and research group "Paleoclimatology and Global Change" from the UCM (Spain). We thank the facilities and permissions given by the Junta de Castilla y León (Spain) for accessing and working in the Ojo Guareña Natural Monument. The collaboration of the Grupo Espeleológico Edelweiss (Exma. Dip. Prov. Burgos) is also greatly acknowledged. Our gratitude to the anonymous reviewer of the manuscript.

References

Baldini J, McDermott F, Fairchild I (2006) Spatial variability in cave drip water hydrochemistry: implications for stalagmite paleoclimate records. Chem Geol 235:390–404

Cruz JA, Martín-Chivelet J, Marin-Roldán A et al. (2014) Trace elements in speleothems as indicators of past climate and karst hydrochemistry: a case study from Kaite Cave (N Spain). This volume

Cruz FW Jr, Burns SJ, Jercinovic M et al (2007) Evidence of rainfall variations in southern Brazil from trace element ratios (Mg/Ca and Sr/Ca) in a Late Pleistocene stalagmite. Geochim Cosmochim Acta 71:2250–2263

Domínguez-Villar D, Wang X, Cheng H et al (2008) A high-resolution late Holocene speleothem record from Kaite Cave, northern Spain: $\delta^{18}O$ variability and possible causes. Quat Int 187:40–51

Fairchild IJ, Borsato A, Tooth AF et al (2000) Controls on trace elements (Sr-Mg) compositions of carbonate waters: implications for speleothem climatic records. Chem Geol 166:255–269

Fairchild IJ, Baker A, Borsato A et al (2001) Annual to sub-annual resolution of multiple trace-element trends in speleothems. J Geol Soc Lond 158:831–841

Hellstrom JC, McCulloch MT (2000) Multi-proxy constraints on the climatic significance of trace element records from a New Zealand speleothem. Earth Planet Sci Lett 179(2):287–297

Huang Y, Fairchild IJ, Borsato A et al (2001) Seasonal variations in Sr, Mg and P in modern speleothems (Grotta di Ernesto, Italy). Chem Geol 175:429–448

Huang Y, Fairchild IJ (2001) Partitioning of Sr^{2+} and Mg^{2+} into calcite under karst analogue experimental conditions. Geochim et Cosmochim Acta 65:47–62

Johnson KR, Hu C, Belshaw NS et al (2006) Seasonal trace-element and stable-isotope variations in a Chinese speleothem: the potential for high-resolution paleomonsoon reconstruction. Earth Planet Sci Lett 244(1–2):394–407

Martín-Chivelet J, Muñoz-García B, Domínguez-Villar D et al (2006) Comparative analysis of stalagmites from two caves of northern spain. Implications for holocene paleoclimate studies. Geol Belgica 9(3-4):323–335

Martín-Chivelet J, Muñoz-García MB, Edwards RL et al (2011) Land surface temperature changes in Northern Iberia since 4000 yr BP, based on δ13C of speleothems. Global Planet Change 77:1–12

Martín-Merino MA (1986) Descripción preliminar del karst de Ojo Guareña. Kaite 4–5:53–72

Roberts MS, Smart PL, Baker A (1998) Annual trace element variations in a Holocene speleothem. Earth Planet Sci Lett 154:237–246

Sinclair DJ, Banner JL, Taylor FW et al (2012) Magnesium and Strontium systematics in tropical speleothems from Western Pacific. Chem Geol 295:1–17

Tooth A, Fairchild I (2003) Soil and karst aquifer hydrological controls on the geochemical evolution of speleothem forming drip waters, Crag Cave, southwest Ireland. J Hydrol 273:51–68

Turrero MJ, Garralón A, Martín-Chivelet J et al (2004) Seasonal changes in the chemistry of drip waters in Kaite Cave (N Spain). In: Wanty RB, Seal RS (eds) Proceedings of the twelveth international symposium on Water-Rock Interaction, vol 2. Balkema Publishers, Rotterdam, pp 1407–1410

Turrero MJ, Garralón A, Gómez P et al (2007) Geochemical evolution of drip-water and present-growing calcite at Kaite cave (N Spain). In: Bullen TD, Wang YX (eds) Proceedings of the twelfth international symposium on Water-Rock Interaction. Taylor & Francis Group, London, pp 1407–1410

Treble PC, Shelley JMG, Chappell J (2003) Comparison of high resolution sub-annual records of trace elements in a modern (1911–1992) speleothem with instrumental climate data from southwest Australia. Earth Plan Sci Lett 216:141–153

Tremaine DM, Froelich PN (2013) Speleothem trace element signatures: A hydrologic geochemical study of modern cave dripwaters and farmed calcite. Geochim Cosmochim Acta 121:522–545

Verheyden S, Keppens E, Fairchild IJ et al (2000) Mg, Sr and Sr isotope geochemistry of a Belgian Holocene speleothem: implications for paleoclimate reconstructions. Chem Geol 169(1–2):131–144

Striped Karren on Snake Mountain above Kunming (Yunnan, China)

M. Knez, L. Hong and T. Slabe

Abstract Unique striped karren give a special landscape stamp to the extensive top of Snake Mountain above Kunming. Lithomorphogenetic research reveals the manner of the formation and development of this karren on vertical rock strata in transition from limestone to dolomite and their formation from subsoil karren to karren shaped by exposure to rainwater.

1 Introduction

Located in the north of Kunming, the capital city of Yunnan Province, China, Snake Mountain stretches north to south across the Kunming Basin. Starting from Matou Mountain in north, it ends around 10 km to the south at the Tiefengan temple, which is the boundary of the Wuhua and Xishan districts. The highest peak of Snake mountain is 2,365 m above sea level, 470 m higher than the bottom of the basin.

The study area has a subtropical plateau monsoon climate characterised by no extremely hot or cold weather but distinct wet and dry seasons. The average annual temperature is 14 °C. The average monthly temperature is 7.3°–9.3 °C in winter and 18°–19 °C in summer. The average annual precipitation is 1,027 mm, of which 85 % falls during the June to October wet season; the rainfall in the November to May dry season amounts to only 15 % of the total precipitation.

M. Knez (✉) · T. Slabe
Research Centre of the Slovenian Academy of Sciences and Arts,
Karst Research Institute, Titov Trg 2, SI-6230 Postojna, Slovenia
e-mail: knez@zrc-sazu.si

T. Slabe
e-mail: slabe@zrc-sazu.si

M. Knez · L. Hong · T. Slabe
Yunnan International Karst Environmental Laboratory, Xueyun Rd. 5,
650223 Kunming, China
e-mail: hongliu@ynu.edu.cn

B. Andreo et al. (eds.), *Hydrogeological and Environmental Investigations in Karst Systems*, Environmental Earth Sciences 1,
DOI 10.1007/978-3-642-17435-3_66

Dissection by unique karren has given the landscape of the extensive top of Snake Mountain above Kunming a special stamp. The geological foundation with vertical beds of limestone transforming into dolomite and the formation of rock forms from subsoil rock forms to karren exposed to rainwater fostered the development of the unique shape of this karst phenomenon.

Snake Mountain is a popular excursion point above Yunnan's capital Kunming. It is being arranged for tourism and knowledge about its interesting karst phenomena is the foundation for the modern development of tourism, as much for landscaping as for the content of the tourist attraction.

2 The Shape of Karren

The elongated top of Snake Mountain (Fig. 1a, b, c) stretches along vertical beds of rock. The thicker and more resistant strata protrude several metres from the sediment covering the top. Rows of individual and clustered rock teeth and smaller rock pillars developed from them.

The vertical strata are broken by intersecting vertical and horizontal fissures that subsoil formation has widened into cracks. The density of the vertical fissures dictates the size of the rock pillars and teeth while the distinctness of the horizontal fissures influences their shape. The denser the network of vertical fissures is, the thinner the pillars are, and flat tops often form along denser horizontal fissures (Gabrovšek et al. 1998; Knez and Slabe 2010; Knez et al. 2011). This results in a diversity of rock pillars and teeth that ranges from thinner to stubby pillars with pointed or flat-tops dissected by subsoil rock forms. The latter tops are oblong along strata. Rocky strips of karren that formed mainly on more resistant dolomitized limestone are higher with emphasized relief.

The karren originally formed under fine-grained sediments and was later reshaped by rainwater. The manner of their formation and development can be traced in the rock relief created by the rock forms indenting the surface of the rock teeth and pillars.

3 Lithostratigrafic and Complexometric Features of Rock

Macroscopic description. Alternating outcrops of dolomitized limestone (Fig. 2a) comprises the basic geological feature of the surface of Snake Mountain. The strong diagenetic transformation of the limestone is an important general characteristic of the rock, marked by its recrystallization and dolomitization and considerable secondary porosity in places as well as a high percentage of carbonate (Table 1). In places entire complexes of strata are dolomitized, while elsewhere

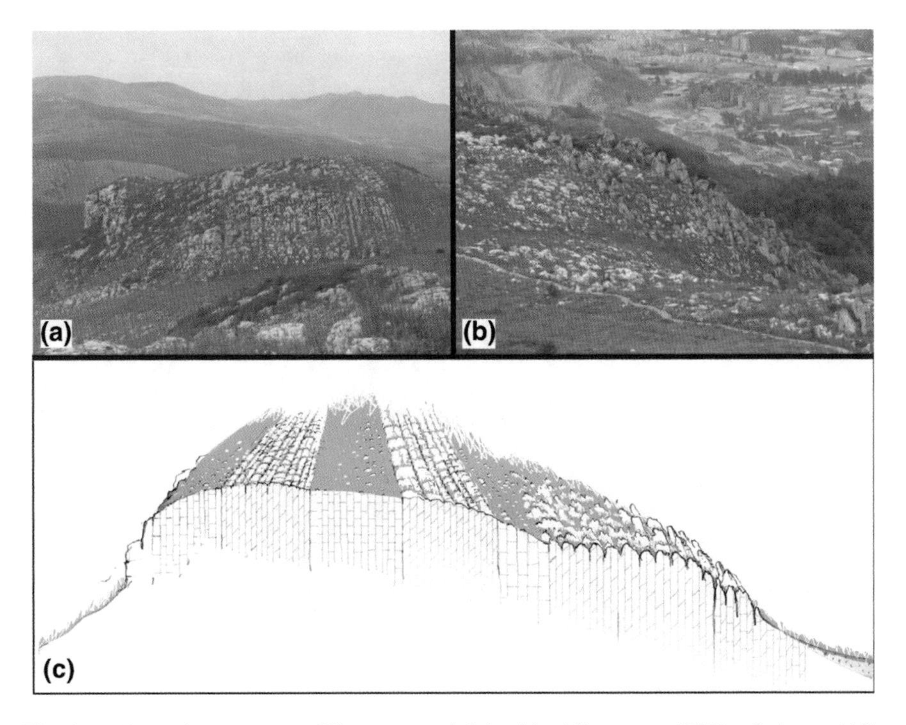

Fig. 1 **a** Alternating outcrops of limestone and dolomitized limestone. Width of view, middle photo, is 150 m. **b** Slopes of Snake Mountain with rock pillars with pointed or flat-tops. Width of view, middle photo, is 150 m. **c** Cross-section of the Snake Mountain

dolomitized rock is found inside more or less frequent lenses in the limestone. The volume of dolomitized lenses ranges from a few cm^3 to many dm^3. In most cases, the borders between the basic rock and the dolomitized areas are sharp. This textural characteristic of light grey (N5, N6, N7) to black (N3, N4) and solid black (N2) and homogenous limestone prevails across the entire area.

Due to their convexity and greater roughness, the dolomite lenses on the surface of the rock are consequently heavily overgrown with dark grey algae and appear as dark grey to black spots against the lighter bedrock (Sample 1B, Table 1). In the major part of the studied profile, the percentage of non-dolomitized limestone surfaces is only slightly smaller than the percentage of dolomitized areas; the degree of dolomitization changes laterally but never quite disappears. Rock with such properties limits the possibility for the development of more minute subsoil features as well as the development of rock relief forms carved by rainwater.

The beds lie concordantly in a sub-vertical position in the 200°–20°direction. On the surface, the dolomitized limestone is morphologically more pronounced and groups of limestone beds are less. In some areas, mainly on rounded hilltops, dolomite beds protrude on the surface and are denuded, while limestone beds are covered with weathered debris and overgrown with grass or shrubbery.

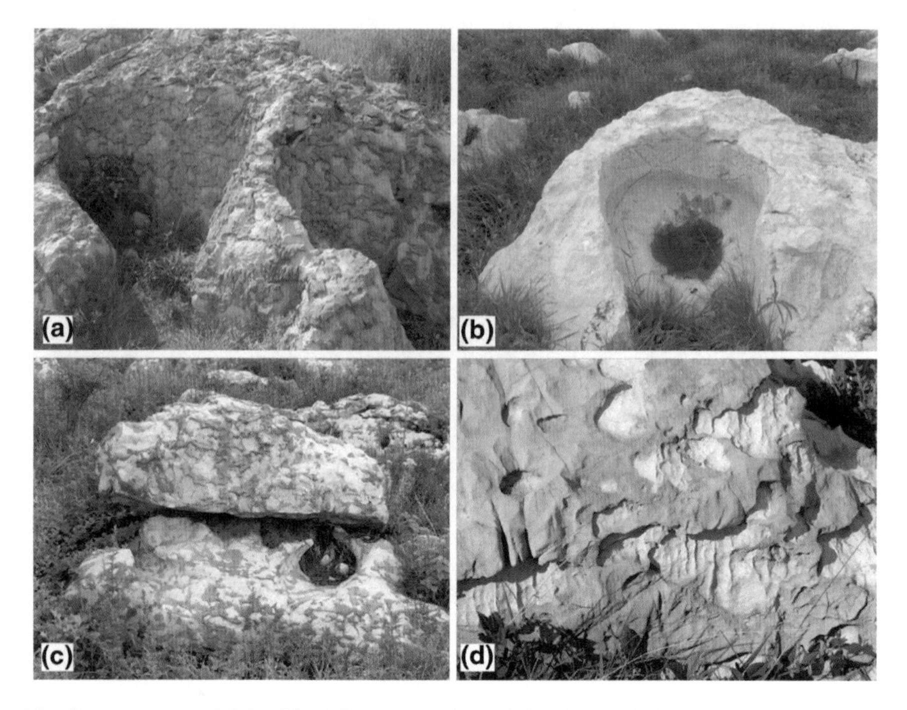

Fig. 2 **a** Outcrop of dolomitized limestone, width of view is 2 m. **b** Subsoil pan developing in funnel-like notch, width of view is 1.5 m. **c** Subsoil tube, width of view is 2 m. **d** Shaping of the dolomite and limestone (rain flutes and wall channels) by rain, width of view is 0.5 m

The rock is fissured and broken in all directions and numerous sub-horizontal fissures are particularly distinct.

The profile is about 250 m thick.

Microscopic description From 12 rock samples, 23 microscopic thin sections were prepared and examined in transmitted light. Prior to the microscopic examination, half of each sample was dyed in alizarin red dye (1,2-dihydroxy-anthraquinone, known also as Mordant Red 11, Evamy and Sherman 1962). Combining the observations with the results of the complexometric titration analysis, we were able to determine the properties of the rock.

The microscopic description begins at the top of the hill in biomicrite to bio-microsparite limestone which alternate to the bottom of profile at the edge of the valley with more or less dolomitized limestone beds.

Particles of various bioclasts in the limestone dominate among the allochems: uniserial, biserial, and fusulinid foraminifera, mollusks, gastropods, calcareous algae, and other unidentifiable particles of organisms (Table 1). In some areas, pellets occur. Also observed were individual damaged small stem parts of echinoderms. Fossil remains constitute around 25 % of the rock. Individual intraclasts also occur among the allochems. Micrite to microsparite fills most of the space between grains. Typical roundness, sorting or fabric of grains in the rock was not

Table 1 Complexometric titration analysis of rock samples

Rock sample	CaO (%)	MgO (%)	Dolomite (%)	Total carbonate (%)	Calcite (%)	CaO/ MgO	Insoluble residue (%)
1A	55.07	0.81	3.69	99.97	96.28	67.99	0.03
1B	38.52	14.71	67.68	99.78	32.10	2.60	0.22
1C1	36.84	15.72	71.92	98.62	26.70	2.34	1.38
1C2	51.51	0.24	1.11	97.29	96.18	227.12	2.71
1D1	55.32	0.26	1.29	99.27	0.98	212.77	0.73
1D2	40.36	12.42	56.80	97.84	41.04	3.24	2.16
2	31.63	18.91	86.31	96.00	9.69	1.73	4.00
3A	55.07	0.32	1.47	98.95	97.48	172.09	1.65
3B	31.06	21.33	96.08	100.00	5.92	14.56	0.00
4	55.01	0.28	1.11	98.76	97.65	196.46	1.24
5	53.50	1.85	8.48	99.34	90.86	28.92	0.66
6A	49.07	5.28	24.15	98.64	74.47	9.29	1.38
6B	31.40	19.31	88.33	96.43	8.10	1.63	3.57
7	0.16	55.24	97.69	98.42	0.73	0.002	1.58
8	53.89	1.45	6.64	99.21	92.57	37.16	0.79
9	53.11	0.16	0.74	95.12	94.38	331.93	4.88
10	54.56	0.12	0.55	97.63	97.08	454.66	2.37
11	54.67	0.24	1.11	98.07	97.96	227.79	1.93
12	29.83	19.83	90.73	94.71	3.98	1.50	5.29

observed anywhere in the profile. Inside the limestone component between the cavity structures, we can mention individual fenestrae, filled with late diagenetic dolomite. The crystals of dolomite are mostly idiomorphic euhedral. In the fenestrae, dolomite crystals in most cases form a transition from xenotopic to idiotopic texture. Among the primary porosity types, we observed mostly mouldic and occasionally intercrystalline porosity. Calcite veins occur only rarely. They are filled with drusy calcite spar. Secondary porosity was not observed.

Lenses of dolomitized limestone from several tens of cm^3 to many dm^3 in size occur irregularly distributed inside limestone beds. In places the density and size of dolomitized areas changes laterally as well as vertically. The original limestone is strongly late diagenetically dolomitized (Table 1). Limestone of the wackestone to packstone type displays barely identifiable fragments of fossil remains, parts of which are often mutually microscopically shifted along tiny fissures. Mostly, idiomorphic to hypidiomorphic grains of dolomite dominate in this section of the rock and build a dolosparite type of grainstone. The grains of dolomite are well bound. Calcite fills the rare 2–10 mm large more or less spherical fenestrae. Secondary intercrystalline porosity changes laterally; in general it is estimated at a few percent. The rock is criss-crossed with tiny calcite veins running in different directions.

In diagenetically completely dolomitized limestone, the original structure of the limestone has been erased completely. Dolosparite of the grainstone type laterally transforms from xenotopic to idiotopic dolomite mosaic texture. Where the xenotopic texture of dolomite dominates, the crystals have anhedral shapes. The spaces between larger rhombic crystals are filled by smaller dolomite crystals of anhedral shapes. The grains of dolomite are well bound everywhere. Well-shaped euhedral crystals comprise around 40 %. Intercrystalline porosity is estimated at around 3 %. Some euhedral rhombic crystals were coloured with alizarin red dye. According to the complexometric analysis, dedolomitization occurred to a smaller degree. No calcite veins were observed there.

Complexometric analysis Because macroscopically in some parts of the rock, we could not be certain whether the rock is dolomitized or to what extent it is dolomitized, we performed several complexometric titration analyses (Engelhardt et al. 1964) on selected samples at the laboratory of the Karst Research Institute (Samples 1C1, 1C2, 1D1, 1D2, 3A, 3B, 6A, 6B, Table 1). The analyses were performed using the standard method. Twenty samples of the rock or sections of the rock were analysed.

In general, the results of the analyses indicate a high total content of carbonate, except for the beds where samples 2, 9 and 12 were taken (Table 1).

With one exception, all the samples exceed 95 % total carbonate with 11 samples exceeding 98 % and six exceeding 99 %. About 2 % insoluble residue was found in six samples and another six samples were very pure with less than 1 % insoluble residue.

The calcite content in the samples varies from 47 % to almost 98 %, with an average value between 96 and 97 %. In dolomitized limestone, the content of dolomite is relatively not very high. The highest content was found in Sample 12, followed by Sample 2.

4 Rock Relief

4.1 Subsoil Rock Forms

Subsoil rock forms can be found on all types of rock. Subsoil cups are the most frequent and distinctive forms and dominate on the flat-tops of rock teeth; subsoil channels are found on their walls, and subsoil tubes are found along bedding planes and fissures.

Subsoil cups are large, a metre or more in diameter, and can be simple or composed, where several cups merge into one. On limestone rock, their shapes are regular and circular, and at the edges of tops, they are semicircular (Fig. 2b); however, on dolomitized rock, their shapes are irregular and their edges jagged.

Subsoil channels through which water drains are most often found below open cups at the edges of tops. Some channels develop into vertical funnel-like notches (Fig. 2a). Due to the gradual lowering of the sediment that covers and surrounds the rock, larger subsoil cups have several stories (Fig. 2b) and their walls are dissected by longitudinal notches which reflect a long-term level of the soil that once filled them. Their bottoms are covered by silt or are bare rock in which solution pans have formed.

Subsoil channels are another distinct subsoil rock form. Larger ones are a metre or more in diameter. Water runs along subsoil channels widening the cracks between the rock teeth at the contact with the sediment downwards along vertical bedding planes and fissures. The largest channels form below subsoil notches or funnel-like mouths. They have semicircular or omega-shaped cross-sections. Larger volumes of water from the top of the teeth collect in them. Smaller channels whose diameters reach a few decimeters are found under more or less horizontal tops and can stand side by side with a shared ridge between them. Water flows down them, trickling relatively evenly down the entire wall. Subsoil channels criss-cross even the wider gently sloping tops of rock teeth that in places are covered by soil. The water shaping them permeates through the sediment and soil to the rock. Their shape resembles an inverted omega sign. They are filled with soil or are empty, the latter being partly reshaped by rainwater. The gently sloping tops of wider rock teeth are often dissected by subsoil channels arranged in a star shape. The water here runs off in all directions toward conducting bedding planes and fissures. At the edges, the subsoil channels dissecting the tops continue into funnel-shaped notches and then in wall channels on the walls.

The subsoil scallops also found on vertical walls bear witness to the often even creeping of water along the permeable contact between the rock and the sediment surrounding it.

Subsoil tubes form along bedding planes and fissures. They also develop in the rock below horizontal fissures and therefore have cross-sections resembling an inverted omega sign (Fig. 2c). This reflects the relatively good permeability of the contact between the rock and the sediment covering it and the distinct percolation of water downwards. Subsoil karren often contains subsoil tubes that can develop into an anastomotic network in the rock above a fissure, which indicates a local phreatic zone and a poorly permeable contact, although this is not often the case here.

4.2 Rock Forms Carved by Rain

Rain flutes occur only on evenly composed limestone and on more or less dolomitized rock with dolomite nodules protruding from the rock surface only on smaller sections of limestone (Fig. 2d). They are not found on dolomite. The surface of dolomite is dissected in accordance with the composition of the rock and its fissuring with slightly pronounced traces of water trickling down the walls where small flute-like notches occur along tiny vertical fissures.

Rain pits are found on the more or less horizontal tops of limestone rock teeth or on limestone sections of dolomitized rock; they rarely appear on dolomite, where small notches develop along tiny fissures.

Networks of rain scallops are found on overhanging surfaces.

Solution pans typically develop from subsoil cups, in subsoil channels, and at the bottom of funnel-like notches that criss-cross the tops. On dolomite rock, they have irregular shapes and jagged edges.

5 Conclusions

The pronounced tectonic action of this area caused the strata to stand upright and become vertically and horizontally fissured, contributing to the formation of the top of Snake Mountain. This rich geological history is revealed by its cover of thick layers of fine-grained alluvial sediment. Precipitation water permeates through the sediment, corroding the rock below it and shaping subsoil teeth in the process. With the rapid washing away of the sediment, especially along the vertical bedding planes as well as along vertical and horizontal fissures, rock teeth in strips appeared on the surface and rock pillars developed from the most resistant rock. The density of the network of vertical fissures dictated the extent of the rock teeth while horizontal fissures, especially if there are many, frequently fostered flat-tops. The rock relief reveals a complete collection of the karren rock forms that develop from subsoil karren into karren shaped by rainwater. The rock forms can be classified into subsoil and rain carved as well as composite rock forms that are divided into partly subsoil and partly "rain" and rain-transformed subsoil forms. It also reveals the manner of formation of the entire spectrum of rock from limestone to dolomite, the diverse composition and resistance to corrosion of the rock, and the fissuring that indicates the decisive importance of lithological and tectonic characteristics in the varied formation of karren under specific conditions. Larger subsoil rock forms occur on all types of rock while smaller rock forms such as rain flutes only occur on evenly composed limestone or limestone sections of dolomitized rock. The wider tops of rock teeth that often formed along horizontal fissures are criss-crossed by subsoil cups and networks of subsoil channels.

The rock relief reveals a recent and relatively rapid denudation of the surface and formation of karren from subsoil forms, since such forms still dominate; it is understandable that the higher sections have been transformed by rain to a somewhat greater extent. Even before the denudation, the distinct vertical percolation of water dominated in the subsoil formation of the karst; gently sloping subsoil hollows were carved downwards that generally show no traces of paragenesis. Denudation can be linked to the removal of vegetation.

Acknowledgments Research was included in UNESCO IGCP project No. 598.

References

Engelhardt W, Füchtbauer H, Müller G (1964) Sediment-Petrologie, Methoden der Sediment-Untersuchung, Teil 1.- E. Schweizerbart́sche Verlagsbuchhandlung (Nägele u. Obermiller), pp 303, Stuttgart

Evamy BD, Sherman DJ (1962) The application of chemical staining techniques to the study of diagenesis of limestones. Proc Geol Soc London 1599:102–103

Gabrovšek F et al (eds) (1998) South China Karst I.- Založba ZRC, 247 p, Ljubljana

Knez M et al (eds) (2011) South China Karst II.- Založba ZRC, 237 p, Ljubljana

Knez M, Slabe T (2010) Karren of mushroom mountain (Junzi Shan) in the Eastern Yunnan Ridge, Yunnan, China: karstological and tourist attraction. Acta Geol Sinica (Engl Ed) 84/2:424–431

Influence of the Rivers on Speleogenesis Combining KARSYS Approach and Cave Levels. Picos de Europa, Spain

D. Ballesteros, A. Malard, P.-Y. Jeannin, M. Jiménez-Sánchez, J. García-Sansegundo, M. Meléndez-Asensio and G. Sendra

Abstract The influence of rivers on speleogenesis is studied analyzing the cave levels located in the underground drainage areas related to two fluvial basins. Cave levels are analyzed through their vertical distribution profiles. The underground limits of the fluvial basins are defined using a 3D geometric model of the karst aquifer established according to the KARSYS approach. The aim of this work is to analyze the influence of the rivers on cave evolution using cave morphology. The study area corresponds to the Western and Central massifs of Picos de Europa (Northern Spain), with 214 km of cave conduits up to 1.6 km vertical range. As a result, we established two sequences of development of the cave levels related to the differences of the incision rate of the Cares and Dobra Rivers, and the partial capture of the Western Massif by the Cares River.

D. Ballesteros (✉) · M. Jiménez-Sánchez · J. García-Sansegundo
Department of Geology, University of Oviedo, C/Arias de Velasco S/N, 33005 Oviedo, Spain
e-mail: ballesteros@geol.uniovi.es

M. Jiménez-Sánchez
e-mail: mjimenez@geol.uniovi.es

J. García-Sansegundo
e-mail: j.g.sansegundo@geol.uniovi.es

A. Malard · P.-Y. Jeannin
Swiss Institute for Speleology and Karst Studies (SISKA), Rue de La Serre 68, La Chaux-de-Fonds, Switzerland
e-mail: info@isska.ch

P.-Y. Jeannin
e-mail: info@isska.es

M. Meléndez-Asensio
Geological Survey of Spain, C/Matemático Pedrayes 25, 33005 Oviedo, Spain
e-mail: m.melendez@igme.es

G. Sendra
SQM, Los Militares 4290, Piso 8, Las Condes, Santiago, Chile
e-mail: gemma.sendralopez@gmail.com

B. Andreo et al. (eds.), *Hydrogeological and Environmental Investigations in Karst Systems*, Environmental Earth Sciences 1,
DOI 10.1007/978-3-642-17435-3_67

1 Introduction

The speleogenesis of the karst massifs is usually conditioned by the evolution of the fluvial network. Many studies relate together the fluvial incision, the drop of the base level, the descent of the water table, and the development of cave levels and vadose shafts (Häuselmann et al. 2007; Piccini 2011). The karst massifs may entail two or more fluvial basins which history can be diverse, producing differences on cave development. These differences are not well established due to the lack of knowledge on the underground limits between adjacent fluvial basins. In order to solve this uncertainty, the KARSYS approach (Jeannin et al. 2013) allows us to define these boundaries based on a 3D geometric model of the karst aquifer. On the other hand, the influence of the fluvial evolution on speleogenesis can be studied analyzing the vertical distribution of the cave conduits (Filipponi et al. 2009). The cave levels are indicators of the fluvial incision stages since they represent the past positions of the water table, related to the base level of the fluvial basins (Audra and Palmer 2013). The relationships between cave levels and river evolution can be approached classifying the caves levels according to the underground drainage areas defined by KARSYS and, later, comparing the cave levels between them on the basis of their vertical distribution profiles. The aim of this work is to analyze the capture of caves by fluvial streams as a result of the evolution of the large fluvial basin. The methodology of work is based on the hypothesis that the cavities were originated from the current underground basin. This allows us to compare initially the cave levels and the limits of the underground basin in order to establish their relationships and highlight fluvial captures and differences between basins.

2 Setting

The study area (20 × 25 km) corresponds to the Western and Central massifs of the Picos de Europa mountains (Northern Spain), which reach a maximum altitude at 2,648 m a.s.l. (Fig. 1). This area involves an alpine karst with few hundreds of kilometers of cave conduits, exceeding 500 m depth in 29 shafts. The fluvial network shows a low drainage density and includes eight rivers organized in two major basins. One of them is the basin of the Dobra River, ranging from 350 (base level) to 1,450 m a.s.l, which includes Hunhumia, Pomperi, and La Beyera tributaries and flows to the NW. The second basin corresponds to the Cares River, with altitudes ranging from 100 (base level) to 1,050 m a.s.l. (in the study area), which mainly flows to the NE, and it comprises the Casaño and Duje tributaries. All of these rivers are incised, carving fluvial canyons up to 2,000 m high as the Cares Gorge.

The study area is mainly formed by 2,000 m of Carboniferous limestone and, secondarily, Ordovician quartzite and Carboniferous and Permian-Mesozoic shale

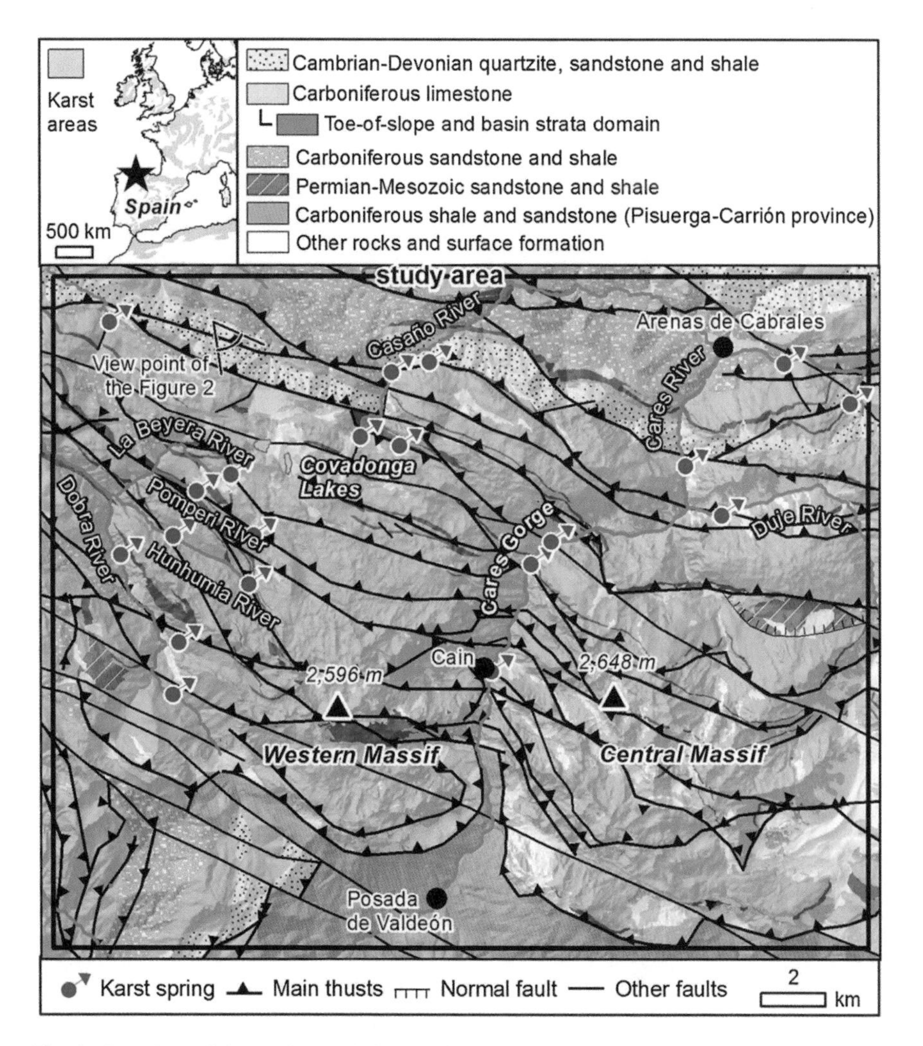

Fig. 1 Overview of the study area (after Merino-Tomé et al. 2013a, b)

and sandstone (Fig. 1). Carboniferous limestone is placed in the core of Picos de Europa and includes an interbedded strata domain named “toe-of-slope and basin” (Bahamonde et al. 2007). This strata domain comprises alternations of limestone, chert, and shale. The geological structure includes a Variscan thrusts imbricate system that was modified during the Alpine Cycle (Merino-Tomé et al. 2009). In the South of Picos de Europa, the detachment level of the Variscan system dips 30–45° to the North and is developed below the Carboniferous limestone. The footwall of the detachment level is formed by the Carboniferous sandstone and shale that crops out to the South of these mountains.

3 Methodology

The methodology includes: (1) collection of existing speleological, hydrogeological, and geological data; (2) 3D reconstruction of 214 km surveys of cave conduits; (3) inventory of 20 karst springs; (4) identification of hydrogeological units classifying them as aquifers or aquicludes (impervious rocks) based on their lithology and the presence of caves and springs; (5) elaboration of a 3D geological model; (6) elaboration of the hydrogeological 3D model at low water stage by KARSYS (Jeannin et al. 2013) combining the geological model with the position of the main karst springs and assuming a set of hydraulic principles in karst aquifers; (7) delineation of the springs systems catchments based on the hydrogeological model and their validation by previous dye-tracings; (8) grouping of the cave conduits by catchment areas and elaboration of their vertical distribution profiles with their absolute elevation (m a.s.l) (Filipponi et al. 2009); and (9) definition of cave levels and the graphical comparison of their length versus their vertical distribution.

4 Results

A zoom from the elaborated hydrogeological model at low water stage is displayed in Fig. 2, showing the defined five hydrogeological units. The units 1, 4, and 5 are formed by quartzite, shale, and compact and clayey sandstone and are considered as aquicludes due to their low permeability; the unit 2 entails limestone and forms the main karst aquifer; the unit 3, formed by limestone, chert, and shale, is interpreted in this work as an aquiclude due to the presence of springs along the contact with the unit 2. The karst aquifer is strongly divided into, at least, 19 groundwater bodies (phreatic zones) laying lie as terraces from 140 to 1,425 m a.s.l. These groundwater bodies are compartmentalized by the units 1, 3, 4, and 5 and their elevation decreases step by step toward the North and to the position of the Cares River, displaying a geometry in "terraces". The units 1, 3, and 4 represent the lateral and lower boundaries of the groundwater bodies, whereas the unit 5 usually marks their lower limit. Most of the groundwater bodies are unconfined, with flow directions to the NW, SE, and sometimes, to the North. The hydrogeological model includes 42 cave siphons documented by speleological works which are sited where the conduits appear saturated by water. The siphons are usually perched between 100 and 200 m over the supposed level of groundwater bodies. In some areas placed in the core of the massifs, groups of neighbor siphons of different caves are placed at the same altitude and perched up to 650 m above the inferred groundwater bodies. This data suggests that the geometry of the aquifer in these areas could present more compartments than the divisions that can be recognized.

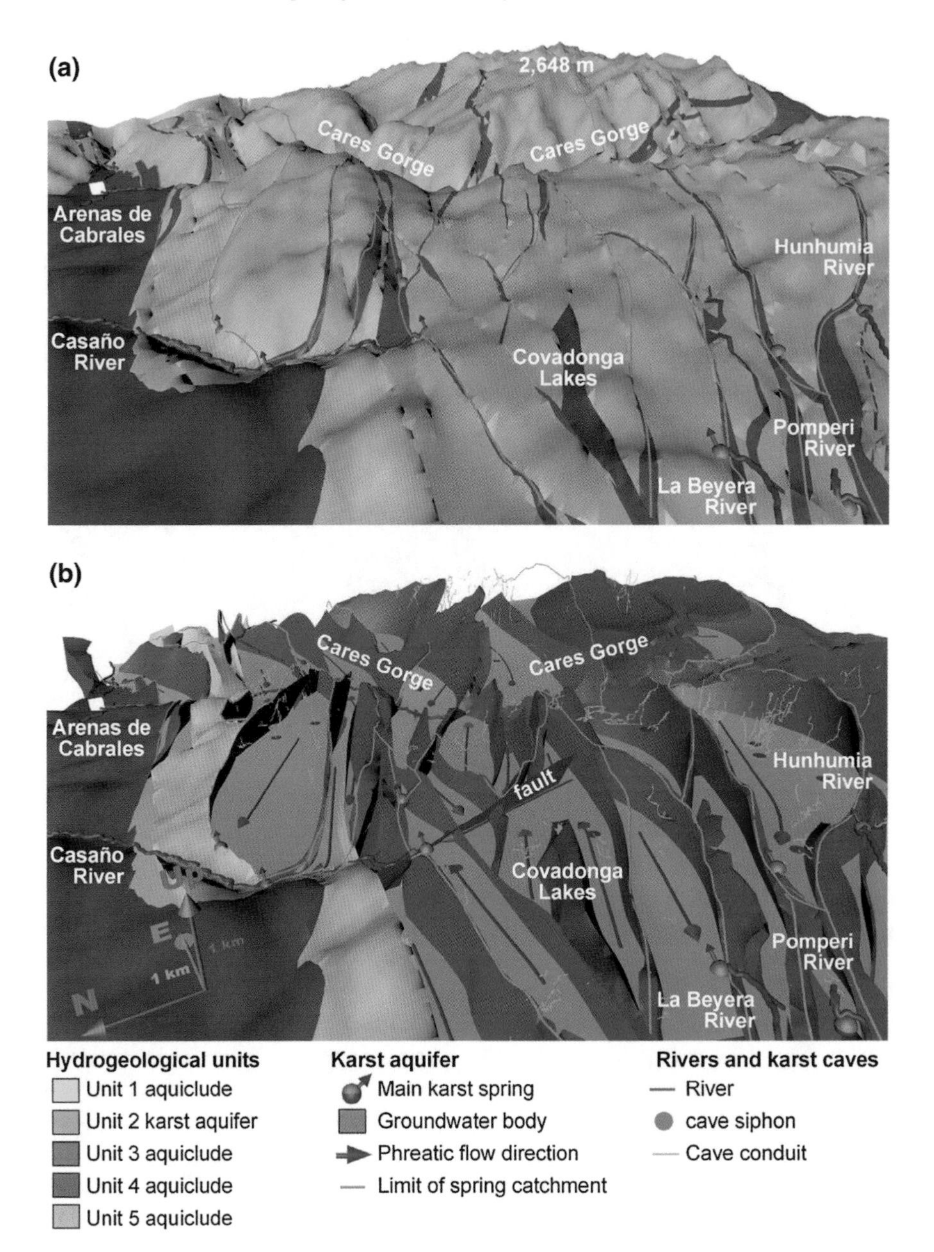

Fig. 2 **a** Zoom on the hydrogeological model showing the limits of the main karst springs catchment. The position of the point of view is displayed in Fig. 1. **b** Same view excluding the volumes of limestone (karst aquifer) in order to display the groundwater bodies (saturated aquifer), phreatic flow directions, and documented caves conduits and siphons

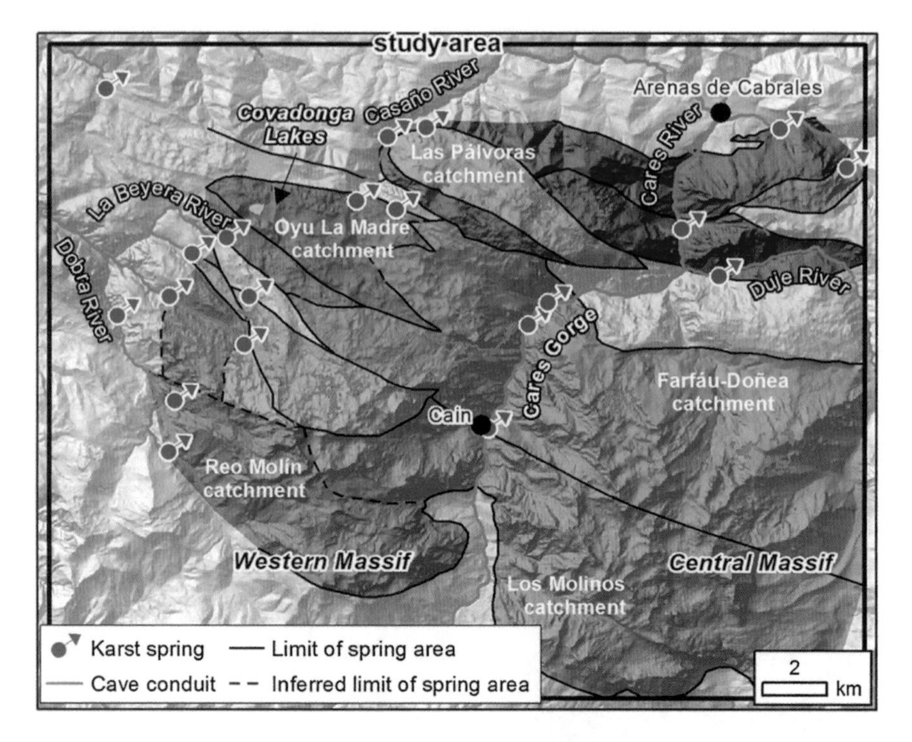

Fig. 3 Cave conduits and spring catchments delineated on the basis of the hydrogeological model, showing the name of the biggest areas. Cave data are provided from speleologists

The hydrogeological model allows us to delineate 17 catchment areas of springs (named "spring catchment" in this work) which limits are shown in Fig. 3. The geometry of the spring catchment is designed for low-flow conditions and is conform to tracer-tests results performed for low water stage. Their extension ranges from 1 to 86 km^2 and their geometry presents high differences from one area to another. Eight spring catchments (the bigger ones) are drained by springs related with the Cares Basin and seven of them flow to the Sella Basin. The base levels of the spring catchments are marked by the altitude of the main permanent springs, although some conduits can be placed under the base level. The base levels of the spring catchments associated to the Dobra basin are sited varying from 1,050 to 1,450 m a.s.l. and the base levels related to the Cares basin are located between 140 and 440 m a.s.l., although one of them, Oyu La Madre spring, lies at 835 m a.s.l.

The hydrogeological model involves 214 km of cave passages, usually showing cavities with length ranging from 1 to 4 km and depth exceeding 100 m. The caves are formed by horizontal and vertical conduits. Previous works (Ballesteros et al. 2011; Smart 1984) evidence that vertical conduits are vadose shafts while horizontal conduits correspond to vadose canyons or phreatic and epiphreatic conduits. The vertical distribution of the conduits density highlights 24 altitudes where the

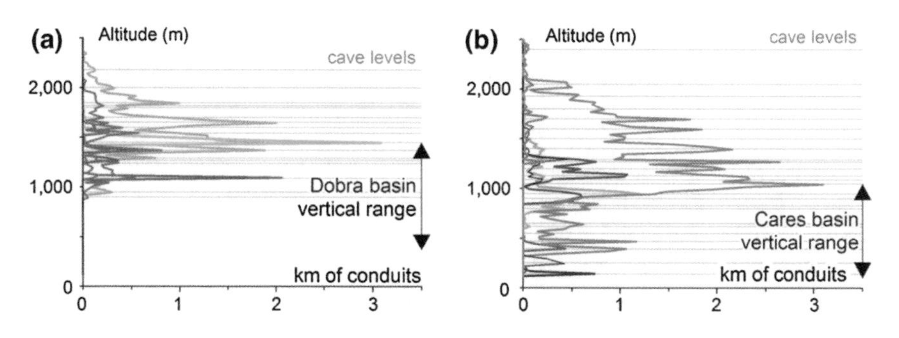

Fig. 4 **a** Vertical distribution of conduits density of the spring catchments related to Dobra basin. **b** Vertical distribution of conduits density of the spring catchments related to Cares basin

abundance of conduits is higher and where phreatic and epiphreatic morphologies are identified by previous works (Ballesteros et al. 2011; Smart 1984; Senior 1987 and others) and by the author. These altitudes define 24 cave levels in the entire area of study ranging from 140 to 2,400 m a.s.l. Most of the cave levels can be correlated in several spring catchments although the abundance of conduits is different from one level to another. Phreatic and epiphreatic features were identified in most of these cave levels in previous works or by the authors.

The comparison of the altimetric distribution profiles of the spring catchments related to Dobra and Cares basins are shown in Fig. 4. The cave levels of both basins are similar over the 1,090 m a.s.l., but below this elevation spring catchments of the Dobra basin do not present cavities in the study area. In this way, two sequences of cave levels can be established. The first sequence includes 13 cave levels defined over 1,090 m a.s.l. and the second sequence involves 11 cave levels, only in the Cares basin, at less than 1,090 m altitude.

5 Discussion and Conclusions

The hydrogeological model shows that the karst aquifer is strongly compartmentalized in, at least, 19 groundwater bodies related to caves with few kilometers length. These data evidence that the cave evolution should be related to the evolution of the aquifer, controlled by the fluvial network. The idea is supported by the three points: (1) the geometry of the spring catchments and the position of the rivers suggest that small western and northern spring catchments and their caves are being captured by the Cares River, increasing those areas of Farfáu-Doñea and Los Molinos springs; (2) the cave level sequence is related to the two fluvial basin and the lowering of the base level, allowing us to define two phases of caves evolution; and (3) fluvial network acts as regional base level of the karst systems, the organization of the caves reflects the evolution of the fluvial network. In the first phase, the first sequence of cave levels (2,400–1,090 m a.s.l.) was developed at a regional scale during the incision of the Cares and Dobra Rivers. In the second phase, a second sequence of caves levels (1,090–140 m a.s.l.) was originated in the Central Massif and the West and North of the Western Massif related to the incision of the Cares River. During this second phase, the incision of Dobra River was continuous, and probably, unknown cave levels and spring areas from the Western Massif were captured by the Cares River and deep shafts were developed intercepting previous cave levels in both massifs. The differences in the evolution of the Cares and Dobra basins are unknown and, perhaps, can be related to different erosion rates, differences about the lithology of the bedrock, the effects of the glaciations, or other reasons that must be considered for future research in the area.

Acknowledgments This research has been funded through the GEOCAVE project (MAGRAMA-580/12, OAPN), a two fellowship granted to D. Ballesteros by the Asturias regional Government: the Severo Ochoa Program (FICYT) and a grant of the Campus of International Excellence Development Plan (University of Oviedo-Department of Education). We acknowledge the support and help of the Picos de Europa National Park, Ó. Merino-Tomé, *Bureau de Recherches Géologiques et Minières* Catalogue of Cavities of León and the following cavers and clubs: AD Cuasacas, Expeditions to Castil, S Wroclaw, GE Diañu Burlón, GE Gorfolí, GE Matallana, GE Polifemo, GEMA, GS Matese, GS Doubs, HP Savez, IE Valenciano, L'Esperteyu CEC, LUSS, OUCC, CADE, SAR d' Ixelles, SCOF, SCS Matese, SEB Escar, SIE CE Áliga, SIS CE Terrassa, Soc. Suisse Speleol., WCC, YUCPC, Llambrión Project, CDG, GERSOP, GES,C. Puch (STD-BAT), J. Alonso, J. Sánchez (CES Alfa), B. Hivert (AS Charentaise) and V. Ferrer.

References

Audra P, Palmer AN (2013) The vertical dimension of karst: controls of vertical cave pattern. In: Shroder JF (ed) Treatise on geomorphology, vol 6. Academic Press, San Diego, pp 186–206

Bahamonde JR, Merino-Tomé O, Heredia N (2007) A Pennsylvanian microbial boundstone-dominated carbonate shelf in a distal foreland margin (Picos de Europa Province, NW Spain). Sediment Geol 198:167–193

Ballesteros D, Jiménez-Sánchez M, García-Sansegundo J, Giralt S (2011) Geological methods applied to speleogenetical research in vertical caves: the example of Torca Teyera shaft (Picos de Europa, Northern Spain). Carbonates Evaporites 26:29–40

Filipponi M, Jeannin P-Y, Tacher L (2009) Evidence of inception horizons in karst conduit networks. Geomorphology 106:86–99

Häuselmann P, Granger DE, Jeannin P-Y, Lauritzen S-E (2007) Abrupt glacial valley incision at 0.8 Ma dated from cave deposits in Switzerland. Geology 35:143–146

Jeannin P-Y, Eichenberger U, Sinreich M, Vouillamoz J, Malard A, Weber E (2013) KARSYS: a pragmatic approach to karst hydrogeological system conceptualisation. Assessment of groundwater reserves and resources in Switzerland. Environ Earth Sci 69:999–1013

Merino-Tomé O, Bahamonde JR, Colmenero JR, Heredia N, Villa E, Farias P (2009) Emplacement of the Cuera and Picos de Europa imbricate system at the core of the Iberian-Armorican arc (Cantabrian zone, north Spain): new precisions concerning the timing of arc closure. Geol Soc Am Bull 121:729–751

Merino-Tomé O, Suárez Rodríguez A, Alonso J (2013a) Mapa Geológico Digital continuo E. 1: 50.000, Zona Cantábrica (Zona-1000). [WWW Document]. GEODE. Mapa Geológico Digit Contin España SIGECO-IGME. http://cuarzo.igme.es/sigeco/default.htm

Merino-Tomé O, Suárez Rodríguez A, Alonso J, González Menéndez L, Heredia N, Marcos A (2013b) Mapa Geológico Digital continuo E. 1:50.000, Principado de Asturias (Zonas: 1100-1000-1600) [WWW Document]. GEODE. Mapa Geológico Digit Contin España, SIGECO-IGME. http://cuarzo.igme.es/sigeco/default.htm

Piccini L (2011) Speleogenesis in highly geodynamic contexts: the quaternary evolution of Monte Corchia multi-level karst system (Alpi Apuane, Italy). Geomorphology 134:49–61

Senior KJ (1987) Geology and speleogenesis of the M2 cave system, Western Massif, Picos de Europa, Northern Spain. Cave Sci 14:93–103

Smart P (1984) The geology, geomorphology and speleogenesis in the eastern massifs, Picos de Europa,Spain. Cave Sci 11(4):238–245

Smart PL (1986) Origin and development of glacio-karst closed depressions in the Picos de Europa, Spain. Zeitschrift für Geomorphol 30:423–443

Geodiversity of a Tropical Karst Zone in South-East Mexico

P. Fragoso-Servón, A. Pereira, O. Frausto and F. Bautista

Abstract The Yucatan Peninsula in south-eastern Mexico is the largest karstic area of the country; however, there are very few studies on the geodiversity of the region. The objective of this study is the identification and qualification of the diversity of elements of the relief, considering geomorphology, geology, hydrology and soil properties as components of geodiversity. To calculate the geodiversity, a simple additive model of thematic diversity was used, following Jačková and Romportl (2008). The geodiversity in the east of the Yucatan Peninsula manifests as a north to south banding. The geodiversity of coastal portion presents values ranging from medium to very high, in the central portion it ranges from medium to very low and in the western portion, where are the oldest geological formations, have high and very high values.

1 Introduction

The term geodiversity was first used in 1993 to describe the variety of nature not alive (highlighted, soils, climate, water bodies and rock). Gray (2008b) defines the term geodiversity as the natural diversity of traits of geological, geomorphological and soils, including their relationships, properties, interpretations and systems. In recent years at the international level this concept has gained acceptance in a large number of countries as a new geological paradigm (Gray 2008a).

P. Fragoso-Servón (✉) · A. Pereira · O. Frausto
Boulevard Bahía s/n Esquina Ignacio Comonfort, Universidad de Quintana Roo, Col. del Bosque, C.P. 77019 Chetumal, Quintana Roo, México
e-mail: pfragoso@uqroo.edu.mx

F. Bautista
Centro de Investigaciones en Geografía Ambiental, Universidad Nacional Autónoma de México, Morelia, Michoacán, México

B. Andreo et al. (eds.), *Hydrogeological and Environmental Investigations in Karst Systems*, Environmental Earth Sciences 1,
DOI 10.1007/978-3-642-17435-3_68

The methods for the evaluation and assessment of geodiversity are still rare; some seek to establish objective indicators through qualitative or quantitative scales with the use of various models and indexes. Some of the techniques have been derived from their equivalents for the study of biodiversity (Pike 2000; Ibàñez and García 2002; Xavier-da-Silva and de Carvalho-Filho 2004; Cantú et al. 2004; Benito-Calvo et al. 2009; Zwoliński and Stachowiak 2012).

One of the challenges in the study of geodiversity is to give an explanation of those places that show a certain homogeneity (lithologic, structural and geomorphological) to medium scales, but however show a remarkable diversity of traits in detail, as shown at the karstic massif (Priego et al. 2003).

In Mexico, there are no studies on the geodiversity of one of the most extensive karst massifs of the country as it is the Yucatan Peninsula. However, there are some descriptions of its geology and geomorphology at medium and small scales that reported a relative homogeneity of the area (Lugo et al. 1992; Bautista et al. 2011).

The objective of this study is the identification and qualification of the diversity of elements of the relief, geological, hydrological and soil as components of the geodiversity of the east of the Yucatan Peninsula on the basis of a zoning derived from the relative heights of geoforms and publicly available information according to the methodology of (Jačková and Romportl 2008).

2 Study Area

The Yucatan Peninsula in the south-east of the country is a large karstic platform whose eastern portion is located as the area of study, between 17° 49′ and 21° 36′ north latitude and 86° 44′ and 89° 24′ 52″ west longitude. The altitude of the area ranges from 0 m in the coastal area to the east and north up to a maximum altitude of 380 m above sea level in the south-west.

The dominant climate is warm humid with summer rainfall. The high amount of precipitation and the high capacity of infiltration cause the greatest amount of water, (near to 80 %) to move to basement level and the remaining 20 % is distributed between what the vegetation intercepts, surface runoff and the direct recruitment of the water bodies: flood zones, lagoons and cenotes (University of Quintana Roo 2004; according to the standard method used nationwide to calculate aquifer recharge). The reduced topographic slope contributes to the surface runoffs being few, low-flow and very short-stroke.

3 Materials and Methods

For the present work we use the wealth of attributes per unit of study; the units were defined by the analysis of the relative heights of terrain formations. We used the 80 topographic charts of the INEGI (2005) covering the area of study in vector format. From the synthetic chart a Digital Elevation Model (DEM) was built by interpolation in raster format; (Takagi and Shibasaki 1996; Yue et al. 2007) with horizontal resolution 1:50000 and vertical resolution of 10 m. This DEM was used to make the measurement of the relative heights of the land formations with the technique proposed by Priego et al. (2010). The resulting representation in m/km^2 was used to build a zoning of the territory as a base for the allocation of the various attributes. To calculate the geodiversity we used an additive model for the categories of thematic diversity considering four aspects related to the heterogeneity of the space: Geology, relief, soils and hydrology according to the methodology of Jačková and Romportl (2008).

Geology was simplified to the count of the different geological structures that are reported within each of the areas defined based on the relief. Pedological diversity was processed in a similar way, using the primary groups of soil or of greater dominance as identifiers.

The diversity of the relief was defined based on an analysis of the relative heights of the land formations for each unit. The hydrography was worked out on the basis of the existing records in the cartography of the INEGI (2005) bodies of water and depressions with permanent flood, the hierarchy was based on the types and number of bodies of water or depressions flooded by each of the areas defined by the relief.

The richness of each of the above components is divided into five classes one being the lowest that represents the less diverse and five the highest (Table 1).

The total Geodiversity (G) was calculated by adding the geologic (Dg), pedological (Ds), hydrological (Dh) and relief (Dr) diversities.

The full range of geodiversities was divided into five classes with the Jenks algorithm (Jenks 1977), which generates the classification shown in Table 1.

All the processes of identification, quantification or attribute mapping was carried out in a geographic information system managed with ArcGIS® 9.3.

4 Results

4.1 Geological Diversity

In the northern portion with lower height (Fig. 1a) there is low geological diversity; this large area is limited to the coast by a strip that goes from 1,000 to 1,500 m wide at its southern part up to 20 km in the north, where there are proliferate deposits of the Quaternary. The areas with very high diversities are mostly located in the southern and western parts, at the oldest geological formations corresponding to different ages of the tertiary period, which are combined with deposits of the quaternary mainly fluvial and lacustrine in the valleys and lower parts of the basins (University of Quintana Roo 2004). In coastal areas there is a mix of spaces with deposits of the quaternary like coastal, lake and swamps that render to high values of geologic diversity.

Table 1 Categorizations of the diversity of attributes of the territory (thematic diversity)

Category	Score	Attribute
Geology		
Very high	5	More than 4 geological formations in the unit
High	4	4 Geological formations in the unit
Media	3	3 Geological formations in the unit
Low	2	2 Geological formations in the unit
Very low	1	1 Geological formation in the unit
Pedology		
Very high	5	8–9 Groups of soil in the unit
High	4	6–7 Groups of soil in the unit
Media	3	4–5 Groups of soil in the unit
Low	2	2–3 Groups of soil in the unit
Very low	1	1 Group of soil in the unit
Relative height		
Very high	5	Relative height of 60–250 m/km^2
High	4	Relative height of 30–60 m/km^2
Media	3	Relative height of 15–30 m/km^2
Low	2	Relative height of 2.5–15 m/km^2
Very low	1	Relative height < 2.5 m/km^2
Hydrology		
Very high	5	More than 4 types of body of water in the unit
High	4	4 Types of body of water in the unit
Media	3	3 Types of body of water in the unit
Low	2	2 Types of body of water in the unit
Very low	1	0–1 type of body of water in the unit

4.2 Pedological Diversity

In the area, 12 soil groups are present; Fig. 1b shows the zones with the five categories of diversity and stresses the central-west and south with very high values. In these spaces the dominance of the Leptosols is offset with interleaving surfaces of Gleysols, Luvisols, Phaeozems and Vertisols. These areas coincide with that in which the altimetric diversities, geological and hydrological are medium and high. However, Gleysols are slightly concentrated in the lower portions of the structures where the altimetric diversity decreases towards plains with undulations and gullies. The other set of spaces that express very high diversity of

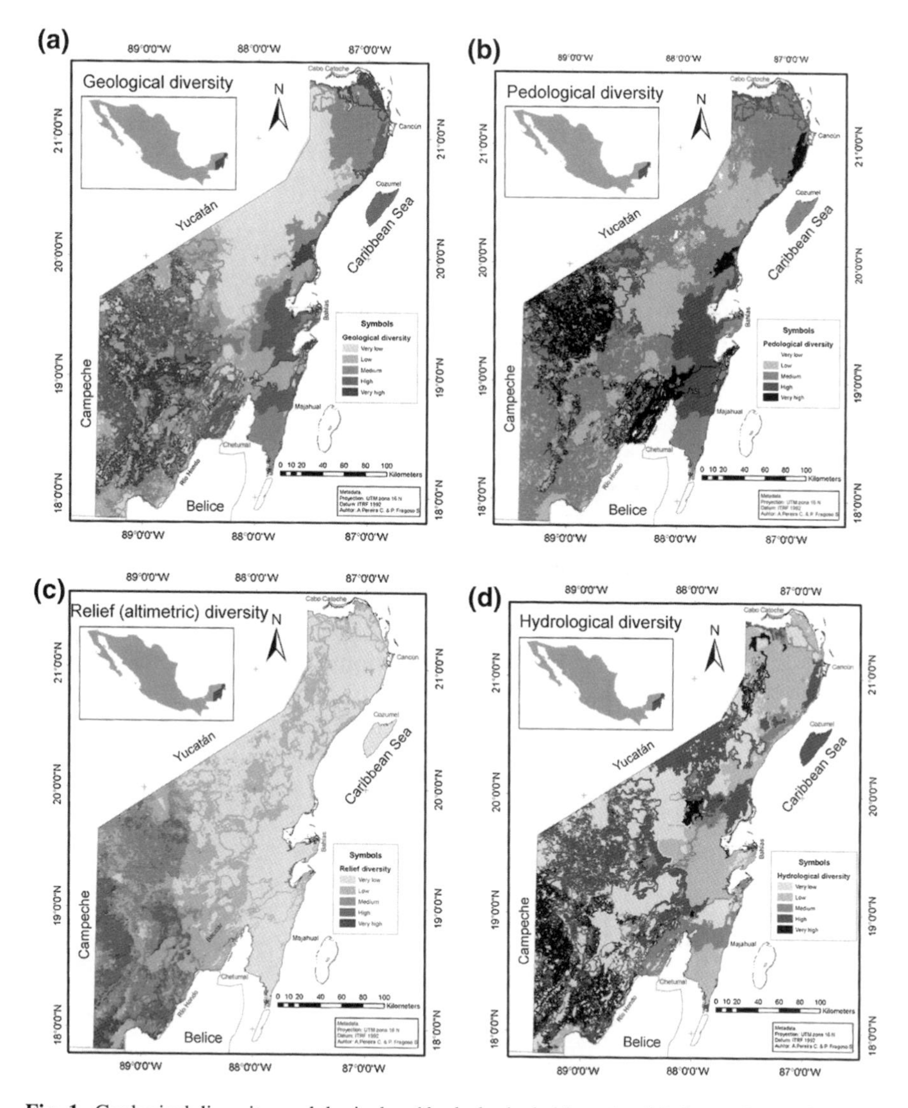

Fig. 1 Geological diversity, pedological and hydrological altimetric of Quintana Roo. **a** Geologic diversity based on number of types of surface materials by ages. **b** Pedological diversity based on number of types of soil per unit. **c** Altimetric diversity or energy of the relief on the basis of the relative heights of the forms of terrain. **d** Hydrological diversity based on number and type of water bodies and depressions with permanent flood

soils is associated with water bodies in the south–south-east of the State and its system of effluents that branches off to the coast. There are two areas with low diversity of soils in the center and a little more north in the study area.

4.3 Diversity of Relative Heights

In Fig. 1c can be seen the diversity of relative heights for the study area, it is apparent that it is distributed in the form of parallel strips from the coastal area with very low energy (category 1), increasing towards the west and south to the extreme south-west the highest values were recorded. The intermediate transitional area range from medium to high, there are alternating spaces of low, medium and even high diversity coexisting. Excels in this distribution a series of narrow stripes that correspond to small spaces in which the combination of landforms increases the geodiversity and highlight areas of transition between those with relatively homogeneous diversity.

4.4 Diversity of the Surface Hydrology

As regards surface water bodies, Fig. 1d shows that near 50 % of the territory is located within categories of high and very high diversity. Such areas are distributed mainly in the west, central, northern and in some coastal areas, where there is an abundance of water bodies characteristic of karst areas such as the so-called cenotes and lagoons, the vast majority of which are connected to the groundwater table. In the central part of the State and to the north, in the plains are large areas with low diversities matching with geologic formations from the tertiary barely altered and with areas of permanent or semi-permanent floods such as those found west of the bays.

4.5 Geodiversity

The geodiversity of the east portion of the Peninsula shows three large areas (Fig. 2).

The first of them, the region with the greatest geodiversity in the west, is the oldest geological area of the State; here the relief with elevations over 70 m above sea level promotes the formation of areas of erosion and accumulation that allow the development of several floors, accumulation areas and runoff of water.

This distribution emphasizes the area west and south of the State where there are discontinuous spaces in which the dominance of geodiversities is high and very high. Some valleys with medium geodiversities and intermountain plains which geodiversity is low, due to the existence of strong discontinuities in the local groups both from the point of view of the morphometric, geological, hydrological and Pedological attribute.

The intermediate zone between the area of altitude higher than 70 m above sea level and that with less than 20 m above sea level presents values of geodiversity

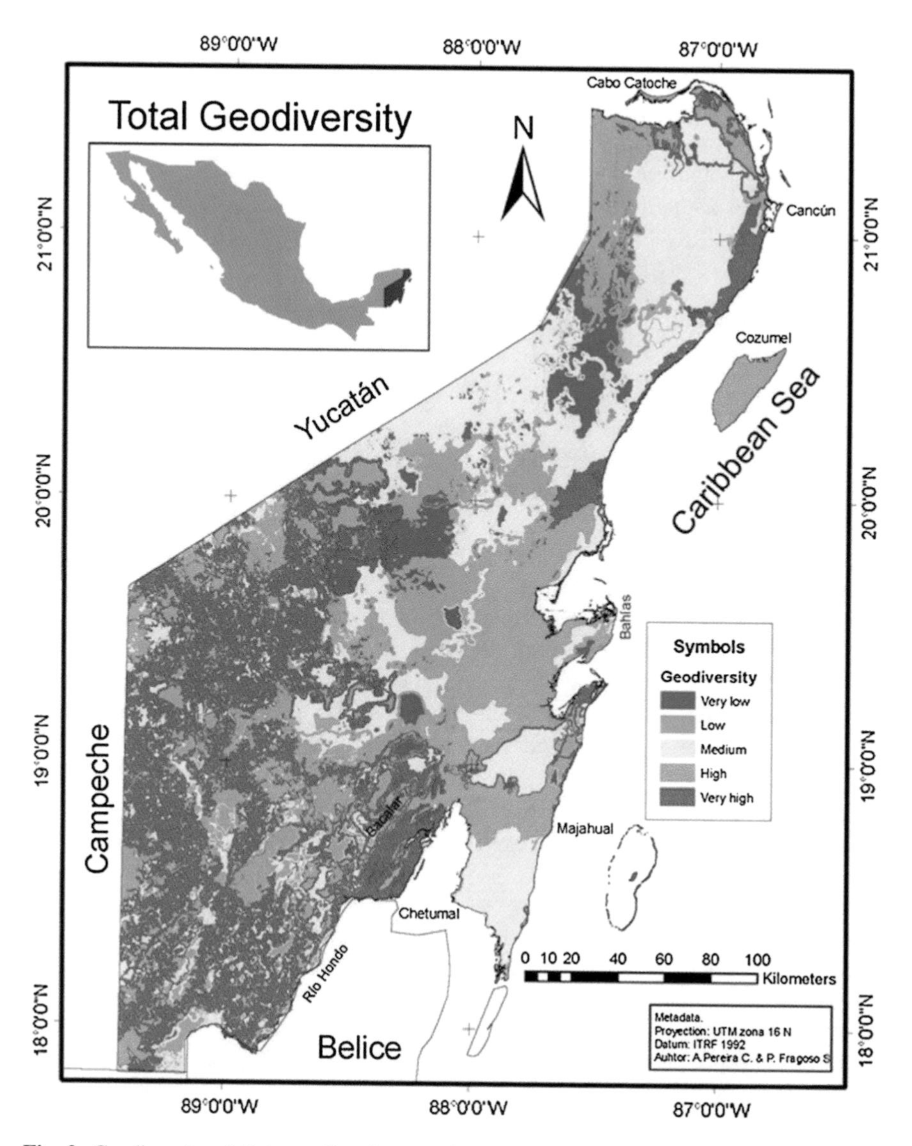

Fig. 2 Geodiversity of Quintana Roo in accordance with the model G = Dg + of + Dr + Dh

from low to very low. This homogeneous zone is located on the geological formation of the Pliocene, sufficiently far away from the coast to not have a strong influence on her and with a gradient such that it will allow the water to drip and not accumulate and thus there is no such abundance of water bodies exposed.

The area to the east, in spite of being the area with the smaller heights, is an area where again there is geodiversity medium to very high. In this area, there are a large number of bodies of water formed by the dissolution of the limestone rock and the shallow depth of the groundwater table. The same dissolution of the rock allows the formation of negative forms of stress in which the accumulation of materials enables the development of the soils with differentiated characteristics making it an area rich in alluvial and lacustrine deposits. In areas close to the coast lacustrine deposits, swamp and coastal environments develop the transition sea-land as the wetland areas where there are soils and characteristic vegetation, all this contributes to high geodiversity of the area of lower height.

5 Discussion

The use of the term geodiversity is increasing everyday, but the methodologies for evaluation and assessment are still limited. So far, most studies use qualitative methods in which the delimitation of the geomorphological units and the inventory of the components of the geodiversity (geology, soils, relief and hydrology) are the starting points for analysis.

The energy of the karstic Yucatan Peninsula relief is scarce compared with other areas of the country or the world, in which the volcanic formations generated structures and wildest landscapes. However, on the one hand they are not comparing two objects of the same nature, the Peninsula is a large deposit of sedimentary platform character and on the other hand, the comparisons are made in terms of classification criteria for national or continental scales and not to regional scales.

The work of description of the geodiversity of the Polish Carpathians considered that a structure with more than 50 m of relative height is of very high geodiversity Zwolinski (2010). In the present work, this value is defined as high geodiversity because in the area we find recognizable landforms with heights of more than 100 m.

The techniques employed were efficient to handle the attributes used. The morphometric technique used succeeded to highlight both strong changes in the geological formations and small variations within them despite the low energy of the relief.

The results obtained allow us to ensure that a study with detailed information could provide important insight about the spatial distribution of biological diversity. Also for the identification of areas of high species richness; for the design of protected natural areas; or definition of conservation strategies; (Priego et al. 2003; Parks and Mulligan 2010); and for the identification of environmental or ecosystem services (Gray 2011). Such detailed information could allow the use of more advanced diversity indexes, among other techniques to define geodiversity.

These results clearly show that geodiversity fosters the diversity of environments; however, to the extent that we know the details of the highlighted attributes at larger scales can foster our knowledge and understanding of the great diversity of environments in the karst of the portion east of the Yucatan Peninsula. For this reason, it is necessary to ensure the completion of studies in detail to enable the recognition, identification and registration of the geodiversity that translates into environmental heterogeneity and diversity of species.

6 Conclusions

In the east of the Yucatan Peninsula, there are differences other than altimetric ones that have allowed the development of a wide range of geodiversity.

The methodology of (Jačková and Romportl 2008) used as a model of simple additive categories of thematic diversity works for the karst features in the area.

Despite the fact that richness counts were useful to estimate geodiversities, more quantitative approaches will give more precise results and a deeper insight into the phenomenon with the same traits as geology, soil types, altimetric units and hydrology as defined by Gray (2008b) and Kozlowski (2004).

This work is the first for the area in which it was reported the altitudinal floors zoning to substantiate that the apparent homogeneity of the relief is not such and to support that in such large karst plains there is a wide diversity of environments product of the morphometric and structural heterogeneity of the territory.

There are areas with very high to very low geodiversity in well-defined spaces and major contributors whose attributes are equally well-defined. We identified three major areas: the area to the west with values of geodiversity medium to very high, an intermediate zone in the central part in where the values of geodiversity range from medium to very low, and a third area, in the part of lower height and towards the coastal zone where the geodiversity has medium to very high values.

References

Bautista F, Palacio-Aponte G, Quintana P, Zinck JA (2011) Spatial distribution and development of soils in tropical karst areas from the Peninsula of Yucatan, Mexico. Geomorphology 135:308–321. doi:10.1016/j.geomorph.2011.02.014

Benito-Calvo A, Pérez-González A, Magri O, Meza P (2009) Assessing regional geodiversity: the Iberian Peninsula. Earth Surf Process Landf 34:1433–1445

Cantú C, Gerald Wright R, Michael Scott J, Strand E (2004) Assessment of current and proposed nature reserves of Mexico based on their capacity to protect geophysical features and biodiversity. Biol Conserv 115:411–417

Gray M (2008a) Geodiversity: developing the paradigm. Proc Geol Assoc 119:287–298

Gray M (2008b) Geoheritage 1. Geodiversity: a new paradigm for valuing and conserving geoheritage. Geosci Can 35:51–59

Gray M (2011) Other nature: geodiversity and geosystem services. Env Conserv 38:271–274

Ibàñez Martí JJ, García Alvarez A (2002) Diversidad: biodiversidad edáfica y geodiversidad. Edafología 9:329–385

INEGI (2005) Cartas Topográficas 1:50000

Jačková K, Romportl D (2008) The relationship between geodiversity and habitat richness in Šumava national park and Křivoklátsko Pla (Czech Republic): a quantitative analysis approach. J Landsc Ecol 1:23–38

Jenks G (1977) Optimal data classification for choropleth maps. University of Kansas Department of Geography, Kansas

Kozłowski S (2004) Geodiversity. The concept and scope of geodiversity. Przegląd Geol 52:833–883

Lugo HJ, Aceves JF, Espinoza R (1992) Rasgos geomorfológicos mayores de la Península de Yucatán. Inst Geogr 10:143–150

Parks KE, Mulligan M (2010) On the relationship between a resource based measure of geodiversity and broad scale biodiversity patterns. Biodivers Conserv 19:2751–2766

Pike RJ (2000) Geomorphometry-diversity in quantitative surface analysis. Prog Phys Geogr 24:1–20

Priego Santander Á, Bocco G, Mendoza ME, Garrido A (2010) Propuesta para la generación semiautomatizada de unidades de paisajes, 1st edn. Instituto Nacional de Ecología, México

Priego Santander Á, Moreno Cassasola P, Palacio Prieto JL et al (2003) Relación entre la heterogeneidad del paisaje y la riqueza de especies de flora en cuencas costeras del estado de Veracruz, México. Investig Geográficas Bol Inst Geogr 52:31–52

Takagi M, Shibasaki R (1996) An interpolation method for continental DEM generation using small scale contour maps. Int Arch Photogramm Remote Sens 31:847–852

Universidad de Quintana Roo (2004) Programa Estatal de Ordenamiento Territorial 3987.

Xavier-da-Silva J, de Carvalho-Filho LM (2004) Geodiversity: Some simple geoprocessing indicators to support environmental biodiversity studies. Dir Mag 1–4. http://www.directionsmag.com/article.php?article_id=473

Yue T-X, Du Z-P, Song D-J, Gong Y (2007) A new method of surface modeling and its application to DEM construction. Geomorphology 91:161–172

Zwoliński Z (2010) The routine of landform geodiversity map design for the Polish Carpathian Mts. Landf Anal 11:77–85

Zwoliński Z, Stachowiak J (2012) Geodiversity map of the Tatra National Park for geotourism. Quaest Geogr 31:99–107

Geoheritage and Geodiversity Evaluation of Endokarst Landscapes: The Picos de Europa National Park, North Spain

D. Ballesteros, M. Jiménez-Sánchez, M.J. Domínguez-Cuesta, J. García-Sansegundo and M. Meléndez-Asensio

Abstract The endokarst presents a spectacular Geoheritage involving many singular features with thousands to millions years in age. The Picos de Europa National Park has one of the most important karst landscapes in the world. The endokarst of Picos de Europa shows high natural, scientific, and cultural values mainly related to cave features, with a spectacular vertical development (14 % of known world caves deeper than 1 km), the presence of geomorphological and sedimentary records related to the Quaternary evolution of the Cantabrian Mountain Range, and the traditional and sport uses of the cavities. The aim of this work is to inventory and to evaluate the Geoheritage and Geodiversity of the endokarst of Picos de Europa National Park combining speleological data, field work, and geomorphological mapping from nine selected caves. As a result, the Picos de Europa Geodiverstiy features show a high variability: 75 different natural features have been recognized with a density ranging from 0.3 to 1.1 different features per cm^2 of cave area.

D. Ballesteros (✉) · M. Jiménez-Sánchez · M.J. Domínguez-Cuesta · J. García-Sansegundo
Department of Geology, University of Oviedo, c/Arias de Velasco s/n, 33005 Oviedo, Spain
e-mail: ballesteros@geol.uniovi.es

M. Jiménez-Sánchez
e-mail: mjimenez@geol.uniovi.es

M.J. Domínguez-Cuesta
e-mail: mjdominguez@geol.uniovi.es

J. García-Sansegundo
e-mail: j.g.sansegundo@geol.uniovi.es

M. Meléndez-Asensio
Geological Survey of Spain, C/Matemático Pedrayes 25, 33005 Oviedo, Spain
e-mail: m.melendez@igme.es

B. Andreo et al. (eds.), *Hydrogeological and Environmental Investigations in Karst Systems*, Environmental Earth Sciences 1,
DOI 10.1007/978-3-642-17435-3_69

1 Introduction

The endokarst encloses a spectacular Geoheritage hidden in many kilometers of underground conduits, involving many singular morphologies and deposits preserved since thousands to millions years ago (Ali et al. 2008). In some countries, as Slovenia, its subsurface reaches 50.4 % of the natural features (Environmental Agency of the Republic of Slovenia, in Erhartic 2010). The evaluation of cave Geoheritage is difficult due to the methodological constraints of cave working. However, some cavities are considerably documented by speleological work, which can be combined with geomorphological techniques in order to evaluate the endokarst Geoheritage. The Picos de Europa National Park is considered an international reference in karstology, speleology, and Geoheritage (Ballesteros et al. 2013), and was declared Global Geosite in 2007 by Geological Survey of Spain (Carcavilla et al. 2007) because the Park includes 14 % of the known world caves deeper than 1 km. The karst of Picos de Europa is included in the "carbonate and evaposite karst system of Iberian Peninsula and the Balearic islands" declared by the Government of Spain (Law 42/2007, Natural Heritage and Biodiversity) as Spanish geological frameworks with international relevance (Durán and Robledo Ardila 2009). The Spanish-funded GEOCAVE research project is being developed in Picos de Europa in order to characterize the endokarst from a geomorphological and geochronological point of view, proposing and validating new methodologies adapted to these environments. In this way, the aim of this work is to evaluate the Geoheritage and Geodiversity of the endokarst of Picos de Europa combining speleological data, field work, and geomorphological mapping from nine selected caves.

2 Setting

The Picos de Europa National Park (4°58′ W, 43°15′ N) is located in the Cantabrian Range, North of Spain and has a 647 km^2 surface. Most of the Park corresponds to the Picos de Europa mountains, divided into three massifs: the Western, Central, and Eastern massifs (Fig. 1). Initially, the Covadonga Mountain National Park (including 169 km^2 of the Western Massif) was declared in 1918 by its spectacular landscape and historic significance. In 1995, the National Park area was extended to the East, including also both Central and Eastern Picos de Europa Massifs, to become the Picos de Europa National Park. Some areas of the Park were also declared Regional Park (1994), Special Protection Area (2000), Reserve of the Biosphere (2003), and Site of Community Importance (ES0000003I; 2004). Moreover, four caves were considered as Natural Monument by the regional Government of Asturias for their geological and hydrogeological interest.

The Picos de Europa landscape is mainly formed by a calcareous rough relief up to 2,650 m above sea level (a.s.l.) involving an alpine karst, relict glacial forms,

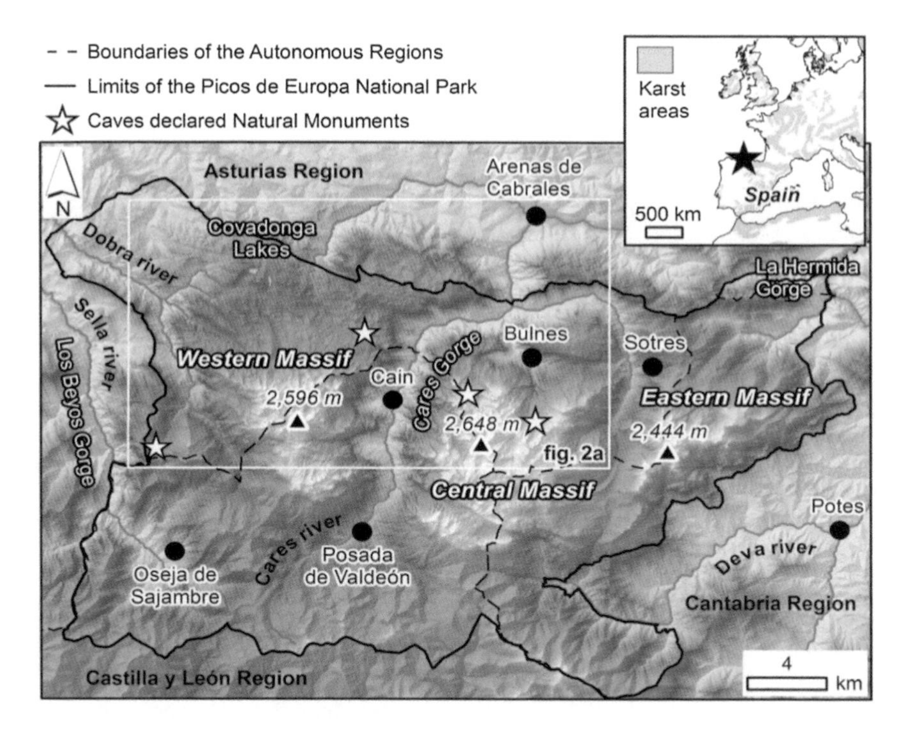

Fig. 1 Setting of the Picos de Picos de Europa National Park

deep fluvial canyons, and few ices patches with a high value as Geoheritage (Fuertes-Gutiérrez and Fernández-Martínez 2012; Serrano et al. 2011). Picos de Europa includes hundreds of kilometers of caves with a great vertical development up to 1,589 m (Margaliano et al. 1998). The Geoheritage of some caves was studied analyzing the use values and management based on speleological data or evaluating the natural value of several ice caves (e. g., Serrano and González-Trueba 2005).

3 Methodology

The methodology, based on previous works (Jiménez-Sánchez et al. 2006, 2014; Ballesteros et al. 2011) includes: (1) collection of cave data from 247 speleological works and 46 scientific articles; (2) elaboration of a catalog of caves in ArcGIS 9.3; (3) 3d modeling of the caves based on speleological cave surveys; (4) collection of cave information by field work into 128 cavities; (5) elaboration of catalogs of natural and cultural features; (6) geomorphological mapping of nine selected caves (14.3 km length), including also geological, hydrogeological, and paleontological features; (7) evaluation of the natural, scientific, and cultural values of the caves; (8) definition and calculation of three indexes for a preliminary evaluation of cave Geoheritage: (a) Cave Geoheritage Area Index (GhAI), defined as the percentage of the area occupied by the entire features divided by the cave area, excluding the features that represent the conduits themselves, (b) Feature Area Index (FAI), defined as the area occupied by each group of features divided by the cave area, and (c) Cave Geodiversity Index (GdvI) (feature/cm^2), defined as the number of features divided by the cave area, expressed in cm^2. The values of these indexes calculated in nine caves have been extrapolated to the whole of the Picos de Europa cavities.

4 Results

The catalog of caves and 3D modeling of their conduits allows us to analyze the spatial cave distribution and their geometry. The catalog of caves includes 3,648 cavities located in the national park and surroundings. These caves involve 355 km of conduits, with 4 to 12 km of new passages discovered by national and foreign speleologists. The area and volume of the whole of the caves is estimated in 885 km^2 and 0.02 km^3, respectively. About 82 % of the caves of Picos de Europa are small cavities less than 50 m length and 10 m depth, and the remaining 18 % are caves containing 83 % of the conduits. Figure 2 shows the main caves from the NW of Picos de Europa. Regarding to their length, 112 cavities with more than 1 km length are documented, including 14 shafts larger than 5 km. Respect to their vertical range, 235 cavities are from 100 to 500 m in depth, 55 caves are between 500 and 1,000 m and 14 shafts are deeper than 1 km.

The caves of Picos de Europa are formed by phreatic and epiphreatic conduits (47 %), vadose canyons (42 %), vadose shafts (10 %) and *soutirage* conduits, paragenetic canyons and gravity-enlarged passages (1 %). The evolution of the cavities began, at least, in the Middle Pleistocene and was conditioned by the drop of the water table, the strong compartmentalize of karst aquifers and the geological structure (Ballesteros et al. 2011; Fernández-Gibert et al. 2000; Senior 1987; Smart 1984). The cavities show several cave levels related to the ancient position of the water table, and include records of speleothems and fluvial deposits. These records have been studied in few works, which have established links between the erosion of the Permian-Mesozoic rocks and the origin of the fluvial deposits (Fernández-Gibert et al. 2000; Smart 1986).

The catalogs of natural and cultural heritage features of the caves are elaborated based on field work in 128 caves (Fig. 2). The catalog of natural features includes 75 elements classified in five groups: geomorphological (46 features), geological (19), hydrogeological (8), and paleontological (2) features. Geomorphological features are divided in fluviokarst features (13 features), speleothems (19), gravity forms (5 features), and snow-ice forms (9); geological features include geological underground recognized elements, as tectonic and sedimentary features (4), lithological features (5), mineralogical features (carbonates, detrital minerals, and ore

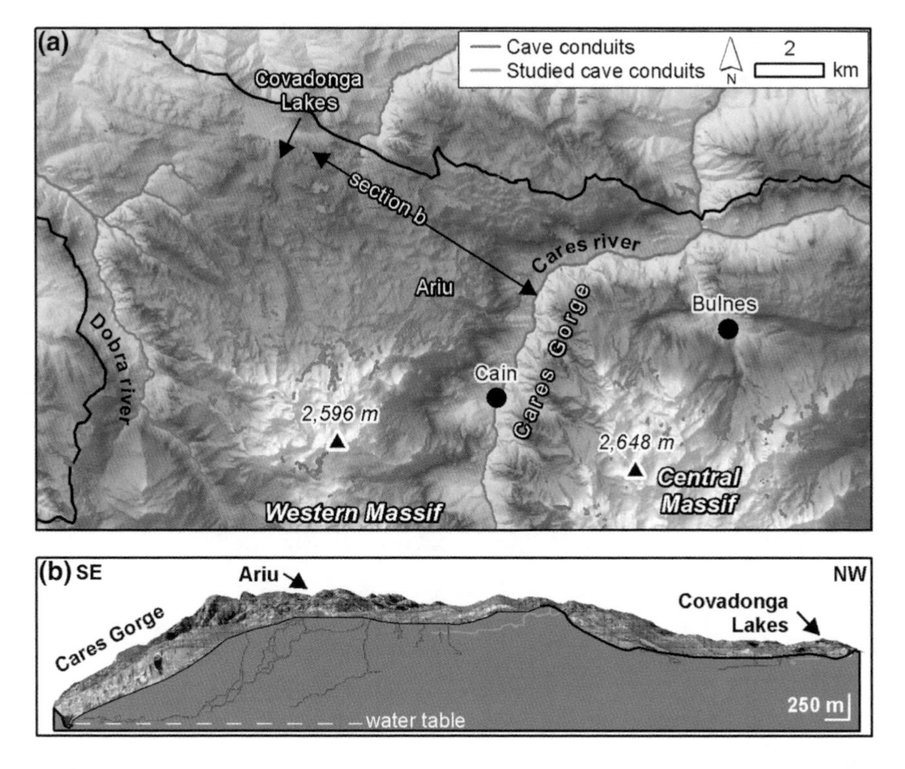

Fig. 2 **a** Cave conduits of the North of the Western and Central massifs, highlighting the studied caves in this work. The position of the picture is shown in Fig. 1. **b** 3D model of the karst massif with the main caves. Their position is shown in section (**a**). Cave data are courtesy of the Speleologists included into the acknowledgments

deposits (8), and paleokarst related features (2). Finally, hydrogeological features involve three zones of karst aquifer and five hydrological elements of the caves (e.g., siphons, lakes, and rivers), and paleontological features include bones and other remains. The catalog of cultural features involves 10 features: one of them includes archeological remains and the remaining nine features correspond to traditional uses of the caves: livestock farming, cheese aging, mining, water supply, and hydropower (Fig. 3).

Figure 4 shows a detail of a cave geomorphological map, displaying the position of speleothems, fluvial and gravity deposits in the cavity survey. The map is complemented by two interpreted photographs. Based on geomorphological maps of nine caves, three indexes have been calculated, providing an initial evaluation of the cave Geoheritage. GhEI and FEI inform about the area of the cave features: GhEI ranges from 22 to 80 % of the cave, being Geomorphological FEI 20–80 %, Geological FEI 4–5 %, Hydrogeological FEI 0–3 %, and Paleontological FEI 0–0.1 %. GdvI, which depicts the density of natural features inside the caves, ranges from 0.3 to 1.1 features/cm^2.

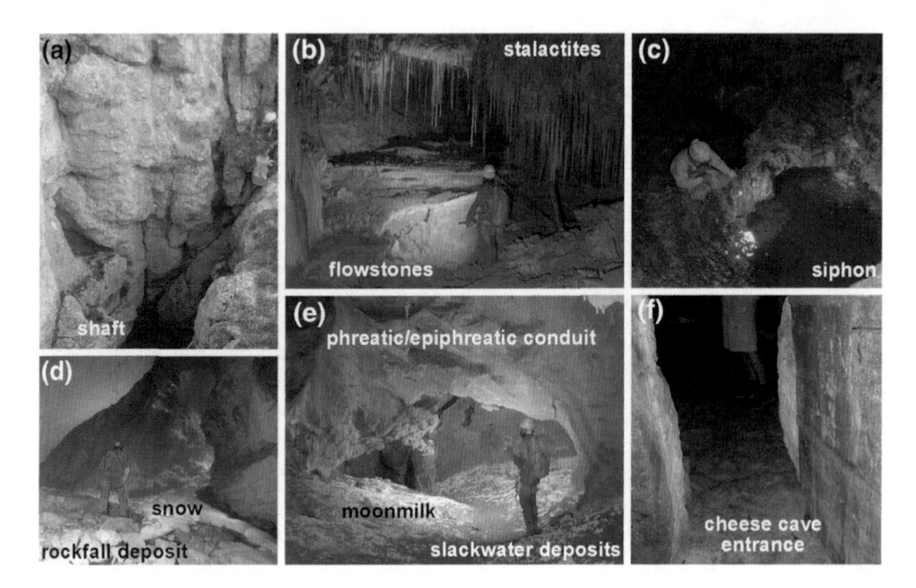

Fig. 3 **a** Cave entrance. **b** Stalactites and stalagmites in a gallery. **c** Cave siphon, with the conduit completely occupied by water. **d** Snow and gravity deposit close to a cave entrance. **e** Slackwater deposits and moonmilk in a phreatic/epiphreatic conduit. **f** Door of a cave used for aging cheese with Certificate of Origin

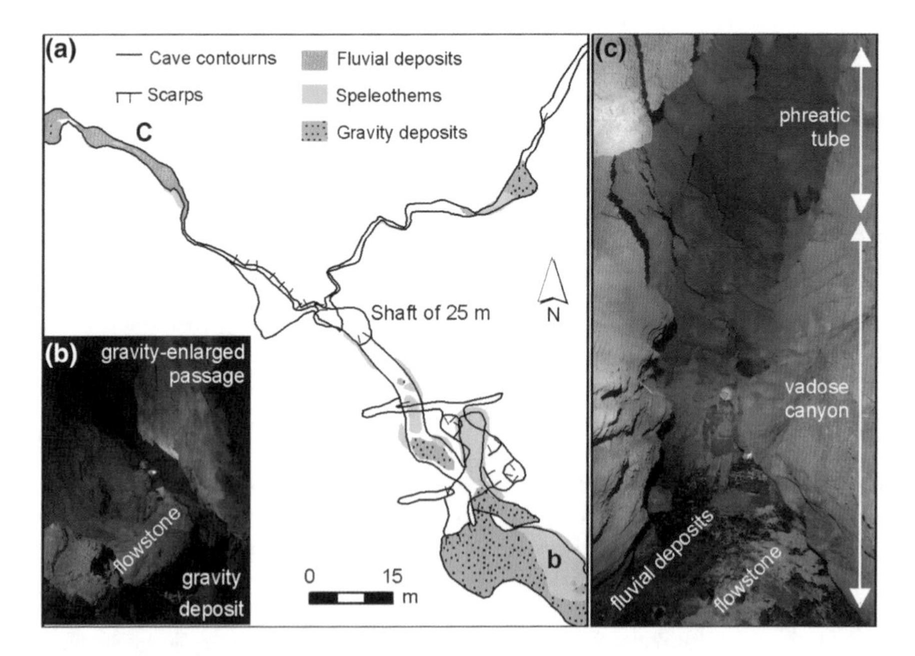

Fig. 4 **a** Zoom of the geomorphological map of Pozu Llucia shaft. **b** and **c** pictures from selected passages of section (**a**)

5 Discussion

The results evidence that the caves of Picos de Europa present high natural, scientific, and cultural values. The natural value corresponds to the singularity and spectacular vertical development of the caves and a very high Geodiversity of cave features. Picos de Europa shows a high number of deep caves, including 69 shafts deeper than 500 m; this area can only be compared with few places, as the Kanin Mountain (Slovenia) or the Arabika Massif (Abkhazia, Western Caucasus). The natural value of the caves is also related to the presence and area of their features; 22 to 80 % of the cave area is covered by natural features, showing strong variations from one cave to another. Most of them are geomorphological features and includes mainly shafts and canyons without deposits and galleries with some speleothems and fluvial deposits. Their Geodiversity value is high, identifying 75 different features inside the caves, with 0.3 to 1.1 different features per cm^2 of cave.

The scientific value is related to cave records that allow us to study the evolution of the caves, the karst massif and the global evolution of the Cantabrian Range. The scientific studies evidence that the caves are originated since, at least, Middle Pleistocene, related to mountain uplift and fluvial incision with the development of deep gorges. Moreover, the caves contribute to the hydrogeological knowledge of the karst aquifer and preserve deposits belonging to the Permian-Mesozoic rocks that previously covered the karstified limestone (Fernández-Gibert et al. 2000; Smart 1986).

The cultural value is related with the specific uses of the cave and the singularity of their names used by the stockbreeder. The uses include traditional customs, as the livestock farming, the water collection, the elaboration of five types of cheese with Certificated of Origin, and sport uses by the speleologists. Nowadays, many cavers from Europe and other countries organize expeditions to Picos de Europa to discover and survey new caves and conduits.

6 Conclusion

The Geoheritage and Geodiversity of the endokarst can be evaluated combining the catalog of caves, 3D models of their conduits, catalogs of natural and cultural features, cave geomorphological mapping and three quantitative indexes: the GhAI, Cave Feature Index and GdvI. The GhAI and Cave feature Area indexes allows us to estimate the relative area of the features of the underground Geoheritage, while the GdvI provide a precise quantification of the different cave features. The application of this methodology to the endokarst of the Picos de Europa National Park evidences its high natural, scientific, and cultural interest related to a high Geodiversity, represented by 75 natural features in the caves with a density varying from 0.3 to 1.1 features/cm^2.

Acknowledgments This research has been funded through the GEOCAVE project (MAGRA-MA-580/12, OAPN) and fellowship granted to D. Ballesteros by the Severo Ochoa Program (FICYT, Asturias regional Government). We acknowledge the support of the Picos de Europa National Park and Federación d'Espeleoloxía del Principáu d'Asturies. Cave survey data is from: AD Cuasacas, Expeditions to Castil, S Wroclaw, GE Diañu Burlón, GE Gorfolí, GE Matallana, GE Polifemo, AD GEMA, GS Matese, GS Doubs, HP Savez, IE Valenciano, L'Esperteyu CEC, LUSS, OUCC, CADE, SAR d'Ixelles, SCAL, SCOF, GE GET, ERE, SCS Matese, SEB Escar, SIE CE Áliga, SIS CE Terrassa, Soc. Suisse Speleol., WCC, YUCPC, CDG, GE Aude, GE Llubí, ANEM, GES, Llambrión Project, GERSOP, C. Puch (STD-BAT), J. Alonso, J. Sánchez (CES Alfa), B. Hivert (AS Charentaise) and V. Ferrer.

References

Ali CA, Mohamed KR, Komoo I (2008) Geoheritage of Pulau Balambangan, Kudat Sabah. Bull Geol Soc Malays 54:91–95

Ballesteros D, Jiménez-Sánchez M, García-Sansegundo J (2013) Patrimonio geológico subterráneo en espacios naturales protegidos: caracterización geomorfológica preliminar de sistemas kársticos profundos en el Parque Nacional de los Picos de Europa (España). In: Vega J, Salazar A, Díaz-Martínez E, Marchán C (eds) Patrimonio Geológico, Un Recurso Para El Desarrollo. Publicaciones del Instituto Geológico y Minero de España, Madrid, pp 361–370

Ballesteros D, Jiménez-Sánchez M, García-Sansegundo J, Giralt S (2011) Geological methods applied to speleogenetical research in vertical caves: the example of Torca Teyera shaft (Picos de Europa, Northern Spain). Carbonates Evaporites 26:29–40

Carcavilla L, López-Martínez J, Durán JJ (2007) Patrimonio geológico y geodiversidad: investigación, conservación, gestión y relación con los espacios naturales protegidos, Serie Cuadernos del Museo Geominero. Instituto Geológico y Minero de España, Madrid

Durán JJ, Robledo Ardila P (2009) Carbonate and evaporite karst systems of the Iberian Peninsula and the Baleatic Islands. In: García-Cortés Á (ed) Spanish geological frameworks and geosites. Instituto Geológico y Minero de España, Madrid, pp 200–214

Erhartic B (2010) Conserving geoheritage in Slovenia through geomorphosite mapping. In: Regolini-Bissig G, Reynard E (eds) Géovisions n°35, mapping geoheritage. Institut de géographie, Lausanne, pp 47–63

Fernández-Gibert E, Calaforra JM, Rossi C (2000) Speleogenesis in the Picos de Europa Massif, Northern Spain. In: Klimchouk A, Ford D, Palmer A, Dreybrodt W (eds) Speleogenesis: evolution of karst aquifers. National Speleological Society, Huntsville, pp 352–357

Fuertes-Gutiérrez I, Fernández-Martínez E (2012) Mapping geosites for geoheritage management: a methodological proposal for the regional park of Picos de Europa (León, Spain). Environ Manage 50:789–806

Jiménez-Sánchez M, Aranburu A, Domínguez-Cuesta, Martos E (2006) Cuevas prehistóricas como Patrimonio Geológico en Asturias: métodos de trabajo en la cueva de Tito Bustillo. Trab Geol 26:163–174

Jiménez-Sánchez M, Domínguez-Cuesta MJ, Aranburu A, Martos E (2011) Quantitative indexes based on geomorphologic features: a tool for evaluating human impact on natural and cultural heritage in caves. J Cult Herit 12:270–278

Jiménez-Sánchez M, Ballesteros D, Rodríguez-Rodríguez L, Domínguez-Cuesta MJ (2014) The Picos de Europa National and Regional Parks. In: Gutiérrez F, Gutiérrez F (eds) Landscape and Landforms of Spain. Springer, Dordrecht, pp 155–163

Margaliano D, Muñoz J, Estévez JA (1998) −1.589 m Récord de España en la Torca del Cerro del Cuevón. Subterránea 10:20–29

Serrano E, González-Trueba J (2005) Assessment of geomorphosites in natural protected areas: the Picos de Europa National Park (Spain). Géomorphol Reli Process Environ 3:197–208

Serrano E, González-Trueba JJ, Sanjosé JJ, Del Río LM (2011) Ice patch origin, evolution and dynamics in a temperate high mountain environment: the Jou Negro, Picos De Europa (Nw Spain). Geogr Ann Ser A Phys Geogr 93:57–70

Senior KJ (1987) Geology and Speleogenesis of the M2 Cave Systems, Western Massif, Picos de Europa, Northern Spain. Cave Sci 14:93–103

Smart PL (1984) The geology, geomorphology and speleogenesis in the eastern massifs, Picos de Europa Spain. Cave Sci 11:238–245

Smart PL (1986) Origin and development of glacio-karst closed depressions in the Picos de Europa Spain. Z Geomorphol 30:423–443

Hydrogeochemical and Isotopic Characterization of Karstic Endorheic Estaña Lakes (Huesca, Spain)

C. Pérez Bielsa, L.J. Lambán Jiménez and P. Martínez Santos

Abstract The study area is located in the Marginal Pyrenean Sierras, which include the Hydrogeological Unit Estopiñán Syncline, consisting of two permeable levels: Upper Cretaceous and Eocene, and Hydrogeological Unit Estaña where endorheic and karstic Estaña Lakes are located. This work introduces the first results obtained from the hydrogeochemical and isotopic (δO^{18}, δH^2 and H^3) precipitacion and groundwater description, made to determine the general hydrological and hydrogeological of Estaña Lakes. This description allowed to differentiate the groundwater main chemical groups and to deduce qualitatively the dominating chemical processes. A large regional chemical variability is observed, mainly attributed to the different permeable levels as well as to mixing processes. The Estaña's Big Lake presents a marked calcium sulphate composition, constant either in depth and time. Multiparametric profiles have been done (EC, T, pH, DO, $\delta^{18}O$) confirming its monomictic character, sufering a process of stratification in the water between March and October and with a termocline about 6 m of depth.

Keywords Hydrogeology • Hydrochemistry • Isotopes • Estaña lakes • Estopiñán syncline

C. Pérez Bielsa (✉)
ADIEGO HNOS. S.A, Zaragoza, Spain
e-mail: cristinaperezbielsa@gmail.com

L.J. Lambán Jiménez
Spanish Geological Survey (IGME), Madrid, Spain
e-mail: javier.lamban@igme.es

P. Martínez Santos
Geology Department, Universidad Complutense de Madrid, Madrid, Spain
e-mail: pemartin@ucm.es

B. Andreo et al. (eds.), *Hydrogeological and Environmental Investigations in Karst Systems*, Environmental Earth Sciences 1,
DOI 10.1007/978-3-642-17435-3_70

1 Introduction

From the time water infiltrates into the ground until it reaches the discharge zone, there is close interaction between this and the materials it goes through. This interaction is defined by a series of chemical processes that change its composition (Appelo and Postma 2005; Garcia-Vera 1994). The identification of these processes allows to complete the hydrogeological characterization of the aquifer and reach a deeper understanding of the dynamics of groundwater flow (Vallejos 1997; Kohfahl et. al 2008).

The study area, known as Estaña Lakes, is located in Southern Pyrenees (North-Eastern Spain) (Fig. 1). It is included in the Natura 2000 European Network of Nature Protection Areas as a "Site of Community Importance". Despite its environmental recognition, its hydrogeology and hydrogeologic boundaries have not previously been analysed. The Estaña Lakes are a set of natural water ponds of karstic origin with perennial freshwater. They are located inside a closed basin (surface water does not flow out and eventually does not reach the ocean, it leaves the basin only by infiltration or evaporation), where carbonates and evaporites are dominant, and in a tectonically complex area without subsurface information. The goal of this work is to determine the general hydrological and hydrogeological functioning of Estaña Lakes based on the hydrogeochemical characterization.

2 Study Area

The Estaña Lakes system is located at the Sierras Marginales, one of the youngest thrust sheets of the Southern Pyrenees in Northern Spain (Martínez-Peña and Pocoví 1988). The study area is close to the zone that links the Sierras Exteriores and Sierras Marginales, at the western end of the South Pyrenean Central Unit (SPCU) (Séguret 1975) (Fig. 1). The Sierras Marginales are composed of Middle and Upper Triassic, Upper Cretaceous and Palaeogene rocks with a total thickness ranging from 300 to more than 2,000 m (Millán 2000).

The main structure of the study area is the Estopiñán syncline, located to the SW of the Estaña Lakes system. It is interpreted as a structure associated with an oblique ramp joining the Sierras Marginales, with a higher displacement towards the foreland, with the Sierras Exteriores (Martínez-Peña and Pocoví 1988). The Estopiñán syncline forms a broad hanging-wall syncline located over the flat part of a basal thrust. It is surrounded by a wide band of Middle and Upper Triassic rocks, forming a broken group of blocks of Mid-Triassic carbonates (Muschelkalk facies) within the Mid- and Upper-Triassic lutites and evaporites (Muschelkalk and Keuper facies) (Millán 2000), where Estaña Lakes are located.

The Estaña Lakes system is formed by two endorrheic main lakes of karstic origin (López-Vicente 2007), with different ecological conditions and water chemistry. The smaller one (Fig. 2) has a circular shape with a diameter of about

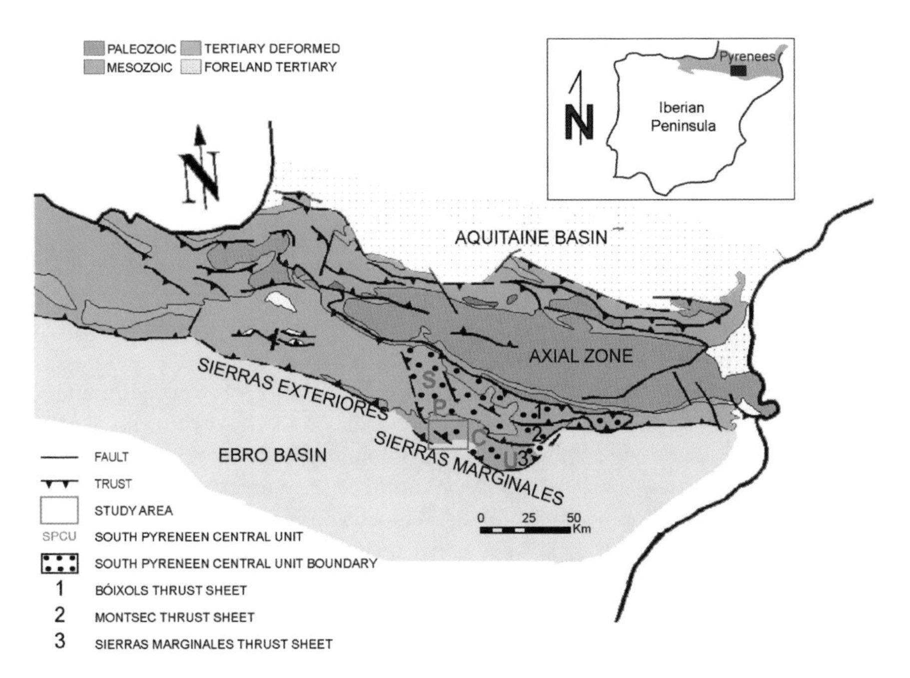

Fig. 1 Geological sketch of the South–Central Pyrenees. The study area is located in the South Pyrenean Central Unit (SPCU) to the North of Ebro basin. Modified from Pueyo et al. (1999)

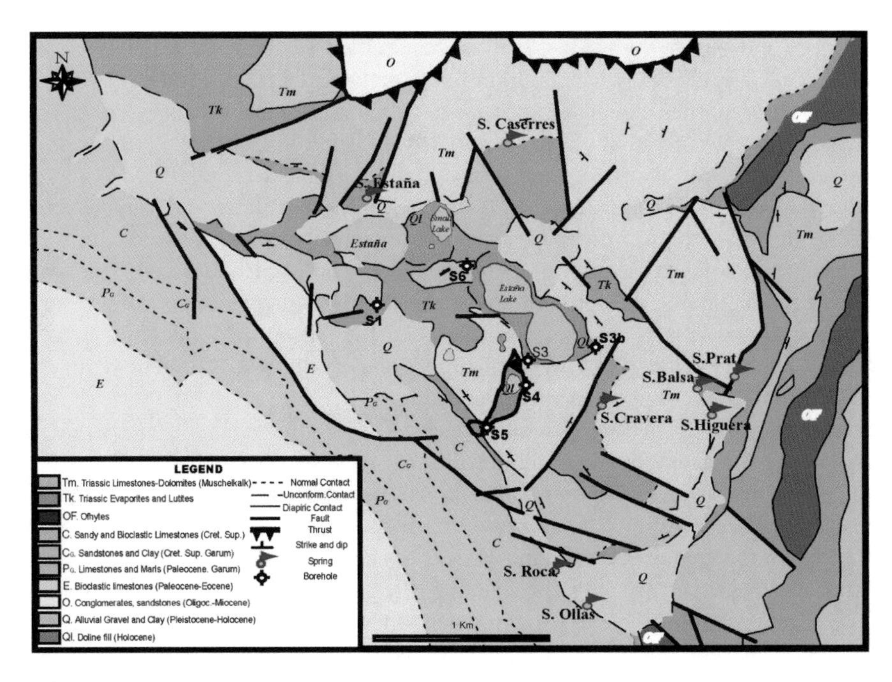

Fig. 2 Detailed geological map in the surroundings of Estaña Lakes (Modified from Pérez Bielsa et al. 2008). MRS, ERT and boreholes distribution are shown. It also presents the most important discharges of the area

200 m and a maximum depth of 7 m. The larger one, 250 m to the southeast of the smaller one and aligned with it, forms an uvala (sinkholes coalescence) with a sill on its central part that separates two depressions of about 360 and 325 m in diameter and maximum depths of 12 and 20 m respectively (Ávila et al. 1984). These lakes have been interpreted as collapse dolines (sinkholes) associated with the dissolution of the underlying Triassic carbonates and evaporites (Sancho-Marcén 1988). Up to 15 m of sediments have accumulated in the deepest portion of the lake (Morellón et al. 2008). Historical records document the existence of higher lake levels in the past (Riera et al. 2004) and sedimentological studies indicate five different depositional environments, showing abrupt and large hydrological changes during the last 21 kyr (Morellón et al. 2008).

The aquifer that has been identified in the Estaña Lakes system and surrounding area corresponds to the Middle Triassic carbonates known as the Muschelkalk facies. They are described as limestones and dolomites 60–80 m thick divided into two sections. The lower part is a highly dolomitized grey limestone with massive-to-tabular appearance, sometimes with cargneules and intensely karstified ("Gray Limestone" of Calvet et al. 2004). The upper section is made of dark or grey well-stratified micritic limestone to fine-grained dolomite, with finely laminated layers and occasional remains of bivalves. All the strata are affected by a very penetrative fracture system orthogonal to the bedding that shows evidence of water circulation (karstification fillings and red patina). These Mid-Triassic carbonated rocks are highly deformed and appear in isolated blocks of about 1 hm in surface area. The contact with the Mid- and Upper-Triassic evaporites and clays is usually a fault.

The Mid- and Upper- Triassic lutites and evaporites (Mid-Muschelkalk and Keuper facies) have been identified as the aquitard of the Estaña Lakes system (Pérez Bielsa et al. 2012). The high similarity between the Mid- and Upper Triassic lutites and evaporites has led to confusion as to which is which, causing stratigraphic and tectonic interpretation errors (Virgili 1958). The Middle Muschelkalk is dominated by red clays and laminated to nodulated gypsum, with a total thickness of more than 100 m. The presence of sandstone layers is distinctive. Keuper facies, with a thickness up to 250 m, are mostly gypsum and clays, interbedded with some marls and carbonates. A lower and upper member can be distinguished. The lower member is made of marls or carbonated clays with limestone intercalations, mud cracks and white gypsum that correspond to the Yesos de Canelles formation of Salvany and Bastida (2004). The upper section consists of red clay and gypsum of massive and messy appearance corresponding to "Yesos del Boix" formation of Salvany and Bastida (2004). Within the Keuper are a large number of massive dark disconnected ophitic bodies (alkaline volcanic rocks) without an apparent structure or in some cases with a pillow structure, probably located on top of the Keuper facies (Lago and Pocoví 1982). The existence of several points of steady discharge associated with the contact between the Mid-Triassic carbonates and Mid- and Upper-Triassic evaporites and clays for example, the Prat, Estaña and Caserres springs (Fig. 2), suggests that the Triassic carbonates are water saturated, working as a local aquifer discharging into the lakes (Pérez Bielsa et al. 2008). All groundwaters and lake waters analyzed have a calcium sulphate composition, which is consistent with this hypothesis. Thus, the geophysical characterization of this lake system was designed to check it.

3 Hydrogeochemical and Isotopic Characterization

3.1 *Hydrogeochemical and Isotopic Characterization of Precipitation*

Chemical analysis of rainwater samples collected next to the Estaña Big Lake have been studied. These correspond to the period 2008–2009, in which a total of 16 samples were analyzed (Fig. 3). The obtained data show a significant compositional variability at a spatial and at a temporal level, being less mineralized waters in winter, as expected. Hydrogeochemical facies vary between calcium bicarbonated and calcium sulphated waters, consistent with the outcropping materials in the recharge area.

For isotopic characterization of precipitation, $\delta^{18}O$–$\delta^{2}H$ relationship was analyzed and compared to VSMOW ‰ (Fig. 4), showing a large variability in composition, being lighter waters from January to April. All samples were adjusted to

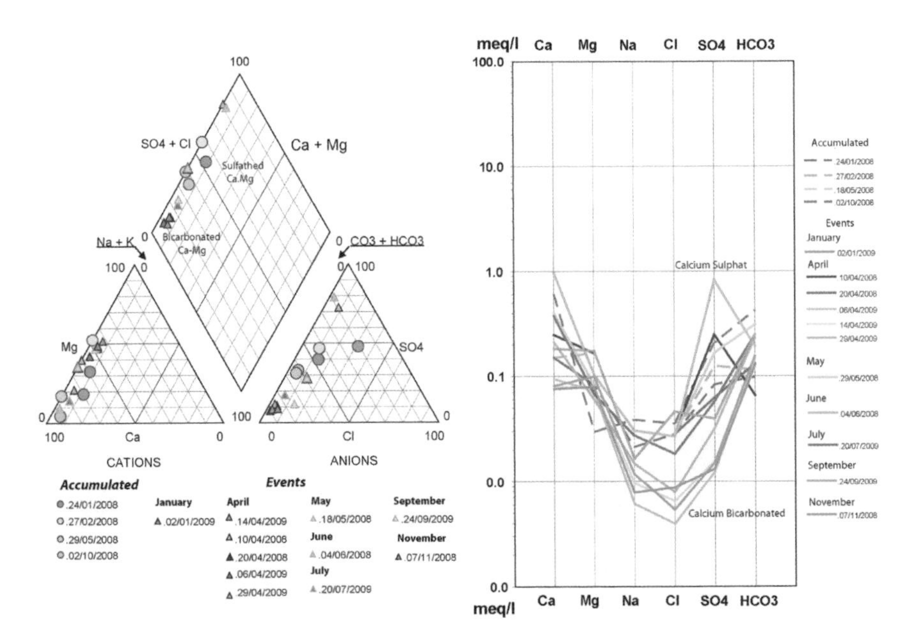

Fig. 3 Piper and Schoeller diagrams of precipitation that has been sampled in the surrounding area of Estaña Big Lake (January 2008 and September 2009). Modified from Pérez Bielsa (2013)

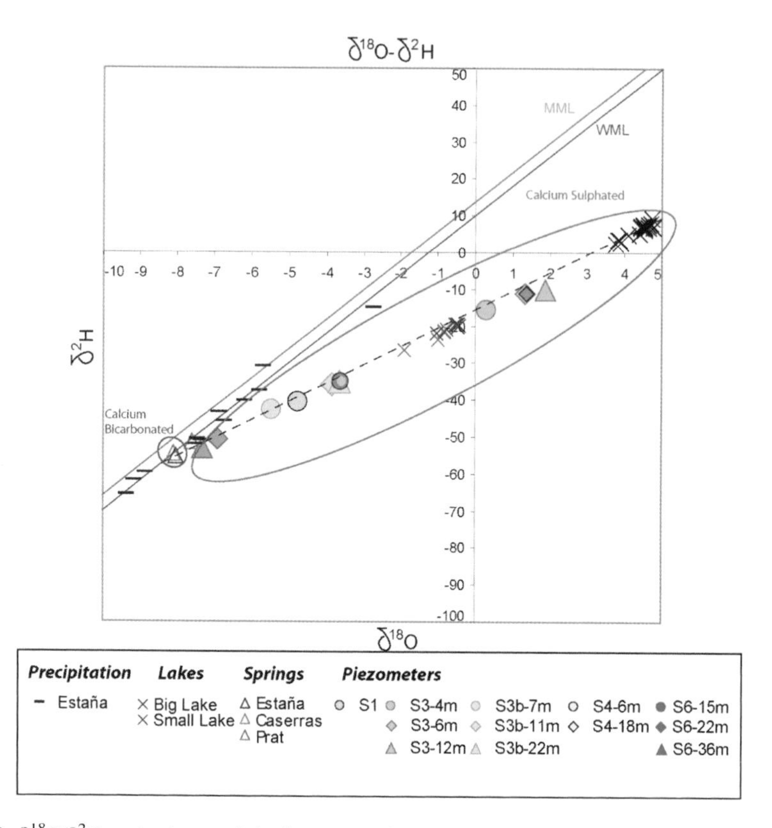

Fig. 4 $\delta^{18}O/\delta^{2}H$ ratio in precipitation, groundwater (springs and piezometers) and lakes. (WML = World Meteoric Line; MML = Mediterranean Meteoric Line). Modified from Pérez Bielsa (2013)

World Meteoric Average Line. Values of *d* range from 10 to 14 ‰, indicating an Atlantic predominantly origin of rain fronts, and a lesser proportion of Western Mediterranean.

3.2 Hydrogeochemical and Isotopic Characterization of Groundwater and Lakes

Available groundwater points in Estaña Lakes surrounding area are six springs and five piezometers. Previous available information was gathered to characterize the chemical and isotopic groundwater composition. Additionally, several sampling campaigns and multiparametric profiles were carried out between July 2007 and January 2010. The spatial distribution of the sampled points is shown in Fig. 2.

From a hydrogeochemical point of view, three main facies are identified: calcium-magnesium bicarbonated, calcium sulphated and sodium chlorinated (Fig. 5).

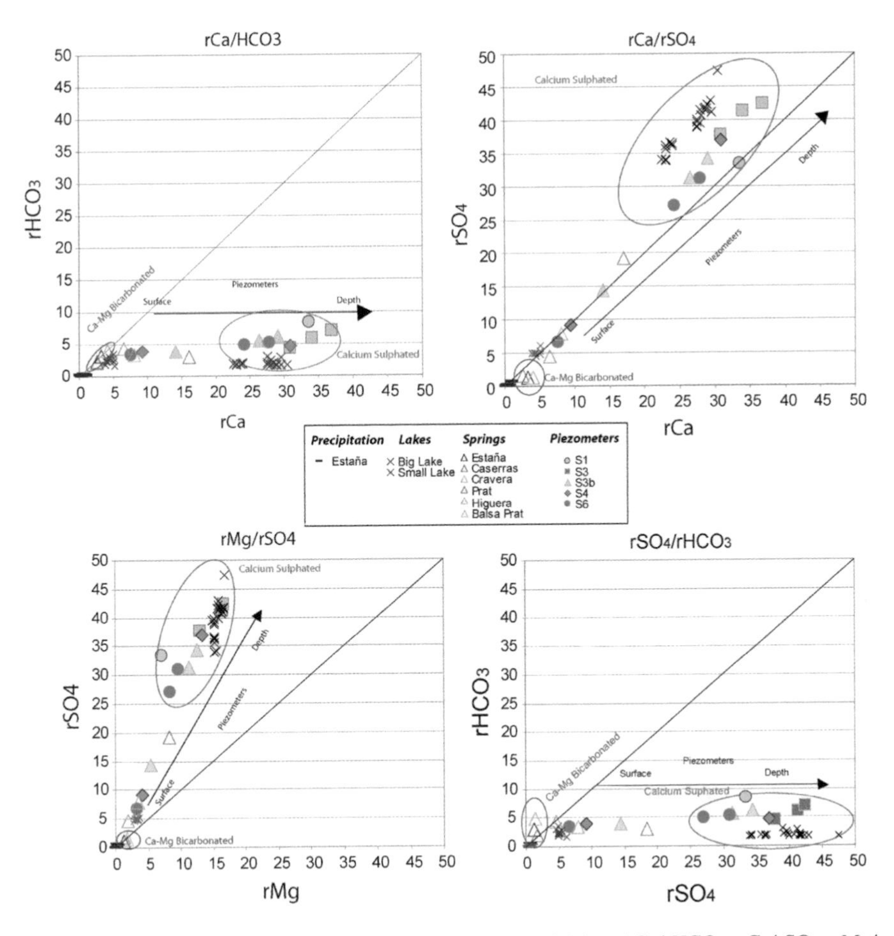

Fig. 5 Main ionic ratios of precipitation, groundwater and lakes ($rCa/rHCO_3$, rCa/rSO_4, rMg/rSO_4, $rSO_4/rHCO_3$). Modified from Pérez Bielsa (2013)

Caserras, Estaña and Cravera springs present calcium-magnesium bicarbonated waters with low-grade mineralization. Furthermore, all piezometers present calcium sulphated waters. Prat Spring water is sodium chlorinated, while Higuera Spring has an intermediate composition between calcium suphated and sodium chlorinated.

In general, at all points with several analyses an important temporal homogeneity is observed. Water type 1 (calcium -magnesium bicarbonated) is characterized by an $rCa/rHCO_3$ ratio next to 1, with an HCO_3^-opposite to SO_4^{2-}, indicating a dominant process of carbonates dissolution (calcite). The rCa/rSO_4 ratio is around 3 in Cravera and Estaña springs and it is 1, 5 in Caserras spring. It is also observed an enrichment of Ca^{2+} opposite to Mg^{2+} and $SO_4{}^{2-}$ (Fig. 5).

Water type 2 (calcium sulphated) is characterized by a relationship $rCa/rSO_4 \approx 1$, indicating a gypsum and/or anhydrite dissolution process. Relative enrichment in $SO_4^{2-}$is observed by increasing depth in the piezometers and by approaching to Estaña Big Lake, so as a Ca^{2+} versus HCO_3^-enrichment (Fig. 5). SO_4^{2-}rising is directly related to an increase in Mg^{2+} and Ca^{2+} content, indicating a possible dedolomitisation process (maximum in Estaña Big Lake) associated with the presence of carbonates and gypsum. SO_4^{2-}can be related to Triassic evaporitic materials dissolution processes, while Mg^{2+} content is associated with the limestone-dolomite aquifer (Pérez Bielsa 2013). Variations in this ion may be caused by changes in the solubility of calcite and/or dolomite, or conditioned by the dissolution of gypsum and/or anhydrite (Hanshaw and Back 1979; Back et al. 1983; Plummer et al. 1990).

The large variability observed in the piezometers hydrogeochemical composition between types 1 and 2 (calcium -magnesium bicarbonated and calcium sulphated), shows significant mixing processes according to the crossed materials (Muschelkalk carbonated facies and evaporite-clay facies). The $\delta^{18}O$–δ^2H relationship study in all samples corresponding to precipitation, lakes and groundwater (springs and piezometers) as well as its comparison with the Global Average Meteoric Linea ($\delta^2H = 8\ \delta^{18}O + 10$) and the Western Mediterranean Meteoric Line ($\delta^2H = 8\ \delta^{18}O + 14$, Jiménez-Martínez and Custodio 2008) allows to highlight the following results; Spring samples correspond with the Global Average Meteoric Line.

In the case of Estaña Lakes, its waters are isotopically heavier, in which an isotopic fractionation by free surface evaporation has been identified, largest at Big Lake than in Small Lake. Samples taken in the piezometers have an intermediate composition between springs and Estaña Lakes. In addition, except in piezometer S6, deeper samples are often more isotopically enriched. All samples from the piezometers are aligned along a slope equation of 4.87, seeming to show the influence of isotopic fractionation by free surface evaporation. Finally, the obtained ranges for tritium in springs (5, 4 to 8, 4), piezometers (4 to 5, 3), Small Lake (7, 3 to 8, 2) and Big Lake (5, 2 to 5, 7), are consistent with the values of local precipitation, between 3 and 7 TU. This fact seems to indicate, in general, the existence of fast groundwater flow with relatively short transit, which seems to be consistent with the karst nature of the aquifer (Pérez Bielsa 2013).

A calcium sulphated composition in Big Lake has been observed, which remains constant in depth and in time. Small Lake has a less mineralized calcium bicarbonated-sulphated composition. The multiparametric profiles made in the summer in this lake confirm its monomictic nature (Avila et al. 1984), with a stratification between the months of March to October and a thermocline around 6 m reflected in a sharp drop in temperature. Electrical conductivity (EC) stay without variation along the entire profile, whereas both temperature and pH suffer a sharp decrease from 6 m. Dissolved oxygen and Eh stay also without variation until 6 m, where an increase in both parameters is observed, to decrease sharply again from about 10 m deep. It is important to outline the existence of anoxic conditions at depth.

4 Conclusions

The whole interpretation of chemical analysis corresponding to precipitation, lakes and groundwater, reveals the existence of two predominant water types: (1) calcium-magnesium bicarbonated waters with low grade mineralization (EC = 64–596 μS/cm) and (2) calcium sulphated waters with high grade mineralization (1135–4970 EC = μS/cm). The remaining waters have an intermediate composition between these extremes. The dominant hydrogeochemical processes from a qualitative point of view are dissolution of carbonates, gypsum and/or anhydrite and halite. Environmental isotopes in groundwater show some enrichment in $\delta^{18}O$ y $\delta^{2}H$ in relation to World Average Meteoric Line and Western Mediterranean Line, as indicated by *d* content minor than 10 ‰. This may be due to pre-infiltration evaporation processes. Furthermore, the hydrogeochemical and isotopic study of precipitation water reveals an important spatial and temporal heterogeneity with a predominantly Atlantic origin, as well as a possible preferential recharge between January and May. The tritium obtained ranges in springs, piezometers and lakes are consistent with the values of local precipitation, suggesting recent recharge waters, with relatively fast transit through the aquifer. Performed profiles in Estaña Big Lake reveal a chemical temporally homogeneity as well as by depth, while the profiles made in summer, during the period of stratification of the water column, large variations in the dissolved oxygen content and in Eh are observed. In addition, the low mineralization of water in lakes and the no increase of salt concentration in depth, suggests the existence of an important renovation of water in the lakes associated with groundwater.

Acknowledgments This research was undertaken within the framework of the project "Funcionamiento hidrológico de humedales relacionados con las aguas subterráneas en la Cuenca del Ebro (2007–2010)" (Hydrological functioning of wetlands related to groundwater in Ebro Basin), funded by the Instituto Geológico y Minero de España (IGME, Spanish Geological Survey). The authors would like to thank the Instituto Pirenaico de Ecología (IPE-CSIC, Pyrenean Institute of Ecology) for their collaboration. Meteorological data have been provided by the Spanish Meteorological Agency (AEMET) and isotopic rainfall data by Applied Techniques Study Center of "Centro de Estudios y Experimentación de Obras Públicas" (CEDEX).

References

Appelo CAJ, Postma D (2005) Geochemistry, groundwater and pollution, 2nd edn. A. Balkema, Rotterdam, 536p

Ávila A, Burrel A, Domingo A, Fernández E, Godall J, Llopart JM (1984) Limnología del Lago Grande de Estanya (Huesca). Oecología Aquatica 7:3–24

Back W, Hanshaw BB, Plummer LN, Rahn PH, Rightmire CT, Rubin M (1983) Process and rate of dedolomitisation: mass transfer and 14C dating in a regional carbonate aquifer. Geol Soc Am Bull 94:1414–1429

Calvet F, Anglada E, Salvany JM (2004) El Triásico de los Pirineos. In: Vera JA (ed) Geología de España. SGEIGME, Madrid, pp 272–273

García–Vera MA (1994) Hidrogeología de zonas endorreicas en climas semiáridos: aplicación a los Monegros, Zaragoza. Ph.D. Thesis, Universidad Politécnica de Cataluña, Spain, 297pp

Hanshaw BB, Back W (1979) Major geochemical processes in the evolution of carbonate aquifer systems. J Hydrol 43:287–321

Kohfahl K, Sprenger C, Benavente Herrera J, Meyer B, Chacón Fernández, Pekdeger A (2008) Recharge sources and hydrogeochemical evolution of groundwater in semiarid and karstic environments: a field study in the Granada Basin (Southern Spain). Appl Geochem 23(2008):846–862

Jiménez-Martínez J, Custodio E (2008) El exceso de deuterio en la lluvia y en la recarga a los acuíferos en el área circum-mediterránea y en la costa mediterránea española. Boletín Geológico y Minero. Monográfico Iberoamericano. AIH-IGME-UNESCO. 21p

Lago M, Pocovi A (1982) Nota preliminar sobre la presencia de estructuras fluidales en las ofitas del área de Estopiñán (provincia de Huesca). Acta Geológica Hispánica 17(4):227–233

López-Vicente M (2007) Erosión y redistribución del suelo en agrosistemas mediterráneoas: Modelización predictiva mediante SIG y validación con 137Cs (Cuenca de Estaña, Pirineo Central). Ph.D. Thesis, Univ. Zaragoza-CSIC, Spain, 212pp

Martínez-Peña MB, Pocoví A (1988) El amortiguamiento frontal de la estructura de la cobertera sur-pirenaica y su relación con el anticlinal de Barbastro-Balaguer. Acta Geológica Hispánica 23:81–94

Millán H (2000) Estructura y cinemática del frente de cabalgamiento surpirenaico. Sierras Exteriores Aragonesas. Ph.d. Thesis, Univ. Zaragoza, Spain

Morellón M, Valero-Garcés B, Moreno A, Anselmetti F, Ariztegui D, Schnellmanns M, Moreno A, Mata P, Rico M and Corella JP (2008) Late Quaternary deposition and facies model for kasrtic Lake Estanya (North-Eastern Spain). Sedimentology. doi:10.1111/j.1365-3091.2008.01044.x

Pérez Bielsa C, Ramajo J, Lambán LJ (2008) Marco geológico e hidrogeológico de las Lagunas de Estaña (Huesca, España). VII Congreso Geológico de España, Las Palmas de Gran Canaria. Geo-temas 10, 2008. P 252

Pérez Bielsa C, Lambán J, Plata JL, Rubio FM, Soto R (2012) Characterization of a karstic aquifer using magnetic resonance sounding (MRS) and electrical resistivity tomography (ERT): a case-study of Estaña Lakes (Northern Spain). Hydrogeol J 20:1045–1059

Pérez Bielsa C (2013). Funcionamiento hidrogeológico de un humedal hipogénico de origen kárstico en las Sierras Marginales Pirenaicas: Las Lagunas de Estaña (Huesca). Ph.D. Thesis, Univ. Complutense de Madrid, Spain, 420pp

Plummer LN, Busby JF, Lee RW, Hanshaw BB (1990) Geochemical modeling in the Madison Aquifer in parts of Montana Wyoming and South Dakota. Water Resour 26:1981–2014

Pueyo-Morer EL, Millán H, Pocoví A, Parés JM (1999) Cinemática rotacional del cabalgamiento basal surpirenáico en las Sierras Exteriores Aragonesas: Datos magnetotectónicos. Acta Geológica Hispánica 32(3–4):119–138

Riera S, Wansard R, Juliá R (2004) 2000-year environmental history of a karstic lake in the Mediterranean Pre-Pyrenees: the Estanya lakes (Spain). Catena 55:293–324

Salvany JM, Bastida J (2004) Análisis litoestratigráfico del Keuper surpirenaico Central. Rev Soc Geol 17:3–26

Sancho Marcén C (1988) El Polje de (Saganta Sierras Exteriores pirenaicas. prov. de Huesca). Cuaternario y Geomorfología 2(1–4):107–113

Seguret M (1975) Etude tectonique des nappes et series decollées de la partie centrale du versant sud des Pyrénées. Caractére sedymentaire role de la compression et de la gravité. Ph.D. thesis, Université des Sciences et Techniques du Languedoc (USTELA), France, Sér. Geol. Struct. no 2, 155pp

Vallejos A (1997) Caracterización Hidrogeoquímica de la recarga de los acuíferos del campo de Dalias a partir de la Sierra de Gador (Almería). Ph.D. Thesis, Universidad de Granada, Spain, 257pp

Printed by Printforce, the Netherlands